Advanced Data Analytics for Power Systems

Experts in data analytics and power engineering present techniques addressing the needs of modern power systems, covering theory and applications related to power system reliability, efficiency, and security. With topics spanning large-scale and distributed optimization, statistical learning, big data analytics, graph theory, and game theory, this is an essential resource for graduate students and researchers in academia and industry with backgrounds in power systems engineering, applied mathematics, and computer science.

Ali Tajer is an associate professor of electrical, computer, and systems engineering at Rensselaer Polytechnic Institute.

Samir M. Perlaza is a chargé de recherche at INRIA, Centre de Recherche de Sophia Antipolis – Méditerranée, France.

H. Vincent Poor is the Michael Henry Strater University Professor of Electrical Engineering at Princeton University.

"There are only a few industries that generate an equally large amount of data with a comparable variety, and societal importance. Data analytics is thus rightfully at the heart of modern power systems operations and planning. Focusing on applications in power systems, this book gives an excellent account of recent developments and of the broad range of algorithms and tools in the area of data analytics, as well as of the applications of these tools for solving challenging problems from a novel angle. Covering a wide range of fundamental problems, from state estimation to load scheduling and anomaly detection, the book is not only an excellent source of inspiration, but can also serve as an extensive reference for the gamut of operational problems faced in the power industry."

György Dán, KTH Royal Institute of Technology

Advanced Data Analytics for Power Systems

Edited by

ALI TAJER
Rensselaer Polytechnic Institute, Troy, NY

SAMIR M. PERLAZA
INRIA, Sophia Antipolis, France

H. VINCENT POOR
Princeton University, Princeton, NJ

CAMBRIDGE
UNIVERSITY PRESS

University Printing House, Cambridge CB2 8BS, United Kingdom

One Liberty Plaza, 20th Floor, New York, NY 10006, USA

477 Williamstown Road, Port Melbourne, VIC 3207, Australia

314–321, 3rd Floor, Plot 3, Splendor Forum, Jasola District Centre, New Delhi – 110025, India

79 Anson Road, #06–04/06, Singapore 079906

Cambridge University Press is part of the University of Cambridge.

It furthers the University's mission by disseminating knowledge in the pursuit of education, learning, and research at the highest international levels of excellence.

www.cambridge.org
Information on this title: www.cambridge.org/9781108494755
DOI: 10.1017/9781108859806

© Cambridge University Press 2021

This publication is in copyright. Subject to statutory exception and to the provisions of relevant collective licensing agreements, no reproduction of any part may take place without the written permission of Cambridge University Press.

First published 2021

Printed in the United Kingdom by TJ Books Ltd, Padstow Cornwall

A catalogue record for this publication is available from the British Library.

Library of Congress Cataloging-in-Publication Data
Names: Tajer, Ali, editor. | Perlaza, Samir M., editor. | Poor, H. Vincent, editor.
Title: Advanced data analytics for power systems / edited by Ali Tajer, Rensselaer Polytechnic
   Institute, New York , Samir M. Perlaza, INRIA, H. Vincent Poor, Princeton University, New Jersey.
Description: Cambridge, United Kingdom ; New York, NY, USA : Cambridge University Press, 2020. |
   Includes bibliographical references and index.
Identifiers: LCCN 2020013357 (print) | LCCN 2020013358 (ebook) | ISBN 9781108494755 (hardback) |
   ISBN 9781108859806 (epub)
Subjects: LCSH: Electric power systems–Mathematical models.
Classification: LCC TK1001 .A36 2020 (print) | LCC TK1001 (ebook) | DDC 621.3101/5118–dc23
LC record available at https://lccn.loc.gov/2020013357
LC ebook record available at https://lccn.loc.gov/2020013358

ISBN 978-1-108-49475-5 Hardback

Cambridge University Press has no responsibility for the persistence or accuracy of URLs for external or third-party internet websites referred to in this publication and does not guarantee that any content on such websites is, or will remain, accurate or appropriate.

Contents

List of Contributors		page xvi
Preface		xxi

Part I Statistical Learning — 1

1 Learning Power Grid Topologies — 3
Guido Cavraro, Vassilis Kekatos, Liang Zhang, and Georgios B. Giannakis

- 1.1 Introduction — 3
- 1.2 Grid Modeling — 4
- 1.3 Topology Detection Using Smart Meter Data — 5
  - 1.3.1 A Maximum Likelihood Approach — 5
  - 1.3.2 A Maximum a posteriori Probability Approach — 7
  - 1.3.3 Numerical Tests on Topology Detection Using Smart Meter Data — 8
- 1.4 Topology Detection Using Partial Correlations — 9
  - 1.4.1 Nonlinear Partial Correlations — 9
  - 1.4.2 A Frank–Wolfe–Based Solver — 11
  - 1.4.3 Numerical Tests on Topology Detection Using Partial Correlations — 12
- 1.5 Grid Probing for Topology Learning — 13
- 1.6 Identifiability Analysis of Grid Probing — 14
  - 1.6.1 Topology Identifiability with Complete Voltage Data — 15
  - 1.6.2 Topology Identifiability with Partial Voltage Data — 17
- 1.7 Graph Algorithms for Topology Identification Using Probing — 18
  - 1.7.1 Topology Identification with Complete Voltage Data — 18
  - 1.7.2 Topology Identification with Partial Voltage Data — 19
  - 1.7.3 Graph Algorithms Operating under Noisy Data — 20
- 1.8 Topology Identification Using Probing through Convex Relaxation — 21
- 1.9 Numerical Tests on Topology Identification Using Probing — 23
- 1.10 Conclusion — 24

2 Probabilistic Forecasting of Power System and Market Operations — 28
Yuting Ji, Lang Tong, and Weisi Deng

- 2.1 Introduction — 28
  - 2.1.1 Related Work — 29

| | | 2.1.2 | Summary and Organization | 31 |

	2.2	Probabilistic Forecasting Using Monte Carlo Simulations	31
	2.3	Models of System and Market Operations	32
		2.3.1 A Multiparametric Model for Real-Time System and Market Operations	33
		2.3.2 Single-Period Economic Dispatch	34
		2.3.3 Multiperiod Economic Dispatch with Ramping Products	35
		2.3.4 Energy and Reserve Co-optimization	37
	2.4	Structural Solutions of Multiparametric Program	39
		2.4.1 Critical Region and Its Geometric Structure	39
		2.4.2 A Dictionary Structure of MLP/MQP Solutions	40
	2.5	Probabilistic Forecasting through Online Dictionary Learning	41
		2.5.1 Complexity of Probabilistic Forecasting	42
		2.5.2 An Online Dictionary Learning Approach to Probabilistic Forecasting	42
	2.6	Numerical Simulations	44
		2.6.1 General Setup	44
		2.6.2 The 3120-Bus System	45
		2.6.3 The IEEE 118-Bus System	46
	2.7	Conclusion	49

3 Deep Learning in Power Systems — 52
Yue Zhao and Baosen Zhang

3.1	Introduction	52
3.2	Energy Data Analytics with Deep Learning: Scenario Generation	53
	3.2.1 Scenario Generation Using Generative Adversarial Networks	54
	3.2.2 GANs with Wasserstein Distance	54
	3.2.3 Training GANs	56
	3.2.4 Examples	57
	3.2.5 Quality of Generated Scenarios	59
3.3	Power System Monitoring with Deep Learning: Real-Time Inference	61
	3.3.1 Model, Problems, and Challenges	62
	3.3.2 A Machine Learning–Based Monitoring	64
	3.3.3 Case Study 1: Multi-line Outage Identification	66
	3.3.4 Case Study 2: Voltage Stability Margin Estimation	68
3.4	Conclusion	71

4 Estimating the System State and Network Model Errors — 74
Ali Abur, Murat Göl, and Yuzhang Lin

4.1	Introduction	74
4.2	Power System State Estimation Using SCADA Measurements	75
4.3	Network Observability	76
4.4	Bad Data Processing	77
4.5	Power System State Estimation in the Presence of PMUs	78

		4.5.1	Phasor Measurement–Based State Estimation	79
		4.5.2	Observability and Criticality Analyses in the Presence of PMUs	81
		4.5.3	Hybrid State Estimation	82
		4.5.4	PMU–Based Three-Phase State Estimation	85
	4.6	Detection of Network Parameter Errors		88
		4.6.1	Detection, Identification, and Correction of Parameter Errors	88
		4.6.2	Detectability and Identifiability of Parameter Errors	91
		4.6.3	Use of Multiple Measurement Scans	92
		4.6.4	Robust State Estimation against Parameter and Measurement Errors	94
	4.7	Conclusion		95

Part II Data-Driven Anomaly Detection — 99

5 Quickest Detection and Isolation of Transmission Line Outages — 101
Venugopal V. Veeravalli and Alejandro Dominguez-Garcia

- 5.1 Introduction — 101
- 5.2 Quickest Change Detection Background — 102
  - 5.2.1 Shewhart Test — 104
  - 5.2.2 CuSum Test — 104
  - 5.2.3 Optimization Criteria and Optimality of CuSum — 105
  - 5.2.4 Incompletely Specified Observation Models — 106
  - 5.2.5 QCD under Transient Dynamics — 107
- 5.3 Power System Model — 108
  - 5.3.1 Pre-outage Model — 108
  - 5.3.2 Instantaneous Change during Outage — 110
  - 5.3.3 Post-outage — 111
  - 5.3.4 Measurement Model — 111
- 5.4 Line Outage Detection Using QCD — 112
  - 5.4.1 Meanshift Test — 112
  - 5.4.2 Generalized Shewhart Test — 113
  - 5.4.3 Generalized CuSum Test — 114
  - 5.4.4 Generalized Dynamic CuSum Test — 114
- 5.5 Line Outage Identification — 115
- 5.6 Numerical Results — 116
  - 5.6.1 Line Statistic Evolution — 116
  - 5.6.2 Delay Performance — 118
  - 5.6.3 Probability of False Isolation — 120
- 5.7 Conclusion — 121

6 Active Sensing for Quickest Anomaly Detection — 124
Ali Tajer and Javad Heydari

- 6.1 Anomaly Detection — 124
- 6.2 Need for Agile Detection of Anomalies — 125

6.3	Introduction to Active Sensing		126
6.4	Active Sensing in Networks		128
6.5	Modeling Active Sensing in Power Systems		128
	6.5.1	Sensor Measurement Model	129
	6.5.2	Anomalous Event Model	130
6.6	Decision Rules and Algorithm		130
	6.6.1	Bus Selection Rule	132
	6.6.2	Termination Rule and Localization Decision	134
6.7	Case Study		137
	6.7.1	Adaptivity Gain	137
	6.7.2	Multiple Outage	138
	6.7.3	Tradeoff among Performance Measures	139
	6.7.4	Scalability and Complexity	140

7 Random Matrix Theory for Analyzing Spatio-Temporal Data 144
Robert Qiu, Xing He, Lei Chu, and Xin Shi

7.1	Data-Driven View of Grid Operation and Its Matrix Form		144
	7.1.1	Fundamental Rules for Grid Network Operation	144
	7.1.2	Jacobian Matrix	146
	7.1.3	Power Flow Analysis and Theoretical Calculation	147
7.2	Random Matrix Theory: An Analytical Tool for Large-Scale Power Grids		147
	7.2.1	When the Power Data Meets Large Random Matrices	148
	7.2.2	Some Fundamental Results	149
	7.2.3	On the Analytical Methods	150
	7.2.4	Universality Principle of RMT	150
	7.2.5	Random Matrix Tools for Power Systems	151
	7.2.6	Linear Eigenvalue Statistics and Its Central Limit Theorem	151
	7.2.7	Matrices Concatenation Operation	153
	7.2.8	Data Observation from Power Systems	153
	7.2.9	Empirical Spectrum Distribution of the Online Monitoring Data	154
	7.2.10	Residual Formulation and Discussion	155
7.3	Applications of RMT on Power Grids		156
	7.3.1	Anomaly Detection Based on Factor Model	156
	7.3.2	More Discussions about the Proposed Approach	159
	7.3.3	Case Studies	160
	7.3.4	Invisible Units Detection and Estimation	167
	7.3.5	Estimation of Jacobian Matrix	171

8 Graph-Theoretic Analysis of Power Grid Robustness 175
Dorcas Ofori-Boateng, Asim Kumer Dey, Yulia R. Gel, and H. Vincent Poor

8.1	Introduction	175
8.2	Power Grid Network and Its Robustness	176
8.3	Failure Variations	177

	8.4	Robustness Metrics	177
		8.4.1 Conventional Robustness Metrics under Failures	178
		8.4.2 Network Motifs as a Local Robustness Measure	179
		8.4.3 Persistent Homology as a Local Robustness Measure	180
		8.4.4 Statistical Inference for Local Higher-Order Topological Features of a Network	183
	8.5	Analysis of Real-World Power Grid Networks	184
		8.5.1 Data Description	184
		8.5.2 Statistical Significance of Motifs and Betti Numbers	185
		8.5.3 Comparison of Robustness Metrics under Failure Scenarios	186
	8.6	Conclusion	188
	A	Appendix Resilience analysis – additional results	189
		A.1 Robustness metrics versus attack strategies: *Romania* & *Poland*	189
		A.2 Attack strategies versus robustness metrics: *Romania* & *Poland*	190

Part III Data Quality, Integrity, and Privacy — 195

9 Data-Injection Attacks — 197
Iñaki Esnaola, Samir M. Perlaza, and Ke Sun

	9.1	Introduction	197
	9.2	System Model	198
		9.2.1 Bayesian State Estimation	198
		9.2.2 Deterministic Attack Model	199
		9.2.3 Attack Detection	200
	9.3	Centralized Deterministic Attacks	201
		9.3.1 Minimum Probability of Detection Attacks	201
		9.3.2 Maximum Distortion Attacks	204
	9.4	Decentralized Deterministic Attacks	205
		9.4.1 Game Formulation	206
		9.4.2 Achievability of an NE	208
		9.4.3 Cardinality of the Set of NEs	208
	9.5	Information-Theoretic Attacks	210
		9.5.1 Random Attack Model	211
		9.5.2 Information-Theoretic Setting	212
		9.5.3 Generalized Stealth Attacks	213
		9.5.4 Probability of Detection of Generalized Stealth Attacks	215
		9.5.5 Numerical Evaluation of Stealth Attacks	219
	9.6	Attack Construction with Estimated State Variable Statistics	222
		9.6.1 Learning the Second-Order Statistics of the State Variables	222
		9.6.2 Ergodic Stealth Attack Performance	223
	9.7	Conclusion	227

10	**Smart Meter Data Privacy**	230
	Giulio Giaconi, Deniz Gündüz, and H. Vincent Poor	
	10.1 The SG Revolution	230
	10.2 ALM Techniques	231
	10.3 SM Privacy Concerns and Privacy-Preserving Techniques	232
	10.4 SMDM Techniques	234
	10.4.1 Data Aggregation Techniques	234
	10.4.2 Data Obfuscation Techniques	236
	10.4.3 Data Anonymization Techniques	237
	10.4.4 Data-Sharing Prevention Techniques	238
	10.4.5 Data-Downsampling Techniques	238
	10.5 UDS Techniques	238
	10.5.1 Heuristic Privacy Measures: Variations in the Grid Load Profile	241
	10.5.2 Theoretical Guarantees on SM Privacy	248
	10.6 Conclusion	255
11	**Data Quality and Privacy Enhancement**	261
	Meng Wang and Joe H. Chow	
	11.1 Introduction	261
	11.2 Low Dimensionality of PMU Data	262
	11.3 Missing PMU Data Recovery	263
	11.4 Bad Data Correction	267
	11.5 Location of Corrupted Devices	271
	11.6 Extensions in Data Recovery and Correction	273
	11.6.1 Data Recovery Using Time-Varying Low-Dimensional Structures to Characterize Long-Range Temporal Correlations	273
	11.6.2 Data Recovery under Multiple Disturbances	274
	11.6.3 Online Data Recovery	275
	11.7 Data Recovery from Privacy-Preserving Measurements	275
	11.8 Conclusion	278

Part IV Signal Processing 283

12	**A Data Analytics Perspective of Fundamental Power Grid Analysis Techniques**	285
	Danilo P. Mandic, Sithan Kanna, Yili Xia, and Anthony G. Constantinides	
	12.1 Introduction	285
	12.2 Problem Formulation	287
	12.3 Background: Tracking the Instantaneous Frequency	287
	12.3.1 Rate of Change of the Voltage Phase Angle	287
	12.3.2 Fixed Frequency Demodulation	288
	12.4 Frequency Estimation from Voltage Phase Angles versus Frequency Demodulation	288
	12.5 Smart DFT-Based Frequency Estimation	291
	12.6 Maximum Likelihood Frequency Estimation	292

	12.7 Real-World Case Studies	293
	12.8 Meeting the Needs of Unbalanced Smart Grids: Smart Clarke and Park Transforms	296
	12.9 Conclusion	308

13 Graph Signal Processing for the Power Grid   312
Anna Scaglione, Raksha Ramakrishna, and Mahdi Jamei

	13.1 Preliminaries	312
	13.2 Graph Signal Processing for the Power Grid	314
	13.3 Complex-Valued Graph Signal Processing	316
	13.3.1 Time-Vertex Frequency Analysis	317
	13.4 Voltage Measurements as Graph Signals	318
	13.4.1 Coefficients of the Grid Graph Filter	319
	13.5 Sampling and Compression for Grid-Graph Signals	321
	13.5.1 Dimensionality Reduction of PMU Data	322
	13.5.2 Sampling and Recovery of PMU Grid-Graph Signals	325
	13.6 The Role of the Topology on Grid-GSP Data	328
	13.6.1 Fault Localization Using Undersampled Grid-Graph Signals	329
	13.6.2 Identification of Community Structure in the Electrical Grid	333
	13.7 Conclusion	336

14 A Sparse Representation Approach for Anomaly Identification   340
Hao Zhu and Chen Chen

	14.1 Introduction	340
	14.2 Power Grid Modeling	341
	14.3 Sparse Representations of Anomaly Events	342
	14.3.1 Power Line Outages	343
	14.3.2 Fault Events	344
	14.3.3 Phase Imbalance	345
	14.4 Efficient Solvers Using Compressed Sensing	346
	14.4.1 Greedy Anomaly Identification Using OMP	347
	14.4.2 Lassoing Anomalies Using CD	349
	14.5 Meter Placement	351
	14.5.1 A Greedy Search Method	352
	14.6 Uncertainty Quantification	354
	14.6.1 Optimal Decision Rules	355
	14.6.2 Effects of PMU Uncertainty on Bayesian Risk	356
	14.7 Conclusion	358

Part V Large-Scale Optimization   361

15 Uncertainty-Aware Power Systems Operation   363
Daniel Bienstock

	15.1 Introduction: Power Systems Operations in Flux	363

	15.1.1 Power Engineering in Steady State	364
	15.1.2 Power-Flow Problems	367
	15.1.3 Three-Level Grid Control	368
	15.1.4 Responding to Changing Conditions	371
15.2	Safe DC-OPF	374
	15.2.1 Specific Formulations	378
	15.2.2 The Gaussian Case	381
	15.2.3 Numerical Algorithmics	383
	15.2.4 Two-Tailed versus Single-Tailed Models	388
	15.2.5 Gaussian Data or Not?	389
	15.2.6 Robust Optimization	390
	15.2.7 The AC Case	391
15.3	Data-Driven Models and Streaming Covariance Estimation	392
15.4	Tools from Financial Analytics: Variance-Aware Models, Sharpe Ratios, Value-at-Risk	394
15.5	Conclusion	395

16 Distributed Optimization for Power and Energy Systems 400
Emiliano Dall'Anese and Nikolaos Gatsis

16.1	Introduction	400
16.2	General Problem Setup and Motivating Applications	401
	16.2.1 Notation and Assumptions	403
	16.2.2 Optimization of Power Distribution Systems	404
	16.2.3 Optimization of Power Transmission Systems	406
	16.2.4 Demand-Side Management	406
	16.2.5 State Estimation	407
	16.2.6 Optimization of Wind Farms	408
16.3	Dual Methods for Constrained Convex Optimization	408
	16.3.1 Dual Method	409
	16.3.2 Further Reading: Cutting Plane and Bundle Methods	412
16.4	Model-Based and Measurement-Based Primal-Dual Methods	413
	16.4.1 Model-Based Regularized Gradient Methods	414
	16.4.2 Measurement-Based Gradient Methods	415
	16.4.3 Further Reading	418
16.5	Consensus-Based Distributed Optimization	418
	16.5.1 ADMM-Based Distributed Algorithms	421
	16.5.2 A More General Formulation	422
	16.5.3 Further Reading	424
16.6	Further Reading	424
	16.6.1 Distributed Methods for Nonconvex Optimization	424
	16.6.2 Lyapunov Optimization	425
	16.6.3 Online Measurement-Based Optimization of Power Systems	425

17	**Distributed Load Management**	431
	Changhong Zhao, Vijay Gupta, and Ufuk Topcu	
	17.1 Distributed Charging Protocols for Electric Vehicles	432
	17.2 An Online-Learning–Based Implementation	440
	17.3 Distributed Feedback Control of Networked Loads	446

18	**Analytical Models for Emerging Energy Storage Applications**	455
	I. Safak Bayram and Michael Devetsikiotis	
	18.1 Energy Storage Systems Landscape	456
	18.1.1 Energy Storage Classification	456
	18.1.2 Energy Storage Applications	459
	18.2 Comparing Energy Storage Solutions	461
	18.2.1 Technology Lock-In	462
	18.3 Analytical Problems in Energy Storage Systems	462
	18.4 Case Study: Sizing On-Site Storage Units in a Fast Plug-In Electric Vehicle Charging Station	464
	18.4.1 Stochastic Model	465
	18.4.2 Numerical Examples	467
	18.5 Case Study: Community Energy Storage System Sizing	468
	18.5.1 Energy Storage Operational Dynamics	470
	18.5.2 Storage Sizing for Single Customer ($K = 1, N = 1$)	473
	18.5.3 Storage Sizing for Multiuser Case ($K = 1, N > 1$)	474
	18.5.4 Storage Sizing for the General Case ($K > 1, N > 1$)	475
	18.5.5 Numerical Results	477
	18.6 Conclusion	478

Part VI Game Theory 481

19	**Distributed Power Consumption Scheduling**	483
	Samson Lasaulce, Olivier Beaude, and Mauricio González	
	19.1 Introduction	483
	19.2 When Consumption Profiles Have to Be Rectangular and Forecasting Errors Are Ignored	484
	19.2.1 System Model	484
	19.2.2 Power Consumption Scheduling Game: Formulation and Analysis	487
	19.2.3 A Distributed Power Consumption Scheduling Algorithm	489
	19.2.4 Numerical Illustration	490
	19.3 When Consumption Profiles Are Arbitrary and Forecasting Errors Are Ignored	491
	19.3.1 Introduction to a Dynamic Framework	492
	19.3.2 Distributed Strategies in This Dynamical Setting	492
	19.3.3 Numerical Illustration	494

	19.4 When Forecasting Errors Are Accounted	496
	19.4.1 Markov Decision Process Modeling	497
	19.4.2 Relaxed Version of MDP Modeling	499
	19.4.3 Iterative MDP Solution Methodology	500
20	**Electric Vehicles and Mean-Field**	**504**
	Dario Bauso and Toru Namerikawa	
	20.1 Introduction	504
	20.1.1 Highlights of Contributions	505
	20.1.2 Literature Overview	505
	20.1.3 Notation	506
	20.2 Modeling EVs in Large Numbers	507
	20.2.1 Forecasting Based on Holt's Model	508
	20.2.2 Receding Horizon	509
	20.2.3 From Mean-Field Coupling to Linear Quadratic Problem	511
	20.3 Main Results	513
	20.3.1 Receding Horizon and Stability	513
	20.4 Numerical Studies	514
	20.5 Discussion	518
	20.6 Conclusion	519
21	**Prosumer Behavior: Decision Making with Bounded Horizon**	**524**
	Mohsen Rajabpour, Arnold Glass, Robert Mulligan, and Narayan B. Mandayam	
	21.1 Introduction	524
	21.2 Experimental Design: Simulation of an Energy Market	527
	21.3 Modeling Approaches	530
	21.3.1 Modeling Prosumer Behavior Using Expected Utility Theory	530
	21.3.2 Bounded Horizon Model of Prosumer Behavior	533
	21.4 Data Fitting, Results, and Analysis	536
	21.4.1 Data Fitting with Mean Deviation	537
	21.4.2 Data Fitting with Proportional Deviation	539
	21.5 Conclusion	541
22	**Storage Allocation for Price Volatility Management in Electricity Markets**	**545**
	Amin Masoumzadeh, Ehsan Nekouei, and Tansu Alpcan	
	22.1 Introduction	546
	22.1.1 Motivation	546
	22.1.2 Contributions	548
	22.2 Related Works	549
	22.3 System Model	550
	22.3.1 Inverse Demand (Price) Function	550
	22.3.2 Upper-Level Problem	552
	22.3.3 Lower-Level Problem	552

22.4	Solution Approach	557
	22.4.1 Game-Theoretic Analysis of the Lower-Level Problem	557
	22.4.2 The Equivalent Single-Level Optimization Problem	559
22.5	Case Study and Simulation Results	560
	22.5.1 One-Region Model Simulations in South Australia	561
	22.5.2 Two-region Model Simulations in South Australia and Victoria	565
22.6	Conclusion	566

Index   571

Contributors

Ali Abur
Northeastern University, Boston, MA, USA

Tansu Alpcan
University of Melbourne, Melbourne, Australia

Dario Bauso
University of Groningen, Groningen, the Netherlands; and University of Palermo, Palermo, Italy

I. Safak Bayram
University of Strathclyde, Glasgow, UK

Olivier Beaude
Électricité de France (EDF), Alaiseau, France

Daniel Bienstock
Columbia University, New York, NY, USA

Guido Cavraro
National Renewable Energy Laboratory, Golden, CO, USA

Chen Chen
Argonne National Laboratory, Lemont, IL, USA

Joe H. Chow
Rensselaer Polytechnic Institute, Troy, NY, USA

Lei Chu
Shanghai Jiaotong University, Shanghai, China

Anthony G. Constantinides
Imperial College London, London, UK

Emiliano Dall'Anese
University of Colorado Boulder, Boulder, CO, USA

Weisi Deng
China Southern Power Grid Co., Ltd, Guangzhou, China

Michael Devetsikiotis
University of New Mexico, Albuquerque, NM, USA

Asim Kumer Dey
University of Texas at Dallas, Richardson, TX, USA

Alejandro Dominguez-Garcia
University of Illinois Urbana-Champaign, Urbana, IL, USA

Iñaki Esnaola
University of Sheffield, Sheffield, UK

Nikolaos Gatsis
University of Texas at San Antonio, San Antoni, TX, USA

Yulia R. Gel
University of Texas at Dallas, Richardson, TX, USA

Giulio Giaconi
BT Labs, Adastral Park, Martlesham Heath, Ipswich, Suffolk, UK

Georgios B. Giannakis
University of Minnesota, Minneapolis, MN, USA

Arnold Glass
Rutgers University, Piscataway, NJ, USA

Murat Göl
Middle East Technical University, Ankara, Turkey

Mauricio González
Université Paris-Saclay, Cachan, France

Deniz Gündüz
Imperial College London, London, UK

Vijay Gupta
University of Notre Dame, Notre Dame, IN, USA

List of Contributors

Xing He
Shanghai Jiaotong University, Shanghai, China

Javad Heydari
LG Electronics, Santa Clara, CA, USA

Mahdi Jamei
Invenia Labs, Cambridge, UK

Yuting Ji
Stanford University, Stanford, CA, USA

Sithan Kanna
Imperial College London, London, UK

Vassilis Kekatos
Virginia Tech, Blacksburg, VA, USA

Samson Lasaulce
CNRS, CentraleSupélec, Gif-sur-Yvette, France

Yuzhang Lin
University of Massachusetts-Lowell, Lowell, MA, USA

Narayan B. Mandayam
Rutgers University, North Brunswick, NJ, USA

Danilo P. Mandic
Imperial College London, London, UK

Amin Masoumzadeh
AGL Energy, Melbourne, VIC, Australia

Robert Mulligan
The Kohl Group, Inc., Parsippany, NJ, USA

Toru Namerikawa
Keio University, Yokohama, Japan

Ehsan Nekouei
City University of Hong Kong, Kowloon Tong, Hong Kong

Dorcas Ofori-Boateng
University of Texas at Dallas, Richardson, TX, USA

Samir M. Perlaza
INRIA, Sophia Antipolis, France

H. Vincent Poor
Princeton University, Princeton, NJ, USA

Robert Qiu
Tennessee Technological University, Cookeville, TN, USA

Mohsen Rajabpour
Rutgers University, North Brunswick, NJ, USA

Raksha Ramakrishna
Arizona State University, Tempe, AZ, USA

Anna Scaglione
Arizona State University, Tempe, AZ, USA

Xin Shi
Shanghai Jiaotong University, Shanghai, China

Ke Sun
University of Sheffield, Sheffield, UK

Ali Tajer
Rensselaer Polytechnic Institute, Troy, NY, USA

Lang Tong
Cornell University, Ithaca, NY, USA

Ufuk Topcu
University of Texas at Austin, Austin, TX, USA

Venugopal V. Veeravalli
University of Illinois Urbana-Champaign, Urbana, IL, USA

Meng Wang
Rensselaer Polytechnic Institute, Troy, NY, USA

Yili Xia
Southeast University, Nanjing, China

Baosen Zhang
University of Washington, Seattle, WA, USA

Liang Zhang
University of Minnesota, Minneapolis, MN, USA

Changhong Zhao
National Renewable Energy Laboratory, Golden, CO, USA

Yue Zhao
Stony Brook University, Stony Brook, NY, USA

Hao Zhu
University of Texas at Austin, Austin, TX, USA

Preface

The existing power grids, being recognized as one of the most significant engineering accomplishments, work exceptionally well for the purposes they have been designed to achieve. Enabled by advances in sensing, computation, and communications, power grids are rapidly growing in scale, inter-connectivity, and complexity. Major paradigm shifts in power grids include departing producer-controlled structures and transforming to more decentralized and consumer-interactive ones, being more distributed in electricity generation, enhancing the coupling between the physical and cyber layers, and operating in more variable and stochastic conditions. Driven by these emerging needs, power grids are anticipated to be complex and smart networked platforms in which massive volumes of high-dimensional and complex data are routinely generated and processed for various monitoring, control, inferential, and dispatch purposes.

There has been growing recent interest in developing data analysis tools for designing or evaluating various operations and functions in power systems. Due to the complex nature of power systems, often the existing theories and methodologies cannot be directly borrowed, and there is a critical need for concurrently advancing the theories that are driven by the needs for analyzing various aspects of power systems. This has led to new research domains that lie at the intersection of applied mathematics and engineering. The research in these domains is often conducted by researchers who have expertise in developing theoretical foundations in data analytics, and at the same time are domain experts in power systems analysis. Some of these domains include large-scale and distributed optimization, statistical learning, high-dimensional signal processing, high-dimensional probability theory, and game theory.

Analyzing large-scale and complex data constitutes a pivotal role in the operations of modern power systems. The primary purpose of this book is to prepare a collection of the data analytics tools that prominent researchers have identified as the key tools that have critical roles in various aspects of power systems' reliability, efficiency, and security. Different chapters discuss the state of the art in different and complementary theoretical tools with in-depth discussions on their applications in power systems.

The focus of this book is at the interface of data analytics and modern power systems. While there is extensive literature on power systems, that on *modern* systems is rather limited. Furthermore, there is an explosive amount of literature being developed on data analytics by different scientific communities. Most of these techniques are being applied to various technological domains (including power systems) as they are being developed.

There is a growing need for having a coherent collection of topics that can serve as a main reference for researchers in power systems analysis.

This book brings together experts in both data analytics and power systems domains. The shared purpose in all the contributed chapters is maintaining a balance between introducing and discussing foundational and theoretical tools, as well as their applications to the engineering-level power system problems. These chapters, categorically, fall under the following six broad topics.

Part I: Statistical Learning: The first part introduces cutting-edge learning techniques and their applications to power systems operations. It covers topics on topology learning, system forecasting and market operations, deep learning, and real-time monitoring.

Part II: Data-Driven Anomaly Detection: The second part is focused on statistical inference techniques and their applications to agile and reliable detection of anomalous events in power systems. This part includes topics on change-point detection theory, active (control) sensing, random matrix theory, and graph-theoretic modeling of grid resilience.

Part III: Data Quality, Integrity, and Privacy: The third part covers challenges pertinent to data reliability. This part includes topics on data integrity attacks and counter-measures, information-theoretic analysis of cyber attacks, and data-dimension reduction methodologies for enhancing data quality and privacy.

Part IV: Signal Processing: The fourth part discusses modern signal processing techniques and their applications to power system analysis. Specifically, it covers topics on graph signal processing, Fourier analysis of power system data, and compressive sensing.

Part V: Large-Scale Optimization: The fifth part encompasses topics on large-scale power flow optimization when facing system uncertainties, distributed power flow optimization, load management, storage planning and optimization, and optimization techniques involved in integrating renewable resources and electric vehicles.

Part VI: Game Theory: Finally, the sixth part focuses on the interactions of the different decision makers that are involved in the generation, transport, distribution, and consumption of energy using tools from game theory, mean fields, and prospect theory. This part includes the analysis of energy-storage, large populations of electrical vehicles, consumer behavior, and distributed power scheduling.

This book is designed to be primarily used as a reference by graduate students, academic researchers, and industrial researchers with backgrounds in electrical engineering, power systems engineering, computer science, and applied mathematics. While the primary emphasis is on the theoretical foundations, all the chapters address specific challenges in designing, operating, protecting, and controlling power systems.

Part I

Statistical Learning

1 Learning Power Grid Topologies

Guido Cavraro, Vassilis Kekatos, Liang Zhang, and Georgios B. Giannakis

1.1 Introduction

To perform any meaningful grid optimization task, distribution utilities need to know the topology and line impedances of their grid assets. One may distinguish two major topology learning tasks: (i) In *topology detection*, the utility knows the existing line infrastructure and impedances, and wants to find which lines are energized; and (ii) in *topology identification*, the utility aims at identifying both the connectivity and line impedances; hence, it is a harder task.

Grid topology learning oftentimes relies on second-order statistics from smart meter data [1–3]. Nonetheless, sample statistics converge to their ensemble values only after a large amount of grid data has been collected. Detecting which lines are energized can be posed as a maximum likelihood detection task [4]; sparse linear regression [5, 6]; or as a spanning tree recovery task using the notion of graph cycles [7]. Line impedances are estimated using a total least-squares fit in [8]. In [9], deep neural networks are trained to detect which lines are energized; nevertheless, the data set feeding the classifiers may not be available in distribution grids. A Kron-reduced admittance matrix is recovered using a low rank-plus-sparse decomposition in [10], though the deployment of micro-phasor measurement units presumed there occurs at a slower pace in distribution grids.

This chapter presents a gamut of statistical tools for learning the topology of distribution grids. Toward this end, utilities could rely on smart meter data polled from customers' sites. In addition to passively collecting data, this chapter puts forth an active data acquisition paradigm that we term *grid probing using smart inverters*. The rest of the chapter is organized as follows. Section 1.2 reviews an approximate grid model. Assuming smart meter data, Section 1.3 poses topology detection as a statistical learning task. The method of partial correlations is extended to the nonlinear setting to detect meshed grid topologies in Section 1.4. Section 1.5 puts forth the novel data acquisition paradigm of grid probing through smart inverters. Section 1.6 provides conditions on inverter placement to ensure topology identifiability. Once specific blocks of the inverse Laplacian matrix have been recovered from probing data, a radial grid topology can be identified using the graph algorithms of Section 1.7. Because these algorithms may become impractical under low signal-to-noise ratios, the grid topology can be identified using the convex relaxation approach of Section 1.8. The different algorithmic alternatives for topology identifiability using probing data are tested in Section 1.9. The chapter is concluded in Section 1.10.

Regarding notation, lower- (upper-) case boldface letters denote column vectors (matrices), and calligraphic letters are reserved for sets. Vectors $\mathbf{0}$, $\mathbf{1}$, and $\mathbf{e}_n$ denote, respectively, the all-zero, all-one, and the n-th canonical vectors. The superscript T stands for transposition and $Tr(\mathbf{X})$ is the trace of $\mathbf{X}$. The operator $dg(\mathbf{x})$ returns a diagonal matrix with $\mathbf{x}$ on its main diagonal. For $\mathbf{W} \succ \mathbf{0}$, define the norm $\|\mathbf{X}\|_\mathbf{W}^2 := Tr(\mathbf{X}^\top \mathbf{W}\mathbf{X}) = \|\mathbf{W}^{1/2}\mathbf{X}\|_F^2$, where $\|\mathbf{X}\|_F$ is the Frobenius norm of $\mathbf{X}$. The pseudo-norm $\|\mathbf{X}\|_{0,\text{off}}$ counts the number of nonzero off-diagonal entries of $\mathbf{X}$.

1.2 Grid Modeling

We build upon an approximate distribution grid model briefly reviewed next. A radial single-phase grid having $N + 1$ buses can be represented by a tree graph $\mathcal{G} = (\mathcal{N}_o, \mathcal{L})$, whose nodes $\mathcal{N}_o := \{0, \ldots, N\}$ correspond to buses and its edges $\mathcal{L}$ to lines. The tree is rooted at the substation indexed by $n = 0$. Let $p_n + jq_n$ be the complex power injection and $v_n e^{j\phi_n}$ the complex voltage phasor at bus $n \in \mathcal{N}$. Collect the voltage magnitudes and phases and power injections of buses in $\mathcal{N}$ in the vectors $\mathbf{v}$, $\boldsymbol{\phi}$, $\mathbf{p}$, and $\mathbf{q}$, respectively. The impedance of line $\ell : (m, n) \in \mathcal{L}$ is denoted by $r_\ell + jx_\ell$ or $r_{mn} + jx_{mn}$, depending on the context. The grid connectivity is captured by the branch-bus incidence matrix $\tilde{\mathbf{A}} \in \{0, \pm 1\}^{L \times (N+1)}$; which can be partitioned into the first and the rest of its columns as $\tilde{\mathbf{A}} = [\mathbf{a}_0 \ \mathbf{A}]$. For a radial grid ($L = N$), the *reduced incidence matrix* $\mathbf{A}$ is square and invertible [11].

The complex power injections are nonlinear functions of voltages. Nonetheless, the power flow equations are oftentimes linearized at the flat voltage profile of $v_n = 1$ and $\phi_n = 0$ for all n, to yield the linearized grid model [12–14]

$$p_n = \sum_{(n,m) \in \mathcal{L}} g_{nm}(v_n - v_m) + b_{nm}(\phi_n - \phi_m) \tag{1.1a}$$

$$q_n = \sum_{(n,m) \in \mathcal{L}} b_{nm}(v_n - v_m) - g_{nm}(\phi_n - \phi_m). \tag{1.1b}$$

where $g_{nm} + jb_{nm} := 1/(r_{nm} + jx_{nm})$ is the admittance of line (n, m). The model in (1.1) constitutes a system of $2N$ linear equations of $2N$ unknowns, namely the voltage magnitudes and phases. The system can be inverted to obtain

$$\begin{bmatrix} \mathbf{v} \\ \boldsymbol{\phi} \end{bmatrix} = \begin{bmatrix} \mathbf{R} & \mathbf{X} \\ \mathbf{X} & -\mathbf{R} \end{bmatrix} \begin{bmatrix} \mathbf{p} \\ \mathbf{q} \end{bmatrix} + \begin{bmatrix} \mathbf{1} \\ \mathbf{0} \end{bmatrix}. \tag{1.2}$$

where the $N \times N$ matrices $\mathbf{R}$ and $\mathbf{X}$ are defined as

$$\mathbf{R} := (\mathbf{A}^\top dg^{-1}(\mathbf{r})\mathbf{A})^{-1} \quad \text{and} \quad \mathbf{X} := (\mathbf{A}^\top dg^{-1}(\mathbf{x})\mathbf{A})^{-1}. \tag{1.3}$$

It is worth mentioning that the widely used *linearized distribution flow* (LDF) model originally proposed by [12], involves the squared voltage magnitudes rather than the voltage magnitudes in (1.2). For this reason, the matrices $\mathbf{R}$ and $\mathbf{X}$ appearing in LDF take the values of (1.3) divided by a factor of 2; see also [13]. Note finally that, different from (1.2), the LDF model does not provide a linearized model to approximate voltage angles by complex power injections.

The topology learning schemes presented later in this chapter do not consider voltage phases. This is because, despite the recent developments of phasor measurement unity (PMU) for distribution grids, smart meters report to utilities only voltage magnitudes, henceforth simply referred to as voltages.

Voltage samples are collected at a sampling period of T_s and indexed by $t = 0, \ldots, T$. Upon applying (1.2) over two consecutive voltage samples, the changes in voltages caused by changes in power injections $\tilde{\mathbf{p}}_t := \mathbf{p}_t - \mathbf{p}_{t-1}$ and $\tilde{\mathbf{q}}_t := \mathbf{q}_t - \mathbf{q}_{t-1}$ can be modeled as

$$\tilde{\mathbf{v}}_t := \mathbf{v}_t - \mathbf{v}_{t-1} = \mathbf{R}\tilde{\mathbf{p}}_t + \mathbf{X}\tilde{\mathbf{q}}_t + \mathbf{n}_t \tag{1.4}$$

where $\mathbf{n}_t$ captures measurement noise; the approximation error introduced by the linearized grid model; and unmodeled load dynamics. There are two advantages of using the differential model of (1.4) over (1.2): When dealing with smart meter data, the operator may be observing some or none of the entries of $(\mathbf{p}_t, \mathbf{q}_t)$, so then one can only exploit the second-order moments of $(\tilde{\mathbf{p}}_t, \tilde{\mathbf{q}}_t)$ using blind signal processing techniques (see Section 1.3). With active data collection, power injections can be intentionally perturbed over short intervals, so that $(\tilde{\mathbf{p}}_t, \tilde{\mathbf{q}}_t)$ take zero entries for nonactuated buses and known nonzero entries for actuated buses. Then, topology learning can be cast as a system identification task; see Section 1.5.

1.3 Topology Detection Using Smart Meter Data

Feeders are built with redundancy in line infrastructure. This redundancy improves system reliability against failures or during scheduled maintenance, while grids are reconfigured for loss minimization [12]: at each time, not all existing lines in $\mathcal{L}$ are energized. We will be considering a period of operation where the subset of energized lines $\mathcal{E} \subset \mathcal{L}$ with $|\mathcal{E}| = L_e$ remains constant, yet unknown.

1.3.1 A Maximum Likelihood Approach

To capture the status of each line, introduce variable b_ℓ for line ℓ taking values $b_\ell = 1$ if line is energized ($\ell \in \mathcal{L}_e$); and $b_\ell = 0$ otherwise. Collect the b_ℓ's in the binary L_e-length vector $\mathbf{b}$. To verify grid topologies using smart meter data, parameterize $(\mathbf{R}, \mathbf{X})$ from (1.3) as

$$\mathbf{R}(\mathbf{b}) = \left(\sum_{\ell \in \mathcal{L}} \frac{b_\ell}{r_\ell} \mathbf{a}_\ell \mathbf{a}_\ell^\top\right)^{-1} \quad \text{and} \quad \mathbf{X}(\mathbf{b}) = \left(\sum_{\ell \in \mathcal{L}} \frac{b_\ell}{x_\ell} \mathbf{a}_\ell \mathbf{a}_\ell^\top\right)^{-1} \tag{1.5}$$

where $\mathbf{a}_\ell^\top$ is the ℓ-th row of $\mathbf{A}$. By slightly abusing notation, matrix $\mathbf{A}$ here has been augmented to include both energized and nonenergized lines. Under this representation, verifying the grid topology entails finding $\mathbf{b}$ from grid data.

The utility collects voltage readings, but power injections are described only through their first- and second-order moments. Smart meter data comprise voltages and (re)active powers. Nonetheless, a smart meter monitors a single household, which may not necessarily correspond to a bus. The nodes of low-voltage grids are typically mapped to pole transformers, each one serving 5–10 residential costumers. Hence, power readings from meters may not be useful. Interpreting powers and voltages as system inputs and outputs, respectively, topology detection has to be posed as a blind system identification problem where power injections (system inputs) are characterized only statistically. To this end, we postulate the ensuing statistical model; see also [2, 14, 15].

ASSUMPTION 1.1 *Differential injection data* $(\tilde{\mathbf{p}}_t, \tilde{\mathbf{q}}_t)$ *are zero-mean random vectors with covariance matrices* $\boldsymbol{\Sigma}_p := \mathbb{E}[\tilde{\mathbf{p}}_t \tilde{\mathbf{p}}_t^\top]$; $\boldsymbol{\Sigma}_q := \mathbb{E}[\tilde{\mathbf{q}}_t \tilde{\mathbf{q}}_t^\top]$; *and* $\boldsymbol{\Sigma}_{pq} := \mathbb{E}[\tilde{\mathbf{p}}_t \tilde{\mathbf{q}}_t^\top]$. *Noise* $\mathbf{n}_t$ *is a zero-mean independent identically distributed (iid) Gaussian random vector with covariance matrix* $\sigma_n^2 \mathbf{I}_N$.

Under Assumption 1.1 and from (1.4)–(1.5), the differential voltages $\{\tilde{\mathbf{v}}_t\}$ (termed *voltage data* for brevity) are zero-mean with covariance matrix parameterized as

$$\boldsymbol{\Sigma}(\mathbf{b}) = \mathbf{R}(\mathbf{b})\boldsymbol{\Sigma}_p \mathbf{R}(\mathbf{b}) + \mathbf{X}(\mathbf{b})\boldsymbol{\Sigma}_q \mathbf{X}(\mathbf{b}) + \mathbf{R}(\mathbf{b})\boldsymbol{\Sigma}_{pq}\mathbf{X}(\mathbf{b}) + \mathbf{X}(\mathbf{b})\boldsymbol{\Sigma}_{pq}^\top \mathbf{R}(\mathbf{b}) + \sigma_n^2 \mathbf{I}_N.$$

We postulate that the probability density function (pdf) of each random vector $\tilde{\mathbf{v}}_t$ converges asymptotically in N to a multivariate Gaussian pdf, even if $(\tilde{\mathbf{p}}_t, \tilde{\mathbf{q}}_t)$ are not Gaussian; this stems from contemporary variants of the central limit theorem as detailed in [16]. Statistical tests on actual data validate this assumption [16]. Thus, the pdf of $\tilde{\mathbf{v}}_t$ for each t can be approximated as

$$p(\tilde{\mathbf{v}}_t; \mathbf{b}) = \frac{|\boldsymbol{\Sigma}(\mathbf{b})|^{-1/2}}{(2\pi)^{N/2}} \exp\left(-\frac{1}{2}\tilde{\mathbf{v}}_t^\top \boldsymbol{\Sigma}^{-1}(\mathbf{b})\tilde{\mathbf{v}}_t\right).$$

To fully characterize the collected voltage data $\{\tilde{\mathbf{v}}_t\}_{t=1}^T$, their joint pdf should be provided. It has been demonstrated that voltage data are relatively uncorrelated across time, especially for sampling periods larger than $T_s = 5$ min; see [16, figure 3]. Due to Gaussianity, uncorrelatedness implies independence. Therefore, the joint pdf for the entire voltage data set becomes

$$p(\{\tilde{\mathbf{v}}_t\}_{t=1}^T; \mathbf{b}) = \prod_{t=1}^T p(\tilde{\mathbf{v}}_t; \mathbf{b}) = \frac{|\boldsymbol{\Sigma}(\mathbf{b})|^{-T/2}}{(2\pi)^{NT/2}} \exp\left(-\frac{1}{2}\sum_{t=1}^T \tilde{\mathbf{v}}_t^\top \boldsymbol{\Sigma}^{-1}(\mathbf{b})\tilde{\mathbf{v}}_t\right). \quad (1.6)$$

From the preceding modeling, *topology detection* amounts to finding the subset $\mathcal{E}$ given: the line infrastructure, that is $\{r_\ell, x_\ell, \mathbf{a}_\ell\}_{\ell \in \mathcal{L}}$; the covariance matrices $\boldsymbol{\Sigma}_p, \boldsymbol{\Sigma}_q, \boldsymbol{\Sigma}_{pq}$; and voltage data $\{\tilde{\mathbf{v}}_t\}_{t=1}^T$. Upon observing $\{\tilde{\mathbf{v}}_t\}_{t=1}^T$, function (1.6) becomes the likelihood function of the line indicator vector $\mathbf{b}$. After ignoring constant terms and adopting a maximum likelihood (ML) approach, vector $\mathbf{b}$ can be found as the minimizer of the negative log-likelihood function

$$\hat{\mathbf{b}} := \arg\min_{\mathbf{b}} \left\{ f(\mathbf{b}) : \mathbf{b} \in \{0,1\}^L, \mathbf{1}^\top \mathbf{b} = N \right\} \quad (1.7)$$

where $f(\mathbf{b}) := \log |\boldsymbol{\Sigma}(\mathbf{b})| + Tr(\boldsymbol{\Sigma}^{-1}(\mathbf{b})\hat{\boldsymbol{\Sigma}})$; the operator $|\cdot|$ is the matrix determinant; and $\hat{\boldsymbol{\Sigma}} := \frac{1}{T}\sum_{t=1}^{T} \tilde{\mathbf{v}}_t \tilde{\mathbf{v}}_t^\top$ is the sample covariance of voltage data. The second summand in $f(\mathbf{b})$ aggregates the information from data, while the first one acts as a regularizer guarding $\boldsymbol{\Sigma}(\mathbf{b})$ within the positive definite matrix cone and away from singularity [17]. The constraint $\mathbf{1}^\top \mathbf{b} = N$ ensures a tree structure.

Solving (1.7) is nontrivial due to the binary variables and the nonlinear objective. A lower bound can be obtained by solving the box relaxation of (1.7)

$$\check{\mathbf{b}} := \arg\min_{\mathbf{b}} \left\{ f(\mathbf{b}) : \mathbf{b} \in [0,1]^L, \mathbf{1}^\top \mathbf{b} = N \right\}. \quad (1.8)$$

Because $f(\mathbf{b})$ is nonconvex in general, one may only be able to find a local minimum of (1.8). Consistent with the properties of the MLE, the true indicator vector minimizes (1.7) and (1.8), when the number of data T grows to infinity.

PROPOSITION 1.1 ([16]) *Let $\mathbf{b}_o$ be the true line indicator vector. If the sample covariance $\hat{\boldsymbol{\Sigma}}$ has converged to the ensemble covariance $\boldsymbol{\Sigma}(\mathbf{b}_o)$, then $\mathbf{b}_o$ is a stationary point of $f(\mathbf{b})$ and global minimizer for (1.7) and (1.8).*

To obtain a feasible $\mathbf{b}$, one may apply a heuristic on $\check{\mathbf{b}}$, such as selecting the lines corresponding to the L largest entries of $\check{\mathbf{b}}$; or finding the minimum spanning tree on a graph having $\check{\mathbf{b}}$ as edge weights. Obviously, $f^\star \leq f(\hat{\mathbf{b}})$ and $f(\check{\mathbf{b}}) \leq f(\hat{\mathbf{b}})$. Even though $\hat{\mathbf{b}}$ yields reasonable detection performance (see Section 1.9), vector $\check{\mathbf{b}}$ may not be a global minimizer of (1.8).

The issues with the nonconvexity of $f(b)$ can be alleviated by resorting to two simplifying assumptions: a1) the resistance-to-reactance ratios are identical or $r_\ell / x_\ell = \alpha$ for all lines $\ell \in \mathcal{L}$; and a2) the noise term n_t is negligible. In this case, the voltage data model of (1.4) simplifies as $\tilde{\mathbf{v}}_t = \alpha \mathbf{X}\tilde{\mathbf{p}}_t + \mathbf{X}\tilde{\mathbf{q}}_t$, and the ensemble covariance of voltage data becomes

$$\boldsymbol{\Sigma}(\mathbf{b}) := \mathbf{X}(\mathbf{b})(\alpha^2 \boldsymbol{\Sigma}_p + \boldsymbol{\Sigma}_q + \alpha(\boldsymbol{\Sigma}_{pq} + \boldsymbol{\Sigma}_{pq}^\top))\mathbf{X}(\mathbf{b}).$$

Under a1)–a2), the original negative log-likelihood is surrogated by

$$\tilde{f}(\mathbf{b}) := -2\log |\mathbf{X}^{-1}(\mathbf{b})| + Tr\left(\mathbf{X}^{-1}(\mathbf{b})\boldsymbol{\Sigma}_\alpha^{-1}\mathbf{X}^{-1}(\mathbf{b})\hat{\boldsymbol{\Sigma}}^{-1}\right).$$

Interestingly, function $\tilde{f}(\mathbf{b})$ is convex and so (1.8) becomes a convex program [16].

1.3.2 A Maximum a posteriori Probability Approach

In meshed grids, the utility may not know the exact number of energized lines L. However, prior information on individual lines being energized could be known through the generalized state estimator or current readings on transformers. To cope with such scenarios, a maximum a posteriori probability (MAP) approach can be adopted. The indicator b_ℓ for line ℓ is modeled as a Bernoulli random variable with given mean $\mathbb{E}[b_\ell] = \pi_\ell$. The prior probability mass function (pmf) for b_ℓ can be expressed as $\Pr(b_\ell) = \pi_\ell^{b_\ell}(1-\pi_\ell)^{1-b_\ell}$. To derive a tractable statistical model, lines are assumed to be energized independently. Then, the joint pmf for **b** is $\Pr(\mathbf{b}) = \prod_{\ell \in \mathcal{L}_e} \Pr(b_\ell)$ up to a normalization constant and then

$$-\log \Pr(\mathbf{b}) = \sum_{\ell \in \mathcal{L}_e} b_\ell \beta_\ell - \log(1 - \pi_\ell) \quad (1.9)$$

where $\beta_\ell := \log\left(\frac{1-\pi_\ell}{\pi_\ell}\right)$ for all $\ell \in \mathcal{L}_e$. The MAP estimate for **b** is defined as the maximizer of the posterior $p(\mathbf{b}|\{\tilde{\mathbf{v}}_t\}_{t=1}^T)$. From Bayes's rule, the latter is proportional to the product $p(\{\tilde{\mathbf{v}}_t\}_{t=1}^T; \mathbf{b}) \Pr(\mathbf{b})$, so that the MAP estimate can be found by minimizing

$$-\log p(\mathbf{b}|\{\tilde{\mathbf{v}}_t\}_{t=1}^T) = -\log p(\{\tilde{\mathbf{v}}_t\}_{t=1}^T; \mathbf{b}) - \log \Pr(\mathbf{b}).$$

Collecting β_ℓ's in $\boldsymbol{\beta}$ and ignoring constants, the latter leads to the problem

$$\mathbf{b}_{\text{MAP}} := \arg \min_{\mathbf{b} \in \{0,1\}^L} \frac{T}{2} f(\mathbf{b}) + \boldsymbol{\beta}^\top \mathbf{b} \quad (1.10)$$

Contrasted to (1.7), problem (1.10) leverages prior information: if line ℓ is likely to be energized, then $\pi_\ell > 1/2$ and $\beta_\ell < 0$. If line ℓ is known to be energized, then $\pi_\ell = 1$ and $\beta_\ell = -\infty$, thus forcing the ℓ-th entry of $\mathbf{b}_{\text{MAP}}$ to one. No prior information on line ℓ means $\pi_\ell = 1/2$ and $\beta_\ell = 0$. If $\pi_\ell = \pi_o$ for $\ell \in \mathcal{L}_e$, then $\boldsymbol{\beta}^\top \mathbf{b} = \pi_o \mathbf{1}^\top \mathbf{b}$, and (1.10) takes a Lagrangian form of (1.7). Similar to (1.7), a box relaxation of (1.10) can be pursued.

1.3.3 Numerical Tests on Topology Detection Using Smart Meter Data

The schemes were validated using the modified version of the IEEE 37-bus system depicted in Figure 1.1 (a), in which additional lines were added. Data collected from the Pecan Street project were used as power profiles [18]. Voltages were obtained through a power flow solver and the measurement noise was modeled as zero-mean Gaussian with a 3-sigma deviation matching 0.5% of the actual value [16]. The ML task (1.8) was solved by a projected gradient (PGD) scheme over 50 Monte Carlo runs and a sampling period of $T_s = 5$ min. The actual topology was randomly chosen. The solutions $\check{\mathbf{b}}$ were projected onto the feasible set using randomization by treating $\check{\mathbf{b}}$ as the mean of a multivariate Bernoulli distribution [16]. The MAP approach (1.10) was tested using 50 Monte Carlo runs and solved by a PGD scheme too. The prior probabilities π_ℓ were set to 0.5 for switches and 0.9 for lines. The entries of the solution were truncated to binary upon thresholding. The line error probabilities for the two problems are depicted in Figure 1.1 (b). It is worth emphasizing that reliable line detection can be obtained even for $T < N$, when matrix $\hat{\boldsymbol{\Sigma}}$ is *singular*. Figure 1.1 (c) tests the effect of T_s on the line status error probability achieved by the nonconvex ML task of (1.8). The performance improves as T_s and the total collection time increase.

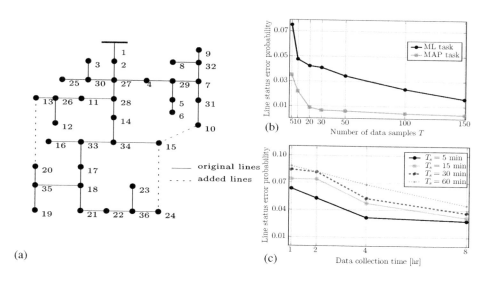

Figure 1.1 (a) Line structure for the IEEE 37-bus feeder. (b) Line error probability for the ML and the MAP task. (c) Line error probability for the ML task (1.10). (c) Effect of T_s on the line status error probability for the ML task.

1.4 Topology Detection Using Partial Correlations

Collect the data associated with bus n for $t = 1, \ldots, T$, into vector $\mathbf{x}_n := [x_n[1] \ldots x_n[T]]^\top$. For power network topology inference, vector $\mathbf{x}_n$ may collect voltage readings across time [2]. Given $\{\mathbf{x}_n\}_{n \in \mathcal{N}}$, we would like to recover the edge set $\mathcal{L}$. To this end, we adopt the method of Partial Correlations (PC), which is effective in capturing unmediated linear influence between nodes [2, 19, 20]. Consider two buses $\{n, m\}$ and define the error vector $\boldsymbol{\epsilon}_{n|m} = \mathbf{x}_n - \hat{\mathbf{x}}_{n|m}$, where $\hat{\mathbf{x}}_{n|m}$ is the estimate of $\mathbf{x}_n$ based on $\{\mathbf{x}_i\}$ for $i \notin \{n, m\}$. The *empirical PC coefficient* between $\mathbf{x}_n$ and $\mathbf{x}_m$ is given by [20]

$$\rho_{nm} := \frac{(\boldsymbol{\epsilon}_{n|m} - \bar{\boldsymbol{\epsilon}}_{n|m})^\top (\boldsymbol{\epsilon}_{m|n} - \bar{\boldsymbol{\epsilon}}_{m|n})}{\|\boldsymbol{\epsilon}_{n|m} - \bar{\boldsymbol{\epsilon}}_{n|m}\|_2 \cdot \|\boldsymbol{\epsilon}_{m|n} - \bar{\boldsymbol{\epsilon}}_{m|n}\|_2} \tag{1.11}$$

where $\bar{\boldsymbol{\epsilon}}_{n|m} := \frac{1}{T} \boldsymbol{\epsilon}_{n|m}^\top \mathbf{1}_T \mathbf{1}_T$. Having computed the PC coefficients ρ_{nm}'s for all pairs $(n, m) \in \mathcal{N} \times \mathcal{N}$, determining whether node m is connected with n entails a hypothesis test: an edge between m and n is declared present if $|\rho_{nm}| \geq \tau > 0$, where τ trades off the relative true positive for the false positive decisions.

1.4.1 Nonlinear Partial Correlations

The PC method assumes that $\hat{\mathbf{x}}_{n|m}$ is a linear function of $\{\mathbf{x}_i\}_{i \notin \{n,m\}}$. It is worth mentioning here that partial correlation coefficients are preferred over ordinary correlation coefficients because they can reveal direct (nonmediated) links rather than mediated ones, e.g., [20]. Nonetheless, the dependence of $\mathbf{x}_n$ on $\{\mathbf{x}_i\}_{i \notin \{n,m\}}$ may be nonlinear.

To accommodate that, an ℓ_2-norm-based multikernel partial correlation–based approach has been studied in [19], which works as follows. Let vector $\chi_{nm}[t]$ collect the data associated with time t and all buses except for $\{n,m\}$. By replacing $\chi_{nm}[t]$ with its lifted image using a feature map $\phi_{nm}[t]$, a nonlinear data generation model can be postulated as

$$x_n[t] = \langle \phi_{nm}[t], \beta_{nm} \rangle + e_n[t] \tag{1.12}$$

where β_{nm} is a parameter vector to learn and $e_n[t]$ captures modeling inaccuracies. Along the lines of ridge regression, β_{nm} can be estimated as

$$\hat{\beta}_{nm} := \arg\min_{\beta} \quad \frac{C}{N}\|\xi_n\|_2^2 + \frac{1}{2}\|\beta\|_2^2 \tag{1.13a}$$

$$\text{s.to} \quad \xi_n = \mathbf{x}_n - \mathbf{\Phi}_{nm}^\top \beta \tag{1.13b}$$

where matrix $\mathbf{\Phi}_{nm}$ has $\phi_{nm}[t]$ for $t = 1, \ldots, T$ as its columns, and $C \geq 0$ is a given constant. Because $\phi_{nm}[t]$ has high (potentially infinite) dimension, the dual of (1.13), which has only T variables, will be used henceforth. Specifically, the dual of (1.13) can be succinctly written as [21]

$$\max_{\alpha} -\mu \alpha^\top \alpha + 2\alpha^\top \mathbf{x}_n - \alpha^\top \mathbf{K}_{nm} \alpha \tag{1.14}$$

where $\alpha \in \mathbb{R}^T$ denotes the Lagrange multiplier associated with (1.13b); constant $\mu := N/(2C)$; and $\mathbf{K}_{nm} := \mathbf{\Phi}_{nm}^\top \mathbf{\Phi}_{nm}$. The maximizer of (1.14) can be found in closed form as $\hat{\alpha}_n = (\mathbf{K}_{nm} + \mu \mathbf{I}_N)^{-1} \mathbf{x}_n$, and so $\hat{\mathbf{x}}_{n|m}$ is obtained as

$$\hat{\mathbf{x}}_{n|m} = \mathbf{\Phi}_{nm}^\top \hat{\beta}_{nm} = \mathbf{\Phi}_{nm}^\top \mathbf{\Phi}_{nm} \hat{\alpha}_n = \mathbf{K}_{nm}(\mathbf{K}_{nm} + \mu \mathbf{I}_N)^{-1} \mathbf{x}_n. \tag{1.15}$$

The latter entails computing inner products between high-dimensional feature vectors of the form $\langle \phi_{nm}[t], \phi_{nm}[t'] \rangle$. Fortunately, such a costly computation can be reduced by invoking the so-called kernel trick [22], which allows computing the wanted inner products in (1.15) by evaluating a kernel function $\kappa\left(\phi_{nm}[t], \phi_{nm}[t']\right)$ for all pairs (t, t').

The accuracy of the estimates in (1.15) depends on the selected kernel function $\kappa(\cdot, \cdot)$ [22]. To choose a suitable kernel, multikernel ridge regression (MKRR) is invoked here [23], which seeks $\kappa(\cdot, \cdot)$ as a conic combination of user-defined kernel functions; that is, $\kappa(\cdot, \cdot) := \sum_{i=1}^{M} \theta_i \kappa_i(\cdot, \cdot)$. The coefficients $\{\theta_i \geq 0\}_{i=1}^{M}$ can be deciphered from data by solving [23]

$$\theta^* := \arg\min_{\theta \in \Theta_p} \max_{\alpha \in \mathbb{R}^N} -\mu \alpha^\top \alpha + 2\alpha^\top \mathbf{x}_n - \sum_{i=1}^{M} \theta_i \alpha^\top \mathbf{K}_{nm}^i \alpha \tag{1.16}$$

where the (t, t')-th entry of the kernel matrix is $\kappa_i\left(\chi_{nm}[t], \chi_{nm}[t']\right)$, and the constraint set Θ_p is defined as $\Theta_p := \{\theta \in \mathbb{R}^M \mid \theta \geq \mathbf{0}, \|\theta\|_p \leq \Lambda\}$ with $p \geq 1$ and $\Lambda > 0$ is a preselected constant. Define the optimal value of the inner optimization (maximization over α) as $F(\theta)$. Then, the problem in (1.16) can be compactly expressed as

$$\theta^* := \arg\min_{\theta \in \Theta_p} F(\theta) \tag{1.17}$$

where $F(\boldsymbol{\theta}) := \mathbf{x}_n^\top \left(\mu \mathbf{I} + \sum_{m=1}^M \theta_m \mathbf{K}_{nm}^i\right)^{-1} \mathbf{x}_n$. Upon obtaining $\boldsymbol{\theta}^*$ and $\mathbf{K}_{nm} = \sum_{m=1}^M \theta_m^* \mathbf{K}_{nm}^i$, we get from (1.15)

$$\hat{\mathbf{x}}_{nm} = \mathbf{K}_{nm} \left(\mu \mathbf{I} + \mathbf{K}_{nm}\right)^{-1} \mathbf{x}_n. \tag{1.18}$$

This formulation generalizes the nonlinear estimator in [19] beyond $p = 2$. This generalization is well motivated because $\|\boldsymbol{\epsilon}_{n|m}\|_2$ can be reduced for $p \neq 2$; see [24].

Returning to the grid topology inference task at hand, let vector $\mathbf{x}_n$ collect the voltage angles $\phi_n[t]$ for $t = 1, \ldots, T$. Given data $\{\mathbf{x}_n\}_{n \in \mathcal{N}}$, the goal is to infer the connectivity between buses. For each bus pair (n,m), we follow the ensuing procedure. Having selected candidate kernel functions $\{\kappa_i\}_{i=1}^M$, one first forms the kernel matrices $\{\mathbf{K}_{nm}^i\}_{i=1}^M$, and learns the best kernel combination $\mathbf{K}_{nm} = \sum_{i=1}^M \theta_i^* \mathbf{K}_{nm}^i$ by solving (1.17). The next step is to obtain the nonlinear estimators $\hat{\mathbf{x}}_{n|m} = \mathbf{K}_{nm}(\mu \mathbf{I} + \mathbf{K}_{nm})^{-1}\mathbf{x}_n$, and likewise for $\hat{\mathbf{x}}_{m|n}$. The PC coefficient ρ_{nm} is found from (1.11), and an edge between buses n and m is claimed to be present if $|\rho_{nm}| > \tau$. This pairwise hypotheses test is repeated for all pairs (n,m). We next present an algorithm for solving (1.17).

1.4.2 A Frank–Wolfe–Based Solver

Though the ℓ_p-norm-based MKRR (1.17) can lead to improved estimation accuracy, solving it may not be easy unless $p = 1$ or 2; see [24, 25]. For this reason, we put forward an efficient solver by leveraging the Frank–Wolfe or conditional gradient method [26]. The latter algorithm targets the convex problem

$$\mathbf{y}^* \in \arg\min_{\mathbf{y} \in \mathcal{Y}} F(\mathbf{y}) \tag{1.19}$$

where F is differentiable and $\mathcal{Y}$ is compact. The Frank–Wolfe solver starts with an arbitrary point $\mathbf{y}^0$, and iterates between the updates [27]

$$\mathbf{s}^k \in \arg\min_{\mathbf{s} \in \mathcal{Y}} \mathbf{s}^\top \nabla F(\mathbf{y}^k) \tag{1.20a}$$

$$\mathbf{y}^{k+1} := \mathbf{y}^k + \eta_k(\mathbf{s}^k - \mathbf{y}^k) \tag{1.20b}$$

where $\eta_k = 2/(k+2)$. The iterates $\{\mathbf{y}^k\}$ remain feasible for all k because $\eta^0 = 1$, $\mathbf{y}^1 = \mathbf{s}^0 \in \mathcal{Y}$, and $\mathbf{s}^k \in \mathcal{Y}$; see [26].

Because the cost in (1.17) is convex and differentiable, and set Θ_p is convex and compact, problem (1.17) complies with the form in (1.19). The negative gradient of $F(\boldsymbol{\theta})$ can be computed as

$$-\nabla F(\boldsymbol{\theta}) = [\hat{\boldsymbol{\alpha}}^\top \mathbf{K}_{nm}^1 \hat{\boldsymbol{\alpha}} \ \cdots \ \hat{\boldsymbol{\alpha}}^\top \mathbf{K}_{nm}^M \hat{\boldsymbol{\alpha}}]^\top \tag{1.21}$$

where $\hat{\boldsymbol{\alpha}}$ depends on $\boldsymbol{\theta}$ through

$$\hat{\boldsymbol{\alpha}} = \left(\mu \mathbf{I} + \sum_{i=1}^M \theta_i \mathbf{K}_{mn}^i\right)^{-1} \mathbf{x}_n. \tag{1.22}$$

Because $\mathbf{K}^i_{nm} \succeq \mathbf{0}$ for all i, it follows that $\mathbf{g}^k := -\nabla F(\boldsymbol{\theta}^k) \geq \mathbf{0}$ for all $\boldsymbol{\theta}^k \in \Theta_p$. Applying (1.20a) to (1.17) yields

$$\mathbf{s}^k \in \arg\max_{\mathbf{s} \in \Theta_p} \mathbf{s}^\top \mathbf{g}^k. \tag{1.23}$$

By Hölder's inequality and for all $\mathbf{s} \in \Theta_p$, it holds that $\mathbf{s}^\top \mathbf{g}^k \leq \|\mathbf{s}\|_p \|\mathbf{g}^k\|_q \leq \Lambda \|\mathbf{g}^k\|_q$ for any (p,q) with $1/p + 1/q = 1$. Because $\mathbf{g}^k = -\nabla F(\boldsymbol{\theta}^k) \geq \mathbf{0}$, it can be deduced that the solution to (1.23) satisfies the previous inequalities with equalities, and so

$$s^k_i = \Lambda \frac{(g^k_i)^{q-1}}{\|\mathbf{g}^k\|_q^{q-1}}, \quad i = 1, \ldots, M. \tag{1.24}$$

The Frank–Wolfe solver of (1.17) is summarized next: the algorithm is initialized at $\boldsymbol{\theta}^0 = \mathbf{0}$. At iteration k, $\hat{\boldsymbol{\alpha}}^k$ is found from (1.22); the negative gradient $\mathbf{g}^k$ is updated from (1.21); the direction $\mathbf{s}^k$ is found from (1.24); and the sought vector is updated as $\boldsymbol{\theta}^{k+1} = \boldsymbol{\theta}^k + \eta_k(\mathbf{s}^k - \boldsymbol{\theta}^k)$. The algorithm converges to $\boldsymbol{\theta}^*$ at sublinear rate, that is $F(\boldsymbol{\theta}^k) - F(\boldsymbol{\theta}^*) \leq \mathcal{O}(1/k)$; see [26].

1.4.3 Numerical Tests on Topology Detection Using Partial Correlations

The performance of the proposed method was evaluated based on voltage angles from the IEEE 14-bus benchmark system, using real load data from [28]. Specifically, the first 10-day loads of zones 114 were normalized to match the scale of active demands in the benchmark, and then corrupted by noise. Voltage angle measurements $\phi_n[t]$ across $T = 240$ times were found by solving the AC power flow equations. We employed a total of $M = 20$ kernels to form the dictionary, which consists of 10 polynomial kernels of orders varying by 1 from 1 to 10, as well as 10 Gaussian kernels with variances distributed uniformly from 0.5 to 5. The regularization coefficients in (1.16) were set as $\mu = 1$ and $\Lambda = 3$.

The first test assesses the convergence and computational performance of the Frank–Wolfe solver. To serve as a benchmark, problem (1.17) is first reformulated as a semidefinite program (SDP) and solved by SeDuMi [29]; see [25]. The left panel of Figure 1.2 depicts the evolution of the relative error $(F(\boldsymbol{\theta}^k) - F(\boldsymbol{\theta}^*))/F(\boldsymbol{\theta}^*)$ of (1.17) with $p = 1.5$ and $p = 2$, where $\{\mathbf{x}_n\}_{n=1}^2$ and $\{\mathbf{x}_n\}_{n=5}^{14}$ were used to predict $\mathbf{x}_3$. The solver converged roughly at rate $\mathcal{O}(1/k)$.

We next tested the topology recovery performance of the proposed scheme for $p = 1.5$ and $p = 2$, against the linear PC- and concentration matrix–based methods [30]. The right panel of Figure 1.2 depicts the obtained empirical receiver operating characteristics (ROC). For our scheme and its linear PC counterpart, the ROC curves were obtained using $|\rho_{nm}|$'s as test statistics. For the concentration matrix–based method, entries of the negative concentration matrix were used as test statistics. The area under the curve for our scheme with $p = 1.5$ and $p = 2$, the linear PC-based, and the concentration matrix–based methods were 0.755, 0.743, 0.646, and 0.604, accordingly. The results demonstrate the improved recovery performance of the novel scheme, and the advantage of selecting $p \neq 2$.

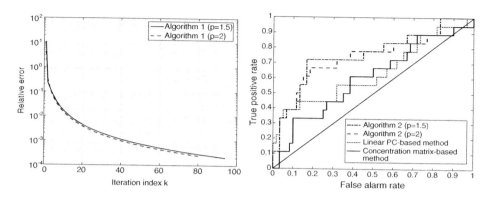

Figure 1.2 Topology recovery for the IEEE 14-bus system. *Left*: convergence speed for the devised Frank–Wolfe iterations; *right*: ROC curves for line detection.

1.5 Grid Probing for Topology Learning

Existing topology processing schemes rely on passively collected smart meter data. We next put forth an active data acquisition approach to topology learning. The idea is to leverage the actuation and sensing capabilities of smart inverters. An inverter can be commanded to shed solar generation, (dis)-charge a battery, or change its power factor within milliseconds. The distribution feeder as an electric circuit responds within a second and reaches a different steady-state voltage profile. This enables a new data collection paradigm, where the operator purposefully probes a grid by changing inverter injections and measuring the circuit response to identify the grid topology. Rather than processing smart meter data on a 15-minute basis, probing actively senses voltages on a per-second basis.

The buses hosting controllable inverters are collected in $\mathcal{P} \subseteq \mathcal{N}$ with $P := |\mathcal{P}|$. Consider the probing action at time t. Each bus $m \in \mathcal{P}$ perturbs its active injection by $\delta_m(t)$. Vector $\boldsymbol{\delta}(t)$ collects all inverter perturbations at time t. The incurred perturbation in voltage magnitudes is expressed from (1.4) as

$$\tilde{\mathbf{v}}(t) = \mathbf{R}\mathbf{I}_\mathcal{P}\boldsymbol{\delta}(t) + \boldsymbol{\epsilon}(t) \tag{1.25}$$

where the $N \times P$ matrix $\mathbf{I}_\mathcal{P}$ collects the canonical vectors associated with $\mathcal{P}$ and basically selects the rows of $\mathbf{R}$ related to inverters. Vector $\boldsymbol{\epsilon}(t)$ captures measurement noise, modeling errors, and voltage deviations attributed to possible load variations during probing.

The grid is perturbed over T probing periods. Stacking the probing actions $\{\boldsymbol{\delta}(t)\}_{t=1}^T$, the measured voltage deviations $\{\tilde{\mathbf{v}}(t)\}_{t=1}^T$, and the error terms $\{\boldsymbol{\epsilon}(t)\}_{t=1}^T$ as columns of matrices $\boldsymbol{\Delta}$, $\tilde{\mathbf{V}}$, and $\mathbf{E}$ accordingly, yields

$$\tilde{\mathbf{V}} = \mathbf{R}\mathbf{I}_\mathcal{P}\boldsymbol{\Delta} + \mathbf{E}. \tag{1.26}$$

Define the *weighted Laplacian* matrix $\tilde{\Theta} = \tilde{\mathbf{A}}^\top dg^{-1}(\mathbf{r})\tilde{\mathbf{A}}$ and the *reduced weighted Laplacian* Θ as

$$\Theta := \mathbf{R}^{-1} = (\mathbf{A}^\top dg(\mathbf{r})\mathbf{A})^{-1}. \tag{1.27}$$

The grid topology is equivalently captured by the reduced weighted Laplacian and the nonreduced Laplacian matrix $\tilde{\Theta} = \tilde{\mathbf{A}}^\top dg^{-1}(\mathbf{r})\tilde{\mathbf{A}}$. To see this, note that the two matrices are related through the linear mapping $\boldsymbol{\Phi} : \mathbb{S}^N \to \mathbb{S}^{N+1}$ as

$$\tilde{\Theta} = \boldsymbol{\Phi}(\Theta) := \begin{bmatrix} \mathbf{1}^\top \Theta \mathbf{1} & -\mathbf{1}^\top \Theta \\ -\Theta \mathbf{1} & \Theta \end{bmatrix}. \tag{1.28}$$

Topology identification can be now posed as the system identification task of finding Θ given $(\tilde{\mathbf{V}}, \boldsymbol{\Delta})$ from (1.26). This is a major advantage over the blind schemes of Section 1.3 and [1, 2]. Albeit we have so far considered perturbing active power injections, the developments carry over to reactive ones too.

1.6 Identifiability Analysis of Grid Probing

This section studies whether the actual topology can be uniquely recovered by probing the buses in $\mathcal{P}$ [31, 32]. We first review some graph theory background. Let $\mathcal{G} = (\mathcal{N}, \mathcal{L})$ be an undirected tree graph, where $\mathcal{N}$ is the set of nodes and $\mathcal{L}$ the set of edges $\mathcal{L} := \{(m,n) : m,n \in \mathcal{N}\}$. A tree is termed rooted if one of its nodes, henceforth indexed by 0, is designated as the root.

In a tree graph, a *path* is the unique sequence of edges connecting two nodes. The set of nodes adjacent to the edges forming the path between nodes n and m will be denoted by $\mathcal{P}_{n,m}$. The nodes belonging to $\mathcal{A}_m := \mathcal{P}_{0,m}$ are termed the *ancestors* of node m. If $n \in \mathcal{A}_m$, then m is a *descendant* of node n. Reversely, if $n \in \mathcal{A}_m$, m is a *descendant* of node n. The descendants of node m comprise the set $\mathcal{D}_m$. By convention, $m \in \mathcal{A}_m$ and $m \in \mathcal{D}_m$. If $n \in \mathcal{A}_m$ and $(m,n) \in \mathcal{E}$, node n is the *parent* of m. A node without descendants is called a *leaf* or *terminal* node. Leaf nodes are collected in the set $\mathcal{F}$, while nonleaf nodes will be termed *internal* nodes; see Figure 1.3. The *depth* d_m of node m is defined as the number of its ancestors, i.e., $d_m := |\mathcal{A}_m|$. The depth of the entire tree is $d_\mathcal{G} := \max_{m \in \mathcal{N}} d_m$. If $n \in \mathcal{A}_m$ and $d_n = k$, node n is the unique k-depth ancestor of node m and will be denoted by α_m^k for $k = 0, \ldots, d_m$. Let also $\mathcal{T}_m^k$ denote the subset of the nodes belonging to the subtree of $\mathcal{G}$ rooted at the k-depth node m and containing all the descendants of m. Finally, the *k-th level set*[1] of node m is defined as [31, 32]

$$\mathcal{N}_m^k := \begin{cases} \mathcal{D}_{\alpha_m^k} \setminus \mathcal{D}_{\alpha_m^{k+1}} & , k = 0, \ldots, d_m - 1 \\ \mathcal{D}_m & , k = d_m. \end{cases} \tag{1.29}$$

---

[1] The notion of level sets has been used in [33] to derive a meter placement strategy for detecting which switches are energized.

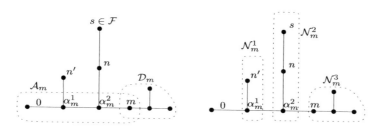

Figure 1.3 Node s is a leaf node, while m is an internal node of the left graph. The ancestor ($\mathcal{A}_m$) and descendant ($\mathcal{D}_m$) sets for node m are also shown. The level set $\mathcal{N}_m^1$ consists of α_m^1 and the subtrees rooted at α_m^1 excluding the subtree containing m.

The concept of the level set is illustrated in Figure 1.3. In essence, the level set $\mathcal{N}_m^k$ consists of node α_m^k and all the subtrees rooted at α_m^k excluding the one containing node m.

The entries of the reduced Laplacian inverses ($\mathbf{R}, \mathbf{X}$) can also be defined exploiting graphical quantities. If $R_{m,n}$ is the (m,n)-th entry of $\mathbf{R}$, then [14]

$$R_{mn} = \sum_{\substack{\ell=(c,d)\in\mathcal{L}\\ c,d\in\mathcal{A}_m\cap\mathcal{A}_n}} r_\ell. \tag{1.30}$$

The entry $R_{m,n}$ can be equivalently interpreted as the voltage drop between the substation and bus m when a unitary active power is injected as bus n and the remaining buses are unloaded. Through this interpretation, the entries of $\mathbf{R}$ relate to the levels sets in $\mathcal{G}$ as follows.

LEMMA 1.1 ([31])   *Let m, n, s be nodes in a radial grid.*

(i) *if $m \in \mathcal{F}$, then $R_{mm} > R_{nm}$ for all $n \neq m$;*
(ii) *$n, s \in \mathcal{N}_m^k$ for a k if and only if $R_{nm} = R_{sm}$; and*
(iii) *if $n' \in \mathcal{N}_m^{k-1}$, $s \in \mathcal{N}_m^k$, then $R_{sm} = R_{mn'} + r_{\alpha_m^{k-1}, \alpha_m^k}$.*

We next provide conditions on inverter placement under which the topology of a radial grid can be identified. We consider the case in which voltages are collected at all buses (complete data), or only at probed buses (partial data).

1.6.1   Topology Identifiability with Complete Voltage Data

As customary in identifiability analysis, probing data are considered noiseless ($\mathbf{E} = \mathbf{0}$), so that (1.26) yields

$$\tilde{\mathbf{V}} = \mathbf{R}_\mathcal{P} \Delta \tag{1.31}$$

where $\mathbf{R}_\mathcal{P} := \mathbf{R}\mathbf{I}_\mathcal{P}$. It is convenient to design $\Delta \in \mathbb{R}^{C\times T}$ to be full row-rank. For example, one can set $T = C$ and $\Delta_1 := dg(\{\delta_m\})$ with $\delta_m \neq 0$ for all $m \in \mathcal{C}$. A diagonal Δ requires simple synchronization because each $m \in \mathcal{C}$ probes the grid at different time slots. Another practical choice is $\Delta_2 := dg(\{\delta_m\})\otimes[+1 \ -1]$, where $\otimes$ is the Kronecker product. In this case, $T = 2C$ and each inverter induces two probing actions: it first drops its generation to zero, yielding a perturbation of δ_m, and then resumes generation at its nominal value, thus incurring a perturbation of $-\delta_m$. The process is repeated over all C inverters.

The task of topology identification can be split into three stages:

s1) Finding $\mathbf{R}_\mathcal{P}$ from (1.31). If $\mathbf{\Delta}$ is full row-rank, then matrix $\mathbf{R}_\mathcal{P}$ can be uniquely recovered as $\tilde{\mathbf{V}}\mathbf{\Delta}^+$, where $\mathbf{\Delta}^+$ is the pseudo-inverse of $\mathbf{\Delta}$. Under this setup, probing for $T = P$ times suffices to find $\mathbf{R}_\mathcal{P}$.

s2) Recovering the level sets for all buses in $\mathcal{P}$. Let $\mathbf{r}_m$ be the m-th column of $\mathbf{R}$. Using Lemma 1.1, we can recover the level sets for each bus $m \in \mathcal{P}$ by grouping together indices associated with the same entries of $\mathbf{r}_m$; see Lemma 1.1(ii). The depth of each level set can be found by ranking the unique values of $\mathbf{r}_m$ in increasing order; see Lemma 1.1(iii).

s3) Recover the grid topology and resistances given the level sets for all $m \in \mathcal{P}$.

The true Θ may not be identifiable from (1.31). A sufficient condition guaranteeing identifiability is provided next.

THEOREM 1.1 *Given probing data $(\tilde{\mathbf{V}}, \mathbf{\Delta})$ where rank $(\mathbf{\Delta}) = P$, the resistive network topology is identifiable if the grid is probed at all leaf nodes, that is $\mathcal{F} \subseteq \mathcal{P}$, and voltage data are collected at all nodes.*

Theorem 1.1 establishes that the topology is identifiable if the grid is probed at all leaf nodes and voltages are collected at all nodes. Under this setup, one needs at least $T = |\mathcal{F}|$, which can be significantly smaller than N. When not all leaf nodes are probed, a portion of the network may still be identifiable. Let $\mathcal{P}_F$ be the set collecting the probing buses who have only regular nodes as descendants, i.e., $n \in \mathcal{P}_F$ if $\mathcal{D}_N \cap \mathcal{P} = \emptyset$. Consider graph $\mathcal{G}'$ obtained from $\mathcal{G}$ by removing the descendants of buses in $\mathcal{P}_F$. Theorem 1.1 ensures that the topology of $\mathcal{G}'$ can be recovered if $\mathcal{G}'$ is a tree whose leaf nodes are exactly nodes in $\mathcal{P}_F$. That is, $\mathcal{G}$ can be reconstructed up to the descendants of nodes $\mathcal{P}_F$, see Figure 1.4.

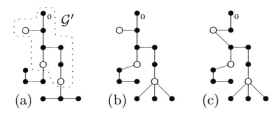

Figure 1.4 The white (black) nodes represent probing (nonprobing) nodes. The dashed portion of the network represent graph $\mathcal{G}'$. Panel (a) represents the actual network. Panel (b) reports one of the networks that can be identified when the three white nodes are chosen as probing buses. Note that the topology is correctly identified down to the probing buses. Panel (c) shows one of the erroneous topologies recovered if not all leaf nodes of $\mathcal{G}'$ are probed.

1.6.2 Topology Identifiability with Partial Voltage Data

Section 1.6.1 assumed that voltages are collected at all buses. This may be unrealistic in grids where the operator can only access the probed buses of $\mathcal{P}$. This section considers probing under the next more realistic setup.

ASSUMPTION 1.2 *Voltage differences are metered only in $\mathcal{P}$.*

Under Assumption 1.2, the probing model (1.26) becomes

$$\tilde{\mathbf{V}} = \mathbf{R}_{\mathcal{P}\mathcal{P}}\boldsymbol{\Delta} \tag{1.32}$$

where now $\tilde{\mathbf{V}}$ is of dimension $P \times T$ and $\mathbf{R}_{\mathcal{P}\mathcal{P}} := \mathbf{I}_{\mathcal{P}}^{\top}\mathbf{R}\mathbf{I}_{\mathcal{P}}$ is obtained from $\mathbf{R}$ upon maintaining only the rows and columns in $\mathcal{P}$. Similar to (1.31), $\mathbf{R}_{\mathcal{P}\mathcal{P}}$ is identifiable if $\boldsymbol{\Delta}$ is full row-rank. This is the equivalent of stage *s1)* in Section 1.5 under the partial data setup.

Toward the equivalent of stage *s2)*, because column $\mathbf{r}_m$ is partially observed, only the *metered level sets* of node $m \in \mathcal{P}$ defined as [32]

$$\mathcal{M}_m^k(\mathbf{w}) = \mathcal{N}_m^h(\mathbf{w}) \cap \mathcal{O}, \tag{1.33}$$

can be found, where $\mathcal{N}_m^h(\mathbf{w})$ is the k-th level set having at least one observed node, see Figure 1.5. The metered level sets for node m can be obtained by grouping the indices associated with the same values in the observed subvector of $\mathbf{r}_m$. Albeit the topology cannot be fully recovered based on $\mathcal{M}_m^k$'s, one can recover a *reduced grid* relying on the concept of internally identifiable nodes; see Figure 1.6. The set $\mathcal{I} \subset \mathcal{N}$ of *internally identifiable* nodes consists of all buses in $\mathcal{G}$ having at least two children with each of one of them being the ancestor of a probing bus.

The reduced grid induced by $\mathcal{P}$ is defined as the graph $\mathcal{G}^r := (\mathcal{N}^r, \mathcal{L}^r)$ with

- node set $\mathcal{N}^r := \mathcal{P} \cup \mathcal{I}$;
- $\ell = (m, n) \in \mathcal{L}^r$ if $m, n \in \mathcal{N}^r$ and all other nodes on the path from m to n in $\mathcal{G}$ do not belong to $\mathcal{N}^r$; and
- line $\ell = (m, n) \in \mathcal{L}^r$ resistance equals the effective resistance between m and n in $\mathcal{G}$, i.e., the sum of resistances across the $m - n$ path [34].

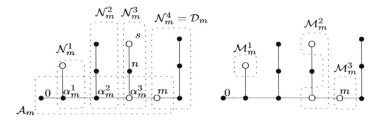

Figure 1.5 White nodes represent metered nodes. The level sets and the metered level sets of node m are reported in the left panel and in the right panel, respectively.

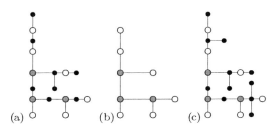

Figure 1.6 (a) The original feeder; (b) its reduced equivalent; and (c) another feeder with the same $\mathbf{R}_{\mathcal{PP}}$. White nodes are probed; black and gray are not. Gray nodes are internally identifiable nodes comprising $\mathcal{I}$.

Let $\mathbf{R}^r$ be the inverse reduced Laplacian associated with $\mathcal{G}^r$. From the properties of effective resistances, it holds [34]

$$\mathbf{R}^r_{\mathcal{PP}} = \mathbf{R}_{\mathcal{PP}}. \qquad (1.34)$$

Heed the grid $\mathcal{G}^r$ is not the only electric grid sharing $\mathbf{R}_{\mathcal{PP}}$ as the top-left block of its $\mathbf{R}$ matrix with $\mathcal{G}$: the (meshed) Kron reduction of $\mathcal{G}$ given $\mathcal{P}$; and even grids having extra nodes relative to $\mathcal{N}$ can yield the same $\mathbf{R}_{\mathcal{PP}}$; see Figure 1.6. However, $\mathcal{G}^r$ features desirable properties: i) it is radial; ii) it satisfies (1.34) with the minimal number of nodes; and iii) its resistances correspond to the effective resistances of $\mathcal{G}$. Furthermore, $\mathcal{G}^r$ conveys all the information needed to solve an optimal power flow task [35]. The next result provides a sufficient condition for identifying $\mathcal{G}^r$.

THEOREM 1.2 *Let the tree* $\mathcal{G} = (\mathcal{N}, \mathcal{L})$ *represent a distribution grid, and let Assumption 1.2 hold. Given probing data* $(\tilde{\mathbf{V}}, \mathbf{\Lambda})$, *where* $\text{rank}(\mathbf{\Lambda}) = P$, *the resistive network topology of its reduced graph* $\mathcal{G}^r = (\mathcal{N}^r, \mathcal{L}^r)$ *is identifiable if the grid is probed at all leaf nodes, that is* $\mathcal{F} \subseteq \mathcal{P}$.

1.7 Graph Algorithms for Topology Identification Using Probing

This section presents graph algorithms for identifying grid topologies using probing data under the following assumption.

ASSUMPTION 1.3 *All leaf nodes are probed, that is* $\mathcal{F} \subseteq \mathcal{P}$.

Note that although Assumption 1.3 ensures topology identifiability, it does not provide a solution for *s3*), which will be devised next.

1.7.1 Topology Identification with Complete Voltage Data

A recursive algorithm for topology identification is presented next [32]. The input to the recursion is a depth k and a maximal subset of probing nodes $\mathcal{P}^k_n$ having the *same* $(k-1)$-depth and k-depth ancestors and the level sets $\mathcal{N}^k_m$ for all $m \in \mathcal{P}^k_n$. The $(k-1)$-depth ancestor α_n^{k-1} is known. The k-depth ancestor n is known to exist, yet it is unknown for now. The recursion proceeds in three steps.

The *first step* finds n as the unique intersection of the sets $\mathcal{N}_m^k$ for all $m \in \mathcal{P}_n^k$.

At the *second step*, node n is connected to node α_n^{k-1}. Since $n = \alpha_m^k \in \mathcal{N}_m^k$ and $\alpha_n^{k-1} = \alpha_m^{k-1} \in \mathcal{N}_m^{k-1}$, the resistance of line (n, α_n^{k-1}) is given by

$$r_{(\alpha_n^{k-1}, n)} = r_{(\alpha_m^{k-1}, \alpha_m^k)} = R_{\alpha_m^k m} - R_{\alpha_m^{k-1} m} \qquad (1.35)$$

for any $m \in \mathcal{P}_n^k$; see Lemma 1.1-(iii).

The *third step* partitions $\mathcal{P}_n^k \setminus \{n\}$ into subsets of buses sharing the same $(k+1)$-depth ancestor. The buses forming one of these partitions $\mathcal{P}_s^{k+1}$ have the same k-depth and $(k+1)$-depth ancestors. Node n was found to be the k-depth ancestor. The $(k+1)$-depth ancestor is known to exist and is assigned the symbol s. The value of s is found by invoking the recursion with new inputs the depth $(k+1)$, the set of buses $\mathcal{P}_s^{k+1}$ along with their $(k+1)$-depth level sets, and their common k-depth ancestor (node n).

Algorithm 1.1 Topology Identification with Complete Data

Require: $\mathcal{N}$, $\{\mathcal{N}_m^k\}_{k=0}^{d_m}$ for all $m \in \mathcal{P}$.
1: Run Root&Branch$(\mathcal{P}, \emptyset, 0)$.
Ensure: Radial grid and line resistances over $\mathcal{N}$.
Function Root&Branch$(\mathcal{P}_n^k, \alpha_n^{k-1}, k)$
1: Identify the node n as the common k-depth ancestor for all buses in $\mathcal{P}_n^k$.
2: **if** $k > 0$, **then**
3:     Connect node n to α_n^{k-1} with the resistance of (1.35).
4: **end if**
5: **if** $\mathcal{P}_n^k \setminus \{n\} \neq \emptyset$, **then**
6:     Partition $\mathcal{P}_n^k \setminus \{n\}$ into groups of buses $\{\mathcal{P}_s^{k+1}\}$ having identical k-depth level sets.
7:     Run Root&Branch$(\mathcal{P}_s^{k+1}, n, k+1)$ for all s.
8: **end if**

To *initialize* the recursion, set $\mathcal{P}_n^0 = \mathcal{P}$ since every probing bus has the substation as 0-depth ancestor. At $k = 0$, the second step is skipped as the substation does not have any ancestors to connect. The recursion *terminates* when $\mathcal{P}_n^k$ is a singleton $\{m\}$. In this case, the first step identifies m as node n; the second step links m to its known ancestor α_m^{k-1}; and the third step has no partition to accomplish. The recursion is tabulated as Alg. 1.1.

1.7.2 Topology Identification with Partial Voltage Data

A three-step recursion operating on metered rather than ordinary level sets and aiming at reconstructing the reduced grid is detailed next [32]. Suppose we are given the set of probing nodes $\mathcal{P}_n^k$ having the same $(k-1)$-depth and k-depth ancestors (known and unknown, respectively), along with their k-depth metered level sets.

At the *first step*, if there exists a node $m \in \mathcal{P}_n^k$ such that $\mathcal{M}_m^k = \mathcal{P}_n^k$, then the k-depth ancestor n is set as m. Otherwise, a non-probing node is added and assigned to be the k-depth ancestor.

At the *second step*, node $n = \alpha_m^k$ is connected to node $\alpha_n^{k-1} = \alpha_m^{k-1}$. The line resistance can be found through the modified version of (1.35)

$$r_{(\alpha_m^{k-1}, \alpha_m^k)} = R_{\alpha_m^k m} - R_{\alpha_m^{k-1} m} = R_{s'm} - R_{sm}. \tag{1.36}$$

At the *third step*, the set $\mathcal{P}_n^k \setminus \{n\}$ is partitioned into subsets of buses having the same $(k+1)$-depth ancestor, by comparing their k-depth metered level sets.

The recursion is tabulated as Alg. 1.2. It is initialized at $k = 1$, since the substation is not probed and $\mathcal{M}_m^0$ does not exist; and is terminated as in Section 1.6.

Algorithm 1.2 Topology Recovery with Partial Data

Require: $\mathcal{M}, \{\mathcal{M}_m^k\}_{k=1}^{d_m}$ for all $m \in \mathcal{P}$.
1: Run Root&Branch-P($\mathcal{P}, \emptyset, 1$).
Ensure: Reduced grid $\mathcal{G}^r$ and resistances over $\mathcal{L}^r$.
Function Root&Branch-P($\mathcal{P}_n^k, \alpha_n^{k-1}, k$)
1: **if** $\exists$ node n such that $\mathcal{M}_n^k = \mathcal{P}_n^k$, **then**
2:     Set n as the parent node of subtree $\mathcal{T}_n^k$.
3: **else**
4:     Add node $n \in \mathcal{I}$ and set it as the root of $\mathcal{T}_n^k$.
5: **end if**
6: **if** $k > 1$, **then**
7:     Connect n to α_n^{k-1} via a line with resistance (1.36).
8: **end if**
9: **if** $\mathcal{P}_n^k \setminus \{n\} \neq \emptyset$, **then**
10:     Partition $\mathcal{P}_n^k \setminus \{n\}$ into groups of buses $\{\mathcal{P}_s^{k+1}\}$ having identical k-depth metered level sets.
11:     Run Root&Branch-P($\mathcal{P}_s^{k+1}, n, k+1$) for all s.
12: **end if**

1.7.3 Graph Algorithms Operating under Noisy Data

When measurements are corrupted by noise and loads are time varying, a least-squares estimate of $\mathbf{R}_\mathcal{P}$ can be found as

$$\hat{\mathbf{R}}_\mathcal{P} := \arg\min_{\Theta} \|\tilde{\mathbf{V}} - \Theta \Delta\|_F^2 = \tilde{\mathbf{V}} \Delta^+. \tag{1.37}$$

To facilitate its statistical characterization and implementation, assume the probing protocol Δ_2. The m-th column of $\mathbf{R}_\mathcal{P}$ can be found as the scaled sample mean of voltage differences obtained when node m was probed

$$\hat{\mathbf{r}}_m = \sum_{t \in \mathcal{T}_m} \frac{1}{(-1)^{t-1} \delta_m T_m}, \quad \mathcal{T}_m := \left\{ \sum_{\tau=1}^{m-1} T_\tau + 1, \ldots, \sum_{\tau=1}^{m} T_\tau \right\} \tilde{\mathbf{v}}(t). \tag{1.38}$$

Let Assumption 1.1 hold and, for simplicity, let $\{\tilde{\mathbf{p}}(t), \tilde{\mathbf{q}}(t), \mathbf{n}(t)\}$ have diagonal covariances $\sigma_p^2 \mathbf{I}$, $\sigma_q^2 \mathbf{I}$, and $\sigma_n^2 \mathbf{I}$. Hence, the error vector $\epsilon(t)$ is zero-mean with covariance $\Sigma_\epsilon := \sigma_p^2 \mathbf{R}^2 + \sigma_q^2 \mathbf{X}^2 + \sigma_n^2 \mathbf{I}$ and the estimate $\hat{\mathbf{r}}_m$ can be approximated as zero-mean Gaussian with covariance $\frac{1}{\delta_m^2 T_m} \Sigma_\epsilon$. By increasing T_m and/or δ_m, $\hat{\mathbf{r}}_m$ goes arbitrarily close to $\mathbf{r}_m$ and their distance can be bounded probabilistically using Σ_ϵ. Note however, that Σ_ϵ depends on the unknown $(\mathbf{R}, \mathbf{X})$. To resolve this issue, suppose the spectral radii $\rho(\mathbf{R})$ and $\rho(\mathbf{X})$, and the variances $(\sigma_p^2, \sigma_q^2, \sigma_w^2)$ are known; see [35] for upper bounds. Then, it holds that $\rho(\Sigma_\epsilon) \leq \sigma^2$, where $\sigma^2 := \sigma_p^2 \rho^2(\mathbf{R}) + \sigma_q^2 \rho^2(\mathbf{X}) + \sigma_n^2$. The standard Gaussian concentration inequality bounds the deviation of the n-th entry of $\hat{\mathbf{r}}_m$ from its actual value as $\Pr\left(|\hat{R}_{nm} - R_{nm}| \geq \frac{4\sigma}{\delta_m \sqrt{T_m}} \right) \leq \pi_0 := 6 \cdot 10^{-5}$.

For s2), no two noisy entries of $\hat{\mathbf{r}}_m$ will be identical almost surely. The entries will be concentrated around their actual values. To identify groups of similar values, sort the entries of $\hat{\mathbf{r}}_m$ in increasing order, and take the differences of the successive sorted entries. Lemma 1.1-(iii) guarantees that the minimum difference between the entries of $\mathbf{r}_m$ is larger or equal to the smallest line resistance $r_{\min}$. Hence, if all estimates were confined within $|\hat{R}_{nm} - R_{nm}| \leq r_{\min}/4$, a difference of sorted $\hat{R}_{nm}$'s larger than $r_{\min}/2$ would pinpoint the boundary between two bus groups. In practice, if the operator knows $r_{\min}$ and selects (T_m, δ_m) so that

$$\delta_m \sqrt{T_m} \geq 16\sigma / r_{\min} \tag{1.39}$$

then $|\hat{R}_{nm} - R_{nm}| \leq r_{\min}/4$ will be satisfied with probability larger than 99.95%. Taking the union bound, the probability of recovering all level sets is larger than $1 - N^2 \pi_0$. The argument carries over to $\mathbf{R}_{\mathcal{PP}}$ under the partial data setup.

1.8 Topology Identification Using Probing through Convex Relaxation

Section 1.7 first estimated a part of $\mathbf{R}$, and then run a graph algorithm to identify the grid topology. Because the latter algorithm may become impractical for low signal-to-noise ratios, this section estimates Θ directly from probing data. From (1.3), it follows that $\Theta \succ \mathbf{0}$, since $\mathbf{A}$ is non-singular [36]. The off-diagonal terms of Θ are non-positive, while the diagonal ones are positive. Also, since the first column of $\tilde{\Theta}$ has non-positive off-diagonal entries and $\tilde{\Theta} \mathbf{1} = \mathbf{0}$, it also follows that $\Theta \mathbf{1} \geq \mathbf{0}$. Finally, the grid operator may know that two specific buses are definitely connected, e.g., through flow sensors or line status indicators. To model known line statuses, introduce $\tilde{\Gamma} \in \mathbb{S}^{N+1}$ with $\tilde{\Gamma}_{mn} = 0$ if line (m,n) is known to be non-energized; and $\tilde{\Gamma}_{mn} = 1$ if there is no prior information for line (m,n). If there is no information for any line, then $\tilde{\Gamma} = \mathbf{1}\mathbf{1}^\top$. Based on $\tilde{\Gamma}$, define the set

$$\mathcal{S}(\tilde{\Gamma}) := \left\{ \Theta : \begin{array}{l} \Theta_{mn} \leq 0, \text{ if } \tilde{\Gamma}_{mn} = 1 \\ \Theta_{mn} = 0, \text{ if } \tilde{\Gamma}_{mn} = 0 \end{array} m, n \in \mathcal{N}, m \neq n \right\}.$$

The set $\mathcal{S}(\tilde{\Gamma})$ ignores possible prior information on lines fed directly by the substation. This information is encoded on the zero-th column of $\tilde{\Gamma}$. In particular, if $\tilde{\Gamma}_{0n} = 1$, then $\tilde{\Theta}_{0n} \leq 0$ and $\sum_{m=1}^{N} \Theta_{mn} \geq 0$. Otherwise, it holds that $\tilde{\Theta}_{0n} = \sum_{m=1}^{N} \Theta_{mn} = 0$. The two properties pertaining to lines directly connected to the substation are captured by the set

$$\mathcal{S}_0(\tilde{\Gamma}) := \left\{ \Theta : \begin{array}{l} e_n^\top \Theta \mathbf{1} \geq 0, \text{ if } \tilde{\Gamma}_{0n} = 1 \\ e_n^\top \Theta \mathbf{1} = 0, \text{ if } \tilde{\Gamma}_{0n} = 0 \end{array}, n \in \mathcal{N} \right\}.$$

Summarizing, the set of admissible reduced Laplacian matrices for arbitrary graphs with prior edge information $\tilde{\Gamma}$ is

$$\mathcal{C} := \left\{ \Theta : \Theta \in \mathcal{S}(\tilde{\Gamma}) \cap \mathcal{S}_0(\tilde{\Gamma}), \Theta = \Theta^\top \right\}. \quad (1.40)$$

By invoking the Gershgorin's disc theorem, it can be shown that $\Theta \succeq \mathbf{0}$ for all $\Theta \in \mathcal{C}$, that is $\mathcal{C} \subseteq \mathbb{S}^+$. The reduced Laplacian matrices included in $\mathcal{C}$ correspond to possibly meshed and/or disconnected graphs. Enforcing two additional properties can render Θ a proper reduced Laplacian of a tree: First, matrix $\Phi(\Theta)$ should have exactly $2N$ non-zero off-diagonal entries. Second, matrix Θ should be strictly positive definite since the graph is connected. Then, define the set

$$\mathcal{T} := \left\{ \Theta : \|\Phi(\Theta)\|_{0, \text{off}} = 2N, \Theta \succ \mathbf{0} \right\}. \quad (1.41)$$

Back to topology identification using (1.26), matrix Θ can be estimated via a (weighted) least-squares (LS) fit of the probing data under Laplacian constraints

$$\min_{\Theta \in \mathcal{C} \cap \mathcal{T}} f(\Theta) := \frac{1}{2} \|\Theta \tilde{\mathbf{V}} - \mathbf{I}_{\mathcal{C}} \Delta\|_{\mathbf{W}}^2. \quad (1.42)$$

Albeit its objective and set $\mathcal{C}$ are convex, the optimization in (1.42) is challenging because $\mathcal{T}$ is non-convex and open. To arrive at a practical solution, we surrogate $\mathcal{T}$ by adding two penalties in the objective of (1.42). Heed that the property $\Theta \succ \mathbf{0}$ is equivalent to enforcing a finite lower bound on $\log |\Theta|$. On the other hand, the non-convex pseudo-norm $\|\Phi(\Theta)\|_{0, \text{off}}$ can be relaxed by its convex envelope $\|\Phi(\Theta)\|_{1, \text{off}} := \sum_{m, n \neq m} |\Phi(\Theta)_{mn}|$; see also [37–39] for related approaches aiming to recover sparse inverse covariance or Laplacian matrices. By defining $\Pi := \mathbf{I} + \mathbf{1}\mathbf{1}^\top$, $\|\Phi(\Theta)\|_{1, \text{off}}$ can be rewritten as

$$\|\Phi(\Theta)\|_{1, \text{off}} = Tr(\Theta \Pi)$$

Upon dualizing the constraints comprising $\mathcal{T}$, (1.42) can be convexified as

$$\hat{\Theta} := \arg \min_{\Theta \in \mathcal{C}} \frac{1}{2} \|\Theta \tilde{\mathbf{V}} - \mathbf{I}_{\mathcal{C}} \Delta\|_{\mathbf{W}}^2 + \lambda Tr(\Theta \Pi) - \mu \log |\Theta| \quad (1.43)$$

where $\lambda, \mu > 0$ are tunable parameters. Since the minimizer of (1.43) does not necessarily belong to $\mathcal{T}$, one may invoke heuristics to convert it to the reduced Laplacian of a tree graph. As suggested in [2], a Laplacian $\tilde{\Theta}$ belonging to $\mathcal{T}$ can be found by running a minimum spanning tree algorithm for the graph defined by $\Phi(\hat{\Theta})$. Problem (1.43) can be solved using the alternating direction method of multipliers (ADMM).

Before closing this section, two comments are in order: *i)* beyond identification, probing can be used for topology detection as well [31]; and *ii)* Instead of the convex relaxation proposed here, topology identification using probing can be solved *exactly* via the mixed-integer linear program (MILP) formulation of [40].

1.9 Numerical Tests on Topology Identification Using Probing

Our probing algorithms were validated on the IEEE 37-bus feeder, see also Section 1.3.3. Figures 1.7(a) and 1.7(b) show the actual and reduced topologies. Probing buses were equipped with inverters having the same rating as the related load. Loads were generated by adding a zero-mean Gaussian variation to the benchmark data, with standard deviation 0.067 times the average of nominal loads. Voltages were obtained via a power flow solver, and then corrupted by zero-mean Gaussian noise with 3σ deviation of 0.01% per unit (pu). Probing actions were performed using the probing protocol Δ_2. The algorithms were tested through 200 Monte Carlo tests. At every run, the actual topology was randomly drawn.

Firstly, the graph algorithms of Section 1.7 were tested. For the 37-bus feeder, $r_{\min} = 0.0014$ pu. From the rated δ_m's; the $r_{\min}$; and (1.39), the number of probing actions was set as $T_m = 90$. In the partial data case, the smallest effective resistance was 0.0021 pu, yielding $T_m = 39$. Level sets obtained using the procedure described in Sec. 1.7.3 were given as inputs to Alg. 1.1 and 1.2. Table 1.1 demonstrates that the error probability (EP) in topology recovery and the mean percentage error (MPE) of line resistances decay as T_m increases.

Secondly, the identification problem in (1.42) was solved for $\lambda = 5 \cdot 10^{-3}$, $\mu = 1$, and $\mathbf{W} = \mathbf{I}$. No prior information on line statuses was assumed ($\tilde{\boldsymbol{\Gamma}} = \mathbf{1}\mathbf{1}^\top$). Kruskal's algorithm was used to obtain a Laplacian matrix $\tilde{\Theta}$ corresponding to a radial grid. The

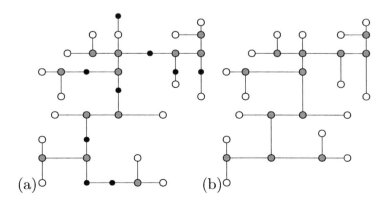

Figure 1.7 Line infrastructure for the IEEE 37-bus feeder benchmark.

Table 1.1 Numerical tests under complete and partial noisy data

T_m	1	10	40	90	1	5	20	39
EP [%]	98.5	55.3	3.1	0.2	97.2	45.8	18.9	0.1
MPE [%]	35.1	32.5	30.9	28.5	97.2	45.8	18.9	0.1
		Alg. 1.1				Alg. 1.2		

Table 1.2 Probability of detecting a wrong topology via (1.42)

T_m	1	2	5	10
Topology error prob. [%]	5.1	3.9	3.7	2.6

connectivity of $\tilde{\Theta}$ was compared against the actual one. The average number of line status errors, including both the energized lines not detected and the non-energized lines detected, are reported in Table 1.2.

1.10 Conclusion

This chapter has put forth a grid topology learning toolbox. If the operator collects data passively, grid topology can be identified using ML, MAP, or partial correlation-based schemes. To accelerate learning, the operator may perturb inverter injections and collect voltage responses to infer the grid topology. We have provided conditions under which the grid topology can be correctly identified by probing. Computationally, under high SNR setups, the topology can be revealed using a recursive graph algorithm. For lower SNR setups and complete data, a convex relaxation approach has been suggested instead. The presented works motivate several interesting questions such as extending these results to multiphase setups; using the tap changes of voltage regulators and capacitor banks as a means for active topology learning; and combining actively and passively collected grid data. Additional directions include active semi-blind approaches along the lines of passive ones dealt with for a general graph in [41].

Acknowledgment

V. Kekatos was supported in part by NSF grant 1751085. L. Zhang and G. B. Giannakis were supported in part by NSF grants 1423316, 1442686, 1508993, and 1509040; and by the Laboratory Directed Research and Development Program at the NREL.

References

[1] D. Deka, M. Chertkov, and S. Backhaus, "Structure learning in power distribution networks," *IEEE Trans. Control Netw. Syst.* vol. 5, no. 3, pp. 1061–1074, Sept. 2018.

[2] S. Bolognani, N. Bof, D. Michelotti, R. Muraro, and L. Schenato, "Identification of power distribution network topology via voltage correlation analysis," in *Proc. IEEE Conf. on Decision and Control, Florence, Italy*, pp. 1659–1664, Dec. 2013.

[3] S. Park, D. Deka, and M. Chertkov, "Exact topology and parameter estimation in distribution grids with minimal observability," in *Proc. Power Syst. Comput. Conf.*, Dublin, Ireland, June 2018.

[4] Y. Sharon, A. M. Annaswamy, A. L. Motto, and A. Chakraborty, "Topology identification in distribution network with limited measurements," in *Proc. IEEE Conf. on Innovative Smart Grid Technologies*, Washington, DC, Jan. 2012.

[5] V. Kekatos and G. B. Giannakis, "Joint power system state estimation and breaker status identification," in *Proc. North American Power Symposium*, Urbana-Champaign, IL, Sept. 2012.

[6] H. Zhu and G. B. Giannakis, "Sparse overcomplete representations for efficient identification of power line outages," *IEEE Trans. Power Syst.*, vol. 27, no. 4, pp. 2215–2224, Nov. 2012.

[7] R. A. Sevlian, Y. Zhao, R. Rajagopal, A. Goldsmith, and H. V. Poor, "Outage detection using load and line flow measurements in power distribution systems," *IEEE Trans. Power Syst.*, vol. 33, no. 2, pp. 2053–2069, Mar. 2018.

[8] J. Yu, Y. Weng, and R. Rajagopal, "PaToPa: A data-driven parameter and topology joint estimation framework in distribution grids," *IEEE Trans. Power Syst.*, vol. 33, no. 4, pp. 4335–4347, July 2017.

[9] Y. Zhao, J. Chen, and H. V. Poor, "A Learning-to-Infer Method for Real-Time Power Grid Multi-Line Outage Identification," vol. 11, no. 1, pp. 555–564, Jan 2020.

[10] O. Ardakanian, Y. Yuan, V. Wong, R. Dobbe, S. Low, A. von Meier, and C. J. Tomlin, "On identification of distribution grids," *IEEE Trans. Control Net. Syst.*, vol. 6, no. 3, Sept. 2019, pp. 950–960.

[11] C. Godsil and G. Royle, *Algebraic Graph Theory*. New York, NY: Springer, 2001.

[12] M. Baran and F. Wu, "Network reconfiguration in distribution systems for loss reduction and load balancing," *IEEE Trans. Power Del.* vol. 4, no. 2, pp. 1401–1407, Apr. 1989.

[13] S. Bolognani and F. Dorfler, "Fast power system analysis via implicit linearization of the power flow manifold," in *Proc. Allerton Conf.*, Allerton, IL, pp. 402–409, Sept. 2015.

[14] D. Deka, S. Backhaus, and M. Chertkov, "Learning topology of distribution grids using only terminal node measurements," *IEEE International Conference on Smart Grid Communications (SmartGridComm)*, Sydney, NSW, Australia, pp. 205–211.

[15] G. Cavraro and R. Arghandeh, "Power distribution network topology detection with time-series signature verification method," *IEEE Trans. Power Syst.*, vol. 33, no. 4, pp. 3500–3509, July 2018.

[16] G. Cavraro, V. Kekatos, and S. Veeramachaneni, "Voltage analytics for power distribution network topology verification," *IEEE Trans. Smart Grid.*, vol. 10, no. 1, pp. 1058–1067, Jan. 2019.

[17] V. Kekatos, G. Giannakis, and R. Baldick, "Online energy price matrix factorization for power grid topology tracking," *IEEE Trans. Smart Grid.*, vol. 7, no. 7, pp. 1239–1248, May 2016.

[18] (2013) Pecan Street Inc. Dataport, https://dataport.pecanstreet.org/

[19] G. V. Karanikolas, G. B. Giannakis, K. Slavakis, and R. M. Leahy, "Multi-kernel based nonlinear models for connectivity identification of brain networks," in *Proc. IEEE Intl. Conf. on Acoustics, Speech, and Signal Process.*, Shanghai, China, pp. 6315–6319, Mar. 2016.

[20] G. B. Giannakis, Y. Shen, and G. V. Karanikolas, "Topology identification and learning over graphs: Accounting for nonlinearities and dynamics," *Proc IEEE.*, vol. 106, no. 5, pp. 787–807, May 2018.

[21] C. Saunders, A. Gammerman, and V. Vovk, "Ridge regression learning algorithm in dual variables," in *Proc. Intl. Conf. Machine Learning*, Madison, WI, pp. 515–521, July 1998.

[22] J. Shawe-Taylor and N. Cristianini, *Kernel Methods for Pattern Analysis*. New York, NY: Cambridge University Press, 2004.

[23] G. R. G. Lanckriet, N. Cristianini, P. Bartlett, L. E. Ghaoui, and M. I. Jordan, "Learning the kernel matrix with semidefinite programming," *J. Mach. Lear. Res.*, vol. 5, pp. 27–72, Jan. 2004.

[24] M. Kloft, U. Brefeld, P. Laskov, K.-R. Müller, A. Zien, and S. Sonnenburg, "Efficient and accurate ℓ_p-norm multiple kernel learning," in *Neural Inf. Process. Syst.*, Vancouver, Canada, pp. 1-9, Dec. 2009.

[25] L. Zhang, D. Romero, and G. B. Giannakis, "Fast convergent algorithms for multi-kernel regression," in *IEEE Wkshp. on Statistical Signal Process.*, Mallorca, Spain, pp. 1-4, June 2016.

[26] M. Jaggi, "Revisiting Frank–Wolfe: Projection-free sparse convex optimization," in *Intl. Conf. on Machine Learning*, vol. 28, no. 1, Atlanta, GA, pp. 427–435, June 2013.

[27] L. Zhang, G. Wang, D. Romero, and G. B. Giannakis, "Randomized block Frank–Wolfe for convergent large-scale learning," *IEEE Trans. Signal Process.* vol. 65, no. 24, pp. 6448–6461, Sept. 2017.

[28] Kaggle. (2012) Global energy forecasting competition, www.kaggle.com/c/global-energy-forecasting-competition-2012-load-forecasting/data

[29] J. F. Sturm, "Using SeDuMi 1.02, a MATLAB toolbox for optimization over symmetric cones," *Optim. Method Softw.*, vol. 11, no. 1–4, pp. 625–653, Jan. 1999.

[30] D. Deka, S. Talukdar, M. Chertkov, and M. Salapaka. (2017) Topology estimation in bulk power grids: Guarantees on exact recovery, https://arxiv.org/abs/1707.01596

[31] G. Cavraro and V. Kekatos, "Inverter probing for power distribution network topology processing," *IEEE Trans. Control Net. Syst.*, vol. 6, no. 3, pp. 980–992, Sept. 2019.

[32] ——, "Graph algorithms for topology identification using power grid probing," *IEEE Control Syst. Lett.*, vol. 2, no. 4, pp. 689–694, Oct. 2018.

[33] G. Cavraro, A. Bernstein, V. Kekatos, and Y. Zhang, "Real-time identifiability of power distribution network topologies with limited monitoring," *IEEE Control Syst. Lett.*, vol. 4, no. 2, pp. 325-330, Apr. 2020.

[34] F. Dorfler and F. Bullo, "Kron reduction of graphs with applications to electrical networks," *IEEE Trans. Circuits Syst I.*, vol. 60, no. 1, pp. 150–163, Jan. 2013.

[35] S. Bolognani, G. Cavraro, and S. Zampieri. "A distributed feedback control approach to the optimal reactive power flow problem," in *Control of Cyber-Physical Systems, Danielle C. Tarraf, ed.* pp. 259–277, Heidelberg: Springer International Publishing, 2013.

[36] V. Kekatos, L. Zhang, G. B. Giannakis, and R. Baldick, "Voltage regulation algorithms for multiphase power distribution grids," *IEEE Trans. Power Syst.* vol. 31, no. 5, pp. 3913–3923, Sept. 2016.

[37] J. Friedman, T. Hastie, and R. Tibshirani, "Sparse inverse covariance estimation with the graphical lasso," *Biostatistics*, vol. 9, no. 3, pp. 432–441, Dec. 2008.

[38] V. Kekatos, G. B. Giannakis, and R. Baldick, "Grid topology identification using electricity prices," in *Proc. IEEE Power & Energy Society General Meeting,* Washington, DC, July 2014, pp. 1–5.

[39] H. E. Egilmez, E. Pavez, and A. Ortega, "Graph learning from data under Laplacian and structural constraints," *IEEE J. Sel. Topics Signal Process,* vol. 11, no. 6, pp. 825–841, Sept. 2017.

[40] S. Taheri, V. Kekatos, and G. Cavraro, "An MILP approach for distribution grid topology identification using inverter probing," in *IEEE PowerTech*, Milan, Italy, pp. 1–5, June 2019.

[41] V. N. Ioannidis, Y. Shen, and G. B. Giannakis, "Semi-blind inference of topologies and dynamical processes over graphs," *IEEE Trans. Signal Process,* vol. 67, no. 9, pp. 2263–2274, Apr. 2019.

2 Probabilistic Forecasting of Power System and Market Operations

Yuting Ji, Lang Tong, and Weisi Deng

2.1 Introduction

The increasing penetration of renewable resources has changed the operational characteristics of power systems and electricity markets. For instance, the net load[1] profile in some areas now follows a so-called duck curve [1] where a steep down-ramp during the hours when a large amount of solar power is injected into the network is followed by a sharp up-ramp when the solar power drops in late afternoon hours.

While the duck curve phenomenon represents an *average* net load behavior, it is the highly stochastic and spatial-temporal dependent ramp events that are most challenging to system operators. In such new operation regimes, the ability to adapt to changing environments and manage risks arising from complex scenarios of contingencies is essential. For this reason, there is a need for informative and actionable characterization of the overall operation uncertainty regarding an extended planning horizon, which models interdependencies among power flows, congestions, reserves, and locational marginal prices (LMPs). To this end, probabilistic forecasts defined by the conditional probability distributions of these variables of interest serve as sufficient statistics for operational decisions.

Some system operators are providing real-time price forecasts currently. The Electric Reliability Council of Texas [2], for example, offers one-hour ahead real-time LMP forecasts, updated every 5 minutes. Such forecasts signal potential shortage or oversupply caused by the anticipated fluctuation of renewable generation or the likelihood of network congestions. The Alberta Electric System Operator [3] provides two short-term price forecasts with prediction horizons of 2 hours and 6 hours, respectively. The Australia Energy Market Operator [4] provides 5-minute ahead to 40-hour ahead spot price and load forecasts.

Most LMP forecasts, especially those provided by system operators, are *point forecasts* that predict directly future LMP values. Often they are generated by substituting actual realizations of stochastic load and generation by their expected (thus deterministic) trajectory in the calculation of future generation and price quantities. Such *certainty equivalent* approaches amount to approximating the expectation of a function of random

---

[1] The net load is defined as the total electrical demand plus interchange minus the renewable generation. The interchange refers to the total scheduled delivery and receipt of power and energy of the neighboring areas.

variables by the function of the expectation of random variables; they can lead to gross mischaracterization of the behavior of the system operation and systematic error in forecasting.

In this chapter we consider the problem of *online probabilistic forecasting* in real-time wholesale electricity market from an operator perspective. Specifically, we are interested in using real-time supervisory control and data acquisition (SCADA) and Power Management Unit (PMU) measurements along with operating conditions known to the operator to produce short-term probabilistic forecasts of nodal prices, power flows, generation dispatches, operation reserves, and discrete events such as congestions and occurrences of contingencies. Unlike point forecasting, which predicts the realization of a random variable, say, the LMP at a particular bus 1-hour ahead, probabilistic forecasting produces the probability distribution of the random variable 1-hour ahead conditional on the real-time measurements obtained at the time of prediction.

For an operator, probabilistic forecasting provides essential information for evaluating operational risks and making contingency plans. It is only with probabilistic forecasts that stochastic optimization becomes applicable for unit commitment and economic dispatch problems [5]. For the market participants, however, probabilistic forecasts of prices are essential signals to elicit additional generation resources when the probability of high prices is high and to curtail otherwise. Indeed, in the absence of a day-ahead market, continuing price and load forecasts by the independent market operator that cover 30 minutes and longer forecasting horizons play an essential role in the Australian national electricity market [6].

2.1.1 Related Work

There is an extensive literature on point forecasting techniques from perspectives of external market participants. See [7, 8] for a review. These techniques typically do not assume having access to detailed operating conditions. Results focusing on probabilistic forecasting by an operator are relatively sparse. Here we highlight some of the relevant results in the literature applicable to probabilistic forecasting by a system operator.

A central challenge of obtaining probabilistic forecasting is estimating the (joint or marginal) conditional probability distributions of future system variables. One approach is to approximate the conditional distributions based on point estimates. For instance, a probabilistic LMP forecasting approach is proposed in [9] based on attaching a Gaussian distribution to a point estimate. While the technique can be used to generalize point forecasting methods to probabilistic ones, the Gaussian distribution is, in general, a poor approximation as there are typically multiple distribution modalities caused by different realizations of binding constraints in the optimization. The authors of [10, 11] approximate the probabilistic distribution of LMP using higher order moments and cumulants. These methods are based on representing the probability distribution as an infinite series involving moments or cumulants. In practice, computing or estimating higher order moments and cumulants are challenging; lower order approximations are necessary.

A more direct approach is to estimate the conditional distributions directly. Other than some simple cases, however, probabilistic forecasting in a large complex system can only be obtained by Monte Carlo simulations in which conditional distributions are estimated from sample paths generated either according to the underlying system model or directly from measurements and historical data. In this context, the problem of probabilistic forecasting is essentially the same as online Monte Carlo simulations. To this end, there is a premium on reducing computation costs.

The idea of using simulation techniques for LMP forecasting was first proposed in [12], although probabilistic forecasting was not considered. Min et al. proposed in [13] a direct implementation of Monte Carlo simulations to obtain short-term forecasting of transmission congestion. For M Monte Carlo runs over a T-period forecasting horizon, the computation cost is dominated by the computation of $M \times T$ direct current optimal power flow (DCOPF) solutions that are used to generate the necessary statistics. For a large-scale system with a significant amount of random generations and loads, such computation costs may be too high to be used for online forecasting.

A similar approach based on a nonhomogeneous Markov chain modeling of real-time LMP was proposed in [14]. The Markov chain technique exploits the discrete nature of LMP distributions and obtains LMP forecasts by the product of transition matrices of LMP states. Estimating the transition probabilities, however, requires thousands to millions of Monte Carlo simulations, thus requiring approximately the same number of DCOPF computations.

A significant reduction of simulation cost is achieved by exploiting the structure of economic dispatch by which random variables (such as stochastic generations) enter the optimal power flow (OPF) problem. In [15], a multiparametric programming formulation is introduced where random generations and demands are modeled as the right-hand side parameters in the constraints of the DCOPF problem. From the parametric linear/quadratic programming theory, the (conditional) probability distributions of LMP and power flows, given the current system state, reduce to the conditional probabilities that realizations of the random demand and generation fall into one of the *finite* number of *critical regions* in the parameter space. The reduction of the modeling complexity from high-dimensional continuous random variables of stochastic generations and loads to a discrete random variable with a finite number of realizations represents a fundamental step toward a scalable solution to probabilistic forecasting.

Although the multiparametric programming approach in [16] reduces the modeling complexity to a finite number of critical regions in the parameter space, the cardinality of the set of critical regions grows exponentially with the number of constraints in the DCOPF problem. For a large power system with many constraints, the cost of characterizing these critical regions, even if performed offline, has an exponentially growing complexity in computation and storage.

In [17], an online dictionary learning (ODL) approach is proposed. The main idea is to avoid computing and store all critical regions ahead of time, instead using online learning to acquire and update sequentially a dictionary that captures the parametric structure of DCOPF solutions. In particular, each entry of the dictionary corresponds to an *observed* critical region within which a sample of random generation/demand has been realized. A new entry of the dictionary is produced only when the realization of the renewable generation and demand falls in a new critical region that has not been recorded in the dictionary. This approach allows solution recall directly from the dictionary without solving a new DCOPF. Because renewable generation and load processes are physical processes, they tend to be concentrated around the mean trajectory. As a result, despite that there is potentially an exponentially large number of potential entries in the dictionary, only a tiny fraction of the dictionary entries are observed in the simulation process.

2.1.2 Summary and Organization

This article highlights and extends recent advances in simulation-based probabilistic forecasting techniques [16, 17]. We focus on simulation-based approaches for two reasons. First, simulation-based forecasting is so far the only type of techniques for large complex power systems that can yield *consistent estimates* of conditional probability distributions. Second, although computation costs of such methods have long been recognized as the primary barrier for their applications in large systems, novel computation and machine learning techniques and new computation resources such as cloud computing and graphical processing units (GPUs) will lead to a scalable online forecasting platform for system operators.

This chapter is organized as follows. Section 2.2 presents a general simulation-based approach to probabilistic forecasting. Examples of real-time operation models are provided in Section 2.3. Section 2.4 summarizes key structural results on multiparametric programming for developing the probabilistic forecasting algorithm. Details of the proposed forecasting approach are presented in Section 2.5, followed by case studies in Section 2.6 and concluding remarks in Section 2.7.

2.2 Probabilistic Forecasting Using Monte Carlo Simulations

We highlight in this section a general approach to probabilistic forecasting based on Monte Carlo simulation. *Probabilistic forecast*, in contrast to point forecast, aims to provide (marginal or joint) probability distributions of future quantities of interest such as load, power flow, reserve, and LMP.

Two critical components of simulation-based probabilistic forecasting are (i) a model that captures the physical characteristics of the system and (ii) a probability model that characterizes the underlying random phenomena in the system. For power system and market operations, the former includes a physical model of the underlying power system and an algorithmic model of the decision process for unit commitment, economic dispatch, and LMP. The latter comes from probabilistic load and generation forecasts and probability models for contingencies.

Figure 2.1 illustrates a general schematic of simulation-based probabilistic forecasting, similar to that in [12]. The main engine of the probabilistic forecasting is a

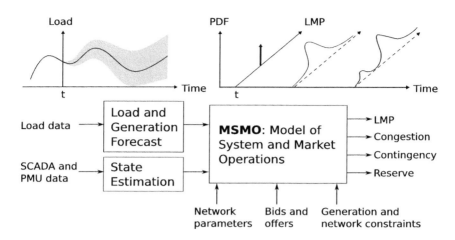

Figure 2.1 Schematics of simulation-based probabilistic forecasting.

computational model of real-time system and market operations (MSMO). Details of MSMO are the subject of Section 2.3.

One set of inputs to MSMO are the exogenous *random processes* from real-time measurements from a SCADA system or PMUs that define the system state S_t at the time of forecasting t. Another set of inputs are the probabilistic forecasts of generations and loads. Shown in Figure 2.1 is an illustration of probabilistic load forecast at time t where the gray area represents possible load trajectories of the future. Probabilistic forecasting of load and renewable generation has been studied extensively. See review articles [18–20] and references therein.

MSMO also imports a set of *deterministic parameters* that characterize the network condition (topology and electrical parameters), bids and offers from market participants, and operation constraints such as generation and line capacities. Using a physical model of the power system, the MSMO simulates the actual system operations from generated scenarios using probabilistic load and generation forecasts. Statistics of variables of interests such as LMPs, congestion patterns, and levels of the reserve are collected and presented to the operator.

The output of MSMO is the set of the conditional probability distributions of the variables of interest. In particular, given the system state $S_t = s$ estimated from PMU/SCADA data, MSMO outputs the conditional probability distribution $f_{t+T|t}$ of LMP at time $t + T$ in the form of a histogram or parametric distributions learned from LMP samples generated by MSMO.

2.3 Models of System and Market Operations

Most wholesale electricity markets consist of day-ahead and real-time markets. The day-ahead market enables market participants to commit to buy or sell wholesale electricity one day ahead of the operation. The day-ahead market is only binding in that the market participants are obligated to pay or be paid for the amount cleared in the day-ahead market at the day-ahead clearing price. The amount of energy cleared in the day-ahead market is not related to the actual power delivery in realtime.

The real-time system operation determines the actual power delivery based on the actual demand and supply. The real-time market, in contrast to the day-ahead market, is a *balancing market* that prices only the differences between day-ahead purchases and the actual delivery in realtime.

Modern power systems and electricity markets are highly complex. Here we present a stylized parametric model for real-time system and market operations, which captures essential features of the optimal economic dispatch, energy-reserve co-optimization, and related LMP computations.

2.3.1   A Multiparametric Model for Real-Time System and Market Operations

We defer to later sections for the detailed specification of several real-time operations. Here we present a generic multiparametric optimization model for real-time system and market operations.

Consider a multiparametric linear or quadratic program (MLP/MQP) of the following form

$$\minimize_{x} z(x) \text{ subject to } Ax \leq b + E\theta \qquad (y) \qquad (2.1)$$

where x is the decision variable typically representing the dispatch of generation or flexible load; $z(\cdot)$ the overall cost function; the inequalities are the constraints of generation and network; and y the vector of dual variables from which the energy prices are calculated. Special to this optimization is the parameter θ that captures the realized exogenous stochastic generations and demands.

The real-time system is operated in discrete periods, each of duration, say, 1 to 5 minutes. For a 24-hour operation, let $\theta_t, t = 1, 2, \ldots, T$ be the sequence of realized stochastic demands and (renewable) generations. The *single-period operation model* is to solve a sequence of optimizations of the form (2.1) for each realized θ_t to determine the optimal generation dispatch x_t^* and related LMP y_t^*. When θ_t's are drawn repeatedly from a probability distribution based on load and generation forecasts, we obtain samples of dispatch levels and LMPs from which their distributions can be estimated. Sections 2.3.2 and 2.3.4 describe techniques for solving economic dispatch and calculating LMPs under the single-period operation model.

In practice, there are the so-called *ramping constraints* for generators on how much the generation level can change from one interval to the next. This means that, given a sequence of realized demand θ_t, obtaining optimal dispatch independently one interval at a time according to the single-period operation model may violate the ramping constraints, leading to an infeasible dispatch sequence.

One approach to deal with significant ramping events is to call up the reserve in cases of shortage, which is a costly proposition. A more economic approach is to schedule generations based on a *multiperiod operation* model in which generation levels of the entire operation horizon are considered jointly. For instance, at time t, if the future load and stochastic generation can be forecasted perfectly as $\theta = (\theta_t, \theta_{t+1}, \ldots, \theta_{t+T'})$, the problem of jointly determining generation levels $x^* = (x_t^*, x_{t+1}^*, \ldots, x_{t+T'}^*)$ can be solved from (2.1).

In practice, one does not have the perfect prediction of $\theta_{t+1}, \ldots \theta_{t+T'}$, so the problem of determining the optimal sequence of dispatch becomes one of multiperiod stochastic (or robust) optimizations, for which the computation complexity is substantially higher. Practical solutions based on certainty equivalent heuristics or model predictive control (MPC) are used in practice. We present one such approach in Section 2.3.3.

We should note that, unlike computing (and forecasting) LMP under the single-period operation model, pricing dispatch under the multiperiod operation model is highly nontrivial and is an area of active research. See [21] and references therein.

2.3.2 Single-Period Economic Dispatch

Here we consider the problem of economic dispatch in the energy-only real-time market under the single-period operation model. The system operator sets generation adjustments by solving a so-called DCOPF problem in which the one-step-ahead real-time demand is balanced subject to system constraints.

For simplicity, we assume that each bus has a generator and a load. The single-period DCOPF problem solves, in each period t, for the optimal generation dispatch vector g_t^* given the forecast demand d_t in period t subject to generation and network constraints from the following optimization:

$$\underset{g}{\text{minimize}} \quad C_g(g) \tag{2.2}$$

$$\text{subject to} \quad \mathbf{1}^\top (g - d_t) = 0 \quad (\lambda_t) \tag{2.3}$$

$$S(g - d_t) \leq F^{\max} \quad (\mu_t) \tag{2.4}$$

$$G^{\min} \leq g \leq G^{\max} \tag{2.5}$$

$$\hat{g}_{t-1} - R^{\max} \leq g \leq \hat{g}_{t-1} + R^{\max} \tag{2.6}$$

where
- $C_g(\cdot)$ real-time generation cost function;
- $\mathbf{1}$ the vector of all ones;
- d_t vector of net load forecast at time t;
- g vector of ex-ante dispatch at time t;
- $\hat{g}_{t-1}$ vector of generation estimate at time $t-1$;
- d_t vector of one-step net load forecast at time t;
- $F^{\max}$ vector of transmission capacities;
- $G^{\max}$ vector of maximum generator capacities;
- $G^{\min}$ vector of minimum generator capacities;
- $R^{\max}$ vector of ramp limits;
- S power transfer distribution factor matrix;
- λ_t shadow price for the energy balance constraint at time t;
- μ_t shadow prices for transmission constraints at time t.

In the preceding optimization, the first equality constraint (2.3) represents the power balance dictated by the Kirchhoff law, the second (2.4) on the maximum power transfer over each branch, and the third (2.5) on the maximum and minimum restrictions on generations. The last constraint (2.6) is on the up and down ramping capabilities of generators from the previously set dispatch levels.

The preceding model clearly is an instance of the general parametric program defined in (2.1). In this model, the generation costs can be linear or quadratic. The real-time LMP π_{it} at bus i and time t is defined by the marginal cost of demand d_{it} at that bus. In other words, the LMP is the total cost increase induced by an ϵ increase of demand d_{it}. In the limit,

$$\pi_{it} = \frac{\partial}{\partial d_{it}} C(g^*(d_t)).$$

By the envelope theorem, the LMP vector π_t can be computed from the dual variable y^* as

$$\pi_t = \lambda_t^* \mathbf{1} - S^\top \mu_t^*, \qquad (2.7)$$

where the first term corresponds to the sum of energy prices λ^* and weighted sum of congestion prices μ^*. Note that the ith entry of μ^* corresponds to the constraint associated with the ith branch of the system. Thus, we have $\mu_i^* > 0$ only if the ith branch is *congested,* i.e., the power flow constraint on branch i is binding.

2.3.3 Multiperiod Economic Dispatch with Ramping Products

With increasing levels of variable energy resources and behind-the-meter generation, the operational challenge of ramping capability becomes more prominent. ISOs [22, 23] are adopting market-based "ramp products" to address the operational challenges of maintaining the power balance in the real-time dispatch. Here we present a multiperiod economic dispatch model [24] based on the so-called *flexible ramping product* (FRP) [22] recently adopted in the California Independent System Operator.

Given the load forecast $\bar{d}_t$ for the next T period, the following optimization, again in the general form of (2.1), produces a sequence of dispatch levels (g_t^*) and up and down ramping levels (r_t^{up}, r_t^{down}):

$$\min_{\{g_t, r_t^{up}, r_t^{down}\}} \sum_{t=t_0+1}^{t_0+T} C_g(g_t) + C_r^{up}\left(r_t^{up}\right) + C_r^{down}\left(r_t^{down}\right) \qquad (2.8)$$

$$\text{subject to } \forall t = t_0 + 1, \ldots, t_0 + T, \qquad (2.9)$$

$$\mathbf{1}^\top (g_t - \bar{d}_t) = 0, \qquad (\lambda_t) \qquad (2.10)$$

$$S(g_t - \bar{d}_t) \le F^{\max}, \qquad (\mu_t) \qquad (2.11)$$

$$g_t + r_t^{up} \le G^{\max}, \qquad (2.12)$$

$$g_t - r_t^{down} \ge G^{\min}, \qquad (2.13)$$

$$g_t - g_{t-1} + r_{t-1}^{up} + r_t^{up} \le R^{\max}, \qquad (2.14)$$

$$g_t - g_{t-1} + r_{t-1}^{down} + r_t^{down} \le R^{\max}, \qquad (2.15)$$

$$\mathbf{1}^\top r_t^{\text{up}} \geq R_t^{\text{up}}, \qquad (\alpha_t) \qquad (2.16)$$

$$\mathbf{1}^\top r_t^{\text{down}} \geq R_t^{\text{down}}, \qquad (\beta_t) \qquad (2.17)$$

$$r_t^{\text{up}} \geq 0, r_t^{\text{down}} \geq 0. \qquad (2.18)$$

where

$C^g(\cdot)$	energy generation cost function;
$C^{\text{up}}(\cdot)$	cost function to provide upward ramping;
$C^{\text{down}}(\cdot)$	cost function to provide downward ramping;
$\bar{d}_t$	vector of net load forecast at time t conditioning on the system state at time t_0;
g_t	vector of generation at time t;
r_t^{up}	vector of upward flexible ramping capacity at time t;
r_t^{down}	vector of downward flexible ramping capacity at time t;
F^{max}	vector of transmission capacities;
G^{max}	vector of maximum generator capacities;
G^{min}	vector of minimum generator capacities;
S	shift factor matrix;
R^{max}	vector of ramp limits;
R_t^{up}	upward ramping requirement of the overall system at time t;
R_t^{down}	downward ramping requirement of the overall system at time t.

This multiperiod economic dispatch is performed at time t_0 to minimize the overall system cost, consisting of energy generation cost and generation ramping cost, over time steps $t = t_0 + 1, t_0 + 2, \ldots, t_0 + T$. The constraints of power balance (2.10) and branch flow capacity (2.11) are the same as in the single-period economic dispatch. Constraints (2.12)–(2.15) correspond to the capacity of generations and ramping constraints. The rest of constraints (2.16)–(2.18) enforce the flexible ramping products to satisfy the ramping requirements.

Ramping requirements R_t^{up} and R_t^{down} will ensure that there is sufficient ramping capability available to meet the predicted net load. In practice, ISO uses the historical forecast error to calculate the distribution of ramping needs. The last constraint is the risk-limiting constraint, which implies that the system operator needs to meet the actual demand at all times with probability at least p.

One way to determine the ramping requirements is as follows. The values of R_t^{up} and R_t^{down} are chosen to achieve confidence level of p with respect to the point prediction of load $\bar{d}_t$ at time t:

$$\mathcal{P}\left[\mathbf{1}^\top \bar{d}_t - R_t^{\text{down}} \leq \mathbf{1}^\top d_t \leq \mathbf{1}^\top \bar{d}_t + R_t^{\text{up}}\right] \geq p \qquad (2.19)$$

where d_t is the actual load at time t.

Because the sequence of demand forecasts ($\bar{d}_t$) is never perfect, the further ahead of the forecast, the higher the forecast error, only the dispatch g_t^* is implemented in reality. The dispatch at time $t_0 + 1$ is determined by solving the preceding optimization upon receiving the updated forecasts. Such type of sequential decision processes follows MPC.

The MPC approach seeks to perform the economic dispatch for time steps $t = t_0 + 1, t_0 + 2, \ldots, t_0 + T$ under the condition that ramping capacity needs to be reserved for steps $t = t_0 + 1, t_0 + 2, \ldots, t_0 + T - 1$. Ramping capacity for time $t = t_0$ has been reserved in the previous time step, hence, there are no variables to be determined. Note that the load predictions are updated as time goes by. Hence, only the energy dispatch profile for $t = t_0$, i.e., g_{t_0}'s, and the flexible ramping requirements for $t = t_0 + 1$, i.e., $r_{t_0+1}^{\text{down}}$ and $r_{t_0+1}^{\text{up}}$, will be applied.

If the single-period pricing model is used, by the envelop theorem, the energy LMP vector π_t at time t is given by

$$\pi_t = \mathbf{1}\lambda_t^* + S^\top \mu_t^*. \tag{2.20}$$

The clearing prices for upward ramping and downward ramping at time t are α_t^* and β_t^*, respectively.

2.3.4 Energy and Reserve Co-optimization

In the joint energy and reserve market, dispatch and reserve are jointly determined using a linear program that minimizes the overall cost subject to operating constraints. In the co-optimized energy and reserve market, systemwide and locational reserve constraints are enforced by the market operator to procure enough reserves to cover the first and the second contingency events. We adopt the co-optimization model in [25] as follows:

$$\underset{g,r,s}{\text{minimize}} \quad \sum_i \left(c_i^g g_i + \sum_j c_{ij}^r r_{ij} \right) + \sum_u c_u^p s_u^l + \sum_v c_v^p s_v^s \tag{2.21}$$

$$\text{subject to} \quad \mathbf{1}^\top (g - d) = 0 \qquad (\lambda) \tag{2.22}$$

$$S(g - d) \leq F^{\max} \qquad (\mu) \tag{2.23}$$

$$\sum_{ij} \delta_{iju}^l r_{ij} + (I_u^{\max} - S_u^{\text{int}}(g - d)) + s_u^l \geq Q_u^l, \forall u \qquad (\alpha_u) \tag{2.24}$$

$$\sum_{ij} \delta_{ijv}^s r_{ij} + s_v^s \geq Q_v^s, \forall v \qquad (\beta_v) \tag{2.25}$$

$$G_i^{\min} \leq g_i + \sum_j r_{ij} \leq G_i^{\max}, \forall i \tag{2.26}$$

$$R^{\text{down}} \leq g - \hat{g}_{t-1} \leq R^{\text{up}} \tag{2.27}$$

$$0 \leq r \leq R^{\max} \tag{2.28}$$

$$s_u^l, s_v^s \geq 0, \forall u, v \tag{2.29}$$

where
- i index of buses;
- j index of reserve types, 10-min spinning, 10-min nonspinning, or 30-min operating reserve;
- u/v index of locational/systemwide reserve constraints;
- k index of transmission constraints;

d_i	net load at bus i;
g_i	dispatch of online generator at bus i;
r_{ij}	generation reserve of type j on bus i;
s^l/s^s	local/system reserve deficit of constraint;
c_i^g	cost for generation at bus i;
c_{ij}^r	cost for reserve type j at bus i;
$c_{u/v}^p$	penalty for reserve deficit of constraint u/v;
$I_u^{\max}$	interface flow limit for locational reserve constraint u;
$F^{\max}$	vector of transmission capacities;
Q^l/Q^s	locational/systemwide reserve requirement of constraint;
S	shift factor matrix for transmission lines;
S^{int}	shift factor matrix for interface flows;
δ_{iju}^l	binary value that is 1 when reserve j on bus i belongs to locational reserve constraint u;
δ_{ijv}^s	binary value that is 1 when reserve j on bus i belongs to systemwide reserve constraint v;
$G_i^{\max/\min}$	max/min generation capacity for generator at bus i;
$R^{\text{up/down}}$	vector of upward/downward ramp limits;
$R^{\max}$	vector of reserve capacities.

In this model, the real-time dispatch problem is formulated as a linear program with the objective to maximize the social surplus, subject to real-time operating constraints and the physical characteristics of resources. Energy balance constraint (2.22) and transmission constraint (2.23) are the same as in the economic dispatch for the energy-only market. The systemwide reserve requirement constraint (2.25) is enforced for the market operator to procure enough reserves to cover the first contingency events. A locational reserve constraint (2.24) is used to cover the second contingency event caused by the loss of a generator or a second line in a local area. Therefore, the unloaded tie-line capacity ($I_u^{\max} - S_u^{\text{int}}(g - d)$), as well as the reserve provided by units located in the local reserve zone, can be utilized to cover the local contingency or reserve requirement within 30 min. Note that the interface flow is calculated the same way as transmission line flow using the shift factor matrix S^{int}. Constraints (2.26)–(2.27) correspond to the generation capacity and ramping limits, respectively.

Based on the envelop theorem, the energy price vector π for all buses and the reserve clearing price ρ_{ij} of each reserve product j at bus i are defined as

$$\pi = \lambda^* \mathbf{1} - S^\top \mu^* + (S^{\text{int}})^\top \alpha^* \qquad (2.30)$$

$$\rho_{ij} = \sum_u \alpha_u^* \delta_{iju}^l + \sum_v \beta_v^* \delta_{ijv}^s \qquad (2.31)$$

where $\lambda^*, \mu^*, \alpha^*$, and β^* are the optimal values of Lagrangian multipliers.

Note again that the energy-reserve co-optimization model is also of the form of parametric program of (2.1) with parameter $\theta = (d_t, \hat{g}_{t-1})$ that is realized prior to the co-optimization.

2.4 Structural Solutions of Multiparametric Program

We have seen in the previous sections that many real-time market operations can be modeled as a general form of the multiparametric linear or quadratic program defined in (2.1). Here we present several key structural results on multiparametric linear/quadratic programming that we use to develop our approach. See [26–28] for multiparametric programming for more comprehensive expositions.

2.4.1 Critical Region and Its Geometric Structure

Recall the multiparametric (linear/quadratic) program defined earlier in (2.1)

$$\underset{x}{\text{minimize }} z(x) \text{ subject to } Ax \le b + E\theta \quad (y)$$

where the decision variable x is in $\mathcal{R}^n$, the cost function $z(x)$ is linear or quadratic, and the parameter θ is in a bounded subspace $\Theta \subset \mathcal{R}^m$.

Let the optimal primal solution be $x^*(\theta)$, the associated dual solution $y^*(\theta)$, and the value of optimization $z^*(\theta)$. We will assume that the MLP/MQP is not (primal or dual) degenerate[2] for all parameter values. Approaches for the degeneracy cases are presented in [28].

The solution structure of (2.1) is built upon the notion of *critical region partition*. There are several definitions for critical region. Here we adopt the definition from [28] under the primal/dual nondegeneracy assumption.

DEFINITION 2.1 *A critical region $\mathcal{C}$ is defined as the set of all parameters such that for every pair of parameters $\theta, \theta' \in \mathcal{C}$, their respective solutions $x^*(\theta)$ and $x^*(\theta')$ of (2.1) have the same active/inactive constraints.*

To gain a geometric insight into this definition and its implication, suppose that, for some θ_o, the constraint of (2.1) is not binding at the optimal solution, i.e., $Ax^*(\theta_o) < b + E\theta_o$. Then the constraint (2.1) is not binding for all θ's in a small enough neighborhood of θ_o. Thus, this small neighborhood belongs to the critical region $\mathcal{C}_o$ in which all linear inequality constraints of (2.1) are not binding at their own optima. We can then expand this neighborhood until one of the linear inequalities becomes binding. That particular linear equality (the newly binding inequality constraint) defines a boundary of $\mathcal{C}_0$ in the form of a hyperplane.

Conceptually, the preceding process defines the critical region $\mathcal{C}_0$ of the form of a polyhedron. For a quadratic objective function, because all linear inequality constraints are not binding for all $\theta \in \mathcal{C}_0$, the solution of (2.1) must all be of the form $x^*(\theta) = f_0$.

Similarly, if we take θ_1, for which only one linear inequality constraint is binding at $x^*(\theta_1)$, we can then obtain a different polyhedral critical region $\mathcal{C}_1$ containing θ_1, for which the same constraint binding condition holds. By ignoring all nonbinding constraints and eliminating one of the variables, say x_1 in the equality constraint, we

---

[2] For a given θ, (2.1) is said to be primal degenerate if there exists an optimal solution $x^*(\theta)$ such that the number of active constraints is greater than the dimension of x. By dual degeneracy we mean that the dual problem of (2.1) is primal degenerate.

can solve an unconstrained quadratic optimization with θ only appears in the linear term. Thus, for all $\theta \in \mathcal{C}_1$, the optimal solution is of a parametric affine form

$$x^*(\theta) = F_1 \theta + f_1 \text{ for all } \theta \in \mathcal{C}_1,$$

where F_1 and f_1 are independent of $\theta \in \mathcal{C}_1$.

The significance of the preceding parametric form of the solution is that, once we have (F_1, f_1), we no longer need to solve the optimization whenever $\theta \in \mathcal{C}_1$.

It turns out that every θ is in one and only one critical region. Because there is only a finite number of binding constraint patterns, the parameter space Θ is partitioned into a finite number of critical regions. In particular, within critical region $\mathcal{C}_i$, the optimal solution is of the form $x^*(\theta) = F_i \theta + f_i$ for all $\theta \in \mathcal{C}_i$.

2.4.2 A Dictionary Structure of MLP/MQP Solutions

We now make the intuitive arguments given in the preceding text precise both mathematically and computationally. In particular, we are interested in not only the existence of a set of a finite number critical regions $\{\mathcal{C}_i\}$ that partitions the parameter space but also the computation procedure to obtain these critical regions and their associated functional forms of the primal and dual solutions.

The following theorem summarizes the theoretical and computational basis for the proposed probabilistic forecasting approach.

THEOREM 2.1 ([16, 17]) *Consider (2.1) with cost function $z(x) = c^\top x$ for MLP and $z(x) = \frac{1}{2} x^\top Q x$ for MQP where Q is positive definite. Given a parameter θ_0 and the solution of the parametric program $x^*(\theta_0)$, let $\tilde{A}, \tilde{E}$ and $\tilde{b}$, be, respectively, the submatrices of A, E, and subvector of b corresponding to the active constraints. Let $\bar{A}, \bar{E},$ and $\bar{b}$ be similarly defined for the inactive constraints. Assume that (2.1) is neither primal nor dual degenerate.*

(1) *For the MLP, the critical region $\mathcal{C}_0$ that contains θ_0 is given by*

$$\mathcal{C}_0 = \{\theta | (\bar{A} \tilde{A}^{-1} \tilde{E} - \bar{E}) \theta < \bar{b} - \bar{A} \tilde{A}^{-1} \tilde{b} \}. \tag{2.32}$$

And for any $\theta \in \mathcal{C}_0$, the primal and dual solutions are given by, respectively,

$$x^*(\theta) = \tilde{A}^{-1}(\tilde{b} + \tilde{E}\theta), \quad y^*(\theta) = y^*(\theta_0).$$

(2) *For the MQP, the critical region $\mathcal{C}_0$ that contains θ_0 is given by*

$$\mathcal{C}_0 = \{\theta | \theta \in \mathcal{P}_p \bigcap \mathcal{P}_d\}, \tag{2.33}$$

where $\mathcal{P}_p$ and $\mathcal{P}_d$ are polyhedra defined by

$$\mathcal{P}_p = \{\theta | \bar{A} H^{-1} \tilde{A}^\top (\tilde{A} Q^{-1} \tilde{A}^\top)^{-1} (\tilde{b} + \tilde{E}\theta) - \bar{b} - \bar{E}\theta < \mathbf{0}\},$$
$$\mathcal{P}_d = \{\theta | (\tilde{A} Q^{-1} \tilde{A}^\top)^{-1} (\tilde{b} + \tilde{E}\theta) \leq \mathbf{0}\}.$$

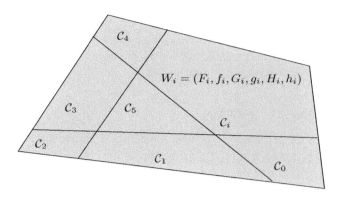

Figure 2.2 Illustration of the geometry of the solution structure on the parameter space Θ.

And for any $\theta \in \mathcal{C}_0$, the primal and dual solutions are given by

$$x^*(\theta) = H^{-1}\tilde{A}^\top(\tilde{A}H^{-1}\tilde{A}^\top)^{-1}(\tilde{b}+\tilde{E}\theta), \tag{2.34}$$

$$y_i^*(\theta) = \begin{cases} 0 & \text{the } i\text{th constraint is inactive} \\ -e_i^\top(\tilde{A}H^{-1}\tilde{A}^\top)^{-1}(\tilde{b}+\tilde{E}\theta) & \text{otherwise} \end{cases} \tag{2.35}$$

where e_i is the unit vector with ith entry equal to one and zero otherwise.

Figure 2.2 illustrates the geometric structure of MLP/MQP on the parameter space Θ partitioned by a finite number of polyhedral critical regions $\{\mathcal{C}_i\}$. We then attach each critical region $\mathcal{C}_i$ a unique signature, which we shall refer to as a *word* $W_i = (F_i, f_i, G_i, g_i, H_i, h_i)$, that defines completely the primal/dual solutions and the critical region. Specifically, the primal and dual solutions for $\theta \in \mathcal{C}_i$ is given by, respectively,

$$x^*(\theta) = F_i\theta + f_i, \ y^*(\theta) = G_i\theta + g_i, \tag{2.36}$$

and the critical $\mathcal{C}_i$ is defined by $\mathcal{C}_i = \{\theta | H_i\theta + h_i \leq 0\}$.

Thus, the complete solution of the MLP/MQP can be viewed as a *dictionary* in which each word of the dictionary defines the solution from within a critical region. In the next section, we present an online learning approach that learns the dictionary adaptively.

2.5 Probabilistic Forecasting through Online Dictionary Learning

In this section, we present an online learning approach to probabilistic forecasting aimed at achieving computation scalability for large networks and a high degree of accuracy that requires a large number of Monte Carlo runs. To this end, we first examine the computation costs of probabilistic forecasting, which motivates the development of an online dictionary learning solution.

In developing a tractable forecasting solution, we emulate the process by which a child learns how to speak a language. A natural way is to acquire words as she encounters them. Along the process, she accumulates a set of words that are most useful for her.

Even if a word was known to her earlier but is forgotten later, she can relearn the word, making the words less likely to forget. Among all the words in the language dictionary, only a small fraction of them are sufficient for all practical purposes, and the set of words that are most useful may change dynamically.

2.5.1 Complexity of Probabilistic Forecasting

As outlined in Section 2.2, the engine of the probabilistic forecast is the Monte Carlo simulations defined by MSMO. Thus, the main computation cost comes from repeatedly solving the optimal dispatch problem defined by the multiparametric program (2.1) for random samples generated from load and generation forecasts.

Consider a system of N buses. The structural results in Theorem 2.1 allows us to solve the linear or quadratic program (LP/QP) defined in (2.1) no more than K_N times where K_N is the number of critical regions. For fixed N, the complexity in terms of the number of LP/QP calls to achieve arbitrarily accurate probabilistic forecasting is finite. Let M be the number of Monte Carlo simulations needed to generate probabilistic forecasts. To achieve a consistent estimate of the conditional probability distributions, we require $M \to \infty$.

For fixed N, as M increases, the computation cost C_M per Monte Carlo run is bounded by

$$C_M \leq \lim_{M \to \infty} \frac{c_1 N^2 + c_2 K_N N^\alpha}{M} = c_1 N^2$$

where the first term in the numerator corresponds to cases when solving LP/QP are unnecessary and the second term for the cases when LP/QP has to be solved (only once per critical region) by a polynomial algorithm. This means that, the overall complexity per-Monte Carlo run is dominated by that of matrix-vector multiplication.

Note, however, that the worst-case complexity scaling with respect to the number of buses is dominated by $K_N N^\alpha$ with K_N potentially exponential with respect to N. The average cases, however, are likely to be far less pessimistic. In practice, typically, the number of binding constraints does not grow linearly with N, and K_N does not grow exponentially.

2.5.2 An Online Dictionary Learning Approach to Probabilistic Forecasting

We leverage the dictionary structure of the MLP/MQP to develop an online learning approach[3] in which critical regions are learned sequentially to overcome the curse of dimensionality and memory explosion. To this end, we also allow words previously learned, but rarely used, to be forgotten. It is the combination of sequential learning and dynamic management of remembered words that makes the simulation-based probabilistic forecasting scalable for large systems.

---

[3] Widely used in the signal processing community, dictionary learning refers to acquiring a dictionary of signal bases to represent a rich class of signals using words (atoms) in the dictionary [29, 30].

Given that, in the worst case, there may be exponentially (in N) a large number of critical regions for an N-bus system, obtaining analytical characterizations of all critical regions and the associated solution functions can be prohibitive. If, however, we are interested not in the worst case, but in the likely realizations of the stochastic load and generation parameter θ_t, not all critical regions need to be characterized. In fact, because θ_t represents the physical processes of load and generation, it is more likely that θ_t concentrates around its mean. As a result, each realization of θ_t may encounter a few critical regions.

A skeleton algorithm of the online dictionary approach is given in Algorithm 2.1. We assume that at time t, we have acquired a dictionary $\mathcal{D}_t = \{W_i, i = 1, \ldots, K_t\}$ with K_t entries, each corresponds to a critical region that has been learned from the past. Upon receiving a realization of random load and generation θ_t, the algorithm checks if θ_t belongs to a critical region whose word representation is already in the dictionary. This mounts to search for a word $W_i = (F_i, f_i, G_i, g_i, H_i, h_i)$ such that $H_i \theta_t + h_i \le 0$. If yes, the primal solution is given by the affine mapping defined by F_i and f_i and dual solution by G_i and g_i. Otherwise, we need to solve (2.1) and obtain a new word W according to Theorem 2.1.

The main intuition of the dictionary learning approach is that the parameter process θ_t represents the physical processes of aggregated load and generation. Such processes tend concentrates around its mean. As a result, each realization of θ_t may encounter a few critical regions. This intuition is confirmed in our experiments discussed in Section 2.6 where for a 3,000 bus system with 1 million Monte Carlo runs, less than 20 critical regions cover more than 99% of cases. Thus a dictionary of 20 words is mostly adequate for representing the level of randomness in the system.

Algorithm 2.1 Online Dictionary Learning

1: **Input:** the mean trajectory $\{\bar{d}_1, \bar{d}_2, \ldots, \bar{d}_T\}$ of load and associated (forecast) distributions $\{\mathcal{F}_1, \mathcal{F}_2, \ldots, \mathcal{F}_T\}$.
2: **Initialization:** compute the initial critical region dictionary $\mathcal{C}_0$ from the mean load trajectory.
3: **for** $m = 1, \ldots, M$ **do**
4:     **for** $t = 1, \ldots, T$ **do**
5:         Generate sample d_t^m and let $\theta_t^m = (d_t^m, g_{t-1}^m)$.
6:         Search $\mathcal{C}_{t-1}^m$ for critical region $C(\theta_t^m)$.
7:         **if** $C(\theta_t^m) \in \mathcal{C}_{t-1}^m$ **then**
8:             Compute g_t^m from the affine mapping $g^*_{C(\theta_t^m)}(\theta)$.
9:         **else**
10:            Solve g_t^m from DC-OPF (2.2–2.6) using θ_m^t, compute $C(\theta_t^m)$, and update $\mathcal{C}_t^m = \mathcal{C}_{t-1}^m \cup \{C(\theta_t^m)\}$.
11:         **end if**
12:     **end for**
13: **end for**
14: **Output:** the critical region dictionary $\mathcal{C}_T^M$.

Algorithmically, to limit the total memory required for the dictionary and the search space, a word-forgetting mechanism may be implemented by introducing a "timer" on each critical region.

Specifically, whenever the critical region is visited, its timer is reset to 1. Otherwise, the timer is reduced by a fraction of η, and the word is removed when the clock is below ϵ. Optimizing η allows us to control the memory size and speed up the search.

2.6 Numerical Simulations

We present in this section two sets of simulation results. The first compares the computation cost of the proposed method with that of direct Monte Carlo simulations. To this end, we used the 3210 "Polish network" [31]. The second set of simulations focuses on probabilistic forecasting. With this example, we aim to demonstrate the capability of the proposed method in providing joint and marginal distributions of LMPs and power flows, a useful feature not available in existing forecasting methods.

2.6.1 General Setup

We selected the "duck curve" [1] as the expected net load profile as shown in Figure 2.3. We were particularly interested in three scenarios: Scenario 1 represented a time ($T = 55$) when the net load was held steady at the midrange. Scenario 2 ($T = 142$) was when the net load was on a downward ramp due to the increase of solar power. Scenario 3 ($T = 240$) was at a time when the net load was at a sharp upward ramp. The three scenarios represented different operating conditions and different levels of randomness.

The net load – the load offset by renewable generation – was distributed throughout the network. A renewable generation connected to a bus, say a wind farm, was modeled

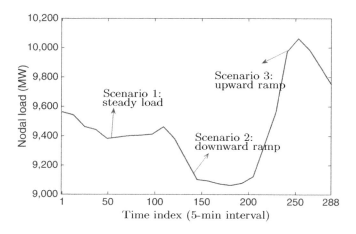

Figure 2.3 The "duck curve" of net load over the different time of the day.

as a Gaussian random variable $\mathcal{N}(\mu,(\eta\mu)^2)$ with mean μ and standard deviation $\eta\mu$. Similar models were used for load forecasts.

Given a forecasting or simulation horizon T, the real-time economic dispatch model was a sequence of optimizations with one DCOPF in each 5-minute interval. In this model, the benchmark technique solved a series of single-period DCOPF models with ramp constraints that coupled the DCOPF solution at time t with that at time $t-1$. Computationally, the simulation was carried out in a MATLAB environment with yalmip toolbox and IBM CPLEX on a laptop with an Intel Core i7-4790 at 3.6 GHz and 32 GB memory.

2.6.2 The 3120-Bus System

The 3120-bus system (Polish network) defined by MATPOWER [31] was used to compare the computation cost of the proposed method with direct Monte Carlo simulation [13]. The network had 3,120 buses, 3,693 branches, 505 thermal units, 2,277 loads, and 30 wind farms. Twenty of the wind farms were located at PV buses and the rest at PQ buses. For the 505 thermal units, each unit had upper and lower generation limits as well as a ramp constraint. Ten transmission lines (1, 2, 5, 6, 7, 8, 9, 21, 36, 37) had capacity limits of 275 MW.

The net load profile used in this simulation was the "duck curve" over a 24-hour simulation horizon. The total load was at the level of 27,000 MW during morning peak load hours with 10% of renewables distributed across 30 wind farms. One large wind farm had rated capacity of 200 MW, 20 midsize wind farms at the rated capacity of 150 MW, and 9 small wind farms at 50–80 MW. Wind farm i produced Gaussian distributed renewable power $\mathcal{N}(\mu_i,(0.03\mu_i)^2)$.

The left panel of Figure 2.4 shows the comparison of the computation cost between the proposed approach and the benchmark technique [13]. The two methods obtained identical forecasts, but ODL had roughly three orders of magnitude reduction in the number of DCOPF computations required in the simulation. This saving came from the

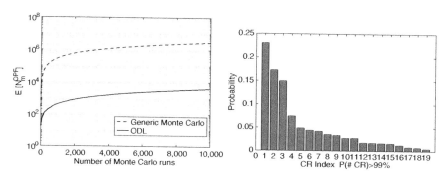

Figure 2.4 Left: The expected number of OPF computations versus the total number of Monte Carlo simulations. Right: The distribution of the critical regions observed for the proposed method.

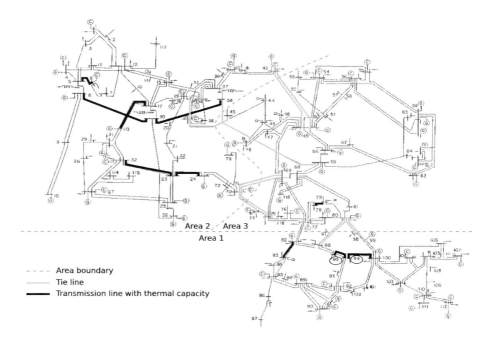

Figure 2.5 The diagram of IEEE 118-bus system.

fact that only 3,989 critical regions appeared in about 2.88 million random parameter samples. In fact, as shown in the right panel of Figure 2.4, 19 out of 3,989 critical regions represented 99% of all the observed critical regions.

2.6.3 The IEEE 118-Bus System

The performance of the proposed algorithm was tested on the IEEE 118-bus system [31] shown in Figure 2.5. The system was partitioned into three areas where the dashed lines show the boundaries and the maximum capacity of 175 MW was imposed on 10 transmission lines, which are highlighted by the thick black lines. The system included 54 thermal units with ramp limits, 186 branches, and 91 load buses, all of which were connected with Gaussian distributed load with standard deviation at the level of $\eta = 0.15\%$ of its mean. The mean trajectory of the net load again followed the "duck curve" with simulation/forecasting horizon of 24 hours. Three scenarios were tested, each included 1,000 Monte Carlo runs to generate required statistics.

Scenario 1: Steady Load

The first scenario was the forecast of LMP at $t = 55$. This was a case in which the system operated in a steady load regime where the load did not have a significant change. Figure 2.6 shows some of the distributions obtained by the proposed technique. The top left panel shows the average LMP at all buses where the average LMPs were relatively flat with the largest LMP difference appearing between bus 94 and bus 95. The top right

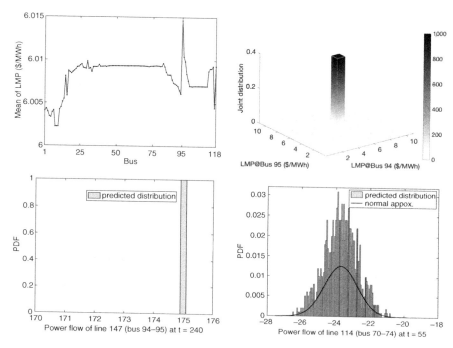

Figure 2.6 Top left: The expected LMPs at all buses. Top right: Joint LMP distribution at buses 94–95. Bottom left: Power flow distribution on line 147. Bottom right: Power flow distribution on line 114.

panel shows the joint LMP distribution at bus 95 and 94. It was apparent that the joint distribution of LMP at these two buses was concentrated at a single point mass, which corresponded to the case that all realizations of the random demand fell in the same critical region. The bottom left panel shows the power flow distribution at line 147 connecting bus 94 and 95. As expected, line 147 was congested. The bottom right panel shows the power flow distribution of line 114, which was one of the tie lines connecting areas 2 and 3. The distribution of power flow exhibited a single mode Gaussian-like shape.

Scenario 2: Downward Ramp

The second scenario at $t = 142$ involved a downward ramp. This was a case in which the load crossed the boundaries of multiple critical regions. In Figure 2.7, the top left panel shows the joint probability distribution of LMP at buses 94 and 95, indicating that the LMPs at these two buses had two possible realizations, one showing small LMP difference with a high probability, the other a bigger price difference with a low probability. The top right panel shows the power flow distribution on the line connecting bus 94 and 95. It was apparent that the line was congested with nonzero but relatively small probability, which gave rise to the more significant price difference between these two buses. The bottom panels show the power flow distributions on tie lines 115 and 153. In both cases, the power flow distribution had three modes, showing little resemblance of Gaussian distributions.

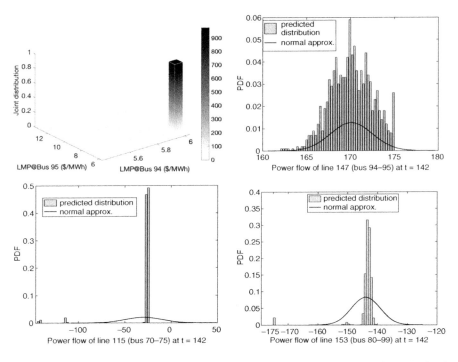

Figure 2.7 Top left: Joint LMP distribution at buses 94–95. Top right: Power flow distribution on line 147. Bottom left: Power flow distribution on line 115. Bottom right: Power flow distribution on line 153.

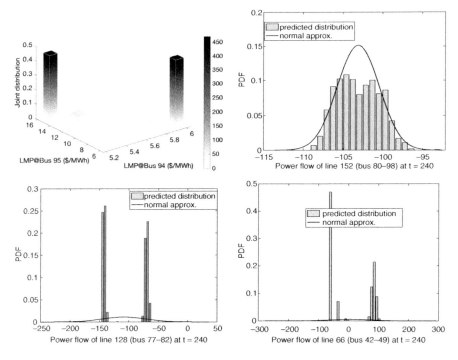

Figure 2.8 Top left: Joint LMP distribution at buses 94–95. Top right: Power flow distribution on line 152. Bottom left: Power flow distribution on line 128. Bottom right: Power flow distribution on line 152.

Scenario 3: Steep Upward Ramp

The third scenario at $t = 240$ involved a steep upward ramp at high load levels. This was also a case in which the random load crossed the boundaries of multiple critical regions. In Figure 2.8, the top left panel indicates 4 possible LMP realizations at buses 94 and 95. There is nearly 50% probability that the LMPs of buses 94 and 95 had a significant disparity, and the other 50% that these LMPs were roughly the same. The power flow on tie line 152 had a Gaussian-like distribution shown in the top right panel whereas tie line 128 had a power flow distribution spread in four different levels shown in the bottom left panel. It is especially worthy of pointing out, from the bottom right panel, that the power flow on line 66 had opposite directions.

2.7 Conclusion

We present in this chapter a new methodology of online probabilistic forecasting and simulation of the electricity market. The main innovation is the use of online dictionary learning to obtain the solution structure of parametric DCOPF sequentially. The resulting benefits are the significant reduction of computation costs and the ability to adapt to changing operating conditions. Numerical simulations show that, although the total number of critical regions associated with the parametric DCOPF is very large, only a tiny fraction of critical regions appear in a large number of Monte Carlo runs. These insights highlight the potential of further reducing both computation costs and storage requirements.

Acknowledgment

This work is supported in part by the National Science Foundation under Awards 1809830 and 1816397.

References

[1] California Independent System Operator, "What the duck curve tells us about managing a green grid," www.caiso.com/Documents/FlexibleResourcesHelpRenewables_FastFacts.pdf, accessed March 13, 2019.

[2] Electric Reliability Council of Texas, "Ercot launches wholesale pricing forecast tool," www.ercot.com/news/press_releases/show/26244, accessed March 13, 2019.

[3] Alberta Electric System Operator, "Price, supply and demand information," www.aeso.ca/market/market-and-system-reporting/, accessed March 13, 2019.

[4] Australian Energy Market Operator, "Data dashboard: Forecast spot price (pre-dispatch)," https://aemo.com.au/Electricity/National-Electricity-Market-NEM/Data-dashboard, accessed March 13, 2019.

[5] Q. P. Zheng, J. Wang, and A. L. Liu, "Stochastic optimization for unit commitment – a review," *IEEE Transactions on Power Systems*, vol. 30, no. 4, pp. 1913–1924, 2015.

[6] Independent Market Operator, "Wholesale electricity market design summary," www.aemo.com.au/-/media/Files/PDF/wem-design-summary-v1-4-24-october-2012.pdf, 2012, accessed March 13, 2019.

[7] G. Li, J. Lawarree, and C.-C. Liu, "State-of-the-art of electricity price forecasting in a grid environment," in *Handbook of Power Systems II*. Berlin and Heidelberg: Springer, pp. 161–187, 2010.

[8] R. Weron, "Electricity price forecasting: A review of the state-of-the-art with a look into the future," *International Journal of Forecasting*, vol. 30, no. 4, pp. 1030–1081, 2014.

[9] R. Bo and F. Li, "Probabilistic LMP forecasting considering load uncertainty," *IEEE Transactions on Power Systems*, vol. 24, no. 3, pp. 1279–1289, 2009.

[10] A. Rahimi Nezhad, G. Mokhtari, M. Davari, S. Hosseinian, and G. Gharehpetian, "A new high accuracy method for calculation of LMP as a random variable," in *International Conference on Electric Power and Energy Conversion Systems*, pp. 1–5, 2009.

[11] M. Davari, F. Toorani, H. Nafisi, M. Abedi, and G. Gharepatian, "Determination of mean and variance of LMP using probabilistic DCOPF and T-PEM," *2nd International Power and Energy Conference*, pp. 1280–1283, 2008.

[12] J. Bastian, J. Zhu, V. Banunarayanan, and R. Mukerji, "Forecasting energy prices in a competitive market," *IEEE Computer Applications in Power*, vol. 12, no. 3, pp. 40–45, 1999.

[13] L. Min, S. T. Lee, P. Zhang, V. Rose, and J. Cole, "Short-term probabilistic transmission congestion forecasting," in *Proc. of the 3rd International Conference on Electric Utility Deregulation and Restructuring and Power Technologies*, pp. 764–770, 2008.

[14] Y. Ji, J. Kim, R. J. Thomas, and L. Tong, "Forecasting real-time locational marginal price: A state space approach," in *Proc. of the 47th Asilomar Conference on Signals, Systems, and Computers*, pp. 379–383, 2013.

[15] Y. Ji, R. J. Thomas, and L. Tong, "Probabilistic forecast of real-time LMP via multiparametric programming," in *Proc. of the 48th Hawaii International Conference on System Sciences*, pp. 2549–2556, 2015.

[16] ——, "Probabilistic forecasting of real-time LMP and network congestion," *IEEE Transactions on Power Systems*, vol. 32, no. 2, pp. 831–841, 2017.

[17] W. Deng, Y. Ji, and L. Tong, "Probabilistic forecasting and simulation of electricity markets via online dictionary learning," in *Proceedings of the 50th Annual Hawaii International Conference on System Sciences*, pp. 3204-3210, 2017.

[18] T. Hong and S. Fan, "Probabilistic electric load forecasting: A tutorial review," *International Journal of Forecasting*, vol. 32, no. 3, pp. 914–938, 2016.

[19] Y. Zhang, J. Wang, and X. Wang, "Review on probabilistic forecasting of wind power generation," *Renewable and Sustainable Energy Reviews*, vol. 32, pp. 255–270, 2014.

[20] D. W. Van der Meer, J. Widén, and J. Munkhammar, "Review on probabilistic forecasting of photovoltaic power production and electricity consumption," *Renewable and Sustainable Energy Reviews*, vol. 81, pp. 1484–1512, 2018.

[21] Y. Guo and L. Tong, "Pricing multi-period dispatch under uncertainty," in *Proceedings of the 56th Annual Allerton Conference on Communication, Control, and Computing (Allerton)*, pp. 341–345, 2018.

[22] California Independent System Operator, "Flexible ramping product," www.caiso.com/informed/Pages/StakeholderProcesses/CompletedClosedStakeholderInitiatives/FlexibleRampingProduct.aspx, accessed March 3, 2019.

[23] N. Navid and G. Rosenwald, "Ramp capability product design for MISO markets," *Market Development and Analysis*, https://cdn.misoenergy.org/Ramp%20Product%20Conceptual%20Design%20Whitepaper271170.pdf, 2013; accessed May 17, 2020.

[24] C. Wu, G. Hug, and S. Kar, "Risk-limiting economic dispatch for electricity markets with flexible ramping products," *IEEE Transactions on Power Systems*, vol. 31, no. 3, pp. 1990–2003, 2016.

[25] T. Zheng and E. Litvinov, "Ex post pricing in the co-optimized energy and reserve market," *IEEE Transactions on Power Systems*, vol. 21, no. 4, pp. 1528–1538, 2006.

[26] T. Gal and J. Nedoma, "Multiparametric linear programming," *Management Science*, vol. 18, pp. 406–442, 1972.

[27] A. Bemporad, M. Morari, V. Dua, and E. N. Pistikopoulos, "The explicit linear quadratic regulator for constrained systems," *Journal of Automatica*, vol. 38, no. 1, pp. 3–20, 2002.

[28] F. Borrelli, A. Bemporad, and M. Morari, *Predictive Control for Linear and Hybrid Systems*. New York: Cambridge University Press, 2017.

[29] K. Kreutz-Delgado, J. F. Murray, B. D. Rao, K. Engan, T.-W. Lee, and T. J. Sejnowski, "Dictionary learning algorithms for sparse representation," *Neural Computation*, vol. 15, no. 2, pp. 349–396, 2003.

[30] R. Rubinstein, A. M. Bruckstein, and M. Elad, "Dictionaries for sparse representation modeling," *Proceedings of the IEEE*, vol. 98, no. 6, pp. 1045–1057, 2010.

[31] R. D. Zimmerman, C. E. Murillo-Sánchez, and R. J. Thomas, "MATPOWER: Steady-state operations, planning, and analysis tools for power systems research and education," *IEEE Transactions on Power Systems*, vol. 26, no. 1, pp. 12–19, 2011, https://matpower.org/; accessed May 17, 2020.

3 Deep Learning in Power Systems

Yue Zhao and Baosen Zhang

3.1 Introduction

Deep learning (DL) has seen tremendous recent successes in many areas of artificial intelligence. It has since sparked great interests in its potential use in power systems. However, success from using DL in power systems has not been straightforward. Even with the continuing proliferation of data collected in the power systems from, for example, Phasor Measurement Units (PMUs) and smart meters, how to effectively use the data, especially with DL techniques, remains a widely open problem. Firstly, the data collected for power systems are typically poorly labeled. Even for data with labels, there is often a mismatch between the labels needed for an intended task and the labels that are available. Secondly, power systems operate under normal conditions most of the time. Therefore, while it is natural for DL to be used for predicting interesting events in the grid, there is a bottleneck because of the extreme label asymmetry in any measured data. Thirdly, some forecasting tasks are fundamentally difficult. For example, no matter how complicated a DL algorithm is used, predicting wind generation in future hours or days is still limited by the fundamental uncertainty of weather. The lack and asymmetry of labels and the fundamental unpredictability make it nontrivial to effectively apply DL to power systems problems.

Recognizing these limitations of applying DL to power systems, this chapter offers two classes of problems in energy data analytics and power system monitoring with approaches that overcome the previously mentioned limitations.

In energy data analytics, we focus on deriving values from time series of renewable power generation (e.g., wind and solar). Instead of trying to forecast a single possible trajectory of future renewable power realization, the focus here is to provide *a realistic set of scenarios* of renewable energies. As such, (a) the ample data from historical renewable power generation are used in an unsupervised fashion, and, moreover, (b) the fundamental unpredictability is not a limiting factor for such tasks. Deep Generative Adversarial Networks (GANs) are employed in learning the probability distribution of renewable generation, and the power of DL is fully exploited. This is in contrast to conventional data generation methods that typically require (a) a probabilistic model to be specified, (b) historical data to be used to fit the parameters of the model, and (c) new scenarios to be generated by sampling the distribution. In such approaches, both the model fitting and the probabilistic sampling process are challenging. In addition, it is difficult to even specify the distribution for high-dimensional distributions that are not jointly Gaussian. GANs sidestep these difficulties by directly learning a neural network that transforms "noise" to data without specifying the particular parametric form of the data distribution. Because of the representation capability of deep neural networks, even complex spatial and temporal patterns can be learned directly from historical data.

In power system monitoring, the fundamental problem is an *inference* one: based on real-time sensor measurements, quantities of interests (e.g., system states and component statuses) need to be estimated. Instead of taking a purely data-driven approach and training a predictor from historical data, the approach here is to *exploit the information from both the data and the physical power system model*. In particular, unlike in artificial intelligence (AI) areas such as computer vision and natural language processing, power system possesses the unique advantage of having a well-understood and engineered physical model. Instead of deriving information from the physical model directly, which can be technically very challenging, we present an "indirect" approach that can fully exploit the power of DL: First, *labeled* data are generated through simulation in an unlimited fashion based on the underlying physical model. Second, a predictor is trained based on the generated data set, and captures the information embedded in the physical model. Notably, with model-based data generation, (a) there is no issue of lack of labels because labeled data can be cheaply generated, and (b) in principle infinitely more diverse situations can be captured in the data sets than those recorded in historical data, overcoming the issue of label asymmetry. These set up perfect conditions for the power of DL to be unleashed.

This chapter is organized as follows. Section 3.2 presents deep learning–based methods for energy data analytics, with a focus on renewable energy scenario generation. A learning framework using deep GAN is described. Numerical experiments performed with real-world wind and solar generation data are provided. Section 3.3 presents deep learning–based methods for power system monitoring, including tasks for situational awareness and preventive analysis. A Learning-to-Infer framework is described: based on model-based data generation, predictors are trained that approximate the theoretically optimal ones. Two case studies, multi-line outage detection and voltage stability margin estimation, are provided to illustrate the methodology. Concluding remarks are made in Section 3.4.

3.2 Energy Data Analytics with Deep Learning: Scenario Generation

High levels of renewable penetration pose challenges in the operation, scheduling, and planning of power systems. Because renewables are intermittent and stochastic, accurately modeling the uncertainties in them is key to overcoming these challenges. One widely used approach to capture the uncertainties in renewable resources is by using a set of time-series scenarios [1]. By using a set of possible power generation scenarios, renewables producers and system operators are able to make decisions that take uncertainties into account, such as stochastic economic dispatch/unit commitment, optimal operation of wind and storage systems, and trading strategies (e.g., see [2–4] and the references within).

Despite the tremendous advances recently achieved, scenario generation remains a challenging problem. The dynamic and time-varying nature of weather, the nonlinear and bounded power conversion processes, and the complex spatial and temporal interactions make model-based approaches difficult to apply and hard to scale, especially when multiple renewable power plants are considered. These models are typically constructed based on statistical assumptions that may not hold or are difficult to test in practice (e.g., forecast errors are Gaussian), and sampling from high-dimensional distributions (e.g., non-Gaussian) is also nontrivial [1]. In addition, some of these methods depend on certain probabilistic forecasts as inputs, which may limit the diversity of the generated scenarios and under-explore the overall variability of renewable resources.

3.2.1 Scenario Generation Using Generative Adversarial Networks

To overcome these difficulties, the recently popularized machine learning concept of GANs [5] is employed to fulfill the task of scenario generation [6]. Generative models have become a research frontier in computer vision and machine learning in general, with the promise of utilizing large volumes of unlabeled training data. There are two key benefits of applying such a class of methods. The first is that they can directly generate new scenarios based on historical data, without explicitly specifying a model or fitting probability distributions. The second is that they use unsupervised learning, avoiding a cumbersome manual labeling process that is sometimes impossible for large data sets.

`Network Architecture` The intuition behind GANs is to leverage the power of deep neural networks (DNNs) to both express complex nonlinear relationships (the generator) as well as classify complex signals (the discriminator). The key insight of a GAN is to set up a minimax two-player game between the generator DNN and the discriminator DNN (thus the use of "adversarial" in the name). During each training epoch, the generator updates its weights to generate "fake" samples trying to "fool" the discriminator network, while the discriminator tries to tell the difference between true historical samples and generated samples. In theory, at reaching the Nash equilibrium, the optimal solution of GANs will provide us a generator that can exactly recover the distribution of the real data so that the discriminator would be unable to tell whether a sample came from the generator or from the historical training data. At this point, generated scenarios are indistinguishable from real historical data, and are thus as realistic as possible. Figure 3.1 shows the general architecture of a GAN's training procedure under our specific setting.

3.2.2 GANs with Wasserstein Distance

The architecture of GANs we use is shown in Figure 3.1. Assume observations x_j^t for times $t \in T$ of renewable power are available for each power plant j, $j = 1, \ldots, N$. Let the true distribution of the observation be denoted by $\mathbb{P}_X$, which is of course unknown and hard to model. Suppose we have access to a group of noise vector input z under a known distribution $Z \sim \mathbb{P}_Z$ that is easily sampled from (e.g., jointly Gaussian). Our goal is to transform a sample z drawn from $\mathbb{P}_Z$ such that it follows $\mathbb{P}_X$ (without ever

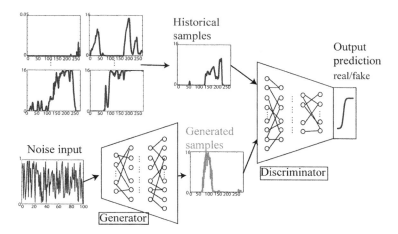

Figure 3.1 The architecture for GANs used for wind scenario generation. The input to the generator is noise that comes from an easily sampled distribution (e.g., Gaussian), and the generator transforms the noise through the DNN with the goal of making its output to have the same characteristics as the real historical data. The discriminator has two sources of inputs, the "real" historical data and the generator data, and tries to distinguish between them. After training is completed, the generator can produce samples with the same distribution as the real data, without efforts to explicitly model this distribution.

learning $\mathbb{P}_X$ explicitly). This is accomplished by simultaneously training two deep neural networks: the generator network and the discriminator network. Let G denote the generator function parametrized by $\beta^{(G)}$, which we write as $G(\cdot; \beta^{(G)})$; let D denote the discriminator function parametrized by $\beta^{(D)}$, which we write as $D(\cdot; \beta^{(D)})$. Here, $\beta^{(G)}$ and $\beta^{(D)}$ are the weights of two neural networks, respectively. For convenience, we sometimes suppress the symbol β.

Generator: During the training process, the generator is trained to take a batch of inputs and, by taking a series of up-sampling operations by neurons of different functions, to output realistic scenarios. Suppose that Z is a random variable with distribution $\mathbb{P}_Z$. Then $G(Z; \beta^{(G)})$ is a new random variable, whose distribution we denote as $\mathbb{P}_G$.

Discriminator: The discriminator is trained simultaneously with the generator. It takes input samples either coming from real historical data or coming from a generator, and by taking a series of operations of down-sampling using another deep neural network, it outputs a continuous value p_{real} that measures to what extent the input samples belong to $\mathbb{P}_X$. The discriminator can be expressed as

$$p_{real} = D(x; \beta^{(D)}) \tag{3.1}$$

where x may come from $\mathbb{P}_{data}$ or $\mathbb{P}_Z$. The discriminator is trained to learn to distinguish between $\mathbb{P}_X$ from $\mathbb{P}_G$, and thus to maximize the difference between $\mathbb{E}[D(X)]$ (real data) and $\mathbb{E}[D(G(Z))]$ (generated data).

With the objectives for discriminator and generator defined, we need to formulate loss function L_G for generator and L_D for discriminator to train them (i.e., update neural networks' weights based on the losses). To set up the game between G and D

so that they can be trained simultaneously, we also need to construct a game's value function $V(G, D)$. During training, a batch of samples drawn with distribution $\mathbb{P}_Z$ are fed into the generator. At the same time, a batch of real historical samples are fed into the discriminator. A small L_G shall reflect the generated samples are as realistic as possible from the discriminator's perspective, for example, the generated scenarios are looking like historical scenarios for the discriminator. Similarly, a small L_D indicates the discriminator is good at telling the difference between generated scenarios and historical scenarios, which reflect there is a large difference between $\mathbb{P}_G$ and $\mathbb{P}_X$. Following this guideline and the loss defined in [7], we can write L_D and L_G as follows:

$$L_G = -\mathbb{E}_Z[D(G(Z))] \tag{3.2}$$

$$L_D = -\mathbb{E}_X[D(X)] + \mathbb{E}_Z[D(G(Z))]. \tag{3.3}$$

Because a large discriminator output means the sample is more realistic, the generator will try to minimize the expectation of $-D(G(\cdot))$ by varying G (for a given D), resulting in the loss function in (3.2). On the other hand, for a given G, the discriminator wants to minimize the expectation of $D(G(\cdot))$, and at the same time maximize the score of real historical data. This gives the loss function in (3.3). Note the functions D and G are parametrized by the weights of the neural networks.

We then combine (3.2) and (3.3) to form a two-player minimax game with the value function $V(G, D)$:

$$\min_{\beta^{(G)}} \max_{\beta^{(D)}} V(G, D) = \mathbb{E}_X[D(X)] - \mathbb{E}_Z[D(G(Z))] \tag{3.4}$$

where $V(G, D)$ is the negative of L_D.

At the early stage of training, G just generates scenario samples $G(z)$ totally different from samples in $\mathbb{P}_X$, and discriminator can reject these samples with high confidence. In that case, L_D is small, and L_G, $V(G, D)$ are both large. The generator gradually learns to generate samples that could let D output high confidence to be true, while the discriminator is also trained to distinguish these newly fed generated samples from G. As training moves on and goes near to the optimal solution, G is able to generate samples that look as realistic as real data with a small L_G value, while D is unable to distinguish $G(z)$ from $\mathbb{P}_X$ with large L_D. Eventually, we are able to learn an unsupervised representation of the probability distribution of renewables scenarios from the output of G.

More formally, the minimax objective (3.4) of the game can be interpreted as the dual of the so-called Wasserstein distance (Earth–Mover distance). This distance measures the effort (or "cost") needed to transport the probability distribution of X to the probability distribution of Y. The connection to GANs comes from the fact that we are precisely trying to get two random variables, $\mathbb{P}_X(D(X))$ and $\mathbb{P}_Z(D(G(X)))$, to be close to each other.

3.2.3 Training GANs

In our setup, $D(x; \beta^{(D)})$ and $G(z; \beta^{(G)})$ are both differentiable functions that contain different neural network layers composed of multilayer perceptron, convolution, normalization, max-pooling, and Rectified Linear Units (ReLU). Thus, we can use standard training methods (e.g., gradient descent) on these two networks to optimize their performances. Training (see Alg. 3.1) is implemented in a batch updating style, while a learning rate self-adjustable gradient descent algorithm *RMSProp* is applied for weights updates in both discriminator and generator neural networks. Clipping is also applied to constrain $D(x;\beta^{(D)})$ to satisfy certain technical conditions as well as prevent gradients explosion.

Algorithm 3.1 Conditional GANs for Scenario Generation

Require: Learning rate α, clipping parameter c, batch size m, number of iterations for discriminator per generator iteration n_{discri}
Require: Initial weights $\beta^{(D)}$ for discriminator and $\beta^{(G)}$ for generator
  while $\beta^{(D)}$ has not converged **do**
    for $t = 0, \ldots, n_{discri}$ **do**
      # Update parameter for Discriminator
      Sample batch from historical data:
      $\{(x^{(i)}, y^{(i)})\}_{i=1}^{m} \; \mathbb{P}_X$
      Sample batch from Gaussian distribution:
      $\{(z^{(i)}, y^{(i)})\}_{i=1}^{m} from \; \mathbb{P}_Z$
      Update discriminator nets using gradient descent:
      $g_{\beta^{(D)}} \leftarrow \nabla_{\beta^{(D)}}[-\frac{1}{m}\sum_{i=1}^{m} D(x^{(i)}|y^{(i)}) + \frac{1}{m}\sum_{i=1}^{m} D(G(z^{(i)}|y^{(i)}))]$
      $\beta^{(D)} \leftarrow \beta^{(D)} - \alpha \cdot RMSProp(\beta^{(D)}, g_{\beta^{(D)}})$
      $\beta^{(D)} \leftarrow clip(w, -c, c)$
    end for
    # Update parameter for Generator
    Update generator nets using gradient descent:
    $g_{\beta^{(G)}} \leftarrow \nabla_{\beta^{(G)}} \frac{1}{m}\sum_{i=1}^{m} D(G(z^{(i)}|y^{(i)}))$
    $\beta^{(G)} \leftarrow \beta^{(G)} - \alpha \cdot RMSProp(\beta^{(G)}, g_{\beta^{(G)}})$
  end while

3.2.4 Examples

In this section we illustrate the algorithm on several different setups for wind and solar scenario generation. We first show that the generated scenarios are visually indistinguishable from real historical samples, then we show that they also exhibit the same statistical properties [8].

We build training and validation data sets using power generation data from NREL Wind[1] and Solar[2] Integration Datasets. The original data has resolution of 5 minutes. We choose 24 wind farms and 32 solar power plants located in the State of Washington to

---
[1] www.nrel.gov/grid/wind-integration-data.html
[2] www.nrel.gov/grid/sind-toolkit.html

use as the training and validating data sets. We shuffle the daily samples and use 80% of them as the training data, and the remaining 20% as the testing data sets. Along with the wind read measurements, we also collect the corresponding 24-hour-ahead forecast data, which is later used for conditional generation based on forecasts error. The 10% and 90% quantile forecasts are also available for Gaussian copula method setup. All these power generation sites are of geographical proximity that exhibit correlated (although not completely similar) stochastic behaviors. Our method can easily handle joint generation of scenarios across multiple locations by using historical data from these locations as inputs with no changes to the algorithm. Thus, the spatiotemporal relationships are learned automatically.

The architecture of our deep convolutional neural network is inspired by the architecture of DCGAN and Wasserstein GAN [7, 9]. The generator G includes 2 deconvolutional layers with stride size of 2×2 to firstly up-sample the input noise z, while the discriminator D includes two convolutional layers with stride size of 2×2 to down-sample a scenario x. The generator starts with fully connected multilayer perceptron for upsampling. The discriminator has a reversed architecture with a single sigmoid output. We observe two convolution layers are adequate to represent the daily dynamics for the training set, and are efficient for training. Details of the generator and discriminator model parameters are listed in the Table 3.1.

All models in this chapter are trained using RmsProp optimizer with a mini-batch size of 32. All weights for neurons in neural networks were initialized from a centered normal distribution with standard deviation of 0.02. Batch normalization is adopted before every layer except the input layer to stabilize learning by normalizing the input of every layer to have zero mean and unit variance. With exception of the output layer, ReLU activation is used in the generator and Leaky-ReLU activation is used in the discriminator. In all experiments, n_{discri} was set to 4, so that the models were training alternatively between four steps of optimizing D and one step of G. We observed model convergence in the loss for discriminator in all the group of experiments. Once the discriminator has converged to similar outputs value for $D(G(z))$ and $D(x)$, the generator was able to generate realistic power generation samples.

Table 3.1 The GANs model structure. MLP denotes the multilayer perceptron followed by number of neurons; Conv/Conv_transpose denotes the convolutional/deconvolutional layers followed by number of filters; Sigmoid is used to constrain the discriminator's output in [0,1].

	Generator G	**Discriminator** D
Input	100	24*24
Layer 1	MLP, 2048	Conv, 64
Layer 2	MLP, 1024	Conv, 128
Layer 3	MLP, 128	MLP, 1024
Layer 4	Conv_transpose, 128	MLP, 128
Layer 5	Conv_transpose, 64	

We also set up Gaussian copula method for scenario generation to compare the result with proposed method [10]. In order to capture the interdependence structure, we recursively estimated the Gaussian copula $\Sigma \in \mathbb{R}^{d \times d}$ based on d-dimension historical power generation observations $\{\mathbf{x_j}\}$, $j = 1, \ldots, N$ for N sites of interests. Then with a normal random number generator with zero mean and covariance matrix Σ, we are able to draw a group of scenarios (after passing through the Copula function).

3.2.5 Quality of Generated Scenarios

We first trained the model to validate that GANs can generate scenario with diurnal patterns. We then fed the trained generator with 2,500 noise vectors Z drawn from the predefined Gaussian distribution $z \sim \mathbb{P}_Z$. Some generated samples are shown in Figure 3.2 with comparison to some samples from the validation set. We see that the generated scenarios closely resemble scenarios from the validation set, which were not used in the training of the GANs. Next, we show that generated scenarios have two important properties:

1. *Mode Diversity*: The diversity of modes variation are well captured in the generated scenarios. For example, the scenarios exhibit hallmark characteristics of renewable generation profiles: for example, large peaks, diurnal variations, fast ramps in power. For instance, in the third column in Figure 3.2, the validating and generated sample both include sharp changes in its power. Using a traditional model-based approach to capture all these characteristics would be challenging, and may require significant manual effort.
2. *Statistical Resemblance*: We also verify that generated scenarios have the same statistical properties as the historical data. For the original and generated samples

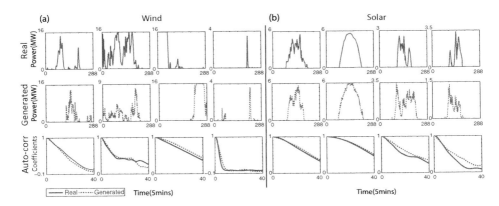

Figure 3.2 Selected samples from our validation sets (top) versus generated samples from our trained GANs (middle) for both wind and solar groups. The pair of samples are selected using Euclidean distance-based search. Without using these validation samples in training, our GANs are able to generate samples with similar behaviors and exhibit a diverse range of patterns. The autocorrelation plots (bottom) also verify generated samples' ability to capture the correct time-correlations.

shown in Figure 3.2, we first calculate and compare sample autocorrelation coefficients $R(\tau)$ with respect to time-lag τ:

$$R(\tau) = \frac{E[(S_t - \mu)(S_{t+\tau} - \mu)]}{\sigma^2} \qquad (3.5)$$

where S represents the stochastic time-series of either generated samples or historical samples with mean μ and variance σ. Autocorrelation represents the temporal correlation at a renewable resource, and capture the correct temporal behavior is of critical importance to power system operations. The bottom rows of Figure 3.2 verify that the real-generated pair has very similar autocorrelation coefficients.

In addition to comparing the stochastic behaviors in single time-series, in Figure 3.3 we show the cumulative distribution function (CDF) of historical validating samples and GANs-generated samples. We find that the two CDFs nearly lie on top of each other. This indicates the capability of GANs to generate samples with the correct marginal distributions.

PSD evaluates the spectral energy distribution that would be found per unit time. To verify the periodic component and temporal correlation of each individual

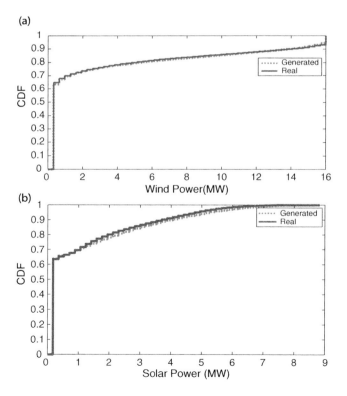

Figure 3.3 The top figure (a) hows the marginal CDFs of generated and historical scenarios of wind and solar. The two CDFs are nearly identical. The bottom figure (b) shows the power spectral density (PSD) plots for both wind and solar scenario generation.

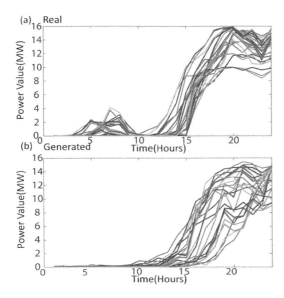

Figure 3.4 A group of one-day training (top) and generated (bottom) wind farm power output. The latter behaves similarly to the former both spatially and temporally.

scenario, we calculate the PSD ranging from 6 days to 2 hours for the validation set and generated scenarios. In Figure 3.3 we plot the results for wind scenarios and solar scenarios, respectively. We observe that for both cases, generated scenarios closely follow the overall shape of historical observations coming from the validation set.

Instead of feeding a batch of sample vectors $x^{(i)}$ representing a single site's diurnal generation profile, we can also feed GANs with a real data matrix $\{x^{(i)}\}$ of size $N \times T$, where N denotes the total number of generation sites, while T denotes the total number of timesteps for each scenario. Here we choose $N = 24$, $T = 24$ with a resolution of 1 hour. A group of real scenarios $\{x^{(i)}\}$ and generated scenarios $\{G(z^{(i)})\}$ for the 24 wind farms are plotted in Figure 3.4. By visual inspection we find that the generated scenarios retain both the spatial and temporal correlations in the historical data (again, not seen in the training stage).

3.3 Power System Monitoring with Deep Learning: Real-Time Inference

As a critical infrastructure that supports all sectors of society, our power system's efficient, reliable, and secure operation is crucially dependent on effective monitoring of the system. The real-time information on the current grid status is not only key for optimizing resources to economically maintain normal system operation but also crucial for identifying current and potential problems that may lead to blackouts. With increasing penetration of distributed energy resources (DERs), such as wind and solar energies and electric vehicles (EVs), power systems become even more dynamic. Faster power system monitoring that offers more accurate actionable information is needed.

The good news is that more and more real-time data are now collected by pervasive sensors deployed throughout power systems. However, while it is widely recognized that the proliferation of data offers great potential for grid monitoring, there are significant challenges toward extracting real-time actionable information from these data. In fact, it must be recognized that some of the essential grid prognosis tasks are fundamentally hard. For example, inferring complex scenarios of anomalies and predicting grid instability and cascading failures, all crucial for reliable operation of the grid, are fundamentally difficult due to the high dimensionality of the problems and the nonlinearity of the physical models of power systems. This section of the chapter introduces a new learning-based monitoring paradigm for overcoming these difficulties, exploiting the information from both the sensor data and the underlying physical model.

3.3.1 Model, Problems, and Challenges

Power systems are typically monitored in real time by a variety of sensors, including Phasor Measurement Units (PMUs) and sensors in Supervisory Control and Data Acquisition (SCADA) systems. The measured quantities (e.g., voltage, current, power flow) satisfy the underlying AC power flow equations:

$$P_m = V_m \sum_{n=1}^{N} V_n \left(G_{mn}\cos(\theta_m - \theta_n) + B_{mn}\sin(\theta_m - \theta_n) \right),$$

$$Q_m = V_m \sum_{n=1}^{N} V_n \left(G_{mn}\sin(\theta_m - \theta_n) - B_{mn}\cos(\theta_m - \theta_n) \right), \quad (3.6)$$

The vector of all measurements at a time instant t is denoted by,

$$\boldsymbol{y}(t) = \boldsymbol{h}(\boldsymbol{x}(t), \boldsymbol{s}(t); \boldsymbol{\alpha}) + \boldsymbol{v}(t), \quad (3.7)$$

where (a) $\boldsymbol{x}(t)$ represents the states of the system, i.e., the nodal complex voltages; (b) $\boldsymbol{s}(t)$ represents the component statuses, i.e., the on-off binary variables of whether each system component (among, e.g., transmission lines, generators, transformers) is in failure or not; and (c) $\boldsymbol{\alpha}$ represents the system component parameters, which typically stay constant for relatively long time scales. The function $\boldsymbol{h}(\cdot)$ represents the *noiseless* measurements, which is determined by the power flow equations (3.6), and $\boldsymbol{v}(t)$ is the measurement noise. In this section, we focus on semi-steady states of power systems, and hence drop the time index t in the notations.

Based on the observed measurements, the tasks of power system monitoring can be cast as computing a *desired function* $f(\boldsymbol{s}, \boldsymbol{x}, \boldsymbol{y}; \boldsymbol{\alpha})$. The goal of computing $f(\cdot)$ can be categorized into two types, (a) situational awareness, where we estimate the *current* values of certain variables in the system, and (b) preventive analysis, where we infer what *changes* of certain variables in the system can lead to system-wide failures, and thus provide an estimate of the *robustness* of the system against potential disruptions. Examples of situational awareness include

- Power system state estimation, where $f(\cdot) = x$, i.e., the nodal voltages, and
- Component outage detection, where $f(\cdot) = s$, i.e., the component statuses.

Examples of preventive analysis include

- Voltage stability analysis, where $f(\cdot)$ is the voltage stability margin, and
- Contingency analysis, where $f(\cdot)$ is whether the system is $N - k$ secure.

As such, computing $f(s, x, y; \alpha)$ based on y (cf.(3.7)) for situational awareness is an *inference* problem. As the focus of this section is power system monitoring in real time, we assume the knowledge of the long-term system component parameters α, and hence drop α in the notations later.

Model-Based Approaches and Limitations

The inference problems from y to $f(\cdot)$ have traditionally been solved by model-based approaches because power systems have well-defined physical models (cf. (3.6) and (3.7)).

For situational awareness, state estimation can be formulated as solving a nonlinear least square problem, assuming s is known, [11]

$$\min_{x} \| y - h(x,s) \|^2. \tag{3.8}$$

Joint component outage detection and state estimation can be formulated as a mixed continuous (x) and discrete (s) search problem.

$$\min_{x,s} \| y - h(x,s) \|^2. \tag{3.9}$$

In both problems, the observation model h (implied by the power flow model (3.6)) plays a key role in developing practical algorithms for solving the problems. For preventive analysis, computing the voltage stability margin can be formulated as finding the minimum disturbance in the system input (power injection) such that there is no feasible (or acceptable) voltage solution from the power flow equations (3.6). A popular approach to approximate this margin is to compute, as a surrogate, the smallest eigenvalue of the Jacobian from solving the power flow equations [12, 13]. Checking $N - k$ security is another combinatorial search problem, where one needs to check if the system can continue to operate under every combination of k-component outages, by re-solving the power flow equations.

However, such physical model-based approaches for power system monitoring face several notable limitations. First, the nonlinear nature of the power flow models (3.6) often leads to optimization problems that are non-convex and computationally hard to solve (e.g., (3.8)). Moreover, inference of discrete variables (e.g., component statuses, $N-k$ insecurity) is fundamentally hard due to the combinatorial complexity (e.g., (3.9)). More recent model-based approaches to solve these problems include convex relaxation, sparsity enforcement, and graphical models [14–16].

3.3.2 A Machine Learning–Based Monitoring

As opposed to model-based approaches, what we will describe next is a machine learning–based paradigm that both takes advantage of the power of deep learning and exploits the information from the physical models.

We begin with a general statistical model. Consider the system states x, component statuses s, and sensor measurements y are random variables that follow some joint probability distribution

$$p(x,s,y) = p(x,s)p(y|x,s). \tag{3.10}$$

Notably, the conditional probability $p(y|x,s)$ is fully determined by the observation model (3.7), which is again determined by the power flow equations (3.6) and the noise distribution.

Generative Model (3.10) represents a generative model [17] with which (a) x, s are generated according to a prior distribution $p(x,s)$, and (b) the noise-free measurements $h(\cdot)$ are also computed by solving the power flow equations (3.6), based on which the conditional probability distribution of the noisy measurements y, $p(y|x,s)$, is derived.

The goal of inferring some desired function $f(x,s,y)$ from observing y can be formulated as providing a *predictor* function $F(y)$ that, given inputs y, outputs values that are "closest" to $f(x,s,y)$. For example, in the case of state estimation, $f(x,s,y) = x$, and the goal is to find $F(y)$ that predicts the ground truth x in the closest sense.

Rigorously, the inference problem can be formulated as follows,

$$\min_{F \in \mathcal{F}} \mathbb{E}_{x,s,y} \left[L\left(f(x,s,y), F(y)\right) \right] \tag{3.11}$$

where (a) $\mathcal{F}$ is some class of functions from which a predictor $F(\cdot)$ is chosen, and (b) $L(f, F)$ is some loss function that characterizes the distance between f and F. We note that (3.11) is, in general, an infinite dimensional optimization problem. In practice, we typically restrict ourselves to some *parameterized* class of functions, so that (3.11) becomes an optimization over a set of parameters, and is finite dimensional.

Importantly, when $\mathcal{F}$ contains all possible functions, the optimal solution of (3.11), denoted by F^*, minimizes the *expected loss* among all possible predictor functions. We call this unconstrained optimal predictor $F^*(\cdot)$ the *target function*. For example, if $L(f, F) = (f - F)^2$, i.e., the squared error loss (assuming the desired function $f(\cdot)$ takes numeric values), it can be shown that the target function is the conditional expectation,

$$F^*(y) = \mathbb{E}_{x,s|y} \left[f(x,s,y) | y \right]. \tag{3.12}$$

In this case, the goal of inference is to compute this conditional expectation. Recall that the joint distribution $p(x,s,y)$ is in part determined by the nonlinear power flow equations (3.6). This makes computing the conditional expectation (3.12) difficult even for f as simple as $f = x$ as in state estimation. The physical model-based approaches such as (3.8) are precisely exploiting the role of power flow equations in $p(x,s,y)$ to simplify the computation of (3.12). More generally, computing (3.12) faces the challenges discussed in the previous section, which significantly limit the effectiveness of physical model-based approaches.

A Learning-to-Infer Apprach

There are two fundamental difficulties in performing the optimal prediction using the target function $F^*(y)$:

- Finding $F^*(y)$ can be very difficult (beyond the special case of $L(\cdot)$ being the squared error loss), and
- Even if $F^*(y)$ is conceptually known (see, e.g., (3.12)), it is often computationally intractable to be evaluated.

To address these two issues, we begin with an idea from variational inference [18]: we would like to find a predictor function $F(y)$ so that

- $F(y)$ matches the target function $F^*(y)$ as closely as possible, and
- $F(y)$ takes a form that allows fast evaluation of its value.

If such an $F(y)$ is found, it provides fast and near-optimal prediction of the desired function $f(x, s, y)$ based on any observed y. In fact, this procedure can precisely be formulated as in (3.11), where the class of functions $\mathcal{F}$ is set so that any $F \in \mathcal{F}$ allows fast evaluation of $F(y)$.

Solving the optimization problem (3.11), however, requires evaluating expectations over all the variables x, s and y. This is unfortunately extremely difficult due to (a) the nonlinear power flow equations embedded in the joint distribution, and (b) the exponential complexity in summing out the discrete variables s.

To overcome this difficulty, the key step forward is to transform the variational inference problem into a learning problem [19], by approximating the expectation with the empirical mean over a large number of Monte Carlo samples, generated according to (ideally) $p(x, s, y)$ (cf. (3.10)). We denote the Monte Carlo samples by $\{x^i, s^i, y^i; i = 1, \ldots, I\}$. Accordingly, (3.11) is approximated by the following,

$$\min_{F \in \mathcal{F}} \frac{1}{I} \sum_{i=1}^{I} L\big(f(x^i, s^i, y^i), F(y^i)\big). \qquad (3.13)$$

In practice, we typically set $\mathcal{F}$ to be a parameterized set of functions, with the parameters denoted by β. We thus have the following problem,

$$\min_{\beta} \frac{1}{I} \sum_{i=1}^{I} L\big(f(x^i, s^i, y^i), F_\beta(y^i)\big). \qquad (3.14)$$

With a data set $\{x^i, s^i, y^i\}$ generated using Monte Carlo simulations, (3.14) can then be solved as a deterministic optimization problem. The optimal solution of (3.14) approaches that of the original problem (3.11) as $I \to \infty$.

Notably, the problem (3.13) is an empirical risk minimization problem in machine learning, as it trains a discriminative predictor $F_\beta(y)$ with a data set $\{x^i, s^i, y^i\}$ generated from a generative model $p(x, s, y)$. As a result of this offline learning/training process (3.14), a near-optimal predictor $F_{\beta^*}(y)$ is obtained. Table 3.2 summarizes this "Learning-to-Infer" approach.

Table 3.2 The Learning-to-Infer Method

Offline computation:
1. Generate a data set $\{x^i, s^i, y^i\}$ using Monte Carlo simulations with the power flow and sensor models.
2. Select a parametrized function class $\mathcal{F} = \{F_\beta(y)\}$.
3. Train the function parameters β with $\{x^i, s^i, y^i\}$ using (3.14).

Online inference (in real time):
1. Collect instant measurements y from the system.
2. Compute the prediction $F_{\beta^*}(y)$.

3.3.3 Case Study 1: Multi-line Outage Identification

In this section, we demonstrate how machine learning–based monitoring applies to a situational awareness task: line outage identification. In power networks, transmission line outages, if not rapidly identified and contained, can quickly escalate to cascading failures. Real-time line outage identification is thus essential to all network control decisions for mitigating failures. In particular, because the first few line outages may have already escaped the operators' attention, the ability to identify in real time the network topology with an *arbitrary* number of line outages becomes critical to prevent system collapse.

System Model

We consider a power system with N buses, and its *baseline topology* (i.e., the network topology when there is no line outage) with M lines. The binary variable s_m denotes the status of a line m. $s = [s_1, \ldots, s_M]^T$ thus represents the actual topology of the network. We denote the real and reactive power injections at all the buses by $P, Q \in \mathbb{R}^N$, and the voltage magnitudes and phase angles by $V, \theta \in \mathbb{R}^N$. In particular, given the network topology s and a set of controlled input values $\{P, Q^{in}, V^{in}\}$ (where Q^{in} and V^{in} consist of some subsets of Q and V, respectively), the remaining values of $\{Q, V, \theta\}$ can be determined by solving (3.6).

Learning to Infer Line Outages

We are interested in identifying the post-outage network topology s in real time based on instant measurements y collected in the power system. Ideally, we would like to compute the posterior conditional probabilities $p(s|y), \forall s$. However, as there are up to 2^M possibilities for s, even *listing* the probabilities $p(s|y), \forall s$ has an exponential complexity.

Instead, we focus on computing the posterior *marginal* conditional probabilities $p(s_m|y), m = 1, \ldots, M$. Note that the posterior marginals are characterized by just M numbers, $\mathbb{P}(s_m = 1|y), m = 1, \ldots, M$, as opposed to $2^M - 1$ numbers required for characterizing $p(s|y)$. Although listing the posterior marginals $p(s_m|y)$ are tractable, computing them, however, still remains intractable. In particular, even with $p(s|y)$ given, summing out all $s_k, k \neq m$, to obtain $p(s_m|\mathbf{y})$ still requires exponential computational complexity [20].

We now employ the machine learning–based framework in Section 3.3.2. The desired function $f(\cdot)$ corresponds to the set of conditional marginals $p(s_m|\mathbf{y}), \forall m$, and $F_\beta(\mathbf{y})$ corresponds to *approximate* conditional marginals, denoted by $q_\beta(s_m|\mathbf{y}), \forall m$. The objective of (3.11) corresponds to minimizing some distance between $p(s_m|\mathbf{y})$ and $q_\beta(s_m|\mathbf{y})$, for which we choose the Kullback–Leibler (KL) divergence: $\forall m$,

$$D(p\|q_\beta)_m \triangleq \sum_{s_m} p(s_m|\mathbf{y}) \log \frac{p(s_m|\mathbf{y})}{q_\beta(s_m|\mathbf{y})}. \tag{3.15}$$

The expected loss minimization (3.11) becomes (with some simple algebra,) $\forall m$,

$$\max_\beta \mathbb{E}_{s_m, \mathbf{y}} \left[\log q_\beta(s_m|\mathbf{y}) \right]. \tag{3.16}$$

With a generated data set $\{\mathbf{x}^i, \mathbf{s}^i, \mathbf{y}^i\}$, the learning problem (3.14) becomes, $\forall m$,

$$\max_\beta \frac{1}{I} \sum_{i=1}^{I} \log q_\beta(s_m^i|\mathbf{y}^i), \tag{3.17}$$

which is equivalent to finding the *maximum likelihood estimation* of the predictor function parameters β.

Numerical Experiments

An experiment of the Learning-to-Infer method on identifying line outages in the IEEE 300 bus system is given in the following text.

Data Set Generation As opposed to considering only a small number of simultaneous line outages, we generate the line statuses $\{s_m\}$ using independent and identically distributed (IID) Bernoulli random variables, so that the average numbers of line outages is 11.6. To capture a wide range of system states, the bus voltage phase angles θ are generated as IID uniformly distributed random variables in $[0, 0.2\pi]$. $2.2M$ data samples are generated, of which subsets will be used for training, validation, and testing. All the generated multi-line outages are distinct from each other. As a result, these generated data sets can very well evaluate the *generalizability* of the trained classifiers, as all data samples in the test set have post-outage topologies *unseen* in the training set.

Training Setup Neural networks are employed as the parametric functions $q_\beta(s_m|\mathbf{y}), \forall m$: given the input data $\mathbf{y}$, the logistic functions in the output layer of the neural network will produce the probabilities $q_\beta(s_m|\mathbf{y})$ (based on which binary identification decisions are then made). In particular, we employ a neural network architecture that allows classifiers for different lines to *share features*: Instead of training M separate neural networks each with one node in its output layer, we train *one* neural network whose output layer consists of M nodes each predicting a different line's status. Specifically, three-layer (i.e., one hidden layer) fully connected neural networks and Rectified Linear Units (ReLUs) are employed.

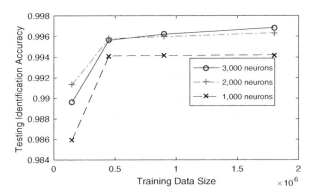

Figure 3.5 Effect of function model size and sample size, IEEE 300 bus system.

Testing Results We vary the size of the hidden layer of the neural network as well as the training data size, and evaluate the learned predictors. The testing results are plotted in Figure 3.5. It is observed that the best performance is achieved with $1.8M$ data and with 3,000 neurons. Further increasing the data size or the neural network size would see much diminished returns. Specifically, an outage identification accuracy of 0.997 is achieved. This translates to that, out of on average 11.6 simultaneous line outages, only one line status on average would be misidentified by the learned predictor.

3.3.4 Case Study 2: Voltage Stability Margin Estimation

In this section, we demonstrate how machine learning–based monitoring applies to a preventive analysis task: voltage stability margin estimation. Voltage collapse has been one of the major causes for large-scale blackouts [21]. With renewable energies increasingly integrated into power systems, their volatilities need to be handled in system operations, pushing the system even closer to its physical capacity. As a result, determining the system stability margin in real time is greatly valuable for system operators to maintain a safe operating margin.

Computing the power system voltage stability margin amounts to finding the distance from the current operating point to the boundary of the stability region. A major limitation of the existing approaches is that they are primarily for exploring the voltage stability limit *along a specific loading direction*, with the continuation power flow (CPF) method [22] as a classic tool of this kind. However, finding the voltage stability margin of an operating point requires searching over a very large number of directions to identify the "worst-case" loading direction, along which the current operating point is *closest* to voltage instability. Sidestepping the problem of finding the worst-case loading direction, another approach is to approximate the voltage stability margin using easily computable metrics, notably the smallest singular value (SSV) of the Jacobian matrix [12, 13] computed from power flow algorithms. Such metrics, however, are only approximate.

System Model

A power profile S, that is, the real and reactive power injections P_k and Q_k of all the buses k[3], can either induce a voltage collapse or not. A voltage collapse means that, given S, the power flow equations (3.6) do not admit a feasible voltage solution [23]. In this case, we call this power profile S unstable. A power system's voltage stability region is defined to be the set of all power profiles that do not induce voltage collapse, denoted by $\mathcal{C}$. $\mathcal{C}^c$ is thus the voltage *instability* region.

The voltage stability margin at an operating point S is the distance of S to the voltage instability region $\mathcal{C}^c$,

$$dist\left(S, \mathcal{C}^c\right) \triangleq \min_{S' \in \mathcal{C}^c} \|S - S'\|_2. \tag{3.18}$$

In other words, the voltage stability margin captures the *minimum* power injection perturbation that, if applied, would lead to a voltage collapse.

Learning to Infer Stability Margin with Transfer Learning

To solve (3.18), one needs to sample a large number of directions (a.k.a. loading directions), and search along each of them starting from S by running CPF. This does not work for *real-time* voltage stability margin inference, especially in medium- to large-scale power systems where S can have hundreds to thousands of dimensions.

We now employ the machine learning–based framework in Section 3.3.2. The desired function $f(\cdot)$ is $dist\,(S, \mathcal{C}^c)$ (cf. (3.18)), and $F_\beta(S)$ corresponds to approximate voltage stability margins. A squared error loss is employed, i.e., $L(\cdot) = (dist\,(S, \mathcal{C}^c) - F_\beta(S))^2$.

However, one key challenge arises: as described above, computing the "label," i.e., $dist\,(S, \mathcal{C}^c)$, even for just one simulated operating point S would consume a considerable time. As such, to construct a relatively large data set, even to be done offline, is not practical, and yet is crucial for effective training especially for high-dimensional inference as needed in power systems of reasonable sizes. This difficulty practically limits our ability to use direct supervised learning for training a margin predictor, simply because it is too time consuming to generate a sufficiently large labeled data set.

To overcome this challenge, we observe the following fact: for an operating point, while computing its voltage stability margin by searching is very time consuming (e.g., minutes), verifying *whether it is stable or not*, nonetheless, is very fast (e.g., milliseconds). Thus, within similar time limits, one can generate a data set of *[operating point, binary stability label]* with a size many orders of magnitude larger than a data set of *[operating point, voltage stability margin label]*. As such, while it is infeasible to generate a margin-labeled data set sufficiently large to capture the high-dimensional boundary of the voltage stability region $\mathcal{C}$, it is feasible to generate a sufficiently large binary stability-labeled data set that does so. The problem is, however, training on a data set with only the binary stability labels does not offer us a predictor that outputs stability margins.

---

[3] More generally, one can consider a subset of buses of interests, which can also include PV buses in addition to PQ buses.

Transfer Learning The key step forward is to use *transfer learning* to jointly exploit both the information embedded in a large binary stability-labeled data set and that in a small margin-labeled data set, with the end goal of obtaining an accurate voltage stability margin predictor [24]. In particular, we (a) train a neural network–based binary classifier from a large binary stability-labeled data set, (b) take the trained hidden layer of the NN as a feature extractor, with the hope that it implicitly captures sufficient information of the boundary of $\mathcal{C}$, and (c) add an additional layer of NN to fine-tune based on only a small margin-labeled data set. In a sense, we transfer the knowledge learned in the binary classifier in order to make it tractable to learn a margin predictor based on only a small data set with stability margin labels.

Numerical Experiments

An experiment of the Transfer Learning method on voltage stability margin inference in the IEEE 118 bus system is given in the following text.

Data Set Generation To construct a *binary stability-labeled data set*, we find a large number of pairs of stable and unstable operating points (S_i, S_o) *near the boundary* of $\mathcal{C}$. We find such boundary sample pairs by running CPF starting from the origin along different directions toward different target points. To generate the target points, we sample the power factor for each bus with uniform distribution $U[0.4, 1]$. We record labeled data pairs as $(S_i, 1)$ and $(S_o, 0)$, where 1 indicates voltage stability and 0 indicates instability. A total of $200K$ data points are generated. To construct a *stability margin-labeled data set*, for any two feasible operating points S_1 and S_2 generated in the preceding text, we shrink them by a random factor. We then apply the CPF algorithm with a starting point S_1, searching along the direction of $S_2 - S_1$ to find the distance to $\mathcal{C}^c$ along this particular direction. For one S_1, we search along 7,000 directions, each time with a different S_2, and pick the minimum distance as the approximate margin. We repeat this procedure for 1,000 different S_1 and generate $1K$ data points.

Training Setup Based on the $200K$ data set of *[operating point S, binary stability label 0/1]*, we train a neural network classifier $\hat{g}(S)$ with one hidden layer and ReLU activation [25]. Based on the only $1K$ data set of *[operating point S, stability margin $dist\,(S,\mathcal{C}^c)]$*, we first import the weights of the hidden layer from the pre-trained binary classifier $\hat{g}(S)$ as a feature extractor, and then add another hidden layer for regression on the $1K$ margin-labeled data set to train an overall margin predictor $\hat{d}(S)$.

Testing Results As a benchmark method, we employ a widely used voltage stability margin approximator – the smallest singular value (SSV) of the Jacobian matrix from running power flow algorithms. The SSV provides us a measure of how close the Jacobian is to singularity, implying voltage instability. We evaluate the SSVs as the predictors to fit the $1K$ margin-labeled data set. The resulting MSE is 15,876.

In evaluating the transfer learning method, we use 64 neurons in the hidden layer for the classifier $\hat{g}(S)$. Next, we transfer the hidden layer of the trained classifier $\hat{g}(S)$ to learn a stability margin predictor $\hat{d}(S)$ based on the $1K$ margin-labeled data set: 800 data are for training and 200 for testing. We use 256 neurons for the newly added layer in the predictor $\hat{d}(S)$. The trained predictor $\hat{d}(S)$ achieves a testing MSE of 1,624.

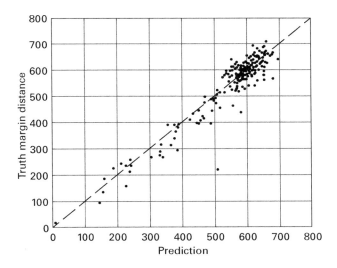

Figure 3.6 Scatterplot from using the proposed method.

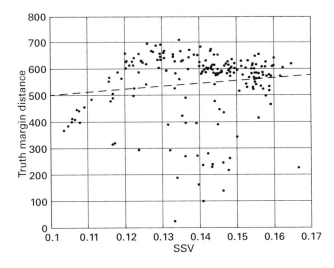

Figure 3.7 Scatterplot from using Jacobian's SSVs.

The scatterplots with the transfer learning method and with the benchmark method using Jacobian's SSV are shown in Figures 3.6 and 3.7, respectively. It is clear both from the MSEs and the scatterplots that the proposed method significantly outperforms the benchmark using Jacobian's SSV.

3.4 Conclusion

Effective use of Deep Learning techniques relies on the availability of massive data sets for dedicated tasks. Data sets collected for power systems are, however, often not well labeled. To break through this limitation, this chapter presents two classes of problems with approaches that successfully exploited the power of deep learning.

The first class of problems is scenario generation based on historical data, with a focus on renewable power generation. The historical renewable generation data are used for learning in an unsupervised fashion. GANs are designed for generating realistic scenarios. The method not only generates scenarios with great fidelity but also has significant advantages over existing methods in its flexibility, offered by deep learning.

The second class of problems is real-time power system monitoring. A crucial property of a power system is that it possess a well-defined and engineered physical model. A Learning-to-Infer method is presented that exploits the information embedded in the physical model using a data-driven approach. This approach first generates massive labeled data from physical-model–based simulations, and then learn predictors from the generated data where the power of deep learning is exploited. Two case studies, multi-line outage identification and voltage stability margin estimation, are presented to illustrate the methodology.

Deep learning has the potential to dramatically improve the situational awareness in power systems, especially for a grid with significant penetration of renewable resources in which handling uncertainties becomes key. This is not without challenges, however, because the data has to be integrated with physical laws. Therefore, adapting and designing new neural network structures specifically for power system applications would be a fruitful area of research.

Lastly, deep learning methods based on model-based simulated data also have great potential in offering powerful tools for power system operations. Prominent examples include solving computational challenging problems such as security constrained optimal power flow (OPF) and unit commitment (UC). It is of great interest to overcome the fundamental computational complexity in these problems with the help of deep learning.

References

[1] J. M. Morales, R. Minguez, and A. J. Conejo, "A methodology to generate statistically dependent wind speed scenarios," *Applied Energy*, vol. 87, no. 3, pp. 843–855, 2010.

[2] W. B. Powell and S. Meisel, "Tutorial on stochastic optimization in energy part I: Modeling and policies," *IEEE Transactions on Power Systems*, vol. 31, no. 2, pp. 1459–1467, 2016.

[3] C. Zhao and Y. Guan, "Unified stochastic and robust unit commitment," *IEEE Transactions on Power Systems*, vol. 28, no. 3, pp. 3353–3361, 2013.

[4] Y. Wang, Y. Liu, and D. S. Kirschen, "Scenario reduction with submodular optimization," *IEEE Transactions on Power Systems*, vol. 32, no. 3, pp. 2479–2480, 2017.

[5] I. Goodfellow, J. Pouget-Abadie, M. Mirza, B. Xu, D. Warde-Farley, S. Ozair, A. Courville, and Y. Bengio, "Generative adversarial nets," in *Advances in Neural Information Processing Systems*, Montréal, Quebec, Canada, pp. 2672–2680, Dec. 2014.

[6] Y. Chen, Y. Wang, D. Kirschen, and B. Zhang, "Model-free renewable scenario generation using generative adversarial networks," *IEEE Transactions on Power Systems*, vol. 33, no. 3, pp. 3265–3275, May 2018.

[7] M. Arjovsky, S. Chintala, and L. Bottou, "Wasserstein GAN," *arXiv preprint arXiv:1701.07875*, 2017.

[8] P. Pinson and R. Girard, "Evaluating the quality of scenarios of short-term wind power generation," *Applied Energy*, vol. 96, pp. 12–20, 2012.

[9] A. Radford, L. Metz, and S. Chintala, "Unsupervised representation learning with deep convolutional generative adversarial networks," *arXiv:1511.06434*, 2015.

[10] P. Pinson, H. Madsen, H. A. Nielsen, G. Papaefthymiou, and B. Klöckl, "From probabilistic forecasts to statistical scenarios of short-term wind power production," *Wind Energy*, vol. 12, no. 1, pp. 51–62, 2009.

[11] A. Abur and A. Gomez-Exposito, *Power System State Estimation: Theory and Implementation*. New York: Marcel Dekker, 2004.

[12] A. Tiranuchit et al., "Towards a computationally feasible on-line voltage instability index," *IEEE Transactions on Power Systems*, vol. 3, no. 2, pp. 669–675, 1988.

[13] P.-A. Lof, T. Smed, G. Andersson, and D. Hill, "Fast calculation of a voltage stability index," *IEEE Transactions on Power Systems*, vol. 7, no. 1, pp. 54–64, 1992.

[14] V. Kekatos and G. B. Giannakis, "Joint power system state estimation and breaker status identification," *Proc. North American Power Symposium (NAPS)*, Champaign, IL, pp. 1 to 6, Sept. 2012.

[15] H. Zhu and G. B. Giannakis, "Sparse overcomplete representations for efficient identification of power line outages," *IEEE Transactions on Power Systems*, vol. 27, no. 4, pp. 2215–2224, Nov. 2012.

[16] J. Chen, Y. Zhao, A. Goldsmith, and H. V. Poor, "Line outage detection in power transmission networks via message passing algorithms," in *Proc. 48th Asilomar Conference on Signals, Systems and Computers*, Pacific Grove, CA, pp. 350–354, 2014.

[17] C. M. Bishop, *Pattern Recognition and Machine Learning*. New York: Springer, 2006.

[18] D. Koller and N. Friedman, *Probabilistic Graphical Models: Principles and Techniques*. Cambridge, MA: MIT Press, 2009.

[19] Y. Zhao, J. Chen, and H. V. Poor, "A Learning-to-Infer method for real-time power grid multi-line outage identification," *IEEE Transactions on Smart Grid*, vol. 11, no. 1, pp. 555–564, Jan. 2020.

[20] M. Mezard and A. Montanari, *Information, Physics, and Computation*. Oxford: Oxford University Press, 2009.

[21] G. Andersson et al., "Causes of the 2003 major grid blackouts in North America and Europe, and recommended means to improve system dynamic performance," *IEEE Transactions on Power Systems*, vol. 20, no. 4, pp. 1922–1928, 2005.

[22] V. Ajjarapu and C. Christy, "The continuation power flow: A tool for steady state voltage stability analysis," *IEEE Transactions on Power Systems*, vol. 7, pp. 416–423, 1992.

[23] J. D. Glover, M. Sarma, and T. Overbye, *Power System Analysis & Design*. Stamford, CT: Cengage Learning, 2011.

[24] J. Li, Y. Zhao, Y.-h. Lee, and S.-J. Kim, "Learning to infer voltage stability margin using transfer learning," in *Proc. IEEE Data Science Workshop (DSW)*, Minnesota, MN, 2019.

[25] V. Nair and G. E. Hinton, "Rectified linear units improve restricted Boltzmann machines," in *Proceedings of the 27th International Conference on Machine Learning*, Haifa, Israel, pp. 807–814, 2010.

4 Estimating the System State and Network Model Errors

Ali Abur, Murat Göl, and Yuzhang Lin

4.1 Introduction

This chapter provides an overview of the methods used for monitoring large interconnected power systems to ensure their secure operation. Monitoring involves both the system states, i.e., voltage phasors at all system buses, and network model parameters. Power systems have been commonly monitored by measurements collected at transmission substations by the Supervisory Control and Data Acquisition (SCADA) systems. These measurements are referred as SCADA measurements and comprise of bus voltage magnitude, bus real and reactive power injections, branch real and reactive power flows, and branch current magnitude (ampere). In some systems, taps and/or phase shifts associated with load tap changing (LTC) transformers are also measured. In the last couple of decades, synchronized voltage and current phasor measurements that are provided by phasor measurement units (PMUs) have been added to the available set of measurements from substations. SCADA and PMU measurements are not compatible due to the significant difference in their update rates and the fact that PMU measurements are synchronized while SCADA measurements are not.

Given the slowly changing nature of system loads and generation, it has been sufficient to scan SCADA measurements every few seconds. These scanned set of measurements are then processed to obtain the best estimates of bus voltage phasors (or the system states) corresponding to that scan. This is referred as static state estimation because it implicitly assumes that the system is in a pseudo steady-state operating condition that is characterized by the estimated system states. If one is interested in tracking the system states during power system transients caused by switching, sudden load or generation changes, short circuits, etc. then dynamic states associated with generators and loads will have to be incorporated in the estimation problem formulation. A dynamic state estimator will have to be used for monitoring power systems during such transients. This chapter will only review static state estimation that is used to monitor power systems during pseudo steady-state operation.

To simplify the terminology, the rest of the chapter will simply use term state estimation to mean static state estimation. Given the substantially different sets of PMU and SCADA measurements, depending on their numbers and availability state estimation can be carried out using strictly PMU, SCADA, or a mix of both types of measurements. When using a measurement set with mixed type of measurements, several practical implementation questions need to be addressed such as how to reconcile the different scan rates between SCADA and PMU measurements, how to select measurement weights, how often to execute state estimation particularly when there are relatively few PMU measurements, etc. Another important consideration is the choice of an estimation method that will be computationally fast to keep up with the scan rates of the used measurements and also capable of handling gross errors in measurements. Finally, the estimator should be able to not only detect but also distinguish between gross errors in measurements and network parameters.

4.2 Power System State Estimation Using SCADA Measurements

The idea of using redundant SCADA measurements to estimate the state of the power system was first introduced by F. C. Schweppe in [1–3]. Schweppe's problem formulation uses the following assumptions:

- Power systems operate under balanced three phase conditions and, therefore, they can be represented by their positive sequence models.
- Measurements contain errors that can be assumed to have a Gaussian distribution with zero mean and known variances under normal operating conditions. Occasional gross errors may appear in a small subset of measurements and they need to be detected and removed to avoid biased state estimates.
- Network topology and parameters are perfectly known.
- Measurement errors are independent, yielding a diagonal error covariance matrix.

The following equation can be written for a set of m measurements represented by the vector z for a system with n states:

$$z = h(x) + e \tag{4.1}$$

where:
$h^T = [h_1(x), h_2(x), \ldots, h_m(x)]$
$h_i(x)$ is the nonlinear function relating measurement z_i to the state vector x
$x^T = [x_1, x_2, \ldots, x_n]$ is the system state vector
$e^T = [e_1, e_2, \ldots, e_m]$ is the vector of measurement errors.

Following the preceding assumptions, the statistical properties of the measurement errors will be:

- $E(e_i) = 0, \quad i = 1, \ldots, m$.
- $E[e_i e_j] = 0$.
- $Cov(e) = E[e \cdot e^T] = R = \text{diag}\{\sigma_1^2, \sigma_2^2, \ldots, \sigma_m^2\}$.

The standard deviation σ_i of each measurement i is assigned a value reflecting its assumed accuracy.

The most commonly used method for power system state estimation has so far been the weighted least squares (WLS) where the estimator minimizes the following objective function:

$$J(x) = \sum_{i=1}^{m}(z_i - h_i(x))^2/R_{ii}$$
$$= [z - h(x)]^T R^{-1}[z - h(x)] \quad (4.2)$$

$J(x)$ provides a measure of how well the measurements satisfy their estimated values that are obtained by substituting the estimated state in the nonlinear measurement equations. This implicitly assumes that the nonlinear measurement function $h(x)$ is perfectly known, i.e. the network topology and all model parameters used in formulating $h(x)$ are accurate. This assumption will be relaxed later in the chapter and methods of detecting errors in model parameters will be presented.

To find the optimal solution $\hat{x}$ that minimizes the objective function $J(\hat{x})$, one needs to solve the following nonlinear set of equations:

$$\frac{\partial J(x)}{\partial x} = -H^T(x)R^{-1}[z - h(x)] = 0 \quad (4.3)$$

where $H(x) = \left[\frac{\partial h(x)}{\partial x}\right]$

Equation 4.3 can be solved using the following iterative scheme:

$$\hat{x}^{k+1} = \hat{x}^k + [G(\hat{x}^k)]^{-1} \cdot H^T(\hat{x}^k) \cdot R^{-1} \cdot (z - h(\hat{x}^k))$$

where k is the iteration index,

$\hat{x}^k$ is the state estimate at iteration k,

$G(\hat{x}^k) = H^T(\hat{x}^k) \cdot R^{-1} \cdot H(\hat{x}^k)$.

Note that in the preceding iterative scheme, there is no need to invert the matrix $G(\hat{x})$. Instead, it can be factorized using the Cholesky method (as for obervable systems it is symmetric and positive definite) and the lower and upper triangular factors can be used to solve for the change in the estimated state $\Delta\hat{x}(k+1) = \hat{x}^{k+1} - \hat{x}^k$:

$$[G(\hat{x}^k)]\Delta\hat{x}^{k+1} = H^T(\hat{x}^k)R^{-1}[z - h(\hat{x}^k)] \quad (4.4)$$

Solution for equation 4.4 exists provided that the matrix $G(\hat{x})$, which is also referred as the "gain" matrix, is nonsingular. This condition will be satisfied when the system is observable, i.e. when there are sufficient number of properly placed measurements in the system. Hence, network observability needs to be checked prior to executing state estimation.

4.3 Network Observability

Network observability analysis is an integral part of state estimation because measurements may become temporarily unavailable due to communication system failures, meter or instrument transformer errors, or reconfiguration of substation topology and, as a result, the system may become unobservable and state estimation cannot be carried out. In such cases, the operators need to identify the observable subsystems (or islands) within the overall system and consider using the so called pseudo-measurements to merge these observable islands and restore full network observability. The types of pseudo-measurements used for this purpose include bus load forecasts and generator scheduled outputs. Because these values are in general not accurate, they are assigned relatively low weights compared to SCADA measurements.

Network model parameters or the system state do not have any effect on observability of a given system. Network observability can be analyzed strictly based on the measurement configuration and network topology. Topological methods [4, 5] can be used to determine if a network is observable and, if not, to determine the observable islands. Alternatively, methods based on numerical analysis of the gain matrix [6, 7] exist to identify observable islands. Details of implementing both types of methods can be found in these references. Irrespective of the method used, if a system is found to be unobservable, pseudo-measurements need to be used to merge existing observable islands. The branches connecting observable islands are referred as "unobservable branches" because their flows cannot be determined. Pseudo-measurements are selected such that they are incident to these unobservable branches.

Power system measurements are classified into two mutually exclusive categories, namely the "critical" and "redundant" measurements. If the removal of a measurement from the measurement set causes the system to become unobservable, then that measurement will be referred as a "critical" measurement. All measurements that are not critical are referred as "redundant" measurements. As will be discussed later in this chapter, critical measurements represent vulnerability points for bad data processing because errors in critical measurements cannot be detected. Hence, a well-designed measurement system should have a minimum number of critical measurements.

4.4 Bad Data Processing

One of the important functions of any state estimator is to detect, identify, and remove measurements with gross errors. The most commonly used state estimators in power systems are based on the weighted least squares (WLS) method, which is known to be not robust against such gross errors in measurements. In fact, even a single measurement with a gross error can significantly bias the estimated state or cause divergence of the state estimation algorithm. Hence, WLS state estimators typically use a postestimation procedure to detect and remove measurements with gross errors. Assuming that the WLS state estimator converges to a solution, the value of the objective function $J(x)$ in equation 4.2 can be shown to follow an approximate χ^2 distribution and thus can be checked against a threshold. This threshold can be selected based on the desired confidence level (e.g., 95%), and degrees of freedom (difference between the total number of measurements and system states). This is referred as the χ^2 test for bad data detection. Once bad data are detected, their identification and subsequent removal is accomplished by applying the largest normalized residual (LNR) test after the state estimator converges and measurement residuals are computed. Hence, both of these tests can only be applied if the state estimator converges that may not always be the case.

Measurement residuals can be expressed in terms of the unknown measurement errors, e using the residual sensitivity matrix S:

$$r = z - \hat{z} = Se \tag{4.5}$$

It can be shown that [8]:

$$S = I - HG^{-1}H^T R^{-1} \tag{4.6}$$

Taking the covariance of both sides in equation 4.5 and substituting equation 4.6 for S, residual covariance matrix Ω can be readily derived as:

$$Cov(r) = \Omega = E[rr^T] = S \cdot R$$

LNR test is based on the analysis of measurement residuals. To decide whether or not a measurement is sufficiently accurate to be safely used in state estimation, the expected value of its corresponding residual is compared with the computed value. This derivation assumes a normal distribution for the measurement residuals with zero mean and a covariance matrix of Ω as derived in the preceding text, thus:

$$r \sim N(0, \Omega)$$

Note that large off-diagonal elements of the residual covariance matrix point to strongly interacting measurements. The LNR test involves an iterative process in which absolute value of each residual is first normalized by using the square root of the corresponding diagonal entry in Ω and then these are sorted. If the largest normalized residual is above a detection threshold ϵ, then the corresponding measurement is suspected of bad data and removed from the measurement set. The choice of ϵ is usually based on 99.9% confidence level that yields a value of 3.0. Iterative process is terminated when the largest normalized residual falls below the detection threshold.

4.5 Power System State Estimation in the Presence of PMUs

PMUs, which provide synchronized voltage and current phasor measurements, were invented in 1988 by Phadke and Thorp at Virginia Polytechnic Institute and State University [9]. However, their widespread deployment in power systems was not before the Northeast Blackout of 2003. The final report of the U.S.-Canada Power System Outage Task Force on the blackout recommended the use of PMUs to provide real-time wide-area grid visibility [10]. Later, the U.S. Department of Energy sponsored multiple well-funded projects to deploy PMUs in power systems, and as a result, the number of PMUs installed in the U.S. power grid has rapidly increased, such that while there were 166 installed PMUs in 2010, this number exceeded 2,500 in 2017. As the number of PMUs increased in power systems, it became inevitable to incorporate those devices in power system state estimation. In this section, an overview of this incorporation will be presented.

4.5.1 Phasor Measurement–Based State Estimation

PMUs measure the voltage and current phasors in a synchronized manner thanks to the time-stamps provided by the GPS. Those synchrophasor measurements are linearly related to system states as shown in equation 4.7 and equation 4.8, where the real and imaginary parts of bus voltage phasors are the system states.

$$V_k^m = Re\{V_k\} + jIm\{V_k\} + e_V \tag{4.7}$$

$$I_{ij}^m = Re\{I_{ij}\} + jIm\{I_{ij}\} + e_I \tag{4.8}$$

$$Re\{I_{ij}\} = G_{ij}(Re\{V_i\} - Re\{V_j\}) - B_{ij}(Im\{V_i\} - Im\{V_j\}) - B_{ii}Im\{V_i\} \tag{4.9}$$

$$Im\{I_{ij}\} = G_{ij}(Im\{V_i\} - Im\{V_j\}) + B_{ij}(Re\{V_i\} - Re\{V_j\}) + B_{ii}Re\{V_i\} \tag{4.10}$$

where,
$G_{ij} + jB_{ij}$ is the series admittance of the branch connecting buses i and j,
B_{ii} is the shunt admittance at bus-i.
The superscript m is used to indicate that the considered quantity is a measurement.
This linear relation can be expressed in compact form as follows.

$$z = Hx + e \tag{4.11}$$

where,
$H(2m \times 2n)$ is the measurement Jacobian, which is independent of system states,
$x = [x_1^r \ x_2^r \ \ldots \ x_n^r \ x_1^i \ x_2^i \ \ldots \ x_n^i]^T (2n \times 1)$ is the system state vector,
x_j^r and x_j^i are the real and imaginary parts of j-th state, respectively,
$z = [z_1^r \ z_2^r \ \ldots \ z_m^r \ z_1^i \ z_2^i \ \ldots \ z_m^i]^T (2m \times 1)$ is the measurement vector, and
z_i^r and z_i^i are the real and imaginary parts of ith measurement, respectively.

In the presence of sufficient PMU measurements to make the system observable, WLS estimation will become noniterative and fast, thanks to the linearity between the PMU measurements and the system states. Consider a system with n buses, and measured by m phasor measurements. As measurements are linearly related to the system states, the estimated states can be computed directly as follows using WLS.

$$\hat{x} = (H^T R^{-1} H)^{-1} H^T R^{-1} z \tag{4.12}$$

where,
$R(2m \times 2m)$ is the diagonal measurement covariance matrix.

As seen in equation 4.12, WLS gives the solution without any iteration, and hence remains as the most computationally efficient estimator for power system applications. Despite its computational performance; however, WLS is vulnerable to bad data. Therefore, a measurement postprocessing is required for bad data analysis. It is known that the bad-data analysis and removal has a significant computational burden, as it requires recomputation of the system states if bad data is encountered in the measurement set.

Considering that the bad data analysis should be carried out every time a PMU measurement set is received (up to 60 times a second), use of a robust state estimator becomes an effective alternative.

The Least Absolute Value (LAV) estimator is known to be robust against bad data, such that the LAV estimator rejects bad data automatically, provided that necessary measurement redundancy exists [11]. Its computational performance in power system state estimation is not very good in the presence of power measurements. However, if the measurement set consists of solely PMU measurements, the computational burden of the LAV estimator becomes superior to that of WLS estimator followed by the bad data processing [12]. The LAV estimation problem can be written as follows, for the measurement set defined in equation 4.11.

$$\min \quad c^T |r| \quad (4.13)$$
$$\text{s.t.} \quad r = z - H\hat{x} \quad (4.14)$$

where,
$c(2m \times 1)$ is the weight vector,
$r = [r_1^r \ r_2^r \ \ldots \ r_m^r \ r_1^i \ r_2^i \ \ldots \ r_m^i]^T (2m \times 1)$ is the measurement residual vector,
$r_i^r = z_i^r - \hat{z}_i^r$ and $r_i^i = z_i^i - \hat{z}_i^i$ are the real and imaginary parts of ith measurement residual, respectively.
$\hat{x}_i^r$ and $\hat{x}_i^i$ are the real and imaginary parts of ith estimated measurement, respectively.

The LAV estimation problem can be formulated as a linear programming problem, which can be solved using efficient linear programming solvers. The rearranged equations and defined new strictly nonnegative variables are given here.

$$\min \quad c^T y \quad (4.15)$$
$$\text{s.t.} \quad My = z \quad (4.16)$$
$$y \geq 0 \quad (4.17)$$
$$c^T = [0_n \ 1_m] \quad (4.18)$$
$$y = [\hat{x}_a \ \hat{x}_b \ u \ v] \quad (4.19)$$
$$M = [H \ -H \ I \ -I] \quad (4.20)$$

where,
$0_n (1 \times 2n)$ is the vector of '0's,
$1_m (1 \times 2m)$ is the vector of '1's,
$\hat{x} = \hat{x}_a^T - \hat{x}_b^T$
$r = u^T - v^T$

The bad data rejection property of LAV is at risk in the presence of leverage measurements. A leverage point can be defined as an observation (x_k, y_k) where x_k lies far away from bulk of the observed x_i's in the sample space [13]. In power systems, corresponding row of a leverage measurement in H is an outlier, such that entries of those rows are either very large or very small compared to values in the remaining rows. Leverage measurements have very small residuals, and hence they force the estimator to closely satisfy their values even when these values are not correct. If a leverage measurement carries bad data, it is likely that estimates of the LAV estimator will be biased. In a power system there are few conditions that lead to presence of leverage measurements:

- An injection measurement at a bus incident to large number of branches.
- An injection measurement placed at a bus incident to branches with very different impedances.
- Flow measurements on the lines with impedances, which are very different from the rest of the lines.
- Using a very large weight for a specific measurement.

The PMU-only LAV estimation is a special case, where only the third condition may cause presence of leverage measurements, as PMUs provide bus voltage and current flow phasor measurements. Considering that, one can eliminate the leveraging effect of the measurement by scaling both sides of equation 4.11, thanks to the linearity between the PMU measurements and system states. The following example illustrates scaling, where y_{km} is the self-admittance of the line between buses k and m.

$$\begin{bmatrix} V_1 \\ V_2 \\ I_{12} \\ I_{23} \end{bmatrix} = \begin{bmatrix} 1 & 0 & 0 \\ 0 & 1 & 0 \\ y_{12} & -y_{12} & 0 \\ 0 & y_{23} & -y_{23} \end{bmatrix} \begin{bmatrix} \hat{V}_1 \\ \hat{V}_2 \\ \hat{V}_3 \end{bmatrix}$$

Once the current measurements are scaled, the following relation can be obtained. Note that charging susceptances of the lines are ignored for simplicity. If charging susceptances are included, one may scale the corresponding relation simply using the largest number in the corresponding row of H as the scaling factor.

$$\begin{bmatrix} V_1 \\ V_2 \\ I_{12}/y_{12} \\ I_{23}/y_{23} \end{bmatrix} = \begin{bmatrix} 1 & 0 & 0 \\ 0 & 1 & 0 \\ 1 & -1 & 0 \\ 0 & 1 & -1 \end{bmatrix} \begin{bmatrix} \hat{V}_1 \\ \hat{V}_2 \\ \hat{V}_3 \end{bmatrix}$$

4.5.2 Observability and Criticality Analyses in the Presence of PMUs

According to the numerical observability analysis method in the presence of only SCADA measurements [6, 7], a system will be observable if the rank of the gain matrix, G, $(H^T R^{-1} H)$, is $n - 1$, where n is the number of the buses because one of the bus phase angles will be chosen as the reference. However, necessity for an angle reference bus is eliminated in the presence of at least one PMU because all PMU measurements will be synchronized using the Global Positioning Satellite (GPS) system's time-stamps. Therefore, for an observable power system with PMUs, rank of G should be equal to n.

The numerical observability analysis method can detect observable islands in the presence of PMUs as well as conventional SCADA measurements. In the conventional approach, zero pivots of G are determined, and among those n_z zero pivots, $n_z - 1$

of them are used to identify the observable islands. However, even if a single PMU exists, all the zero pivots must be considered. Once the zero pivots are determined, the unobservable branches of the system can be identified as done in the conventional approach.

To conduct observability analysis, the decoupled Jacobian H should be formed. If power measurements are also considered, the decoupling should be performed with respect to voltage magnitudes and phase angles. This decoupling strategy must be slightly modified because the current phasor measurements cannot be decoupled with respect to voltage magnitudes and phase angles unlike the power measurements. PMUs will measure the voltage phasor at a bus and one or more current phasors incident to the same bus. Therefore, one can represent the active and reactive power flow on line $k - m$ in terms of voltage phasor at bus-k and current phasor from bus-k to m as follows:

$$V_k I_{km}^* = P_{km} + jQ_{km} \tag{4.21}$$

Based on equation 4.21, a power flow measurement can be used in lieu of a current phasor measurement for observability analysis. Because voltage phasor measurements can be decoupled trivially, H can be used for system observability and measurement criticality analyses. In both analyses, each current phasor measurement should be represented as a power flow measurement, and a "1" should be added to the column corresponding to voltage phasor measurement bus.

Before going any further, new definitions should be introduced for observability analysis in the presence of PMUs. Observable islands will be classified into two groups: anchored and floating observable islands, depending on whether voltage and current phase angles inside the observable island can be calculated synchronized to GPS or not, respectively. Power flows among the anchored observable islands can be computed, even in the absence of any flow measurements because the phase angles of these observable islands are already GPS synchronized. Note that, all the anchored observable islands constitute a single observable island.

To detect the critical measurements of the power systems with PMU measurements, the sensitivity matrix S should be computed as described in equation 4.6. The decoupled Jacobian H, defined in this section should be utilized for this purpose.

4.5.3 Hybrid State Estimation

While the number of installed PMUs is rapidly increasing, it is unlikely that power systems will be monitored using solely PMUs rather than existing SCADA in the medium term. Even if full system observability based on PMUs is obtained, considering the potential information that conventional SCADA systems provide, it will not be advisable to disregard them altogether. Therefore, use of a hybrid state estimator appears to be an attractive option for taking advantage of existing PMU measurements in state estimation.

This section seeks answers to the implementation questions raised in the introduction of this chapter. Considering the different update rates of PMUs (up to 60 times a second) and conventional SCADA system (once in every few seconds), it can be said that between two consecutive SCADA scans, PMU measurements will continue refreshing and the system will become unobservable. Therefore, one has to partially utilize SCADA measurements at those intermediate instants between SCADA update instants to satisfy full system observability.

In this section, two hybrid state estimation methods will be presented. The difference between those two methods is how they handle absence of SCADA measurements between two consecutive SCADA updates. Both estimators use WLS estimator when both SCADA and PMU measurements are updated because of its computationally supremacy. Note that, SCADA measurements are related to system states in a nonlinear manner, hence computation times of the robust state estimators may not be compatible with the refresh rate of PMU measurements.

During the instants when only PMU measurement updates are received, the first hybrid state estimator switches to a modified measurement equation and use a LAV-based state estimator. Note that, the LAV estimator uses just the required number of measurements to compute the final state estimates. Therefore, the LAV estimator will use PMU measurements thanks to their higher accuracy (higher weights) and a minimum number pseudo-measurements, which are generated based on the SCADA measurements that have not been updated.

To improve the computational speed of the LAV estimator in the presence of SCADA measurements, a first order linear approximation will be applied to the measurement equations. Assuming quasisteady state operation, the following relation can be used to generate pseudo-measurements.

$$\Delta z_k = H \Delta x_k + e \tag{4.22}$$

where,
$H(2m \times 2n)$ is the Jacobian matrix,
$\Delta z_k = z_k - h(x_k)$,
$\Delta x_k = x_{k+1} - x_k$,
$z_k = [z_{SCADA,k}^T \quad z_{PMU,k}^T]^T$ is the vector of measurements at time instant k,
$z_{SCADA,k}$ is the vector of updated SCADA measurements (pseudo-measurements) at time instant k,
$z_{PMU,k}$ is the vector of refreshed PMU measurements (actual measurements) at time instant k,
x_k is the vector of estimates of states at time instant k.

Based on equation 4.22, one can generate pseudo-measurements (update SCADA measurements) as below each time PMU measurements are refreshed and estimation is repeated, until a new set of SCADA measurements is received.

$$z_{SCADA,k+1} = h_{SCADA}(x_k) \tag{4.23}$$

where,
$h_{SCADA}(.)$ represents the nonlinear relations between SCADA measurements and system states.

Solving equation 4.22 using LAV estimator will yield state updates. The state estimates can be found adding those values to the previous state estimates. Note that, in case of an abrupt change in the operating point, the estimation results may become partly biased because of the use of pseudo-SCADA measurement updates.

The second hybrid state estimator, which is a forecast aided estimator, uses pseudo-injection measurements calculated based on historical data, rather than using a linearized measurement transition. Load demands are recorded regularly by the Transmission System Operators for operational and planning purposes. Those recorded values may be actual measurements or calculated using the state estimates. The recorded data is also employed by the conventional state estimators as pseudo-measurements. The presented estimator also makes use of those pseudo-measurements.

Note that electric power demand is affected by many factors such as economics, climate, and demographic properties. Therefore, the most recent 60-day-long data is utilized to generate those pseudo-injection measurements, to reflect the effects of those properties automatically. To keep the demand model updated considering the changes in those factors, the 60-day-long data set is updated daily, by replacing the data corresponding to the oldest day with the data recorded on the most recent day.

It is assumed that the load demands have a Gaussian distribution. Therefore, the pseudo-measurements can be modeled with an expected value (mean) and a standard deviation. Mean and standard deviation (square root of the variance) of the demand of each load point for each day of the week is calculated separately as follows.

$$P_{d,t}^{ave} = \frac{1}{N_d} \sum_{i=1}^{N_d} P_{d,t,i} \qquad (4.24)$$

$$\sigma_{d,t}^2 = \frac{1}{N_d} \sum_{i=1}^{N_d} (P_{d,t,i} - P_{d,t}^{ave})^2 \qquad (4.25)$$

where,

$P_{d,t}^{ave}$ is the expected value of active power demand at time instant-t for day-d of week. $P_{d,t,i}$ is the measured active power value at instant-t on the ith day-d within the most recent 60 days. Same relations can be written for reactive power demands.

The pseudo-measurements are used once SCADA measurements do not refresh. Because the system will lack required number of measurements, pseudo-measurements will be used to restore observability for the system that has only PMU updates. The anchored observable island can be identified by grouping the sending and receiving end buses of the current phasor measurements, provided that each PMU provides a voltage phasor measurement as well as current phasor measurements. Each one of the remaining buses constitutes a floating observable island. Note that a pseudo-measurement is a forecasted/predicted value, and hence may carry error.

The pseudo-measurement placement problem aims to determine the minimum number of pseudo-measurements with the highest accuracy (lowest standard deviation) that will restore observability. The minimum number consideration ensures that each placed pseudo-measurement is a critical measurement, which avoids propagation of the error associated with the pseudo-measurements among the state estimates. The optimal pseudo-measurement problem can be solved using the method defined in [14], which also considers the standard deviations of the pseudo-measurements.

Note that, even if the pseudo-measurements with minimum standard deviations are selected, the real demand values may vary significantly in practical operation. Therefore, rather than using those pseudo-measurements directly, one should update the values with respect to the available observations. To update the pseudo-measurements Kalman Filter based very short-term load forecasting methods can be employed [15].

4.5.4 PMU–Based Three-Phase State Estimation

Renewable energy source penetration to the power grid requires real-time monitoring of those sources because of their uncertain intermittent generation characteristics. Therefore, modeling of distribution systems in state estimation problem became a necessity to accurately monitor those sources. Considering that distribution systems commonly contain unbalanced loads, one should utilize the complete three-phase model of the power system to estimate the three-phase bus voltages as the states. Although, this approach can handle unbalanced loads, it brings a significant computational burden because of the increased problem size.

Traditionally, symmetrical components have been used in the analysis of the unbalanced operation in power systems, as they provide three decoupled networks to be analyzed, namely the positive, negative, and zero sequence networks. Although, use of the sequence networks is very common in power system analysis, the transformation unfortunately fails to decompose the state estimation problem because of the nonlinearity of the power balance equations. However, thanks to the linearity between the synchrophasor measurements and the system states, the state estimation problem can be decoupled using symmetrical components.

Consider the three-phase measurement model given here.

$$z^{3\phi} = H^{3\phi} V + e \qquad (4.26)$$

where,
$z^{3\phi}(3m \times 1)$ is the three-phase phasor measurement vector,
$H^{3\phi}(3m \times 3n)$ is the measurement Jacobian,
m is the number of three phase sets of phasor measurements,
$V(3n \times 1)$ is the three-phase system state vector,
n is the number buses, and
$e(3m \times 1)$ is the measurement error vector.

Let T be the inverse of symmetrical components transformation matrix, and T_Z be a $3m \times 3m$ block diagonal square matrix having 3×3 modal transformation matrices, T, on its diagonal, such that;

$$T = \frac{1}{3} \begin{bmatrix} 1 & 1 & 1 \\ 1 & e^{j\frac{2}{3}} & e^{-j\frac{2}{3}} \\ 1 & e^{-j\frac{2}{3}} & e^{j\frac{2}{3}} \end{bmatrix} \qquad (4.27)$$

$$T_Z = \begin{bmatrix} T & 0 & \cdots & 0 \\ 0 & T & & \vdots \\ \vdots & & \ddots & 0 \\ 0 & \cdots & 0 & T \end{bmatrix} \quad (4.28)$$

Then multiplying both sides of equation 4.26 by T_Z from the left will yield the following.

$$T_Z z^{3\phi} = T_Z H^{3\phi} V + T_Z e \quad (4.29)$$

$$z_M^{3\phi} = H_M^{3\phi} V + e_M \quad (4.30)$$

where,
$z_M^{3\phi} = T_Z z^{3\phi}$,
$H_M^{3\phi} = T_Z H^{3\phi} T_V H$,
$V = T_V V_M$
$T_V = T_Z^{-1}$
$e_M = T_Z e$

The explicit representation of equation 4.30 is given in the following text, which yields three decoupled equations shown in equation 4.32 and equation 4.33.

$$\begin{bmatrix} z_0 \\ z_r^+ \\ z_r^- \end{bmatrix} = \begin{bmatrix} H_0 & 0 & 0 \\ 0 & H_r & 0 \\ 0 & 0 & H_r \end{bmatrix} \begin{bmatrix} V_0 \\ V_r^+ \\ V_r^- \end{bmatrix} + \begin{bmatrix} e_0 \\ e_r^+ \\ e_r^- \end{bmatrix} \quad (4.31)$$

$$z_0 = H_0 V_0 + e_0 \quad (4.32)$$

$$z_r^{+/-} = H_r V_r^{+/-} + e_r^{+/-} \quad (4.33)$$

where,
A_0 is the zero sequence component of A, and
$A_r^{+/-}$, represents the positive/negative sequence components of A.

The presented transformation, however, cannot decouple the current phasor measurements on untransposed lines. Consider the admittance matrix of an untransposed line with a horizontal conductor configuration. The phase b conductor lies in the middle of the other phase conductors.

$$\overline{Y} = \begin{bmatrix} Y_S & M_1 & M_2 \\ M_1 & Y_S & M_1 \\ M_2 & M_1 & Y_S \end{bmatrix} \quad (4.34)$$

where,
Y_S indicates self admittance of the line, and
M represents mutual admittance between the phase conductors.

Considering equation 4.34, one can write the following relation between the current phasor measurement and bus voltages of bus-1 and bus-2.

$$\overline{I} = \overline{Y}(\overline{V}_1 - \overline{V}_2) \quad (4.35)$$

where,
$$\overline{I} = \begin{bmatrix} I_{1-2}^a \\ I_{1-2}^b \\ I_{1-2}^c \end{bmatrix}, \overline{V}_1 = \begin{bmatrix} V_1^a \\ V_1^b \\ V_1^c \end{bmatrix}, \text{ and } \overline{V}_2 = \begin{bmatrix} V_2^a \\ V_2^b \\ V_2^c \end{bmatrix}.$$

If the symmetrical components transformation is applied to equation 4.35, the following relation will be obtained.

$$\overline{I}_M = H_Y(\overline{V}_{M,1} - \overline{V}_{M,2}) \tag{4.36}$$

In equation 4.36, H_Y is not diagonal, hence equation 4.36 cannot be decoupled. However, the relation obtained through multiplying both sides of equation 4.35 by the inverse of admittance matrix will yield a relation that can be decoupled. The resulting relation is shown in the following text. In equation 4.37, current phasor measurements are represented as voltage differences, and thus decoupling can be achieved trivially.

$$\begin{bmatrix} Y_S & M_1 & M_2 \\ M_1 & Y_S & M_1 \\ M_2 & M_1 & Y_S \end{bmatrix}^{-1} \overline{I} = \begin{bmatrix} 1 & 0 & 0 \\ 0 & 1 & 0 \\ 0 & 0 & 1 \end{bmatrix} (\overline{V}_{M,1} - \overline{V}_{M,2}) \tag{4.37}$$

Once the charging susceptances are considered; however, the complete decoupling vanishes. Consider the following relation in the presence of the charging susceptances.

$$\overline{I} = \overline{Y}(\overline{V}_1 - \overline{V}_2) + \overline{B}\overline{V}_1 \tag{4.38}$$

where,
$\overline{Y}$ and $\overline{B}$ are the series and shunt admittance matrices of the corresponding untransposed line, respectively. If equation 4.38 is scaled using $\overline{Y}^{-1}$, the following relation will be obtained.

$$\overline{Y}^{-1}\overline{I} = \overline{Y}^{-1}\overline{Y}(\overline{V}_1 - \overline{V}_2) + \overline{Y}^{-1}\overline{B}\overline{V}_1 \tag{4.39}$$

$$\overline{Y}^{-1}\overline{I} = (\overline{V}_1 - \overline{V}_2) + \overline{Y}^{-1}\overline{B}\overline{V}_1 \tag{4.40}$$

where,

$$\overline{Y}^{-1}\overline{B} = \begin{bmatrix} Y_S & M_1 & M_2 \\ M_1 & Y_S & M_1 \\ M_2 & M_1 & Y_S \end{bmatrix}^{-1} \begin{bmatrix} B & 0 & 0 \\ 0 & B & 0 \\ 0 & 0 & B \end{bmatrix}^{-1} = \begin{bmatrix} BZ_S & BZ_{M1} & BZ_{M2} \\ BZ_{M1} & BZ_S & BZ_{M1} \\ BZ_{M2} & BZ_{M1} & BZ_S \end{bmatrix}$$

It is obvious that the off-diagonal entries of the product of $\overline{Y}^{-1}\overline{B}$ are very small compared to the diagonal entries. Moreover, once the symmetrical components transformation is applied to this multiplication, the obtained off-diagonal entries, which are the problems of the decoupling, will be even smaller. Therefore, decoupling the sequence components independent of those resulting diagonal entries will yield a very small error in estimation results.

4.6 Detection of Network Parameter Errors

Power system models are never perfect. In practice, model parameter errors can be widely present due to various reasons including inaccurate manufacturing data, human entry error, failure to update status of devices, change of ambient conditions, and even malicious cyber attacks. If not successfully identified and corrected, they may exert wide impacts on many online and offline model-based applications in system operation.

State estimation can be extended as an effective tool for handling parameter errors in power network models. Historically, two types of methods have been proposed to address parameter errors. The first type of methods estimate parameter errors using sensitivity analysis between measurement residuals and parameter errors [16–19]. The second type of methods include the parameters of interest into the state vector, and estimate them together with the state variables. They are referred to as augmented state estimation approaches [20–23]. These methods are unable to resolve the following major challenges:

1. Differentiation between parameter and measurement errors. In state estimation, when the network model is not consistent with the measurements, it is important to find out whether the error comes from the model, or comes from the measurements.

2. Selection of a suspect set of parameters to be estimated. To avoid observability issues, a subset of parameters needs to be selected from the large parameter database of a system.

This section will present methods that effectively and systematically address the challenges mentioned in the preceding text.

4.6.1 Detection, Identification, and Correction of Parameter Errors

Consider the measurement equation that explicitly considers parameter errors:

$$z = h(x, p_e) + e \tag{4.41}$$

where p_e is the parameter error vector, i.e., the difference between the parameter vector p and the true parameter vector p_t:

$$p_e = p - p_t \tag{4.42}$$

Assuming initially that the parameters are errorfree, a modified version of the WLS state estimator can be given as

$$\begin{aligned} \min \quad & J(x, p_e) = \tfrac{1}{2} r(x, p_e)^T R^{-1} r(x, p_e) \\ \text{s.t.} \quad & p_e = 0 \end{aligned} \tag{4.43}$$

where R is the covariance matrix associated with measurement errors, and $r(x, p_e)$ is the residual vector given by

$$r(x, p_e) = z - h(x, p_e) \tag{4.44}$$

Note that problem 4.43 is equivalent to the conventional formulation of the WLS state estimator, and can be solved in exactly the same way.

The Lagrange multipliers associated with the equality constraints regarding the parameter errors can be recovered after equation 4.43 is solved. The Lagrangian of equation 4.43 can be written as

$$L(x, p_e, \lambda) = \frac{1}{2} r^T R^{-1} r - \lambda^T p_e \tag{4.45}$$

At the solution point, the first-order necessary condition will be satisfied:

$$\frac{\partial L}{\partial p} = H_p^T R^{-1} r + \lambda = 0 \tag{4.46}$$

where H_p is the Jacobian matrix of the measurement function h with respect to the parameter vector p or the parameter error vector p_e. The Lagrange multiplier vector can therefore be evaluated as

$$\lambda = -H_p^T R^{-1} r \tag{4.47}$$

The Lagrange multipliers show the reduction brought to the objective function with incremental relaxation of the corresponding equality constraints. In other words, they show the degrees of inconsistency between the model and the measurements caused by imposing the assumption of the corresponding parameters being error free. Therefore, they contain useful information for detection and identification of parameter errors. To better understand the relationship between Lagrange multipliers and parameter errors, further derivations are needed. In [24], it is shown that using the linearized measurement equation, both the measurement residuals and Lagrange multipliers are linear combinations of measurement errors and parameter errors:

$$\begin{aligned} r &= Se - SH_p p_e \\ &= Se + B p_e \end{aligned} \tag{4.48}$$

$$\begin{aligned} \lambda &= H_p^T R^{-1} S H_p p_e - H_p^T R^{-1} S e \\ &= \Lambda p_e + A e \end{aligned} \tag{4.49}$$

where

$$A = -H_p^T R^{-1} S \tag{4.50}$$

$$\Lambda = H_p^T R^{-1} S H_p \tag{4.51}$$

$$B = -S H_p \tag{4.52}$$

$$S = I - K \tag{4.53}$$

$$K = H G^{-1} H^T R^{-1} \tag{4.54}$$

$$G = H^T R^{-1} H \tag{4.55}$$

Equation 4.49 can be used to study the statistical properties of the Lagrange multipliers. Under the assumption of Gaussian measurement noise with zero mean and covariance matrix R, the expected values of the Lagrange multipliers can be evaluated as

$$E(\lambda) = E(\Lambda p_e) + E(-H_p^T R^{-1} S e) = \Lambda p_e \tag{4.56}$$

while the covariance matrix of the Lagrange multipliers can be evaluated as

$$\begin{aligned}
\text{cov}(\lambda) &= \text{cov}\big(H_p^T R^{-1} S H_p p_e + H_p^T R^{-1} S e\big) \\
&= \text{cov}\big(H_p^T R^{-1} S e\big) \\
&= H_p^T R^{-1} (SR) R^{-T} H_p \\
&= H_p^T R^{-1} S H_p = \Lambda
\end{aligned} \qquad (4.57)$$

It is interesting to note that Λ is both the sensitivity matrix linking parameter errors to Lagrange multipliers, and the covariance matrix of Lagrange multipliers.

The normalized Lagrange multiplier (NLM) associated with the ith parameter can be defined as

$$\lambda_i^N = \frac{\lambda_i}{\sqrt{\Lambda_{ii}}} \qquad (4.58)$$

It is clear from equations 4.56, 4.57, and 4.58 that in the absence of errors the probability distribution of the Lagrange multiplier will be given by

$$H_0: \quad \lambda_i^N \sim N(0, 1) \qquad (4.59)$$

while in the presence of errors the probability distribution will be given by

$$H_1: \quad \lambda_i^N \sim N\big(\sqrt{\Lambda_{ii}}\, p_{e,i}, 1\big) \qquad (4.60)$$

Therefore, parameter error detection can be formulated as a standard hypothesis testing problem, with H_0 being the null hypothesis, and H_1 being the alternative hypothesis. For a given probability of false alarm α, a positive threshold t can be set up for parameter error detection:

$$\big|\lambda_i^N\big| \geq \Phi^{-1}(1 - \alpha/2) = t \qquad (4.61)$$

where Φ is the probability density function of a standard normal distribution. Typically, t is set equal to 3.0, which corresponds to a false alarm probability of $\alpha = 0.0026$.

Once parameter errors are detected, they need to be identified and corrected. Moreover, parameter and measurement errors need to be jointly identified and differentiated because they may both be present for the same measurement scan.

To this end, the NLMs associated with parameters, and the NRs associated with measurements, are jointly used. The "largest normalized variable" criterion is applied: the largest variable among all the NLMs and NRs is found, and the corresponding parameter or measurement is identified as in error. If the largest normalized variable is a NLM (NR), then a parameter (measurement) error will be identified:

$$\arg\max_{i,j} \big\{\big|\lambda_i^N\big|, \big|r_j^N\big|\big\} \qquad (4.62)$$

The theoretical foundation (theorems and proofs) of the "largest normalized variable" criterion is given in [24].

Finally, the detected and identified parameter/measurement error needs to be corrected. The correction can be preformed using the derived relationship in equations 4.48 and 4.49 [25].

For an identified parameter error, the correction can be performed by

$$p_{corr,i} = p_{bad,i} - \lambda_i/\Lambda_{ii} \tag{4.63}$$

For an identified measurement error, the correction can be performed by

$$z_{corr,i} = z_{bad,i} - r_i/S_{ii} \tag{4.64}$$

Note that because the "largest normalized variable" criterion is used, errors have to be identified and corrected one at a time. After the correction of an identified error, WLS state estimation should be repeated to search for the next error. This procedure is summarized here.

Step 1. Execute WLS state estimation by solving problem 4.43.

Step 2. Compute NLMs for all network parameters and NRs for all measurements.

Step 3. Detect an error using equation 4.61, i.e., by determining whether any normalized variable has a greater absolute value than the threshold t.

Step 4. If no error is detected, go to step 7; otherwise, go to step 5.

Step 5. Identify an error using equation 4.61, i.e., by selecting the parameter/measurement corresponding to the largest normalized variable.

Step 6. If the identified error is a parameter error, correct it using equation 4.63; otherwise, correct it using equation 4.63. After the correction, go to step 1.

Step 7. Report the state estimate and all the identified erroneous parameters/measurements along with their corrected values.

4.6.2 Detectability and Identifiability of Parameter Errors

The capability of parameter error processing methods is always limited by the redundancy of measurements. The measurements need to provide sufficient information for error detection and identification in addition to estimating the state variables. The concepts regarding the detectability and identifiability of parameter are presented here.

DEFINITION 4.1 *If Λ matrix in equation 4.51 has a null column, the corresponding parameter will be defined as critical.*

THEOREM 4.1 *Errors in critical parameters are not detectable.*

DEFINITION 4.2 *If two columns of Λ in equation 4.51 are linearly dependent, then the corresponding parameters will be defined as a critical parameter pair. In a similar vein, if a column of Λ in equation 4.51 and a column of A in equation 4.50 are linearly dependent, the corresponding parameter and measurement will be defined as a critical parameter-measurement pair.*

THEOREM 4.2 *Errors in a critical pair are not identifiable.*

DEFINITION 4.3 *Assume that $k = p+q$, then p parameters and q measurements will be defined as a critical k-tuple, if the corresponding p columns of Λ defined by equation 4.51 and q columns of A defined by equation 4.50 are linearly dependent.*

THEOREM 4.3 *Errors in a critical parameter-measurement k-tuple are not identifiable.*

The detailed proof of the preceding theorems can be found in [24].

When a parameter is critical, the associated Lagrange mutiplier and its variance will both be zero, hence the NLM cannot be evaluated. When two parameters (or one parameter and one measurement) consist of a critical pair, their NLMs (or NLM and NR) can be evaluated, but will always be identifical. Therefore, it is impossible to differentiate their errors from one another based on the largest normalized variable criterion. Similar situations can be found in a critical k-tuple.

It should be noted that NLM/NR-based method's failure to detect or identify errors in the previously mentioned situations is not due to a limitation of the method but a limitation of the existing measurement configuration. Such cases can thus be avoided by strategic meter placement to increase the local measurement redundancy.

4.6.3 Use of Multiple Measurement Scans

When determining the threshold for parameter error detection, equation 4.61 is used to keep the probability of false alarms at a reasonably low level. However, it does not guarantee that truly existing errors will be detected reliably. In fact, the probability of missing a existing parameter error can be evaluated as

$$\beta = P(|\lambda_i^N| < t) = \Phi(t - \sqrt{\Lambda_{ii}}|p_{e,i}|) \tag{4.65}$$

from which it can be seen that if $\sqrt{\Lambda_{ii}}$ is small, even if a substantial error $|p_{e,i}|$ is present, a significantly large probably of missing this error will result. Therefore, $\sqrt{\Lambda_{ii}}$ can be viewed as the sensitivity of λ_i^N to $p_{e,i}$. Unfortunately, this sensitivity varies widely for different parameters in a power network, mainly influenced by factors including network topology, measurement configuration, measurement accuracy, operating point, parameter type, and parameter value.

For parameters with low NLM sensitivity, potential errors of concern may not be reliably detected. To address this problem, multiple measurement scans can be used. If meters are properly calibrated, measurement noise will obey statistical distribution with zero means. Meanwhile, parameter errors can be assumed constant across different scans because they do not vary or vary slowly over time. When multiple measurement scans are used, the impact of parameter errors accumulates while measurement noise cancels out, which is helpful for improving the probability of error detection.

Suppose there are s measurement scans. Writing the measurement equation of the qth scan as

$$z^{(q)} = h(x^{(q)}, p_e) + e^{(q)} \tag{4.66}$$

The WLS state estimation problem is given by

$$\min \quad J\left(x^{(1)}, \ldots, x^{(q)}, \ldots, x^{(s)}, p_e\right)$$
$$= \frac{1}{2} \sum_{q=1}^{s} \left(z^{(q)} - h\left(x^{(q)}, p_e\right)\right)^T R^{-1} \left(z^{(q)} - h\left(x^{(q)}, p_e\right)\right) \tag{4.67}$$
$$\text{s.t.} \quad p_e = 0$$

Similar to the case of a single measurement scan, the Lagrange multiplier vector can be obtained by [26]:

$$\lambda = -\sum_{q=1}^{s}\left[H_p^{(q)}\right]^T R^{-1} r^{(q)} = \sum_{q=1}^{s} \lambda^{(q)} \qquad (4.68)$$

where $\lambda^{(q)}$ is the Lagrange multiplier vector obtained when only the qth scan is involved in state estimation.

Assuming Gaussian white measurement noise, it can be shown that the covariance matrix of the Lagrange multipliers can be derived as [26]:

$$\Lambda = \sum_{q=1}^{s}\left[H_p^{(q)}\right]^T R^{-1} S^{(q)} H_p^{(q)} = \sum_{q=1}^{s} \Lambda^{(q)} \qquad (4.69)$$

With the Lagrange multiplier vector and the associated covariance matrix available, the NLMs can be obtained using equation 4.58.

Using the linearized measurement model, the relationship between Lagrange multipliers and parameter errors can be derived in the similar vein as the single scan case. Assuming that the operating point does not change drastically between measurement scans, the measurement Jacobian can also be assumed unchanged. Based on this approximation, one can denote $\Lambda^s = \Lambda^{(q)}$ for all considered scans. It is shown in [26] that the null and alternative hypotheses can be approximately given by

$$H_0: \quad \lambda_i^N \sim N(0, 1) \qquad (4.70)$$

$$H_1: \quad \lambda_i^N \sim N\left(\sqrt{s\Lambda_{ii}^s} p_{e, i}, 1\right) \qquad (4.71)$$

Consequently, the probability of missing a parameter error can be evaluated by

$$\beta = P(|\lambda_i^N| < t) = \Phi\left(t - \sqrt{s\Lambda_{ii}^s}|p_{e,i}|\right) \qquad (4.72)$$

Obviously, the sensitivity of the NLM, $\sqrt{s\Lambda_{ii}^s}$, now becomes a function of the number of measurement scans, s. When the number of scans increases, the sensitivity will also increase, which leads to a lower probability of missing the parameter error.

Because the increase of measurement scans requires more data and computational resources, the number of measurement scans to be used can be customized based on the need of parameter error detection. Given a probability of missing error, β, and a minimum magnitude of error to be detected, the required number of measurement scans can be evaluated by

$$s = \frac{\left(t - \Phi^{-1}(\beta)\right)^2}{p_{e, \min}^2 \Lambda_{ii}^s} \qquad (4.73)$$

To keep both the probability of false alarm α and the probability of missing parameter errors β at very low levels, it is recommended to choose $t = -\Phi^{-1}(\alpha/2) = 3.0$, and

$\Phi^{-1}(\beta) = -3.0$. They correspond to $\alpha = 0.0026$ and $\beta = 0.0013$, respectively. In this case, the required number of measurement scans can be evaluated by

$$s = \frac{(3.0-(-3.0))^2}{p_{e,\min}^2 \Lambda_{ii}^s} = \frac{36}{p_{e,\min}^2 \Lambda_{ii}^s} \tag{4.74}$$

4.6.4 Robust State Estimation against Parameter and Measurement Errors

In the previous subsections, parameter errors are handled using the NLMs calculated based on the output of the WLS state estimator. In this subsection, another alternative for addressing the parameter error issues using robust estimation will be introduced.

The concept of "robustness" in power system state estimation is not new [27–30]. Under the conventional context, a state estimation approach is said to be robust if its solution is not sensitive to a finite number of errors in redundant measurements. However, this concept does not take into account errors in the power system model. In the presence of model parameter errors, conventional robust state estimators can still be biased and, furthermore, good measurements can be mistakenly filtered out as bad data. Therefore, the concept of "robustness" should be extended to model parameter errors.

As one of the well-documented robust alternatives, the LAV state estimation is formulated as:

$$\begin{aligned} \min_{\Delta x} \quad & \|r\|_1 \\ \text{s.t.} \quad & H_x \Delta x + r = \Delta z \end{aligned} \tag{4.75}$$

As an L_1 optimization problem, its solution has the property of automatically selecting a subset of measurements equal to the number of state variables and having them strictly satisfied, while rejecting the rest of measurements that will have nonzero residuals. Due to this desirable feature, occasional bad data can be automatically rejected and do not bias the estimated state.

To account for parameter errors, it is assumed that substantial parameter errors are not common in practice (i.e., majority of parameters are sufficiency accurate). Then, an extended version referred as the Extended Least Absolute Value (ELAV) state estimator is developed [31]:

$$\begin{aligned} \min_{\Delta x, \Delta p} \quad & \|r\|_1 + \|\Delta p\|_1 \\ \text{s.t.} \quad & H_x \Delta x + H_p \Delta p + r = \Delta z \end{aligned} \tag{4.76}$$

In the equality constraint, the parameter update (error) vector Δp is included. In this case, all state variables and network parameters in a system are estimated simultaneously. As is mentioned at the beginning of this section, simultaneous estimation of all state variables and network parameters will bring up observability issues. The problem may become undetermined due to a large number of unknown variables. In the ELAV formulation, the sparse nature of the parameter error vector is exploited. The L_1 norm of the parameter error vector is used to augment the objective function. Using this regularization term, a unique solution to the problem can be obtained, with the parameter error vector being a sparse one. The erroneous parameters that do not well explain the measurements can be automatically selected, with the nonzero error estimate also obtained, while the error estimates for the correct parameters will remain zero. This desirable feature is the result of the selection property of L_1 regularization, which has been studied and applied extensively in the past [32–34].

In summary, the advantages of the ELAV state estimator include:
1. Robust against measurement errors;
2. Robust against parameter errors;
3. Handles all parameters without observability issues;
4. Automatically identifies (selection) erroneous parameters.

In actual implementation, due to the fact that parameter values may not be all in the same order of magnitude among themselves as well as the measurements, scaling will be needed to achieve satisfactory results:

$$\begin{array}{cl} \min_{\Delta x, \Delta p} & \|r\|_1 + \|V\Delta p\|_1 \\ \text{s.t.} & H_x \Delta x + H_p \Delta p + r = \Delta z \end{array} \quad (4.77)$$

where $V = diag(v_1, v_2, \ldots, v_u)$ (u being the number of parameters). One possible scaling method is given as follows:

$$v_i = mean\left\{|H_{p,ij}|, \ H_{p,ij} \neq 0\right\} \quad (4.78)$$

Finally, it is worth mentioning that the ELAV state estimation problem can be readily converted into a linear programming problem. Similar to the conversion performed for the conventional LAV state estimation problem, problem 4.77 can be shown equivalent to the following problem:

$$\begin{array}{cl} \min & c^T \left(r^+ + r^-\right) + v_d^T \left(\Delta p^+ + \Delta p^-\right) \\ \text{s.t.} & H_x \Delta x^+ - H_x \Delta x^- + H_p \Delta p^+ - H_p \Delta p^- + r^+ - r^- = 0 \\ & \Delta x^+, \Delta x^-, \Delta p^+, \Delta p^-, r^+, r^- \geq 0 \end{array} \quad (4.79)$$

where

$$v_d = (v_1, v_2, \ldots, v_u)^T \quad (4.80)$$

After solving problem 4.79, the original variables can be recovered by

$$r = r^+ - r^- \quad (4.81)$$

$$\Delta x = \Delta x^+ - \Delta x^- \quad (4.82)$$

$$\Delta p = \Delta p^+ + \Delta p^- \quad (4.83)$$

Therefore, well-developed highly efficient linear programming algorithms can be readily used to solve this problem.

4.7 Conclusion

Secure and efficient operation of power systems require continuous monitoring of the system state. This chapter provides an overview of the methods and tools used in monitoring the system state and its network parameters. Two different types of measurements are available in today's power grids. One set is provided by the SCADA system and they are scanned at a slower rate and not synchronized. The other set is synchronized and provided by the phasor measurement units at a much higher scan rate. The chapter presents both the methods that are in common use today as well as some of the more recently developed ones that take advantage of the synchronized measurements. In addition, it describes recent results that facilitate simultaneous detection, identification, and correction of errors not only in the analog measurements but also network parameters.

References

[1] F. C. Schweppe and J. Wildes, "Power system static-state estimation, part II: Approximate model," *IEEE Transactions on Power Apparatus and Systems*, vol. PAS-89, pp. 120–125, 1970.

[2] F. C. Schweppe and D. B. Rom, "Power system static-state estimation, part I: Exact model," *IEEE Transactions on Power Apparatus and Systems*, vol. PAS-89, pp. 125–130, 1970.

[3] F. C. Schweppe, "Power system static-state estimation, part III: Implementation," *IEEE Transactions on Power Apparatus and Systems*, vol. PAS-89, pp. 130–135, 1970.

[4] G. R. Krumpholz, K. A. Clements, and P. W. Davis, "Power system observability: A practical algorithm using network topology," *IEEE Transactions on Power Apparatus and Systems*, vol. PAS-99, pp. 1534–1542, 1980.

[5] K. A. Clements, G. R. Krumpholz, and P. W. Davis, "Power system state estimation with measurement deficiency: An observability/measurement placement algorithm," *IEEE Transactions on Power Apparatus and Systems*, vol. PAS-102, pp. 2012–2020, 1983.

[6] A. Monticelli and F. F. Wu, "Network observability: Theory," *IEEE Transactions on Power Apparatus and Systems*, vol. PAS-104, pp. 1042–1048, 1985.

[7] ——, "Network observability: Identification of observable islands and measurement placement," *IEEE Transactions on Power Apparatus and Systems*, vol. PAS-104, pp. 1035–1041, 1985.

[8] A. Monticelli and A. Garcia, "Reliable bad data processing for real-time state estimation," *IEEE Transactions on Power Apparatus and Systems*, vol. PAS-102, no. 5, pp. 1126–1139, May 1983.

[9] A. G. Phadke, "Synchronized phasor measurements: A historical overview," *IEEE PES Transmission and Distribution Conference and Exhibition 2002: Asia Pacific*, vol. 1, pp. 476–479, 2002.

[10] U.S.-Canada Power System Outage Task Force, *Final Report on the August 14, 2003 Blackout in the United States and Canada: Causes and Recommendations*. U.S.-Canada Power System Outage Task Force, 2004.

[11] W. W. Kotiuga and M. Vidyasagar, "Bad data rejection properties of weighted least absolute value technique applied to static state estimation," *IEEE Transactions on Power Apparatus and Systems*, vol. PAS-101, pp. 844–851, 1982.

[12] M. Gol and A. Abur, "LAV based robust state estimation for systems measured by PMUs," *IEEE Transactions on Smart Grid*, vol. 5, pp. 1808–1814, 2014.

[13] P. J. Rousseeuw and A. A. Leroy, *Robust Regression and Outlier Detection*. New York: John Wiley & Sons, 2004.

[14] B. Donmez and A. Abur, "A computationally efficient method to place critical measurements," *IEEE Transactions on Power Systems*, vol. 26, pp. 924–931, 2011.

[15] R. E. Kalman, "A new approach to linear filtering and prediction problems," *ASME Transactions Journal of Basic Engineering*, vol. 82, pp. 35–45, 1960.

[16] D. L. Fletcher and W. O. Stadlin, "Transformer tap position estimation," *IEEE Transactions on Power Apparatus and Systems*, vol. PAS-102, no. 11, pp. 3680–3686, Nov. 1983.

[17] W. H. E. Liu, F. F. Wu, and S. M. Lun, "Estimation of parameter errors from measurement residuals in state estimation (power systems)," *IEEE Transactions on Power Systems*, vol. 7, no. 1, pp. 81–89, Feb. 1992.

[18] W. H. E. Liu and S.-L. Lim, "Parameter error identification and estimation in power system state estimation," *IEEE Transactions on Power Systems*, vol. 10, no. 1, pp. 200–209, Feb. 1995.

[19] M. R. M. Castillo, J. B. A. London, N. G. Bretas, S. Lefebvre, J. Prevost, and B. Lambert, "Offline detection, identification, and correction of branch parameter errors based on several measurement snapshots," *IEEE Transactions on Power Systems*, vol. 26, no. 2, pp. 870–877, May 2011.

[20] E. Handschin and E. Kliokys, "Transformer tap position estimation and bad data detection using dynamic signal modelling," *IEEE Transactions on Power Systems*, vol. 10, no. 2, pp. 810–817, May 1995.

[21] I. W. Slutsker, S. Mokhtari, and K. A. Clements, "Real time recursive parameter estimation in energy management systems," *IEEE Transactions on Power Systems*, vol. 11, no. 3, pp. 1393–1399, Aug. 1996.

[22] O. Alsac, N. Vempati, B. Stott, and A. Monticelli, "Generalized state estimation," *IEEE Transactions on Power Systems*, vol. 13, no. 3, pp. 1069–1075, Aug. 1998.

[23] N. Petra, C. G. Petra, Z. Zhang, E. M. Constantinescu, and M. Anitescu, "A Bayesian approach for parameter estimation with uncertainty for dynamic power systems," *IEEE Transactions on Power Systems*, vol. 32, no. 4, pp. 2735–2743, July 2017.

[24] Y. Lin and A. Abur, "A new framework for detection and identification of network parameter errors," *IEEE Transactions on Smart Grid*, vol. 9, no. 3, pp. 1698–1706, May 2018.

[25] ——, "Fast correction of network parameter errors," *IEEE Transactions on Power Systems*, vol. 33, no. 1, pp. 1095–1096, Jan. 2018.

[26] ——, "Enhancing network parameter error detection and correction via multiple measurement scans," *IEEE Transactions on Power Systems*, vol. 32, no. 3, pp. 2417–2425, May 2017.

[27] H. M. Merrill and F. C. Schweppe, "Bad data suppression in power system static state estimation," *IEEE Transactions on Power Apparatus and Systems*, vol. PAS-90, no. 6, pp. 2718–2725, Nov. 1971.

[28] G. He, S. Dong, J. Qi, and Y. Wang, "Robust state estimator based on maximum normal measurement rate," *IEEE Transactions on Power Systems*, vol. 26, no. 4, pp. 2058–2065, Nov. 2011.

[29] W. Xu, M. Wang, J. F. Cai, and A. Tang, "Sparse error correction from nonlinear measurements with applications in bad data detection for power networks," *IEEE Transactions on Signal Processing*, vol. 61, no. 24, pp. 6175–6187, Dec. 2013.

[30] Y. Weng, R. Negi, C. Faloutsos, and M. D. IliÄǦ, "Robust data-driven state estimation for smart grid," *IEEE Transactions on Smart Grid*, vol. 8, no. 4, pp. 1956–1967, July 2017.

[31] Y. Lin and A. Abur, "Robust state estimation against measurement and network parameter errors," *IEEE Transactions on Power Systems*, vol. 33, no. 5, pp. 4751–4759, Sept. 2018.

[32] E. J. Candes and T. Tao, "Decoding by linear programming," *IEEE Transactions on Information Theory*, vol. 51, no. 12, pp. 4203–4215, Dec. 2005.
[33] H. Wang, G. Li, and G. Jiang, "Robust regression shrinkage and consistent variable selection through the lad-lasso," *Journal of Business and Economic Statistics*, vol. 25, no. 3, pp. 347–355, 2007.
[34] E. J. Candes and P. A. Randall, "Highly robust error correction by convex programming," *IEEE Transactions on Information Theory*, vol. 54, no. 7, pp. 2829–2840, July 2008.

Part II

Data-Driven Anomaly Detection

5 Quickest Detection and Isolation of Transmission Line Outages

Venugopal V. Veeravalli and Alejandro Dominguez-Garcia

5.1 Introduction

Many online decision-making tools used by the operators of a power transmission system, e.g., the state estimator, and contingency analysis and generation re-dispatch procedures, rely on a system model obtained offline assuming a certain system network topology. If the network topology changes due to a transmission line outage, the power system model needs to be updated; otherwise, the aforementioned decision-making tools may provide an inaccurate picture of the system state. For example, in the 2011 San Diego blackout, the system operators were not able to determine that certain transmission lines were overloaded because the system model was not updated following a transmission line outage [1]. The lack of situational awareness in this case prevented system operators from determining the next critical contingency and led to a cascading failure. Similarly, during the 2003 U.S. Northeastern blackout, system operators were unaware of the loss of key transmission lines and as a result they did not initiate the necessary remedial actions schemes that could have prevented the cascading failure that ultimately led to the blackout [2].

In light of the preceding discussion, it is clear that there is a need for developing online techniques that enable detection and identification of transmission line outages in a timely fashion. In this chapter, we address this problem and discuss methods for online detection and identification of line outages in transmission networks recently developed by the authors, their students, and their postdoctoral research associates [3–7]; the foundations of the methods lie in the theory of quickest change detection (QCD) (see, e.g., [8–10]). In QCD, a decision maker observes a sequence of random variables, the probability distribution of which changes abruptly due to some event that occurs at an unknown time. The objective is then to detect this change in distribution as quickly as possible, subject to false alarm constraints.

Our line outage detection method relies on the fact that in transmission networks the incremental changes in the voltage phase angle at each bus of the network can be approximated by a linear combination of incremental changes in the power demand at all load buses. The coefficients of this linear combination can be easily obtained from the network admittance matrix (the sparsity pattern of which depends on the network topology) and some time-varying coefficients capturing the response of generators in the system to small changes in active power demand. An outage in any line will result in a change in the system admittance matrix. As such, the coefficients of the linear combination relating active power demands and voltage phase angles will change after a line outage. By assuming that the incremental changes in the power demand at load buses can be modeled as independent zero-mean Gaussian random variables, one can approximately obtain the probability distribution of the bus voltage angle incremental changes before the occurrence of any line outage, in a short transient period immediately after a line outage (governed by aforementioned coefficients capturing generator responses), and after all transient phenomena have died down and the system settles into a new steady state. Then, by using bus voltage phase angle measurements provided by phasor measurements units (PMUs), one can obtain a random sequence of incremental changes in the voltage phase angle measurements in real time, and feed it to a QCD-based algorithm to detect a change in the probability distribution of this random sequence.

Earlier algorithms for topology change detection in the literature include those based on state estimation [11, 12], and those that mimic system operator decisions [13]. More recent techniques have utilized measurements provided by PMUs for line outage detection [14–16]. All these works rely on the voltage phase angle difference between two sets of measurements obtained before and after the outage occurrence, and attempt to identify the line outage by using hypothesis testing [14], sparse vector estimation [15], and mixed-integer nonlinear optimization [16]. However, these techniques are "one-shot" in the sense that they do not exploit the persistent nature of line outages and do not incorporate the transient behavior immediately after the outage occurrence. In addition, the work in [17] proposes a method to identify line outages using statistical classifiers where a maximum likelihood estimation is performed on the PMU data. This work considers the system transient response after the occurrence of a line outage by comparing data obtained offline using simulation against actual data. However, this method requires the exact time the line outage occurs to be known before executing the algorithm, i.e., it assumes line outage detection has already been accomplished by some other method, whereas the methods discussed in this chapter provide both detection and identification.

The remainder of this chapter is organized as follows. In Section 5.2, we provide some background on the theoretical foundations of QCD. In Section 5.3 we describe the power system model adopted in this work, and introduce the statistical model describing the voltage phase angle before, during, and after the occurrence of an outage. In Section 5.4, we describe several QCD-based line outage detection algorithms. Section 5.5 discusses how to identify line outages using QCD-based algorithms. In Section 5.6, we illustrate and compare the performance of the QCD-based line outage detection and identification algorithms discussed in Section 5.4 using numerical case studies on the IEEE 118-bus test system. Some concluding remarks are provided in Section 5.7.

5.2 Quickest Change Detection Background

In the basic quickest change detection (QCD) problem, we have a sequence of observations $\{X[k], k \geq 1\}$, which initially have certain distribution described by some probability density function (pdf), and at some point in time γ, due to some event, the distribution of the observations changes. We restrict our attention to models where the observations are independent and identically distributed (i.i.d.) in the pre- and postchange regimes because this is a good approximation for the line outage detection problem as we will see in Section 5.3. Thus

$$X[k] \sim \text{i.i.d. with pdf } f_0 \text{ for } k < \gamma,$$
$$X[k] \sim \text{i.i.d. with pdf } f_1 \text{ for } k \geq \gamma. \quad (5.1)$$

We assume that f_0 and f_1 are known but that γ is unknown. We call γ the *change point*.

The goal in QCD is to detect the change in the distribution as quickly as possible (after it happens) subject to constraints on the rate of false alarms. A QCD procedure is constructed through a stopping time[1] τ on the observation sequence $\{X[k], k \geq 1\}$, with the change being declared at the stopping time. To have quick detection, we need to minimize some metric that captures the delay in detection $(\tau - \gamma)^+$ (where we use the notation $x^+ = \max\{x, 0\}$). Clearly, selecting $\tau = 1$ always minimizes this quantity, but if $\gamma > 1$, we end up with a false alarm event. The QCD problem is to find a stopping time τ that results in an optimal tradeoff between the detection delay and false alarm rate.

Before we discuss specific algorithms for the QCD problem, we introduce the Kullback–Leibler (KL) divergence, which is an information-theoretic measure of the discrepancy between two probability distributions, f and g, defined as:

$$D(f \| g) := \mathbb{E}_f \left[\log \frac{f(X)}{g(X)} \right] = \int f(x) \log \frac{f(x)}{g(x)} dx. \quad (5.2)$$

It is easy to show that $D(f \| g) \geq 0$, with equality if and only if $f = g$. The KL divergence plays an important role both in developing good QCD algorithms and in characterizing their performance.

We next introduce the likelihood ratio of the observations

$$L(X[k]) = \frac{f_1(X[k])}{f_0(X[k])}.$$

It is well known that the likelihood ratio plays a key role in the construction of good detection algorithms in general, not just for the QCD problem being discussed here [18].

A fundamental fact that is used in the construction of good QCD algorithms is that the mean of the log-likelihood ratio $\log L(X[k])$ in the prechange regime, i.e., for $k < \gamma$ is given by

$$\mathbb{E}_{f_0}[\log L(X[k])] = -D(f_0 \| f_1) < 0$$

and in the postchange regime, $k \geq \gamma$, is given by

$$\mathbb{E}_{f_1}[\log L(X[k])] = -D(f_1 \| f_0) > 0.$$

---

[1] See, e.g., [10] for the definition of a stopping time.

5.2.1 Shewhart Test

A simple way to use the properties of the log-likelihood ratio $\log L(X[k])$ was first proposed by Shewhart [19], in which a statistic based on the current observation is compared with a threshold to make a decision about the change. That is, the Shewhart test is defined as

$$\tau_S = \inf\{k \geq 1 : \log L(X[k]) > b\} \tag{5.3}$$

where b is chosen to meet a desired false alarm constraint. Shewhart's test is widely employed in practice due to its simplicity; however, significant gain in performance can be achieved by making use of past observations (in addition to the current observation) to make the decision about the change.

5.2.2 CuSum Test

In [20], Page proposed an algorithm that uses past observations, which he called the Cumulative Sum (CuSum) algorithm. The idea behind the CuSum algorithm is based on the behavior of the cumulative log-likelihood ratio sequence:

$$S[k] = \sum_{j=1}^{n} \log L(X[j]).$$

Before the change occurs, the statistic $S[k]$ has a negative "drift" due to the fact that $\mathbb{E}_0[\log L(X[j])] < 0$ and diverges toward $-\infty$. At the change point, the drift of the statistic $S[k]$ changes from negative to positive due to the fact that $\mathbb{E}_1[\log L(X[j])] > 0$, and beyond the change point $S[k]$ starts growing toward ∞. Therefore $S[k]$ roughly attains a minimum at the change point. The CuSum algorithm is constructed to detect this change in drift, and stop the first time the growth of $S[k]$ after change in the drift is large enough (to avoid false alarms).

Specifically, the stopping time for the CuSum algorithm is defined as

$$\tau_C = \inf\{k \geq 1 : W[k] \geq b\} \tag{5.4}$$

where

$$W[k] = S[k] - \min_{0 \leq j \leq k} S[j]$$
$$= \max_{0 \leq j \leq k} S[k] - S[j]$$
$$= \max_{0 \leq j \leq k} \sum_{\ell=j+1}^{k} \log L(X[\ell]) = \max_{1 \leq j \leq k+1} \sum_{\ell=j}^{k} \log L(X[\ell]) \tag{5.5}$$

with the understanding that $\sum_{\ell=k+1}^{k} \log L(X[\ell]) = 0$. It is easily shown that $W[k]$ can be computed iteratively as follows:

$$W[k] = (W[k-1] + \log L(X[k]))^+, \quad W[0] = 0. \tag{5.6}$$

This recursion is useful from the viewpoint of implementing the CuSum test. Note that the threshold b in (5.4) is chosen to meet a desired false alarm constraint.

5.2.3 Optimization Criteria and Optimality of CuSum

Without any prior information about the change point, a reasonable measure of false alarms is the Mean Time to False Alarm (MTFA):

$$\text{MTFA}(\tau) = \mathbb{E}_\infty[\tau], \tag{5.7}$$

where $\mathbb{E}_\infty$ is the expectation with respect to the probability measure when the change never occurs. Finding a uniformly powerful test that minimizes the delay over all possible values of γ subject to a false alarm constraint, say $\text{MTFA}(\tau) \geq \beta$, is generally not possible. Therefore, it is more appropriate to study the quickest change detection problem in what is known as a minimax setting in this case. There are two important minimax problem formulations, one due to Lorden [21] and the other due to Pollak [22].

In Lorden's formulation, the objective is to minimize the supremum of the average delay conditioned on the worst possible realizations, subject to a constraint on the false alarm rate. In particular, we define the Worst-Case Average Detection Delay (WADD) as follows:

$$\text{WADD}(\tau) = \sup_{\gamma \geq 1} \operatorname{ess\,sup} \mathbb{E}_\gamma \left[(\tau - \gamma + 1)^+ | X[1], \ldots, X[\gamma - 1] \right] \tag{5.8}$$

where $\mathbb{E}_\gamma$ denotes the expectation with respect to the probability measure when the change occurs at time γ. We have the following problem formulation.

PROBLEM 5.1 (Lorden) *Minimize* $\text{WADD}(\tau)$ *subject to* $\text{MTFA}(\tau) \geq \beta$.

Lorden showed that the CuSum algorithm (5.4) is asymptotically optimal for Problem 5.1 as $\beta \to \infty$. It was later shown in [23] that an algorithm that is equivalent to the CuSum algorithm is actually exactly optimal for Problem 5.1. Although the CuSum algorithm enjoys such a strong optimality property under Lorden's formulation, it can be argued that WADD is a somewhat pessimistic measure of delay. A less pessimistic way to measure the delay, the Conditional Average Detection Delay (CADD), was suggested by Pollak [22]:

$$\text{CADD}(\tau) = \sup_{\gamma \geq 1} \mathbb{E}_\gamma[\tau - \gamma | \tau \geq \gamma] \tag{5.9}$$

for all stopping times τ for which the expectation is well defined. It can be shown that

$$\text{CADD}(\tau) \leq \text{WADD}(\tau). \tag{5.10}$$

We then have the following problem formulation.

PROBLEM 5.2 (Pollak) *Minimize* $\text{CADD}(\tau)$ *subject to* $\text{MTFA}(\tau) \geq \beta$.

It can be shown that the CuSum algorithm (5.4) is asymptotically optimal for Problem 5.2 as $\beta \to \infty$ (see, e.g., [10]).

5.2.4 Incompletely Specified Observation Models

Thus far we have assumed that pre- and postchange distributions, i.e., f_0 and f_1 are completely specified. For the line outage problem that is of interest in this chapter, while it reasonable to assume that the prechange distribution is completely specified, the postchange distribution can only be assumed to come from a parametric family of distributions, with an unknown parameter a that depends on the line that goes into outage. The observation model of (5.1) becomes:

$$X[k] \sim \text{i.i.d. with pdf } f_0 \text{ for } k < \gamma,$$
$$X[k] \sim \text{i.i.d. with pdf } f_a \text{ for } k \geq \gamma, a \in \mathcal{A} \quad (5.11)$$

where $\mathcal{A}$ is the set of all possible values that the parameter can take.

A powerful approach to constructing a good test for the observation model of (5.11) is the *generalized likelihood ratio* (GLR) approach [24, 25], where at any time step, all the past observations are used to first obtain a maximum likelihood estimate of the postchange parameter a, and then the postchange distribution corresponding to this estimate of a is used to compute the test statistic. In particular, the GLR-CuSum test can be constructed as:

$$\tau_C^{\text{GLR}} = \inf\{k \geq 1 : W_G[k] \geq b\} \quad (5.12)$$

where

$$W_G[k] = \max_{0 \leq j \leq k} \max_{a \in \mathcal{A}} \sum_{\ell=j+1}^{k} \log L_a(X[\ell]) \quad (5.13)$$

with

$$L_a(X[\ell]) = \frac{f_a(X[\ell])}{f_0(X[\ell])}.$$

Note that for general $\mathcal{A}$, the test statistic $W_G[k]$ cannot be computed recursively as in (5.6). However, when the cardinality of $\mathcal{A}$ is finite, which will be the case for the line outage problem, we can implement τ_C^{GLR} using a recursive test statistic as follows. First, we swap the maxima in (5.13) to obtain:

$$W_G[k] = \max_{a \in \mathcal{A}} \max_{0 \leq j \leq k} \sum_{\ell=j+1}^{k} \log L_a(X[\ell]) \quad (5.14)$$

Then if we define

$$W_a[k] = \max_{0 \leq j \leq k} \sum_{\ell=j+1}^{k} \log L_a(X[\ell]),$$

we can compute $W_a[k]$ recursively as:

$$W_a[k] = (W_a[k-1] + \log L_a(X[k]))^+, \quad W_a[0] = 0$$

and we have

$$\tau_C^{\text{GLR}} = \inf\left\{k \geq 1 : \max_{a \in \mathcal{A}} W_a[k] \geq b\right\} \quad (5.15)$$

where b is chosen to meet a desired false alarm constraint. The GLR-CuSum test can be interpreted as running $|\mathcal{A}|$ CuSum tests in parallel, each corresponding to a different value of a, and stopping at the first time any one of the CuSum tests crosses the threshold b.

5.2.5 QCD under Transient Dynamics

Another variant of the basic QCD problem that is relevant to the line outage detection problem is one in which the change from the initial distribution to the final persistent distribution does not happen instantaneously, but after a series of transient phases. The observations within the different phases are generated by different distributions. The resulting observation model is then given by:

$$\begin{aligned} X[k] &\sim \text{i.i.d. with pdf } f_0 \text{ for } k < \gamma_0 \\ X[k] &\sim \text{i.i.d. with pdf } f_i \text{ for } \gamma_{i-1} \leq k < \gamma_i, \quad 1 \leq i \leq T-1 \\ X[k] &\sim \text{i.i.d. with pdf } f_T \text{ for } k \geq \gamma_{T-1} \end{aligned} \quad (5.16)$$

i.e., there are $T-1$ transient phases before the observations finally settle down to the persistent postchange distribution f_T. The change times $\gamma_i, 0 \leq i \leq T-1$ are assumed be unknown, and the goal is to choose a stopping time τ in such a way as to the detect the change at γ_0 as quickly as possible subject to false alarm constraints.

A dynamic generalized likelihood approach can be used to construct a test for this problem [7, 26], and this approach leads to the following Dynamic CuSum (D-CuSum) Test:

$$\tau_D = \inf\{k \geq 1 : W_D[k] \geq b\}. \quad (5.17)$$

Denoting

$$L_i(X[k]) = \frac{f_i(X[k])}{f_0(X[k])}$$

the statistic $W_D[k]$ can be computed recursively through the following nested iterations. For $i = 1, 2, \ldots, T$, we first compute iteratively:

$$\Omega_i[k] = \max\{\Omega_1[k-1], \Omega_2[k-1], \ldots, \Omega_i[k-1], 0\} + \log L_i(X[k])$$

with $\Omega_i[0] = 0$, for $i = 1, 2, \ldots, T$. Then, we compute

$$W_D[k] = \max\{\Omega_1[k], \Omega_2[k], \ldots, \Omega_T[k], 0\}. \quad (5.18)$$

It is established in [26] that the D-CuSum algorithm is asymptotically optimum as the mean time to false alarm goes to infinity. We will see in Section 5.4 that the line outage detection problem has both the transient and composite aspects, and therefore a combination of the GLR and dynamic CuSum algorithms will be used in constructing the most effective algorithm for line outage detection.

5.3 Power System Model

We consider a power system with N buses and L transmission lines, respectively, indexed by the elements in $\mathcal{V} = \{1, \ldots, N\}$ and $\mathcal{L} = \{1, \ldots, L\}$. Let $V_i(t)$ and $\theta_i(t)$, respectively, denote the voltage magnitude and phase angle of bus i at time t. Similarly, let $P_i(t)$ and $Q_i(t)$, respectively, denote the net active and reactive power injection into bus i at time t. Then, the quasi steady-state behavior of the system can be described by the power flow equations (see, e.g., [27]), which for bus i can be written as:

$$P_i(t) = \sum_{k=1}^{N} V_i(t) V_k(t) \Big(G_{ik} \cos \big(\theta_i(t) - \theta_k(t)\big) + B_{ik} \sin \big(\theta_i(t) - \theta_k(t)\big) \Big), \quad (5.19)$$

$$Q_i(t) = \sum_{k=1}^{N} V_i(t) V_k(t) \Big(G_{ik} \sin \big(\theta_i(t) - \theta_k(t)\big) - B_{ik} \cos \big(\theta_i(t) - \theta_k(t)\big) \Big) \quad (5.20)$$

where G_{ik} and B_{ik} are, respectively, the real and imaginary part of the (i,k) entry of the network admittance matrix. In the remainder we make the following two assumptions:

A1 Line outages are persistent, i.e., once a line (say line ℓ) is unintentionally tripped offline, its terminals will remain open until it is detected that the line is in such condition.

A2 Any single line outage will not cause the system to break into two electrically isolated islands, i.e., the underlying graph describing the power system network remains connected.

5.3.1 Pre-outage Model

Let Δt [s] denote the PMU sample time and define the kth measurement sample of active and reactive power injections at bus i as

$$P_i[k] := P_i(k\Delta t)$$

and

$$Q_i[k] := Q_i(k\Delta t),$$

respectively. Similarly, define the kth voltage magnitude and phase angle measurement sample at bus i as

$$V_i[k] := V_i(k\Delta t)$$

and

$$\theta_i[k] := \theta_i(k\Delta t),$$

respectively. In addition, define the differences in bus i's voltage magnitudes and phase angles of between consecutive sampling times $k\Delta t$ and $(k+1)\Delta t$ as

$$\Delta V_i[k] := V_i[k+1] - V_i[k]$$

and

$$\Delta \theta_i[k] := \theta_i[k+1] - \theta_i[k],$$

respectively. Similarly, define the differences in the active and reactive power injections into bus i between two consecutive sampling times as

$$\Delta P_i[k] = P_i[k+1] - P_i[k]$$

and

$$\Delta Q_i[k] = Q_i[k+1] - Q_i[k],$$

respectively.

Next, we assume the DC power flow assumptions (see, e.g., [27]) hold, namely

D1 flat voltage profile, i.e., $V_i(t) \approx 1$ p.u. for all $i \in \mathcal{V}$;
D2 negligible line resistances, i.e., $G_{ik} \approx 0$ for all $i,k \in \mathcal{V}$; and
D3 small phase angle differences, i.e.,

$$\cos\left(\theta_i(t) - \theta_k(t)\right) \approx 1$$

and

$$\sin\left(\theta_i(t) - \theta_k(t)\right) \approx \theta_i - \theta_k$$

for all i,k for which $B_{ik} > 0$.

Define $\Delta \tilde{P}[k] = \left[\Delta P_1[k], \ldots, \Delta P_N[k]\right]^\top$ and $\Delta \tilde{\theta}[k] = \left[\Delta \theta_1[k], \ldots, \Delta \theta_N[k]\right]^\top$. Then, by linearizing (5.19) around $(\theta_i[k], V_i[k], P_i[k], Q_i[k])$, $i = 1, \ldots, N$, we obtain

$$\Delta \tilde{P}[k] = \tilde{H}_0 \Delta \tilde{\theta}[k] \tag{5.21}$$

where $\tilde{H}_0 = [B_{ik}] \in \mathbb{R}^{N \times N}$. Without loss of generality, assume that bus 1 is the "slack" bus, i.e., it behaves as a voltage source with known magnitude and angle, which we take to be zero. Then, after omitting the equation corresponding to the slack bus, the relationship between variations in voltage phase angles and the variations in the active power injection is given by:

$$\Delta P[k] \approx H_0 \Delta \theta[k] \tag{5.22}$$

where $\Delta P[k]$ and $\Delta \theta[k] \in \mathbb{R}^{(N-1)}$ are respectively obtained by removing the first entry from $\Delta \tilde{P}[k]$ and $\Delta \tilde{\theta}[k]$, and $H_0 \in \mathbb{R}^{(N-1) \times (N-1)}$ is obtained by removing the first row and first column from the matrix $\tilde{H}_0$.

Next, assume that the buses are indexed in such a way that generator buses appear first in $\Delta P[k]$. Then, we can partition $\Delta P[k]$ as follows:

$$\Delta P[k] = \begin{bmatrix} \Delta P^g[k] \\ \Delta P^d[k] \end{bmatrix}$$

where $\Delta P^d[k] \in \mathbb{R}^{N_d}$ and $\Delta P^g[k] \in \mathbb{R}^{N_g}$, $N_d + N_g = N - 1$, respectively denote the changes in the load demand vector and the power generation vector at time instant k. Now, to capture the response of the generators in the system to random load demand fluctuations, we assume that

$$\Delta P^g[k] = B(t)\Delta P^d[k] \tag{5.23}$$

where $B(t) = [B_{ij}(t)]$ captures the response of the power injected at generation bus i to a change in demand at load bus j at instant k. The $B_{ij}(t)$'s depend on generator inertia coefficients and other control parameters, including droop coefficients and the automatic generation control (AGC) participation factors. In this work, we approximate $B(t)$ by assuming it can only take values in a finite set $\{B_0, B_1, \ldots, B_T\}$, where each B_i captures the generator response during some period $[t_i, t_{i+1})$. Let $B(t) = B_0$ during the pre-outage period and define $M_0 := H_0^{-1}$. Then, we can substitute (5.23) into (5.22) to obtain a pre-outage relation between the changes in the voltage angles and the active power demand at the load buses as follows:

$$\begin{aligned}
\Delta \theta[k] &\approx M_0 \Delta P[k] \\
&= M_0 \begin{bmatrix} \Delta P^g[k] \\ \Delta P^d[k] \end{bmatrix} \\
&= [M_0^{(1)} \; M_0^{(2)}] \begin{bmatrix} B_0 \Delta P^d[k] \\ \Delta P^d[k] \end{bmatrix} \tag{5.24} \\
&= (M_0^{(1)} B_0 + M_0^{(2)}) \Delta P^d[k] \\
&= \tilde{M}_0 \Delta P^d[k]
\end{aligned}$$

where $\tilde{M}_0 = M_0^{(1)} B_0 + M_0^{(2)}$.

5.3.2 Instantaneous Change during Outage

At the time an outage occurs, $t = t_f$, there is an instantaneous change in the mean of the voltage phase angle measurements that affects only one incremental sample, namely, $\Delta \theta[\gamma_0] = \theta[\gamma_0 + 1] - \theta[\gamma_0]$, with γ_0 such that $\Delta t \gamma_0 \leq t_f < \Delta t(\gamma_0 + 1)$, i.e., $\theta[\gamma_0]$ is obtained immediately prior to the outage, whereas $\theta[\gamma_0 + 1]$ is obtained immediately after the outage. Assume that the outaged line ℓ connects buses m and n. Then, the effect of an outage in line ℓ can be modeled with a power injection of $P_\ell[\gamma_0]$ at bus m and $-P_\ell[\gamma_0]$ at bus n, where $P_\ell[\gamma_0]$ is the pre-outage line flow across line ℓ from m to n. Following a similar approach as the one in [4], the relation between the voltage phase angle incremental change at the instant of outage, $\Delta\theta[\gamma_0]$, and the variations in the active power flow can be expressed as:

$$\Delta\theta[\gamma_0] \approx M_0\Delta P[\gamma_0] - P_\ell[\gamma_0 + 1]M_0 r_\ell \quad (5.25)$$

where $r_\ell \in \mathbb{R}^{N-1}$ is a vector with the $(m-1)$th entry equal to 1, the $(n-1)$th entry equal to -1, and all other entries equal to 0. Then, by using (5.23) together with (5.25), we obtain

$$\Delta\theta[\gamma_0] \approx \tilde{M}_0\Delta P^d[\gamma_0] - P_\ell[\gamma_0 + 1]M_0 r_\ell. \quad (5.26)$$

5.3.3 Post-outage

Following a line outage, the power system undergoes a transient response governed by $B_i, i = 1, 2, \ldots, T-1$ until a new quasi-steady state is reached, in which $B(t)$ settles to B_T. For example, immediately after the outage occurs, the power system is dominated by the inertial response of the generators, which is immediately followed by the governor response, and then followed by the response of the automatic generation control system response. As a result of the line outage, the system topology changes, which manifests itself in the matrix H_0. This change in the matrix H_0 resulting from the outage can be expressed as the sum of the pre-outage matrix and a rank-one perturbation matrix, ΔH_ℓ, i.e., $H_\ell = H_0 + \Delta H_\ell$. Then, by letting $M_\ell := H_\ell^{-1} = [M_\ell^{(1)} \; M_\ell^{(2)}]$, and proceeding in the same manner as the pre-outage model of (5.24), we obtain the post-outage relation between the changes in the voltage angles and the active power demand as:

$$\Delta\theta[k] \approx \tilde{M}_{\ell,i}\Delta P^d[k], \quad \gamma_{i-1} \le k < \gamma_i \quad (5.27)$$

where $\tilde{M}_{\ell,i} = M_\ell^{(1)} B_i + M_\ell^{(2)}, i = 1, 2, \ldots, T$.

5.3.4 Measurement Model

We assume that the voltage phase angles are measured at only a subset of the load buses, and denote this reduced measurement set by $\hat{\theta}[k]$. Assume that we choose $p \le N_d$ locations to deploy the PMUs. Then, there are $\binom{N_d}{p}$ possible locations to place the PMUs. Here, we assume that the PMU locations are fixed; in general, the problem of optimal PMU placement is NP-hard but one can resort to heuristics to obtain suboptimal placement policies (see, e.g., [6]).

Let

$$\tilde{M} = \begin{cases} \tilde{M}_0, & \text{if } 1 \le k < \gamma_0, \\ \vdots \\ \tilde{M}_{\ell,T}, & \text{if } k \ge \gamma_T. \end{cases} \quad (5.28)$$

Then, the absence of a PMU at bus i corresponds to removing the ith row of $\tilde{M}$. Thus, let $\hat{M} \in \mathbb{R}^{p \times N_d}$ be the matrix obtained by removing $N-p-1$ rows from $\tilde{M}$. Therefore, we can relate $\hat{M}$ to $\tilde{M}$ in (5.28) as follows:

$$\hat{M} = C\tilde{M} \quad (5.29)$$

where $C \in \mathbb{R}^{p \times (N-1)}$ is a matrix of 1's and 0's that appropriately selects the rows of $\tilde{M}$. Accordingly, the increments in the phase angle can be expressed as follows:

$$\Delta \hat{\theta}[k] \approx \hat{M} \Delta P^d[k]. \tag{5.30}$$

Because small variations in the active power injections at the load buses, $\Delta P^d[k]$, can be attributed to random fluctuations in electricity consumption, we may model the $\Delta P^d[k]$'s as i.i.d. random vectors. Then, by using the Central Limit Theorem [28], we can argue that each $\Delta P^d[k]$ is a Gaussian vector, i.e., $\Delta P^d[k] \sim \mathcal{N}(0, \Lambda)$, where Λ is a covariance matrix (note that we make the additional assumption that the elements in $\Delta P^d[k]$ are roughly independent). Because $\Delta \hat{\theta}[k]$ depends on $\Delta P^d[k]$ through the linear relationship given in (5.30), we have that:

$$\Delta \hat{\theta}[k] \sim \begin{cases} f_0 := \mathcal{N}(0, \hat{M}_0 \Lambda \hat{M}_0^\top), \text{ if } 1 \leq k < \gamma_0, \\ f_\ell^{(0)} := \mathcal{N}(\mu_\ell, \hat{M}_0 \Lambda \hat{M}_0^\top), \text{ if } k = \gamma_0, \\ f_\ell^{(1)} := \mathcal{N}(0, \hat{M}_{\ell,1} \Lambda \hat{M}_{\ell,1}^\top), \text{ if } \gamma_1 \leq k < \gamma_2, \\ \vdots \\ f_\ell^{(i)} := \mathcal{N}(0, \hat{M}_{\ell,i} \Lambda \hat{M}_{\ell,i}^\top), \text{ if } \gamma_i \leq k < \gamma_{i+1}, \\ \vdots \\ f_\ell^{(T)} := \mathcal{N}(0, \hat{M}_{\ell,T} \Lambda \hat{M}_{\ell,T}^\top), \text{ if } \gamma_T \leq k \end{cases} \tag{5.31}$$

where $\mu_\ell := -P_\ell[\gamma+1] C M_0 r_\ell$ is the instantaneous meanshift and $\gamma_1 = \gamma_0 + 1$. It is important to note that for $\mathcal{N}(0, \hat{M} \Lambda \hat{M}^\top)$ to be a nondegenerate pdf, its covariance matrix, $\hat{M} \Lambda \hat{M}^\top$, must be full rank. We enforce this by ensuring that the number of PMUs placed, p, is less than or equal to the number of load buses, N_d, and that they are deployed at nodes such that the measured voltage phase angles are independent. The matrices $\hat{M}$ are known based on the system topology following a line outage and Λ can be estimated from historical data.

5.4 Line Outage Detection Using QCD

In the line outage detection problem setting, the goal is to detect the outage in line ℓ as quickly as possible subject to false alarm constraints. The outage induces a change in the statistical characteristics of the observed sequence $\{\Delta \hat{\theta}[k]\}_{k \geq 1}$. The aim is to design stopping rules that detect this change using the QCD methodology discussed in Section 5.2.

We begin by introducing the Meanshift test, a detection scheme that can be shown to be equivalent to that proposed in [14] (see also [16] and [15]) for detecting line outages.

5.4.1 Meanshift Test

The Meanshift test is a "oneshot" detection scheme like the Shewhart test of (5.3), i.e., only the most recent observation is used to calculate the test statistics. An additional simplification is that only a single log-likelihood ratio between the distribution of the observations at the changepoint and before the changepoint is used to construct the test. As seen in (5.31), there is a mean shift at exactly the time of outage γ_0, and only this mean shift is exploited in constructing the test statistic:

$$W_\ell^M[k] = \log \frac{f_\ell^{(0)}(\Delta\hat{\theta}[k])}{f_0(\Delta\hat{\theta}[k])}. \tag{5.32}$$

Because an outage can occur at any line, we use a generalized test structure as in the GLR-CuSum test described in Section 5.2.4, i.e., we take the maximum $W_\ell^M[k]$ over ℓ to form the test statistic. Consequently, the stopping time for the Meanshift test is given by

$$\tau_M = \inf\left\{k \geq 1 : \max_{\ell \in \mathcal{L}} W_\ell^M[k] > b\right\} \tag{5.33}$$

with $b > 0$ chosen to meet the false alarm constraint.

In addition to the fact that it is a "one-shot" detection scheme, the Meanshift test has the drawback that the details of the postchange statistics, the transient dynamics and the persistent distribution in (5.31), are not incorporated in the test statistic. As a result, the log-likelihood ratio used in the test statistic does not match the true distribution of the observations after the changepoint γ_0. For example, during the first transient period ($\gamma_0 < k \leq \gamma_1$), the expected value of the test statistic could be negative as

$$\mathbb{E}_\ell^{(1)}\left[\log \frac{f_\ell^{(0)}(\Delta\hat{\theta}[k])}{f_0(\Delta\hat{\theta}[k])}\right] = D(f_\ell^{(1)} \| f_0) - D(f_\ell^{(1)} \| f_\ell^{(0)}) \tag{5.34}$$

where $\mathbb{E}_\ell^{(1)}$ denotes the expectation under distribution $f_\ell^{(1)}$.

5.4.2 Generalized Shewhart Test

We can improve on the Meanshift test by incorporating all the postchange statistics in (5.31) into a Shewhart type "one-shot" statistic. In particular, the test statistic corresponding to outage in line ℓ is given by:

$$W_\ell^S[k] = \max_{i \in \{0,1,\ldots,T\}} \left\{\log \frac{f_\ell^{(i)}(\Delta\hat{\theta}[k])}{f_0(\Delta\hat{\theta}[k])}\right\} \tag{5.35}$$

and the corresponding generalized Shewhart test is given by:

$$\tau_S = \inf\left\{k \geq 1 : \max_{\ell \in \mathcal{L}} W_\ell^S[k] > b\right\} \tag{5.36}$$

with $b > 0$ chosen to meet the false alarm constraint. While we can expect the generalized Shewhart test to perform better than the Meanshift test, it is still a "one-shot" test that does not use the past information in the observations.

5.4.3 Generalized CuSum Test

We now explore the incorporation of past information in constructing the test statistic as in the CuSum test described in Section 5.2.2. We first make the approximation that the transition between pre- and postoutage periods is not characterized by any transient behavior other than the mean shift that occurs at the instant of outage. In particular, if line ℓ goes into outage, we make the approximation that the distribution of the observations goes from f_0 to $f_\ell^{(0)}$ at the time of outage, and then directly to $f_\ell^{(T)}$, the persistent change distribution for line ℓ. Then, we construct a slight generalization of the CuSum test, in which the test statistic corresponding to line ℓ being in outage incorporates the mean shift term in the following manner:

$$W_\ell^C[k] = \max\left\{ W_\ell^C[k-1] + \log \frac{f_\ell^{(T)}(\Delta\hat{\theta}[k])}{f_0(\Delta\hat{\theta}[k])},\ \log \frac{f_\ell^{(0)}(\Delta\hat{\theta}[k])}{f_0(\Delta\hat{\theta}[k])}, 0 \right\} \quad (5.37)$$

with $W_\ell^C[0] = 0$ for all $\ell \in \mathcal{L}$.

Although the preceding statistic does not take any transient dynamics into consideration, it is a good approximation when: (i) the transient distributions and the final postchange distribution are "similar," e.g., when the KL divergence between $f_\ell^{(i)}$ and $f_\ell^{(T)}$ is small, for $i = 1, 2, \ldots, T-1$, or (ii) when the expected detection delay is large so that the algorithm stops well into the persistent phase of the change.

Because the line that is in outage is not known *a priori*, we again use the GLR approach of Section 5.2.4 to construct the following generalized test, which we refer to as the G-CuSum test:

$$\tau_C = \inf\left\{ k \geq 1 : \max_{\ell \in \mathcal{L}} W_\ell^C[k] > b \right\} \quad (5.38)$$

with $b > 0$ chosen to meet the false alarm constraint.

5.4.4 Generalized Dynamic CuSum Test

We now construct a test that takes into account all the details of the statistical model of (5.31), the mean shift at the changepoint, the transient dynamics after the changepoint, and the final persistent change. We use the approach described in Section 5.2.5 to form the following test statistic for line ℓ:

$$W_\ell^D[k] = \max\left\{ \Omega_\ell^{(0)}[k], \ldots, \Omega_\ell^{(T)}[k], 0 \right\} \quad (5.39)$$

where

$$\Omega_\ell^{(i)}[k] = \max\{\Omega_\ell^{(i)}[k-1], \Omega_\ell^{(i-1)}[k-1]\} + \log \frac{f_\ell^{(i)}(\Delta\hat{\theta}[k])}{f_0(\Delta\hat{\theta}[k])} \quad (5.40)$$

for $i \in \{1, \ldots, T\}$,

$$\Omega_\ell^{(0)}[k] := \log \frac{f_\ell^{(0)}(\Delta\hat{\theta}[k])}{f_0(\Delta\hat{\theta}[k])}$$

and $\Omega_\ell^{(i)}[0] := 0$, for all $\ell \in \mathcal{L}$ and all i. We refer to the corresponding generalized test as the Generalized Dynamic CuSum (G-D-CuSum) test, which is given by

$$\tau_D = \inf\left\{k \geq 1 : \max_{\ell \in \mathcal{L}} W_\ell^D[k] > b\right\} \tag{5.41}$$

with $b > 0$ chosen to meet the false alarm constraint.

5.5 Line Outage Identification

The line outage detection algorithms described in Section 5.4 can also be used to identify the line that has been affected by the outage. A natural strategy that exploits the generalized likelihood nature of the test statistics would be to declare the affected line as the one corresponding to the largest statistic, i.e.,

$$\hat{\ell} = \arg\max_{j \in \mathcal{L}} W_j[\tau] \tag{5.42}$$

where W_ℓ denotes the test statistic corresponding to line ℓ for the given test (Meanshift, Shewhart, G-CuSum, G-D-CuSum). However, algorithms based on maximum of line statistics that perform well for line outage detection may not perform well at line outage isolation. The reason is that the statistics for other lines may also increase following a line outage, even if the statistic corresponding to the line in outage has the largest drift. In particular, the drift (average increase) in the test statistic for line ℓ when line j is in outage is given by:

$$\mathbb{E}_{f_j}\left[\log\frac{f_\ell(\Delta\hat{\theta}[k])}{f_0(\Delta\hat{\theta}[k])}\right] = D(f_j \| f_0) - D(f_j \| f_\ell). \tag{5.43}$$

If $D(f_j \| f_\ell)$ is small compared to $D(f_j \| f_0)$, then the test statistic for line ℓ will have a positive drift when line j is in outage, and it is possible that the statistic for line ℓ is larger than that of line j when the threshold is crossed; see Figure 5.2(b) for an example where line 36 is in outage, but the statistic for line 37 crosses the threshold first.

We could adopt a joint approach for detection and isolation, where we stop only when we are confident that the false isolation probability is small, as in the matrix CuSum test [9]. However, if the KL divergence between the postchange distributions for some pairs of lines are small as in the example discussed in the previous paragraph, then good isolation will come at the cost of unacceptably long detection delays.

Another approach to obtaining good isolation error probabilities is to allow for the possibility that when an outage is declared by the detection algorithm, more than one line can be checked by the system operator. Then, we can create a precomputed ranked list of candidate lines that should be checked when an outage in a particular line is declared by the algorithm. This is similar to the notion of list decoding in digital communications (see, e.g., [29]). Then, false isolation occurs only if list of lines checked does not contain true outaged line.

We can create the ranked list based on the drifts of the line statistics given in (5.43). In particular, for the line ℓ statistic, we create the ranked list based on the smallest values of $D(f_j \| f_\ell)$ because the j's correspond to the lines that are most likely to the true outaged line if the line ℓ statistic crosses the threshold when the change is declared. If we constrain the ranked list to have r elements, then we can define the ranked list corresponding to line ℓ as:

$$\mathcal{R}_\ell = \{\ell, j_2, \ldots, j_r\}. \tag{5.44}$$

Note that line ℓ has to be in $\mathcal{R}_\ell$ because $D(f_\ell \| f_\ell) = 0$. The remaining elements in $\mathcal{R}_\ell$, $j_2, \ldots, j_r$ are the indices of the lines with $D(f_j \| f_\ell)$ values that are successively larger.

To quantify the performance of our algorithm with respect to its ability to identify the outaged line accurately, we define the probability of false isolation (PFI). When line ℓ is in outage, a false isolation event occurs when line ℓ is not included in the ranked list $\mathcal{R}_{\hat{\ell}}$, with $\hat{\ell}$ chosen according to (5.42). Therefore, we can define the PFI when line ℓ is outaged as:

$$\text{PFI}_\ell(\tau) = \mathbb{P}\{\ell \notin \mathcal{R}_{\hat{\ell}} | \text{line } \ell \text{ in outage}\}. \tag{5.45}$$

The size r of the ranked list should be chosen to optimize the tradeoff between PFI and number of lines that need to be checked after an outage detection has occurred. In particular, larger ranked lists lead to lower PFI, but to a larger set of possibly outaged lines to check.

5.6 Numerical Results

In this section, we verify the effectiveness of the algorithm in (5.39)–(5.41) for detecting line outages in the IEEE 118-bus test system (see [30] for the model data). To simulate the dynamics of the system following a line outage, we utilize the Power System Toolbox (PST) [31], which includes detailed models of the generator dynamics and control. For simplicity, in (5.31) we only consider one transient period after the line outage occurrence, i.e., $T = 2$ with a duration of 100 samples. As discussed earlier, we assume the changes in load demands to be zero-mean independent Gaussian random variables with variance 0.03. We also assume that the PMUs provide 30 measurements per second and choose to place them at all load buses.

5.6.1 Line Statistic Evolution

We simulate two different line outages occurring at $k = 10$; the results are shown in Figure 5.1. In one case, the detection takes place during the transient period, whereas in the other case the detection takes places after the transient period has ended. Figure 5.1(a) shows some typical progressions of $W_{180}[k]$ for the various line outage detection schemes discussed earlier with a detection threshold of $b = 120$. As shown, a line outage can be declared after 50 samples when the G-D-CuSum stream crosses the threshold. Also as shown in the figure, the other algorithms incur a much larger detection delay. The progress of $W_{32}[k]$ for the different algorithms for an outage in line 32 is shown

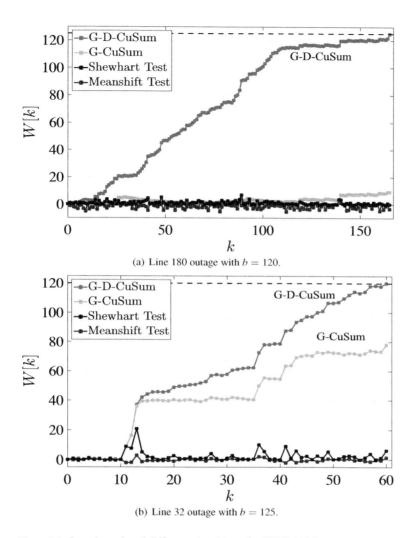

Figure 5.1 Sample paths of different algorithms for IEEE 118-bus test system.

in Figure 5.1(b), where one can see that for the detection threshold, $b = 125$, the G-D-CuSum detects an outage 156 samples after its occurrence. In this case, the detection occurs after the transient dynamics have ended. From the plots, we can conclude that the G-D-CuSum algorithm has a smaller detection delay than the G-CuSum algorithm even though the line outage is detected after the transient period has finished at approximately $k = 110$. Also note that the G-CuSum algorithm has a large delay because its statistic does not grow during the transient period.

Figure 5.2 shows the evolution of the G-D-CuSum statistic for different lines and two different outages in lines 180 and 36 occurring at time $k = 10$. For an outage in line 180, as shown in Figure 5.2(a), $W_{180}[k]$ grows faster than that of other line statistics, and an

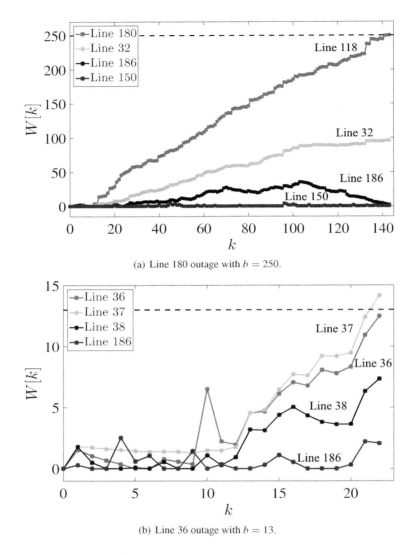

(a) Line 180 outage with $b = 250$.

(b) Line 36 outage with $b = 13$.

Figure 5.2 Sample paths of the G-D-CuSum algorithm for IEEE 118-bus test system.

an outage is declared after 135 samples, when $W_{180}[k]$ crosses the detection threshold, $b = 250$. Next, we illustrate the occurrence of a false isolation event. Specifically, we simulate an outage in line 36; however, in Figure 5.2(b) we see that $W_{37}[k]$ crosses the detection threshold, $b = 13$, before $W_{36}[k]$ does.

5.6.2 Delay Performance

We performed Monte Carlo simulations for outages in lines 36, 180, and 104, and computed detection delay versus MTFA results for the G-D-CuSum, the Meanshift test,

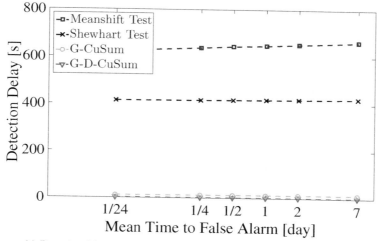

(a) Detection delay versus mean time to false alarm for different algorithms.

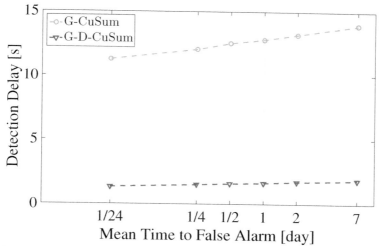

(b) Detection delay versus mean time to false alarm for G-D-CuSum and G-CuSum algorithms.

Figure 5.3 Monte Carlo simulation results for an outage in line 36 for IEEE 118-bus test system.

the Shewhart test, and the G-CuSum algorithm; their performance for an outage in line 36 – measured as the worst delay of all three outaged cases – is compared in Figure 5.3. From the results in Figure 5.3(a), we can conclude that the G-D-CuSum algorithm achieves the lowest detection delay among all algorithms for a given MTFA. In Figure 5.3(b) we show the performance gain that is achieved when choosing the G-D-CuSum test over the Meanshift and Shewhart tests, and the G-CuSum test. As shown, for an outage in line 36 and a given false alarm rate, the delay achieved with the G-D-CuSum tests is an order of magnitude less than that of other tests.

Table 5.1 Probability of false isolation for IEEE 118-bus test system simulated with a ranked list of length of 1.

$\mathbb{E}_\infty[\tau]$ [day]	1/24	1/4	1/2	1	2	7
Line 36	0.5999	0.5472	0.5266	0.5182	0.5102	0.4911
Line 180	0.0344	0.0237	0.0186	0.0160	0.0148	0.0091
Line 104	$<10^{-6}$	$<10^{-6}$	$<10^{-6}$	$<10^{-6}$	$<10^{-6}$	$<10^{-6}$

Table 5.2 Probability of false isolation for IEEE 118-bus test system simulated with a ranked list of length of 3.

$\mathbb{E}_\infty[\tau]$ [day]	1/24	1/4	1/2	1	2	7
Line 36	0.1477	0.1392	0.1371	0.1303	0.1251	0.1219
Line 180	0.0197	0.0127	0.0112	0.0097	0.0062	0.0049
Line 104	$<10^{-6}$	$<10^{-6}$	$<10^{-6}$	$<10^{-6}$	$<10^{-6}$	$<10^{-6}$

Table 5.3 Probability of false isolation for IEEE 118-bus test system simulated with a ranked list of length of 5.

$\mathbb{E}_\infty[\tau]$ [day]	1/24	1/4	1/2	1	2	7
Line 36	0.0295	0.0203	0.0158	0.0137	0.0108	0.0067
Line 180	0.0097	0.0074	0.0052	0.0041	0.0036	0.0027
Line 104	$<10^{-6}$	$<10^{-6}$	$<10^{-6}$	$<10^{-6}$	$<10^{-6}$	$<10^{-6}$

5.6.3 Probability of False Isolation

Tables 5.1–5.3 contains the PFI versus MTFA for outages in lines 36, 104, and 180. We computed the PFI using the ranked list method discussed in Section 5.5 for a ranked list of fixed length 1, 3, and 5. Table 5.1 contains the PFI versus MTFA results for a ranked lists of length 1; this is equivalent to identifying the outaged line using (5.42), i.e., identifying the line with the highest statistic at the stopping time as outaged. Here it is important to note that while this technique might be efficient to handle some line outages (e.g., outage in line 180 and 104), some other line outages, e.g., an outage in line 36, may lead to large PFI values. As discussed in Section 5.5, this is due to the fact that many line statistics other than the one corresponding to the outaged line grow postoutage. Table 5.2 contains the PFI versus MTFA values for a ranked list of length 3; one can see that in this case the PFI is significantly reduced for line 36. Finally, Table 5.3 contains the PFI versus MTFA for a ranked list of length 5, where one case that in this case the PFI for line 36 is below 5%. Finally, note that the PFI decreases as the MTFA increases; this is due to the fact that larger MTFAs corresponds to larger thresholds, which result in smaller PFI values.

5.7 Conclusion

In this chapter, we discussed algorithms based on the methodology of statistical quickest change detection for detecting and identifying line outages. The algorithms exploit the statistical properties of voltage phase angle measurements obtained from PMUs in realtime and feature a set of statistics that are used to capture each distribution shift that results from a line outage.

The most effective of these algorithms, the G-D-CuSum algorithm, takes a generalized likelihood ratio approach and incorporates the transient dynamics after the outage in the test statistic. We showed that the algorithm outperforms other previously developed line outage detection methods. In particular, we compared the detection delay performance of this QCD-based algorithm against other line outage detection algorithms for line outages simulated on the IEEE 118-bus test system. We used this system to also simulate the delay performance of the G-D-CuSum for different line outage scenarios, as well as to illustrate the use of the algorithm for line outage identification. To achieve effective identification, we used the mechanism of ranked lists, i.e., sets of possibly outaged lines that are highly likely to contain the true outaged line.

The G-D-CuSum algorithm can be further improved by incorporating prior knowledge about the durations of the transients that follow the line outage, but our simulation studies for the IEEE 118-bus test system indicate that the improvement may be marginal. Another topic that is of interest to explore is the optimal placement of limited PMU resources in the network to maximize line outage detection performance. Some initial studies along these lines are presented in [6].

The techniques discussed in this chapter can be easily extended to double-line outage detection (and in general to multiple line outage detection), as in [4], by characterizing the distribution of the observations after a double-line outage. Then, the G-D-CuSum scheme can be used to detect such a change in the distribution. In particular, to handle double-line outages, a test statistic needs to be calculated for every possible pair of lines; however, this might be impractical in large systems due to combinatorial explosion. An alternative is just to focus on the double-line outages that are more likely to occur. These can be characterized from historical data or from simulations by noting that a double-line outage is usually the result of a line tripping offline causing a second line to trip offline shortly after due to the particular loading conditions at the time the first line trips. Then, by taking into account the probability of each line tripping offline in the first place (which can also be determined from historical data for a particular system) and the loading conditions probability distribution, one can conduct Monte Carlo simulations to determine which lines are more likely to trip offline immediately after the occurrence of the first outage. As an alternative to Monte Carlo simulation, one can potentially use so-called line outage distribution factors (LODFs) – linear sensitivities that determine how the power flow on a particular line changes after the occurrence of an outage in another line – to determine which lines may become overloaded, and thus trip offline, following an outage in some particular line. This LODF-based method is less accurate that Monte Carlo simulation but also less computationally expensive.

References

[1] FERC and NERC. Arizona-southern California outages on September 8, 2011: Causes and recommendations [Online]. www.ferc.gov, Apr. 2012.

[2] U.S.-Canada Power System Outage Task Force. Final report on the August 14th blackout in the United States and Canada: Causes and recommendations [Online]. http://energy.gov, Apr. 2004.

[3] T. Banerjee, Y. Chen, A. Domínguez-García, and V. Veeravalli, "Power system line outage detection and identification: A quickest change detection approach," in *Acoustics, Speech and Signal Processing (ICASSP), 2014 IEEE International Conference*, pp. 3450–3454, May 2014.

[4] Y. C. Chen, T. Banerjee, A. D. Domínguez-García, and V. V. Veeravalli, "Quickest line outage detection and identification," *IEEE Trans. Power Syst.*, vol. 31, no. 1, pp. 749–758, Jan. 2016.

[5] G. Rovatsos, X. Jiang, A. D. Domínguez-García, and V. V. Veeravalli, "Comparison of statistical algorithms for power system line outage detection," in *Proc. of the IEEE International Conference on Acoustics, Speech, and Signal Processing*, Apr. 2016.

[6] X. Jiang, Y. C. Chen, V. V. Veeravalli, and A. D. Domínguez-García, "Quickest line outage detection and identification: Measurement placement and system partitioning," in *2017 North American Power Symposium (NAPS)*, pp. 1–6, Sept. 2017.

[7] G. Rovatsos, X. Jiang, A. D. Domínguez-García, and V. V. Veeravalli, "Statistical power system line outage detection under transient dynamics," *IEEE Transactions on Signal Processing*, vol. 65, no. 11, pp. 2787–2797, June 2017.

[8] H. V. Poor and O. Hadjiliadis, *Quickest detection*. Cambridge: Cambridge University Press, 2009.

[9] A. G. Tartakovsky, I. V. Nikiforov, and M. Basseville, *Sequential Analysis: Hypothesis Testing and Change-Point Detection*, ser. Statistics. Boca Raton, FL: CRC Press, 2014.

[10] V. V. Veeravalli and T. Banerjee, *Quickest Change Detection*. Elsevier: E-reference Signal Processing, http://arxiv.org/abs/1210.5552, 2013.

[11] K. A. Clements and P. W. Davis, "Detection and identification of topology errors in electric power systems," *IEEE Trans. Power Syst.*, vol. 3, no. 4, pp. 1748–1753, Nov. 1988.

[12] F. F. Wu and W. E. Liu, "Detection of topology errors by state estimation," *IEEE Trans. Power Syst.*, vol. 4, no. 1, pp. 176–183, Feb. 1989.

[13] N. Singh and H. Glavitsch, "Detection and identification of topological errors in online power system analysis," *IEEE Trans. Power Syst.*, vol. 6, no. 1, pp. 324–331, Feb. 1991.

[14] J. E. Tate and T. J. Overbye, "Line outage detection using phasor angle measurements," *IEEE Trans. Power Syst.*, vol. 23, no. 4, pp. 1644–1652, Nov. 2008.

[15] H. Zhu and G. B. Giannakis, "Sparse overcomplete representations for efficient identification of power line outages," *IEEE Trans. Power Syst.*, vol. 27, no. 4, pp. 2215–2224, Nov. 2012.

[16] G. Feng and A. Abur, "Identification of faults using sparse optimization," in *Proceedings of Communication, Control, and Computing (Allerton Conference)*, pp. 1040–1045, Sept 2014.

[17] M. Garcia, T. Catanach, S. V. Wiel, R. Bent, and E. Lawrence, "Line outage localization using phasor measurement data in transient state," *IEEE Trans. Power Syst.*, vol. 31, no. 4, pp. 3019–3027, July 2016.

[18] P. Moulin and V. Veeravalli, *Statistical Inference for Engineers and Data Scientists*. Cambridge: Cambridge University Press, 2019.

[19] W. A. Shewhart, "The application of statistics as an aid in maintaining quality of a manufactured product," *J. Amer. Statist. Assoc.*, vol. 20, no. 152, pp. 546–548, Dec. 1925.

[20] E. S. Page, "Continuous inspection schemes," *Biometrika*, vol. 41, no. 1/2, pp. 100–115, June 1954.

[21] G. Lorden, "Procedures for reacting to a change in distribution," *Ann. Math. Statist.*, vol. 42, no. 6, pp. 1897–1908 [Online]. http://dx.doi.org/10.1214/aoms/1177693055, Dec. 1971.

[22] M. Pollak, "Optimal detection of a change in distribution," *Ann. Statist.*, vol. 13, no. 1, pp. 206–227, Mar. 1985.

[23] G. V. Moustakides, "Optimal stopping times for detecting changes in distributions," *Ann. Statist.*, vol. 14, no. 4, pp. 1379–1387 [Online]. http://dx.doi.org/10.1214/aos/1176350164, Dec. 1986.

[24] T. L. Lai, "Information bounds and quick detection of parameter changes in stochastic systems," *IEEE Trans. Inf. Theory*, vol. 44, no. 7, pp. 2917–2929, Nov. 1998.

[25] D. Siegmund and E. S. Venkatraman, "Using the generalized likelihood ratio statistic for sequential detection of a change-point," *Ann. Statist.*, vol. 23, no. 1, pp. 255–271, Feb. 1995.

[26] S. Zou, G. Fellouris, and V. Veeravalli, "Quickest change detection under transient dynamics: Theory and asymptotic analysis," *IEEE Trans. Inf. Theory*, vol. 65, no. 3, Mar. 2019.

[27] A. R. Bergen and V. Vittal, *Power Systems Analysis*. Upper Saddle River, NJ: Prentice Hall, 2000.

[28] B. Hajek, *Random Processes for Engineers*. Cambridge: Cambridge University Press, 2015.

[29] P. Elias, "Error-correcting codes for list decoding," *IEEE Trans. Inf. Theory*, vol. 37, no. 1, pp. 5–12, Jan. 1991.

[30] "Power system test case archive" [Online]. www2.ee.washington.edu/research/pstca, Oct. 2012.

[31] J. Chow and K. Cheung, "A toolbox for power system dynamics and control engineering education and research," *IEEE Trans. Power Syst*, vol. 7, no. 4, pp. 1559–1564, Nov. 1992.

6 Active Sensing for Quickest Anomaly Detection

Ali Tajer and Javad Heydari

6.1 Anomaly Detection

Transmission lines are constantly exposed to various kinds of disturbances such as equipment malfunctioning and natural disasters. Due to the highly inter-connected structure of the power grid, an anomalous behavior by a certain operation or process can transcend its realm and can propagate through the ensuing operations or processes. For instance, an anomalous behavior or decisions by the monitoring functions (e.g., state estimators) can also mislead all the ensuing scheduling and control decisions that rely on the monitoring functions.

Because the interconnectivities facilitate propagation of anomalies, quickly detecting and localizing any anomalous behavior in a system as soon as it emerges has a critical role for containing the undesired consequences of the anomalies, expediting the repair of the faulty components, speeding up restoration of the grid, reducing outage time, and improving overall power system reliability. Hence, agile localization of anomalous events is vital for enhancing the overall resiliency of a power system. This is in particular important in the large-scale grids due to their geographical extent and the complexity and volume of their data.

Designing algorithms for detecting anomalous events can be grouped into two broad categories, based on the type of the data the algorithms use. In the first group, the algorithms use voltage and current measurements to estimate the impedance values. Deviation of the impedance values of transmission lines indicates an anomalous event being potentially underway. Hence, a group of algorithms monitor impedance fluctuations for detecting and localizing anomalous events. In this group of algorithms, the availability and reliability of the data has an important impact on the quality of detecting anomlaies as well as the computational complexity of the algorithms involved [1–14].

In the other category, anomalous events are detected by analyzing signature waves that are sent along the transmission lines. Specifically, the traveling durations of these waves indicate the potential presence of anomalies, where they can explicitly determine the distance of the anomalous points from the reference points at which the waves are transmitted [15–21]. This category of algorithms are insensitive to fault types, fault resistance, and source parameters of the system, and are independent of the equipments installed in the network. For the arrival times, feature extraction techniques such as wavelet transform are leveraged to distinguish between the normal signal and the one containing high frequency components.

6.2 Need for Agile Detection of Anomalies

The existing algorithms, irrespectively of their differences, have a prespecified set of sensors or locations of the grid that are being continuously probed for finding the traces of anomalous behaviors. Such designs, subsequently, require that a significant amount of data to be continuously collected and processed. This, in turn, leads to substantial computational complexity, delay in analysis, and cost of using the communication infrastructure. Hence, despite their effectiveness for small networks, a significant majority of the existing approaches become inefficient as the network size and complexity grows.

To circumvent this issue, we develop an *active sensing* approach. The hallmark of this approach is that it consists of a decision mechanism that governs how data should be collected. Specifically, this approach is designed so that it does not collect the data that are unlikely to be informative. This leads to a significant reduction in the amount of the data that is collected from the sensors, and subsequently, considerably lowers the computational complexity and the delay in decision-making. The core idea is that due to the strong interconnectivity in the power systems, a measurement collected from any sensor (e.g., on a bus) is also partially informative about the states of the data collected in the neighboring sensors (e.g., neighboring buses). Motivated by this premise, active sensing will be responsible for dynamically and data-adaptively selecting the sensors (and collecting data from them) that, at the time of data acquisition, are found to be most informative about a potential anomaly emerging or existing in the system.

This chapter proposes an active sensing framework for collecting data in networks. Within this framework, we can design a class of algorithms whose objective is to localize the anomalies by collecting the minimum amount of data possible. The active sensing aspect dynamically focuses the sensing resources on the segments of the data that are deemed to be most informative about the state of the network and its potential anomalous behavior. For instance, when an anomaly starts emerging in a specific group of buses, our algorithms are designed to quickly sense that and place their focus on the data collected from these buses or their close vicinity. The algorithms that can be developed within this framework obtain the measurements sequentially, and progressively update their decisions about the location of the anomaly. The process resumes until the location of the anomaly can be identified with desired reliability. We provide a general theory for active sensing in networked data for the quickest anomaly localization. To also show how the theoretical framework can be adopted by practitioners, we will discuss the theory in the context of line outage detection.

The key technique that facilitates such fast convergence and focus is capitalizing on the strong correlation structures observed among the data collected from different buses. Specifically, the data collected from different sensors (e.g., bus measurements) throughout the system follow a specific correlation structure that depends on the topology of the network and the grid parameters (e.g., impedance values of transmission lines). When an anomalous event occurs, this correlation model deviates from its normal behavior. Detecting the change in the correlation, subsequently, can be used to identify the locations of the anomalous events. This framework is leveraged to minimize the number of measurements required to ensure that all the events can be localized with target reliability through designing a coupled data-acquisition and decision-making process. This leads to minimizing the amount of data required for localizing the fault.

6.3 Introduction to Active Sensing

Active (controlled) sensing is different from the canonical notions of data acquisition (sensing) that have prespecified strategies for data acquisition according to which the data is collected and fed to a decision maker. Specifically, active sensing algorithms assume some notion of *control* over how to collect the data. The critical distinction between active and passive sensing mechanisms is that in the passive approaches, the decision-making and data acquisition are decoupled and are designed as isolated processes. In these methods, the design of the data-acquisition process is also independent of the data, and it is pre-designed to often be optimal for the average of the data. In active sensing algorithms, in contrast, the way that the data is collected is data-adaptive. Furthermore, often, the decision-making and the data acquisition processes are coupled, and their designs are optimized jointly. In these algorithms, data acquisition being data-adaptive allows for collecting the data that is optimal in the very specific system realization instead of being optimal for the average performance.

In many circumstances, active sensing algorithms can provide equally powerful decisions with a considerably lower amount of information compared with the canonical non-active algorithms. In this section, for simplicity, we describe the notion of active sensing in networks in the context of making a decision in a simple dichotomous setting (e.g., binary hypothesis testing). The idea of active sensing, also called *controlled* sensing, was initially developed by Chernoff for binary composite hypothesis testing through incorporating a controlled information gathering process that dynamically decides about taking one of a finite number of possible actions at each time [22]. Depending on the action taken, the statistical model of the data collected will be different.

As an illustrative example, consider the case of testing the validity of a scientific hypothesis. Without having access to any data, of course, it is impossible to examine the validity of the hypothesis. Hence, the first step will be designing a proper experiment that guides the scientist testing the hypothesis toward accepting or rejecting the hypothesis. After an initial process through which some preliminary data is collected, the scientist might have some impression about whether the hypothesis is correct, but for forming a reliable decision she/he might still need stronger evidence. This will often require designing new and more informative experiments. This cycle continues up to the point that the scientist can make a confident decision with a desired level of accuracy.

Such sequential searching and designing proper experiments, often referred to as the *sequential design of experiments* constitute the core principle of active sensing: selecting the experiments is the control action and the experiments are the data-acquisition processes. It is noteworthy that the behavior and model of the data under different experiments are expected to be distinct. Hence, when we decide about a control action,

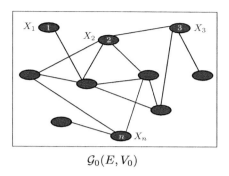 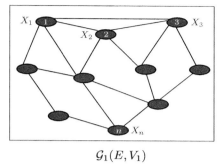

Figure 6.1 Anomaly detection using probabilistic graphical models.

we know that that decision will affect the statistical model of the data that we will be collecting after that action.

To further expand this theory to the anomaly detection objective of interest, we next further explain the implications of this theory when facing decision-making in power systems (e.g., detecting anomalies). For this purpose consider a power system that is operating normally. We abstract this system with an undirected weighted graph $\mathcal{G}_0$ shown in Figure 6.1, in which the nodes represent the buses and the edges represent the transmission lines. At a random instance of time a failure occurs that leads to outage in a few transmission lines. Hence, the topology of the grid captured by $\mathcal{G}_0$ deviated to a different graphical, which we represent by $\mathcal{G}_1$ in Figure 6.1.

A simple approach to monitoring the system and deciding whether it has deviated from the nominal model $\mathcal{G}_0$ is to continually collect data from all the buses, and determine when a change has occurred. Nevertheless, this approach becomes computationally complex, or even prohibitive, as the size of the system (graph dimension) and connections complexity (the number of edges) grow. For this purpose, for continuously monitoring the grid, instead of monitoring all the buses at all times, at each instance in time we select only a subset of the buses and monitor those. Selection of the subsets over time, when done judiciously, allows for rapidly focusing the sensing resources on the neighborhoods that are deemed to contain an anomaly. We comment that the size of the subset of buses directly depends on the computational complexity that we can afford. The more buses that we can select at any give time, the more accurate our decisions become. However, increasing the number of sensors increases the dimension and complexity of the data and, as a result, increases the required computational complexity.

In the context of the line outage detection described earlier, the control actions over time are the selection of a subset of buses from which we collect data. Consequently, an active sensing algorithm in this context is a sequential process over time such that at each time t, it has to make one of the following possible decisions:

- **Exploration:** The information collected up to time t is not conclusive about the potential presence or the location of an anomaly, and we need to identify the set of buses (or sensors in general) to collect measurements from at time $t+1$.

- **Stopping rule:** The information collected provides us with the sufficient confidence to stop probing and exploring, and declare that an anomaly has occurred.
- **Localization:** Based on the information available provides the most accurate decision about the location of the anomaly.

6.4 Active Sensing in Networks

A significant part of the existing literature on active sensing is inspired by the pioneering study of Chernoff on binary hypothesis testing under independent control actions. In this context, Chernoff's rule decides in favor of the control action with the best *immediate* return according to proper information measures. Chernoff's rule, specifically, at each time identifies the most likely true hypothesis based on the collected data, and takes the action that reinforces the decision. Extensions of Chernoff's rule to various settings are studied in [23–30].

Besides the Chernoff rule and its variations, other efforts have also been made to devise alternative strategies admitting certain optimality guarantees. Studies in [31] and [32] offer a strategy that initially takes a number of measurements according to a pre-designated rule to identify the correct hypothesis, after which they select the action that maximizes the information under the identified hypothesis. The study in [33] proposes a heuristic strategy and characterizes the deviation of its average delay from the optimal rule.

Despite their discrepancies in settings and approaches, all the aforementioned studies on controlled sensing assume that the available actions are *independent*, or they follow a first-order stationary Markov process. This is in contrast to the setting of this chapter, in which the correlation structure in the generated data under one hypothesis or both induces co-dependence among the control actions. In this chapter, we provide a sequential sampling strategy for detecting networks, in which the correlation model plays a significant role in forming the sampling decisions. Specifically, the devised selection rule, unlike the Chernoff rule, incorporates the correlation structure into the decision-making by accounting for the impact of each action on the future ones and selecting the one with the largest *expected* information under the most likely true hypothesis. The associated optimality guarantees are established, and the specific results for the particular case of Gaussian distributions are characterized. The gains of the proposed selection rule are also delineated analytically and by numerical evaluations. More details about the optimality properties and the performance analysis can be found in [34–36].

6.5 Modeling Active Sensing in Power Systems

Monitoring power system for detecting various anomalous behavior involves collecting different kinds of measurements from all around the grid. In what follows, we first show how such sensor measurements can be modeled as a Markov network, based on which we next demonstrate that detecting anomalies is equivalent to testing for different correlation structures in those measurements.

6.5.1 Sensor Measurement Model

To formalize the data model for sensor measurements, consider a power system with N buses and L transmission lines, represented as a graph $\mathcal{G}(\mathcal{B}, \mathcal{E})$ where $\mathcal{B}$ and $\mathcal{E} \subseteq \mathcal{B} \times \mathcal{B}$ denote the set of buses and transmission lines, respectively. Each bus is equipped with a sensor that measures the voltage phasor and send it to the monitoring center. If we denote the voltage phasor angle and the injected active power at bus $i \in \mathcal{B}$ by θ_i and p_i, from the DC power flow model we obtain:

$$p_i = \sum_{j \in \mathcal{N}_i} \left(\frac{\theta_i - \theta_j}{x_{ij}} \right), \quad \forall i \in \mathcal{B}, \tag{6.1}$$

where $\mathcal{N}_i$ is the set of buses connected to bus i and x_{ij}, for any $j \in \mathcal{N}_i$, is the reactance of the line connecting bus i to bus j. Collecting all the equations in (6.1) and defining $\boldsymbol{p} \triangleq [p_1, \ldots, p_N]^T$ and $\boldsymbol{\theta} \triangleq [\theta_1, \ldots, \theta_N]^T$, we can rewrite (6.1) in the matrix form as follows:

$$\boldsymbol{p} = \boldsymbol{H} \cdot \boldsymbol{\theta}, \tag{6.2}$$

where $\boldsymbol{H} \in \mathbb{R}^{N \times N}$ is the weighted Laplacian matrix of the grid defined as

$$H[ij] = \begin{cases} \sum_{(i,\ell) \in \mathcal{E}} \frac{1}{x_{i\ell}} & \text{if } i = j \\ -\frac{1}{x_{ij}} & \text{if } (i, j) \in \mathcal{E} \\ 0 & \text{Otherwise} \end{cases} . \tag{6.3}$$

Furthermore, from (6.1) it follows that θ_i can be represented as

$$\theta_i = \sum_{j \in \mathcal{N}_i} r_{ij} \theta_j + \beta_i p_i, \tag{6.4}$$

where we have defined

$$\beta_i \triangleq \left(\sum_{(i,j) \in \mathcal{E}} \frac{1}{x_{ij}} \right)^{-1}, \quad \text{and} \quad r_{ij} \triangleq \frac{\beta_i}{x_{ij}}. \tag{6.5}$$

Injected power at different buses $\boldsymbol{p}$ are influenced by various sources of uncertainty in the system, such as the random disturbances and varying load profiles, and they can be modeled as independent random variables [37, 38]. Hence, given θ_j for $j \in \mathcal{N}_i$, from (6.4) we can conclude that θ_i is independent of the rest of the measurements, and therefore satisfies the global Markov property.

To overcome the rank deficiency of matrix $\boldsymbol{H}$, which makes (6.2) under-determined, one bus is selected as the reference, and the rest are measured with respect to it. To reflect this in (6.2) we remove the row and column corresponding to the reference bus, so that the remaining $(N-1) \times (N-1)$ matrix $\boldsymbol{H}$ is full-rank. Then, we can solve (6.2) according to

$$\boldsymbol{\theta} = \boldsymbol{H}^{-1}\boldsymbol{p}. \tag{6.6}$$

If the distribution of the injected power remains constant over time, any anomaly that changes the topology of the grid or the impedance of one of the transmission lines alters matrix $\boldsymbol{H}$, and as a result, changes the joint distribution of voltage phasor angles $\boldsymbol{\theta}$. Therefore, $\boldsymbol{\theta}$ can be used for detection and localization of anomalous events.

6.5.2 Anomalous Event Model

Any anomaly in the grid that changes the distribution of $\boldsymbol{\theta}$ through $\boldsymbol{H}$ can be detected by testing the joint distribution of $\boldsymbol{\theta}$. To formally model anomalous events, we define $\mathcal{L} \triangleq \{1, \ldots, L\}$ as the set of transmission lines in the grid, and $\mathcal{R} \triangleq \{R_1, \ldots, R_M\}$ as the set of anomalies that we are interested in detecting and localizing, where M is the number of events that should be localized in the quickest fashion. R_k for $k \in \mathcal{L}$ represents one anomalous event, where $R_k \subseteq \mathcal{L}$ contains the indices of the lines experiencing anomaly under event $k \in \{1, \ldots, M\}$. The normal operation mode is denoted by $R_0 = \{\}$.

Under any anomalous event R_k, we denote the connectivity graph of the grid and the reactance of the line connecting buses i and j by $\mathcal{G}_k(\mathcal{B}, \mathcal{E}_k)$ and x_{ij}^k, respectively. We also define the Laplacian matrix $\boldsymbol{H}_k$ corresponding to $\mathcal{G}_k(\mathcal{B}, \mathcal{E}_k)$ accordingly:

$$\boldsymbol{H}_k[ij] \triangleq \begin{cases} \sum_{(i,\ell) \in \mathcal{E}_k} \frac{1}{x_{i\ell}^k} & \text{if } i = j \\ -\frac{1}{x_{ij}^k} & \text{if } (i,j) \in \mathcal{E}_k \\ 0 & \text{Otherwise} \end{cases}. \tag{6.7}$$

Hence, anomaly detection can be abstracted as the following multi-hypothesis testing problem:

$$\mathsf{H}_k: \quad \boldsymbol{\theta} = \boldsymbol{H}_k^{-1}\boldsymbol{p}, \quad \text{for } k \in \{0, \ldots, M\}. \tag{6.8}$$

Because the Laplacian matrix changes under each anomalous event, $\boldsymbol{\theta}$ follows a distinct joint distribution governed by the associated topology and line reactances of the network. In practice, detecting anomalies by collecting measurements from all the buses is costly due to the prohibitive sensing, communication, storage, and processing costs. In the next section, we show that active sensing comes to the rescue by devising a data-adaptive decision-making framework that can form arbitrarily reliable decisions about the state of the grid with *minimal* number of measurements.

6.6 Decision Rules and Algorithm

In the active sensing paradigm, the anomaly detection is performed as follows: the process starts by collecting ℓ measurements from a subset of buses. Then, the collected measurements are leveraged to update the decision about the presence or absence of an anomaly and guiding the future data collection. For instance, if the belief is that no conclusive decision can be made, then ℓ new buses are selected and their measurements are collected. The selection of new buses is guided by the previously collected measurements in a data-adaptive fashion. The process continues until time $\tau \in \mathbb{N}$, as the stopping time of the process, at which point it terminates, and a decision about the underlying event is formed.

The motivation behind this data-acquisition and decision-making mechanism is that when a specific anomaly occurs, different buses are affected with varying degrees. For instance, the measurements generated at the buses closer to a line in outage are affected more than those of a remote bus. Such discrepancies among the level of information provided by different buses are leveraged to guide the data-collection process toward the buses that are expected to be more informative about the underlying events.

The dynamic decisions about the buses to be observed is denoted by the selection function $\psi(t) \triangleq [\psi(t,1), \ldots, \psi(t,\ell)]$, which captures the indices of ℓ buses to be measured at time $t \in \{1, \ldots, \tau\}$. The measurements collected at time t from $\psi(t)$ are denoted by $\theta(t) \triangleq [\theta(t,1), \ldots, \theta(t,\ell)]$ where $\theta(t,i)$ is the measurement collected from bus $\psi(t,i)$. Accordingly, we denote the vector of observed buses and their corresponding measurements up to time t by ψ_t and θ_t, respectively, i.e.,

$$\psi_t \triangleq [\psi(1), \ldots, \psi(t)]^T \tag{6.9}$$

$$\text{and } \theta_t \triangleq [\theta(1), \ldots, \theta(t)]^T. \tag{6.10}$$

When the data collection terminates, a decision $\delta \in \mathcal{R}$ about the true hypothesis based on the observed data is formed.

In any anomaly detection problem, we are interested in fast and reliable decisions. The reliability of the final decision is captured by the decision error probability, i.e.,

$$\mathsf{P_e} = \mathbb{P}(\delta \neq \mathsf{T}) = \sum_{i=0}^{M} \mathbb{P}(\mathsf{T}=i) \sum_{j \neq i} \mathbb{P}(\delta = j \mid \mathsf{T} = i), \tag{6.11}$$

where T is the true hypothesis. Hence, by controlling the reliability by constraining the probability of erroneous decisions below a threshold $\beta \in (0,1)$ and minimizing the average delay in reaching a reliable decision, the optimal sampling strategy is obtained as the solution to the following optimization problem:

$$\underset{\tau, \delta, \psi_\tau}{\text{minimize}} \; \mathbb{E}\{\tau\} \quad \text{subject to} \quad \mathsf{P_e} \leq \beta. \tag{6.12}$$

Here, the dynamic bus selection rule, the stopping time, and the final decision rule are optimized jointly. In the remainder of this chapter, we limit the total number of lines that can be concurrently anomalous to be less than $\eta_{\max}$ and propose a data-adaptive algorithm for anomaly detection and localization.

6.6.1 Bus Selection Rule

Active sensing works on the premise that different sensing actions have distinct information content. In power systems, the amount of information obtained from the measurement of one bus under various anomalous events are different. Hence, it is of paramount importance to devise a bus selection strategy that can adapt to different outage events to strike an optimal balance between the decision quality and the quickness of the process, as formalized in (6.12). Inspired by techniques from controlled sensing, and specifically the Chernoff rule [22], we propose a bus selection rule that applies maximum likelihood decision to identify the most likely event, based on which it selects buses that are most informative about that event. While it is shown that Chernoff rule is asymptotically optimal under some conditions, it loses such optimality property when the actions are correlated [36], as is the case in power systems. The proposed bus selection rule addresses this issue by incorporating the correlation structure of the data into the information measures assigned to each bus. The new information measures account for the impact of future decisions on the current ones. Specifically, we apply an *exhaustive* search over all buses selected in the future, which can have prohibitive complexity, especially as the grid size grows. We show that the new measure achieves the asymptotic optimality property similar to that of the Chernoff rule for correlated actions. Next, we leverage the Markovian structure of the measurements to limit the search domain to the neighbors of each bus, which in turn, reduces the computational complexity significantly.

To prove the optimality properties, consider a binary setting, i.e., $\mathcal{R} = \{R_0, R_1\}$ with $\ell = 1$. Under R_0, assume that the measurements form a Gauss-Markov random field (GMRF) with mean $\bar{\boldsymbol{\theta}}$, which represents the empirical average of $\boldsymbol{\theta}$ based on the historical data, and covariance matrix $(\boldsymbol{I} - \boldsymbol{Q}_0)$, where the elements of $\boldsymbol{Q}_0 = [r_{ij}]$ are defined in (6.5). Under R_1, however, assume that the measurements form an alternative GMRF with a different covariance matrix $\boldsymbol{Q}_1$. For simplicity in notations we assume $\boldsymbol{Q}_1 = \boldsymbol{I}$, which will also be useful when dealing with the general setting. To characterize the information measure of bus i at time t, we define

$$M_i^0(t, \mathcal{S}_t^i) = \frac{1}{2} \sum_{j \in \psi_{t-1}} \log \frac{1}{1 - r_{ij}^2} + r_{ij}^2(\Delta\theta_j^2 - 1)$$

$$+ \frac{1}{2|\mathcal{S}_t^i|} \sum_{j \in \mathcal{S}_t^i} \log \frac{1}{1 - r_{ij}^2}, \qquad (6.13)$$

$$\text{and } M_i^1(t, \mathcal{S}_t^i) = \frac{1}{2} \sum_{j \in \psi_{t-1}} \log(1 - r_{ij}^2) + \frac{r_{ij}^2(\Delta\theta_j^2 + 1)}{1 - r_{ij}^2}$$

$$+ \frac{1}{2|\mathcal{S}_t^i|} \sum_{j \in \mathcal{S}_t^i} \log(1 - r_{ij}^2) + \frac{2r_{ij}^2}{1 - r_{ij}^2}, \qquad (6.14)$$

where we have defined $\Delta\theta_i \triangleq \theta_i - \bar{\theta}_i$ for $i \in \{1, \ldots, N\}$, and set $\mathcal{S}_t^i$ is a subset of unobserved buses prior to time t that contain bus i, i.e.,

$$\mathcal{S}_t^i \subseteq \mathcal{B} \setminus \psi_{t-1} \quad \text{and} \quad i \in \mathcal{S}_t^i. \tag{6.15}$$

With these information measures, when the maximum likelihood decision about the true model at time $(t-1)$ is H_j, at time t we set the bus selection rule according to

$$\psi(t) = \arg\max_{i \notin \psi_{t-1}} \max_{\mathcal{S}_t^i} M_i^j(t, \mathcal{S}_t^i). \tag{6.16}$$

Next, we show that the selection rule given in (6.16) achieves asymptotic optimality as the size of the network grows and the frequency of erroneous decisions tends to zero. This statement is formalized in the following theorem.

THEOREM 6.1 For the quickest anomaly detection and localization problem given in (6.12), the selection functions in (6.16) achieve asymptotic optimality as β approaches zero, i.e., for $j \in \{0, 1\}$

$$\lim_{\beta \to 0} \frac{\inf_{\tau, \delta, \psi_\tau} \mathbb{E}_j\{\tau\}}{\inf_\delta \mathbb{E}_j\{\hat{\tau}\}} = 1, \tag{6.17}$$

where $\hat{\tau}$ is the stopping time when the bus selection rules are given in (6.16).

Proof Refer to [36]. □

The exhaustive search involved in the optimization used in the selection function in (6.16) is computationally prohibitive. However, the Markov properties of the network can be leveraged to reduce the complexity of the search. To formalize this, let us define the subset of the unobserved neighbors of i at time t by $\mathcal{U}_t^i$, i.e.,

$$\mathcal{U}_t^i \triangleq \mathcal{N}_i \setminus \psi_{t-1}. \tag{6.18}$$

We have shown that instead of exhaustively searching over the entire network, it is sufficient to consider a smaller set $\mathcal{U}_t^i$, which is significantly smaller than that of all the unobserved nodes. This observation is formalized in the following theorem.

THEOREM 6.2 At time t, for all valid sequences $\mathcal{S}_t^i$ and for $j \in \{0, 1\}$ we have

$$\arg\max_{i \notin \psi_{t-1}} \max_{\mathcal{S}_t^i} M_i^j(t, \mathcal{S}_t^i) = \arg\max_{i \notin \psi_{t-1}} \max_{\mathcal{S}_t^i \subseteq \mathcal{U}_t^i} M_i^j(t, \mathcal{S}_t^i). \tag{6.19}$$

Proof Refer to [36]. □

Because the connectivity degree of the graph underlying the grids is substantially smaller than the size of the grid (e.g., in the IEEE 118-bus model the degree is 12), the complexity of the proposed bus selection rule is substantially lower than that of the exhaustive search.

Generally, the number of measures assigned to each bus is equal to the number of anomalous events of interest M, which grows as the size of the network increases. To circumvent this issue, we provide some simplification and implementation tips such that the proposed algorithm becomes practical. The main idea is that the metric for each bus under different anomalous events takes relatively similar values. This is primarily due to the fact that each outage event only affects a limited number of buses and, consequently, the effects on the metrics are minor. Therefore, instead of calculating M distinct metrics for each bus, we only compute one metric, denoted by

$$M_i(t) = \frac{1}{2} \sum_{j \in \psi_{t-1}} \log \frac{1}{1-r_{ij}^2} + r_{ij}^2(\Delta \theta_j^2 - 1)$$
$$+ \frac{1}{2|\mathcal{S}_t^i|} \sum_{j \in \mathcal{S}_t^i} \log \frac{1}{1-r_{ij}^2}. \quad (6.20)$$

To further simplify this metric, we offer an alternative two-stage selection rule to place more emphasis on the data. By taking a closer look into the metric in (6.20), we note that the first and third terms are functions of the correlation structure through $\{r_{ij}\}$, while the second term also depends on the data through $\Delta \theta_j$. In the two-stage approach, we first focus on the buses that are already observed, and identify the buses whose measurements have the largest level of deviation from the expected values, i.e., the buses with largest $|\theta_i - \bar{\theta}_i|$. This provides an estimate of the location of the underlying anomaly event, and it is equivalent to maximum likelihood decision about the true hypothesis model. In the second stage, among the neighbors of the buses with larger $|\Delta \theta|$, we identify buses with the largest metric $M_i(t)$. Also, at $t = 1$, data collection is initialized by selecting ℓ buses with the most number of neighbors. The steps of bus selection rule are presented in Algorithm 6.1.

6.6.2 Termination Rule and Localization Decision

To minimize detection delay, the data-collection process is terminated as soon as a sufficiently reliable decision can be made, or equivalently, when the collected data can explain one of the possible events with sufficient confidence. To formalize this, we define the perturbations in the power injection under an anomaly by

$$n \triangleq p - \bar{p}, \quad (6.21)$$

which can be modeled as a zero-mean Gaussian random vector [39]. Replacing p and $\bar{p}$ under event R_k from (6.2), we have

$$H\bar{\theta} + n = H_k \theta, \quad \text{for } k \in \{0, \ldots, M\}. \quad (6.22)$$

On the other hand, we can rewrite H_k as

$$H_k = \sum_{i \in \mathcal{L}} X_k[ii] m_i m_i^T \quad (6.23)$$
$$= M X_k M^T \quad (6.24)$$
$$= H - \sum_{i \in R_k} (X_0[ii] - X_k[ii]) m_i m_i^T, \quad (6.25)$$

where $X_k \in \mathbb{R}^{L \times L}$ is a diagonal matrix defined such that when the ith transmission line connects buses m and n we have $X_k[ii] = \frac{1}{x_{mn}^k}$, x_{mn}^k is the line reactance value under event R_k, and $M \in \mathbb{R}^{N \times L}$ represents the incident matrix of the grid, i.e., the ith column

Algorithm 6.1 Data-adaptive bus selection for anomaly detection

---
1 Set $t = 0$ and $\mathcal{T} = \{1, \ldots, N\}$
2 **For** $i = 1, \ldots, N$ **repeat**
3   $\deg(i) \leftarrow$ Number of buses connected to bus i
4   $M(i) \leftarrow \max_{\mathcal{U} \subseteq \mathcal{N}_i} \frac{1}{|\mathcal{U}|} \sum_{j \in \mathcal{U}} \log \frac{1}{1-r_{ij}^2}$
5 **End for**
6 $\mathcal{T} \leftarrow$ Sorted $\mathcal{T}$ based on decreasing $\deg(\cdot)$
7 $\psi(t) \leftarrow$ First ℓ elements of $\mathcal{T}$
8 **While** stopping criterion is not met **do**
9   Take measurements from buses in $\psi(t)$
10   $S \leftarrow \psi_t$
11   $t \leftarrow t + 1$
12   $\psi(t) \leftarrow \{\}$
13   **While** $|\psi(t)| < \ell$ **do**
14    $i \leftarrow \arg\max_{j \in S} |\theta_j - \bar{\theta}_j|$
15    $\mathcal{N}_i \leftarrow$ Neighbors of i sorted based on decreasing $M(\cdot)$
16   **If** $|\mathcal{N}_i| < \ell - |\psi(t)|$ **then**
17     $\psi(t) \leftarrow \psi(t) \cup \mathcal{N}_i$
18   **Else**
19     $\psi(t) \leftarrow \psi(t) \cup \{\mathcal{N}_i(1), \ldots, \mathcal{N}_i(\ell - |\psi(t)|)\}$
20   **End if**
21    $S \leftarrow S \setminus i$
22   **End while**
23 **End while**
24 Set $\tau = t$
---

of M, denoted by m_i, corresponds to line $i \in \mathcal{L}$ and all its entries are zero except at two locations that specify the buses connected by line i. Combining (6.22)–(6.25) yields

$$H \cdot \Delta\theta = \sum_{i \in R_k} (X_0[ii] - X_k[ii]) m_i m_i^T \theta + n = M s_k + n,$$

where $s_k \in \mathbb{R}^L$ is defined as

$$s_k[i] \triangleq \begin{cases} (X_0[ii] - X_k[ii]) m_i^T \theta & \text{if } i \in R_k \\ 0 & \text{Otherwise} \end{cases}. \quad (6.26)$$

The locations of the non-zero elements of vector s_k is in one-to-one correspondence with the indices of the anomalous lines. Because each anomalous event affects a small fraction of the total number of transmission lines, s_k is a sparse vector, where the location of each non-zero element is the index of an anomalous line. Any off-the-shelf sparse signal recovery algorithm can be applied to find s_k. By further simplifying we obtain

$$\Delta\theta = BMs_k + Bn, \quad (6.27)$$

where we have defined $B \triangleq H^{-1}$. At time t we have

$$\Delta\theta_t = B_t M s_k + B_t n, \tag{6.28}$$

where B_t is the matrix constructed from B by keeping its rows corresponding to set ψ_t. We include a pre-processing whitening stage for the noise vector $B_t n$. For this purpose, suppose that matrix $B_t \Sigma_k^{\frac{1}{2}}$ has the following singular value decomposition (SVD):

$$B_t \Sigma_k^{\frac{1}{2}} = U_k \Lambda_k V_k^T, \tag{6.29}$$

where Σ_k is the covariance matrix of n under event R_k. Then, corresponding to event R_k we have

$$y_k = A_k s_k + \tilde{n}_k, \tag{6.30}$$

where we have defined

$$y_k \triangleq \Lambda_k^{-1} U_k^T \Delta\theta_\tau, \tag{6.31}$$

$$A_k \triangleq V_k^T M, \tag{6.32}$$

$$\text{and} \quad \tilde{n}_k \triangleq \Lambda_k^{-1} U_k^T B_t n. \tag{6.33}$$

We note that $\tilde{n}_k$ is a white noise vector with covariance matrix I. To solve the inverse problem given in (6.30), we use orthogonal matching pursuit (OMP) as a fast sparse recovery algorithm, which is summarized in Algorithm 6.2. The data-acquisition process stops as soon as the collected data θ (or its projection y_k) can be explained by the identified lines in $\mathcal{T}_k$, or equivalently, when the residual vector r_k becomes sufficiently small. In Algorithm 6.2, $a_{k,i}$ is the ith column of matrix A_k, and $A_{k,\mathcal{T}}$ is the matrix

Algorithm 6.2 Termination rule and final decision.

---
1   **Inputs** y_k and A_k for $k \in \{0, \ldots, M\}$
2   Set $r_k = y_k$, $s_k = 0$ and $\mathcal{T}_k = \{\}$
3   **While** $\min_k \|r_k\| > \gamma$ **and** $|\mathcal{T}_1| < \eta_{\max}$ **do**
3   **For** $k = 1, \ldots, M$ **Repeat**
4            $I_k \leftarrow \arg\max_i \dfrac{|a_{k,i}^T r_k|}{|a_{k,i}|}$
4            $\mathcal{T}_k \leftarrow \mathcal{T}_k \cup \{I_k\}$
5            $s_k[\mathcal{T}_k] = (A_{k,\mathcal{T}_k}^T A_{k,\mathcal{T}_k})^{-1} A_{k,\mathcal{T}_k}^T y_k$
6            $r_k = y_k - A_k s_k$
7   **End for**
7   **End while**
8   **If** $\min_k \|r_k\| > \gamma$
9   **Continue** sampling
10   **Else**
11       $i \leftarrow \arg\min_k \|r_k\|$
11   **Stop** sampling and **Return** s_i
12   **End if**
---

composed of a set of columns of matrix A_k indexed in set $\mathcal{T}$. The value of threshold γ controls the balance between quickness and the reliability of the decision and can be calculated based on some historical data or through a comprehensive simulation of power grid under different events.

6.7 Case Study

Detecting and localizing outage events are critical in maintaining a power system and ensuring a reliable power supply. Fast detection of outage events improves the quality of service by decreasing the outage duration. Active sensing is a paradigm that can be leveraged for quick detection and localization of outage anomalies and minimizing the cost of data collection and processing. Hence, in this section, we use outage detection as a case study to evaluate the performance of the devised algorithm.

We generate synthetic data by simulating IEEE standard power systems by using the software toolbox MATPOWER [40]. We compare the results of the active sensing approach with the pre-specified bus selection method as well as full observation of the network in terms of detection accuracy and the number of required measurements. Furthermore, we evaluate the tradeoff among delay, the number of measurements, and decision accuracy. For the pre-specified bus selection rule, to identify the pre-specified set of buses, we fix the number of measurements and identify the buses whose measurements give the best average detection accuracy over all possible outage events. Then, this set is selected as the pre-specified set for all possible outage events.

6.7.1 Adaptivity Gain

The main advantage of active sensing stems from its ability to progressively focus the sensing resources on the areas of the network that are more informative about the underlying outage event. This adaptability to different situations enables our algorithm to detect and localize the lines under outage with the minimum number of measurements to achieve a target reliability level. To quantify the adaptivity gain, we compare the accuracy performance of our proposed algorithm with the prespecified bus selection rule in the 118-bus IEEE standard system when the number of measurements in the prespecified bus selection rule and the average number of measurements in data-adaptive approach are the same. We set $\ell = 5$, $\eta_{\max} = 1$ and consider the set of single line outage events in which the network remains connected. Figure 6.2 compares the detection and localization accuracy performance of both methods when the power of the perturbation noise vector is 1% of the average injected power before any outage. The data-adaptive approach uniformly outperforms the prespecified selection rule due to the exploitation of the correlation structure of the measurements for selecting the most informative buses. Also, it is observed that in the data-adaptive approach the performance gap between 70 measurements and 110 measurements (almost full observation of the network) is marginal, which indicates that by observing nearly half the network we can achieve a performance similar to complete observation.

Table 6.1 Comparison of average running time for the proposed and exhaustive search.

Average Number of Measurements	30	50	70	90
t_{DA}(sec)	0.1242	0.3784	0.6934	1.0138
t_{ES}(sec)	3.0351	6.871	10.0597	12.8147
t_{ES}/t_{DA}	24.4	18.2	14.5	12.6

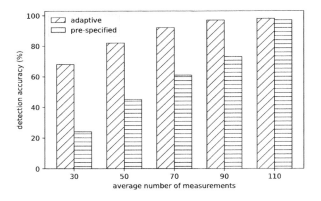

Figure 6.2 Comparison of detection accuracy versus number of measurements for data-adaptive and pre-specified bus selection rules for $\ell = 5$ and 118-bus system.

One of the main advantages of the data-adaptive approach is reducing the computational complexity of the bus selection rule, as established in Theorem 6.2. To quantify the reduction in computation cost, in Table 6.1, we compare the simulation time required for implementing the proposed approach with that of an exhaustive search for finding the most informative buses. We consider single line outage events and denote the average simulation time over all possible events for the exhaustive search and the proposed search by t_{ES} and t_{DA}, respectively. It is observed that for 70 measurements, the algorithm performs close to observing the entire network, data-adaptive collection of measurements is 14.5 times faster.

6.7.2 Multiple Outage

When an outage occurs, if not dealt with promptly, it overloads neighboring lines, which might lead to multiple line outages. Hence, when studying multiple line outages we assume that the lines under outage are in the same locality of the grid. In Figure 6.3, we consider a different number of lines in outage under the same settings as in Figure 6.2. We set the number of measurements in both methods to 70 and also include the results for full observation of the network for comparison. The adaptivity gain, for single and multiple line outage events, is significant and indicates at least a 50% improvement in detection accuracy performance. The performance of the proposed data-adaptive algorithm, as expected, is close to the full observation of the network.

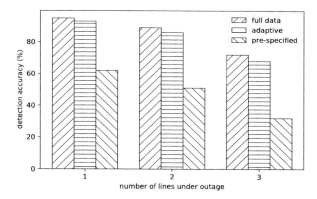

Figure 6.3 Comparison of detection accuracy versus number of lines under outage for data-adaptive and pre-specified bus selection rules.

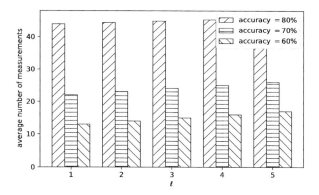

Figure 6.4 Comparison of the average number of measurements versus ℓ for different accuracy levels.

6.7.3 Tradeoff among Performance Measures

We evaluate the interplay among detection accuracy, the average number of measurements, average delay, and ℓ under the single line outage setting. To this end, in Figure 6.4, we compare the number of measurements required to achieve a certain accuracy level for different values of ℓ. Interestingly, as ℓ increases, more measurements are required to achieve the same decision accuracy. This behavior can be attributed to the fact that for larger ℓ we collect more measurements at the same time without incorporating the information of the current time instant. For instance, when $\ell = 1$, each new measurement is collected based on the entire past measurements, while in $\ell = 5$ the information used for all five new measurements are the same. Also, it is observed that when the accuracy improves from 60% to 70%, the number of required measurements is less compared to when it is enhanced from 70% to 80%.

Detection delay is a critical aspect of anomaly detection in power systems. In Figure 6.5, we evaluate the role of ℓ in data-collection delay, which is the number of time

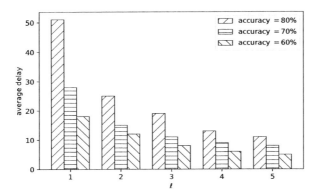

Figure 6.5 Comparison of the average delay versus ℓ for different accuracy levels.

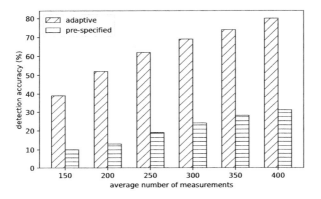

Figure 6.6 Comparison of detection accuracy versus number of measurements for data-adaptive and pre-specified bus selection rules for $\ell = 10$ and 2383-bus system.

steps required to collect all the measurements. To this end, we compare the average delay incurred to achieve a certain accuracy level for various ℓ. It is observed that when ℓ decreases, improving detection accuracy incurs less delay. However, we know from Figure 6.4 that larger ℓ means more number of measurements, which leads to more sensing, communication, storage, and processing cost. Furthermore, Figure 6.5 shows that decreasing ℓ leads to more delay for smaller ℓ and the same accuracy performance.

6.7.4 Scalability and Complexity

The proposed algorithm is designed to scale well to more extensive networks as it progressively identifies the most likely location of an event and collects its data from that locality. To evaluate this behavior, in Figure 6.6 we consider the Polish power system provided by MATPOWER case2383wp casefile which has 2,383 buses and use the settings of Figure 6.2 except that we set $\ell = 10$ and consider the noise-free case. The main observation is that the gap between the proposed algorithm and the pre-specified bus selection rule has increased significantly compared to the smaller network with 118 buses. It indicates that even for large-scale power systems, the data-adaptive selection rule can achieve considerable performance by selecting a subset of buses in the grid, while the pre-specified selection approach performance degrades. In fact, as the grid size grows, the performance gain improves too. This is primarily because larger grids provide more freedom for selecting the buses.

Acknowledgment

This work is supported in part by the National Science Foundation under the CAREER Award EECS-1554482 and grant DMS-1737976.

References

[1] T. Takagi, Y. Yamakoshi, M. Yamaura, R. Kondow, and T. Matsushima, "Development of a new type fault locator using the one-terminal voltage and current data," *IEEE Transactions on Power Apparatus and Systems*, vol. PAS-101, no. 8, pp. 2892–2898, Aug. 1982.

[2] L. Eriksson, M. M. Saha, and G. D. Rockefeller, "An accurate fault locator with compensation for apparent reactance in the fault resistance resulting from remore-end infeed," *IEEE Transactions on Power Apparatus and Systems*, vol. PAS-104, no. 2, pp. 423–436, Feb. 1985.

[3] T. Kawady and J. Stenzel, "A practical fault location approach for double circuit transmission lines using single end data," *IEEE Transactions on Power Delivery*, vol. 18, no. 4, pp. 1166–1173, Oct. 2003.

[4] C. E. de Morais Pereira and L. C. Zanetta, "Fault location in transmission lines using one-terminal post fault voltage data," *IEEE Transactions on Power Delivery*, vol. 19, no. 2, pp. 570–575, Apr. 2004.

[5] M. Kezunovic and B. Perunicic, "Automated transmission line fault analysis using synchronized sampling at two ends," in *Proc. Power Industry Computer Application Conference*, Salt Lake City, UT, pp. 407–413, May 1995.

[6] A. L. Dalcastagne, S. N. Filho, H. H. Zurn, and R. Seara, "An iterative two-terminal fault-location method based on unsynchronized phasors," *IEEE Transactions on Power Delivery*, vol. 23, no. 4, pp. 2318–2329, Oct. 2008.

[7] Y. Liao and N. Kang, "Fault-location algorithms without utilizing line parameters based on the distributed parameter line model," *IEEE Transactions on Power Delivery*, vol. 24, no. 2, pp. 579–584, Apr. 2009.

[8] J. Izykowski, E. Rosolowski, P. Balcerek, M. Fulczyk, and M. M. Saha, "Accurate non-iterative fault-location algorithm utilizing two-end unsynchronized measurements," *IEEE Transactions on Power Delivery*, vol. 26, no. 2, pp. 547–555, Apr. 2011.

[9] C. A. Apostolopoulos and G. N. Korres, "A novel algorithm for locating faults on transposed/untransposed transmission lines without utilizing line parameters," *IEEE Transactions on Power Delivery*, vol. 25, no. 4, pp. 2328–2338, Oct. 2010.

[10] T. Nagasawa, M. Abe, N. Otsuzuki, T. Emura, Y. Jikihara, and M. Takeuchi, "Development of a new fault location algorithm for multi-terminal two parallel transmission lines," in *Proc. IEEE Power Engineering Society*, Dallas, TX, pp. 348–362, Sep. 1991.

[11] D. A. Tziouvaras, J. B. Roberts, and G. Benmouyal, "New multi-ended fault location design for two- or three-terminal lines," in *Proc. International Conference on Developments in Power System Protection*, Amsterdam, the Netherlands, pp. 395–398, Apr. 2001.

[12] G. Manassero, E. C. Senger, R. M. Nakagomi, E. L. Pellini, and E. C. N. Rodrigues, "Fault-location system for multiterminal transmission lines," *IEEE Transactions on Power Delivery*, vol. 25, no. 3, pp. 1418–1426, July 2010.

[13] T. Funabashi, H. Otoguro, Y. Mizuma, L. Dube, and A. Ametani, "Digital fault location for parallel double-circuit multi-terminal transmission lines," *IEEE Transactions on Power Delivery*, vol. 15, no. 2, pp. 531–537, Apr. 2000.

[14] S. M. Brahma, "Fault location scheme for a multi-terminal transmission line using synchronized voltage measurements," *IEEE Transactions on Power Delivery*, vol. 20, no. 2, pp. 1325–1331, Apr. 2005.

[15] P. G. McLaren and S. Rajendra, "Traveling-wave techniques applied to the protection of teed circuits: Multi-phase/Multi-circuit system," *IEEE Transactions on Power Apparatus and Systems*, vol. PAS-104, no. 12, pp. 3551–3557, Dec. 1985.

[16] A. O. Ibe and B. J. Cory, "A traveling wave-based fault locator for two- and three-terminal networks," *IEEE Transactions on Power Delivery*, vol. 1, no. 2, pp. 283–288, Apr. 1986.

[17] F. H. Magnago and A. Abur, "Fault location using wavelets," *IEEE Transactions on Power Delivery*, vol. 13, no. 4, pp. 1475–1480, Oct. 1998.

[18] C. Y. Evrenosoglu and A. Abur, "Traveling wave based fault location for teed circuits," *IEEE Transactions on Power Delivery*, vol. 20, no. 2, pp. 1115–1121, Apr. 2005.

[19] D. Spoor and J. G. Zhu, "Improved single-ended traveling-wave fault-location algorithm based on experience with conventional substation transducers," *IEEE Transactions on Power Delivery*, vol. 21, no. 3, pp. 1714–1720, July 2006.

[20] P. Jafarian and M. Sanaye-Pasand, "A traveling-wave-based protection technique using wavelet/PCA analysis," *IEEE Transactions on Power Delivery*, vol. 25, no. 2, pp. 588–599, Apr. 2010.

[21] M. Korkali, H. Lev-Ari, and A. Abur, "Traveling-wave-based fault-location technique for transmission grids via wide-area synchronized voltage measurements," *IEEE Transactions on Power Systems*, vol. 27, no. 2, pp. 1003–1011, May 2012.

[22] H. Chernoff, "Sequential design of experiments," *The Annals of Mathematical Statistics*, vol. 30, no. 3, pp. 755–770, Sept. 1959.

[23] S. Bessler, "Theory and applications of the sequential design of experiments, k-actions and infinitely many experiments: Part I theory," Department of Statistics, Stanford University, Technical Report, Mar. 1960.

[24] A. E. Albert, "The sequential design of experiments for infinitely many states of nature," *The Annals of Mathematical Statistics*, vol. 32, no. 3, pp. 774–799, Sept. 1961.

[25] G. E. P. Box and W. J. Hill, "Discrimination among mechanistic models," *Technometrics*, vol. 9, no. 1, pp. 57–71, Feb. 1967.

[26] W. J. Blot and D. A. Meeter, "Sequential experimental design procedures," *Journal of the American Statistical Association*, vol. 68, no. 343, pp. 586–593, Sept. 1973.

[27] S. Nitinawarat, G. K. Atia, and V. V. Veeravalli, "Controlled sensing for multihypothesis testing," *IEEE Transactions on Automatic Control*, vol. 58, no. 10, pp. 2451–2464, May 2013.

[28] S. Nitinawarat and V. V. Veeravalli, "Controlled sensing for sequential multihypothesis testing with controlled Markovian observations and non-uniform control cost," *Sequential Analysis*, vol. 34, no. 1, pp. 1–24, Feb. 2015.

[29] K. Cohen and Q. Zhao, "Active hypothesis testing for anomaly detection," *IEEE Transactions on Information Theory*, vol. 61, no. 3, pp. 1432–1450, Mar. 2015.

[30] N. K. Vaidhiyan and R. Sundaresan, "Active search with a cost for switching actions," *arXiv:1505.02358v1*, May 2015.

[31] G. Schwarz, "Asymptotic shapes of Bayes sequential testing regions," *The Annals of Mathematical Statistics*, vol. 33, no. 1, pp. 224–236, Mar. 1962.

[32] J. Kiefer and J. Sacks, "Asymptotically optimum sequential inference and design," *The Annals of Mathematical Statistics*, vol. 34, no. 3, pp. 705–750, Sept. 1963.

[33] S. P. Lalley and G. Lorden, "A control problem arising in the sequential design of experiments," *The Annals of Probability*, vol. 14, no. 1, pp. 136–172, Jan. 1986.

[34] J. Heydari, A. Tajer, and H. Poor, "Quickest detection of Gauss Markov random fields," in *Proc. 53rd Allerton Conference on Communication, Control, and Computing*, Monticello, IL, Oct. 2015.

[35] J. Heydari, A. Tajer, and H. V. Poor, "Quickest detection of Markov networks," in *International Symposium on Information Theory*, Barcelona, Spain, pp. 1341–1345, July 2016.

[36] ———, "Quickest detection of Markov networks," https://arxiv.org/abs/1711.04268, Technical Report 2019.

[37] J. F. Dopazo, O. A. Klitin, and A. M. Sasson, "Stochastic load flows," *IEEE Transactions on Power Apparatus and Systems*, vol. 94, no. 2, pp. 299–309, Mar. 1975.

[38] A. Schellenberg, W. Rosehart, and J. Aguado, "Cumulant-based probabilistic optimal power flow (P-OPF) with Gaussian and Gamma distributions," *IEEE Transactions on Power Systems*, vol. 20, no. 2, pp. 773–781, May 2005.

[39] H. Zhu and G. Giannakis, "Sparse overcomplete representations for efficient identification of power line outages," *IEEE Transactions on Power Systems*, vol. 27, no. 4, pp. 2215–2224, Nov. 2012.

[40] R. D. Zimmerman, C. E. Murillo-Sanchez, and R. J. Thomas, "MATPOWER: Steady-state operations, planning, and analysis tools for power systems research and education," *IEEE Transactions on Power Systems*, vol. 26, no. 1, pp. 12–19, Feb. 2011.

7 Random Matrix Theory for Analyzing Spatio-Temporal Data

Robert Qiu, Xing He, Lei Chu, and Xin Shi

Modern power gird is one of the most complex engineering systems in existence. Data of the grid, however, are not similar to big data such as image data. Power grid data are sampled by various sensors deployed within the network, such as phase measurement units (PMUs). The massive data set is in high-dimensional vector space, and in the form of time series: the temporal variations (T sampling instants) are simultaneously observed together with spatial variations (N grid nodes). The extraction of statistical information, especially temporal-spatial correlations, from the preceding data sets is a challenge that does not meet the prerequisites of most classical mathematical tools. Also, the mining task is inappropriate for supervised training algorithms such as neural networks due to the lack or the asymmetry of the labeled data. Unifying time and space through their ratio $c = T/N$, random matrix theory (RMT) deals with such kind of data mathematically rigorously. Moreover, linear eigenvalue statistics (LESs) built from data matrices follow Gaussian distributions for very general conditions, and other statistical variables are studied due to the latest breakthroughs in probability on the central limit theorems of those LESs.

In this chapter, we firstly model spatial-temporal data sets as random matrix sequences. Secondly, some basic principles of RMT, such as asymptotic spectrum laws, transforms, convergence rate, and free probability, are introduced briefly for a better understanding of the applications of RMT. Lastly, the case studies based on synthetic data and real data are developed to evaluate the performance of the RMT-based schemes in different application scenarios (i.e., state evaluation and situation awareness).

7.1 Data-Driven View of Grid Operation and Its Matrix Form

7.1.1 Fundamental Rules for Grid Network Operation

This part studies the fundamental rules for grid network operation for the purpose of revealing the primary characteristics of the grid operation data model.

For each node i in a grid network, choosing the reference direction as shown in Figure 7.1, Kirchhoff's current law and Ohm's law say that:

$$\dot{I}_i = \sum_{\substack{j=1 \\ j \neq i}}^{n} \dot{I}_j = \sum_{j \neq i} \dot{Y}_{ij} \cdot (\dot{U}_j - \dot{U}_i). \tag{7.1}$$

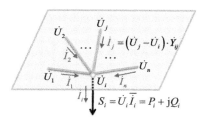

Figure 7.1 Schematic diagram for grid network operation.

where $\dot{Y}_{ij} = G_{ij} + j \cdot B_{ij}$ is the admittance in Cartesian form[1], and $\dot{U}_i = |\dot{U}_i| \angle \theta_i = V_i \angle \theta_i = V_i e^{j\theta_i}$ and $\dot{I}_i = |\dot{I}_i| \angle \phi_i$ are node voltage and node current, respectively, in polar form[2].

For all the nodes of the network, we obtain

$$\begin{bmatrix} \dot{I}_1 \\ \dot{I}_2 \\ \vdots \\ \dot{I}_n \end{bmatrix} = \begin{bmatrix} y_{11} & y_{12} & \cdots & y_{1n} \\ y_{21} & y_{22} & \cdots & y_{2n} \\ \vdots & \vdots & \cdots & \vdots \\ y_{n1} & y_{2n} & \cdots & y_{nn} \end{bmatrix} \begin{bmatrix} \dot{U}_1 \\ \dot{U}_2 \\ \vdots \\ \dot{U}_n \end{bmatrix}$$

or equivalently:

$$\dot{\mathbf{I}} = \dot{\mathbf{y}} \dot{\mathbf{U}} \tag{7.2}$$

where $[y]_{ij} = \begin{cases} \dot{Y}_{ij} & i \neq j \\ -\sum_{k \neq i} \dot{Y}_{ik} & i = j \end{cases}$

$$\dot{\mathbf{S}} = \begin{bmatrix} P_1 + jQ_1 \\ P_2 + jQ_2 \\ \vdots \\ P_n + jQ_n \end{bmatrix} = \dot{\mathbf{U}} \circ \overline{\dot{\mathbf{I}}} = \dot{\mathbf{U}} \circ \overline{\mathbf{Y}\dot{\mathbf{U}}} = \dot{\mathbf{U}} \circ \overline{\mathbf{Y}}\overline{\dot{\mathbf{U}}} \tag{7.3}$$

$$= \sum \dot{U}_{i1} \mathbf{E}_{i1} \circ \sum \overline{y_{jk}} \mathbf{E}_{jk} \sum \overline{\dot{U}_{l1}} \mathbf{E}_{l1}$$

$$= \sum \dot{U}_{i1} \overline{y_{jk}} \overline{\dot{U}_{l1}} \mathbf{E}_{i1} \circ (\mathbf{E}_{jk} \mathbf{E}_{l1}) = \sum \dot{U}_i \overline{y_{ik}} \overline{\dot{U}_k} \mathbf{E}_{i1}$$

where $\circ$ is the Hadamard product[3], and $\mathbf{E}_{ij}$ is the single-entry matrix—1 at (i, j) and 0 elsewhere. It is worth mentioning that $\mathbf{A} = \sum_{i,j} [A]_{ij} \mathbf{E}_{ij}$, abbreviated as $\sum A_{ij} \mathbf{E}_{ij}$.

---

[1] G is the conductance, B is the susceptance, and j is the imaginary unit.
[2] $V_i \angle \theta_i = V_i e^{j\theta_i} = V_i (\cos \theta_i + j \cdot \sin \theta_i)$.
[3] $\mathbf{C} = \mathbf{A} \circ \mathbf{B}, [C]_{ij} = [A]_{ij} [B]_{ij}$.

And thus, for each node i, its active power P and reactive power Q are expressed as:

$$\begin{cases} P_i = V_i \sum_{k \neq i} V_k \left(G_{ik} \cos\theta_{ik} + B_{ik} \sin\theta_{ik} \right) - V_i^2 \sum_{k \neq i} G_{ik} \\ Q_i = V_i \sum_{k \neq i} V_k \left(G_{ik} \sin\theta_{ik} - B_{ik} \cos\theta_{ik} \right) + V_i^2 \sum_{k \neq i} B_{ik} \end{cases} \quad (7.4)$$

Taking account of the ith node-to-ground admittance y_i, we obtain

$$\begin{cases} P_i := P_i - V_i^2 g_i \\ Q_i := P_i + V_i^2 b_i \end{cases} \quad (7.5)$$

7.1.2 Jacobian Matrix

Jacobian matrix is a sparse matrix that results from a sensitivity analysis of the power flow equations. It is the key part of power flow analysis, which is the basis for power system planning and operations. Additionally, the eigenvalues of **J** have long been used as indices of system vulnerability to voltage instabilities [1]. Besides, the sparsity structure of **J** inherently contains the most up-to-date network topology and corresponding parameters. Topology errors have long been cited as a cause of inaccurate state estimation results [2, 3]. Moreover, the Jacobian matrix, in practice, may be out date due to erroneous records, faulty telemetry from remotely monitored circuit breakers, or unexpected operating conditions resulting from unforeseen equipment failure.

Starting from (7.5), the physical power system, abstractly, may be viewed as an analog engine, taking bus voltage magnitudes V and phases θ as *inputs*, and "computing" active and reactive power injection P, Q as *outputs*. Thus, the entries of **J**, i.e. $[J]_{ij}$, are composed of partial derivatives of P and Q with respect to V and θ. **J** consists of four parts H, N, K, L as follows:

$$\begin{cases} H_{ij} = V_i V_j \left(G_{ij} \sin\theta_{ij} - B_{ij} \cos\theta_{ij} \right) - \delta_{ij} \cdot Q_i + \delta_{ij} \cdot V_i^2 b_i \\ N_{ij} = V_i V_j \left(G_{ij} \cos\theta_{ij} + B_{ij} \sin\theta_{ij} \right) + \delta_{ij} \cdot P_i - \delta_{ij} \cdot V_i^2 g_i \\ K_{ij} = -V_i V_j \left(G_{ij} \cos\theta_{ij} + B_{ij} \sin\theta_{ij} \right) + \delta_{ij} \cdot P_i + \delta_{ij} \cdot V_i^2 g_i \\ L_{ij} = V_i V_j \left(G_{ij} \sin\theta_{ij} - B_{ij} \cos\theta_{ij} \right) + \delta_{ij} \cdot Q_i + \delta_{ij} \cdot V_i^2 b_i \end{cases} \quad (7.6)$$

where $H_{ij} = \frac{\partial P_i}{\partial \theta_j}, N_{ij} = \frac{\partial P_i}{\partial V_j} V_j, K_{ij} = \frac{\partial Q_i}{\partial \theta_j}, L_{ij} = \frac{\partial Q_i}{\partial V_j} V_j$, and δ is the Kronecker delta function defined as

$$\delta_{\alpha,\beta} = \begin{cases} 1 & \alpha = \beta \\ 0 & \alpha \neq \beta. \end{cases}$$

In (7.7), f is a differentiable mapping, $f : \mathbf{x} \in \mathbb{R}^{2n-m-2} \to \mathbf{y} \in \mathbb{R}^{2n-m-2}$, and **J** is the Jacobian matrix,

$$\mathbf{y} \triangleq \begin{bmatrix} P_1 \\ \vdots \\ P_{n-1} \\ Q_{m+1} \\ \vdots \\ Q_{n-1} \end{bmatrix} = f \begin{bmatrix} \theta_1 \\ \vdots \\ \theta_{n-1} \\ V_{m+1} \\ \vdots \\ V_{n-1} \end{bmatrix} \triangleq f(\mathbf{x}) \qquad \mathbf{J} = \begin{bmatrix} \frac{\partial y_1}{\partial x_1} & \cdots & \frac{\partial y_1}{\partial x_l} \\ \vdots & \ddots & \vdots \\ \frac{\partial y_l}{\partial x_1} & \cdots & \frac{\partial y_l}{\partial x_l} \end{bmatrix} \qquad (7.7)$$

According to (7.6), $\mathbf{J}$ can be represented by

$$\mathbf{J} = \begin{bmatrix} [\mathbf{H}]_{n-1,n-1} & [\mathbf{N}]_{n-1,n-m-1} \\ [\mathbf{K}]_{n-m-1,n-1} & [\mathbf{L}]_{n-m-1,n-m-1} \end{bmatrix} \qquad (7.8)$$

7.1.3 Power Flow Analysis and Theoretical Calculation

Power flow (PF) analysis is the fundamental and the most heavily used tool for solving many complex power system operation problems, such as fault diagnosis, state estimation, $N-1$ security, optimal power dispatch. PF analysis deals mainly with the calculation of the steady-state voltage magnitude and phase for each network bus, for a given set of variables such as load demand and real power generation, under certain assumptions such as balanced system operation [4]. Based on this information, the network operating conditions, in particular, real and reactive power flows on each branch, power losses, and generator reactive power outputs, can be determined [5]. Thus, the input (output) variables of the PF problem typically fall into three categories:

- Active power P and voltage magnitude V (reactive power Q and voltage angle θ) for each voltage-controlled bus, i.e. PV buses;
- Active power P and reactive power Q (voltage magnitude V and voltage angle θ) for each load bus, i.e. PQ buses;
- Voltage magnitude V and voltage angle θ (active power P and reactive power Q) for reference or slack bus.

Theoretical calculation for PF analysis is model- and assumption-based. That is to say, prior information of the topological parameter, the admittance Y, is required in advance. Consider a power system with n buses among which there are m PV buses, $n-m-1$ PQ buses, and 1 slack bus, the PF problem can be formalized in general as (7.7), which simultaneously solves a set of equations with an equal number of unknowns [6].

7.2 Random Matrix Theory: An Analytical Tool for Large-Scale Power Grids

Reliable operation and intelligent management of electric power systems influence heavily on everyday life. Recently, power companies, scholars, and researchers keep their eyes on utilizing large-scale phase measurement units (PMUs) to improve wide-area monitoring, protection, and control [7–11].

The model-driven based analysis (i.e., eigenvalue analysis) is a classical method, exploiting off-line dynamic properties of power systems. It focuses on the assumed system model and the linear systems control theory. Based on the eigenvalues, eigenvectors, or some participation factors, the system characteristics can be determined (or predicted). However, this hardly meets the demand for efficient monitoring of dynamically changed power systems of a high level of uncertainty [12–16]. Besides, there has been increasing challenges introduced by the renewable energy resources. All the preceding issues affect the suitability and robustness of the traditional model-based approach for revealing the dynamic properties of modern (large-scale) power systems.

In the era of big power-grid data, random matrix theory paves the way for further progress in the promising field of data-driven based analysis for modern power systems.

7.2.1 When the Power Data Meets Large Random Matrices

Random matrix theory, an analytical tool from nuclear and statistical physics, has been shown to be useful in the study of large-size systems such as electric grids [15–20]. For instance, the transmission systems can be represented as large complex networks, with high-voltage lines and the edges connecting nodes (power plants or substations) [20, 21]. Besides, RMT can be used to identify new characteristics of these networks [22–24]. In this section, the basics of RMT are introduced for a better understanding of the principle of large power grids data analysis.

We start by introducing the several representative classes of large random matrices (ensembles): Gaussian unitary ensemble (GUE), Gaussian orthogonal ensemble (GOE), and Laguerre unitary ensemble (LUE), which are useful for data modeling of massive streaming power grids data [16, 18, 19].

DEFINITION 7.1 *An $N \times N$ matrix $\mathbf{A}$ is called a GUE if it satisfies [25]:*

(i) *The entries of $\mathbf{A}$ are i.i.d Gaussian random variables.*
(ii) *For $1 \leq i \leq j \leq N$, $Re(A_{ij})$ and $Re(A_{ij})$, are i.i.d. with distribution $N(0, \frac{1}{2}\sigma^2)$.*
(iii) *For any i, j in $\{1, 2, \ldots, N\}$, $A_{ij} = \bar{A}_{ji}$.*
(iv) *The diagonal entries of $\mathbf{A}$ are real random variable with distribution $N(0, \sigma^2)$.*

For the convenience of analysis, we can write the GUE $\mathbf{A}$ as $\mathbf{A} = \frac{1}{2}(\mathbf{X} + \mathbf{X}^H)$, where $\mathbf{X}$ is a standard Gaussian matrix.

DEFINITION 7.2 *A matrix is a GOE if it is uniformly distributed on the set, $\mathcal{U}(\mathcal{N})$, of $N \times N$ unitary matrices. The density function on $\mathcal{U}(\mathcal{N})$ is denoted by [25, 26]*

$$2^{-N} \pi^{-\frac{1}{2}N(N+1)} \prod_{n=1}^{N} (N-n)!. \tag{7.9}$$

DEFINITION 7.3 *Let $\{X_{ij}\}_{1 \leq i \leq M, 1 \leq j \leq N}$ be i.i.d. Gaussian random variables with $\mathbb{E}(X_{ij}) = 0$ and $\mathbb{E}X_{ij}^2 = \frac{1}{2}(1 + \delta_{ij})$. The so-called Wishart matrix or Laguerre unitary ensemble LUE can be expressed as $\mathbf{A} = \frac{1}{N}\mathbf{X}\mathbf{X}^H$. The p.d.f. of $\mathbf{A}$ for $N \geq M$ is [26]*

$$\frac{\pi^{-M(M-1)/2}}{\det \sum \prod_{i=1}^{M}(N-i)!} \exp[-Tr\{\mathbf{A}\}] \det \mathbf{A}^{N-M}. \quad (7.10)$$

In practice, these random matrices are useful for modeling the data collected from modern power systems. For example, it has shown in previous work [17] that one can use random matrix model (i.e., GOE, LUE) to describe the interconnection of multiple grids and construct a simple model of a distributed grid. Specially, most grids can be characterized by the GOE, an indicator of chaos in many complex systems. Recently, by using RMT for modeling data sampled from the modern power systems and exploiting the related high-order statistics [9, 15, 16, 22, 23], RMT-based methods can tackle the tricky problems in large-scale power grids. Moreover, it has been shown in [20] that the combination of different kinds of random matrices is a choice beneficial to modeling the dynamics of the modern power systems.

7.2.2 Some Fundamental Results

This part introduces some fundamental results from random matrix theory that will be needed in the development of RMT-based algorithms.

For a square random matrix $\mathbf{A} \in \mathbb{C}^{M \times M}$, the resolvent of $\mathbf{A}$ is defined as

$$\mathbf{G}_{\mathbf{A}}(z) = (z\mathbf{I}_M - \mathbf{A})^{-1}. \quad (7.11)$$

The normalized trace of (7.11) gives

$$\mathfrak{g}_{\mathbf{A}}^{M}(z) = \frac{1}{M} \text{tr}\,[\mathbf{G}_{\mathbf{A}}(z)].$$

In the limit of large dimension, one has $\mathfrak{g}_{\mathbf{A}}^{M}(z) \underset{M \to \infty}{\to} \mathfrak{g}_{\mathbf{A}}(z)$, where $\mathfrak{g}_{\mathbf{A}}(z)$ is the *Stieltjes* transform of $\mathbf{A}$. The asymptotic empirical distribution of eigenvalues $F_{\mathbf{A}}(\lambda)$ can be described in terms of its *Stieltjes* transform [22], defined by

$$\mathfrak{g}_{\mathbf{A}}(z) = \int \frac{dF_{\mathbf{A}}(\lambda)}{\lambda - z} = \int \frac{\rho(\lambda)}{\lambda - z} d\lambda, \quad (7.12)$$

In random matrix theory, a central result states that when $\mathbf{A}$ is a LUE, the empirical distribution of the eigenvalues of $\mathbf{A}$ converges almost surely to the so-called Marcenko-Pastur law [27], which can be derived by the *Stieltjes* transform method.

Suggested by problems of interest in power grid, one needs to pay attention to other useful transforms. For example, the *R*-transform, a handy transform that enables the characterization of the limiting eigen-spectra of a sum of free random matrices from their individual limiting eigen-spectra, can be defined as

$$R_{\mathbf{A}}(z) = \mathfrak{g}_{\mathbf{A}}^{-1}(z) - \frac{1}{z}. \quad (7.13)$$

Furthermore, the *R*-transform can be expanded as:

$$R_{\mathbf{A}}(z) = \sum_{l=1}^{\infty} \kappa_l(\mathbf{A}) z^{l-1}, \quad (7.14)$$

where $\kappa_l(\mathbf{A})$ denotes the so-called *free cumulant*, which can be expressed as a function of moments of $\mathbf{A}$. Specially, given the mth moments of $\mathbf{A}$ and using the so-called cumulant formula [28], one can have

$$\varphi(\mathbf{A}^m) = \sum_{l=1}^{m} \kappa_l(\mathbf{A}^m) \sum_{m_1,\ldots,m_l} \varphi(\mathbf{A}^{m_1-1}) \cdots \varphi(\mathbf{A}^{m_l-1}), \qquad (7.15)$$

where, for $k = 1, \ldots, l$, m_k is a nonnegative integer and satisfies $m_1 + \cdots + m_l = m$.

The basic transforms and the related moments (or cumulants) fully characterize the asymptotic properties of random matrices (GUE, GOE, LUE, or combinations), which are critical for modeling and analysis of dynamic power systems. Besides, the random matrix theory basics introduced in the preceding text, despiting their simplicity, can form a strong basis for the reader to be able to appreciate RMT-based methods for engineering problems shown in later sections or the extensions in technical details [26, 29, 30].

7.2.3 On the Analytical Methods

The methods used for RMT-based algorithm design and analysis are mainly segmented into:

(i) *Stieltjes* transform, an analytical method treating asymptotical eigenvalue distribution of large random matrices in a comprehensive framework, which allows us to solve a large range of problems such as estimation of network parameters (i.e., the smallest eigenvalue of the Jacobian matrix, an indicator of the stability of a power system).

(ii) The method of moments, which establishes results on the successive moments of the asymptotic eigenvalues probability distribution.

(iii) *Free probability*, a powerful method that can bridge the analytical tools and the method of moments by derivatives of the *Stieltjes* transform, the *R*-transform, and the *S*-transform. Besides, the $free\ probability$ is a very efficient framework to study limiting distributions of some complex models of large dimensional random matrices, which has been shown to be useful for modeling the dynamics of a modern power network [19, 20, 26].

(iv) The *replica method*, an approach borrowed from physics and the field of statistical mechanics. While classical *Stieltjes* transform approaches used to fail to derive nice closed-form formulas for the asymptotic functionals of the empirical spectral density of the covariance matrix, the $replica$ method often conjecture that these complex expressions of the covariance matrix take the form of Jacobian matrices.

Based on the aforementioned methods, one can develop a solid framework for RMT-based data analysis of a power system. We refer interested readers to monographs [25, 26, 28–30] for more comprehensive studies.

7.2.4 Universality Principle of RMT

The mathematical descriptions previously mentioned assume that the original system has independent (or at least weakly dependent) constituents. What if independence is not a realistic approximation and strong correlations need to be modeled? Is there a universality for strongly correlated models?

Indeed, there is a wealth of literature establishing that many properties of large random matrices do not depend on the details of how their entries are distributed, e.g., many results are universal. For instance, the universality principle, rigorously introduced in [30], shows that the spectrum of Wishart matrices asymptotically approaches the Marcenko–Pastur (M–P) law regardless of the distribution of the individual entries, as long as they are independent, and have zero mean, unit variance, and finite k-th moment, for $k > 2$. Analogous results can be found for other types of matrices (i.e., GUE and GOE). In summary, applying the universality principle, the independence assumption for the data in the modern power system is theoretically valid in the high-dimensional regime and has been proven to be effective in finite-size [9, 15, 16].

7.2.5 Random Matrix Tools for Power Systems

Laws for Spectral Analysis

RMT says that for a Laguerre unitary ensemble (LUE) matrix $\mathbf{A} \in \mathbb{C}^{N \times T}$ ($c = N/T \leq 1$), its empirical spectral density $g_{\mathbf{A}}(x)$ follows the M–P law [31]:

$$g_{\mathbf{A}}(x) = \frac{1}{2\pi c x}\sqrt{(x-a)(b-x)}, x \in [a,b] \tag{7.16}$$

where $a = (1 - \sqrt{c})^2, b = (1 + \sqrt{c})^2$.

7.2.6 Linear Eigenvalue Statistics and Its Central Limit Theorem

The LES τ of an arbitrary matrix $\mathbf{\Gamma} \in \mathbb{C}^{N \times N}$ is defined in [32, 33] using the continuous test function $\varphi : \mathbb{C} \to \mathbb{C}$,

$$\tau_\varphi = \sum_{i=1}^{N} \varphi(\lambda_i) = \text{Tr}\varphi(\mathbf{\Gamma}), \tag{7.17}$$

where the trace of the function of a random matrix is involved.

Law of Large Numbers

The law of large numbers tells us that $N^{-1}\tau_\varphi$ converges in probability to the limit

$$\lim_{N \to \infty} \frac{1}{N} \sum_{i=1}^{N} \varphi(\lambda_i) = \int \varphi(\lambda)\rho(\lambda)\,d\lambda, \tag{7.18}$$

where $\rho(\lambda)$ is the probability density function of λ.

Central Limit Theorem

The central limit theorem (CLT) [33], as the natural second step, aims to study LES fluctuations [26].

THEOREM 7.1 ([33]) *Consider a $N \times T$ random matrix $\mathbf{X}$; $\mathbf{M}$ is the covariance matrix $\mathbf{M} = \frac{1}{N}\mathbf{X}\mathbf{X}^H$. The CLT for $\mathbf{M}$ is given as follows: Let the real valued test function φ*

satisfy condition $\|\varphi\|_{3/2+\varepsilon} < \infty$ $(\varepsilon > 0)$. Then $\tau_\varphi^\circ = \tau_\varphi - \mathbb{E}(\tau_\varphi)$, in the limit $N, T \to \infty$, $c = N/T \leq 1$, converges in the distribution to the Gaussian random variable with zero mean and the variance:

$$\sigma^2(\tau_\varphi) = \frac{2}{c\pi^2} \iint_{-\frac{\pi}{2} < \theta_1, \theta_2 < \frac{\pi}{2}} \psi^2(\theta_1, \theta_2)(1 - \sin\theta_1 \sin\theta_2) d\theta_1 d\theta_2 \qquad (7.19)$$
$$+ \frac{\kappa_4}{\pi^2} \left(\int_{-\frac{\pi}{2}}^{\frac{\pi}{2}} \varphi(\zeta(\theta)) \sin\theta d\theta \right)^2,$$

where $\psi(\theta_1, \theta_2) = \frac{[\varphi(\zeta(\theta))]|_{\theta=\theta_1}^{\theta=\theta_2}}{[\zeta(\theta)]|_{\theta=\theta_1}^{\theta=\theta_2}}$, $[\zeta(\theta)]|_{\theta=\theta_1}^{\theta=\theta_1} = \zeta(\theta_1) - \zeta(\theta_2)$, and $\zeta(\theta) = 1 + 1/c + 2/\sqrt{c}\sin\theta$; $\kappa_4 = \mathbb{E}(X_{ij}^4) - 3$ is the 4-th cumulant of entries of $\mathbf{X}$.

To study the convergence as a function of N, we study the LES instead of the probability distribution of eigenvalues in (7.16). For an arbitrary test function with enough smoothness, the LES τ (see it as a random variable Y) is a positive scalar random variable defined in (7.17). As $N \to \infty$, the asymptotic limit of its expectation, $\mathbb{E}(Y)$, is given in (7.18). As $N \to \infty$, the asymptotic limit of its variance, $\sigma(Y)$, is given in (7.19). These two equations are sufficient to study the scalar random variable Y. This approach can be viewed as a dimensionality reduction. The random data matrix of size $N \times T$ is reduced to a positive scalar random variable Y. This dimension reduction is mathematically rigorous only when $N \to \infty, T \to \infty$ but $\frac{N}{T} \to c$. Experiences demonstrate, however, that moderate values of N and T are accurate enough for our practical purposes.

Change Point Detection Using LES

Change-point detection began with Page's (1954, 1955) classical formulation, which was further developed by Shiryaev (1963) and Lorden (1971) [34]. Change-point detection is the following problem: suppose $X_1, X_2, \ldots, X_m$ are independent observations. For $j \leq M$, they have the distribution F_0; for $j > M$, they have the distribution F_1. The distributions F_1 may be completely specified or may depend on unknown parameters. In the case of a fixed number m of observations, we would like to test the null hypothesis of no change, that $F_0 = F_1$, and to estimate M.

This part formulates the hypothesis testing problem in terms of statistical properties of LES indicators. Theorem 7.1 says that LES τ_φ, in the limit $N, T \to \infty, c = N/T \leq 1$, converges in the distribution to a Gaussian random variable with mean $\mathbb{E}(\tau_\varphi)$ and variance $\sigma(\tau_\varphi)$. Due to the Gaussian property, following a standard procedure, the detection can be modeled as a binary hypothesis testing problem: the normal hypothesis $\mathcal{H}_0$ (no anomaly present) and the abnormal one $\mathcal{H}_1$, denoted by:

$$\begin{aligned} \mathcal{H}_0 &: \left| \frac{\tau_\varphi - \mathbb{E}(\tau_\varphi)}{\sigma(\tau_\varphi)} \right| < \epsilon, \\ \mathcal{H}_1 &: \left| \frac{\tau_\varphi - \mathbb{E}(\tau_\varphi)}{\sigma(\tau_\varphi)} \right| \geq \epsilon, \end{aligned} \qquad (7.20)$$

where ϵ is a threshold value that needs to be preset based on experiences.

7.2.7 Matrices Concatenation Operation

Numerous *causing factors* affect the *system state* in different ways. Sensitivity analysis is a valuable and hot topic. The sampling data are in the form of multiple time-series, and we assume that there are N state variables and M factors. Within a fixed period of interest t_i ($i = 1, \ldots, T$), the sampling data of N state variables consist of matrix $\mathbf{B} \in \mathbb{C}^{N \times T}$ (i.e., *state matrix*), and the factors consist of vector $\mathbf{c}_j^{\mathrm{T}} \in \mathbb{C}^{1 \times T}$ ($j = 1, \ldots, M$) (i.e., *factor vector*). Two matrices with the same length can be put together and thus a new concatenated matrix is formed. In such a way, matrix $\mathbf{A}_j$ is formed by concatenating state matrix $\mathbf{B}$ with factor vector $\mathbf{c}_j^{\mathrm{T}}$.

To balance the proportion (to increase statistic correlation), a factor matrix $\mathbf{C}_j$ is formed by duplicating each factor vector $\mathbf{c}_j^{\mathrm{T}}$ for K times (K is empirically assigned around $0.3 \times N$ [15]), written as

$$\mathbf{C}_j = \begin{bmatrix} \mathbf{c}_j & \mathbf{c}_j & \cdots & \mathbf{c}_j \end{bmatrix}^{\mathrm{T}} \in \mathbb{C}^{K \times T}.$$

Then, white noise is introduced into $\mathbf{C}_j$ to avoid extremely strong cross-correlations. Thus, factor matrix $\mathbf{D}_j$ for factor vector $\mathbf{c}_j^{\mathrm{T}}$ is expressed as

$$\mathbf{D}_j = \mathbf{C}_j + \eta_j \mathbf{R} \quad (j = 1, 2, \ldots, m), \tag{7.21}$$

where η_j is related to signal-to-noise ratio (SNR), and matrix $\mathbf{R}$ is a standard Gaussian random matrix.

In parallel, we construct concatenated matrix $\mathbf{A}_j$ with each factor $\mathbf{c}_j^{\mathrm{T}}$, expressed as

$$\mathbf{A}_j = \begin{bmatrix} \mathbf{B} \\ \mathbf{D}_j \end{bmatrix} \in \mathbb{C}^{(N+K) \times T} \quad (j = 1, 2, \ldots, m). \tag{7.22}$$

Relationships between causing factors $\mathbf{c}_j^{\mathrm{T}}$ and system state $\mathbf{B}$ can be revealed by concatenated matrix $\mathbf{A}_j$. This concatenated model is compatible with different units and different measurements for each variable data that are in the form of rows of $\mathbf{A}_j$, due to the normalization during data preprocessing. It is worth mentioning that some mathematical methods, e.g., interpolation, may be applied to handle sensor data with different sampling rates.

7.2.8 Data Observation from Power Systems

For power systems, voltage magnitudes U and power consumptions P are preferred for the following reasons: (1) they are easily accessible and usually at a high accuracy; (2) they belong to measurement parameters, which are independent from network topology; and (3) our previous work [16] proves that, for certain scenarios, different types of streaming data, such as V and I, may have similar statistical properties in high-dimensional space. Similar to (7.22), $\mathbf{F}_j$ is formed as

$$\mathbf{F}_j = \begin{bmatrix} \mathbf{U} \\ \mathbf{P}_j^{(\Sigma)} \end{bmatrix} \in \mathbb{C}^{(N+K) \times T} \quad (j = 1, 2, \ldots, N). \tag{7.23}$$

The state matrix $\mathbf{U} \in \mathbb{C}^{N \times T}$ consists of voltage magnitudes $[U]_{j,t} (j = 1, \ldots, N, t = 1, \ldots, T)$, and the j-th factor matrix $\mathbf{P}_j^{(\Sigma)} \in \mathbb{C}^{K \times T}$ consists of active power consumptions $[P_j^{(\Sigma)}]_{k,t} (j = 1, \ldots, N, k = 1, \ldots, K, t = 1, \ldots, T)$ according to (7.21).

7.2.9 Empirical Spectrum Distribution of the Online Monitoring Data

We apply the M–P law for the online monitoring data from a distribution network. Assume matrix $\mathbf{R}$ is the three-phase voltage measurements from 77 monitoring devices installed on the low voltage side of distribution transformers within one feeder. The data were sampled every 15 minutes and the sampling time was 7 days. We convert $\mathbf{R}$ into the standard form $\hat{\mathbf{R}}$ through

$$\hat{r}_{ij} = (r_{ij} - \mu(\mathbf{r}_i)) \times \frac{\sigma(\hat{\mathbf{r}}_i)}{\sigma(\mathbf{r}_i)} + \mu(\hat{\mathbf{r}}_i), \tag{7.24}$$

where $\mathbf{r}_i = (r_{i1}, r_{i2}, \ldots)$, $\mu(\hat{\mathbf{r}}_i) = 0$, and $\sigma(\hat{\mathbf{r}}_i) = 1$. The covariance matrix of $\hat{\mathbf{R}}$ is calculated and the corresponding ESD under both normal and abnormal operating states of the feeder is shown in Figure 7.2.

From Figure 7.2, we can see that the spectrum of the covariance matrix of $\hat{\mathbf{R}}$ (i.e., the rectangle bars) typically exhibits two aspects: bulk and spikes (i.e., the deviating eigenvalues). The bulk arises from random noise or fluctuations and the spikes are mainly caused by large disturbances or anomaly signals. No matter when the feeder operates in normal or abnormal state, the spectrum can not be fitted by the M–P law. To be noted is that the region of the bulk and the size of the spikes are different when the feeder operates in different states. Therefore, the spectrum can not be trivially dissected by using the M–P law and we must consider a new approach instead to analyze the spectrum for assessing the operating states of the feeder more accurately.

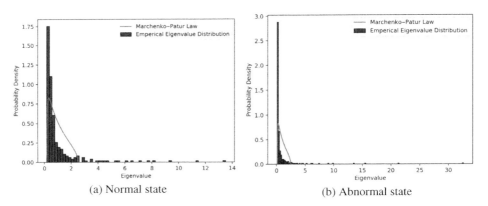

(a) Normal state (b) Abnormal state

Figure 7.2 The ESD of the covariance matrix of $\hat{\mathbf{R}}$ and its comparison with the theoretical M–P law under both normal (a) and abnormal (b) operating conditions of the feeder.

7.2.10 Residual Formulation and Discussion

From subsection 7.2.9, the spectrum of the covariance matrix of $\hat{\mathbf{R}}$ inspires us to decompose the real-world online monitoring data into systematic components (factors) and idiosyncratic noise (residuals). Assume matrix $\mathbf{R}$ is of N measurements and T observations, thus a factor model regarding $\mathbf{R}$ can be written as

$$\mathbf{R} = LF + U, \tag{7.25}$$

where L is an $N \times p$ matrix of factor loadings, F is a $p \times T$ matrix of factors, p is the number of factors, and U is an $N \times T$ matrix of residuals. For the real-world online monitoring data, the ESD of the covariance matrix of the residuals does not fit to the M–P law, no matter how many factors are removed, as is shown in Figure 7.3.

To estimate the spectrum of the real residuals, we connect the estimation of the number of factors to the limiting ESD of the covariance matrix of U. Assume there are cross- and auto-correlated structures in U, then it can be denoted as $U = A_N^{1/2} S B_T^{1/2}$. The covariance matrix of U is written as $\Sigma = \frac{1}{T} U U^T = \frac{1}{T} A_N^{1/2} S B_T S^T A_N^{1/2}$, where S is an $N \times T$ matrix, A_N and B_T are $N \times N$ and $T \times T$ symmetric nonnegative

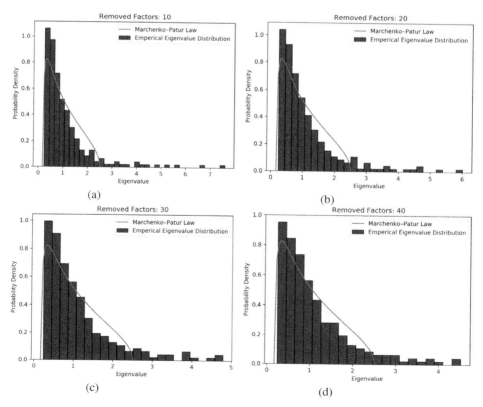

Figure 7.3 No matter how many factors are removed, the ESD of the covariance matrix of the residuals from the real-world online monitoring data does not converge to the M–P law.

definite matrices, respectively representing cross- and auto-covariances. By restricting the structures of A_N and B_T, we can use simple parameter set θ (i.e., $\theta = (\theta_{A_N}, \theta_{B_T})$) to determine them. The objective of our estimation method is to match the eigenvalue distribution of Σ to the ESD of the covariance matrix of residuals from the real-world online monitoring data. The latter is controlled by the number of removing factors (i.e., parameter p), and the former is determined by the parameter set θ. To simplify the modeling for A_N and B_T, referring to Yeo's work [35], we also make two assumptions here.

ASSUMPTION 7.1 *The cross-correlations of the real residual $U^{(p)}$ are effectively eliminated by removing p factors, i.e., $A_N \approx I_{N \times N}$.*

ASSUMPTION 7.2 *The auto-correlations of the model residual U are exponentially decreasing, i.e., $\{B_T\}_{ij} = b^{|i-j|}$, with $|b| < 1$.*

From *Assumption 7.1* and *7.2*, the estimation of cross- and auto-correlation problems is approximated by the number of removing factors p and the autoregressive rate b, respectively. The two parameters effectively characterize the features of the ESD of the covariance matrix of the residuals from the real data: the parameter p controls the range of spikes, and the parameter b reflects the shape variability of the bulk. Combining the analysis in Section 7.2.9, the anomalies in distribution network can be detected by estimating the number of removed factors p and the autoregressive rate b.

7.3 Applications of RMT on Power Grids

7.3.1 Anomaly Detection Based on Factor Model

This part proposes a spatial-temporal correlation analysis approach for anomaly detection in distribution networks. In this section, the estimation method of factor models is described in detail, in which, free random variable techniques proposed in Burda's work [36] are used to calculate the modeled spectral density. Then, specific procedures of our proposed anomaly detection approach are given and characteristics regarding them are further analyzed. Finally, more discussions about our approach are carried out.

Factor Model Estimation

We can estimate p and b by minimizing the distance between the ESD of the covariance matrix of the residuals from the real-world online monitoring data and the limiting eigenvalue density of Σ, which is stated as

$$\{\hat{p}, \hat{b}\} = \arg\min_{p,b} \mathcal{D}(\rho_{real}(p), \rho_{model}(b)), \tag{7.26}$$

where $\rho_{real}(p)$ represents the ESD of the covariance matrix of the residuals constructed by removing p factors from the real data, $\rho_{model}(b)$ is the limiting spectral density of the modeled covariance matrix characterized by parameter b, and $\mathcal{D}$ is the spectral distance measure.

To obtain $\rho_{real}(p)$, we firstly calculate the residuals by removing p largest principal components from the real-world online monitoring data. Considering the high dimensionality of the investigated PMU data, according to work [37], principal components can approximately mimic all true factors. Considering the factor model in Equation 7.25, the p-level residual $\hat{U}^{(p)}$ is calculated by

$$\hat{U}^{(p)} = R - \hat{L}^{(p)} \hat{F}^{(p)}, \qquad (7.27)$$

where $\hat{F}^{(p)}$ is an $p \times T$ matrix composed of p principal components from correlation matrix of R, $\hat{L}^{(p)}$ is an $N \times p$ matrix estimated by multivariate least squares regression, namely

$$\hat{L}^{(p)} = R * inv(\hat{F}^{(p)}), \qquad (7.28)$$

where inv denotes the pseudo-inverse method. We can calculate the covariance matrix of $\hat{U}^{(p)}$ by

$$\Sigma_{real}^{(p)} = \frac{1}{T} \hat{U}^{(p)} \hat{U}^{(p)T}, \qquad (7.29)$$

and $\rho_{real}(p)$ is the bulk spectrum of $\Sigma_{real}^{(p)}$.

Then we calculate $\rho_{model}(b)$ by using free random variable techniques proposed in [38]. For the autoregressive model U, by using the moments' generating function $M \equiv M_{\Sigma_{model}}(z)$ and its inverse relation to N-transform [35], we can derive the following polynomial equation

$$\begin{aligned}&a^4 c^2 M^4 + 2a^2 c(-(1+b^2)z + a^2 c)M^3 + ((1-b^2)^2 z^2 \\ &- 2a^2 c(1+b^2)z + (c^2-1)a^4)M^2 - 2a^4 M - a^4 = 0\end{aligned}, \qquad (7.30)$$

where $a = \sqrt{1-b^2}$ and $c = \frac{N}{T}$. See Appendix for the derivation details. Thus, we can obtain $M_{\Sigma_{model}}(z)$ by solving Equation (7.30). Then we can calculate the Stieltjes transform $G_{\Sigma_{model}}(z)$ through

$$G_{\Sigma_{model}}(z) = \frac{M_{\Sigma_{model}}(z) + 1}{z} \quad for \ |z| \neq 0. \qquad (7.31)$$

The mean spectral density $\rho_{model}(b)$ can be reconstructed from the imaginary part of $G_{\Sigma_{model}}(z)$ as

$$\rho_{model}(b) = -\frac{1}{\pi} \lim_{\varepsilon \to 0^+} \Im G_{\Sigma_{model}}(\lambda + i\varepsilon). \qquad (7.32)$$

The spectral distance measure $\mathcal{D}$ must be sensitive to the information disparity in $\rho_{real}(p)$ and $\rho_{model}(b)$. Here, we use Jensen–Shannon divergence, which is defined as

$$\mathcal{D}(\rho_{real} \| \rho_{model}) = \frac{1}{2} \sum_i \rho_{real}^{(i)} \log \frac{\rho_{real}^{(i)}}{\rho^{(i)}} + \frac{1}{2} \sum_i \rho_{model}^{(i)} \log \frac{\rho_{model}^{(i)}}{\rho^{(i)}}, \qquad (7.33)$$

where $\rho = \frac{\rho_{real} + \rho_{model}}{2}$. We can see that $\mathcal{D}(\rho_{real} || \rho_{model})$ becomes smaller as ρ_{real} approaches ρ_{model}, and vice versa. Therefore, we can match $\rho_{model}(b)$ to $\rho_{real}(p)$ by minimizing $\mathcal{D}$, through which the optimal parameter set $(\hat{p}, \hat{b})$ is obtained.

Spatial-Temporal Correlation Analysis Approach for Anomaly Detection
We know that the number of removed factors p and the autoregressive rate b can be used to indicate the variations of spatial and temporal correlation of the original data. Based on the estimated parameter $\hat{p}$, we design a statistical indicator for the top $\hat{p}$ eigenvalues of the covariance matrix of the online monitoring data to measure the spatial correlation, which is defined as

$$\mathcal{N}_\phi = \sum_{i=1}^{\hat{p}} \phi(\lambda_i), \tag{7.34}$$

where $\lambda_1 > \lambda_2 > \cdots > \lambda_{\hat{p}}$, and $\phi(\cdot)$ is a test function that makes a linear or nonlinear mapping for the eigenvalues λ_i. Details about the test function can be found in our previous work [39]. As an indicator to measure the spatial correlation of the data, $\mathcal{N}_\phi$ is more accurate and robust than the estimated number of removed factors $\hat{p}$ because the latter is susceptible to the weak factors caused by some normal fluctuations or disturbances.

Meanwhile, the estimated parameter $\hat{b}$ is directly used to measure the temporal correlation of the real data. It can effectively emulate the variation of the temporal correlation of the data, and provide an insight into system dynamics. What is worth mentioning is, if the residual processes of the real online monitoring data are not auto-correlated, $\hat{b}$ will be far different from the true value.

In real applications, we can move a certain length window on the collected data set $\mathbf{D}$ at continuous sampling times and the last sampling time is the current time, which enables us to track the variations of spatial-temporal correlations of the online monitoring data in real-time. For example, at the sampling time t_j, the obtained raw data matrix $\mathbf{R}(t_j) \in \mathbb{R}^{N \times T}$ is formulated by

$$\mathbf{R}(t_j) = (\mathbf{d}(t_{j-T+1}), \mathbf{d}(t_{j-T+2}), \ldots, \mathbf{d}(t_j)), \tag{7.35}$$

where $\mathbf{d}(t_k) = (d_1, d_2, \ldots, d_N)^H$ for $t_{j-T+1} \leq t_k \leq t_j$ is the sampling data at time t_k. Thus, $\mathcal{N}_\phi$ and $\hat{b}$ are produced for each sampling time.

Based on the works, an anomaly detection approach based on spatial-temporal correlation analysis is designed. The fundamental steps are given as follows. Steps 4–8 are conducted for calculating the ESD of the covariance matrix of the real residuals and steps 9 and 10 are for calculating the limiting spectral density of the built autoregressive model, and the spectral distance of them are calculated and saved in each iteration shown in step 11. We can obtain the optimal parameter set by getting those corresponding the minimum spectral distance for each sampling time in step 12. During the preceding steps, $\mathcal{N}_\phi$ and $\hat{b}$ are calculated as indicators to assess the system states.

The anomaly detection approach proposed is driven by the online monitoring data in distribution networks, and based on high-dimensional statistical theories. It reveals

Steps of Spatial-temporal Correlation Analysis for Anomaly Detection in Distribution Networks

1. For each feeder, construct a spatial-temporal data set **D** by arranging three-phase voltage measurements of all public transformers within the feeder in a series of time.
2. At each sampling time t_j:
3.   Obtain the original data matrix $\mathbf{R}(t_j) \in \mathbb{R}^{N \times T}$ by moving a $N \times T$ window on **D**;
4.   For the number of removing factors $p = 1, 2, \ldots$
5.   Get the real residual $\hat{U}^{(p)}(t_j)$ through Equation (7.27);
6.   Normalize $\hat{U}^{(p)}(t_j)$ into the standard form through Equation (7.24);
7.   Calculate the covariance matrix of the real residual, i.e., $\Sigma_{real}^{(p)}(t_j)$;
8.   Obtain $\rho_{real}^{(p)}(t_j)$ through getting the bulk spectrum of $\Sigma_{real}^{(p)}(t_j)$;
9.   For the autoregressive rate $b \sim U[0, 1]$
10.   Obtain $\rho_{model}^{(b)}(t_j)$ through Equations (7.30), (7.31), and (7.32);
11.   Calculate the spectral distance $\mathcal{D}(\rho_{real}^{(p)}(t_j)||\rho_{model}^{(b)}(t_j))$ through Equation (7.33) and save them;
12. Obtain the optimal parameter set $(\hat{p}(t_j), \hat{b}(t_j))$ through Equation (7.26);
13. Calculate the spatial indicator $\mathcal{N}_\phi(t_j)$ through Equation (7.34);
14. Draw the $\mathcal{N}_\phi - t$ curve and $\hat{b} - t$ curve for each feeder in a series of time t to realize anomaly detection.

the variations of spatial-temporal correlations of the input data when anomalies happen and can detect the weak anomalies occurring in the system through controlling both the number of factors p and the autoregressive rate b. Compared with traditional model-based methods, our approach is purely driven by data and does not require too much knowledge about the complex topology of the complex distribution network. It is theoretically robust against small random fluctuations and measuring errors in the system, which can help to improve anomaly detection accuracy and reduce the potential false detection probability. What's more, our proposed approach is practical for real-time anomaly detection by moving a certain length window method.

7.3.2 More Discussions about the Proposed Approach

In Section 7.2.10, we assume that the cross-correlations of the real residuals can be effectively eliminated by removing p factors. However, for the real-world online monitoring data in a distribution network, whether this assumption holds is questionable. Meanwhile, the factor model estimation method in Section 7.3.1 is suitable for large-dimensional data matrices in theory. However, in practice, the dimensions of the online monitoring data for some feeders in a distribution network are often moderate, such as hundreds or less. Here, we will check how well the built autoregressive model can fit the real residuals, the result of which are shown in Figure 7.4.

Figures 7.4(a) and 7.4(b), respectively, show the fitted result of our built autoregressive model to the real residuals under both normal and abnormal operating states. We

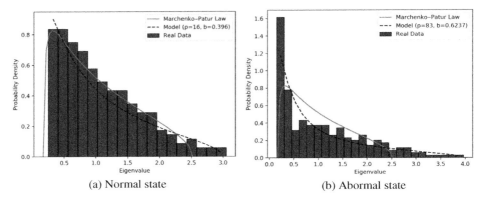

Figure 7.4 Fit of the built autoregressive model to the real data. The data were sampled through the online monitoring devices installed on the low voltage side of the public transformers within one feeder. It was sampled when the feeder operated under both normal and abnormal states, respectively, sampling 672 times. The dimension of the data is 231. The built model with estimated $\hat{p}$ and $\hat{b}$ fits the residuals of the real data very well. For comparison, M–P law for the real residual matrix is plotted.

can see that, with optimal estimated parameter set $(\hat{p}, \hat{b})$, our built model can fit the real residuals well no matter whether the feeder runs in normal or abnormal states. In contrast, the M–P law does not fit the real residuals. The well-fitted result validates our assumption for the real data in Section 7.2.10 and proves the feasibility of our approach for analyzing the online monitoring data. Furthermore, it is noted that the estimated $\hat{p}$ and $\hat{b}$ are different when the feeder runs in different states, which explains why they can be used as basic indicators to detect anomalies in distribution networks.

7.3.3 Case Studies

In this section, our proposed anomaly detection approach is validated through using both the synthetic data from the IEEE 118-bus test system and real-world online monitoring data in a distribution network. Four cases under different scenarios are designed: (1) The first three cases, leveraging the synthetic data, test the effectiveness of our approach for anomaly detection and analyze the implications of parameter p and b. (2) The last case, using the real-world online monitoring data, validates our approach and compares it with the spectrum analysis approach using M–P law in the work [16].

Case Study with Synthetic Data

The synthetic data were sampled from the simulation results of the standard IEEE 118-bus test system, with a sampling rate of 50 Hz. During the simulations, a sudden change of the active load at one bus was considered an anomaly signal and little white noise was introduced to represent random fluctuations and system errors.

(1) Case Study on the Effectiveness of Our Approach for Anomaly Detection: In this case, the synthetic data set contains 118 voltage measurement variables with sampling

Table 7.1 Assumed signals for load of bus 20 in case 1.

Bus	Sampling Time	Active Power (MW)
20	$t_s \in \{1, 2, \ldots, 500\}$	20
	$t_s \in \{501, 502, \ldots, 1000\}$	80
Others	$t_s \in \{1, 2, \ldots, 1000\}$	Unchanged

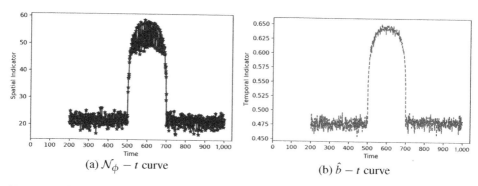

(a) $\mathcal{N}_\phi - t$ curve

(b) $\hat{b} - t$ curve

Figure 7.5 Effectiveness of our approach for anomaly detection in case 1.

1,000 times. To test the effectiveness of the proposed anomaly detection approach, an assumed anomaly signal was set for bus 20 and others stayed unchanged, which is shown in Table 7.1. In the experiment, the size of the split-window was 118×200 and a little random autoregressive noise with a decaying rate $b = 0.5$ was introduced into each split-window to represent the auto-correlations of the residuals. The experiment was repeated for 20 times and the results were averaged. Here, we chose the likelihood radio function (LRF) as the test function.

Figures 7.5(a) and 7.5(b) show the anomaly detection result respectively from the perspective of space and time. It is noted that the $\mathcal{N}_\phi - t$ curve and $\hat{b} - t$ curve begin at $t_s = 200$ because the initial split-window includes 199 times of historical sampling and the present sampling data. It is also important to note that the index number starts from 0 in *Python*. The detection processes are shown as follows:

I. During $t_s = 200 - 500$, $\mathcal{N}_\phi$ and $\hat{b}$ remains almost constant, which means the system operates in normal state and the spatial-temporal correlations of the data stay almost unchanged. As is shown in Figure 7.6(a), the ESD of the covariance matrix of the residuals can be fitted well by the built model with $\hat{p} = 6, \hat{b} = 0.4752$.

II. From $t_s = 501 - 700$, $\mathcal{N}_\phi$ and $\hat{b}$ change dramatically, which denotes an anomaly signal that occurs at $t_s = 501$, where the spatial-temporal correlations of the data begin to change. The curves are almost inverted U-shape and the delay lag of the anomaly signal to the spatial-temporal indicators is equal to the split-window's width. Figure 7.6(b) shows the ESD of the covariance matrix of the residual fits the probability density function (PDF) of our built model with $\hat{p} = 1, \hat{b} = 0.6435$.

III. At $t_s = 701$, $\mathcal{N}_\phi$ and $\hat{b}$ recover to their initial values and remain constant afterward, which indicates the system has returned to normal.

Table 7.2 Assumed signals for active load of bus 20, 30, and 60 in case 2.

Bus	Sampling Time	Active Power(MW)
20	$t_s \in \{1, 2, \ldots, 500\}$	20
	$t_s \in \{501, 502, \ldots, 1000\}$	80
30	$t_s \in \{1, 2, \ldots, 550\}$	20
	$t_s \in \{551, 552, \ldots, 1000\}$	80
60	$t_s \in \{1, 2, \ldots, 600\}$	20
	$t_s \in \{601, 602, \ldots, 1000\}$	80
Others	$t_s \in \{1, 2, \ldots, 1000\}$	Unchanged

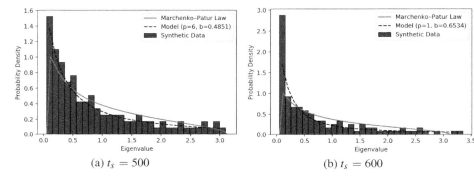

(a) $t_s = 500$

(b) $t_s = 600$

Figure 7.6 The ESD of the covariance matrix of the residuals can be fitted well by our built model with estimated $\hat{p}$ and $\hat{b}$. The optimal parameters estimated are different when the system operates in different states.

From the preceding analyzes, we can conclude that, by exploring the variations of the spatial-temporal correlations of the data, our proposed approach is able to detect anomaly signal effectively. It is noted that the estimated $\hat{p}$ and $\hat{b}$ are different when the system operates in different states. In the following cases, we will discuss their implications.

(2) Case Study on the Implication of p: In our experiments, we found the estimated optimal parameter $\hat{p}$ is related to the system states. In this case, we will further discuss what drives it. To validate the relations of p and the number of anomaly events occurred in the system, different number of assumed anomaly signals were set, which is shown in Table 7.2. All the other parameters were set the same as in case (1) and the experiment was also repeated for 20 times with results being averaged.

The generated $\hat{p} - t$ curve by continuously moving windows is shown in Figure 7.7. Interpretations of $\hat{p}$ are stated as follows:

I. From $t_s = 200$ to $t_s = 500$, the estimated $\hat{p}$ remains almost at 7. During our experiment, we observed that no strong factors appeared during this period. The most likely explanation is that our approach is sensitive to weak factors caused by some normal fluctuations or disturbances and can identify them effectively.

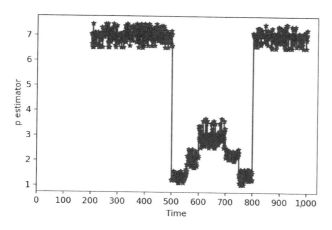

Figure 7.7 $\hat{p} - t$ curve.

II. From $t_s = 501$ to $t_s = 550$, one strong factor is observed during our experiment and the estimated $\hat{p}$ is between 1 and 2. This indicates that one anomaly event occurred, which is consistent with our assumed result in Table 7.2. Then, from $t_s = 551$ to $t_s = 600$, the number of estimated factors is between 2 and 3, which is also consistent with the number of assumed anomaly signals contained in the window. A similar analysis result can be concluded during $t_s \in \{601, 602, \ldots, 650\}$.

III. From $t_s = 651$ to $t_s = 800$, the value of $\hat{p}$ decreases 1 every other 50 sampling times. This is because the width of our moving window is 200 and the number of anomaly events contained in the window decreases 1 every 50 sampling times with its movement.

IV. At $t_s = 801$, the value of $\hat{p}$ returns to nearly 7 and remains afterward, which indicates the system has returned to the normal state.

From the preceding analyses, we can conclude that p is related to the number of anomaly events occurring in the system. Theoretically, the estimated $\hat{p}$ is equal to the number of anomaly events. However, it is noted that our approach is able to identify several weak factors even if the system operates in normal state.

(3) Case Study on the Implication of b: From Case (1), we see that the parameter b can be used to reflect the system states from the perspective of time. In this case, we will further discuss the implication of b and what drives it. To test the relations of b and the degree of abnormality in the system, assumed anomaly signals with different scale degrees were set for bus 20, which is shown in Table 7.3. All the other parameters were set the same as in Case (1). The experiment was also repeated 20 times with results being averaged.

The $\hat{b} - t$ curve generated by continuously moving windows is shown in Figure 7.8. Interpretations of $\hat{b}$ are described as follows:

I. From $t_s = 200$ to $t_s = 500$, the estimated $\hat{b}$ remains almost constant, which indicates the system operates in normal state.

II. From $t_s = 501$ to $t_s = 700$, the $\hat{b} - t$ curves are almost inverted U-shaped and reach their global maximum at $t_s = 600$. It is noted that the estimated $\hat{b}$ corresponding

Table 7.3 Assumed signals with different scale degrees set for bus 20 in case 3.

Bus	Sampling Time	Active Power(MW)
20	$t_s \in \{1, 2, \ldots, 500\}$ $t_s \in \{501, 502, \ldots, 1000\}$	20 60/90/120
Others	$t_s = 1 \sim 1000$	Unchanged

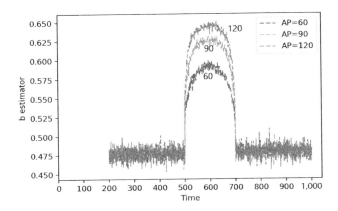

Figure 7.8 $\hat{b} - t$ curve.

to the assumed signal of the active power from 20 to 120 has the largest maximum value and that from 20 to 60 has the smallest value. It shows that the higher the abnormality degree is, the larger the estimated value of $\hat{b}$ is.

III. At $t_s = 701$, the estimated $\hat{b}$ recover to their initial values and remain constant afterward, which indicate that the system has returned to the normal state.

From the preceding analyses, it can be concluded that b is related to the degree of abnormality in the system. The higher the abnormality degree of the system is, the greater the deviation of the estimated $\hat{b}$ from its normal standard is. It reflects essential information on the system movement, which explains why it can be used as a temporal indicator to measure the system state in our approach.

Case Study with Real-World Online Monitoring Data

In this subsection, the online monitoring data obtained from a real-world distribution network is used to test our proposed approach. The feeder in the distribution network consists of different levels of branch lines and substations with 77 distribution transformers in total. On the low voltage side of each distribution transformer, one online monitoring device is installed, through which we can obtain many types of measurement variables such as three-phase voltages, three-phase currents, active load. The data were sampled every 15 minutes and the sampling time was from 2017/3/15 00:00:00 to 2017/4/6 23:45:00. Here, we chose three-phase voltages as the elements to formulate the data set of size 231×2208.

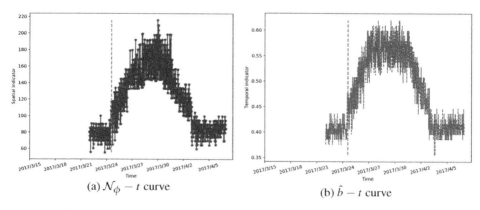

Figure 7.9 Effectiveness of our approach for anomaly detection in distribution networks.

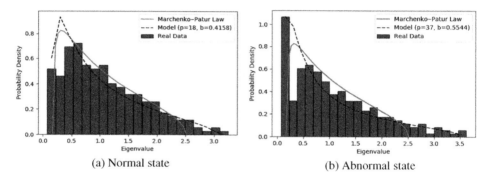

Figure 7.10 The ESD of the covariance matrix of the residuals from the real-world online monitoring data can be fitted very well by our built model with the estimated $\hat{p}$ and $\hat{b}$, while can not be fitted by the M–P law. The optimal parameters estimated are different when the feeder runs in different states.

In the experiment, we used a 231×672 split-window, and moved it at each sampling time, which enabled us to track the variation of the spatial-temporal correlation of the data. With moving windows, the generated spatial-temporal indicators for each sampling time is shown in Figure 7.9. In the figure, the vertical dashed line marks the beginning time of the anomalies. It is noted that the indicator curves begin at 2017/3/7 23:45:00 because our initial spit-window includes 7 days' sampling time.

From the *spatial/temporal indicator* $-t$ curves, we can realize anomaly detection for the feeder as follows:

I. From 2017/3/15 00:00:00 to 2017/3/24 08:45:00, the values of the spatial-temporal indicators remain almost constant, which indicates the feeder operates in normal state. As is shown in Figure 7.10(a), the ESD of covariance matrix of residuals can be fitted well by the built model with $\hat{p} = 18, \hat{b} = 0.4158$ when the feeder runs in normal state, but it does not fit the M–P law.

II. At 2017/3/24 09:00:00, the values of the spatial-temporal indicators begin to change dramatically, which indicate anomaly signals occurring and the operating state of the feeder beginning to change. Comparing with the recorded anomaly time, we can see that the anomaly can be detected earlier than the recorded time. Furthermore, from 2017/3/24 09:00:00 to 2017/4/2 22:30:00, the *spatial/temporal indicator* $- t$ curves are almost inverted U-shaped, which is consistent with our simulation result in Section 7.3.3. Figure 7.10(b) shows, in abnormal state, the ESD of covariance matrix of residuals can be fitted well by our built model with $\hat{p} = 37, \hat{b} = 0.5544$.

III. From 2017/4/2 22:45:00, the calculated spatial-temporal indicators recover their initial values and remain afterward, which indicates the feeder has returned to the normal state.

The preceding analyses show that our approach is capable of detecting anomalies in an early phase in distribution networks. In real-time analysis, we may not observe the completely inverted U-shape curve and we can use the variance radio of the *spatial/temporal indicator* $- t$ curves as the basis to judge whether an anomaly is occurring. It is noted that the delay lag of the anomaly signals to the spatial-temporal indicators is not equal to the split-window's width, which is not in accord with our simulation results. The reason is that the anomaly events in the real data usually last for a period and we can estimate that through subtracting the split-window's width from the width of the inverted "U."

Furthermore, we compare our approach with the spectrum analysis (SPA) method using the M–P law. SPA has been well studied in the work [16]. The difference of SPA to our approach is that it uses the M–P law for analyzing the spectrum of the covariance matrix of the data and calculates a statistical indicator for it. To compare the anomaly detection performances of the two approaches, we normalized the calculated indicators into [0,1], which is shown in Figure 7.11.

From the figure, we can see that both our approach ($SI/TI - t$ curves) and SPA are capable of detecting the anomalies effectively. However, the $SI/TI - t$ curves have a

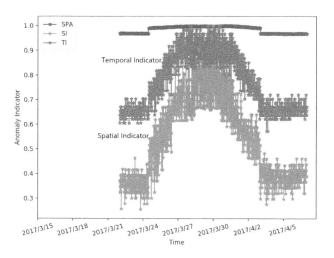

Figure 7.11 Comparison of anomaly detection performance between our approach and SPA using the M–P law.

higher variance radio than SPA using the M–P law, which indicates our approach is more sensitive to the anomalies in the distribution network. The reason is that the spectrum from the real data is complex and it is not accurate to trivially dissect it by using the M–P law adopted in SPA. In contrast, our approach uses factor model for the spectrum analysis of the real data, and it estimates the factor model by controlling both the number of factors and the autoregressive rate of the residuals, making it capable of tracking the data behavior more accurately.

7.3.4 Invisible Units Detection and Estimation

Problem Formulation

Invisible units mainly refer to small-scale units that are not monitored, and thus are not visible to utilities, e.g., small-scale distributed units like unauthorized roof-top photovoltaics, and plug-and-play units like electric vehicles. Integration of invisible units into power systems could significantly affect the way in which a distribution grid is planned and operated. Lack of visibility may result in incorrect planning and operation of power systems and, even worse, damage to system equipment such as transformers, voltage regulators, and customer appliances. In a highly invisible units penetration environment, utilities face technical problems related to overvoltage, frequency control, back feeding flow, and other issues such as a rapid decrease in revenue. Prosumers like EVs also bring many unknowns and risks that need to be identified and managed [40].

This part attempts to conduct invisible units detection and estimation in a non-omniscient distribution network. More precisely, we aim to obtain load/generator ingredients and their weights at the node level.

Firstly, the load/generator ingredients are discussed. Customers are divided into two categories: typical load pattern units (TLPs) and uncertain load pattern units (ULPs).

1. The TLPs operate according to a well-defined profile, which can be denoted as vectors $\mathbf{p}_i^{(T)}$. For instance, the load pattern of streetlights, which are turned on at 18:00 and turned off at 6:00, is modeled as

$$\mathbf{p}^{(\text{Lamp})}(t) = \begin{cases} 1 & t \in [00:00, 06:00] \cup [18:00, 24:00] \\ 0 & t \in [06:00, 18:00] \end{cases}.$$

   If the sampling interval is 6 hours, $\mathbf{p}^{(\text{Lamp})} = [1, 0, 0, 1]$.

2. The ULPs are denoted as vectors $\mathbf{p}_j^{(U)}$, and can be further divided into three categories–completely random behavior, invisible behavior, and fraudulent behavior. Our previous works [16, 41] have already distinguished the completely random behavior from the others using random matrix tools. This part focuses on detection and estimation of the invisible and the fraudulent behavior. The invisible behavior often causes a chain reaction and has an impact on numerous parameters. For instance, unauthorized residential PV installation changes the power flow of grid network. The fraudulent behavior, however, often only causes parameter deviation alone. For example, some metering errors or cyberattacks might merely change value of power consumption P of some node without affecting voltage U.

Considering the weights, we propose to study a general model for each node:

$$\mathbf{p}^{(\Sigma)} = a_1 \mathbf{p}_1^{(T)} + \cdots + a_n \mathbf{p}_n^{(T)} + b_1 \mathbf{p}_1^{(U)} + \cdots + b_m \mathbf{p}_m^{(U)}, \quad (7.36)$$

where vectors $\mathbf{p}_i^{(T)}$ and $\mathbf{p}_j^{(U)}$ are daily patterns of TLPs and ULPs, with their coefficients (weights) $a_i, b_j, i = 1, \ldots, n, j = 1, \ldots, m$, respectively. Thus, vector $a_i \mathbf{p}_i^{(T)}$ is the daily power usage for the ith TLP, similarly vector $b_j \mathbf{p}_j^{(U)}$ for the j-th ULP.

If all units' pattern and behavior are known in advance, i.e., no $\mathbf{p}_j^{(U)}$ exists, or if ULPs are able to be modeled as $\mathbf{p}_{n+j}^{(T)}$ instead of uncertain $\mathbf{p}_j^{(U)}$, then (7.36) is rewritten as

$$\mathbf{p}^{(\Sigma)} = a_1 \mathbf{p}_1^{(T)} + a_2 \mathbf{p}_2^{(T)} + \cdots + a_{n+m} \mathbf{p}_{n+m}^{(T)}. \quad (7.37)$$

Our first step is to formulate the problem in the form of an expression

$$\arg\min_{a_i} \left(\mathbf{p}^{(N)} - \mathbf{p}^{(L)} - \mathbf{p}^{(\Sigma)} \begin{pmatrix} a_1 & a_2 & \cdots & a_{m+n} \end{pmatrix} \right), \quad (7.38)$$

where vectors $\mathbf{p}^{(N)}$ and $\mathbf{p}^{(L)}$ are power injections and power losses on lines, respectively; they are measurable or calculable. It is worth mentioning that the analysis for reactive power Q may be conducted similarly.

For simplicity, we rewrite (7.38) as

$$\arg\min_{\mathbf{a}} \| \mathbf{P}\mathbf{a} - \mathbf{k} \| \quad (7.39)$$

where $\mathbf{P} = \begin{bmatrix} \mathbf{p}_1^{(T)} & \cdots & \mathbf{p}_{n+m}^{(T)} \end{bmatrix}$, $\mathbf{a} = \begin{bmatrix} a_1 & \cdots & a_{n+m} \end{bmatrix}^T$, and $\mathbf{k} = \mathbf{p}^{(N)} - \mathbf{p}^{(L)}$. The expression in (7.39) is a classical optimization problem, which can be solved using least square methods and the estimate value $\hat{\mathbf{a}}$ is given as

$$\hat{\mathbf{a}} = (\mathbf{P}^T \mathbf{P})^{-1} \mathbf{P}^T \mathbf{k} \quad (7.40)$$

In a modern distribution network, ULPs $b_j \mathbf{p}_j^{(U)}, j = 1, \ldots, m$ are present and their influences need to be considered. This phenomenon violates the prerequisites of most algorithms (e.g., least squares method) and has significant effects on the estimated values of coefficients $a_i, i = 1, \ldots, n$ in (7.36). In most cases, it is reasonable to model $\mathbf{p}_j^{(U)}, j = 1, \ldots, m$ as a step signal. This is the case in which plug-in EVs charge or unauthorized PVs generate during t_a to t_b. *Determining the start point t_a and the end point t_b of this step signal* is at the heart of $\mathbf{p}_j^{(U)}$ modeling. Based on random matrix theory and linear eigenvalue statistics, a statistical, data-driven approach, rather than its deterministic, empirical, or model-based counterpart, is proposed as the solution.

Background of the Simulation Cases
Simulations are based on IEEE 33-bus test system for a distribution network. For Node $j (j = 1, \ldots, 33)$, its gross power usage $\mathbf{p}_j^{(\Sigma)}$ and voltage magnitude $\mathbf{u}_j$ are sampled at a high rate, for example, 9,600 points per day (0.11 Hz). The white noise is added to the power injections as

$$\tilde{y}_{nt} = y_{nt}(1 + \gamma_1 z_1) + \gamma_2 z_2, \quad (7.41)$$

where z_1 and z_2 are two standard Gaussian random variables, i.e. $z_1, z_2 \sim \mathcal{N}(0, 1)$; $\gamma_1 = 0.005, \gamma_2 = 0.02$. In this way, the power flow is obtained using MATPOWER.

Determining the start point and the end point of $\mathbf{p}_i^{(U)}$ is the main focus of this study. To long-standing anomalies without sudden change in the observed data segment, some long-term indicators, such as monthly line loss rate, might be effective.

Fraud Events Detection

Fraud events often cause parameter deviation alone. Suppose that active power values P for each node are at their initial points with fluctuations given as (7.41). From 14:00 to 17:00, some fraud events on Node 6 and Node 14 cause a reduction of 0.005 MW (8.33% of P_6, and 4.17% of P_{14}). The lines with legends *data 1* to *data 33* are for the actual power consumption of Node 1 to Node 33, and the lines with legends *data 34* and *data 35* are for the measured power consumption of Node 14 and Node 6, respectively. Note that due to the fraud events, the data with legends *data 14* and *data 6* are not accessible.

Invisible Units Detection

This subsection, focusing on fraudulent behavior and anomaly power usage, proposes a data-driven solution for the problem – determining the start point and the end point to model the invisible unit $\mathbf{p}_i^{(U)}$ as a step signal. Firstly, the following assumptions are made:

(I) Power usage of Node $j (j = 1, \ldots, 33)$ generally consists of four TLPs and one ULP, denoted as

$$\mathbf{p}_j^{(\Sigma)} = a_{j1}\mathbf{p}_1^{(T)} + \cdots + a_{j4}\mathbf{p}_4^{(T)} + b_{j1}\mathbf{p}_1^{(U)}. \quad (7.42)$$

Daily load profiles of TLPs $(\mathbf{p}_1^{(T)}, \ldots, \mathbf{p}_4^{(T)})$ are set as Figure 7.13. Note that the rectangles mean that the load profile has a dramatic change at this time point, i.e., change point (CP) [42].

(II) Coefficients $a_i, b_j (i = 1, 2, 3, 4; j = 1)$ are set according to the table in Figure 7.12.

(III) Invisible power usage events exist on Node 20 and 31: the periods are 1:00–5:00 and 14:00–20:00, and the percentages are 30% and 50%, respectively.

(IV) Fraud events exist on Node 6, 14, and 27: the periods are 20:00–22:00, 14:00–17:00 and 18:00–19:00, and the percentages are 7%, 8%, and 12%, respectively.

In the assumed complex scenario mentioned previously, the active power P and the voltage U of each node are accessible, respectively. Following (7.23), $\mathbf{F}_j$ is obtained for Node $j (j = 1, \ldots, 36)$. The LES indicators of $\mathbf{F}_j$ are obtained as the $\tau_{T_2} - t$ curves.

These figures give a visualization of LES indicators during a whole day. Particular attention should be paid to those spikes – their corresponding time point and line number hold vital clues about when and where the invisible behavior occurs:

- $\tau_{T_2}(\mathbf{B})$; it is relatively smooth because the system is stable without emergencies such as voltage collapse.

	a_1	a_2	a_3	a_4	b_1		a_1	a_2	a_3	a_4	b_1
1	0.25	0.25	0.25	0.25	0	2	0	0.7	0.1	0.2	0
3	0	0.1	0.8	0.1	0	4	0.05	0.75	0.1	0.1	0
5	0	0.1	0.8	0.1	0	6	0.1	0.2	0.5	0.2	0
7	0.8	0.05	0.1	0.05	0	8	0.85	0.05	0	0.1	0
9	0.1	0.15	0.6	0.15	0	10	0	0.15	0.8	0.05	0
11	0	0.2	0.75	0.05	0	12	0.05	0.1	0.75	0.1	0
13	0.05	0.05	0.85	0.05	0	14	0.7	0.05	0.2	0.05	0
15	0	0.05	0.9	0.05	0	16	0	0	0.95	0.05	0
17	0	0.1	0.8	0.1	0	18	0	0.7	0.1	0.2	0
19	0	0.5	0.1	0.4	0	20	0	0.2	0.2	0.3	0.3
21	0	0.8	0.1	0.1	0	22	0.1	0.75	0	0.15	0
23	0.2	0.6	0	0.2	0	24	0.85	0	0.05	0.1	0
25	0.75	0.1	0.1	0.05	0	26	0.2	0	0.7	0.1	0
27	0.1	0	0.75	0.15	0	28	0.25	0.1	0.6	0.05	0
29	0.8	0.05	0.1	0.05	0	30	0.9	0	0.05	0.05	0
31	0.1	0.1	0.05	0.25	0.5	32	0.9	0	0	0.1	0
33	0.95	0	0	0.05	0						

Figure 7.12 Coefficients of TLPs and ULP of each node.

	p_1	p_2	p_3	p_4	p_{u1}		p_1	p_2	p_3	p_4	p_{u1}
0	88	20	25	100	0	12	94	77	35	0	0
1	87	20	23	100	100	13	86	80	30	0	0
2	88	20	22	100	100	14	86	86	33	0	100
3	100	21	22	100	100	15	88	86	44	0	100
4	96	20	27	100	100	16	85	87	50	100	100
5	100	20	31	100	0	17	87	35	56	100	100
6	98	20	29	0	0	18	88	25	85	100	100
7	97	30	28	0	0	19	85	25	80	100	100
8	88	40	31	0	0	20	84	20	70	100	0
9	82	85	37	0	0	21	83	20	76	100	0
10	82	85	42	0	0	22	86	20	43	100	0
11	95	82	42	0	0	23	88	15	30	100	0

Note: Rectangles mean CP.

Figure 7.13 TLPs, ULPs, and their 24-hour power demands.

- For fraud events, the extreme points are located at $t = 5553, 6854, 7655$, etc. According to Section 7.3.4, this phenomenon **matches Assumption IV** that the CPs are at $t = 14:00(5600 \approx 5553+50), 17:00(6800 \approx 6854 - 50), 1900(7600 \approx 7655 - 50)$, etc., respectively.
- For invisible TLPs, *Spike 1 (X: 1250, Y: 32, Z: 4041)* means the existence of a CP at $3:00(1200 = 1250 - 50)$ of Node 32. According to Figure 7.13, 3:00 is a CP of TLP $p_1^{(T)}$. Thus we deduce that TLP $p_1^{(T)}$ takes a dominant part of Node 32. Similar deductions could be made for Node 25, 24, 30, etc. These deductions **match Assumption II** and are confirmed by Figure 7.12.
- For invisible ULPs, *Spike 2 (X: 450, Y: 31 Z:4626)* means the existence of a CP at $1:00(400=450-50)$ of Node 31. According to Figure 7.13, there is no existence of any TLP ($p_1^{(T)}, p_2^{(T)}, p_3^{(T)}, p_4^{(T)}$) matching CP 1:00. As a result, we artificially build a CP at 1:00 for a new ULP ($p_1^{(U)}$ for this case). Step by step, all the CPs of ULP $p_1^{(U)}$ are obtained as Figure 7.13. Then, the general model (7.42) is tuned into the classical model (7.37) $\mathbf{p}^{(\Sigma)} = a_1 \mathbf{p}_1^{(T)} + a_2 \mathbf{p}_2^{(T)} + a_3 \mathbf{p}_3^{(T)} + a_4 \mathbf{p}_4^{(T)} + a_5 \mathbf{p}_5^{(T)}$.

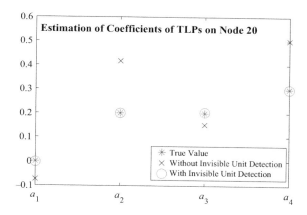

Figure 7.14 Estimate with and without invisible units detection.

Estimation with and without Invisible Units Detection

Existence of invisible units disables classical least squares method. Taking Node 20 for instance, without invisible units detection, we will obtain a bad result shown as Figure 7.14. If the information about the start point and end point of the invisible behavior is acquired, an accurate estimate is produced.

7.3.5 Estimation of Jacobian Matrix

Problem Formula

Jacobian matrix is the key part of power flow analysis. It is a sparse matrix that results from a sensitivity analysis of the power flow equations.

Numerical iteration algorithms and sparse factorization techniques, mainly based on Newton–Raphson or fast-decoupled methods, are employed to approximate the nonlinear PF equations by linearized Jacobian–matrix $\mathbf{J}$. In mathematics, if $\mathbf{p}$ is a point in $\mathbb{R}^n$ and f is differentiable at $\mathbf{p}$, then its derivative is given by $\mathbf{J}_f(\mathbf{p})$. In this case, the linear map described by $\mathbf{J}_f(\mathbf{p})$ is the best linear approximation of f near the point $\mathbf{p}$, in the sense that

$$f(\mathbf{b}) = f(\mathbf{p}) + \mathbf{J}_f(\mathbf{p})(\mathbf{b}-\mathbf{p}) + o(\|\mathbf{b}-\mathbf{p}\|) \tag{7.43}$$

for $\mathbf{b}$ close to $\mathbf{p}$ and where o is the little o-notation (for $\mathbf{b} \to \mathbf{p}$) and $\|\mathbf{b}-\mathbf{p}\|$ is the distance between $\mathbf{b}$ and $\mathbf{p}$.

From (7.43), the linear approximation of the system operating from point $(\mathbf{x}^{(k)}, \mathbf{y}^{(k)})$ to point $(\mathbf{x}^{(k+1)}, \mathbf{y}^{(k+1)})$, the iteration is acquired as follows:

$$\mathbf{x}^{(k+1)} := \mathbf{x}^{(k)} + \mathbf{J}_f^{-1}(\mathbf{x}^{(k)})(\mathbf{y}^{(k+1)} - \mathbf{y}^{(k)}) \tag{7.44}$$

where $:=$ is the assignment symbol in computer science.

The iteration, given in (7.44), depicts how the power system state estimation is carried out. $\mathbf{y}^{(k+1)}$, according to (7.7), is the desired P, Q of the PQ buses and desired P of the

PV buses.[4] $\mathbf{y}^{(k)}$ and $\mathbf{x}^{(k)}$ are measurements available from PMUs. To conduct (7.44), we also need Jacobian matrix $\mathbf{J}$.

Traditionally, $\mathbf{J}$ is computed offline using (7.8) based on a physical model of the network. The model-based approach is not ideal in practice because accurate and up-to-date network topology and relevant parameters (the admittance Y) and operating point $(\mathbf{x}^{(k)}, \mathbf{y}^{(k)})$ are required according to (7.6).

This part aims at a model-free estimation of the Jacobian matrix $\mathbf{J}$. In the model-free mode, the physical model mentioned previously is no longer necessary information, as well as the admittance Y. Moreover, the estimation of the Jacobian matrix $\mathbf{J}$ inherently contains the most up-to-date network topology and corresponding parameters. The data-driven approach could handle the scenario in which system topology and parameter information are wholly unavailable.

Under fairly general conditions, the target of the estimation $\mathbf{J}$ is almost unchanged within a short time, e.g., Δt, due to the stability of the system, or concretely, variable V, θ, Y. During Δt, considering T times observation at time instants t_i, ($i = 1, 2, \ldots, T, t_T - t_1 = \Delta t$), operating points $(\mathbf{x}^{(i)}, \mathbf{y}^{(i)})$ are obtained as (7.7). Defining $\Delta \mathbf{x}^{(k)} \triangleq \mathbf{x}^{(k+1)} - \mathbf{x}^{(k)}$, and similarly $\Delta \mathbf{y}^{(k)} \triangleq \mathbf{y}^{(k+1)} - \mathbf{y}^{(k)}$, from (7.43) we deduce that $\Delta \mathbf{y}^{(k)} = \mathbf{J} \Delta \mathbf{x}^{(k)}$. Because $\mathbf{J}$ keeps nearly constant, the matrix form is written as:

$$\mathbf{B} \approx \mathbf{J}\mathbf{A} \tag{7.45}$$

where $\mathbf{J} \in \mathbb{R}^{N \times N}$ ($N = m + n - 1$), $\mathbf{B} = [\Delta \mathbf{y}^{(1)}, \ldots, \Delta \mathbf{y}^{(T)}] \in \mathbb{R}^{N \times T}$, and $\mathbf{A} = [\Delta \mathbf{x}^{(1)}, \ldots, \Delta \mathbf{x}^{(T)}] \in \mathbb{R}^{N \times T}$. As the fast-sampling of PMUs, the assumption that $T > N$ is made.

It is worth mentioning that sometimes the normalization is required for some purposes. Supposing that $\widehat{\mathbf{B}} = \Lambda_\mathbf{B} \mathbf{B}$, $\widehat{\mathbf{A}} = \Lambda_\mathbf{A} \mathbf{A}$, and $\widehat{\mathbf{B}} \approx \widehat{\mathbf{J}} \widehat{\mathbf{A}}$, it is deduced that $\mathbf{J} = \Lambda_\mathbf{B}^{-1} \widehat{\mathbf{J}} \Lambda_\mathbf{A}$.

References

[1] B. Gao, G. K. Morison, and P. Kundur, "Voltage stability evaluation using modal analysis," *Power Systems IEEE Transactions*, vol. 7, no. 4, pp. 1529–1542, 1992.

[2] R. L. Lugtu, D. F. Hackett, K. C. Liu, and D. D. Might, "Power system state estimation: Detection of topological errors," *IEEE Power Engineering Review*, vol. PER-1, no. 1, pp. 19–19, 1980.

[3] F. F. Wu and W. H. E. Liu, "Detection of topology errors by state estimation," *IEEE Transactions on Power Systems*, vol. 4, no. 2, pp. 50–51, 1989.

[4] A. Gomez-Exposito, A. J. Conejo, and C. Canizares, *Electric Energy Systems: Analysis and Operation*. Boca Raton, FL: CRC Press, 2018.

[5] A. Vaccaro and C. A. Canizares, "A knowledge-based framework for power flow and optimal power flow analyses," *IEEE Transactions on Smart Grid*, vol. 9, no. 1, pp. 230–239, 2017.

---

[4] For PQ buses, neither V and θ are fixed; they are state variables that need to be estimated. For PV buses, V is fixed, and θ are state variables that need to be estimated.

[6] J. Duncan Glover and Mulukutla S. Sarma, *Power System Analysis and Design*. Boston Cengage, 2012.

[7] S. Chakrabarti, E. Kyriakides, T. Bi, D. Cai, and V. Terzija, "Measurements get together," *IEEE Power and Energy Magazine*, vol. 7, no. 1, pp. 41–49, 2009.

[8] L. Luo, H. Bei, J. Chen, G. Sheng, and X. Jiang, "Partial discharge detection and recognition in random matrix theory paradigm," *IEEE Access*, vol. PP, no. 99, pp. 1–1, 2016.

[9] L. Chu, R. C. Qiu, X. He, Z. Ling, and Y. Liu, "Massive streaming PMU data modeling and analytics in smart grid state evaluation based on multiple high-dimensional covariance tests," *IEEE Transactions on Big Data*, vol. 4, no. 1, pp. 2332–7790, 2018.

[10] W. Hou, Z. Ning, G. Lei, and Z. Xu, "Temporal, functional and spatial big data computing framework for large-scale smart grid," *IEEE Transactions on Emerging Topics in Computing*, vol. PP, no. 99, pp. 1–1, 2017.

[11] C. Tu, H. Xi, Z. Shuai, and J. Fei, "Big data issues in smart grid – A review," *Renewable and Sustainable Energy Reviews*, vol. 79, pp. 1099–1107, 2017.

[12] H. Shaker, H. Zareipour, and D. Wood, "A data-driven approach for estimating the power generation of invisible solar sites," *IEEE Transactions on Smart Grid*, vol. 7, no. 5, pp. 2466–2476, 2016.

[13] A. E. Motter, S. A. Myers, M. Anghel, and T. Nishikawa, "Spontaneous synchrony in power-grid networks," *Nature Physics*, vol. 9, no. 3, pp. 191–197, 2013.

[14] L. Wang, H. W. Li, and C. T. Wu, "Stability analysis of an integrated offshore wind and seashore wave farm fed to a power grid using a unified power flow controller," *IEEE Transactions on Power Systems*, vol. 28, no. 3, pp. 2211–2221, 2013.

[15] X. Xu, X. He, Q. Ai, and R. C. Qiu, "A correlation analysis method for power systems based on random matrix theory," *IEEE Transcations on Smart Grids*, vol. 8, no. 4, pp. 1811–1820, 2017.

[16] X. He, Q. Ai, C. Qiu, W. Huang, L. Piao, and H. Liu, "A big data architecture design for smart grids based on random matrix theory," *IEEE Transactions on Smart Grid*, vol. 8, no. 2, pp. 674–686, 2017.

[17] K. Marvel and U. Agvaanluvsan, "Random matrix theory models of electric grid topology," *Physica A: Statistical Mechanics and Its Applications*, vol. 389, no. 24, pp. 5838–5851, 2010.

[18] R. Qiu, X. He, L. Chu, and A. Qian, "Big data analysis of power grid from random matrix theory," "Smarter Energy: From Smart Metering to the Smart Grid, "H. Vincent Poor et al., eds.," Oxford: Wiley, pp. 381–426, 2018.

[19] R. Qiu, L. Chu, X. He, Z. Ling, and H. Liu, "Spatio-temporal big data analysis for smart grids based on random matrix theory," Transportation and Power Grid in Smart Cities: Communication Networks and Services," "Hussein T. Mouftah et al., eds.," Oxford: Wiley, 591–634 2018.

[20] Z. Ling, R. C. Qiu, X. He, and L. Chu, "A novel approach for big data analytics in future grids based on free probability," *ArXiv e-prints*, no. 1612.01089, 2017.

[21] X. He, R. C. Qiu, A. Qian, C. Lei, X. Xu, and Z. Ling, "Designing for situation awareness of future power grids: An indicator system based on linear eigenvalue statistics of large random matrices," *IEEE Access*, vol. 4, pp. 3557–3568, 2017.

[22] X. He, L. Chu, and R. Qiu, "A novel data driven situation awareness approach for future grids: Using large random matrices for big data modeling," *IEEE Access*, vol. 6, no. 1, pp. 13 855–13 865, 2018.

[23] F. Wen, L. Chu, P. Liu and R. Qiu, "A survey on nonconvex regularization based sparse and low-rank recovery in signal processing, statistics, and machine learning," *IEEE Access*, vol. 6, no. 1, pp. 69883–69906, 2018.

[24] F. Xiao and Q. Ai, "Electricity theft detection in smart grid using random matrix theory," *IET Generation Transmission and Distribution*, vol. 12, no. 2, pp. 371–378, 2018.

[25] A. Edelman and N. R. Rao, "Random matrix theory," *Acta Numerica*, vol. 14, no. 14, pp. 233–297, 2005.

[26] R. C. Qiu and P. Antonik, *Smart Grid and Big Data: Theory and Practice*. Hoboken, NJ: John Wiley and Sons, 2015.

[27] V. A. Marchenko and L. A. Pastur, "Distribution of eigenvalues for some sets of random matrices," *Mathematics of the USSR-Sbornik*, vol. 1, no. 1, pp. 507–536, 1967.

[28] "Foundations and Trends in Communications and Information Theory," Random Matrix Theory and Wireless Communications, vol. 1, no. 1, pp. 1–182, 2004

[29] R. Speicher, "Free probability theory," *Jahresbericht der Deutschen Mathematiker-Vereinigung*, vol. 263, no. 10, pp. 1–28, 2009.

[30] T. Tao, V. Vu, and M. Krishnapur, "Random matrices: Universality of ESDS and the circular law," *Annals of Probability*, vol. 38, no. 5, pp. 2023–2065, 2010.

[31] V. A. Marčenko and L. A. Pastur, "Distribution of eigenvalues for some sets of random matrices," *Sbornik: Mathematics*, vol. 1, no. 4, pp. 457–483, 1967.

[32] A. Lytova, L. Pastur et al., "Central limit theorem for linear eigenvalue statistics of random matrices with independent entries," *The Annals of Probability*, vol. 37, no. 5, pp. 1778–1840, 2009.

[33] M. Shcherbina, "Central limit theorem for linear eigenvalue statistics of the Wigner and sample covariance random matrices," *ArXiv e-prints*, no. 1101.3249, Jan. 2011.

[34] D. Siegmund, "Change-points: From sequential detection to biology and back," *Sequential Analysis*, vol. 32, no. 1, pp. 2–14, 2013.

[35] J. Yeo and G. Papanicolaou, "Random matrix approach to estimation of high-dimensional factor models," *ArXiv*, no. 1611.05571, 2016.

[36] Z. Burda, A. Jarosz, M. A. Nowak, and M. Snarska, "A random matrix approach to VARMA processes," *New Journal of Physics*, vol. 12, no. 7, 075036, 2010.

[37] J. H. Stock and M. W. Watson, "Forecasting using principal components from a large number of predictors," *Journal of the American Statistical Association*, vol. 97, no. 460, pp. 1167–1179, 2002.

[38] Z. Burda, A. Jarosz, M. A. Nowak, and M. Snarska, "A random matrix approach to VARMA processes," *New Journal of Physics*, vol. 12, no. 7, p. 075036, 2010.

[39] X. Shi, R. Qiu, X. He, L. Chu, and Z. Ling, "Incipient fault detection and location in distribution networks: A data-driven approach," *ArXiv e-prints*, 2018. http://arxiv.org/pdf/1801.01669.pdf

[40] Y. Parag and B. K. Sovacool, "Electricity market design for the prosumer era," *Nature Energy*, vol. 1, p. 16032, 2016.

[41] X. He, R. C. Qiu, Q. Ai, L. Chu, X. Xu, and Z. Ling, "Designing for situation awareness of future power grids: An indicator system based on linear eigenvalue statistics of large random matrices," *IEEE Access*, vol. 4, pp. 3557–3568, 2016.

[42] X. Zhang and S. Grijalva, "A data-driven approach for detection and estimation of residential pv installations," *IEEE Transactions on Smart Grid*, vol. 7, no. 5, pp. 2477–2485, Sept, 2016.

8 Graph-Theoretic Analysis of Power Grid Robustness

Dorcas Ofori-Boateng, Asim Kumer Dey, Yulia R. Gel, and H. Vincent Poor

8.1 Introduction

The past decade has seen increasing interest in the application of tools developed in the interdisciplinary field of complex network (CN) analysis to improve our understanding of power system behavior. A power grid can be naturally described as a graph where nodes represent units such as transformers, substations, or generators, and edges represent the transmission lines. Numerous CN analytical methods have provided new insights into the fundamental and intrinsic qualities of power system efficiency, vulnerability, and resilience. While there exists no uniquely adopted definition, power grid *resilience* is conventionally understood as the quantification of the ability of the grid network to maintain its functions under component failures from random errors or external causes.

There are generally two main approaches for the study of power system resilience and/or robustness using CN tools. The first approach is based purely on the topological properties of a grid network, and the second approach (a hybrid) aims to incorporate electrical engineering (EE) concepts, such as impedance or blackout size [1], into the CN analysis. Both approaches provide important complementary insights about the hidden mechanisms behind the functionality of power systems, and neither approach can be viewed as a universally preferred method (see [2–7] for a detailed overview).

The most widely explored characteristics of power grid resilience in the CN context are node degree distribution, mean degree, small world properties, and, to a lesser extent, the betweenness centrality metrics – that is, primarily lower-order connectivity features that are investigated at the level of individual nodes and edges [2, 3, 6].

However, a number of recent studies of power system reliability indices and stability estimation suggest that the robustness of power grids is also intrinsically connected to higher-order network features such as network *motifs* [8–11]. Here a motif is defined as a recurrent multinode subgraph pattern such as, e.g., a triangle, star, square, or wheel.

The analysis of the local properties of power grids and their impact on grid functionality becomes even more challenging if we incorporate EE concepts into the grid network structure, especially at a local level. By doing so, the power grid is now viewed as a flow-based network, where edges are weighted, e.g., in terms of the line importance in a given operating state or strength of connections, and nodes are distinguished as, e.g., generators, load buses, and transformers. Recently [12] showed that the local structure of a power grid and its innate underlying geometry play a critical role in mitigating risks due to blackouts and cascading events. Another extension involving the local structure of power flow networks is the recent results on the motif analysis of unweighted power grids and the incorporation of weighted grid networks through the emerging and powerful methodology of *topological data analysis* (TDA) and, in particular, *persistent homology* [13, 14].

While a number of robustness metrics for power grid networks have been developed, the relation between all these robustness metrics, as well as the consistency of the metrics under different failure scenarios remain largely unexplored. In the rest of this chapter, we explore the (in)consistency among power grid robustness metrics under diverse attack strategies. We also evaluate the outcome of the different failure scenarios for all robustness metrics.

8.2 Power Grid Network and Its Robustness

Inherently, grid systems can be viewed as graph structures, with the electrical connections (transmission lines) as edges and the transformers/generators as nodes. We consider a graph $G = (V, E)$ as a model for the grid network, with node set V and set of edges $E \subset V \times V$. Ideally, $e_{uv} \in E$ represents an edge, i.e., a transmission line between nodes u and v, and $v \in V$ represents the node, e.g., generators, transformers, and load buses. We assume that G is *undirected* i.e., for all $e_{uv} \in E$, $e_{uv} \equiv e_{vu}$.

The total number of nodes in G is $n = |V|$ and the number of edges in G is $m = |E|$. Because the topological structure of G does not reflect the functional information about the power grid [3, 4], we also consider an *(edge)-weighted* graph, or a pair (G, ω). Here $\omega : V \times V \mapsto \mathbb{R}_{\geq 0}$ is an *(edge) weight* function such that each edge $e_{uv} \in E$ has a weight ω_{uv}. For technical purposes, we assume $e_{uu} \in E$ and $\omega_{uu} = 0$ for all $u \in V$. Examples of weight functions for power grid networks include, e.g., geographical distances, admittance, average power flow, and electrical conductance [4, 15, 16]. The case of $\omega_{uv} \equiv 1$ for $e_{uv} \in E$ characterizes connectivity of a power grid and corresponds to an unweighted graph G.

Power grid networks are imperiled by many factors, e.g., extreme weather, earthquake, human errors, and terrorist attacks. A variety of metrics have been proposed to assess the robustness of power grids under diverse threats (see, for instance, [17–21]). The *robustness* of a power grid can be defined as the degree to which the network is able to resist a disruptive event and maintain its functions. Power grid *vulnerability*, opposite to the concept of robustness, quantifies how much damage occurs as a consequence of an unexpected perturbation. *Resilience* refers to the ability of a power grid to sustain its functionality after a disruptive event and to modify grid functions and structure to limit the impact of similar hazards in the future. However, resilience is not uniquely defined, and nowadays resilience and robustness are often used interchangeably [22–24], whereas robustness concepts tend to be more rooted in statistical literature. A detailed description of the interdependency of the robustness related terms can be found in [3, 6, 25, 26]. Armed with CN techniques, we explore the emergence of structural robustness and/or resilience of power grid networks.

8.3 Failure Variations

The failure of power systems can be classified into two broad events: *random* and *targeted* [27–30]. Random failures are the consequence of natural events, e.g., hurricane and floods, whereas targeted failures (sometimes referred to as *intentional attacks*) result from intelligent intrusions (i.e., cyberattacks). The nodes (edges) within a random failure are arbitrarily deleted until some predefined condition occurs (like the deletion of a proportion of nodes/edges that fall within a specific climate region), while the nodes (edges) in a targeted failure are removed in decreasing order of importance of a predefined statistic (e.g., degree or betweenness). Following the mentioned failure types, we explore the selection strategies for specific node (edge) removal.

In line with benchmark processes, the node-based attacks will be based on the highest node degree, betweenness centrality, and other quantifiers of small world properties [10, 11, 13, 31–33]. For $a, u, v \in V(G)$ and $e \in E(G)$,

$$ND(v) = \sum_{u \neq v} e_{u,v}$$

refers to the *degree of v*;

$$NS(v) = \sum_{u \neq v} \omega_{u,v}$$

refers to the *strength of v*. Finally,

$$NB(v) = \sum_{a \neq u} \frac{\theta_{a,u}(v)}{\theta_{a,u}}$$

is the *betweenness centrality of v*, where $\theta_{a,u}(v)$ is the cardinality of the shortest paths between a and u through v and $\theta_{a,u}$ is the cardinality of shortest paths between a to u.

Furthermore, the edge-based attacks will be fashioned according to the highest edge betweenness centrality [34–36] and the "raw" edge weights (WH). For $e \in E(G)$,

$$EB(e) = \sum_{a \neq u} \frac{\theta_{a,u}(e)}{\theta_{a,u}}$$

is the *betweenness centrality of e*, where $\theta_{a,u}(e)$ is the cardinality of the shortest paths between a and u that go through e and $\theta_{a,u}$ refers to the cardinality of the shortest paths between a and u.

8.4 Robustness Metrics

Metrics for robustness analysis in power grid systems can be broadly classified into two groups. The first group consists of topology-based performance metrics that quantify grid robustness solely on the premise of network topologies [37]. The second group (flow-based robustness metrics) rates the power grid robustness with a consideration for particle flow over grids [38]. Based on the complex network dynamics of the power grid, we focus on topology-based metrics and extended topological metrics that characterize the local and global geometry of the network.

8.4.1 Conventional Robustness Metrics under Failures

The common topological-based robustness metrics of power grid networks include degree distribution, average path length (APL), diameter (D), clustering coefficient (CC), giant component (GCC), etc. These metrics generally measure stability or the lack of it in the grid network. A network with higher mean degree is considered better connected and is, consequently, likely to be more robust [3, 30, 39]. From small world properties, lower APL and higher CC have been connected with higher robustness [2, 40]. Furthermore, the *percolation limit* states that when the fraction of removed nodes reach a critical limit f_c, the network fragments into many isolated subgraphs. When the nodes are removed randomly, the percolation limit can be defined as $f_c = 1 - (1/\langle k^2 \rangle/\langle k \rangle - 1)$, where, $\langle k \rangle$ and $\langle k^2 \rangle$ are the mean and second moment of the degree distribution [41].

In [28] and [42], the cumulated degree distribution of European grid networks is theorized to follow the exponential distribution. This implied that the probability that a randomly chosen node has degree k or higher follows: $P(K \geq k) = C \cdot \exp(-k/\gamma)$, where C is a normalization constant, k is the node degree, and γ is a characteristic parameter. From this setup, a power grid is considered *robust* if $\gamma < 1.5$ and *fragile* if $\gamma > 1.5$.

The previously mentioned metrics are regularly used to measure the robustness of a networks in a *static* scenario. In robustness under failures, the goal is to evaluate how a power grid network robustness metric behaves when a fraction of nodes (edges) are removed in a simultaneous failure strategy. Two global features – *giant component* (GCC) and *connectivity loss* (LC) have proved relevant to such robustness analysis under different failure strategies [43, 44]. The former relates to the normalized size of the largest connected component [45] and is formally defined as follows:

$$\Delta S' = \frac{S'}{S^0},$$

where S^0 is the largest connected component of the network prior to the failure and S' is the largest connected component of the network after the failure. Ideally, a power grid is more robust at the stages where $\Delta S'$ is largest.

The connectivity loss (LC) measures the average decrease in the number of connected nodes as

$$LC = 1 - \frac{N'}{N},$$

where, N' is the number of connected nodes after the failure. Sometimes we use the measure $1 - LC$, which is the ratio of the number of connected nodes after failure to the number of connected nodes before failure, i.e., $1 - LC = N'/N$.

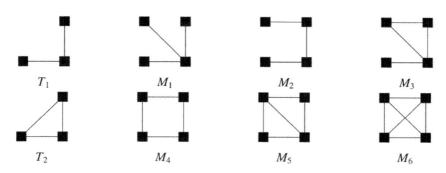

Figure 8.1 3-node and 4-node connected motifs.

8.4.2 Network Motifs as a Local Robustness Measure

The frequently used power grid robustness metrics, described in previous section, mainly explore global topological characteristics, that is, lower-order connectivity features that are investigated at the level of individual nodes and edges. However, recent studies have shown that the robustness of the power grid is also intrinsically connected to higher-order network features such as network *motifs* [8, 9, 46]. A *motif* is broadly defined as a recurrent multinode subgraph pattern, such that a motif $G' = (V', E')$ is an n'-node subgraph of G where $|V'| = n'$. Figure 8.1 shows all 3-node and 4-node connected motifs for an undirected graph.

The existence of motifs in power grid networks is not by chance or random, and such components tend to play relevant connectivity roles. For instance, [5] and [28] found that power system vulnerability tends to increase as the number of {3,4}-node subgraph patterns such as stars and treelike structures increase. In turn, [9] suggests that changing a motif that contains a treelike structure to a cycle can improve grid stability. The dynamics of motifs, in recent decades, have been measured according to their *concentration* in the network [11]. The concentration of an n-node motif type i is the ratio between its number of appearances (N_i) and total number of n-node motifs in the network G:

$$C_i = \frac{N_i}{\sum_j N_j},$$

where $\sum_j N_j$ is the total number of n-node motifs, $i, j = 1, \ldots, m_n$, where m_n is the number of distinct n-node motifs. We focus on a 4-node motif-based performance measure under different failures. More specifically, our goal is to analyze the dynamics of a specific motif concentration under particular type of failure. Furthermore, we evaluate the *reliability* functions of the motifs M_k, $k = 1, 2, \ldots, 6$ under a given attack strategy and combine them to obtain a (joint) measure of the entire network's reliability/survival function.

Under a targeted attack, let A_k be the event that a motif M_k, $k = 1, 2, \ldots, 6$ survives till *time t*, where t refers to the number of attacks. Additionally, the survival/reliability function of M_k written as $MR_k(t) = P_r(A_k) = P_r(L_k > t)$, where L_k is a nonnegative random variable representing the lifetimes, i.e., waiting time until the *death* of individual motifs in M_k. The $MR_k(t)$ can be modeled as an exponential model such that

$$MR_k(t) = \exp\left[-\lambda_k t\right],$$

where $\lambda_k > 0$ [47]. We combine the individual motif reliabilities under the *system-components* framework to obtain the reliability of the entire network, by considering the entire network as a *parallel system* and motifs as its *components* such that

$$MR_s(t) = 1 - \left[\left(\prod_{k=1}^{6} P_r(A_k^c)\right) \min\left\{P_r(A_1^c), \ldots, P_r(A_6^c)\right\}\right]^{\frac{1}{2}},$$

where L_s is is the lifetime of the entire network, $P_r(A_k^c) = 1 - MR_k(t), k = 1, 2, \ldots, 6$ [48–50]. The number of motif deaths at time t can be counted as $D_k = N_k(t) - N_k(t-1)$. The lifetimes of D_k motifs are then considered to be t, i.e., $L_k = t$. After determining lifetimes L_k, we fit the reliability function, $MR_k(t)$ of each M_k motif. Finally, we evaluate reliability of the entire network by combining the $MR_k(t)$ for all motif types.

8.4.3 Persistent Homology as a Local Robustness Measure

While network motifs have a high potential to quantify the geometry of a power grid network and its role in robustness, the motif-based analysis is mainly restricted to unweighted graphs and, therefore, does not allow for straightforward incorporation of relevant power flow characteristics such as series impedance, shunt admittance, and apparent power (Mega Volt Amp, MVA) limits. As an alternative, we study the geometry and topology of weighted power grid networks with the tools of topological data analysis (TDA), and in particular, Betti numbers and persistence diagrams (see [51–58] for a comprehensive overview on TDA).

Consider an (edge)-weighted graph (G, ω) as a representation of a power grid network. For a certain threshold (or scale) $v_j > 0$, we only keep edges with weights $\omega_{uv} \leq v_j$, therefore, we obtain a graph G_j that has the adjacency matrix $A_{uv} = \mathbb{1}_{\omega_{uv} \leq v_j}$. By changing the threshold values $v_1 < v_2 < \cdots < v_n$, we obtain a hierarchical nested sequence of graphs $G_1 \subseteq G_2 \subseteq \cdots \subseteq G_n$, which is called a *network filtration*.

A commonly used filtration is the Vietoris–Rips (VR) complex in which weights ω_{uv} are viewed as some "distance" or similarity measure between nodes u and v. For $v_j > 0$ the VR complex is defined as $VR_j = \{\sigma \subset V | \omega_{uv} \leq v_j \text{ for all } u, v \in \sigma\}$, i.e., VR_j contains all the k-node subsets of $G_j, k = 1, \ldots, K$, which are pairwise connected by an edge as simplices of dimension $k - 1$. A node is a 0-simplex, an edge is a 1-simplex, a triangle is a 2-simplex, etc. In each filtration of the VR complex, $(VR_1 \subseteq VR_2 \subseteq \cdots \subseteq VR_n)$, we detect topological summaries called *Betti numbers*, $\beta_p, p \in \mathbb{Z}^+$. For a given simplicial complex, Betti-0 (β_0) reports the number of connected components, Betti-1 (β_1) is the number of one-dimensional holes, Betti-2 (β_2) reports the number of two-dimensional holes, and so on [57–59]. Figure 8.2 shows the Betti numbers of a point, circle, and torus, where the point has a single connected component; the circle

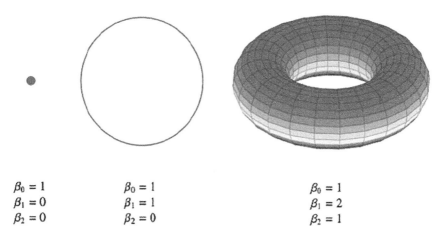

$\beta_0 = 1$
$\beta_1 = 0$
$\beta_2 = 0$

$\beta_0 = 1$
$\beta_1 = 1$
$\beta_2 = 0$

$\beta_0 = 1$
$\beta_1 = 2$
$\beta_2 = 1$

Figure 8.2 Betti number estimation for a point (left), circle (center), and torus (right).

has a single connected component and one-dimensional hole (loop); and the torus has one connected component, two loops, and a single two-dimensional hole.

Figure 8.3 shows the construction outline of a VR complex for a data cloud with 100 points. Notice that some holes that appear in a lower threshold/radius ($v_j = 0.10$) disappear in larger thresholds ($v_j = 0.15, 0.20$). In summary, the objective is to detect features that are long-lived (or *persistent*) over varying thresholds v_j. Such persistent features likely characterize the structural functionality of the power grid network.

Over recent decades, a number of methods have been developed to identify persistent topological features (e.g., persistence barcodes and persistence diagrams). Persistence barcodes capture the birth and death times of each topological feature as a bar. In the persistence diagram, each topological feature is represented by a point in the Cartesian (x, y)-coordinate system, where the x-coordinate represents the birth time and y-coordinate represents the death time in filtration. Features with a longer life span, i.e., a stronger persistence, are those points that are far from the main diagonal [60] of the plane.

As described in Section 8.3, we intend to analyze the evolution of these topological summaries, e.g., Betti numbers β_p, $p \in \mathbb{Z}^+$ under particular type of failure(s). In particular, we calculate the (normalized) change in the Betti numbers as the grid network is attacked as

$$\Delta \beta_k = \frac{\|\vec{\beta}_k^0 - \vec{\beta}_k^p\|_2}{\|\vec{\beta}_k^0\|_2},$$

where $k \in \{0, 1\}$, $\|.\|_2$ denotes the Euclidean distance, p is the fraction of removed nodes, $\vec{\beta}_k^0 = \{\beta_{k1}^0, \ldots, \beta_{kn}^0\}$ and $\vec{\beta}_k^p = \{\beta_{k1}^p, \ldots, \beta_{kn}^p\}$ are the sequences of Betti numbers of a power grid network before and after the failure, respectively [13].

Additionally, we can compare two power grid networks (G_0, ω_0) and (G_p, ω_p), which refer to the before and after failure(s) grid networks, respectively, by comparing their

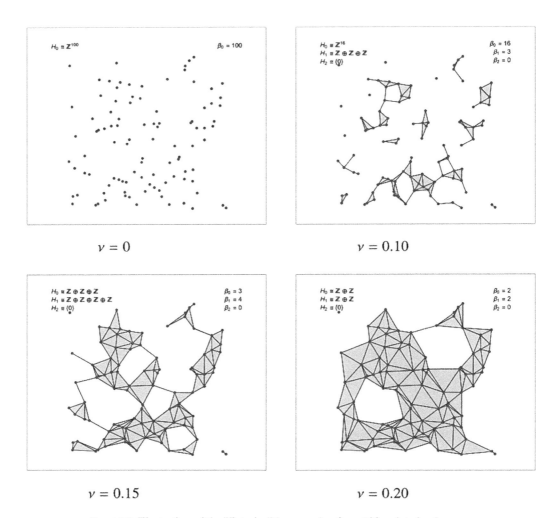

Figure 8.3 Illustration of the Vietoris–Rips complex for a 100-point cloud.

associated persistence diagrams D_0 and D_p, respectively [52, 53]. A commonly used distance measure is the r-th Wasserstein distance, which can be defined as

$$W_r(D_1, D_2) = \inf_{\eta} \left(\sum_{x \in D_1} \|x - \eta(x)\|_\infty^r \right)^{1/r},$$

where η ranges over all bijections between D_0 and D_p, $\|z\|_\infty = \max_i |z_i|$ and $r \geq 1$ [55].

Broadly, $\Delta \beta_k$ and $W_r(D_0, D_p)$ quantify the change in local geometry of the network under failure(s). However, we will use $1 - \Delta \beta_k$-metric, $k \in \{0, 1\}$ and $1 - W_2(D_0, D_\prime)$-metric as the metrics that quantify structural network robustness.

Algorithm 8.1 provides a pseudocode for calculating the different robustness metrics under a degree centrality-based attack. This framework can easily be adapted to the other types of failure strategies we described in Section 8.3.

Algorithm 8.1 Degree of centrality-based attack tolerance of a power grid network

Input : Power grid $G = (V, E, \omega)$.
Output: Network performance metrics under failure.

1. S^0: largest connected component of the network before the failure
2. $\vec{\beta}_k^{or} = \{\beta_{k1}^{or}, \ldots, \beta_{kn}^{or}\}, k \in \{0, 1\}$ - Betti numbers from VR filtration before the failure
3. D_0 : persistence diagram from VR filtration before the failure
4. O_v : degree centrality of node v. Calculate $O_v, \forall v \in V$
5. $H(G) \leftarrow$ sorted V by O_v (descending)
6. **for** $t = 1$ **to** $|H(G)|$ **do**
7. $\quad V = V \setminus H(t)$
   $\quad E = E \setminus \{(x, y) \in E : x = H(t) \text{ or } y = H(t)\}$
8. $\quad$ Calculate the connectivity loss $LC[t]$
9. $\quad$ Calculate largest connected component, S' and then $\Delta S'[t] = S'/S^0$
10. $\quad$ Count motifs, N_i for $i = 1, \ldots, m_n$, their concentration, $C_i[t] = N_i / \sum_j N_j$ and then their reliability $MR_k[t]$
11. $\quad$ Conduct VR complex filtration and calculate Betti numbers, $\vec{\beta}_k = \{\beta_{k1}, \ldots, \beta_{kn}\}$
12. $\quad$ Calculate $\Delta \beta_k[t] = \|\vec{\beta}_k^{or} - \vec{\beta}_k\|_2 / \|\vec{\beta}_k^{or}\|_2$
13. $\quad$ Calculate persistence diagram D_p from VR complex filtration
14. $\quad$ Calculate Wasserstein distance, $W_2[t] = W_2(D_0, D_p)$
15. **end**;

8.4.4 Statistical Inference for Local Higher-Order Topological Features of a Network

Although higher-order networks structures such as motifs and persistent topological summaries appear to play a critical role in understanding the functionality of power grid networks, their metrics are mostly based on deterministic approaches and do not account for data variability and uncertainty estimation. Recently, a number of methods have been proposed for inference on network motif counts, particularly in biological networks [61–63]. In turn, statistical inference for topological features, e.g., Betti numbers and persistence diagrams, has attracted continuing attention (see, for instance, [64–67] and references therein).

In this section we evaluate the statistical significance of such higher-order local structures (i.e., motifs or Betti numbers) in power grid networks. We assume that the observed network data is drawn from a collection of networks, i.e., population, following some unknown probability distribution. In turn, our approach for the statistical inference of higher-order local features in a power grid network is based on simulations, where the observed appearance of motifs and Betti numbers in the power grid networks are compared to their corresponding quantity in a randomized network. Let N_{obs} be the observed number of a particular motif or Betti number in the power grid network and N_{rand} be its appearance in the randomized network. The significance of the higher-order structure is assessed by calculating the probability $P = Pr(N_{rand} \geq N_{obs})$. If the probability is less than a cutoff value, e.g., 0.05, we consider the motif or Betti number to be statistically significant in the power grid network and, hence, imply that the respective higher-order topological feature likely serves a specific function in the network.

The probability, P, can be determined by Z-score defined by

$$Z = \frac{N_{obs} - N_m}{s},$$

where N_m is the mean number of N_{rand} in B replicated randomized networks and s is the corresponding sample standard deviation.

Because the distributions of higher-order local structures in a network are highly skewed, the use of the standard Z- and t-quantiles will be inappropriate. Hence, we compare the observed motif (or Betti number) Z-scores with an estimation of the empirical p-value obtained from the parametric bootstrap of a randomized network as a reference.

8.5 Analysis of Real-World Power Grid Networks

In Section 8.4 we described a number of metrics that quantify global and local robustness of a power grid network. Moreover, multiple failure types have been defined in Section 8.3 according to the current trends in power grid robustness analysis. However, little is known on how the various robustness metrics respond to different failures and which metrics are more appropriate under a particular scenario. In the rest of this section, we evaluate robustness of two real-world power grids under different attack strategies and assess the sensitivity and consistency of the robustness metrics under the attacks.

8.5.1 Data Description

We assess the performance of the global and local robustness metrics on the power grid networks for two European countries (Romania and Poland). The data is obtained from the Union for the Coordination of the Transmission of Electricity, UCTE (www.ucte.org). The edge weight function for each network is defined as the inverse of electrical conductance [68].

Table 8.1 presents a few topological properties for both power grid networks: number of nodes (n), number of edges (m), average node degree $\langle k \rangle \pm$ standard deviation (SD), percolation limit (f_c), average shortest path length (APL), diameter (D), and expected degree distribution (γ).

The Polish power grid is larger and has higher $\langle k \rangle$, that is, on average the nodes of Polish power grid have a higher number of connections. Moreover, the Polish grid has higher APL and D. In turn, the Romanian power grid has higher f_c, and so the

Table 8.1 Network summaries for two real-world power systems.

Country	n	m	$\langle k \rangle \pm$ SD	f_c	APL	D	γ
Poland	162	212	2.617 ± 1.521	0.599	6.943	57.129	1.641
Romania	106	136	2.566 ± 1.543	0.600	5.521	36.142	1.418

Romanian power grid is likely to be more robust than Polish power grid under random failures. Base on the conventional metric that γ parameter is less than 1.5, the Romanian power grid network will be classified as robust ($\gamma = 1.418$) while the Polish power grid network will be classified as fragile, because γ is greater than 1.5 ($\gamma = 1.641$).

8.5.2 Statistical Significance of Motifs and Betti Numbers

We assess the significance of different motifs and the Betti-1 (β_1) for the Poland power grid network based on z-scores. Following the approach outlined in Section 8.4.4, we start from specifying an appropriate random graph model as a reference model for the power grid network. The commonly used random graph models for power grid networks include small-world model, configuration model, preferential attachment (PA) model, Erdos–Renyi model [69, 70]. In our study, we simulate the randomized network according to the following two models: (a) configuration model, that is, a random graph with the same degree sequence as the observed power grid, and (b) Erdos–Renyi model [71]. Each graph is generated to have the same number of nodes and nearly the same number of edges.

Table 8.2 summarizes the Z-scores for the different higher-order local structures that appear in Poland power grid network. The results show that for both configuration model and Erdos–Renyi model, the p-value of T_2, M_3, M_4, and M_5 are less than 0.05, and hence these topological features are significant in the Poland power grid network. That is, the appearance of T_2, M_3, M_4, and M_5 motifs in Poland grid is significantly higher than that of the random graph models. In contrast, motifs T_1, M_1, M_3, and β_1 are not statistically significant. We also observe similar patterns on statistical significance of the motifs T_2, M_3, M_4, and M_5 in the Romanian power grid.

Remarkably, all significant motifs in the two power grid networks are of a detour type, i.e., contain one more *cycle* [9]. At the same time, treelike *dead end* motifs (i.e., T_1, M_1, M_3) appear insignificant. Hence, these findings echo the deterministic approaches on motif analysis by [5, 9], and we can conclude that there likely exists some nonlinear relationship between the detour motifs and the power grid network functionality and stability, even after accounting for uncertainty in the data and estimation processes.

Table 8.2 Significance of 3-node motifs (T_1 and T_2), 4-node motifs (V_1, V_2, ..., V_5) and β_1 in the Poland power grid network. N_{rand}, s and P-value are based on 1,000 simulated networks.

	N_{obs}	Configuration Model			Erdos–Renyi Model		
		$N_{rand}(s)$	Z	p-value	$N_{rand}(s)$	Z	p-value
T_1	491	522 (4.605)	−6.66	1.000	539 (22.064)	−2.191	0.989
T_2	9	2 (1.542)	4.27*	0.000	3 (1.702)	3.57*	0.000
M_1	415	495 (13.772)	−5.84	1.000	443(58.104)	−0.49	0.665
M_2	978	1213 (49.768)	−4.73	1.000	1335 (109.073)	−3.27	1.000
M_3	72	20 (13.354)	3.88*	0.000	22 (13.605)	3.70*	0.002
M_4	11	4 (2.118)	3.23*	0.005	6 (2.462)	2.26*	0.013
M_5	2	0.14 (0.434)	4.35*	0.005	0.19 (0.465)	3.9*	0.004
β_1	41	49 (1.472)	−5.14	1.000	61 (3.193)	−6.21	1.000

* Indicates the significance with 0.05 level of significance.

8.5.3 Comparison of Robustness Metrics under Failure Scenarios

We now turn to investigate the robustness of the two European grid networks based on six metrics: giant component ($\Delta S'$), survival probability of the network (MR), ratio of the remaining connected node ($1-LC$), change in Betti numbers ($1-\Delta\beta_0$, $1-\Delta\beta_1$), and change in the Wasserstein distance ($1 - \Delta W_2(D_0, D_p)$). For each measure, we consider attack strategies based on the following network summaries: node degree (ND), node betweenness centrality (NB), node strength (NS), edge betweenness centrality (EB), and edge weights (WH).

Figures 8.4 and A.1 in the Appendix present the six robustness metrics for the Romanian and iPolish power grid networks under different failure scenarios. In each figure, the x- and y-axes represent the fraction of nodes (edges) removed and a specific robustness metric, respectively.

For all six metrics, we observe that the node-based attack strategies (ND, NB, NS) appear to have a higher impact on the robustness of the grid networks than their edge-based counterparts (EB, WH). In particular, the curves for each robustness metric decay faster for node-based attacks as opposed to what happens under the edge-based attacks.

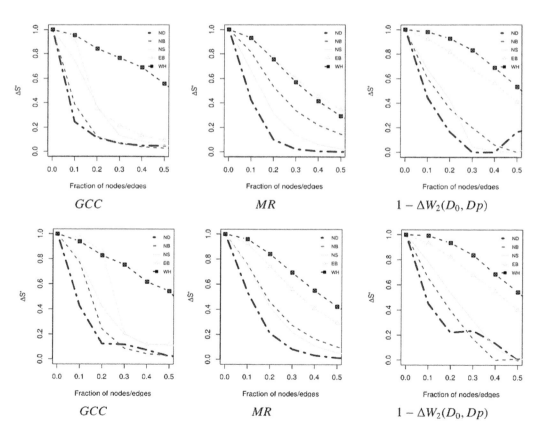

Figure 8.4 Dynamics of robustness metrics under diverse attacks strategies, Romania (first row) and Poland (second row) power grids.

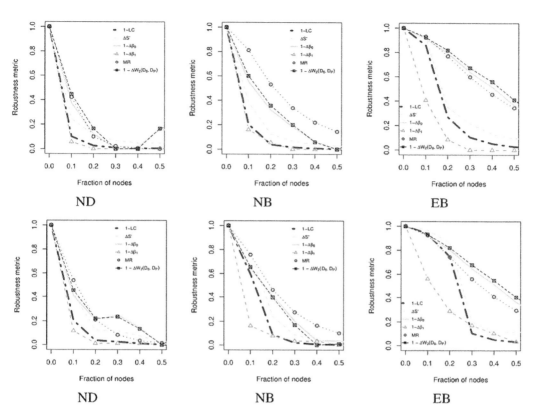

Figure 8.5 Dynamics of attack strategies under diverse robustness metrics, Romania (first row) and Polandro (second row) power grids.

Among the node-based failures, we find that the two power grids are most fragile under ND and least fragile under NS. In turn, the power grids appear to be more robust under the edge weight-based attack (WH), that is, the WH-based attack has least influence on the robustness of the power grid networks.

Moreover, the ranking of the impact from all five attack strategies (ND, NB, NS, EB, WH) are consistent in all six robustness metrics. That is, we find that attack strategy based on node degree (ND) tends to be the most influential form of the targeted attacks and that the edge weight-based attack (WH) is the least influential one. Therefore, we suggest that grid operators consider security measures around and about energy distribution resources (nodes), e.g., power stations and transformers.

To investigate how the robustness metrics behave under particular attack strategies, we study both Figures 8.5 and A.2 (see the Appendix). Among all six robustness metrics, we analyze the dynamics of the local (MR, $1 - \Delta\beta_0$, $1 - \Delta\beta_1$, $1 - \Delta W_2(D_0, D_p)$) and global ($1\text{-}LC, \Delta S'$) robustness metrics under each failure strategy. In each figure, the x- and y-axes represent the fraction of nodes (edges) removed and the robustness metrics, respectively.

In general, we find that the resilience metrics based on the local geometry (with the exception of the $1 - \Delta\beta_1$-metric) of the grid networks tend to be more robust under the five attack strategies. Typically, as the fraction of removed nodes (edges) increases, the curves for the local robustness metric decay steadily. Furthermore, we find that the global geometry of the two power grids appear to be more sensitive under all attack strategies. In particular, we observe that the global robustness metric ($1\text{-}LC, \Delta S'$) curves decay faster and lie below their local counterparts ($MR, 1 - \Delta\beta_0, 1 - \Delta W_2(D_0, D_p)$) under all failure scenarios. These findings on the interplay of local and global power grid network properties suggest that the power grids appear to be highly heterogeneous and global robustness metrics are more sensitive to even moderate system perturbations, thereby, making it more challenging to differentiate multiple system resilience under uncertainties. In turn, analysis of grid resilience and robustness at a local level, using motifs and TDA, enables to segment parts of the grids that are more or less vulnerable to a failure and, as a result, to derive optimal islanding strategies and more efficient micro-grid design. Furthermore, systematic assessment of significant higher-order topological features, e.g., detours and cycles, can be incorporated into design of more resilient new power systems and expansion of the existing grid networks.

8.6 Conclusion

Over recent decades, a considerable number of metrics have been modeled toward assessing the robustness of power systems [10, 11, 13, 21, 30]. However, the joint analysis of these metrics and their systematic response to different failure scenarios remains largely underinvestigated. In this chapter, we have examined the vulnerability of two power grids using six robustness metrics ($\Delta S', MR, 1\text{-}LC, 1 - \Delta\beta_0, 1 - \Delta\beta_1$ and $1 - \Delta W_2(D_0, D_p)$) and five attack strategies (ND, NB, NS, EB, WH). Furthermore, we have analyzed the sensitivity of the targeted power systems under global ($\Delta S', 1 - LC$) and local ($MR, 1 - \Delta\beta_0, 1 - \Delta\beta_1$ and $1 - \Delta W_2(D_0, D_p)$) robustness metrics.

Our findings suggest that the power grid networks appear to exhibit the highest fragility with node-based attacks. As expected, the system topology of the power grids remains more robust under the edge-based attacks. Moreover, we have found that the local geometry of the two power grids appears to be more resilient to all considered node and edge attack strategies, as opposed to the global geometry of power grid networks.

In addition, we have studied the statistical significance of different motifs in power grid networks, based on two benchmark models, namely, the configuration model and Erdos–Renyi model. We have observed that the occurrence of higher connected motifs that contain cycles are highly significant in power grid networks. Hence, while generating synthetic power grid networks, it is necessary to validate how well the synthetic power systems match the real power grids at a local level. Furthermore, to enhance the robustness of new power systems to hazardous scenarios, resilience analysis based on local network topological properties should be incorporated into the design principles of new power grids.

An open issue in this area is the advancement of the proposed robustness approaches based on local higher-order network characteristics to critical infrastructures and multilayer networks [72, 73]. This would allow one to simultaneously assess vulnerability of integrated energy resources such as solar, wind, and hydropower and the cascading effects of power failures on transportation, telecommunication, and other sectors, at both local and global levels.

Of further interest is the investigation of the relationship between local and global robustness metrics based on functional grid properties such as blackout size [1] and power flow analysis [74], relating system resilience at a local level with optimal islanding strategies and microgid design, and evaluating network robustness metrics in dynamic scenarios under uncertainty.

The authors would like to thank Cuneyt Akcora, Murat Kantarcioglu, Jie Zhang, and Binghui Li for assistance with the analysis of the data and stimulating discussion throughout the project.

Ofori-Boateng, Dey, and Gel were partially funded by NSF DMS 1736368, NSF ECCS 1824716, and NSF IIS 1633331. Poor was partially funded by NSF DMS 1736417 and NSF ECCS 1824710.

A Appendix Resilience analysis – additional results

A.1 Robustness metrics versus attack strategies: *Romania* & *Poland*

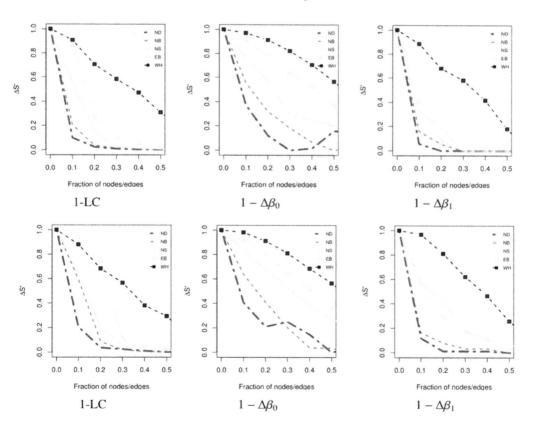

Figure A.1 Dynamics of robustness metrics under diverse attacks strategies, *Romania* (first row) and *Poland* (second row) power grids

A.2 Attack strategies versus robustness metrics: *Romania* & *Poland*

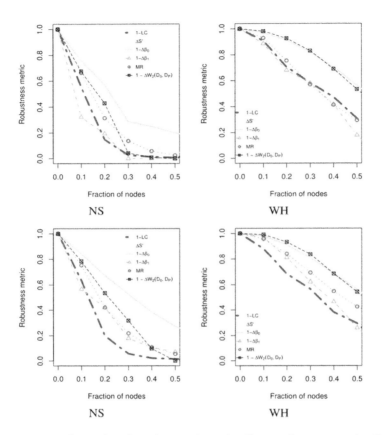

Figure A.2 Dynamics of attack strategies under diverse robustness metrics, *Romania* (first row) and *Poland* (second row) power grids

References

[1] Y. Zhu, J. Yan, Y. Tang, Y. Sun, and H. He, "Resilience analysis of power grids under the sequential attack," *IEEE Transactions on Information Forensics and Security*, vol. 9, no. 12, pp. 2340–2354, 2014.

[2] G. A. Pagani and M. Aiello, "The power grid as a complex network: A survey," *Physica A*, vol. 392, no. 11, pp. 2688–2700, 2013.

[3] L. Cuadra, S. Salcedo-Sanz, J. Del Ser, S. Jiménez-Fernández, and Z. W. Geem, "A critical review of robustness in power grids using complex networks concepts," *Energies*, vol. 8, no. 9, pp. 9211–9265, 2015.

[4] R. J. Sánchez-García, M. Fennelly, S. Norris, N. Wright, G. Niblo, J. Brodzki, and J. W. Bialek, "Hierarchical spectral clustering of power grids," *IEEE Transactions on Power Systems*, vol. 29, no. 5, pp. 2229–2237, 2014.

[5] M. Mureddu, G. Caldarelli, A. Damiano, A. Scala, and H. Meyer-Ortmanns, "Islanding the power grid on the transmission level: Less connections for more security," *Scientific Reports*, vol. 6, p. 34797, 2016.

[6] A. Abedi, L. Gaudard, and F. Romerio, "Review of major approaches to analyze vulnerability in power system," *Reliability Engineering & System Safety*, 2018.

[7] M. Rohden, D. Jung, S. Tamrakar, and S. Kettemann, "Cascading failures in AC electricity grids," *Physical Review E*, vol. 93, p. 032209, 2017.

[8] P. J. Menck, J. Heitzig, J. Kurths, and H. J. Schellnhuber, "How dead ends undermine power grid stability," *Nature Communications*, vol. 5, p. 3969, 2014.

[9] P. Schultz, J. Heitzig, and J. Kurths, "Detours around basin stability in power networks," *New Journal of Physics*, vol. 16, no. 12, p. 125001, 2014.

[10] A. K. Dey, Y. R. Gel, and H. V. Poor, "Motif-based analysis of power grid robustness under attacks," in *2017 IEEE Global Conference on Signal and Information Processing (GlobalSIP)*, pp. 1015–1019, 2017.

[11] ——, "What network motifs tell us about robustness and reliability of complex networks," *Proceedings of the National Academy of Sciences*, 2019.

[12] B. A. Carreras, D. E. Newman, and I. Dobson, "The impact of size and inhomogeneity on power transmission network complex system dynamics," in *Proceedings of the 47th Hawaii International Conference on System Sciences (HICSS), 2014 IEEE*, pp. 2527–2535, 2014.

[13] U. Islambekov, A. K. Dey, Y. R. Gel, and H. V. Poor, "Role of local geometry in robustness of rower grid networks," in *Global Conference on Signal and Information Processing (GlobalSIP), 2018 IEEE*, 2018.

[14] D. Ofori-Boateng, A. K. Dey, Y. R. Gel, B. Li, J. Zhang, and H. V. Poor, "Assessing the resilience of the Texas power grid network," DSW 2019, IEEE Data Science Workshop, 2019.

[15] B. Dörfler and F. Bullo, "Kron reduction of graphs with applications to electrical networks," *IEEE Transactions on Circuits and Systems*, vol. 60, no. 1, pp. 150–163, 2013.

[16] Y. Xu, A. Gurfinkel, and P. Rikvold, "Architecture of the Florida power grid as a complex network," *Physica A*, vol. 401, pp. 130–140, 2014.

[17] J. Wang and L. Rong, "Robustness of the Western United States power grid under edge attack strategies due to cascading failures," *Safety Science*, vol. 49, no. 6, pp. 807–812, 2011.

[18] M. Youssef, C. Scoglio, and S. Pahwa, "Robustness measure for power grids with respect to cascading failures," in *Proceedings of the 2011 International Workshop on Modeling, Analysis, and Control of Complex Networks,* International Teletraffic Congress, pp. 45–49, 2011.

[19] Y. Koç, M. Warnier, R. Kooij, and F. Brazier, "An entropy-based metric to quantify the robustness of power grids against cascading failures," *Safety Science*, vol. 59, pp. 126–134, 2013.

[20] H. Tu, Y. Xia, H. Iu, and X. Chen, "Optimal robustness in power grids from a network science perspective," *IEEE Transactions on Circuits and Systems II: Express Briefs*, vol. 66, no. 1, pp. 126–130, 2019.

[21] F. Hajbani, H. Seyedi, and K. Zare, "Evaluation of power system robustness in order to prevent cascading outages," *Turkish Journal of Electrical Engineering & Computer Sciences*, vol. 27, no. 1, pp. 258–273, 2019.

[22] National Academies of Sciences, Medicine et al., *Enhancing the Resilience of the Nation's Electricity System*. Washington, DC: National Academies Press, 2017.

[23] G. W. Klau and R. Weiskircher, "Robustness and resilience," in *Network Analysis*. Berlin and Heidelberg: Springer, pp. 417–437, 2005.

[24] A. Clark-Ginsberg, "Whats the difference between reliability and resilience?" *Stanford University, Tech. Rep*, 2016.

[25] M. Ouyang, Z. Pan, L. Hong, and L. Zhao, "Correlation analysis of different vulnerability metrics on power grids," *Physica A: Statistical Mechanics and Its Applications*, vol. 396, pp. 204–211, 2014.

[26] M. Panteli and P. Mancarella, "Influence of extreme weather and climate change on the resilience of power systems: Impacts and possible mitigation strategies," *Electric Power Systems Research*, vol. 127, no. 10, 2015.

[27] R. V. Solé, M. Rosas-Casals, B. Corominas-Murtra, and S. Valverde, "Robustness of the European power grids under intentional attack," *Physical Review E*, vol. 77, no. 2, p. 026102, 2008.

[28] M. Rosas-Casals and B. Corominas-Murtra, "Assessing European power grid reliability by means of topological measures," *WIT Transactions on Ecology and the Environment*, vol. 121, pp. 527–537, 2009.

[29] T. C. Gulcu, V. Chatziafratis, Y. Zhang, and O. Yağan, "Attack vulnerability of power systems under an equal load redistribution model," *IEEE/ACM Transactions on Networking*, vol. 26, no. 3, pp. 1306–1319, 2018.

[30] S. Wang, W. Lv, L. Zhao, S. Nie, and H. Stanley, "Structural and functional robustness of networked critical infrastructure systems under different failure scenarios," *Physica A: Statistical Mechanics and its Applications*, vol. 523, pp. 476–487, 2019.

[31] E. Bompard, R. Napoli, and F. Xue, "Analysis of structural vulnerabilities in power transmission grids," *International Journal of Critical Infrastructure Protection*, vol. 2, no. 1–2, pp. 5–12, 2009.

[32] Z. Kong and E. M. Yeh, "Resilience to degree-dependent and cascading node failures in random geometric networks," *IEEE Transactions on Information Theory*, vol. 56, no. 11, pp. 5533–5546, 2010.

[33] S. Sun, Y. Wu, Y. Ma, L. Wang, Z. Gao, and C. Xia, "Impact of degree heterogeneity on attack vulnerability of interdependent networks," *Scientific Reports*, vol. 6, p. 32983, 2016.

[34] M. Girvan and M. E. J. Newman, "Community structure in social and biological networks," *Proceedings of the National Academy of Sciences*, vol. 99, no. 12, pp. 7821–7826, 2002.

[35] M. Arasteh and S. Alizadeh, "A fast divisive community detection algorithm based on edge degree betweenness centrality," *Applied Intelligence*, vol. 49, no. 2, pp. 689–702, 2019.

[36] P. Kumari and A. Singh, "Approximation and updation of betweenness centrality in dynamic complex networks," in *Computational Intelligence: Theories, Applications and Future Directions – Volume I*. Basel: Springer, 2019, pp. 25–37.

[37] M. Rosas-Casals, S. Valverde, and R. V. Solé, "Topological vulnerability of the European power grid under errors and attacks," *International Journal of Bifurcation and Chaos*, vol. 17, no. 7, pp. 2465–2475, 2007.

[38] C. D. Nicholson, K. Barker, and J. E. Ramirez-Marquez, "Flow-based vulnerability measures for network component importance: Experimentation with preparedness planning," *Reliability Engineering and System Safety*, vol. 145, pp. 62–73, 2016.

[39] D. F. Rueda, E. Calle, and J. L. Marzo, "Robustness comparison of 15 real telecommunication networks: Structural and centrality measurements," *Journal of Network and Systems Management*, vol. 25, no. 2, pp. 269–289, Apr. 2017.

[40] C. J. Kim and O. B. Obah, "Vulnerability assessment of power grid using graph topological indices," *International Journal of Emerging Electric Power Systems*, vol. 8, no. 6, pp. 1–15, 2007.

[41] R. Albert and A. Barabási, "Statistical mechanics of complex networks," *Reviews of Modern Physics*, vol. 74, pp. 47–97, Jan. 2002.

[42] R. V. Sóle, M. Rosas-Casals, B. Corominas-Murtra, and S. Valverde, "Robustness of the European power grids under intentional attack," *Physical Review E*, vol. 77, p. 026102, 2008.

[43] F. Salem, M. Jaber, C. Abdallah, O. Mehio, and S. Najem, "A distributed spatiotemporal contingency analysis for the Lebanese power grid," *IEEE Transactions on Computational Social Systems*, vol. 6, no. 1, pp. 162–175, 2019.

[44] L. Lee and P. Hu, "Vulnerability analysis of cascading dynamics in smart grids under load redistribution attacks," *International Journal of Electrical Power & Energy Systems*, vol. 111, pp. 182–190, 2019.

[45] X. Li, R. Wang, Z. Fu, and D. Du, "Cascading failure analysis of power-communication networks considering packets priorities," in *2018 5th IEEE International Conference on Cloud Computing and Intelligence Systems (CCIS)*. IEEE, pp. 993–997, 2019.

[46] N. K. Ahmed, J. Neville, R. A. Rossi, N. Duffield, and T. L. Willke, "Graphlet decomposition: Framework, algorithms, and applications," *Knowledge and Information Systems (KAIS)*, pp. 1–32, 2016.

[47] J. F. Lawless, *Statistical Models and Methods for Lifetime Data*, 2nd ed. New York: John Wiley & Sons, 2003.

[48] N. C. Rasmussen, "Reactor safety study: An assessment of accident risks in U.S. commercial nuclear power plants, wash-1400," *United States Nuclear Regulatory Commission*, Washington, DC., 1975.

[49] M. Rausand and A. Høyland, *System Reliability Theory: Models, Statistical Methods, and Applications, Second Edition*, Wiley Series in Probability and Statistics. New York: Wiley, 2004.

[50] S. LaRocca, J. Johansson, H. Hassel, and S. Guikema, "Topological performance measures as surrogates for physical flow models for risk and vulnerability analysis for electric power systems," *Risk Analysis*, vol. 35, no. 4, pp. 608–623, 2014. https://onlinelibrary.wiley.com/doi/abs/10.1111/risa.12281

[51] H. Edelsbrunner and J. Harer, *Persistent Homology – A Survey*, vol. 453, 2008.

[52] M. Kerber, D. Morozov, and A. Nigmetov, "Geometry helps to compare persistence diagrams," in *Proceedings of the 18th Workshop on Algorithm Engineering and Experiments (ALENEX)*, 2016, pp. 103–112.

[53] N. Otter, M. A. Porter, U. Tillmann, P. Grindrod, and H. A. Harrington, "A roadmap for the computation of persistent homology," *EPJ Data Science*, vol. 6, no. 1, 2017.

[54] A. Patania, F. Vaccarino, and G. Petri, "Topological analysis of data," *EPJ Data Science*, vol. 6, no. 7, 2017.

[55] L. Wasserman, "Topological data analysis," *Annual Review of Statistics and Its Application*, vol. 5, no. 1, pp. 501–532, 2018.

[56] J. F. Cuenca and A. Iske, "Persistent homology for defect detection in non-destructive evaluation of materials," *The e-Journal of Nondestructive Testing*, vol. 21, no. 01, 2016.

[57] G. Carlsson, "Topology and data," *Bulletin of the American Mathematical Society*, vol. 46, no. 2, 2009.

[58] C. Epstein, G. Carlsson, and H. Edelsbrunner, "Topological data analysis," *Inverse Problems*, vol. 27, no. 12, p. 120201, 2011.

[59] A. Zomorodian, "Fast construction of the Vietoris–Rips complex," *Computers and Graphics*, vol. 34, no. 3, pp. 263–271, 2010.

[60] R. Ghrist, "Barcodes: The persistent topology of data," *Bulletin of The American Mathematical Society*, vol. 45, no. 1, pp. 61–75, 2008.

[61] S. Schbath, E. Birmel, J. Daudin, and S. Robin, "Network motifs: Mean and variance for the count," *REVSTAT*, vol. 1, no. 1 2006.

[62] F. Picard, J.-J. Daudin, M. Koskas, S. Schbath, and S. Robin, "Assessing the exceptionality of network motifs," *Journal of Computational Biology*, vol. 15, no. 1, pp. 1–20, 2008.

[63] R. Milo, S. Shen-Orr, S. Itzkovitz, N. Kashtan, D. Chklovskii, and U. Alon, "Network motifs: Simple building blocks of complex networks," *Science*, vol. 298, no. 5594, pp. 824–827, 2002.

[64] M. Kahle and E. Meckes, "Limit the theorems for Betti numbers of random simplicial complexes," *Homology Homotopy and Applications*, vol. 15, no. 1, pp. 343–374, 2013. https://projecteuclid.org:443/euclid.hha/1383943681

[65] P. Bubenik and P. T. Kim, "A statistical approach to persistent homology," *Homology Homotopy and Applications*, vol. 9, no. 2, pp. 337–362, 2007. https://projecteuclid.org:443/euclid.hha/1201127341

[66] B. T. Fasy, F. Lecci, A. Rinaldo, L. Wasserman, S. Balakrishnan, and A. Singh, "Confidence sets for persistence diagrams," *The Annals of Statistics*, vol. 42, no. 6, pp. 2301–2339, 2014. https://doi.org/10.1214/14-AOS1252

[67] S. Agami and R. J. Adler, "Modeling of persistent homology," *arXiv preprint*, 2017.

[68] Y. Xu, A. Gurfinkel, and P. Rikvold, "Architecture of the Florida power grid as a complex network," *Physica A: Statistical Mechanics and Its Applications*, vol. 401, pp. 130–140, 2014.

[69] E. Cotilla-Sanchez, P. D. H. Hines, C. Barrows, and S. Blumsack, "Comparing the topological and electrical structure of the North American electric power infrastructure," *IEEE Systems Journal*, vol. 6, no. 4, pp. 616–626, 2012.

[70] P. Hines, S. Blumsack, E. C. Sanchez, and C. Barrows, "The topological and electrical structure of power grids," in *2010 43rd Hawaii International Conference on System Sciences*, pp. 1–10, 2010.

[71] E. D. Kolaczyk, *Statistical Analysis of Network Data*. Berlin: Springer, 2009.

[72] P. H. J. Nardelli, N. Rubido, C. Wang, M. S. Baptista, C. Pomalaza-Raez, P. Cardieri, and M. Latva-aho, "Models for the modern power grid," *The European Physical Journal Special Topics*, vol. 223, no. 12, pp. 2423–2437, 2014.

[73] A. Aleta and Y. Moreno, "Multilayer networks in a nutshell," *Annual Review of Condensed Matter Physics*, vol. 10, pp. 45–62, 2019.

[74] R. Rocchetta and E. Patelli, "Assessment of power grid vulnerabilities accounting for stochastic loads and model imprecision," *International Journal of Electrical Power & Energy Systems*, vol. 98, pp. 219–232, 2018.

Part III

Data Quality, Integrity, and Privacy

9 Data-Injection Attacks

Iñaki Esnaola, Samir M. Perlaza, and Ke Sun

9.1 Introduction

The pervasive deployment of sensing, monitoring, and data acquisition techniques in modern power systems enables the definition of functionalities and services that leverage accurate and real-time information about the system. This wealth of data supports network operators in the design of advanced control and management techniques that will inevitably change the operation of future power systems. An interesting side-effect of the data collection exercise that is starting to take place in power systems is that the unprecedented data analysis effort is shedding some light on the turbulent dynamics of power systems. While the underlying physical laws governing power systems are well understood, the large scale, distributed structure, and stochastic nature of the generation and consumption processes in the system results in a complex system. The large volumes of data about the state of the system are opening the door to modeling aspirations that were not feasible prior to the arrival of the smart grid paradigm.

The refinement of the models describing the power system operation will undoubtedly provide valuable insight to the network operator. However, that knowledge and the explanatory principles that it uncovers are also subject to be used in a malicious fashion. Access to statistics describing the state of the grid can inform malicious attackers by allowing them to pose the data-injection attack problem [1] within a probabilistic framework [2, 3]. By describing the processes taking place in the grid as a stochastic process, the network operator can incorporate the statistical structure of the state variables in the state estimation procedure by posing it within a Bayesian estimation setting. Similarly, the attacker can exploit the stochastic description of the state variables by incorporating it to the attack construction in the form of prior knowledge about the state variables. Interestingly whether the network operator or the attacker benefit more from adding a stochastic description to the state variables does not have a straightforward answer and depends on the parameters describing the power system.

In this chapter, we review some of the basic attack constructions that exploit a stochastic description of the state variables. We pose the state estimation problem in a Bayesian setting and cast the bad data detection procedure as a Bayesian hypothesis testing problem. This revised detection framework provides the benchmark for the attack detection problem that limits the achievable attack disruption. Indeed, the tradeoff between the impact of the attack, in terms of disruption to the state estimator, and the probability of attack detection is analytically characterized within this Bayesian attack setting. We then generalize the attack construction by considering information-theoretic measures that place fundamental limits to a broad class of detection, estimation, and learning techniques. Because the attack constructions proposed in this chapter rely on the attacker having access to the statistical structure of the random process describing the state variables, we conclude by studying the impact of imperfect statistics on the attack performance. Specifically, we study the attack performance as a function of the size of the training data set that is available to the attacker to estimate the second-order statistics of the state variables.

9.2 System Model

9.2.1 Bayesian State Estimation

We model the state of the system as the vector of n random variables X^n taking values in $\mathbb{R}^n$ with distribution P_{X^n}. The random variable X_i with $i = 1, 2, \ldots, n$, denotes the state variable i of the power system, and therefore, each entry represents a different physical magnitude of the system that the network operator wishes to monitor. The prior knowledge that is available to the network operator is described by the probability distribution P_{X^n}. The knowledge of the distribution is a consequence of the modeling based on historical data acquired by the network operator. Assuming linearized system dynamics with m measurements corrupted by additive white Gaussian noise (AWGN), the measurements are modeled as the vector of m random variables Y^m taking values in $\mathbb{R}^m$ with distribution P_{Y^m} given by

$$Y^m = \mathbf{H} X^n + Z^m, \qquad (9.1)$$

where $\mathbf{H} \in \mathbb{R}^{m \times n}$ is the Jacobian matrix of the linearized system dynamics around a given operating point and $Z^m \sim \mathcal{N}(0, \sigma^2 \mathbf{I})$ is thermal white noise with power spectral density σ^2. Often, the matrix $\mathbf{H}$ is also referred to as the Jacobian measurement matrix. While the operation point of the system induces a dynamic on the Jacobian matrix $\mathbf{H}$, in the following we assume that the timescale over which the operation point changes is small compared to the timescale at which the state estimator operates to produce the estimates. For that reason, in the following we assume that the Jacobian matrix is fixed and the only sources of uncertainty in the observation process originate from the stochasticity of the state variables and the additive noise corrupting the measurements.

The aim of the state estimator is to obtain an estimate $\hat{X}^n$ of the state vector X^n from the system measurements Y^m. In this chapter, we adopt a linear estimation framework resulting in an estimate given by $\hat{X}^n = \mathbf{L} Y^m$, where $\mathbf{L} \in \mathbb{R}^{n \times m}$ is the linear estimation matrix determining the estimation procedure. In the case in which the operator knows the distribution P_{X^n} of the underlying random process governing the state of the network, the estimation is performed by selecting the estimate that minimizes a given error cost function. A common approach is to use the mean square error (MSE) as the error cost function. In this case, the network operator uses an estimator $\mathbf{M}$ that is the unique solution to the following optimization problem:

$$\mathbf{M} = \arg\min_{\mathbf{L}\in\mathbb{R}^{n\times m}} \mathbb{E}\left[\frac{1}{n}\|X^n - \mathbf{L}Y^m\|_2^2\right], \qquad (9.2)$$

where the expectation is taken with respect to X^n and Z^m.

Under the assumption that the network state vector X^n follows an n-dimensional real Gaussian distribution with zero mean and covariance matrix $\Sigma_{XX} \in \mathcal{S}_+^n$, i.e. $X^n \sim \mathcal{N}(\mathbf{0}, \Sigma_{XX})$, the minimum MSE (MMSE) estimate is given by

$$\hat{X}^n \triangleq \mathbb{E}[X^n | Y^m] = \mathbf{M} Y^m, \qquad (9.3)$$

where

$$\mathbf{M} = \Sigma_{XX}\mathbf{H}^\mathsf{T}(\mathbf{H}\Sigma_{XX}\mathbf{H}^\mathsf{T} + \sigma^2 \mathbf{I})^{-1}. \qquad (9.4)$$

9.2.2 Deterministic Attack Model

The aim of the attacker is to corrupt the estimate by altering the measurements. Data-injection attacks alter the measurements available to the operator by adding an attack vector to the measurements. The resulting observation model with the additive attack vector is given by

$$Y_a^m = \mathbf{H}X^n + Z^m + \mathbf{a}, \qquad (9.5)$$

where $\mathbf{a} \in \mathbb{R}^m$ is the attack vector and $Y_a^m \in \mathbb{R}^m$ is the vector containing the compromised measurements [1]. Note that in this formulation, the attack vector does not have a probabilistic structure, i.e. the attack vector is deterministic. The random attack construction is considered later in the chapter.

The intention of the attacker can respond to diverse motivations, and therefore, the attack construction strategy changes depending on the aim of the attacker. In this chapter, we study attacks that aim to maximize the monitoring disruption, i.e., attacks that obstruct the state estimation procedure with the aim of deviating the estimate as much as possible from the true state. In that sense, the attack problem is bound to the cost function used by the state estimator to obtain the estimate, as the attacker aims to maximize it while the estimator aims to minimize it. In the MMSE setting described in the preceding text, it follows that the impact of the attack vector is obtained by noticing that the estimate when the attack vector is present is given by

$$\hat{X}_a^n = \mathbf{M}(\mathbf{H}X^n + Z^m) + \mathbf{M}\mathbf{a}. \qquad (9.6)$$

The term $\mathbf{M}\mathbf{a}$ is referred to as the *Bayesian injection vector* introduced by the attack vector $\mathbf{a}$ and is denoted by

$$\mathbf{c} \triangleq \mathbf{M}\mathbf{a} = \Sigma_{XX}\mathbf{H}^\mathsf{T}(\mathbf{H}\Sigma_{XX}\mathbf{H}^\mathsf{T} + \sigma^2 \mathbf{I})^{-1}\mathbf{a}. \qquad (9.7)$$

The *Bayesian injection vector* is a deterministic vector that corrupts the MMSE estimate of the operator resulting in

$$\hat{X}_a^n = \hat{X}^n + \mathbf{c}, \qquad (9.8)$$

where $\hat{X}^n$ is given in (9.3).

9.2.3 Attack Detection

As a part of the grid management, a network operator systematically attempts to identify measurements that are not deemed of sufficient quality for the state estimator. In practice, this operation can be cast into a hypothesis testing problem with hypotheses

$$\begin{aligned}\mathcal{H}_0 &: \text{There is no attack,} \quad \text{versus} \\ \mathcal{H}_1 &: \text{Measurements are compromised.}\end{aligned} \quad (9.9)$$

Assuming the operator knows the distribution of the state variables, P_{X^n}, and the observation model (9.5), then it can obtain the joint distribution of the measurements and the state variables for both normal operation conditions and the case in which an attack is present, i.e. $P_{X^n Y^m}$ and $P_{X^n Y_a^m}$, respectively.

Under the assumption that the state variables follow a multivariate Gaussian distribution $X^n \sim \mathcal{N}(\mathbf{0}, \mathbf{\Sigma}_{XX})$ it follows that the vector of measurements Y^m follows an m-dimensional real Gaussian random distribution with covariance matrix

$$\mathbf{\Sigma}_{YY} = \mathbf{H}\mathbf{\Sigma}_{XX}\mathbf{H}^\mathsf{T} + \sigma^2 \mathbf{I}, \quad (9.10)$$

and mean $\mathbf{a}$ when there is an attack; or zero mean when there is no attack. Within this setting, the hypotheses described in (9.9) reduces to:

$$\begin{aligned}\mathcal{H}_0 &: Y^m \sim \mathcal{N}(\mathbf{0}, \mathbf{\Sigma}_{YY}), \quad \text{versus} \\ \mathcal{H}_1 &: Y^m \sim \mathcal{N}(\mathbf{a}, \mathbf{\Sigma}_{YY}).\end{aligned} \quad (9.11)$$

A worst-case scenario approach is assumed for the attackers, namely, the operator knows the attack vector, $\mathbf{a}$, used in the attack. However, the operator does not know *a priori* whether the grid is under attack or not, which accounts for the need of an attack detection strategy. That being the case, the optimal detection strategy for the operator is to perform a likelihood ratio test (LRT) $L(\mathbf{y}, \mathbf{a})$ with respect to the measurement vector $\mathbf{y}$. Under the assumption that state variables follow a multivariate Gaussian distribution, the likelihood ratio can be calculated as

$$L(\mathbf{y}, \mathbf{a}) = \frac{f_{\mathcal{N}(\mathbf{0}, \mathbf{\Sigma}_{YY})}(\mathbf{y})}{f_{\mathcal{N}(\mathbf{a}, \mathbf{\Sigma}_{YY})}(\mathbf{y})} = \exp\left(\frac{1}{2}\mathbf{a}^\mathsf{T}\mathbf{\Sigma}_{YY}^{-1}\mathbf{a} - \mathbf{a}^\mathsf{T}\mathbf{\Sigma}_{YY}^{-1}\mathbf{y}\right), \quad (9.12)$$

where $f_{\mathcal{N}(\mu, \Sigma)}$ is the probability density function of a multivariate Gaussian random vector with mean μ and covariance matrix Σ. Therefore, either hypothesis is accepted by evaluating the inequalities

$$L(\mathbf{y}, \mathbf{a}) \underset{\mathcal{H}_1}{\overset{\mathcal{H}_0}{\gtrless}} \tau, \quad (9.13)$$

where $\tau \in [0, \infty)$ is tuned to set the tradeoff between the probability of detection and the probability of false alarm.

9.3 Centralized Deterministic Attacks

This section describes the construction of data-injection attacks in the case in which there is a unique attacker with access to all the measurements on the power system. This scenario is referred to as *centralized attacks* to highlight that there exists a unique entity deciding the data-injection vector $\mathbf{a} \in \mathbb{R}^m$ in (9.5). The difference between the scenario in which there exists a unique attacker or several (competing or cooperating) attackers is subtle and it is treated in Section 9.4.

Let $\mathcal{M} = \{1, \ldots, m\}$ denote the set of all m sensors available to the network operator. A sensor is said to be compromised if the attacker is able to arbitrarily modify its output. Given a total energy budget $E > 0$ at the attacker, the set of all possible attacks that can be injected to the network can be explicitly described by

$$\mathcal{A} = \{\mathbf{a} \in \mathbb{R}^m : \mathbf{a}^\mathsf{T}\mathbf{a} \leqslant E\}. \tag{9.14}$$

9.3.1 Minimum Probability of Detection Attacks

The attacker chooses a vector $\mathbf{a} \in \mathcal{A}$ taking into account the tradeoff between the probability of being detected and the distortion induced by the Bayesian injection vector given by (9.7). However, the choice of a particular data-injection vector is not trivial as the attacker does not have any information about the exact realizations of the vector of state variables $\mathbf{x}$ and the noise vector $\mathbf{z}$. A reasonable assumption on the knowledge of the attacker is to consider that it knows the structure of the power system and, thus, it knows the matrix $\mathbf{H}$. It is also reasonable to assume that it knows the first and second moments of the state variables X^n and noise Z^m as this can be computed from historical data.

Under these knowledge assumptions, the probability that the network operator is unable to detect the attack vector $\mathbf{a}$ is

$$\mathsf{P}_{\mathsf{ND}}(\mathbf{a}) = \mathbb{E}\left[\mathbb{1}_{\{L(Y^m,\mathbf{a})>\tau\}}\right], \tag{9.15}$$

where the expectation is taken over the joint probability distribution of state variables X^n and the AWGN noise vector Z^m, and $\mathbb{1}_{\{\cdot\}}$ denotes the indicator function. Note that under these assumptions, Y^m is a random variable with Gaussian distribution with mean $\mathbf{a}$ and covariance matrix Σ_{YY}. Thus, the probability $\mathsf{P}_{\mathsf{ND}}(\mathbf{a})$ of a vector $\mathbf{a}$ being a successful attack, i.e. a nondetected attack, is given by [4]

$$\mathsf{P}_{\mathsf{ND}}(\mathbf{a}) = \frac{1}{2}\mathsf{erfc}\left(\frac{\frac{1}{2}\mathbf{a}^\mathsf{T}\Sigma_{YY}^{-1}\mathbf{a} + \log \tau}{\sqrt{2\mathbf{a}^\mathsf{T}\Sigma_{YY}^{-1}\mathbf{a}}}\right). \tag{9.16}$$

Often, the knowledge of the threshold τ in (9.13) is not available to the attacker and, thus, it cannot determine the exact probability of not being detected for a given attack vector $\mathbf{a}$. However, the knowledge of whether $\tau > 1$ or $\tau \leqslant 1$ induces different behaviors on the attacker. The following propositions follow immediately from (9.16) and the properties of the complementary error function.

PROPOSITION 9.1 (Case $\tau \leqslant 1$) Let $\tau \leqslant 1$. Then, for all $\mathbf{a} \in \mathcal{A}$, $\mathsf{P}_{\mathsf{ND}}(\mathbf{a}) < \mathsf{P}_{\mathsf{ND}}((0,\ldots,0))$ and the probability $\mathsf{P}_{\mathsf{ND}}(\mathbf{a})$ is monotonically decreasing with $\mathbf{a}^\mathsf{T} \Sigma_{YY}^{-1} \mathbf{a}$.

PROPOSITION 9.2 (Case $\tau > 1$) Let $\tau > 1$ and let also $\Sigma_{YY} = \mathbf{U}_{YY} \Lambda_{YY} \mathbf{U}_{YY}^\mathsf{T}$ be the singular value decomposition of Σ_{YY}, with $\mathbf{U}_{YY}^\mathsf{T} = (\mathbf{u}_{YY,1},\ldots,\mathbf{u}_{YY,m})$ and $\Lambda_{YY} = \mathrm{diag}(\lambda_{YY,1},\ldots,\lambda_{YY,m})$ and $\lambda_{YY,1} \geqslant \lambda_{YY,2} \geqslant \cdots \geqslant \lambda_{YY,m}$. Then, any vector of the form

$$\mathbf{a} = \pm\sqrt{\lambda_{YY,k} 2\log\tau}\,\mathbf{u}_{YY,k}, \qquad (9.17)$$

with $k \in \{1,\ldots,m\}$, is a data-injection attack that satisfies for all $\mathbf{a}' \in \mathbb{R}^m$, $\mathsf{P}_{\mathsf{ND}}(\mathbf{a}') \leqslant \mathsf{P}_{\mathsf{ND}}(\mathbf{a})$.

The proof of Proposition 9.1 and Proposition 9.2 follows.

Proof Let $x = \mathbf{a}^\mathsf{T} \Sigma_{YY}^{-1} \mathbf{a}$ and note that $x > 0$ due to the positive definiteness of Σ_{YY}. Let also the function $g : \mathbb{R} \to \mathbb{R}$ be

$$g(x) = \frac{\frac{1}{2}x + \log\tau}{\sqrt{2x}}. \qquad (9.18)$$

The first derivative of $g(x)$ is

$$g'(x) = \frac{1}{2\sqrt{2x}}\left(\frac{1}{2} - \frac{\log\tau}{x}\right). \qquad (9.19)$$

Note that in the case in which $\log\tau \leqslant 0$ (or $\tau \leqslant 1$), then for all $x \in \mathbb{R}^+$, $g'(x) > 0$ and thus, g is monotonically increasing with x. Because the complementary error function erfc is monotonically decreasing with its argument, the statement of Proposition 9.1 follows and completes its proof. In the case in which $\log\tau \geqslant 0$ (or $\tau > 1$), the solution to $g'(x) = 0$ is $x = 2\log\tau$ and it corresponds to a minimum of the function g. The maximum of $\frac{1}{2}\mathrm{erfc}(g(x))$ occurs at the minimum of $g(x)$ given that erfc is monotonically decreasing with its argument. Hence, the maximum of $\mathsf{P}_{\mathsf{ND}}(\mathbf{a})$ occurs for the attack vectors satisfying:

$$\mathbf{a}^\mathsf{T} \Sigma_{YY}^{-1} \mathbf{a} = 2\log\tau. \qquad (9.20)$$

Solving for $\mathbf{a}$ in (9.20) yields (9.17) and this completes the proof of Proposition 9.2. □

The relevance of Proposition 9.1 is that it states that when $\tau \leqslant 1$, any nonzero data-injection attack vector possesses a nonzero probability of being detected. Indeed, the highest probability $\mathsf{P}_{\mathsf{ND}}(\mathbf{a})$ of not being detected is achieved by the null vector $\mathbf{a} = (0,\ldots,0)$, i.e., there is no attack. Alternatively, when $\tau > 1$ it follows from Proposition 9.2 that there always exists a nonzero vector that possesses maximum probability of not being detected. However, in both cases, it is clear that the corresponding data-injection vectors that induce the highest probability of not being detected are not necessarily the same that inflict the largest damage to the network, i.e., maximize the excess distortion.

From this point of view, the attacker faces the tradeoff between maximizing the excess distortion and minimizing the probability of being detected. Thus, the attack construction can be formulated as an optimization problem in which the solution **a** is a data-injection vector that maximizes the probability $P_{ND}(\mathbf{a})$ of not being detected at the same time that it induces a distortion $\|\mathbf{c}\|_2^2 \geq D_0$ into the estimate, with **c** in (9.7). In the case in which $\tau \leq 1$, it follows from Proposition 9.1 and (9.7) that this problem can be formulated as the following optimization problem:

$$\min_{\mathbf{a} \in \mathcal{A}} \mathbf{a}^\mathsf{T} \Sigma_{YY}^{-1} \mathbf{a} \quad \text{s.t.} \quad \mathbf{a}^\mathsf{T} \Sigma_{YY}^{-1} \mathbf{H} \Sigma_{XX}^2 \mathbf{H}^\mathsf{T} \Sigma_{YY}^{-1} \mathbf{a} \geq D_0. \tag{9.21}$$

The solution to the optimization problem in (9.21) is given by the following theorem.

THEOREM 9.1 *Let* $\mathbf{G} = \Sigma_{YY}^{-\frac{1}{2}} \mathbf{H} \Sigma_{XX}^2 \mathbf{H}^\mathsf{T} \Sigma_{YY}^{-\frac{1}{2}}$ *have a singular value decomposition* $\mathbf{G} = \mathbf{U}_\mathbf{G} \Sigma_\mathbf{G} \mathbf{U}_\mathbf{G}^\mathsf{T}$, *with* $\mathbf{U}_\mathbf{G} = (\mathbf{u}_{G,i}, \ldots, \mathbf{u}_{G,m})$ *a unitary matrix and* $\Sigma_\mathbf{G} = \mathrm{diag}(\lambda_{G,1}, \ldots, \lambda_{G,m})$ *a diagonal matrix with* $\lambda_{G,1} \geq \cdots \geq \lambda_{G,m}$. *Then, if* $\tau \leq 1$, *the attack vector* **a** *that maximizes the probability of not being detected* $P_{ND}(\mathbf{a})$ *while inducing an excess distortion not less than* D_0 *is*

$$\mathbf{a} = \pm \sqrt{\frac{D_0}{\lambda_{G,1}}} \Sigma_{YY}^{\frac{1}{2}} \mathbf{u}_{G,1}. \tag{9.22}$$

Moreover, $P_{ND}(\mathbf{a}) = \frac{1}{2} \mathrm{erfc}\left(\dfrac{\frac{D_0}{2\lambda_{G,1}} + \log \tau}{\sqrt{\frac{2D_0}{\lambda_{G,1}}}} \right)$.

Proof Consider the Lagrangian

$$\mathcal{L}(\mathbf{a}) = \mathbf{a}^\mathsf{T} \Sigma_{YY}^{-1} \mathbf{a} - \gamma \left(\mathbf{a}^\mathsf{T} \Sigma_{YY}^{-1} \mathbf{H} \Sigma_{XX}^2 \mathbf{H}^\mathsf{T} \Sigma_{YY}^{-1} \mathbf{a} - D_0 \right), \tag{9.23}$$

with $\gamma > 0$ a Lagrangian multiplier. Then, the necessary conditions for **a** to be a solution to the optimization problem (9.21) are:

$$\nabla_\mathbf{a} \mathcal{L}(\mathbf{a}) = 2\left(\Sigma_{YY}^{-1} - \gamma \Sigma_{YY}^{-1} \mathbf{H} \Sigma_{XX}^2 \mathbf{H}^\mathsf{T} \Sigma_{YY}^{-1} \right) \mathbf{a} = 0 \tag{9.24}$$

$$\frac{d}{d\gamma} \mathcal{L}(\mathbf{a}) = \mathbf{a}^\mathsf{T} \Sigma_{YY}^{-1} \mathbf{H} \Sigma_{XX}^2 \mathbf{H}^\mathsf{T} \Sigma_{YY}^{-1} \mathbf{a} - D_0 = 0. \tag{9.25}$$

Note that any

$$\mathbf{a}_i = \pm \sqrt{\frac{D_0}{\lambda_{G,i}}} \Sigma_{YY}^{\frac{1}{2}} \mathbf{u}_{G,i} \text{ and} \tag{9.26}$$

$$\gamma_i = \lambda_{G,i}, \text{ with } 1 \leq i \leq \mathrm{rank}(\mathbf{G}), \tag{9.27}$$

satisfy $\gamma_i > 0$ and conditions (9.24) and (9.25). Hence, the set of vectors that satisfy the necessary conditions to be a solution of (9.21) is

$$\left\{ \mathbf{a}_i = \pm \sqrt{\frac{D_0}{\lambda_{G,i}}} \Sigma_{YY}^{\frac{1}{2}} \mathbf{u}_{G,i} : 1 \leq i \leq \mathrm{rank}(\mathbf{G}) \right\}. \tag{9.28}$$

More importantly, any vector $\mathbf{a} \neq \mathbf{a}_i$, with $1 \leq i \leq \text{rank}(\mathbf{G})$, does not satisfy the necessary conditions. Moreover,

$$\mathbf{a}_i^\mathsf{T} \Sigma_{YY}^{-1} \mathbf{a}_i = \frac{D_0}{\lambda_{\mathbf{G},i}} \geq \frac{D_0}{\lambda_{\mathbf{G},1}}. \tag{9.29}$$

Therefore, $\mathbf{a} = \pm \sqrt{\frac{D_0}{\lambda_{\mathbf{G},1}}} \Sigma_{YY}^{\frac{1}{2}} \mathbf{u}_{\mathbf{G},1}$ are the unique solutions to (9.21). This completes the proof. □

Interestingly, the construction of the data-injection attack $\mathbf{a}$ in (9.22) does not require the exact knowledge of τ. That is, only knowing that $\tau \leq 1$ is enough to build the data-injection attack that has the highest probability of not being detected and induces a distortion of at least D_0.

In the case in which $\tau > 1$, it is also possible to find the data-injection attack vector that induces a distortion not less than D_0 and the maximum probability of not being detected. Such a vector is the solution to the following optimization problem.

$$\min_{\mathbf{a} \in \mathcal{A}} \frac{\frac{1}{2} \mathbf{a}^\mathsf{T} \Sigma_{YY}^{-1} \mathbf{a} + \log \tau}{\sqrt{2 \mathbf{a}^\mathsf{T} \Sigma_{YY}^{-1} \mathbf{a}}} \quad \text{s.t.} \quad \mathbf{a}^\mathsf{T} \Sigma_{YY}^{-1} \mathbf{H} \Sigma_{XX}^2 \mathbf{H}^\mathsf{T} \Sigma_{YY}^{-1} \mathbf{a} \geq D_0. \tag{9.30}$$

The solution to the optimization problem in (9.30) is given by the following theorem.

THEOREM 9.2 *Let* $\mathbf{G} = \Sigma_{YY}^{-\frac{1}{2}} \mathbf{H} \Sigma_{XX}^2 \mathbf{H}^\mathsf{T} \Sigma_{YY}^{-\frac{1}{2}}$ *have a singular value decomposition* $\mathbf{G} = \mathbf{U}_\mathbf{G} \Sigma_\mathbf{G} \mathbf{U}_\mathbf{G}^\mathsf{T}$, *with* $\mathbf{U}_\mathbf{G} = (\mathbf{u}_{\mathbf{G},i}, \ldots, \mathbf{u}_{\mathbf{G},m})$ *a unitary matrix and* $\Sigma_\mathbf{G} = \text{diag}(\lambda_{\mathbf{G},1}, \ldots, \lambda_{\mathbf{G},m})$ *a diagonal matrix with* $\lambda_{\mathbf{G},1} \geq \cdots \geq \lambda_{\mathbf{G},m}$. *Then, when* $\tau > 1$, *the attack vector* $\mathbf{a}$ *that maximizes the probability of not being detected* $\mathsf{P}_{\mathsf{ND}}(\mathbf{a})$ *while producing an excess distortion not less than* D_0 *is*

$$\mathbf{a} = \begin{cases} \pm \sqrt{\frac{D_0}{\lambda_{\mathbf{G},k^*}}} \Sigma_{YY}^{\frac{1}{2}} \mathbf{u}_{\mathbf{G},k^*} & \text{if } \frac{D_0}{2 \log \tau \lambda_{\mathbf{G},\text{rank}(\mathbf{G})}} \geq 1, \\ \pm \sqrt{2 \log \tau} \Sigma_{YY}^{\frac{1}{2}} \mathbf{u}_{\mathbf{G},1} & \text{if } \frac{D_0}{2 \log \tau \lambda_{\mathbf{G},\text{rank}(\mathbf{G})}} < 1 \end{cases}$$

with

$$k^* = \arg \min_{k \in \{1,\ldots,\text{rank}(\mathbf{G})\}: \frac{D_0}{\lambda_{\mathbf{G},k}} > 2 \log(\tau)} \frac{D_0}{\lambda_{\mathbf{G},k}}. \tag{9.31}$$

Proof The structure of the proof of Theorem 9.2 is similar to the proof of Theorem 9.1 and is omitted in this chapter. A complete proof can be found in [5]. □

9.3.2 Maximum Distortion Attacks

In the previous subsection, the attacker constructs its data-injection vector $\mathbf{a}$ aiming to maximize the probability of nondetection $\mathsf{P}_{\mathsf{ND}}(\mathbf{a})$ while guaranteeing a minimum distortion. However, this problem has a dual in which the objective is to maximize the distortion $\mathbf{a}^\mathsf{T} \Sigma_{YY}^{-1} \mathbf{H} \Sigma_{XX}^2 \mathbf{H}^\mathsf{T} \Sigma_{YY}^{-1} \mathbf{a}$ while guaranteeing that the probability of not being detected remains always larger than a given threshold $L'_0 \in [0, \frac{1}{2}]$. This problem can be formulated as the following optimization problem:

$$\max_{\mathbf{a} \in \mathcal{A}} \mathbf{a}^T \Sigma_{YY}^{-1} \mathbf{H} \Sigma_{XX}^2 \mathbf{H}^T \Sigma_{YY}^{-1} \mathbf{a} \quad \text{s.t.} \quad \frac{\frac{1}{2}\mathbf{a}^T \Sigma_{YY}^{-1} \mathbf{a} + \log \tau}{\sqrt{2 \mathbf{a}^T \Sigma_{YY}^{-1} \mathbf{a}}} \le L_0, \quad (9.32)$$

with $L_0 = \text{erfc}^{-1}(2L'_0) \in [0, \infty)$.

The solution to the optimization problem in (9.32) is given by the following theorem.

THEOREM 9.3 *Let the matrix* $\mathbf{G} = \Sigma_{YY}^{-\frac{1}{2}} \mathbf{H} \Sigma_{XX}^2 \mathbf{H}^T \Sigma_{YY}^{-\frac{1}{2}}$ *have a singular value decomposition* $\mathbf{U_G} \Sigma_\mathbf{G} \mathbf{U_G}^T$, *with* $\mathbf{U_G} = (\mathbf{u}_{\mathbf{G},i}, \ldots, \mathbf{u}_{\mathbf{G},m})$ *a unitary matrix and* $\Sigma_\mathbf{G} = \text{diag}(\lambda_{\mathbf{G},1}, \ldots, \lambda_{\mathbf{G},m})$ *a diagonal matrix with* $\lambda_{\mathbf{G},1} \ge \cdots \ge \lambda_{\mathbf{G},m}$. *Then, the attack vector* $\mathbf{a}$ *that maximizes the excess distortion* $\mathbf{a}^T \Sigma_{YY}^{-\frac{1}{2}} \mathbf{G} \Sigma_{YY}^{-\frac{1}{2}} \mathbf{a}$ *with a probability of not being detected that does not go below* $L_0 \in [0, \frac{1}{2}]$ *is*

$$\mathbf{a} = \pm \left(\sqrt{2L_0} + \sqrt{2L_0^2 - 2\log \tau} \right) \Sigma_{YY}^{\frac{1}{2}} \mathbf{u}_{\mathbf{G},1}, \quad (9.33)$$

when a solution exists.

Proof The structure of the proof of Theorem 9.3 is similar to the proof of Theorem 9.1 and is omitted in this chapter. A complete proof can be found in [5]. □

9.4 Decentralized Deterministic Attacks

Let $\mathcal{K} = \{1, \ldots, K\}$ be the set of attackers that can potentially perform a data-injection attack on the network, e.g., a decentralized vector attack. Let also $\mathcal{C}_k \subseteq \{1, 2, \ldots, m\}$ be the set of sensors that attacker $k \in \mathcal{K}$ can control. Assume that $\mathcal{C}_1, \ldots, \mathcal{C}_K$ are proper sets and form a partition of the set $\mathcal{M}$ of all sensors. The set $\mathcal{A}_k$ of individual data attack vectors $\mathbf{a}_k = (a_{k,1}, a_{k,2}, \ldots, a_{k,m})$ that can be injected into the network by attacker $k \in \mathcal{K}$ is of the form

$$\mathcal{A}_k = \{\mathbf{a}_k \in \mathbb{R}^m : a_{k,j} = 0 \text{ for all } j \notin \mathcal{C}_k \text{ and } \mathbf{a}_k^T \mathbf{a}_k \le E_k\}. \quad (9.34)$$

The constant $E_k < \infty$ represents the energy budget of attacker k. Let the set of all possible sums of the elements of $\mathcal{A}_i$ and $\mathcal{A}_j$ be denoted by $\mathcal{A}_i \oplus \mathcal{A}_j$. That is, for all $\mathbf{a} \in \mathcal{A}_i \oplus \mathcal{A}_j$, there exists a pair of vectors $(\mathbf{a}_i, \mathbf{a}_j) \in \mathcal{A}_i \times \mathcal{A}_j$ such that $\mathbf{a} = \mathbf{a}_i + \mathbf{a}_j$. Using this notation, let the set of all possible data-injection attacks be denoted by

$$\mathcal{A} = \mathcal{A}_1 \oplus \mathcal{A}_2 \oplus \ldots \oplus \mathcal{A}_K, \quad (9.35)$$

and the set of complementary data-injection attacks with respect to attacker k be denoted by

$$\mathcal{A}_{-k} \stackrel{\Delta}{=} \mathcal{A}_1 \oplus \ldots \oplus \mathcal{A}_{k-1} \oplus \mathcal{A}_{k+1} \oplus \ldots \oplus \mathcal{A}_K. \quad (9.36)$$

Given the individual data-injection vectors $\mathbf{a}_i \in \mathcal{A}_i$, with $i \in \{1, \ldots, K\}$, the global attack vector $\mathbf{a}$ is

$$\mathbf{a} = \sum_{i=1}^{K} \mathbf{a}_k \in \mathcal{A}. \tag{9.37}$$

The aim of attacker k is to corrupt the measurements obtained by the set of meters $\mathcal{C}_k$ by injecting an error vector $\mathbf{a}_k \in \mathcal{A}_k$ that maximizes the damage to the network, e.g., the excess distortion, while avoiding the detection of the global data-injection vector $\mathbf{a}$. Clearly, all attackers have the same interest but they control different sets of measurements, i.e., $\mathcal{C}_i \neq \mathcal{C}_k$, for a any pair $(i,k) \in \mathcal{K}^2$. For modeling this behavior, attackers use the utility function $\phi : \mathbb{R}^m \to \mathbb{R}$, to determine whether a data-injection vector $\mathbf{a}_k \in \mathcal{A}_k$ is more beneficial than another $\mathbf{a}'_k \in \mathcal{A}_k$ given the complementary attack vector

$$\mathbf{a}_{-k} = \sum_{i \in \{1, \ldots, K\} \setminus \{k\}} \mathbf{a}_i \in \mathcal{A}_{-k} \tag{9.38}$$

adopted by all the other attackers. The function ϕ is chosen considering the fact that an attack is said to be successful if it induces a nonzero distortion and it is not detected. Alternatively, if the attack is detected no damage is induced into the network as the operator discards the measurements and no estimation is performed. Hence, given a global attack $\mathbf{a}$, the distortion induced into the measurements is $\mathbb{1}_{\{L(Y_a^m, \mathbf{a}) > \tau\}} \mathbf{x}_a^T \mathbf{x}_a$. However, attackers are not able to know the exact state of the network $\mathbf{x}$ and the realization of the noise $\mathbf{z}$ before launching the attack. Thus, it appears natural to exploit the knowledge of the first and second moments of both the state variables X^n and noise Z^m and to consider as a metric the expected distortion $\phi(\mathbf{a})$ that can be induced by the attack vector $\mathbf{a}$:

$$\phi(\mathbf{a}) = \mathbb{E}\left[\left(\mathbb{1}_{\{L(Y_a^m, \mathbf{a}) > \tau\}} \right) \mathbf{c}^T \mathbf{c} \right], \tag{9.39}$$
$$= \mathsf{P}_{\mathrm{ND}}(\mathbf{a}) \, \mathbf{a}^T \Sigma_{YY}^{-1} \mathbf{H} \Sigma_{XX}^2 \mathbf{H}^T \Sigma_{YY}^{-1}, \tag{9.40}$$

where $\mathbf{c}$ is in (9.7) and the expectation is taken over the distribution of state variables X^n and the noise Z^m. Note that under this assumption of global knowledge, this model considers the worst-case scenario for the network operator. Indeed, the result presented in this section corresponds to the case in which the attackers inflict the most harm onto the state estimator.

9.4.1 Game Formulation

The benefit $\phi(\mathbf{a})$ obtained by attacker k does not only depend on its own data-injection vector $\mathbf{a}_k$ but also on the data-injection vectors $\mathbf{a}_{-k}$ of all the other attackers. This becomes clear from the construction of the global data-injection vector $\mathbf{a}$ in (9.37), the excess distortion $\mathbf{x}_a$ in (9.7), and the probability of not being detected $\mathsf{P}_{\mathrm{ND}}(\mathbf{a})$ in (9.16). Therefore, the interaction of all attackers in the network can be described by a game in normal form

$$\mathcal{G} = \left(\mathcal{K}, \{\mathcal{A}_k\}_{k \in \mathcal{K}}, \phi \right). \tag{9.41}$$

Each attacker is a player in the game $\mathcal{G}$ and it is identified by an index from the set $\mathcal{K}$. The actions player k might adopt are data-injection vectors $\mathbf{a}_k$ in the set $\mathcal{A}_k$ in (9.34). The underlying assumption in the following of this section is that, given a vector of data-injection attacks $\mathbf{a}_{-k}$, player k aims to adopt a data-injection vector $\mathbf{a}_k$ such that the expected excess distortion $\phi(\mathbf{a}_k + \mathbf{a}_{-k})$ is maximized. That is,

$$\mathbf{a}_k \in \mathrm{BR}_k(\mathbf{a}_{-k}), \tag{9.42}$$

where the correspondence $\mathrm{BR}_k : \mathcal{A}_{-k} \to 2^{\mathcal{A}_k}$ is the best response correspondence, i.e.,

$$\mathrm{BR}_k(\mathbf{a}_{-k}) = \arg\max_{\mathbf{a}_k \in \mathcal{A}_k} \phi(\mathbf{a}_k + \mathbf{a}_{-k}). \tag{9.43}$$

The notation $2^{\mathcal{A}_k}$ represents the set of all possible subsets of $\mathcal{A}_k$. Note that $\mathrm{BR}_k(\mathbf{a}_{-k}) \subseteq \mathcal{A}_k$ is the set of data-injection attack vectors that are optimal given that the other attackers have adopted the data-injection vector $\mathbf{a}_{-k}$. In this setting, each attacker tampers with a subset $\mathcal{C}_k$ of all sensors $\mathcal{C} = \{1, 2, \ldots, m\}$, as opposed to the centralized case in which there exists a single attacker that is able to tamper with all the sensors in $\mathcal{C}$.

A game solution that is particularly relevant for this analysis is the NE [6].

DEFINITION 9.1 (Nash Equilibrium) *The data-injection vector* $\mathbf{a}$ *is an NE of the game* $\mathcal{G}$ *if and only if it is a solution of the fix point equation*

$$\mathbf{a} = \mathrm{BR}(\mathbf{a}), \tag{9.44}$$

with $\mathrm{BR} : \mathcal{A} \to 2^{\mathcal{A}}$ *being the global best-response correspondence, i.e.,*

$$\mathrm{BR}(\mathbf{a}) = \mathrm{BR}_1(\mathbf{a}_{-1}) \oplus \ldots \oplus \mathrm{BR}_K(\mathbf{a}_{-K}). \tag{9.45}$$

Essentially, at an NE, attackers obtain the maximum benefit given the data-injection vector adopted by all the other attackers. This implies that an NE is an operating point at which attackers achieve the highest expected distortion induced over the measurements. More importantly, any unilateral deviation from an equilibrium data-injection vector $\mathbf{a}$ does not lead to an improvement of the average excess distortion. Note that this formulation does not say anything about the exact distortion induced by an attack but the average distortion. This is because the attack is chosen with the uncertainty introduced by the state vector X^n and the noise term Z^m.

The following proposition highlights an important property of the game $\mathcal{G}$ in (9.41).

PROPOSITION 9.3 *The game* $\mathcal{G}$ *in (9.41) is a potential game.*

Proof The proof follows immediately from the observation that all the players have the same utility function ϕ [7]. Thus, the function ϕ is a potential of the game $\mathcal{G}$ in (9.41) and any maximum of the potential function is an NE of the game $\mathcal{G}$. □

In general, potential games [7] possess numerous properties that are inherited by the game $\mathcal{G}$ in (9.41). These properties are detailed by the following proposition.

PROPOSITION 9.4 *The game* $\mathcal{G}$ *possesses at least one NE.*

Proof Note that ϕ is continuous in $\mathcal{A}$ and $\mathcal{A}$ is a convex and closed set; therefore, there always exists a maximum of the potential function ϕ in $\mathcal{A}$. Finally, from Lemma 4.3 in [7], it follows that such a maximum corresponds to an NE. $\square$

9.4.2 Achievability of an NE

The attackers are said to play a sequential best response dynamic (BRD) if the attackers can sequentially decide their own data-injection vector $\mathbf{a}_k$ from their sets of best responses following a round-robin (increasing) order. Denote by $\mathbf{a}_k^{(t)} \in \mathcal{A}$ the choice of attacker k during round $t \in \mathbb{N}$ and assume that attackers are able to observe all the other attackers' data-injection vectors. Under these assumptions, the BRD can be defined as follows.

DEFINITION 9.2 (Best Response Dynamics) *The players of the game $\mathcal{G}$ are said to play best response dynamics if there exists a round-robin order of the elements of $\mathcal{K}$ in which at each round $t \in \mathbb{N}$, the following holds:*

$$\mathbf{a}_k^{(t)} \in \mathrm{BR}_k \left(\mathbf{a}_1^{(t)} + \cdots + \mathbf{a}_{k-1}^{(t)} + \mathbf{a}_{k+1}^{(t-1)} + \cdots + \mathbf{a}_K^{(t-1)} \right). \tag{9.46}$$

From the properties of potential games (Lemma 4.2 in [7]), the following proposition follows.

LEMMA 9.1 (Achievability of NE attacks) *Any BRD in the game $\mathcal{G}$ converges to a data-injection attack vector that is an NE.*

The relevance of Lemma 9.1 is that it establishes that if attackers can communicate in at least a round-robin fashion, they are always able to attack the network with a data-injection vector that maximizes the average excess distortion. Note that there might exists several NEs (local maxima of ϕ) and there is no guarantee that attackers will converge to the best NE, i.e., a global maximum of ϕ. It is important to note that under the assumption that there exists a unique maximum, which is not the case for the game $\mathcal{G}$ (see Theorem 9.4), all attackers are able to calculate such a global maximum and no communications is required among the attackers. Nonetheless, the game $\mathcal{G}$ always possesses at least two NEs, which enforces the use of a sequential BRD to converge to an NE.

9.4.3 Cardinality of the Set of NEs

Let $\mathcal{A}_{\mathrm{NE}}$ be the set of all data-injection attacks that form NEs. The following theorem bounds the number of NEs in the game.

THEOREM 9.4 *The cardinality of the set $\mathcal{A}_{NE}$ of NE of the game $\mathcal{G}$ satisfies*

$$2 \leqslant |\mathcal{A}_{NE}| \leqslant C \cdot \mathrm{rank}(\mathbf{H}) \tag{9.47}$$

where $C < \infty$ is a constant that depends on τ.

Proof The lower bound follows from the symmetry of the utility function given in (9.39), i.e. $\phi(\mathbf{a}) = \phi(-\mathbf{a})$, and the existence of at least one NE claimed in Proposition 9.4.

To prove the upper bound the number of stationary points of the utility function is evaluated. This is equivalent to the cardinality of the set

$$\mathcal{S} = \{\mathbf{a} \in \mathbb{R}^m : \nabla_{\mathbf{a}}\phi(\mathbf{a}) = \mathbf{0}\}, \tag{9.48}$$

which satisfies $\mathcal{A}_{NE} \subseteq \mathcal{S}$. Calculating the gradient with respect to the attack vector yields

$$\nabla_{\mathbf{a}}\phi(\mathbf{a}) = \left(\alpha(\mathbf{a})\mathbf{M}^\mathsf{T}\mathbf{M} - \beta(\mathbf{a})\Sigma_{YY}^{-1}\right)\mathbf{a}, \tag{9.49}$$

where

$$\alpha(\mathbf{a}) \stackrel{\Delta}{=} \text{erfc}\left(\frac{1}{\sqrt{2}} \frac{\frac{1}{2}\mathbf{a}^\mathsf{T}\Sigma_{YY}^{-1}\mathbf{a} + \log \tau}{(\mathbf{a}^\mathsf{T}\Sigma_{YY}^{-1}\mathbf{a})^{\frac{1}{2}}}\right) \tag{9.50}$$

and

$$\beta(\mathbf{a}) \stackrel{\Delta}{=} \frac{\mathbf{a}^\mathsf{T}\mathbf{M}^\mathsf{T}\mathbf{M}\mathbf{a}}{\sqrt{2\pi}\mathbf{a}^\mathsf{T}\Sigma_{YY}^{-1}\mathbf{a}}\left(\frac{1}{2} - \frac{\log \tau}{\mathbf{a}^\mathsf{T}\Sigma_{YY}^{-1}\mathbf{a}}\right)\exp\left(-\left(\frac{1}{\sqrt{2}}\frac{\frac{1}{2}\mathbf{a}^\mathsf{T}\Sigma_{YY}^{-1}\mathbf{a} + \log \tau}{(\mathbf{a}^\mathsf{T}\Sigma_{YY}^{-1}\mathbf{a})^{\frac{1}{2}}}\right)^2\right). \tag{9.51}$$

Define $\delta(\mathbf{a}) \stackrel{\Delta}{=} \frac{\beta(\mathbf{a})}{\alpha(\mathbf{a})}$ and note that combining (9.4) with (9.49) gives the following condition for the stationary points:

$$\left(\mathbf{H}\Sigma_{XX}^2\mathbf{H}^\mathsf{T}\Sigma_{YY}^{-1} - \delta(\mathbf{a})\mathbf{I}\right)\mathbf{a} = \mathbf{0}. \tag{9.52}$$

Note that the number of linearly independent attack vectors that are a solution of the linear system in (9.52) is given by

$$R \stackrel{\Delta}{=} \text{rank}\left(\mathbf{H}\Sigma_{XX}^2\mathbf{H}^\mathsf{T}\Sigma_{YY}^{-1}\right) \tag{9.53}$$

$$= \text{rank}(\mathbf{H}). \tag{9.54}$$

where (9.54) follows from the fact that Σ_{XX} and Σ_{YY} are positive definite. Define the eigenvalue decomposition

$$\Sigma_{YY}^{-\frac{1}{2}}\mathbf{H}\Sigma_{XX}^2\mathbf{H}^\mathsf{T}\Sigma_{YY}^{-\frac{1}{2}} = \mathbf{U}\Lambda\mathbf{U}^\mathsf{T} \tag{9.55}$$

where Λ is a diagonal matrix containing the ordered eigenvalues $\{\lambda_i\}_{i=1}^m$ matching the order of of the eigenvectors in $\mathbf{U}$. As a result of (9.53) there are r eigenvalues, λ_k, which are different from zero and $m-r$ diagonal elements of Λ which are zero. Combining this decomposition with some algebraic manipulation, the condition for stationary points in (9.52) can be recast as

$$\Sigma_{YY}^{-\frac{1}{2}} \mathbf{U} (\Lambda - \delta(\mathbf{a})\mathbf{I}) \mathbf{U}^\mathsf{T} \Sigma_{YY}^{-\frac{1}{2}} \mathbf{a} = \mathbf{0}. \qquad (9.56)$$

Let $w \in \mathbb{R}$ be a scaling parameter and observe that the attack vectors that satisfy $\mathbf{a} = w \Sigma_{YY}^{\frac{1}{2}} \mathbf{U} \mathbf{e}_k$ and $\delta(\mathbf{a}) = \lambda_k$ for $k = 1, \ldots, r$ are solutions of (9.56). Note that the critical points associated to zero eigenvalues are not NE. Indeed, the eigenvectors associated to zero eigenvalues yield zero utility. Because the utility function is strictly positive, these critical points are minima of the utility function and can be discarded when counting the number of NE. Therefore, the set in (9.48) can be rewritten based on the condition in (9.56) as

$$\mathcal{S} = \bigcup_{k=1}^{R} \mathcal{S}_k, \qquad (9.57)$$

where

$$\mathcal{S}_k = \{\mathbf{a} \in \mathbb{R}^m : \mathbf{a} = w \Sigma_{YY}^{\frac{1}{2}} \mathbf{U} \mathbf{e}_k \text{ and } \delta(\mathbf{a}) = \lambda_k\}. \qquad (9.58)$$

There are r linearly independent solutions of (9.56) but for each linearly independent solution there can be several scaling parameters, w, which satisfy $\delta(\mathbf{a}) = \lambda_k$. For that reason, $|\mathcal{S}_k|$ is determined by the number of scaling parameters that satisfy $\delta(\mathbf{a}) = \lambda_k$. To that end, define $\delta' : \mathbb{R} \to \mathbb{R}$ as $\delta'(w) \triangleq \delta(w \Sigma_{YY}^{\frac{1}{2}} \mathbf{U} \mathbf{e}_k)$. It is easy to check that $\delta'(w) = \lambda_k$ has a finite number of solutions for $k = 1, \ldots, r$. Hence, for all k there exists a constant C_k such that $|\mathcal{S}_k| \leq C_k$, which yields the upper bound

$$|\mathcal{S}| \leq \sum_{i=1}^{R} |\mathcal{S}_k| \leq \sum_{i=1}^{R} C_k \leq \max_k C_k R. \qquad (9.59)$$

Noticing that the there is a finite number of solutions of $\delta'(w) = \lambda_k$ and that they depend only on τ yields the upper bound. $\square$

9.5 Information-Theoretic Attacks

Modern sensing infrastructure is moving toward increasing the number of measurements that the operator acquires, e.g., phasor measurement units exhibit temporal resolutions in the order of miliseconds while supervisory control and data acquisition (SCADA) systems traditionally operate with a temporal resolution in the order of seconds. As a result, attack constructions that do not change within the same temporal scale at which measurements are reported do not exploit all the *degrees of freedom* that are available to the attacker. Indeed, an attacker can choose to change the attack vector with every measurement vector that is reported to the network operator. However, the deterministic attack construction changes when the Jacobian measurement matrix changes, i.e. with the operation point of the system. Thus, in the deterministic attack case, the attack construction changes at the same rate that the Jacobian measurement matrix changes and, therefore, the dynamics of the state variables define the update cadency of the attack vector.

In this section, we study the case in which the attacker constructs the attack vector as a random process that corrupts the measurements. By equipping the attack vector with a probabilistic structure, we enable the attacker to generate attack vector realizations that corrupt each set of measurements reported to the network operator. The task of the attacker in this case is, therefore, to devise the optimal distribution over the attack vectors. In the following, we pose the attack construction problem within an information-theoretic framework and characterize the attacks that simultaneously minimize the mutual information and the probability of detection.

9.5.1 Random Attack Model

We consider an additive attack model as in (9.5) but with the distinction that the attack vector is modeled by a random process in this case. The resulting vector of compromised measurements is given by

$$Y_A^m = \mathbf{H} X^n + Z^m + A^m, \tag{9.60}$$

where $A^m \in \mathbb{R}^m$ is the vector of random variables introduced by the attacker and $Y_A^m \in \mathbb{R}^m$ is the vector containing the compromised measurements. The attack vector of random variables with P_{A^m} is chosen by the attacker. We assume that the attacker has no access to the realizations of the state variables so the attack vector is independent of the state variables, i.e. it holds that $P_{A^m X^n} = P_{A^m} P_{X^n}$ where $P_{A^m X^n}$ denotes the joint distribution of A^m and X^n.

Similarly to the deterministic attack case, we adopt a multivariate Gaussian framework for the state variables such that $X^n \sim \mathcal{N}(\mathbf{0}, \mathbf{\Sigma}_{XX})$. Moreover, we limit the attack vector distribution to the set of zero-mean multivariate Gaussian distributions, i.e. $A^m \sim \mathcal{N}(\mathbf{0}, \mathbf{\Sigma}_{AA})$ where $\mathbf{\Sigma}_{AA} \in \mathcal{S}_+^m$ is the covariance matrix of the attack distribution. The rationale for choosing a Gaussian distribution for the attack vector follows from the fact that for the measurement model in (9.60) the additive attack distribution that minimizes the mutual information between the vector of state variables and the compromised measurements is Gaussian [8]. As we will see later, minimizing this mutual information is central to the proposed information-theoretic attack construction and indeed one of the objectives of the attacker. Because of the Gaussianity of the attack distribution, the vector of compromised measurements is distributed as

$$Y_A^m \sim \mathcal{N}(\mathbf{0}, \mathbf{\Sigma}_{Y_A Y_A}), \tag{9.61}$$

where $\mathbf{\Sigma}_{Y_A Y_A} = \mathbf{H} \mathbf{\Sigma}_{XX} \mathbf{H}^\mathsf{T} + \sigma^2 \mathbf{I} + \mathbf{\Sigma}_{AA}$ is the covariance matrix of the distribution of the compromised measurements. Note that while in the deterministic attack case, the effect of the attack vector was captured by shifting the mean of the measurement vector, in the random attack case the attack changes the structure of the second-order moments of the measurements. Interestingly, the Gaussian attack construction implies that knowledge of the second-order moments of the state variables and the variance of the AWGN introduced by the measurement process suffices to construct the attack. This assumption significantly reduces the difficulty of devising an optimal attack strategy.

The operator of the power system makes use of the acquired measurements to detect the attack. The detection problem is cast as a hypothesis testing problem with hypotheses

$$\begin{aligned}\mathcal{H}_0 &: Y^m \sim \mathcal{N}(\mathbf{0}, \Sigma_{YY}), \quad \text{versus} \\ \mathcal{H}_1 &: Y^m \sim \mathcal{N}(\mathbf{0}, \Sigma_{Y_A Y_A}).\end{aligned} \qquad (9.62)$$

The null hypothesis $\mathcal{H}_0$ describes the case in which the power system is not compromised, while the alternative hypothesis $\mathcal{H}_1$ describes the case in which the power system is under attack.

Two types of error are considered in this hypothesis testing formulation, Type-I error refers to the probability of a "misdetection" event; while Type-II error refers to the probability of the "false alarm" event. The Neyman–Pearson lemma [9] states that for a fixed probability of Type-I error, the likelihood ratio test achieves the minimum Type-II error when compared with any other test with an equal or smaller Type-I error. Consequently, the LRT is chosen to decide between $\mathcal{H}_0$ and $\mathcal{H}_1$ based on the available measurements. The LRT between $\mathcal{H}_0$ and $\mathcal{H}_1$ takes following form:

$$L(\mathbf{y}) \triangleq \frac{f_{Y_A^m}(\mathbf{y})}{f_{Y^m}(\mathbf{y})} \overset{\mathcal{H}_1}{\underset{\mathcal{H}_0}{\gtrless}} \tau, \qquad (9.63)$$

where $\mathbf{y} \in \mathbb{R}^m$ is a realization of the vector of random variables modeling the measurements, $f_{Y_A^m}$ and f_{Y^m} denote the probability density functions (p.d.f.s) of Y_A^m and Y^m, respectively, and τ is the decision threshold set by the operator to meet a given false alarm constraint.

9.5.2 Information-Theoretic Setting

The aim of the attacker is twofold. Firstly, it aims to disrupt the state estimation process by corrupting the measurements in such a way that the network operator acquires the least amount of knowledge about the state of the system. Secondly, the attacker aspires to remain stealthy and corrupt the measurements without being detected by the network operator. In the following, we describe the information-theoretic measures that provide quantitative metrics for the objectives of the attacker.

The data-integrity of the measurements is measured in terms of the mutual information between the state variables and the measurements. The mutual information between two random variables is a measure of the amount of information that each random variable contains about the other. By adding the attack vector to the measurements, the attacker aims to reduce the mutual information that ultimately results in a loss of information about the state by the network operator. Specifically, the attacker aims to minimize $I(X^n; Y_A^m)$. In view of this, it seems reasonable to consider a Gaussian distribution for the attack vector as the minimum mutual information for the observation model in (9.5) is achieved by additive Gaussian noise.

The probability of attack detection is determined by the detection threshold τ set by the operator for the LRT and the distribution induced by the attack on the vector of compromised measurements. An analytical expression of the probability of attack detection can be described in closed form as a function of the distributions describing the measurements under both hypotheses. However, the expression is involved in general and it is not straightforward to incorporate it into an analytical formulation of the attack construction. For that reason, we instead consider the asymptotic performance of the LRT to evaluate the detection performance of the operator. The Chernoff–Stein lemma [10] characterizes the asymptotic exponent of the probability of detection when the number of observed measurement vectors grows to infinity. In our setting, the Chernoff–Stein lemma states that for any LRT and $\epsilon \in (0, 1/2)$, it holds that

$$\lim_{T \to \infty} \frac{1}{T} \log \beta_T^\epsilon = -D(P_{Y_A^m} \| P_{Y^m}), \tag{9.64}$$

where $D(\cdot \| \cdot)$ is the Kullback–Leibler (KL) divergence, β_T^ϵ is the minimum Type-II error such that the Type-I error α satisfies $\alpha < \epsilon$, and T is the number of m-dimensional measurement vectors that are available for the LRT detection procedure. As a result, minimizing the asymptotic probability of false alarm given an upper bound on the probability of misdetection is equivalent to minimizing $D(P_{Y_A^m} \| P_{Y^m})$, where $P_{Y_A^m}$ and P_{Y^m} denote the probability distributions of Y_A^m and Y^m, respectively.

The purpose of the attacker is to disrupt the normal state estimation procedure by minimizing the information that the operator acquires about the state variables, while guaranteeing that the probability of attack detection is sufficiently small, and therefore, remain stealthy.

9.5.3 Generalized Stealth Attacks

When the two information-theoretic objectives are considered by the attacker, in [11], a stealthy attack construction is proposed by combining two objectives in one cost function, i.e.,

$$I(X^n; Y_A^m) + D(P_{Y_A^m} \| P_{Y^m}) = D(P_{X^n Y_A^m} \| P_{X^n} P_{Y^m}), \tag{9.65}$$

where $P_{X^n Y_A^m}$ is the joint distribution of X^n and Y_A^m. The resulting optimization problem to construct the attack is given by

$$\min_{A^m} D(P_{X^n Y_A^m} \| P_{X^n} P_{Y^m}). \tag{9.66}$$

Therein, it is shown that (9.66) is a convex optimization problem and the covariance matrix of the optimal Gaussian attack is $\Sigma_{AA} = \mathbf{H}\Sigma_{XX}\mathbf{H}^\mathsf{T}$. However, numerical simulations on IEEE test systems show that the attack construction proposed in the preceding text yields large values of probability of detection in practical settings.

To control the probability of attack detection of the attack, the preceding construction is generalized in [12] by introducing a parameter that weights the detection term in the cost function. The resulting optimization problem is given by

$$\min_{A^m} I(X^n; Y_A^m) + \lambda D(P_{Y_A^m} || P_{Y^m}), \qquad (9.67)$$

where $\lambda \geq 1$ governs the weight given to each objective in the cost function. It is interesting to note that for the case in which $\lambda = 1$ the proposed cost function boils down to the effective secrecy proposed in [13] and the attack construction in (9.67) coincides with that in [11]. For $\lambda > 1$, the attacker adopts a conservative approach and prioritizes remaining undetected over minimizing the amount of information acquired by the operator. By increasing the value of λ the attacker decreases the probability of detection at the expense of increasing the amount of information that the operator might acquire using the measurements.

The attack construction in (9.67) is formulated in a general setting. The following propositions particularize the KL divergence and mutual information to our multivariate Gaussian setting.

PROPOSITION 9.5 [10] *The KL divergence between m-dimensional multivariate Gaussian distributions $\mathcal{N}(\mathbf{0}, \Sigma_{Y_A Y_A})$ and $\mathcal{N}(\mathbf{0}, \Sigma_{YY})$ is given by*

$$D(P_{Y_A^m} || P_{Y^m}) = \frac{1}{2}\left(\log \frac{|\Sigma_{YY}|}{|\Sigma_{Y_A Y_A}|} - m + \mathrm{tr}\big(\Sigma_{YY}^{-1} \Sigma_{Y_A Y_A}\big) \right). \qquad (9.68)$$

PROPOSITION 9.6 [10] *The mutual information between the vectors of random variables $X^n \sim \mathcal{N}(\mathbf{0}, \Sigma_{XX})$ and $Y_A^m \sim \mathcal{N}(\mathbf{0}, \Sigma_{Y_A Y_A})$ is given by*

$$I(X^n; Y_A^m) = \frac{1}{2} \log \frac{|\Sigma_{XX}||\Sigma_{Y_A Y_A}|}{|\Sigma|}, \qquad (9.69)$$

where Σ is the covariance matrix of the joint distribution of (X^n, Y_A^m).

Substituting (9.68) and (9.69) in (9.67) we can now pose the Gaussian attack construction as the following optimization problem:

$$\min_{\Sigma_{AA} \in \mathcal{S}_+^m} -(\lambda - 1)\log |\Sigma_{YY} + \Sigma_{AA}| - \log |\Sigma_{AA} + \sigma^2 \mathbf{I}| + \lambda \mathrm{tr}(\Sigma_{YY}^{-1} \Sigma_{AA}). \qquad (9.70)$$

We now proceed to solve the preceding optimization problem. First, note that the optimization domain $\mathcal{S}_+^m$ is a convex set. The following proposition characterizes the convexity of the cost function.

PROPOSITION 9.7 *Let $\lambda \geq 1$. Then the cost function in the optimization problem in (9.70) is convex.*

Proof Note that the term $-\log |\Sigma_{AA} + \sigma^2 \mathbf{I}|$ is a convex function on $\Sigma_{AA} \in \mathcal{S}_+^m$ [14]. Additionally, $-(\lambda - 1)\log |\Sigma_{YY} + \Sigma_{AA}|$ is a convex function on $\Sigma_{AA} \in \mathcal{S}_+^m$ when $\lambda \geq 1$. Because the trace operator is a linear operator and the sum of convex functions is convex, it follows that the cost function in (9.70) is convex on $\Sigma_{AA} \in \mathcal{S}_+^m$. □

THEOREM 9.5 *Let $\lambda \geq 1$. Then the solution to the optimization problem in (9.70) is*

$$\Sigma_{AA}^\star = \frac{1}{\lambda} \mathbf{H} \Sigma_{XX} \mathbf{H}^\mathsf{T}. \qquad (9.71)$$

Proof Denote the cost function in (9.70) by $f(\Sigma_{AA})$. Taking the derivative of the cost function with respect to Σ_{AA} yields

$$\frac{\partial f(\Sigma_{AA})}{\partial \Sigma_{AA}} = -2(\lambda-1)(\Sigma_{YY}+\Sigma_{AA})^{-1} - 2(\Sigma_{AA}+\sigma^2 I)^{-1} + 2\lambda\Sigma_{YY}^{-1} - \lambda\mathrm{diag}(\Sigma_{YY}^{-1})$$
$$+ (\lambda-1)\mathrm{diag}\big((\Sigma_{YY}+\Sigma_{AA})^{-1}\big) + \mathrm{diag}\big((\Sigma_{AA}+\sigma^2 I)^{-1}\big). \qquad (9.72)$$

Note that the only critical point is $\Sigma_{AA}^\star = \frac{1}{\lambda}H\Sigma_{XX}H^\mathsf{T}$. Theorem 9.5 follows immediately from combining this result with Proposition 9.7. □

COROLLARY 9.1 *The mutual information between the vector of state variables and the vector of compromised measurements induced by the optimal attack construction is given by*

$$I(X^n; Y_A^m) = \frac{1}{2}\log\left|H\Sigma_{XX}H^\mathsf{T}\left(\sigma^2 I + \frac{1}{\lambda}H\Sigma_{XX}H^\mathsf{T}\right)^{-1} + I\right|. \qquad (9.73)$$

Theorem 9.5 shows that the generalized stealth attacks share the same structure of the stealth attacks in [11] up to a scaling factor determined by λ. The solution in Theorem 9.5 holds for the case in which $\lambda \geq 1$, and therefore, lacks full generality. However, the case in which $\lambda < 1$ yields unreasonably high probability of detection [11] which indicates that the proposed attack construction is indeed of practical interest in a wide range of state estimation settings.

The resulting attack construction is remarkably simple to implement provided that the information about the system is available to the attacker. Indeed, the attacker only requires access to the linearized Jacobian measurement matrix H and the second-order statistics of the state variables, but the variance of the noise introduced by the sensors is not necessary. To obtain the Jacobian measurement matrix, a malicious attacker needs to know the topology of the grid, the admittances of the branches, and the operation point of the system. The second-order statistics of the state variables however, can be estimated using historical data. In [11] it is shown that the attack construction with a sample covariance matrix of the state variables obtained with historical data is asymptotically optimal when the size of the training data grows to infinity.

It is interesting to note that the mutual information in (9.73) increases monotonically with λ and that it asymptotically converges to $I(X^n; Y^m)$, i.e. the case in which there is no attack. While the evaluation of the mutual information as shown in Corollary 9.1 is straightforward, the computation of the associated probability of detection yields involved expressions that do not provide much insight. For that reason, the probability of detection of optimal attacks is treated in the following section.

9.5.4 Probability of Detection of Generalized Stealth Attacks

The asymptotic probability of detection of the generalized stealth attacks is governed by the KL divergence as described in (9.64). However in the non asymptotic case, determining the probability of detection is difficult, and therefore, choosing a value of λ that provides the desired probability of detection is a challenging task. In this section, we first provide a closed-form expression of the probability of detection by direct evaluation and show that the expression does not provide any practical insight over the choice of λ that achieves the desired detection performance. That being the case, we then provide an upper bound on the probability of detection, which, in turn, provides a lower bound on the value of λ that achieves the desired probability of detection.

Direct Evaluation of the Probability of Detection

Detection based on the LRT with threshold τ yields a probability of detection given by

$$\mathsf{P_D} \stackrel{\Delta}{=} \mathbb{E}\big[\mathbb{1}_{\{L(Y_A^m) \geq \tau\}}\big]. \tag{9.74}$$

The following lemma particularizes the preceding expression to the optimal attack construction described in Section 9.5.3.

LEMMA 9.2 *The probability of detection of the LRT in (9.63) for the attack construction in (9.71) is given by*

$$\mathsf{P_D}(\lambda) = \mathbb{P}\big[(U^p)^\mathsf{T} \Delta U^p \geq \lambda\big(2\log\tau + \log|\mathbf{I} + \lambda^{-1}\Delta|\big)\big], \tag{9.75}$$

where $p = \operatorname{rank}(\mathbf{H}\Sigma_{XX}\mathbf{H}^\mathsf{T})$, $U^p \in \mathbb{R}^p$ *is a vector of p random variables with distribution* $\mathcal{N}(\mathbf{0}, \mathbf{I})$, *and* $\Delta \in \mathbb{R}^{p \times p}$ *is a diagonal matrix with entries given by* $(\Delta)_{i,i} = \lambda_i(\mathbf{H}\Sigma_{XX}\mathbf{H}^\mathsf{T})\lambda_i(\Sigma_{YY}^{-1})$, *where* $\lambda_i(\mathbf{A})$ *with* $i = 1, \ldots, p$ *denotes the ith eigenvalue of matrix* $\mathbf{A}$ *in descending order.*

Proof The probability of detection of the stealth attack is,

$$\mathsf{P_D}(\lambda) = \int_{\mathcal{S}} \mathrm{d}P_{Y_A^m} \tag{9.76}$$

$$= \frac{1}{(2\pi)^{\frac{m}{2}} |\Sigma_{Y_A Y_A}|^{\frac{1}{2}}} \int_{\mathcal{S}} \exp\left\{-\frac{1}{2}\mathbf{y}^\mathsf{T} \Sigma_{Y_A Y_A}^{-1} \mathbf{y}\right\} \mathrm{d}\mathbf{y}, \tag{9.77}$$

where

$$\mathcal{S} = \{\mathbf{y} \in \mathbb{R}^m : L(\mathbf{y}) \geq \tau\}. \tag{9.78}$$

Algebraic manipulation yields the following equivalent description of the integration domain:

$$\mathcal{S} = \{\mathbf{y} \in \mathbb{R}^m : \mathbf{y}^\mathsf{T} \Delta_0 \mathbf{y} \geq 2\log\tau + \log|\mathbf{I} + \Sigma_{AA}\Sigma_{YY}^{-1}|\}, \tag{9.79}$$

with $\Delta_0 \stackrel{\Delta}{=} \Sigma_{YY}^{-1} - \Sigma_{Y_A Y_A}^{-1}$. Let $\Sigma_{YY} = \mathbf{U}_{YY}\Lambda_{YY}\mathbf{U}_{YY}^\mathsf{T}$ where $\Lambda_{YY} \in \mathbb{R}^{m \times m}$ is a diagonal matrix containing the eigenvalues of Σ_{YY} in descending order and $\mathbf{U}_{YY} \in \mathbb{R}^{m \times m}$ is a unitary matrix whose columns are the eigenvectors of Σ_{YY} ordered matching the order of the eigenvalues. Applying the change of variable $\mathbf{y}_1 \stackrel{\Delta}{=} \mathbf{U}_{YY}\mathbf{y}$ in (9.77) results in

$$\mathsf{P_D}(\lambda) = \frac{1}{(2\pi)^{\frac{m}{2}} |\Sigma_{Y_A Y_A}|^{\frac{1}{2}}} \int_{\mathcal{S}_1} \exp\left\{-\frac{1}{2}\mathbf{y}_1^\mathsf{T} \Lambda_{Y_A Y_A}^{-1} \mathbf{y}_1\right\} \mathrm{d}\mathbf{y}_1, \tag{9.80}$$

where $\Lambda_{Y_A Y_A} \in \mathbb{R}^{m \times m}$ denotes the diagonal matrix containing the eigenvalues of $\Sigma_{Y_A Y_A}$ in descending order. Noticing that Σ_{YY}, Σ_{AA}, and $\Sigma_{Y_A Y_A}$ are also diagonalized by $\mathbf{U}_{YY}$, the integration domain $\mathcal{S}_1$ is given by

$$\mathcal{S}_1 = \{\mathbf{y}_1 \in \mathbb{R}^m : \mathbf{y}_1^\mathsf{T} \Delta_1 \mathbf{y}_1 \geq 2\log \tau + \log |\mathbf{I} + \Lambda_{AA} \Lambda_{YY}^{-1}|\}, \qquad (9.81)$$

where $\Delta_1 \triangleq \Lambda_{YY}^{-1} - \Lambda_{Y_A Y_A}^{-1}$ with Λ_{AA} denoting the diagonal matrix containing the eigenvalues of Σ_{AA} in descending order. Further applying the change of variable $\mathbf{y}_2 \triangleq \Lambda_{Y_A Y_A}^{-\frac{1}{2}} \mathbf{y}_1$ in (9.80) results in

$$\mathsf{P}_D(\lambda) = \frac{1}{\sqrt{(2\pi)^m}} \int_{\mathcal{S}_2} \exp\left\{-\frac{1}{2}\mathbf{y}_2^\mathsf{T} \mathbf{y}_2\right\} d\mathbf{y}_2, \qquad (9.82)$$

with the transformed integration domain given by

$$\mathcal{S}_2 = \{\mathbf{y}_2 \in \mathbb{R}^m : \mathbf{y}_2^\mathsf{T} \Delta_2 \mathbf{y}_2 \geq 2\log \tau + \log |\mathbf{I} + \Delta_2|\}, \qquad (9.83)$$

with

$$\Delta_2 \triangleq \Lambda_{AA} \Lambda_{YY}^{-1}. \qquad (9.84)$$

Setting $\Delta \triangleq \lambda \Delta_2$ and noticing that $\mathrm{rank}(\Delta) = \mathrm{rank}(\mathbf{H} \Sigma_{XX} \mathbf{H}^\mathsf{T})$ concludes the proof. □

Notice that the left-hand term $(U^P)^\mathsf{T} \Delta U^P$ in (9.75) is a weighted sum of independent χ^2 distributed random variables with one degree of freedom where the weights are determined by the diagonal entries of Δ that depend on the second-order statistics of the state variables, the Jacobian measurement matrix, and the variance of the noise; i.e. the attacker has no control over this term. The right-hand side contains in addition λ and τ, and therefore, the probability of attack detection is described as a function of the parameter λ. However, characterizing the distribution of the resulting random variable is not practical because there is no closed-form expression for the distribution of a positively weighted sum of independent χ^2 random variables with one degree of freedom [15]. Usually, some moment matching approximation approaches such as the Lindsay–Pilla–Basak method [16] are utilized to solve this problem but the resulting expressions are complex and the relation of the probability of detection with λ is difficult to describe analytically following this course of action. In the following an upper bound on the probability of attack detection is derived. The upper bound is then used to provide a simple lower bound on the value λ that achieves the desired probability of detection.

Upper Bound on the Probability of Detection

The following theorem provides a sufficient condition for λ to achieve a desired probability of attack detection.

THEOREM 9.6 *Let $\tau > 1$ be the decision threshold of the LRT. For any $t > 0$ and $\lambda \geq \max(\lambda^\star(t), 1)$ then the probability of attack detection satisfies*

$$P_D(\lambda) \leq e^{-t}, \qquad (9.85)$$

where $\lambda^*(t)$ is the only positive solution of λ satisfying

$$2\lambda \log \tau - \frac{1}{2\lambda}\text{tr}(\Delta^2) - 2\sqrt{\text{tr}(\Delta^2)t} - 2\|\Delta\|_\infty t = 0, \qquad (9.86)$$

and $\|\cdot\|_\infty$ is the infinity norm.

Proof We start with the result of Lemma 9.2 which gives

$$P_D(\lambda) = \mathbb{P}\left[(U^p)^\mathsf{T} \Delta U^p \geq \lambda \left(2\log \tau + \log \left|\mathbf{I} + \lambda^{-1}\Delta\right|\right)\right]. \qquad (9.87)$$

We now proceed to expand the term $\log \left|\mathbf{I} + \lambda^{-1}\Delta\right|$ using a Taylor series expansion resulting in

$$\log \left|\mathbf{I} + \lambda^{-1}\Delta\right| = \sum_{i=1}^p \log\left(1 + \lambda^{-1}(\Delta)_{i,i}\right) \qquad (9.88)$$

$$= \sum_{i=1}^p \left(\sum_{j=1}^\infty \left(\frac{\left(\lambda^{-1}(\Delta)_{i,i}\right)^{2j-1}}{2j-1} - \frac{\left(\lambda^{-1}(\Delta)_{i,i}\right)^{2j}}{2j}\right)\right). \qquad (9.89)$$

Because $(\Delta)_{i,i} \leq 1$, for $i = 1, \ldots, p$, and $\lambda \geq 1$, then

$$\frac{\left(\lambda^{-1}(\Delta)_{i,i}\right)^{2j-1}}{2j-1} - \frac{\left(\lambda^{-1}(\Delta)_{i,i}\right)^{2j}}{2j} \geq 0, \text{ for } j \in \mathbb{Z}^+. \qquad (9.90)$$

Thus, (9.89) is lower bounded by the second-order Taylor expansion, i.e.,

$$\log |\mathbf{I} + \Delta| \geq \sum_{i=1}^p \left(\lambda^{-1}(\Delta)_{i,i} - \frac{\left(\lambda^{-1}(\Delta)_{i,i}\right)^2}{2}\right) \qquad (9.91)$$

$$= \frac{1}{\lambda}\text{tr}(\Delta) - \frac{1}{2\lambda^2}\text{tr}(\Delta^2). \qquad (9.92)$$

Substituting (9.92) in (9.87) yields

$$P_D(\lambda) \leq \mathbb{P}\left[(U^p)^\mathsf{T} \Delta U^p \geq \text{tr}(\Delta) + 2\lambda \log \tau - \frac{1}{2\lambda}\text{tr}(\Delta^2)\right]. \qquad (9.93)$$

Note that $\mathbb{E}\left[(U^p)^\mathsf{T} \Delta U^p\right] = \text{tr}(\Delta)$, and therefore, evaluating the probability in (9.93) is equivalent to evaluating the probability of $(U^p)^\mathsf{T} \Delta U^p$ deviating $2\lambda \log \tau - \frac{1}{2\lambda}\text{tr}(\Delta^2)$ from the mean. In view of this, the right-hand side in (9.93) is upper bounded by [17, 18]

$$P_D(\lambda) \leq \mathbb{P}\left[(U^p)^\mathsf{T} \Delta U^p \geq \text{tr}(\Delta) + 2\sqrt{\text{tr}(\Delta^2)t} + 2\|\Delta\|_\infty t\right] \leq e^{-t}, \qquad (9.94)$$

for $t > 0$ satisfying

$$2\lambda \log \tau - \frac{1}{2\lambda}\text{tr}(\Delta^2) \geq 2\sqrt{\text{tr}(\Delta^2)t} + 2\|\Delta\|_\infty t. \qquad (9.95)$$

The expression in (9.95) is satisfied with equality for two values of λ, one is strictly negative and the other one is strictly positive denoted by $\lambda^*(t)$, when $\tau > 1$. The result follows by noticing that the left-hand term of (9.95) increases monotonically for $\lambda > 0$ and choosing $\lambda \geq \max(\lambda^\star(t), 1)$. This concludes the proof. □

It is interesting to note that for large values of λ the probability of detection decreases exponentially fast with λ. We will later show in the numerical results that the regime in which the exponentially fast decrease kicks in does not align with the saturation of the mutual information loss induced by the attack.

9.5.5 Numerical Evaluation of Stealth Attacks

We evaluate the performance of stealth attacks in practical state estimation settings. In particular, the IEEE 30-bus and 118-bus test systems are considered in the simulation. In state estimation with linearized dynamics, the Jacobian measurement matrix is determined by the operation point. We assume a DC state estimation scenario [19, 20], and thus, we set the resistances of the branches to 0 and the bus voltage magnitude to 1.0 per unit. Note that in this setting it is sufficient to specify the network topology, the branch reactances, real power flow, and the power injection values to fully characterize the system. Specifically, we use the IEEE test system framework provided by MATPOWER [21]. We choose the bus voltage angle to be the state variables, and use the power injection and the power flows in both directions as the measurements.

As stated in Section 9.5.4, there is no closed-form expression for the distribution of a positively weighted sum of independent χ^2 random variables, which is required to calculate the probability of detection of the generalized stealth attacks as shown in Lemma 9.2. For that reason, we use the Lindsay–Pilla–Basak method [16] and the MOMENTCHI2 R package [22] to numerically evaluate the probability of attack detection.

The covariance matrix of the state variables is modeled as a Toeplitz matrix with exponential decay parameter ρ, where the exponential decay parameter ρ determines the correlation strength between different entries of the state variable vector. The performance of the generalized stealth attack is a function of the weight given to the detection term in the attack construction cost function, i.e. λ, the correlation strength between state variables, i.e. ρ, and the Signal-to-Noise Ratio (SNR) of the power system which is defined as

$$\text{SNR} \triangleq 10 \log_{10} \left(\frac{\text{tr}(\mathbf{H} \boldsymbol{\Sigma}_{XX} \mathbf{H}^\mathsf{T})}{m\sigma^2} \right). \tag{9.96}$$

Figures 9.1 and 9.2 depict the performance of the optimal attack construction for different values of λ and ρ with SNR = 10 dB and SNR = 20 dB, respectively, when $\tau = 2$. As expected, larger values of the parameter λ yield smaller values of the probability of attack detection while increasing the mutual information between the state variables vector and the compromised measurement vector. We observe that the probability of detection decreases approximately linearly for moderate values of $\log \lambda$. However, Theorem 9.6 states that for large values of λ the probability of detection decreases exponentially fast to zero. However, for the range of values of λ in which the decrease of probability of detection is approximately linear, there is no significant

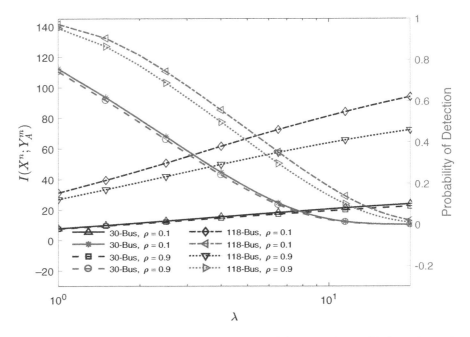

Figure 9.1 Performance of the generalized stealth attack in terms of mutual information and probability of detection for different values of λ and system size when $\rho = 0.1$, $\rho = 0.9$, SNR = 10 dB, and $\tau = 2$.

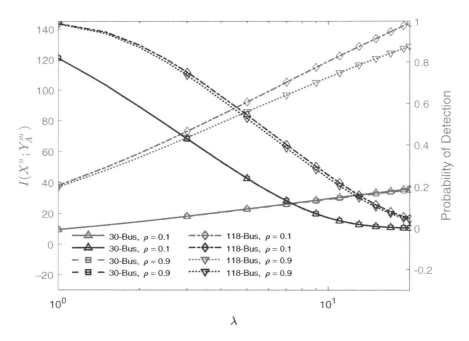

Figure 9.2 Performance of the generalized stealth attack in terms of mutual information and probability of detection for different values of λ and system size when $\rho = 0.1$, $\rho = 0.9$, SNR = 20 dB, and $\tau = 2$.

reduction on the rate of growth of mutual information. In view of this, the attacker needs to choose the value of λ carefully as the convergence of the mutual information to the asymptote $I(X^n;Y^m)$ is slower than that of the probability of detection to zero.

The comparison between the 30-bus and 118-bus systems shows that for the smaller size system the probability of detection decreases faster to zero while the rate of growth of mutual information is smaller than that on the larger system. This suggests that the choice of λ is particularly critical in large size systems as smaller size systems exhibit a more robust attack performance for different values of λ. The effect of the correlation between the state variables is significantly more noticeable for the 118-bus system. While there is a performance gain for the 30-bus system in terms of both mutual information and probability of detection due to the high correlation between the state variables, the improvement is more noteworthy for the 118-bus case. Remarkably, the difference in terms of mutual information between the case in which $\rho = 0.1$ and $\rho = 0.9$ increases as λ increases which indicates that the cost in terms of mutual information of reducing the probability of detection is large in the small values of correlation.

The performance of the upper bound given by Theorem 9.6 on the probability of detection for different values of λ and ρ when $\tau = 2$ and SNR $= 10$ dB is shown in Figure 9.3. Similarly, Figure 9.4 depicts the upper bound with the same parameters

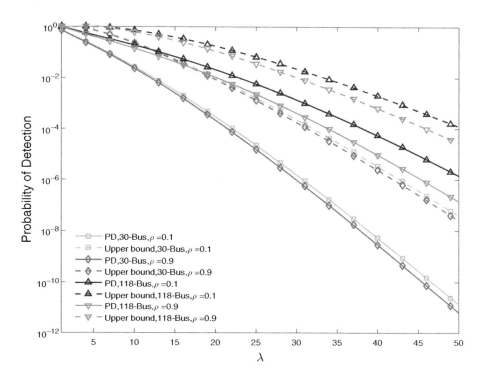

Figure 9.3 Upper bound on probability of detection given in Theorem 9.6 for different values of λ when $\rho = 0.1$ or 0.9, SNR $= 10$ dB, and $\tau = 2$.

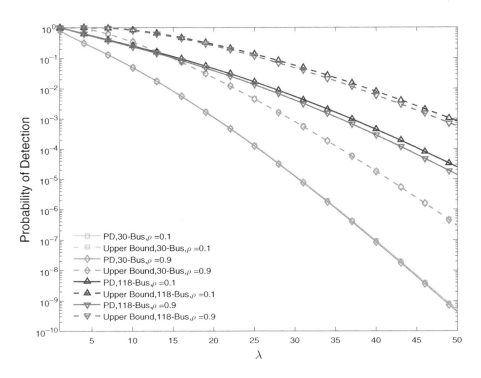

Figure 9.4 Upper bound on probability of detection given in Theorem 9.6 for different values of λ when $\rho = 0.1$ or 0.9, SNR $= 20$ dB, and $\tau = 2$.

but with SNR $= 20$ dB. As shown by Theorem 9.6, the bound decreases exponentially fast for large values of λ. Still, there is a significant gap to the probability of attack detection evaluated numerically. This is partially due to the fact that our bound is based on the concentration inequality in [17] that introduces a gap of more than an order of magnitude. Interestingly, the gap decreases when the value of ρ increases although the change is not significant. More importantly, the bound is tighter for lower values of SNR for both 30-bus and 118-bus systems.

9.6 Attack Construction with Estimated State Variable Statistics

9.6.1 Learning the Second-Order Statistics of the State Variables

The stealth attack construction proposed in the preceding text requires perfect knowledge of the covariance matrix of the state variables and the linearized Jacobian measurement matrix. In [23] the performance of the attack is studied when the second-order statistics are not perfectly known by the attacker but the linearized Jacobian measurement matrix is known. Therein, the partial knowledge is modeled by assuming that the attacker has access to a sample covariance matrix of the state variables. Specifically, the training data consisting of k state variable realizations $\{\mathbf{x}_i\}_{i=1}^{k}$ is available to the attacker.

That being the case, the attacker computes the unbiased estimate of the covariance matrix of the state variables given by

$$\mathbf{S}_{XX} = \frac{1}{k-1} \sum_{i=1}^{k} \mathbf{x}_i \mathbf{x}_i^\mathsf{T}. \quad (9.97)$$

The stealth attack constructed using the sample covariance matrix follows a multivariate Gaussian distribution given by

$$\tilde{A}^m \sim \mathcal{N}(\mathbf{0}, \Sigma_{\tilde{A}\tilde{A}}), \quad (9.98)$$

where $\Sigma_{\tilde{A}\tilde{A}} = \mathbf{H} \mathbf{S}_{XX} \mathbf{H}^\mathsf{T}$.

Because the sample covariance matrix in (9.97) is a random matrix with central Wishart distribution given by

$$\mathbf{S}_{XX} \sim \frac{1}{k-1} W_n(k-1, \Sigma_{XX}), \quad (9.99)$$

the ergodic counterpart of the cost function in (9.65) is defined in terms of the conditional KL divergence given by

$$\mathbb{E}_{\mathbf{S}_{XX}} \left[D \left(P_{X^n Y_A^m | \mathbf{S}_{XX}} \| P_{X^n} P_{Y^m} \right) \right]. \quad (9.100)$$

The ergodic cost function characterizes the expected performance of the attack averaged over the realizations of training data. Note that the performance using the sample covariance matrix is suboptimal [11] and that the ergodic performance converges asymptotically to the optimal attack construction when the size of the training data set increases.

9.6.2 Ergodic Stealth Attack Performance

In this section, we analytically characterize the ergodic attack performance defined in (9.100) by providing an upper bound using random matrix theory tools. Before introducing the upper bound, some auxiliary results on the expected value of the extreme eigenvalues of Wishart random matrices are presented here.

Auxiliary Results in Random Matrix Theory

LEMMA 9.3 *Let $\mathbf{Z}_l$ be an $(k-1) \times l$ matrix whose entries are independent standard normal random variables, then*

$$\mathrm{var}\,(s_{\max}(\mathbf{Z}_l)) \leq 1, \quad (9.101)$$

where $\mathrm{var}\,(\cdot)$ *denotes the variance and* $s_{\max}(\mathbf{Z}_l)$ *is the maximum singular value of* $\mathbf{Z}_l$.

Proof Note that $s_{\max}(\mathbf{Z}_l)$ is a 1-Lipschitz function of matrix $\mathbf{Z}_l$, the maximum singular value of $\mathbf{Z}_l$ is concentrated around the mean [24, Proposition 5.34] given by $\mathbb{E}[s_{\max}(\mathbf{Z}_l)]$. Then for $t \geq 0$, it holds that

$$\mathbb{P}[|s_{\max}(\mathbf{Z}_l) - \mathbb{E}[s_{\max}(\mathbf{Z}_l)]| > t] \leq 2 \exp\{-t^2/2\} \quad (9.102)$$
$$\leq \exp\{1 - t^2/2\}. \quad (9.103)$$

Therefore $s_{\max}(\mathbf{Z}_l)$ is a sub-Gaussian random variable with variance proxy $\sigma_p^2 \leq 1$. The lemma follows from the fact that $\text{var}(s_{\max}(\mathbf{Z}_l)) \leq \sigma_p^2$. □

LEMMA 9.4 *Let $\mathbf{W}_l$ denote a central Wishart random matrix distributed as $\frac{1}{k-1} W_l(k-1, \mathbf{I})$, then the nonasymptotic expected value of the extreme eigenvalues of $\mathbf{W}_l$ is bounded by*

$$\left(1 - \sqrt{l/(k-1)}\right)^2 \leq \mathbb{E}[\lambda_{\min}(\mathbf{W}_l)] \tag{9.104}$$

and

$$\mathbb{E}[\lambda_{\max}(\mathbf{W}_l)] \leq \left(1 + \sqrt{l/(k-1)}\right)^2 + 1/(k-1), \tag{9.105}$$

where $\lambda_{\min}(\mathbf{W}_l)$ and $\lambda_{\max}(\mathbf{W}_l)$ denote the minimum eigenvalue and maximum eigenvalue of $\mathbf{W}_l$, respectively.

Proof Note that [24, Theorem 5.32]

$$\sqrt{k-1} - \sqrt{l} \leq \mathbb{E}[s_{\min}(\mathbf{Z}_l)] \tag{9.106}$$

and

$$\sqrt{k-1} + \sqrt{l} \geq \mathbb{E}[s_{\max}(\mathbf{Z}_l)], \tag{9.107}$$

where $s_{\min}(\mathbf{Z}_l)$ is the minimum singular value of $\mathbf{Z}_l$. Given the fact that $\mathbf{W}_l = \frac{1}{k-1} \mathbf{Z}_l^\mathsf{T} \mathbf{Z}_l$, then it holds that

$$\mathbb{E}[\lambda_{\min}(\mathbf{W}_l)] = \frac{\mathbb{E}[s_{\min}(\mathbf{Z}_l)^2]}{k-1} \geq \frac{\mathbb{E}[s_{\min}(\mathbf{Z}_l)]^2}{k-1} \tag{9.108}$$

and

$$\mathbb{E}[\lambda_{\max}(\mathbf{W}_l)] = \frac{\mathbb{E}[s_{\max}(\mathbf{Z}_l)^2]}{k-1} \leq \frac{\mathbb{E}[s_{\max}(\mathbf{Z}_l)]^2 + 1}{k-1}, \tag{9.109}$$

where (9.109) follows from Lemma 9.3. Combining (9.106) with (9.108), and (9.107) with (9.109), respectively, yields the lemma. □

Recall the cost function describing the attack performance given in (9.100) can be written in terms of the covariance matrix $\mathbf{\Sigma}_{\tilde{A}\tilde{A}}$ in the multivariate Gaussian case with imperfect second-order statistics. The ergodic cost function that results from averaging the cost over the training data yields

$$\mathbb{E}\left[D\left(P_{X^n Y_A^m | S_{XX}} \| P_{X^n} P_{Y^m}\right)\right] = \frac{1}{2}\mathbb{E}[\text{tr}(\mathbf{\Sigma}_{YY}^{-1} \mathbf{\Sigma}_{\tilde{A}\tilde{A}}) - \log |\mathbf{\Sigma}_{\tilde{A}\tilde{A}} + \sigma^2 \mathbf{I}| - \log |\mathbf{\Sigma}_{YY}^{-1}|] \tag{9.110}$$

$$= \frac{1}{2}\left(\text{tr}(\mathbf{\Sigma}_{YY}^{-1} \mathbf{\Sigma}_{AA}^\star) - \log |\mathbf{\Sigma}_{YY}^{-1}| - \mathbb{E}\left[\log|\mathbf{\Sigma}_{\tilde{A}\tilde{A}} + \sigma^2 \mathbf{I}|\right]\right). \tag{9.111}$$

The assessment of the ergodic attack performance boils down to evaluating the last term in (9.110). Closed form expressions for this term are provided in [25] for the same case considered in this section. However, the resulting expressions are involved and are only computable for small dimensional settings. For systems with a large number of dimensions the expressions are computationally prohibitive. To circumvent this challenge we propose a lower bound on the term that yields an upper bound on the ergodic attack performance. Before presenting the main result we provide the following auxiliary convex optimization result.

LEMMA 9.5 Let $\mathbf{W}_p$ denote a central Wishart matrix distributed as $\frac{1}{k-1}W_p(k-1,\mathbf{I})$ and let $\mathbf{B} = \mathrm{diag}(b_1,\ldots,b_p)$ denote a positive definite diagonal matrix. Then

$$\mathbb{E}\left[\log |\mathbf{B} + \mathbf{W}_p^{-1}|\right] \geq \sum_{i=1}^{p} \log\left(b_i + 1/x_i^\star\right), \qquad (9.112)$$

where $x_i^\star$ is the solution to the convex optimization problem given by

$$\min_{\{x_i\}_{i=1}^p} \sum_{i=1}^{p} \log(b_i + 1/x_i) \qquad (9.113)$$

$$\text{s.t.} \quad \sum_{i=1}^{p} x_i = p \qquad (9.114)$$

$$\max(x_i) \leq \left(1 + \sqrt{p/(k-1)}\right)^2 + 1/(k-1) \qquad (9.115)$$

$$\min(x_i) \geq \left(1 - \sqrt{p/(k-1)}\right)^2. \qquad (9.116)$$

Proof Note that

$$\mathbb{E}\left[\log |\mathbf{B} + \mathbf{W}_p^{-1}|\right] = \sum_{i=1}^{p} \mathbb{E}\left[\log\left(b_i + \frac{1}{\lambda_i(\mathbf{W}_p)}\right)\right] \qquad (9.117)$$

$$\geq \sum_{i=1}^{p} \log\left(b_i + \frac{1}{\mathbb{E}[\lambda_i(\mathbf{W}_p)]}\right), \qquad (9.118)$$

where in (9.117), $\lambda_i(\mathbf{W}_p)$ is the ith eigenvalue of $\mathbf{W}_p$ in decreasing order; (9.118) follows from Jensen's inequality due to the convexity of $\log\left(b_i + \frac{1}{x}\right)$ for $x > 0$. Constraint (9.114) follows from the fact that $\mathbb{E}[\mathrm{trace}(\mathbf{W}_p)] = p$, and constraints (9.115) and (9.116) follow from Lemma 9.4. This completes the proof. □

Upper Bound on the Ergodic Stealth Attack Performance

The following theorem provides a lower bound for the last term in (9.110), and therefore, it enables us to upper bound the ergodic stealth attack performance.

THEOREM 9.7 Let $\mathbf{\Sigma}_{\tilde{A}\tilde{A}} = \mathbf{H}\mathbf{S}_{XX}\mathbf{H}^\mathsf{T}$ with $\mathbf{S}_{XX}$ distributed as $\frac{1}{k-1}W_n(k-1,\mathbf{\Sigma}_{XX})$ and denote by $\mathbf{\Lambda}_p = \mathrm{diag}(\lambda_1,\ldots,\lambda_p)$ the diagonal matrix containing the nonzero eigenvalues in decreasing order. Then

$$\mathbb{E}[\log|\Sigma_{\tilde{A}\tilde{A}} + \sigma^2 \mathbf{I}|] \geq \left(\sum_{i=0}^{p-1} \psi(k-1-i) \right) - p \log(k-1) + \sum_{i=1}^{p} \log \left(\frac{\lambda_i}{\sigma^2} + \frac{1}{\lambda_i^\star} \right)$$
$$+ 2m \log \sigma, \qquad (9.119)$$

where $\psi(\cdot)$ is the Euler digamma function, $p = \text{rank}(\mathbf{H}\Sigma_{XX}\mathbf{H}^\mathsf{T})$, and $\{\lambda_i^\star\}_{i=1}^p$ is the solution to the optimization problem given by (9.113) – (9.116) with $b_i = \frac{\lambda_i}{\sigma^2}$, for $i = 1, \ldots, p$.

Proof We proceed by noticing that

$$\mathbb{E}[\log|\Sigma_{\tilde{A}\tilde{A}} + \sigma^2 \mathbf{I}|] = \mathbb{E}\left[\log\left|\frac{1}{(k-1)\sigma^2}\mathbf{Z}_m^\mathsf{T} \Lambda \mathbf{Z}_m + \mathbf{I}\right|\right] + 2m \log \sigma \qquad (9.120)$$

$$= \mathbb{E}\left[\log\left|\frac{\Lambda_p}{\sigma^2} \frac{\mathbf{Z}_p^\mathsf{T} \mathbf{Z}_p}{k-1} + \mathbf{I}\right|\right] + 2m \log \sigma \qquad (9.121)$$

$$= \mathbb{E}\left[\log\left|\frac{\mathbf{Z}_p^\mathsf{T} \mathbf{Z}_p}{k-1}\right| + \log\left|\frac{\Lambda_p}{\sigma^2} + \left(\frac{\mathbf{Z}_p^\mathsf{T} \mathbf{Z}_p}{k-1}\right)^{-1}\right|\right] + 2m \log \sigma$$
$$\qquad (9.122)$$

$$\geq \left(\sum_{i=0}^{p-1} \psi(k-1-i) \right) - p \log(k-1) + \sum_{i=1}^{p} \log \left(\frac{\lambda_i}{\sigma^2} + \frac{1}{\lambda_i^\star} \right)$$
$$+ 2m \log \sigma, \qquad (9.123)$$

where in (9.120), Λ is a diagonal matrix containing the eigenvalues of $\mathbf{H}\Sigma_{XX}\mathbf{H}^\mathsf{T}$ in decreasing order; (9.121) follows from the fact that $p = \text{rank}(\mathbf{H}\Sigma_{XX}\mathbf{H}^\mathsf{T})$; (9.123) follows from [26, Theorem 2.11] and Lemma 9.5. This completes the proof. □

THEOREM 9.8 *The ergodic attack performance given in (9.110) is upper bounded by*

$$\mathbb{E}\left[D\left(P_{X^n Y_A^m | S_{XX}} \| P_{X^n} P_{Y^m}\right)\right] \leq \frac{1}{2}\left(\text{trace }(\Sigma_{YY}^{-1} \Sigma_{AA}^\star) - \log|\Sigma_{YY}^{-1}| - 2m \log \sigma \right.$$
$$- \left(\sum_{i=0}^{p-1} \psi(k-1-i) \right) + p \log(k-1)$$
$$\left. - \sum_{i=1}^{p} \log\left(\frac{\lambda_i}{\sigma^2} + \frac{1}{\lambda_i^\star}\right) \right). \qquad (9.124)$$

Proof The proof follows immediately from combing Theorem 9.7 with (9.110). □

Figure 9.5 depicts the upper bound in Theorem 9.8 as a function of number of samples for $\rho = 0.1$ and $\rho = 0.8$ when SNR $= 20$ dB. Interestingly, the upper bound in Theorem 9.8 is tight for large values of the training data set size for all values of the exponential decay parameter determining the correlation.

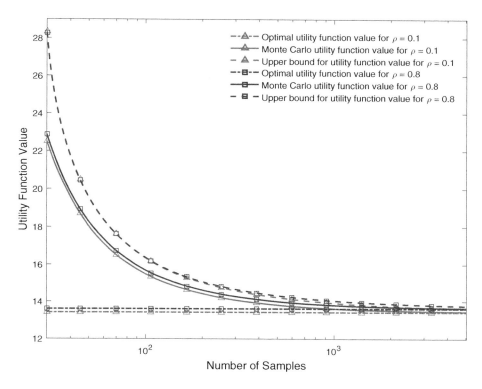

Figure 9.5 Performance of the upper bound in Theorem 9.8 as a function of number of sample for $\rho = 0.1$ and $\rho = 0.8$ when SNR $= 20$ dB.

9.7 Conclusion

We have cast the state estimation problem in a Bayesian setting and shown that the attacker can construct data-injection attacks that exploit prior knowledge about the state variables. In particular, we have focused in multivariate Gaussian random processes to describe the state variables and proposed two attack construction strategies: deterministic attacks and random attacks.

The deterministic attack is specified by the power system and the statistical structure of the state variables. The attack problem is cast as a multiobjective optimization problem in which the attacker aims to simultaneously minimize the MSE distortion induced by the injection vector and the probability of the attack being detected using a LRT. Within this setting, we have characterized the tradeoff between the achievable distortion and probability of detection by deriving optimal centralized attack constructions for a given distortion and probability of detection pair. We have then extended the investigation to decentralized scenarios in which several attackers construct their respective attack without coordination. In this setting, we have posed the interaction between the attackers in a game-theoretic setting. We show that the proposed utility function results in a setting that can be described as a potential game that allows us to claim the existence of an NE and the convergence of BRD to an NE.

The random attack produces different attack vectors for each set of measurements that are reported to the state estimator. The attack vectors are generated by sampling a defined attack vector distribution that yields attack vector realizations to be added to the measurements. The attack aims to disrupt the state estimation process by minimizing the mutual information between the state variables and the altered measurements while minimizing the probability of detection. The rationale for posing the attack construction in information-theoretic terms stems from the fundamental character that information measures grant to the attack vector. By minimizing the mutual information, the attacker limits the performance of a wide range of estimation, detection, and learning options for the operator. We conclude the chapter by analyzing the impact of imperfect second-order statistics about the state variables in the attack performance. In particular, we consider the case in which the attacker has access to a limited set of training state variable observations that are used to produce the sample covariance matrix of the state variables. Using random matrix theory tools we provide an upper bound on the ergodic attack performance.

Acknowledgment

This work was supported in part by the European Commission under Marie Skodowska–Curie Individual Fellowship No. 659316 and in part by the Agence Nationale de la Recherche (ANR, France) under Grant ANR-15-NMED-0009-03 and the China Scholarship Council (CSC, China).

References

[1] Y. Liu, P. Ning, and M. K. Reiter, "False data injection attacks against state estimation in electric power grids," in *Proc. ACM Conf. on Computer and Communications Security*, Chicago, IL, USA, pp. 21–32, Nov. 2009.

[2] O. Kosut, L. Jia, R. J. Thomas, and L. Tong, "Malicious data attacks on the smart grid," *IEEE Trans. Smart Grid*, vol. 2, no. 4, pp. 645–658, Dec. 2011.

[3] I. Esnaola, S. M. Perlaza, H. V. Poor, and O. Kosut, "Maximum distortion attacks in electricity grids," *IEEE Trans. Smart Grid*, vol. 7, no. 4, pp. 2007–2015, Jul. 2016.

[4] H. V. Poor, *An Introduction to Signal Detection and Estimation*. 2nd ed. New York: Springer-Verlag, 1994.

[5] I. Esnaola, S. M. Perlaza, H. V. Poor, and O. Kosut, "Decentralized maximum distortion MMSE attacks in electricity grids," *INRIA, Lyon, Tech. Rep. 466*, Sept. 2015.

[6] J. F. Nash, "Equilibrium points in n-person games," *Proc. National Academy of Sciences of the United States of America*, vol. 36, no. 1, pp. 48–49, Jan. 1950.

[7] D. Monderer and L. S. Shapley, "Potential games," *Games and Economic Behavior*, vol. 14, no. 1, pp. 124–143, May 1996.

[8] I. Shomorony and A. S. Avestimehr, "Worst-case additive noise in wireless networks," *IEEE Trans. Inf. Theory*, vol. 59, no. 6, pp. 3833–3847, Jun. 2013.

[9] J. Neyman and E. S. Pearson, "On the problem of the most efficient tests of statistical hypotheses," in *Breakthroughs in Statistics*, Springer Series in Statistics. New York: Springer pp. 73–108, 1992.

[10] T. M. Cover and J. A. Thomas, *Elements of Information Theory*. Hoboken, NJ: John Wiley & Sons, Nov. 2012.

[11] K. Sun, I. Esnaola, S. M. Perlaza, and H. V. Poor, "Information-theoretic attacks in the smart grid," in *Proc. IEEE Int. Conf. on Smart Grid Commum.*, Dresden, Germany, pp. 455–460, Oct. 2017

[12] K. Sun, I. Esnaola, S. M. Perlaza, and H. V. Poor, "Stealth attacks on the smart grid," *IEEE Trans. Smart Grid*, vol. 11 , no. 2, pp. 1276–1285, Mar. 2020.

[13] J. Hou and G. Kramer, "Effective secrecy: Reliability, confusion and stealth," in *Proc. IEEE Int. Symp. on Information Theory*, Honolulu, HI, USA, pp. 601–605, Jun. 2014.

[14] S. Boyd and L. Vandenberghe, *Convex Optimization*. Cambridge: Cambridge University Press, Mar. 2004.

[15] D. A. Bodenham and N. M. Adams, "A comparison of efficient approximations for a weighted sum of chi-squared random variables," *Stat. Comput.*, vol. 26, no. 4, pp. 917–928, Jul. 2016.

[16] B. G. Lindsay, R. S. Pilla, and P. Basak, "Moment-based approximations of distributions using mixtures: Theory and applications," *Ann. Inst. Stat. Math.*, vol. 52, no. 2, pp. 215–230, Jun. 2000.

[17] B. Laurent and P. Massart, "Adaptive estimation of a quadratic functional by model selection," *Ann. Statist.*, vol. 28, no. 5, pp. 1302–1338, 2000.

[18] D. Hsu, S. Kakade, and T. Zhang, "A tail inequality for quadratic forms of subgaussian random vectors," *Electron. Commun. in Probab.*, vol. 17, no. 52, pp. 1–6, 2012.

[19] A. Abur and A. G. Expósito, *Power System State Estimation: Theory and Implementation*. Boca Raton, FL: CRC Press, Mar. 2004.

[20] J. J. Grainger and W. D. Stevenson, *Power System Analysis*. New York: McGraw-Hill, 1994.

[21] R. D. Zimmerman, C. E. Murillo-Sánchez, and R. J. Thomas, "MATPOWER: Steady-state operations, planning, and analysis tools for power systems research and education," *IEEE Trans. Power Syst.*, vol. 26, no. 1, pp. 12–19, Feb. 2011.

[22] D. Bodenham. *Momentchi2: Moment-Matching Methods for Weighted Sums of Chi-Squared Random Variables*, https://cran.r-project.org/web/packages/momentchi2/index.html, 2016.

[23] K. Sun, I. Esnaola, A. M. Tulino, and H. V. Poor, "Learning requirements for stealth attacks," in *Proc. IEEE International Conference on Acoustics, Speech and Signal Processing*, Brighton, UK, pp. 8102–8106, May 2019.

[24] R. Vershynin, "Introduction to the non-asymptotic analysis of random matrices," in *Compressed Sensing: Theory and Applications*, Y. Eldar and G. Kutyniok, eds. Cambridge: Cambridge University Press, pp. 210–268, 2012.

[25] G. Alfano, A. M. Tulino, A. Lozano, and S. Verdú, "Capacity of MIMO channels with one-sided correlation," in *Proc. IEEE Int. Symp. on Spread Spectrum Techniques and Applications*, Sydney, Australia, Aug. 2004.

[26] A. M. Tulino and S. Verdú, *Random Matrix Theory and Wireless Communications*. Delft: Now Publishers Inc., 2004.

10 Smart Meter Data Privacy

Giulio Giaconi, Deniz Gündüz, and H. Vincent Poor

Smart grids (SGs) promise to deliver dramatic improvements compared to traditional power grids thanks primarily to the large amount of data being exchanged and processed within the grid, which enables the grid to be monitored more accurately and at a much faster pace. The smart meter (SM) is one of the key devices that enable the SG concept by monitoring a household's electricity consumption and reporting it to the utility provider (UP), i.e., the entity that sells energy to customers, or to the distribution system operator (DSO), i.e., the entity that operates and manages the grid, with high accuracy and at a much faster pace compared to traditional meters. However, the very availability of rich and high-frequency household electricity consumption data, which enables a very efficient power grid management, also opens up unprecedented challenges on data security and privacy. To counter these threats, it is necessary to develop techniques that keep SM data private, and, for this reason, SM privacy has become a very active research area. The aim of this chapter is to provide an overview of the most significant privacy-preserving techniques for SM data, highlighting their main benefits and disadvantages.

10.1 The SG Revolution

The SG refers to the set of technologies that have been developed to replace an increasingly ageing power infrastructure. Thanks to an extensive use of information and communication technologies and to the introduction of two-way communication links between the UPs and the end customers, the SG allows for improved system reliability, better quality of power delivered, and more rapid response to outages and thefts. The SG market is rapidly evolving and is fueled by a rapid penetration of SG technologies all over the world, as well as by extensive investments, ranging from US$23.8 billion in 2018 to an estimated US$61.3 billion by 2023, at a compound annual growth rate of 20.9% [1].

Key to the SG development is the installation of SMs at the households' premises, which allow near real-time power consumption information to be recorded and sent to the UPs or to the DSOs. SMs are the crucial elements in the SG revolution, as they send electricity consumption measurements at a much higher resolution and with a higher accuracy compared to traditional meters. SMs provide benefits for all parties in the SG. UPs are able to better understand and control the needs of their customers, as well as adjust electricity price dynamically according to short-term generation and consumption variations, and they are able to communicate this information to the consumers instantly. Additionally, the UPs can generate more accurate bills while reducing the need for back-office rebilling, detect energy theft and outages more rapidly, and implement load-shaping techniques. DSOs are able to reduce operational costs and energy losses; improve grid efficiency, system design, and distributed system state estimation; and better allocate resources to the current demand. Consumers take advantage of SMs to monitor their consumption in near real time, leading to better consumption awareness and energy usage management. Moreover, consumers are able to integrate microgeneration and energy storage devices into the grid, detect failing appliances and waste of energy more quickly, notice expected or unexpected activity, as well as migrate more easily between UPs. The SM market growth demonstrates the value of these technologies, which are expected to reach US$10.4 billion by 2022, with around 88 million SM installations taking place in 2017 [2]. To speed up SM adoption, many countries have introduced legislation that enforces SM installations; for example, most European Union countries need to reach 80% SM penetration rate by 2020 and 100% by 2022 [3].

The very property that allows SMs to deliver a much improved overall performance in the grid management is, however, also a source of concern for the privacy of SM users. In fact, many appliance load monitoring (ALM) techniques have been developed to gain insights into consumer behavior, inferring consumers' habits or preferences, and the number of household occupants. These privacy concerns are echoed by consumer associations and the media, and even delayed the SM rollout in the Netherlands in 2009, which proceeded forward only after the customers were given the possibility to opt out from the SM installation [4]. Privacy concerns are further exacerbated by recent legislation, such as the General Data Protection Regulation (GDPR) in Europe [5], which sets a limit on the collection, use, and processing of personal information. In particular, article 6 of the GDPR clearly states that the user must give explicit consent to processing of her personal information, and such processing should be limited to only a specific purpose.

10.2 ALM Techniques

ALM methods are aimed at monitoring a household's power consumption to achieve a wide range of benefits for the occupants by providing energy consumption data analysis at the appliance level. ALM techniques provide near real-time feedback on the user's power consumption behavior, are able to detect more power-hungry devices, and allow the automation of demand-side management [6]. Recent improvements in artificial intelligence and data communication and sensing technologies have made the SM benefits even more evident. ALM techniques can be divided into nonintrusive and intrusive load monitoring (NILM and ILM, respectively) techniques. While ILM techniques need to monitor power consumption at multiple points in a household [7], NILM techniques aim at recognizing the operation of electric loads within a household without the need to physically monitor each electrical device separately, relying only on aggregate SM measurements. ILM techniques are generally more accurate than NILM ones, however, they are also more invasive and expensive to deploy. For this reason, most of the works analyzing privacy for SM users are focused on NILM techniques, which create the biggest concern from a privacy point of view as they can be run by using a single probe attached to the SM and do not need any device to be physically installed within a target household.

The first NILM prototypes were devised in the eighties by George Hart [8]. Since then, NILM techniques have evolved in various directions, e.g., by considering either low- or high-frequency measurements; by focusing on detecting on/off events by means of either steady-state or transient signatures; or by analyzing the raw SM readings, e.g., by studying their frequency spectra. Additionally, both supervised and unsupervised machine learning models have been used for pattern recognition on SM data. Extensive surveys of NILM techniques are provided in [6], whereas [7] discuss ILM techniques in detail.

10.3 SM Privacy Concerns and Privacy-Preserving Techniques

Adoption of SMs and the use of NILM algorithms generate growing concerns about consumer privacy. An example of a typical power consumption profile along with some detected appliances is illustrated in Figure 10.1. It is noteworthy that such precise and accurate information would be available in principle only if very high-frequency SM readings were available to an attacker. However, even with low-frequency SM readings, the attacker may still be able to gain insights into users' activities and behaviors,

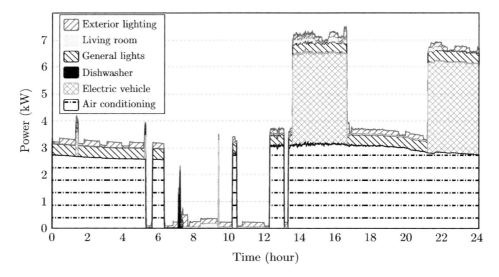

Figure 10.1 A power consumption profile in which the consumption of some appliances is highlighted [9]. (Data retrieved from the Dataport database [10])

determining, for example, a user's presence at home, her religious beliefs, disabilities, and illnesses [11–13]. Moreover, SM privacy risks could be particularly critical for businesses, e.g., factories and data centers, as their power consumption profiles may reveal sensitive information about the state of their businesses to their competitors. Such important privacy concerns in the use of SMs has raised significant public attention and they have been highly debated in the media and by politicians, and, if not properly addressed, they could represent a major roadblock for this multibillion dollar industry.

In the following, we adopt the classification introduced in [9], and divide the privacy-preserving techniques into *SM data manipulation* (SMDM) techniques, which manipulate SM readings before reporting them to the UP, and *user demand shaping* (UDS) techniques, which modify the actual electricity consumption by shaping it by means of physical devices such as renewable energy sources (RESs) or rechargeable batteries (RBs) [9]. The main difference between these sets of techniques is that while the SMDM techniques report corrupted or incomplete electrical consumption data to the UP to preserve user's privacy, the UDS techniques report a fully correct measurement, which is, however, generated by appropriately filtering the original user consumption. Hence, the UDS techniques do not typically suffer from the issue of data mismatch between the UP and the consumer, as opposed to some SMDM techniques, e.g., data obfuscation techniques. SMDM techniques have other notable disadvantages compared to UDS techniques, e.g., an eavesdropper may still be able to measure a user's consumption by installing additional probes outside the target household, hence choosing not to rely exclusively on the SM measurements; or the introduction of trusted third parties (TTPs), considered by many SMDM approaches, which only shifts the problem of trust from the UPs to the TTPs [9]. Finally, the UDS techniques allow the UP to have full visibility of user's consumption data, as opposed to some SMDM techniques, e.g., data aggregation, data anonymization, and data-sharing avoidance techniques. As a result, UDS techniques do not impact the utility of the SG as the actual consumption data is always shared with all the relevant stakeholders.

On the other hand, the major disadvantage of UDS techniques is that they require the presence of a physical device at the household, which can be costly for the user to purchase and install, such as in the case of RESs and RBs. However, such devices are becoming increasingly available [14], thanks to government incentives and decreasing cost of solar panels [15], residential RBs, as well as RBs for electric vehicles [16]. It is noteworthy that sharing RBs and RESs among multiple users, e.g., within the same neighborhood or block of apartments, results in reduced installation and operation costs as well as allowing management of the available energy in a centralized way, leading to a more efficient use of the available resources among multiple users. Other disadvantages of UDS techniques are that the use of physical sources may impact dynamic pricing and demand response, and such interaction has not been properly investigated yet. Finally, the shaping algorithms of the UDS techniques may prevent detecting anomalous consumption patterns.

Many surveys on SM privacy exist to date, each focusing on different aspects and techniques. Within the SMDM techniques, [17] provides an overview of data aggregation techniques, whereas [18] presents an excellent overview of cryptographic techniques and a wide discussion on privacy requirements and privacy legislation. The earlier survey in [19] discusses mostly SMDM techniques, whereas the recent magazine article [9] provides a wide review of UDS techniques. Differently from the previous surveys, the focus of this chapter is to provide an up-to-date technological review of the most significant SMDM and UDS techniques, without focusing on legal and normative aspects.

The following analysis considers the possible compromise of SM data, whereas the SM is assumed to be tamper resistant and trusted as it is equipped with a trusted platform module (TPM) to store cryptographic keys and to perform cryptographic operations. However, it is noteworthy that SMs suffer from physical attacks as well, which can be carried out to manipulate consumption data or to steal energy, and which can lead to devastating effects such as explosions [20].

The remainder of this chapter is organized as follows. SMDM techniques are analyzed in Section 10.4, whereas UDS techniques are discussed in Section 10.5. Conclusions are drawn in Section 10.6.

10.4 SMDM Techniques

The main SMDM techniques are data aggregation, obfuscation, anonymization, data-sharing prevention, and down-sampling. Although these techniques are explained in the following in different paragraphs for simplicity, many of the described works consider various combinations of these techniques. In the following, we denote random variables and their realizations by upper-case and lower-case letters, respectively. Let $X_{i,t}$ denote the total power requested by the appliances in a household i at time t, called the *user load*; and let $Y_{i,t}$, called the *grid load*, denote the electric load that is generated by the application of SMDM techniques to $X_{i,t}$, and which is reported to the UP and the DSO using the SM readings. The objective of the privacy-preserving policies is to keep $X_{i,t}$ private and report only a modified version of it, i.e., $Y_{i,t}$, to the UP. However, in general, the larger the deviation of $Y_{i,t}$ from $X_{i,t}$, the less useful $Y_{i,t}$ is for the UP or the DSO for optimal grid management and correct user billing. For this reason, for these techniques it is often of interest to characterize the tradeoff between privacy and utility, e.g., as studied from an information-theoretic point of view in [21]. We remark here that such a tradeoff is not typically analyzed within the UDS techniques, as the UDS techniques reshape the data by means of physical sources, and report to the UP the power that is actually requested by a household.

10.4.1 Data Aggregation Techniques

Data aggregation techniques typically propose solutions where SM measurements are encrypted and only the aggregate measurement from K different SMs are revealed to the UP. Aggregation may be achieved with the help of a TTP, which has perfect knowledge of all SM readings and sends only the aggregated measurements to the UP, as proposed in [22]. However, considering a TTP only shifts the problem of trust from one entity (UP) to another (TTP) without solving the SM privacy problem. Hence, the most significant data aggregation approaches avoid the presence of a centralized TTP, and propose decentralized approaches where SMs are grouped, e.g., into neighborhoods, and cooperate among themselves to achieve private system operation. Hybrid approaches also exists, where both a TTP and multiple data collectors are considered simultaneously [23].

Aggregation techniques typically require a certification authority that verifies the signatures of single SMs, and the capability of SMs to perform cryptographic operations, e.g., hash functions, symmetric and asymmetric encryption, and pseudorandom number generators, which are performed by a TPM [17]. Homomorphic encryption schemes are often used as they allow the UP to perform operations on the cyphertexts of encrypted messages without the necessity of decrypting the messages first, hence keeping the content of the message private. An encryption scheme is said to be homomorphic over an operation $*$ if $\mathsf{Enc}(m_1) * \mathsf{Enc}(m_2) = \mathsf{Enc}(m_1 * m_2), \forall m_1, m_2 \in M$, where Enc denotes the encryption algorithm and M is the set of all possible messages. Homomorphic encryption schemes are either *partial*, e.g., Paillier or ElGamal, which allow only a certain operation to be performed on the cyphertext, or *full*, which allow all operations to be performed but result in high computational complexity [18].

Paillier homomorphic encryption and additive secret sharing can be used so that the total power consumption is visible to the UP only at a neighborhood level and every SM in the neighborhood knows only a share of the consumption of all the other SMs [24]. To achieve this, each SM divides its readings into random shares that are encrypted using the public keys of other SMs in the neighborhoods. Then, the UP aggregates and encrypts the readings by means of homomorphic encryption and sends the readings back to the SMs whose public keys were used for encryption. Finally, each SM decrypts the shares encrypted using its public key, adds its own share, and sends the aggregated reading back to the UP. Despite its simplicity, this technique is not scalable as the amount of data increases due to the random shares, and the number of homomorphic encryptions is $\mathcal{O}(K^2)$ [17]. Noise $N_{i,t}$ may be added to individual SM readings to obtain $Y_{i,t} = X_{i,t} + N_{i,t}$, where the noise is computed so that it cancels out once all the readings from all the SMs in a neighborhood are aggregated by the UP, i.e., $\sum_{i=1}^{K} Y_{i,t} = \sum_{i=1}^{K} X_{i,t}$ [25]. Alternatively, each SM may output $g_i^{X_{i,t}+N_{i,t}}$, where g_i is the hash of a unique identifier and $N_{i,t}$ is computed so that they cancel out when the readings are aggregated, as proposed in [25], where g_i and N_i are derived by using the Diffie–Hellman key exchange protocol and a bilinear map. However, for the UP to be able to compute the aggregation, it needs to know g_i, $\forall i$, and an approximation of the total consumption, and, moreover, this technique results in $\mathcal{O}(K^2)$ messages, $\mathcal{O}(K)$ modulo multiplications, and $\mathcal{O}(1)$ exponentiations [17]. As SM data is inherently very rich and multidimensional, some techniques can be used to improve the efficiency of homomorphic systems and reduce the computational and communication overhead [26]. A further approach is to allow all SMs in a neighborhood to perform intermediate incremental and distributed aggregation, by constructing an aggregation tree rooted at the UP, and using homomorphic encryption to allow end-to-end secure aggregation so that intermediate aggregations are kept private [27].

Aggregation can also be coupled with differential privacy. A function f is defined to be ϵ-differentially private if, for any data sets D_1 and D_2, where D_1 and D_2 differ in at most a single element, and for all subsets of possible answers $S \subseteq \mathsf{Range}(f)$, the following condition holds: $p(f(D_1) \in S) \leq e^{\epsilon} \cdot p(f(D_2) \in S)$, where p denotes probability. Hence, differentially private functions produce similar outputs for inputs that differ on one element only [28]. A function can be made differentially private by the addition of Laplacian noise $\mathcal{L}(S(f)/\epsilon)$, where $S(f)$ is the global sensitivity of f. A Laplace distribution can be generated by summing independent and identically distributed (i.i.d.) gamma distributed random variables $\mathcal{G}(K,\lambda)$, where λ is a scale parameter for the Laplace distribution. Hence, $\mathcal{L}(\lambda) = \sum_{i=1}^{K}[\mathcal{G}_1(K,\lambda) - \mathcal{G}_2(K,\lambda)]$, where $\mathcal{G}_1(K,\lambda)$ and $\mathcal{G}_2(K,\lambda)$ are drawn independently from the same gamma distribution, i.e., Laplacian noise can be constructed by subtracting gamma distributed random variables. Hence, to achieve a differentially private operation, each SM adds gamma-distributed noise to its readings, encrypt them, and send the encrypted measurement to the UP in the form of $Y_{i,t} = X_{i,t} + \mathcal{G}_1(K,\lambda) - \mathcal{G}_2(K,\lambda)$. Moreover, SM data, corrupted by Laplacian noise and encrypted, can be further aggregated between groups of SMs [29].

Lately, blockchain technology has also been applied to provide privacy to SM users, especially in the context of data aggregation techniques. The use of blockchain technology, with its decentralized infrastructure, removes the need for a TTP, and the aggregator, or *miner*, is chosen directly from the users within a neighborhood. The miner transmits only the neighborhood aggregate consumption, and each user may create multiple pseudonyms to hide her identity [30]. Blockchain has also been considered to provide privacy for users in the context of energy trading [31].

A general issue with data aggregation techniques is the fact that the UP, or the DSO, is prevented from having a clear real-time picture of a single premise's consumption. This can adversely impact the UP in terms of local grid state estimation using SM readings [32], fault detection at the local level, and ability to implement dynamic pricing to mitigate peak demands [33]. Moreover, data aggregation techniques typically suffer from the so-called human-factor-aware attack, whereby an attacker may be able to estimate a user's consumption from the aggregate if she knows for example if the user is, or is not, at home [34]. Cryptographic techniques, heavily used in data aggregation approaches, typically suffer from high computational complexity, key distribution issues and overhead, and poor scalability, which prevent practical applicability in an SM setting where computational and bandwidth resources are limited. Additionally, cryptographic techniques are vulnerable to statistical attacks and power analysis [35].

10.4.2 Data Obfuscation Techniques

Data obfuscation revolves around the introduction of noise in the SM readings, i.e., $Y_{i,t} = X_{i,t} + N_{i,t}$, and many works that propose obfuscation techniques also involve aggregating data. In fact, as described in Section 10.4.1, if noise is properly engineered across multiple SMs, the aggregation at the UP allows the noise to be removed from the sum of the readings so that the UP is able to retrieve the total power consumption correctly. Alternatively, a simpler solution is to add noise to each SM independently of other SMs, e.g., by adding noise with a null expected value so that the expected value of the readings per each pricing period does not change, i.e., $\mathbb{E}[Y_{i,t}] = \mathbb{E}[X_{i,t}]$ and $\text{Var}[Y_{i,t}] = \text{Var}[N_{i,t}]$, as $X_{i,t}$ is not drawn from a random distribution [22]. The UP is able to retrieve an accurate estimate of the aggregate consumption across a group of SMs thanks to the convergence in the central limit theorem. This, however, requires a large number of SMs, which hinders the practical applicability of this technique. More specifically, the number of SMs that are needed is $\left(\frac{w \cdot v \cdot \text{Var}[N_{i,t}]}{d}\right)^2$, where w is the confidence interval width of the UP on the aggregate power consumption, v is the maximum peak power used by the consumer for obfuscation, and d is the allowed average deviation in power consumption at the household [22].

The amount of obfuscation can be determined directly by the UP, and then distributed across multiple SMs with the help of a lead meter. A TTP receives the vector of obfuscated measurements, sums them, and sends them to the UP, which is able to retrieve the correct aggregation value by subtracting the predetermined amount of obfuscation [36]. However, the TTP may represent a bottleneck for network traffic and, if compromised, may lead to the disclosure of the original SM readings. To overcome such security and efficiency issues, multiple TTPs can interact to create obfuscation vectors, which are used by each lead meter in its own subnetwork [33]. The latter approach, which has also been simulated in an IEEE 802.11s wireless mesh network, improves the overall reliability and efficiency but has the obvious disadvantage of requiring multiple TTPs.

Obfuscation techniques deliberately report incorrect readings to the UP, which creates a discrepancy between power production and consumption, and which prevents the UP from quickly reacting to energy outages and thefts and the DSO from properly managing the SG. As an example of the risks involved with obfuscation techniques, consider the optimal power flow problem, i.e., characterizing the best operating levels for electric power plants to meet demands while minimizing operating costs. It has been found that noise injection is positively correlated with the generators' power output, and the locational marginal price on each bus of the grid is mostly influenced by the noise applied at links that are in the bus or immediately adjacent to it [37]. This example shows how injecting noise may have the consequence of undermining the utility of the SG.

10.4.3 Data Anonymization Techniques

Data anonymization is about using pseudonyms to hide SMs' identities. Different pseudonyms for the same SM may be used for various functions, e.g., a pseudonym may be allocated for SM data sent at high frequency, necessary for real-time grid monitoring but more privacy sensitive, whereas another pseudonym may be allocated for SM data sent at low frequency, e.g., for billing purposes, and random time intervals are used to reduce correlation between the use of various pseudonyms [38]. The main problem with this approach is how to link the various pseudonyms to the same SM, which can be trivially achieved by using a TTP [38]. A disadvantage of these techniques is that recent advances in machine learning and anomaly detection lead to techniques that can successfully de-anonymize SM data [39].

10.4.4 Data-Sharing Prevention Techniques

These techniques propose methods to process SM data locally at a household, without the need for the readings to be sent to the UP. Hence, the energy bill is computed directly at the household or on any device trusted by the consumer on the basis of publicly accessible tariffs, while only the final bill is revealed to the UP. The issue of SM data privacy does not arise because user's data never leaves the household, and there is no need for sensitive data to be stored at the UP premises as well. Zero-knowledge proofs [40] are employed so that the UP can verify the integrity of the bill, and SM signatures are used to prove the identity of the sender [41]. As an example, Pedersen commitments can be used in the form $\mathsf{Commit}(x_{i,t}, r_{i,t})$, where $r_{i,t}$ is generated by using known Pedersen generators. These commitments are sent along with the total energy bill over T time slots (TSs) based on the specific time-of-use (ToU) tariff employed, $C = \sum_{t=1}^{T} x_{i,t} c_t$, where c_t is the power cost at TS t [42]. Alternatively, noninteractive zero-knowledge techniques can be used along with Camenisch–Lysyanskaya signatures [43], which can be applied to more complicated nonlinear ToU tariffs, i.e., tariffs that change after exceeding certain consumption thresholds [44].

Data-sharing prevention techniques may solve the basic problem of metering for billing, however, they cannot be applied in more dynamic scenarios in which energy cost changes quickly over short periods based on user demands, or when considering demand-side management and demand response. Also, these techniques do not solve the privacy problem when SM data needs to be necessarily shared for grid management purposes, e.g., with the DSO.

10.4.5 Data-Downsampling Techniques

Alternatively, it is possible to reduce the user load sampling rate, so that the UP receives less frequent SM readings. However, the less frequent the SM readings, the harder it is for the UP (or the DSO) to accomplish its duties. As the SG scenario can be modeled as a closed-loop between the UP and the consumer, whereby the UP reacts to SM readings through demand response, the aim is to minimize the user load sampling rate whilst allowing the closed-loop properties of the system, e.g., safety, stability, and reliability, to hold within acceptable limits [45].

10.5 UDS Techniques

Differently from SMDM techniques, UDS techniques report the actual power requested by the consumer, without any manipulation or the addition of any noise. However, what is reported to the UP is not the original load demand of the user, but rather a version of

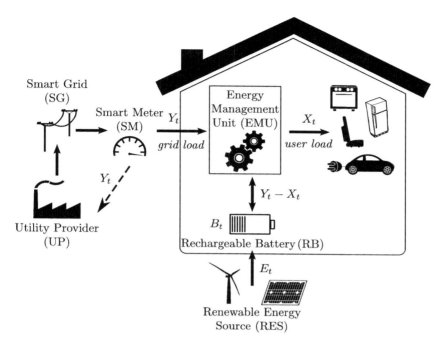

Figure 10.2 Representation of the system model [9]. X_t, Y_t, E_t, and B_t denote the consumer's energy demand, i.e., the user load; the SM readings, i.e., the grid load; the energy produced by the RES; and the level of energy in the RB at TS t, respectively. The meter readings being reported to the UP are shown by the dashed line. The energy management unit (EMU) is the physical or logical unit where the privacy-preserving algorithm resides.

it that is modified by means of the power exchanged with an additional physical device that is present at the household, e.g., an RB or an RES.

In the following we adopt the same discrete-time SM system model of [9], which is represented in Figure 10.2. $X_t \in \mathcal{X}$ and $Y_t \in \mathcal{Y}$ denote the total user load and the grid load at TS t, respectively, where $\mathcal{X}$ and $\mathcal{Y}$ denote the user load and grid load alphabets, respectively. Each TS duration is normalized to unit time, allowing the use of power and energy values interchangeably within a TS. Also, the user and grid loads are assumed to be constant within a TS, hence representing a discrete-time linear approximation of a continuous load profile, whose accuracy can be arbitrarily increased by reducing the TS duration. Because the aim of the UDS techniques is to protect the privacy of customers from the UP, the DSO, and all the SG parties that may be able to access near real-time power consumption information, the TSs in this model do not correspond to the sampling intervals used for transmitting SM measurements to the UP, but rather to the shorter time intervals that are used to request the actual power from the grid [9].

Depending on the user's priorities, part of the demand may not necessarily be satisfied immediately but only by a certain deadline, e.g., fully charging the electric vehicle by 8 a.m., with no specific requirement on the exact time the load needs to take place. Hence, some works explore *load-shifting* techniques that allow part of the user load to be shifted to a later time, appropriately called *elastic demand*, which may be applicable for loads including electric vehicle charging, and dishwasher and clothes washer-dryer cycles. This flexibility allows the consumer to employ *demand response* to increase her privacy as well as to lower the energy cost.

The electricity unit cost at TS t, C_t, can be modeled as a random variable, or in accordance with a specific ToU tariff, and the total cost incurred by a user to purchase Y_t units of power over a time interval of τ_t at the price of C_t is thus given by $\tau_t Y_t C_t$.

Physical Resources: RBs and RESs
Some of the UDS techniques consider an RB for shaping the grid load, so that the difference between the user and the grid load, $X_t - Y_t$, is retrieved from the RB. The amount of energy stored in the RB at TS t is $B_t \in [0, B_{\max}]$, where $B_{\max}$ denotes the maximum RB capacity, while the RB charging and discharging processes may be constrained by the so-called charging and discharging power constraints $\hat{P}_c$ and $\hat{P}_d$, respectively, i.e., $-\hat{P}_c \leq X_t - Y_t \leq \hat{P}_d$, $\forall t$, and additional losses in the battery charging and discharging processes may be taken into account to model a more realistic energy management system. The battery wear and tear due to charging and discharging can also be considered and modeled as a cost variable [46]. Some works also consider a less stringent constraint on the average power that can be retrieved from an RB $\bar{P}$, i.e., $\mathbb{E}\left[\frac{1}{n}\sum_{t=1}^{n}(X_t - Y_t)\right] \leq \bar{P}$. Where an RESs is considered, the renewable energy generated at TS t is denoted by $E_t \in \mathcal{E}$, where $\mathcal{E} = [0, E_{\max}]$ depending on the type of energy source. The amount of energy in the RB at TS $t+1$, B_{t+1}, can be computed on the basis of B_t as

$$B_{t+1} = \min\left\{B_t + E_t - (X_t - Y_t), B_{\max}\right\}. \quad (10.1)$$

Works that characterize theoretical expressions or bounds for the level of privacy achieved in SM systems typically consider the random processes X and E to be Markov or made up of sequences of i.i.d. random variables. Some works also study the scenario in which the UP knows the realizations of the renewable energy process E, which may occur if, for example, the UP has access to additional information from sensors deployed near the household that measure various parameters, e.g., solar or wind power intensity, and if it knows the specifications of the user's renewable energy generator, e.g., model and size of the solar panel. It is noteworthy that RBs and RESs can be used for both privacy protection and cost minimization, and using them jointly greatly increases the potential benefits. For example, from a cost-saving perspective, the user may be able to use the generated renewable energy when electricity is more expensive to buy from the grid, and may even be able to sell surplus energy to the grid.

The Energy Management Policy
The energy management policy (EMP) f, implemented by the EMU, decides on the grid load at any TS t based on the previous values of the user load X^t, renewable energy E^t, level of energy in the battery B^t, and grid load Y^{t-1}, i.e.,

$$f_t : \mathcal{X}^t \times \mathcal{E}^t \times \mathcal{B}^t \times \mathcal{Y}^{t-1} \to \mathcal{Y}, \qquad \forall t, \quad (10.2)$$

where $f \in \mathcal{F}$, and $\mathcal{F}$ denotes the set of feasible policies, i.e., policies that produce grid load values satisfying the RB and RES constraints at any time, as well as the battery update equation (10.1). The EMP is chosen so that it optimizes the user privacy along with other targets, e.g., the cost of energy or the amount of wasted energy, and it has to satisfy the user demand. The EMP in (10.2) can be analyzed either as an *online EMP*, which only considers information available causally right up to the current time to make a decision, or as an *offline EMP*, in which case the future user load values are assumed to be known in a noncausal fashion. Although online algorithms are more realistic and relevant for real-world applications, offline algorithms may still lead to interesting intuition or bounds on the performance, and noncausal knowledge of the electricity price process as well of power consumption for large appliances such as refrigerators, boilers, heating, and electric vehicles may still be considered valid.

A number of privacy measures and techniques have been proposed in the literature, each with its own advantages and drawbacks. In the following, we review the most significant approaches, and distinguish between *heuristic* and *theoretically grounded* techniques [9]. Because NILM algorithms look for sudden changes in the grid load profile $y_t - y_{t-1}$, and assign them to specific electric appliances' on/off events, the so-called *features* [47], heuristic techniques are aimed at minimizing such changes in the grid load. However, because these approaches counter specific NILM techniques, the validity of their privacy guarantees are also limited only against these attacks, and they do not provide theoretical assurances on the amount of privacy that can be achieved. On the contrary, theoretically grounded techniques typically provide a rigorous definition of privacy measure, and characterize ways to achieve privacy providing theoretical guarantees under that measure. However, their practical implementation may be harder to achieve and demonstrate.

10.5.1 Heuristic Privacy Measures: Variations in the Grid Load Profile

Generating a completely flat (equivalently, deterministic) or a completely random (independent of the user load) grid load profile can provide privacy against NILM algorithms. However, this could be achievable in practice only by having a very large RB or a very powerful RES, or by requesting more power than needed from the UP, both options being potentially extremely costly for the consumer. In the following, we describe various EMPs on the basis of the privacy measure or the specific technique being adopted.

Optimization Techniques

A possible solution to reducing the variations in the grid load profile is to set up an appropriate constant *target load profile* W and try to match it over time. The capability of a privacy-preserving algorithm is then measured by how tight this match is, i.e., how small the variance of the grid load Y is around W [48]:

$$\mathcal{V}_T \triangleq \frac{1}{T} \sum_{t=1}^{T} \mathbb{E}\left[(Y_t - W)^2\right], \tag{10.3}$$

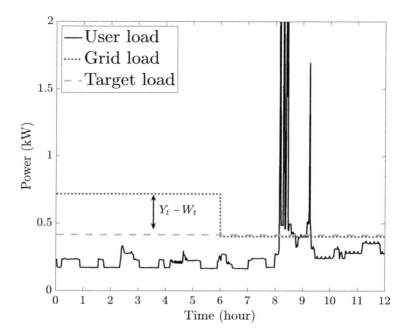

Figure 10.3 Examples of user load, grid load, and target load profiles when considering a constant target load profile [9]. In this figure the "distance" between the grid load and the target load, $Y_t - W_t$, is highlighted. The algorithms presented in this section are aimed at minimizing the average squared distance.

where the expectation is over X_t and Y_t, and $W = \mathbb{E}[X]$ may be considered. In fact, in the limiting scenario in which the target load profile is completely flat this would be equivalent to leaking only the average power consumption to the UP, unless more power than that needed by the consumer has been requested. This scenario is shown in Figure 10.3, where the solid line represents the user load, the dashed line represents the constant target load profile, and the dotted line represents the actual grid load profile. Additionally, also the cost of energy, expressed by the following equation, may need to be minimized:

$$C_T \triangleq \frac{1}{T} \sum_{t=1}^{T} \mathbb{E}\left[C_t Y_t\right]. \tag{10.4}$$

A solution to the joint optimization of Equations (10.3) and (10.4) can be characterized for an offline framework, where the optimal privacy and cost of energy can be found as the points on the Pareto boundary of the convex region formed by all the cost and privacy leakage pairs by solving the following convex optimization problem [48]:

$$\min_{Y_t \geq 0} \sum_{t=1}^{T} \left[(1-\alpha)Y_t C_t + \alpha(Y_t - W)^2\right], \tag{10.5}$$

where $0 \leq \alpha \leq 1$ strikes the tradeoff between privacy and cost of energy, which can be set up by the user. The solution to Equation (10.5) has a water-filling interpretation with a variable *water level* due to to the instantaneous power constraints. When modelling the battery wear and tear caused by charging and discharging the RB, the optimization can be expressed as [46]:

$$\min \frac{1}{T} \sum_{t=1}^{T} \mathbb{E}\left[C_t Y_t + \mathbb{1}_B(t) C_B + \alpha (Y_t - W)^2\right], \qquad (10.6)$$

where $\mathbb{1}_B(t) = 1$ if the battery is charging/discharging at time t, and 0 otherwise and C_B is the battery operating cost due to the battery wear and tear caused by charging and discharging the RB; and the expectation in (10.6) is over the probability distributions of all the involved random variables, i.e., X_t, Y_t, and C_t. The solution to Equation (10.6) has been characterized for an online setting by means of a Lyapunov function with a perturbed weight and by adopting the *drift-plus-penalty* framework, which consists of the simultaneous minimization of a so-called *drift*, i.e., the difference in the level of energy in the RB at successive time instants, and of a *penalty* function, i.e., the optimization target. The solution to this problem leads to a mixed-integer nonlinear program, which can be solved by decomposing the problem into multiple cases and solving each of them separately [46]. With a similar approach, it is possible to constrain the grid load to be within a certain maximum range λ of an average historical load $\bar{Y}$ at any TS, i.e., $\lambda \leq Y(t) - \bar{Y} \leq \lambda$ [49]. In the latter work, load shifting is analyzed to exploit the possibility of shifting nonurgent appliances to improve the privacy-cost tradeoff, and an anomaly detection method is developed to detect attacks on the electricity prices publicized to consumers.

Matching a completely constant target load profile is not feasible most of the time as that would require the user to have a large RB or RES. Additionally, it would conflict with the cost-saving objective as the constant target load completely disregards any ToU tariff. Instead, it would be reasonable to assume that a user would prefer to request more electricity over less expensive TSs compared to more expensive TSs. To allow such flexibility, one can set a piecewise constant target load profile, as shown in Figure 10.4 [50]. Accordingly, the optimization problem can be expressed as

$$\min_{Y_t, W^{(i)}} \sum_{i=1}^{M} \sum_{t=t_c(i-1)}^{t_c(i)-1} \left[\alpha (Y_t - W^{(i)})^2 + (1-\alpha) Y_t C^{(i)}\right], \qquad (10.7)$$

where $C^{(i)}$ and $W^{(i)}$ are the cost of the energy purchased from the UP and the target profile level during the ith price period, respectively, where $1 \leq i \leq M$, M is the total number of price periods during time T, and the ith price period spans from time slot $t_{c(i-1)}$ to $t_{c(i)}$. As expected, considering a piecewise constant target profile allows the system to reach a better privacy-cost tradeoff compared to a constant target profile, as shown in Figure 10.5, and allowing energy to be sold to the grid improves the tradeoff even further [50]. However, it is noteworthy that adopting a piecewise constant target profile introduces an inherent information leakage compared to a constant target load profile that is not fully captured by the tradeoff in Figure 10.5.

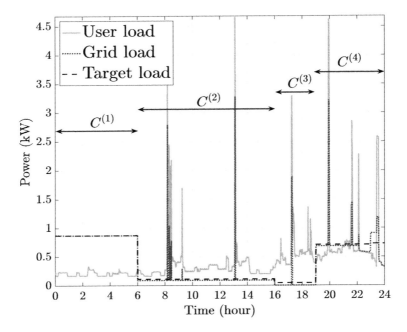

Figure 10.4 Examples of user load, grid load, and target load profiles over the course of a day when considering a piecewise target load profile [9, 50]. The arrows highlight the various price periods. Note that the target assumes a different constant value for each price period. Electricity consumption data retrieved from the UK-Dale data set [51].

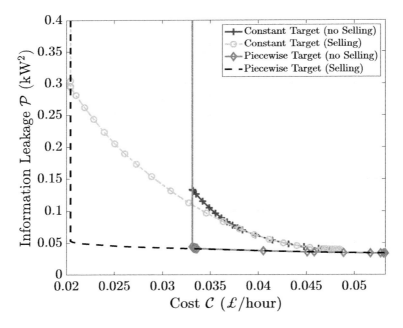

Figure 10.5 Privacy-cost tradeoff when using a Powervault G200-LI-4KWH RB [52] and adopting the strategies characterized in [48] and [50].

The adoption of a piecewise constant target load profile has also been studied in a more realistic scenario, called the *short horizon model* (SHM), in which the consumer's future consumption profile is known to the EMU only for the next H_F TSs, and where a *moving* or *receding horizon* model is considered [53]. Let $\overline{t + H_F} \triangleq \min\{t + H_F, T\}$, and let $\overline{t - H_P} \triangleq \max\{t - H_P, 0\}$. Then, the optimization problem is formulated as

$$\min_{Y_t^{\overline{t+H_F}}, W_t} \alpha \sum_{\tau=\overline{t-H_P}}^{\overline{t+H_F}} (Y_\tau - W_t)^2 + (1-\alpha) \sum_{\tau=t}^{\overline{t+H_F}} Y_\tau C_\tau, \qquad (10.8)$$

which states that at TS t the EMP produces the optimal grid load for the current TS and the prediction horizon $Y_t^{\overline{t+H_F}}$, and the optimal target load for the current time W_t. It is noteworthy that the SM remembers the consumption that occurred during the previous H_P TSs, considered in the term $\sum_{\tau=\overline{t-H_P}}^{t-1}(Y_\tau - W_t)^2$, to ensure a smooth variation of the overall target load profile. Figure 10.6 compares the load profiles of the SHM (Figures 10.6(a) and 10.6(c)) and the offline scenario, called the *long horizon model* (LHM) (Figures 10.6(b) and 10.6(d)) over the course of one day, also including the

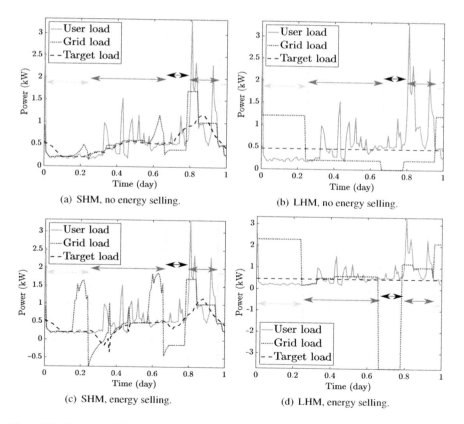

Figure 10.6 Power profiles for the SHM and the LHM scenarios, $\alpha = 0.5$ and $H_F = H_P = 2$ hours [53]. Off-peak, medium and peak price periods for the electricity cost are denoted by increasingly darker arrows.

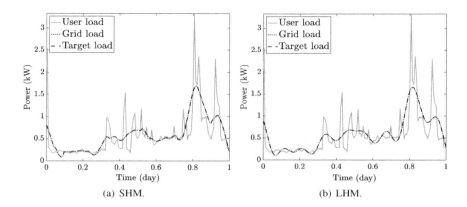

Figure 10.7 Power profiles when considering a filtered target load with cutoff frequency of 0.1 mHz, $\alpha = 1$, and $H_F = H_P = 2$ hours [53].

scenario in which energy can be sold to the grid. The LHM results in a flatter grid load profile compared to the SHM, however, the SHM is also able to flatten the consumption peaks to some extent and the resulting peaks in the grid load are not aligned with the original peaks in the user load [53].

Another target profile that has been considered is a low-pass filtered version of the user load, as high-frequency components in the grid load are more sensitive compared to low-frequency components because they leak more information about a user's activities [54]. The optimization problem for this scenario can be expressed as [53]

$$\min_{Y_t^{t+H_F}} \alpha \sum_{\tau=t}^{\overline{t+H_F}} (Y_\tau - W_\tau)^2 + (1-\alpha) \sum_{\tau=t}^{\overline{t+H_F}} Y_\tau C_\tau, \qquad (10.9)$$

where $W_t, W_{t+1}, \ldots, W_{\overline{t+H_F}}$ are obtained as low-pass filtered versions of the user load. Figure 10.7 shows the power profiles for the SHM and LHM settings and a filtered user load as the target load profile. Compared to the previous scenarios, the almost perfect match between grid and target load profiles in this setting is noteworthy.

Temporal and Spatial Similarities in the Grid Load as Privacy Measures

Let N_a be the total number of electrical appliances in a household, then $X_t = \sum_{a}^{N_a} X_{a,t}$ and $Y_t = \sum_{a}^{N_a} Y_{a,t}$, where $X_{a,t}$ and $Y_{a,t}$ are the user and grid loads generated by the a-th appliance. Other approaches to flattening the grid load are minimizing the difference in the power profile of each single appliance across all other TSs, i.e., the quantity $\sum_{t=1, t \neq t_0}^{T} |Y_{a,t} - Y_{a,t_0}|, \forall a, t_0$; minimizing the difference in the power profile of every combination of two appliances in the same TS, i.e., the quantity $\sum_{a=1}^{N_a-1} \sum_{j=a+1}^{N_a} |Y_{a,t} - Y_{j,t}|, \forall t, a$; and minimizing the difference in the aggregated power profile over consecutive TSs, i.e., the quantity $\sum_{t=1}^{T-1} \left| \sum_{a=1}^{N_a} Y_{a,t+1} - Y_{a,t} \right|$ [55]. In an online optimization framework, the former quantities are computed by estimating the future electricity prices and consumption by means of Monte Carlo simulations, and the optimal solution is characterized through a rolling online stochastic optimization process and by means of *load shifting*.

Markov Decision Process Formulation

The SM privacy problem can be cast as a Markov Decision Process (MDP), where the X and E processes are modeled as Markov processes, and the information leaked about a user is included within the *cost* incurred by the MDP. The *state* of the MDP at time t is typically given by a combination of the energy currently available in the RB B_t, the user demand X_t, and the renewable energy generated E_t, whereas the *action* u_t, expressed by the EMP, is given by the amount of power demanded from the grid, Y_t, and the energy used from the RB and RES, as a function of the current state. The *state transitions* are given by the transitions in the user demand, renewable energy generation, and the battery update equation. The target of an MDP is to determine the policy that can minimize the average, or discounted cost for every possible state, by considering either a finite or an infinite horizon setting. The MDP can be characterized by formulating the optimal Bellman optimality equations [56], which can be solved to obtain the optimal EMP at each state and time instant. One of the prerequisites necessary to formulate a problem as an MDP is to have a cost that is additive over time, i.e., the total cost is computed as the summation of the costs incurred at different TSs. An additive formulation for the SM privacy loss is natural when the privacy loss is expressed as the variance of the user load with respect to a target load, but it is more challenging when considering other measures, e.g., mutual information [9].

When the loss of privacy is measured by the fluctuations of the grid load around a constant target load, and the joint optimization of privacy and cost is considered, the SM problem can be cast as an MDP [57]. Q-learning [58], an iterative algorithm that computes the expected cost for each state-action pair by alternating exploitation and exploration phases, can be adopted when the transition probabilities $p(X_t|X_{t-1})$ and $p(B_t|B_{t-1}, u_t)$ are not known or stochastic, which is typically the case in the SM setting.

Heuristic Algorithms

One intuitive approach to SM privacy is battery charging and discharging algorithms that keep the grid load as constant as possible. For example, the RB could be discharged (charged) when the current user load is larger (smaller) than that at the previous TS, which would hide the higher frequency components of the user load [59]. In [59], the differences between the resulting grid and user load distributions are measured by computing the *empirical relative entropy*, by clustering SM data according to power levels, or by using *cross-correlation* and *regression* procedures, i.e., shifting the grid load in time to reach the point of maximum cross-correlation with the user load and using regression methods to compare the two aligned power profiles [59].

A more advanced method is to consider multiple grid load target values and let the EMP maintain the grid load to be equal to one of these values [60]. In [60] one main target value is considered for the grid load to match, called the *steady-state target*, and high and low *recovery states* are introduced, which are matched by the grid load in case of persistent light or heavy user demand, respectively. When this happens, strategies similar to those employed in [61] are used to modify the steady-state target load to permit the RB to be charged or discharged, and an exponentially weighted moving average of the demand is used to update the steady-state target load to reduce the occurrences of recovery states.

However, these intuitive algorithms suffer from load change recovery attacks that can identify peaks of user demand [62]. The use of a steady-state target load and high- and low-recovery states can be generalized by considering an arbitrary number of steady states, as this is equivalent to considering a larger number of quantization levels for the user load [62]. Such a "stepping" EMP results in an irreversible process because quantization is a "many-to-few" mapping. Let β be the step size that satisfies the RB maximum capacity and power constraints, and let h_t be an integer, so that $y_t = h_t \beta$. The grid load is chosen between the quantization levels that are adjacent to the user load, i.e., $\lceil \frac{x_t}{\beta} \rceil$ and $\lfloor \frac{x_t}{\beta} \rfloor$, where $\lceil \cdot \rceil$ and $\lfloor \cdot \rfloor$ denote the ceiling and floor functions, respectively. Various stepping algorithms are studied in [62]: one that keeps the grid load constant for as long as possible; one that keeps charging (discharging) the RB until it is full (empty); and another that chooses its actions at random. Despite being thoroughly analyzed, it is difficult to determine the levels of privacy these stepping algorithms can achieve, given their heuristic nature. Additionally, heuristic schemes may be based on deterministic schemes, which make them easier to be reverse-engineered.

10.5.2 Theoretical Guarantees on SM Privacy

Above all, being able to provide theoretical guarantees or assurances on the level of privacy that can be achieved in an SM scenario is of utmost importance. Such guarantees should be completely independent of any assumption on the attacker's capabilities, e.g., the NILM algorithms employed or the amount of computational resources available, so that their validity can be absolute. Theoretically grounded methods would also make it easier to compare the level of privacy achieved in various scenarios, e.g., using RBs of various capacities or RESs of various power outputs. To be able to achieve theoretical formulations, these techniques typically assume that the statistics of the user load and renewable energy process are stationary over time and known to the EMU, which is reasonable if these can be learned over a sufficiently long period [63–65]. Additionally, most of the works in this area also develop suboptimal policies that are applied to real power traces, which allow the reader to gain an intuition on the proposed techniques. Finally, the worst-case approach of considering the statistics governing the random processes to be known to the attacker is followed, which further strengthens the privacy guarantees.

Theoretical analysis studies the performance of SM privacy over long time horizons, focusing on the average user information leaked over time and its asymptotic behavior. Because the problem complexity increases with time, one of the challenges of the theoretical analysis is to find "single-letter" expressions for the optimal solutions, which would significantly reduce the complexity. However, the model needs to be simplified, e.g., by considering an i.i.d. or Markov user load or RES generation, to be able to obtain closed-form or single-letter expressions for the information leaked in an SM system.

Mutual Information as a Privacy Measure

The entropy of a random variable X, $H(X)$, measures the uncertainty of its realizations, whereas the mutual information (MI) between random variables X and Y, $I(X;Y)$, measures the amount of information shared between the two random variables and the dependance between them. $I(X;Y)$ ranges between zero, if X and Y are independent, and $H(X) = H(Y)$ if X and Y are completely dependent [66]. Additionally, $I(X;Y)$ can be interpreted as the average reduction in uncertainty of X given the knowledge of Y, hence lending itself perfectly as a measure of the information shared between the user load and the grid load processes X^n and Y^n. For an SM system with only an RB (no RES) and a given EMP f in (10.2), running over n time slots, the average *information leakage rate* $\mathcal{I}_f^n(B_{\max}, \hat{P}_d)$ is defined as [9]

$$\mathcal{I}_f^n(B_{\max}, \hat{P}_d) \triangleq \frac{1}{n} I(X^n; Y^n) = \frac{1}{n} \big[H(X^n) - H(X^n | Y^n) \big], \qquad (10.10)$$

where $0 \leq X_t - Y_t \leq \hat{P}_d$. It is noteworthy that the privacy achieved according to Equation (10.10) depends on the RB capacity $B_{\max}$ and on the discharging peak power constraint $\hat{P}_d$. The minimum information leakage rate, $\mathcal{I}^n(B_{\max}, \hat{P}_d)$, is obtained by minimizing (10.10) over all feasible policies $f \in \mathcal{F}$.

Privacy with an RES Only

Consider first the SM system of Figure 10.2 with an RES but no RB, and without the possibility of selling the generated renewable energy to the UP, to fully analyze the impact of the RES on the SM privacy. Hence, for an i.i.d. user load, the minimum information leakage rate is characterized by the so-called *privacy-power function* $\mathcal{I}(\bar{P}, \hat{P}_d)$, and can be formulated in the following single-letter form:

$$\mathcal{I}(\bar{P}, \hat{P}) = \inf_{p_{Y|X} \in \mathcal{P}} I(X;Y), \qquad (10.11)$$

where $\mathcal{P} \triangleq \{p_{Y|X} : y \in \mathcal{Y}, \mathbb{E}[(X - Y)] \leq \bar{P}, 0 \leq X - Y \leq \hat{P}\}$. If $\mathcal{X}$ is discrete, i.e., X can assume countable values that are multiples of a fixed quantum, the grid load alphabet can be constrained to the user load alphabet without loss of optimality and, because the MI is a convex function of $p_{Y|X} \in \mathcal{P}$, the privacy-power function can be written as a convex optimization problem with linear constraints [67, 68]. Numerical solutions for the optimal conditional distribution can be found using algorithms such as the Blahut-Arimoto (BA) algorithm [66]. When $\mathcal{X}$ is continuous, i.e., X can assume all real values within the limits specified by the constraints, the Shannon lower bound, a computable lower bound on the rate-distortion function widely used in the literature, is shown to be tight for exponential user load distributions [68, 69]. Two interesting observations can be made about the solution to Equation (10.11). First, the EMP that minimizes Equation (10.11) is stochastic and memoryless, that is, the optimal grid load at each time slot is generated randomly using the optimal conditional probability that minimizes (10.11) by considering only the current user load. Secondly, Equation (10.11) has an expression similar to the well-known *rate-distortion function* $R(D)$ in information theory, which characterizes the minimum compression rate R of data, in bits per sample, that is required for a receiver to reconstruct a source sequence within a specified average distortion level D [66]. Shannon computed the following single-letter form for the rate-distortion function for an i.i.d. source $X \in \mathcal{X}$ with distribution p_X, reconstruction alphabet $\hat{\mathcal{X}}$, and distortion function $d(\hat{x}, x)$, where the distortion between sequences X^n and $\hat{X}^n$ is given by $\frac{1}{n} \sum_{i=1}^{n} d(x_i, \hat{x}_i)$:

$$R(D) = \min_{p_{\hat{X}|X}: \sum_{(x,\hat{x})} p_X p_{\hat{X}|X} d(x,\hat{x}) \leq D} I(\hat{X}; X). \qquad (10.12)$$

Hence, tools from rate distortion theory can be used to evaluate Equation (10.11). However, it is important to highlight that there are conceptual differences between the two settings, namely that i) in the SM privacy problem Y^n is the direct output of the encoder rather than the reconstruction at the decoder side and ii) unlike the lossy source encoder, the EMU does not operate over blocks of user load realizations; instead, it operates symbol by symbol, acting instantaneously after receiving the appliance load at each time slot.

An interesting extension to this problem is to consider a multiuser scenario where K users, each equipped with a single SM, share the same RES, and the objective is to jointly minimize the total privacy loss of all consumers [68]. The average information leakage rate has the same expression in (10.10) where X and Y are replaced by $\mathbf{X}_t = [X_{1,t}, \ldots, X_{K,t}]$ and $\mathbf{Y}_t = [Y_{1,t}, \ldots, Y_{K,t}]$ and the privacy-power function has the same expression in (10.11). When the user loads are independent, but not necessarily identically distributed, the optimization problem (ignoring the peak power constraint) can be cast as [68]

$$\mathcal{I}(\bar{P}) = \inf_{\sum_{i=1}^{K} P_i \leq \bar{P}} \sum_{i=1}^{K} \mathcal{I}_{X_i}(P_i), \qquad (10.13)$$

where $\mathcal{I}_{X_i}(\cdot)$ denotes the privacy-power function for the ith user having user load distribution $p_{X_i}(x_i)$. Moreover, it is found that the *reverse water-filling* algorithm determines the optimal allocation of renewable energy for continuous and exponential user loads.

Privacy with an RB Only

In this section an RB only is considered to be present in the SM system, which is thus charged only using the grid. Including an RB in the SM setting complicates significantly the problem as the RB introduces memory in time, and the EMP needs to consider the impact of its decisions not only in the current TS but also in the future.

As discussed in the preceding text, this problem can be cast as an MDP upon determining an additive formulation for the privacy loss. This can be achieved by formulating the optimization problem as follows [70]:

$$L^* \triangleq \min_{f} \frac{1}{n} I(B_1, X^n; Y^n), \qquad (10.14)$$

and by adopting an EMP that decides on the grid load based only on the current user load, level of energy in the RB, and past values of the grid load, which does not lose optimality as the following inequality holds:

$$\frac{1}{n}I(X^n, B_1; Y^n) \geq \frac{1}{n}\sum_{t=1}^{n} I(X_t, B_t; Y_t | Y^{t-1}). \qquad (10.15)$$

Additionally, to avoid an exponential growth in the space of possible conditional distributions in Equation (10.15), the knowledge of Y^{t-1} is summarized into a belief state $p(X_t, B_t | Y^{t-1})$, which is computed recursively and can be interpreted as the belief that the UP has about (X_t, B_t) at TS t, given its past observations, Y^{t-1}. The minimum information leakage rate has been characterized in a single-letter expression for an i.i.d. user load, resulting in an i.i.d. grid load and a memoryless EMP, both for a binary user load [71] and for a generic size for the user load [70, 72, 73].

The level of energy in the RB can be modeled as a *trapdoor channel*, which is a type of unifilar finite state channel, i.e., its output and state at any time depend only on the current input and the previous state, and its state is deterministic given the previous state and the current input and output [74]. Let a certain number of balls, labeled as either "0" or "1," be within the channel. At each TS a new ball is inserted into the channel and an output ball is randomly selected from those within the channel. In an SM context, inserting or removing a ball from the channel represents charging or discharging the RB, respectively. An upper bound on the information leakage rate achieved using this model can be determined by minimizing the information leakage rate over the set of *stable output balls*, i.e., the set of feasible output sequences Y^n that can be extracted from the channel given a certain initial state and an input sequence X^n, and by taking inspiration from codebook construction strategies in [75]. This upper bound is expressed as follows [76]:

$$\frac{1}{n}I(X^n; Y^n) \leq \frac{1}{\lfloor (B_{max} + 1)/X_{max} \rfloor}, \qquad (10.16)$$

where X_{max} is the largest value X can assume. It is also shown in [76] that the average user energy consumption determines the level of achievable privacy.

Above all, it is important to jointly optimize the user's privacy and cost of energy, which allows characterization of the optimal tradeoffs between privacy and cost. Because cost of energy has an immediate additive formulation, it can also be easily embedded within the MDP formulation. Let $C^t = (C_1, \ldots, C_t)$ be the random price sequence over t TSs. Then, user privacy can be defined in the long time horizon as [61]

$$\mathcal{P} \triangleq \lim_{t \to \infty} \frac{H(X^t | Y^t, C^t)}{t}. \qquad (10.17)$$

Two solutions to the problem (10.17) are presented in [61], the most interesting of which proposes a *battery centering approach* aimed at keeping the RB at a medium level of charge so that the EMU is less constrained by the RB or the user load in determining the grid load. Then, the aim is to keep the system in a so-called *hidden state* where the grid load depends only on the current cost of energy but not on the user load or the level of energy in the battery.

Privacy with Both an RES and an RB

The most interesting scenario, as well as the most challenging, is when both an RES and an RB are considered. First, considering either the absence of an RB or the presence of an infinite capacity RB allows us to characterize bounds on the performance of systems with finite capacity RBs. Figure 10.8 shows the minimum information leakage rate with respect to the renewable energy generation rate p_e [77, 78]. When $B_{max} = 0$, the renewable energy that can be used at any TS is limited by the amount of renewable energy generated within that TS, and the privacy performance seriously degrades if the UP further knows the amount of renewable energy generated, as shown in Figure 10.8. The case in which $B_{max} = \infty$ is analogous to the average and peak power-constrained scenario, and no loss of privacy is experienced when the UP knows the exact amount of renewable energy generated. The lower bound is achieved by two different EMPs in [77].

Modeling a finite capacity RB is challenging due to the memory effects, and for this reason single-letter expressions for the general setting are still lacking. Nevertheless, the

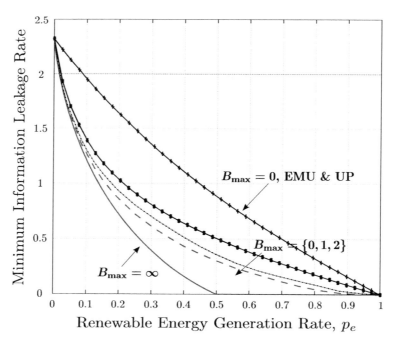

Figure 10.8 Minimum information leakage rate with respect to the renewable energy generation rate p_e with $\mathcal{X} = \mathcal{E} = \mathcal{Y} = \{0, 1, 2, 3, 4\}$ and $B_{max} = \{0, 1, 2, \infty\}$ [77]. The leakage for $B_{max} = \infty$ has been found by setting $\hat{P} = 4$. The curves for $B_{max} = \{0, \infty\}$ are obtained analytically, whereas the curves for the finite battery capacities $B_{max} = \{1, 2\}$ are obtained numerically by considering a suboptimal EMP.

problem may be cast as an MDP by measuring privacy using the MI and by formulating the corresponding Bellman equations [79]. Additionally, the privacy-cost tradeoff may be analyzed as an MDP, as investigated in [80], where a numerical solution focusing on a renewable energy process that recharges the RB fully at random time instances is presented, as well as a lower bound where the user knows noncausally the time when the RES recharges the RB.

Detection Error Probability as a Privacy Measure

In some scenarios the user may want to keep private only specific activities, e.g., the fact that she is eating microwaved food or if there is an active burglar alarm. Considering M possible hypotheses related to the activity that is to be kept private, this problem can be modeled as an M-ary hypothesis test, where $H \in \mathcal{H} = \{h_0, h_1, \ldots h_{M-1}\}$. A binary hypothesis test occurs when $M = 2$, e.g., when answering the question "is the consumer using the oven," and, by convention, the *null hypothesis* h_0 represents the absence of some factor or condition, e.g., "the consumer is not using the oven," while the *alternative hypothesis* h_1 is the complementary condition, e.g., "the consumer is using the oven." Typically, it is assumed that the user load has different statistics under these two hypotheses, i.e., the energy demand at TS t is i.i.d. with $p_{X|h_0}$ (respectively, $p_{X|h_1}$) under hypothesis h_0 (respectively, h_1).

An attacker wishes to determine the best mapping $\hat{H}(\cdot)$ between the grid load and the underlying hypothesis, so that the set of all possible SM readings $\mathcal{Y}^n$ is partitioned into the two disjoint decision regions $\mathcal{A}_0 \triangleq \{y^n | \hat{H}(y^n) = h_0\}$ and $\mathcal{A}_1 \triangleq \{y^n | \hat{H}(y^n) = h_1\}$, corresponding to the subsets of the SM readings for which the UP decides for one of the two hypotheses. When performing a decision, the attacker may incur two types of errors:

- Type I error probability: Make a decision h_1 when h_0 is the true hypothesis (*false positive* or *false alarm*), i.e., $p_I = p_{Y^n|h_1}(\mathcal{A}_0)$;
- Type II error probability: Make a decision h_0 when h_1 is the true hypothesis (*false negative* or *miss*), i.e., $p_{II} = p_{Y^n|h_0}(\mathcal{A}_1)$.

One possible solution to this mapping problem is to perform a Neyman–Pearson test on the grid load, i.e., characterizing the minimum type II error probability $p_{II}^{\min}$ while fixing a maximum type I error probability, and making decisions by thresholding the likelihood ratio $\frac{p_{Y^n|h_0}(y^n|h_0)}{p_{Y^n|h_1}(y^n|h_1)}$. Consider the worst case of an attacker that has perfect knowledge of the EMP employed, the asymptotic regime $n \to \infty$, and, for the sake of simplicity, a memoryless EMP. Then, $p_{II}^{\min}$ is linked to the Kullback–Leibler (KL) divergence $D(\cdot || \cdot)$ by the Chernoff–Stein Lemma [66]:

$$\lim_{n \to \infty} -\frac{\log p_{II}^{\min}}{n} = D(p_{Y|h_0} || p_{Y|h_1}), \quad (10.18)$$

where the KL divergence between two probability distribution functions on X, p_X and q_X, is defined as $D(p_X || q_X) \triangleq \sum_{x \in \mathcal{X}} p_X(x) \log \frac{p_X(x)}{q_X(x)}$ [66]. Hence, to maximize the consumer's privacy the goal of an EMP is to find the optimal grid load distributions, which, given X and the true hypothesis h, minimize the KL divergence in Equation

(10.18), or equivalently, the asymptotic exponential decay rate of $p_{\text{II}}^{\min}$. When considering a constraint on the average RES that can be used, the problem can be cast as

$$\min_{p_{Y|H} \in \mathcal{P}_{Y|H}} D(p_{Y|h_0} || p_{Y|h_1}), \quad (10.19)$$

where $\mathcal{P}_{Y|H}$ is the set of feasible EMPs, i.e., those that satisfy the average RES generation rate $\bar{P}$, so that $\frac{1}{n}\mathbb{E}[\sum_{i=1}^{n} X_i - Y_i | h_j] \leq \bar{P}$, $j = 0, 1$. Asymptotic single-letter expressions for two privacy-preserving EMPs when the probability of type I error is close to 1 are characterized in [81].

Fisher Information as a Privacy Measure

Let θ be a parameter that underpins the distribution of some sample data X. Then, Fisher information (FI) is a statistical measure of the amount of information that X contains about θ. FI can be cast in the SM setting by letting Y^n be the sample data and X^n the parameter underlying the sample data [82]. The FI can be generalized to the multivariate case by the FI matrix, defined as

$$\mathcal{FI}(X^n) = \int_{Y^n \in \mathcal{Y}^n} p(Y^n | X^n) \left[\frac{\partial \log(p(Y^n | X^n))}{\partial X^n} \right] \left[\frac{\partial \log(p(Y^n | X^n))}{\partial X^n} \right]^T dY^n. \quad (10.20)$$

If an *unbiased estimator* is deployed by the attacker, which produces an estimate $\hat{X}^n$ for X^n, then the variance of the estimation error is limited by the Cramér–Rao bound as follows:

$$\mathbb{E}[||X^n - \hat{X}^n(Y^n)||_2^2] \geq \text{Tr}(\mathcal{FI}(X^n)^{-1}), \quad (10.21)$$

where $|| \cdot ||_2^2$ denotes the squared Euclidean norm, and $\text{Tr}(A)$ denotes the trace of a matrix A. Then, to maximize the privacy it is necessary to maximize the right hand side of Equation (10.21). Two settings with an RB can be considered, specifically when the battery charging policy is independent of the user load, and when it is dependent noncausally on the entire user load sequence [82, 83].

Empirical MI as a Privacy Measure

Empirical MI can be used to evaluate numerically the information leakage in an SM system, by considering a "large enough" time interval and sampling the resulting X^n and Y^n sequences [84]. The empirical MI between two sequences x^n and y^n is

$$I(X;Y) \approx -\frac{1}{n} \log p(y^n) - \frac{1}{n} \log p(x^n) + \frac{1}{n} \log p(x^n, y^n), \quad (10.22)$$

where $p(y^n)$, $p(x^n)$ and $p(x^n, y^n)$ are calculated recursively through a sum-product computation. Typically, when using this technique the RB is modeled as a finite state machine, whose transition probabilities are discretized and optimized. A binary RB and an i.i.d. Bernoulli distributed user demand has been studied in [85]. Additionally, the presence of an RES has been included, and the privacy-energy efficiency tradeoff for a binary scenario and equiprobable user load and renewable energy generation processes has been characterized [86]. When the RB and RES are both present, a suboptimal EMP has also been analyzed, which, at each TS, decides among using all of the available energy, half of it, or no energy at all [77, 78]. Empirical MI normalized by the empirical entropy of the user load has also been considered [87]. Although assuming the user load to be i.i.d. allows the problem to be mathematically tractable, this is clearly not the case in reality. To overcome this problem, a feature-dependent first-order Markov process can be considered, where the distribution of the user load at any TS depends on an underlying feature, e.g., time of day, day of week, or, season [88].

Alternatively, $I(X;Y)$ can be approximated by the relative frequency of events (X_t, Y_t) when X and Y are considered to be i.i.d. Such a measure has been considered in [89], where a *model-distribution predictive controller* is employed, which, for each TS t, decides actions for a prediction horizon of duration T, i.e., up to time $t + T$, considering noncausal knowledge of the renewable energy generation process, user load and energy prices, while the EMU's actions, i.e., the energy that is requested from the grid and the RB, are forecast over the prediction horizon. It is noteworthy that considering a small prediction horizon prevents the EMU from fully utilizing the RB capacity, whereas large values for T allow the system to achieve better privacy-cost tradeoffs at the expense of a much higher computational complexity.

10.6 Conclusion

Privacy, a fundamental and inalienable human right, has never been under so much attack and scrutiny as in recent years. Reports of mass surveillance by government agencies as well as private companies have strongly undermined the trust of consumers and the general public. Moreover, the big data and machine learning revolution is also seen as an improved way to profit from consumer's data, which, more often than not, is stored and processed without users' prior authorization and even unbeknownst to them. Privacy in SG is no exception to this debate, as the proliferation of anti-SM movements across the world shows. In fact, UPs, DSOs, and other SG entities may not be incentivized enough in keeping user's data private and in investing in the creation of privacy-preserving technologies. Hence, it is the task of legislators to strengthen privacy guarantees around the use of customer's data by creating new laws that safeguard the consumers' right to privacy, e.g., the GDPR in Europe [5]. However, as these legal initiatives are still limited, it is often up to the research community to investigate and lead the development and the discussion around privacy-preserving techniques for SMs.

To further inspire research and improvements in this domain, in this chapter we have presented a broad overview of privacy-preserving techniques in SMs. We have discussed techniques that manipulate meter readings before sending them to the UP, as well as techniques that adopt physical resources such as RBs or RESs, and we have discussed their main advantages and limitations. We have described theoretically grounded techniques, which shed light on fundamental aspects of the SM privacy problem, as well as more empirical techniques, which have a more immediate practical implementation but tend to provide fewer privacy assurances. Finally, we have also presented various measures for privacy in SMs, which look at the SM problem from various perspectives.

References

[1] MarketsandMarkets, "Smart meters market by type (electric, water, and gas), application (commercial, residential, and industrial), technology (automatic meter reading and advanced metering infrastructure), and by region – global forecasts to 2022," Apr. 2017.

[2] GlobalData, "Smart meters, update 2018 – global market size, competitive landscape, key country analysis, and forecast to 2022," Sept. 2018.

[3] European Union, "Directive 2009/72/EC of the European Parliament and of the council of 13 July 2009 concerning common rules for the internal market in electricity and repealing directive 2003/54/EC," *Official J. European Union*, vol. 52, no. L211, pp. 55–93, Aug. 2009.

[4] C. Cuijpers and B.-J. Koops, "Smart metering and privacy in Europe: Lessons from the Dutch case," *European Data Protection: Coming of Age*, pp. 269–293, Feb. 2012.

[5] The European Parliament and the Council of the European Union, "Regulation (EU) 2016/679 of the European Parliament and of the Council of 27 April 2016 on the protection of natural persons with regard to the processing of personal data and on the free movement of such data, and repealing Directive 95/46/EC (General Data Protection Regulation)," *Official J. European Union*, May 2016.

[6] A. Zoha, A. Gluhak, M. A. Imran, and S. Rajasegarar, "Non-intrusive load monitoring approaches for disaggregated energy sensing: A survey," *Sensors*, vol. 12, no. 12, pp. 16838–16866, 2012.

[7] A. Ridi, C. Gisler, and J. Hennebert, "A survey on intrusive load monitoring for appliance recognition," in *Proc. Int. Conf. Pattern Recognition*, Stockholm, Sweden, pp. 3702–3707, Aug. 2014.

[8] G. Hart, "Prototype nonintrusive appliance load monitor," *MIT Energy Laboratory Technical Report, and Electric Power Research Institute Technical Report*, Sept. 1985.

[9] G. Giaconi, D. Gündüz, and H. V. Poor, "Privacy-aware smart metering: Progress and challenges," *IEEE Signal Process. Mag.*, vol. 35, no. 6, pp. 59–78, Nov. 2018.

[10] "Pecan Street Inc. Dataport," https://dataport.cloud/, accessed Mar. 16, 2019.

[11] A. Prudenzi, "A neuron nets based procedure for identifying domestic appliances pattern-of-use from energy recordings at meter panel," in *Proc. IEEE Power Eng. Soc. Winter Meeting*, vol. 2, New York, pp. 941–946, Jan. 2002.

[12] E. Quinn, "Privacy and the new energy infrastructure," *Social Science. Research Network*, Feb. 2009.

[13] I. Rouf, H. Mustafa, M. Xu, W. Xu, R. Miller, and M. Gruteser, "Neighborhood watch: Security and privacy analysis of automatic meter reading systems," in *Proc. ACM Conf. Comput. Commun. Security*, Raleigh, NC, pp. 462–473, Oct. 2012.

[14] M. Munsell, www.greentechmedia.com/articles/read/led-by-surging-residential-sector-q2-us-energy-storage-deployments-grow-200, Sept. 2018, accessed Feb. 27, 2019.

[15] G. Kavlak, J. McNerney, and J. E. Trancik, "Evaluating the causes of cost reduction in photovoltaic modules," *Energy Policy*, vol. 123, pp. 700–710, 2018.

[16] B. Nykvist and M. Nilsson, "Rapidly falling costs of battery packs for electric vehicles," *Nature Climate Change*, vol. 5, pp. 329–332, 2015.

[17] Z. Erkin, J. R. Troncoso-Pastoriza, R. L. Lagendijk, and F. Perez-Gonzalez, "Privacy-preserving data aggregation in smart metering systems: An overview," *IEEE Signal Process. Mag.*, vol. 30, no. 2, pp. 75–86, Mar. 2013.

[18] M. R. Asghar, G. Dán, D. Miorandi, and I. Chlamtac, "Smart meter data privacy: A survey," *IEEE Commun. Surveys Tutorials*, vol. 19, no. 4, pp. 2820–2835, Fourth Quarter 2017.

[19] S. Finster and I. Baumgart, "Privacy-aware smart metering: A survey," *IEEE Commun. Surveys Tutorials*, vol. 17, no. 2, pp. 1088–1101, Second Quarter 2015.

[20] "Real life stories – Energy theft is all around you," www.stayenergysafe.co.uk/stories/, accessed Feb. 26, 2019.

[21] L. Sankar, S. Rajagopalan, S. Mohajer, and H. V. Poor, "Smart meter privacy: A theoretical framework," *IEEE Trans. Smart Grid*, vol. 4, no. 2, pp. 837–846, June 2013.

[22] J.-M. Bohli, C. Sorge, and O. Ugus, "A privacy model for smart metering," in *Proc. IEEE Int. Conf. Commun.*, Cape Town, South Africa, pp. 1–5, May 2010.

[23] R. Petrlic, "A privacy-preserving concept for smart grids," in *Sicherheit in vernetzten Systemen:18. DFN Workshop*. Books on Demand GmbH, pp. B1–B14, 2010.

[24] F. D. Garcia and B. Jacobs, "Privacy-friendly energy-metering via homomorphic encryption," in *Proc. Int. Conf. Security Trust Manage.*, Athens, Greece, pp. 226–238, Sept. 2010.

[25] K. Kursawe, G. Danezis, and M. Kohlweiss. "Privacy-friendly aggregation for the smart-grid," in *Proc. Int. Symp. Privacy Enhancing Technologies*, Waterloo, Canada, pp. 175–191, July 2011.

[26] R. Lu, X. Liang, X. Li, X. Lin, and X. Shen, "Eppa: An efficient and privacy-preserving aggregation scheme for secure smart grid communications," *IEEE Trans. Parallel Distrib. Syst.*, vol. 23, no. 9, pp. 1621–1631, Sept. 2012.

[27] F. Li, B. Luo, and P. Liu, "Secure and privacy-preserving information aggregation for smart grids," *Int. J. Security and Netw.*, vol. 6, no. 1, pp. 28–39, Apr. 2011.

[28] C. Dwork, F. McSherry, K. Nissim, and A. Smith, "Calibrating noise to sensitivity in private data analysis," in *Proc. Theory of Cryptography Conf.*, New York, pp. 265–284, Mar. 2006.

[29] G. Ács and C. Castelluccia, "I have a dream! (differentially private smart metering)," in *Proc. Int. Workshop Inf. Hiding*, Prague, Czech Republic, May 2011.

[30] Z. Guan, G. Si, X. Zhang, L. Wu, N. Guizani, X. Du, and Y. Ma, "Privacy-preserving and efficient aggregation based on blockchain for power grid communications in smart communities," *IEEE Commun. Mag.*, vol. 56, no. 7, pp. 82–88, July 2018.

[31] N. Z. Aitzhan and D. Svetinovic, "Security and privacy in decentralized energy trading through multi-signatures, blockchain and anonymous messaging streams," *IEEE Trans. Dependable Secure Comput.*, vol. 15, no. 5, pp. 840–852, Sept. 2018.

[32] A. Abdel-Majeed and M. Braun, "Low voltage system state estimation using smart meters," in *Proc. Int. Universities Power Eng. Conf.*, London, pp. 1–6, Sept. 2012.

[33] S. Tonyali, O. Cakmak, K. Akkaya, M. M. E. A. Mahmoud, and I. Guvenc, "Secure data obfuscation scheme to enable privacy-preserving state estimation in smart grid ami networks," *IEEE Internet of Things J.*, vol. 3, no. 5, pp. 709–719, Oct. 2016.

[34] W. Jia, H. Zhu, Z. Cao, X. Dong, and C. Xiao, "Human-factor-aware privacy-preserving aggregation in smart grid," *IEEE Syst. J.*, vol. 8, no. 2, pp. 598–607, June 2014.

[35] S. Mangard, E. Oswald, and T. Popp, *Power Analysis Attacks*. Heidelberg: Springer, 2007.

[36] Y. Kim, E. Ngai, and M. Srivastava, "Cooperative state estimation for preserving privacy of user behaviors in smart grid," in *Proc. IEEE Int. Conf. Smart Grid Commun.*, Brussels, Belgium, pp. 178–183, Oct. 2011.

[37] Z. Yang, P. Cheng, and J. Chen, "Differential-privacy preserving optimal power flow in smart grid," *IET Generation, Transmission Distribution*, vol. 11, no. 15, pp. 3853–3861, Nov. 2017.

[38] C. Efthymiou and G. Kalogridis, "Smart grid privacy via anonymization of smart metering data," in *Proc. IEEE Int. Conf. Smart Grid Commun.*, Gaithersburg, MD, pp. 238–243, Oct. 2010.

[39] M. Jawurek, M. Johns, and K. Rieck, "Smart metering de-pseudonymization," in *Proc. Annual Comput. Security Applicat. Conf.*, Orlando, FL, pp. 227–236, Dec. 2011.

[40] S. Goldwasser, S. Micali, and C. Rackoff, "The knowledge complexity of interactive proof-systems," in *Proc. ACM Symp. Theory Computing*, Providence, RI, pp. 291–304, May 1985.

[41] A. Molina-Markham, P. Shenoy, K. Fu, E. Cecchet, and D. Irwin, "Private memoirs of a smart meter," in *ACM Workshop Embedded Sens. Syst. for Energy-Efficiency Bldg.*, Zurich, Switzerland, pp. 61–66, Nov. 2010.

[42] M. Jawurek, M. Johns, and F. Kerschbaum, "Plug-in privacy for smart metering billing," in *Proc. Int. Symp. Privacy Enhancing Technologies*, Waterloo, Canada, pp. 192–210, July 2011.

[43] J. Camenisch and A. Lysyanskaya, "A signature scheme with efficient protocols," in *Proc. Int. Conf. Security in Commun. Networks*, Amalfi, Italy, pp. 268–289, Sept. 2002.

[44] A. Rial and G. Danezis, "Privacy-preserving smart metering," in *Proc. ACM Workshop Privacy Electron. Soc.*, Chicago, pp. 49–60, Oct. 2011.

[45] A. Cárdenas, S. Amin, and G. A. Schwartz, "Privacy-aware sampling for residential demand response programs," in *Proc. ACM Int. Conf. High Confidence Networked Syst.*, Beijing, China, Apr. 2012.

[46] L. Yang, X. Chen, J. Zhang, and H. V. Poor, "Cost-effective and privacy-preserving energy management for smart meters," *IEEE Trans. Smart Grid*, vol. 6, no. 1, pp. 486–495, Jan. 2015.

[47] A. Zoha, A. Gluhak, M. A. Imran, and S. Rajasegarar, "Non-intrusive load monitoring approaches for disaggregated energy sensing: A survey," *Sensors*, vol. 12, no. 12, pp. 16 838–16 866, 2012.

[48] O. Tan, J. Gómez-Vilardebó, and D. Gündüz, "Privacy-cost trade-offs in demand-side management with storage," *IEEE Trans. Inf. Forens. Security*, vol. 12, no. 6, pp. 1458–1469, June 2017.

[49] J. Wu, J. Liu, X. S. Hu, and Y. Shi, "Privacy protection via appliance scheduling in smart homes," in *Proc. IEEE/ACM Int. Conf. Comput.-Aided Design*, Austin, TX, pp. 1–6, Nov. 2016.

[50] G. Giaconi, D. Gündüz, and H. V. Poor, "Optimal demand-side management for joint privacy-cost optimization with energy storage," in *Proc. IEEE Int. Conf. Smart Grid Commun.*, Dresden, Germany, pp. 265–270, Oct. 2017.

[51] J. Kelly and W. Knottenbelt, "The UK-DALE dataset, domestic appliance-level electricity demand and whole-house demand from five UK homes," *Scientific Data*, vol. 2, no. 150007, Mar. 2015.

[52] Powervault, "Technical specifications," www.powervault.co.uk/downloads/PV_technical-specification_AW-DIGITAL_jan2017.pdf, accessed March 1, 2019.

[53] G. Giaconi, D. Gündüz, and H. V. Poor, "Joint privacy-cost optimization in smart electricity metering systems," *arXiv:1806.09715*, June 2018.

[54] D. Engel and G. Eibl, "Wavelet-based multiresolution smart meter privacy," *IEEE Trans. Smart Grid*, vol. 8, no. 4, pp. 1710–1721, July 2017.

[55] Z. Chen and L. Wu, "Residential appliance DR energy management with electric privacy protection by online stochastic optimization," *IEEE Trans. Smart Grid*, vol. 4, no. 4, pp. 1861–1869, Dec. 2013.

[56] D. P. Bertsekas, *Dynamic Programming and Optimal Control, Vol. II*, 3rd ed. Belmont, MA, Athena Scientific, 2007.

[57] Y. Sun, L. Lampe, and V. W. S. Wong, "Smart meter privacy: Exploiting the potential of household energy storage units," *IEEE Internet Things J.*, vol. 5, no. 1, pp. 69–78, Feb. 2018.

[58] R. S. Sutton and A. G. Barto, *Reinforcement Learning: An Introduction*. Cambridge, MA: MIT Press, 1998.

[59] G. Kalogridis, C. Efthymiou, S. Denic, T. Lewis, and R. Cepeda, "Privacy for smart meters: Towards undetectable appliance load signatures," in *Proc. IEEE Int. Conf. Smart Grid Commun.*, Gaithersburg, MD, pp. 232–237, Oct. 2010.

[60] S. McLaughlin, P. McDaniel, and W. Aiello, "Protecting consumer privacy from electric load monitoring," in *Proc. ACM Conf. Comput. Commun. Security*, Chicago, pp. 87–98, Oct. 2011.

[61] J. Yao and P. Venkitasubramaniam, "The privacy analysis of battery control mechanisms in demand response: Revealing state approach and rate distortion bounds," *IEEE Trans. Smart Grid*, vol. 6, no. 5, pp. 2417–2425, Sept. 2015.

[62] W. Yang, N. Li, Y. Qi, W. Qardaji, S. McLaughlin, and P. McDaniel, "Minimizing private data disclosures in the smart grid," in *Proc. ACM Conf. Comput. Commun. Security*, Raleigh, NC, pp. 415–427, Oct. 2012.

[63] K. Qian, C. Zhou, M. Allan, and Y. Yuan, "Modeling of load demand due to EV battery charging in distribution systems," *IEEE Trans. Power Syst.*, vol. 26, no. 2, pp. 802–810, May 2011.

[64] P. A. Leicester, C. I. Goodier, and P. N. Rowley, "Probabilistic analysis of solar photovoltaic self-consumption using bayesian network models," *IET Renewable Power Generation*, vol. 10, no. 4, pp. 448–455, Mar. 2016.

[65] W. Labeeuw and G. Deconinck, "Residential electrical load model based on mixture model clustering and Markov models," *IEEE Trans. Ind. Informat.*, vol. 9, no. 3, pp. 1561–1569, Aug. 2013.

[66] T. M. Cover and J. A. Thomas, *Elements of Information Theory*. New York: Wiley-Interscience, 1991.

[67] D. Gündüz and J. Gómez-Vilardebó, "Smart meter privacy in the presence of an alternative energy source," in *Proc. IEEE Int. Conf. Commun.*, Budapest, Hungary, pp. 2027–2031, June 2013.

[68] J. Gómez-Vilardebó and D. Gündüz, "Smart meter privacy for multiple users in the presence of an alternative energy source," *IEEE Trans. Inf. Forens. Security*, vol. 10, no. 1, pp. 132–141, Jan. 2015.

[69] J. Gómez-Vilardebó and D. Gündüz, "Privacy of smart meter systems with an alternative energy source," in *Proc. IEEE Int. Symp. Inf. Theory*, Istanbul, Turkey, pp. 2572–2576, July 2013.

[70] S. Li, A. Khisti, and A. Mahajan, "Information-theoretic privacy for smart metering systems with a rechargeable battery," *IEEE Trans. Inf. Theory*, vol. 64, no. 5, pp. 3679–3695, May 2018.

[71] S. Li, A. Khisti, and A. Mahajan, "Structure of optimal privacy-preserving policies in smart-metered systems with a rechargeable battery," in *Proc. IEEE Int. Workshop Signal Process. Adv. Wireless Commun.*, pp. 375–379, June 2015.

[72] ——, "Privacy preserving rechargeable battery policies for smart metering systems," in *Proc. Int. Zurich Seminar Commun.*, Zurich, Switzerland, pp. 121–124, Mar. 2016.

[73] ——, "Privacy-optimal strategies for smart metering systems with a rechargeable battery," in *Proc. American Control Conf.*, Boston, pp. 2080–2085, July 2016.

[74] H. Permuter, P. Cuff, B. Van Roy, and T. Weissman, "Capacity of the trapdoor channel with feedback," *IEEE Trans. Inf. Theory*, vol. 54, no. 7, pp. 3150–3165, July 2008.

[75] R. Ahlswede and A. Kaspi, "Optimal coding strategies for certain permuting channels," *IEEE Trans. Inf. Theory*, vol. 33, no. 3, pp. 310–314, May 1987.

[76] M. Arrieta and I. Esnaola, "Smart meter privacy via the trapdoor channel," in *Proc. IEEE Int. Conf. Smart Grid Commun.*, Dresden, Germany, pp. 277–282, Oct. 2017.

[77] G. Giaconi, D. Gündüz, and H. V. Poor, "Smart meter privacy with renewable energy and an energy storage device," *IEEE Trans. Inf. Forens. Security*, vol. 13, no. 1, pp. 129–142, Jan. 2018.

[78] ——, "Smart meter privacy with an energy harvesting device and instantaneous power constraints," in *Proc. IEEE Int. Conf. Commun.*, London, pp. 7216–7221, June 2015.

[79] G. Giaconi and D. Gündüz, "Smart meter privacy with renewable energy and a finite capacity battery," in *Workshop Sig. Process. Adv. Wireless Commun.*, Edinburgh, July 2016.

[80] E. Erdemir, P. L. Dragotti, and D. Gündüz, "Privacy-cost trade-off in a smart meter system with a renewable energy source and a rechargeable battery," in *Proc. IEEE Int. Conf. Acoust., Speech, Signal Process.*, Brighton, pp. 2687–2691, May 2019.

[81] Z. Li, T. J. Oechtering, and D. Gündüz, "Privacy against a hypothesis testing adversary," *IEEE Trans. Inf. Forens. Security*, vol. 14, no. 6, pp. 1567–1581, June 2019.

[82] F. Farokhi and H. Sandberg, "Fisher information as a measure of privacy: Preserving privacy of households with smart meters using batteries," *IEEE Trans. Smart Grid*, vol. 9, no. 5, Sept. 2017.

[83] ——, "Ensuring privacy with constrained additive noise by minimizing Fisher information," *Automatica*, vol. 99, pp. 275–288, Jan. 2019.

[84] D.-M. Arnold, H.-A. Loeliger, P. Vontobel, A. Kavcic, and W. Zeng, "Simulation-based computation of information rates for channels with memory," *IEEE Trans. Inf. Theory*, vol. 52, no. 8, pp. 3498–3508, Aug. 2006.

[85] D. Varodayan and A. Khisti, "Smart meter privacy using a rechargeable battery: Minimizing the rate of information leakage," in *Proc. IEEE Int. Conf. Acoust., Speech Signal Process.*, Prague, Czech Republic, pp. 1932–1935, May 2011.

[86] O. Tan, D. Gündüz, and H. V. Poor, "Increasing smart meter privacy through energy harvesting and storage devices," *IEEE J. Sel. Areas Commun.*, vol. 31, no. 7, pp. 1331–1341, July 2013.

[87] J. Koo, X. Lin, and S. Bagchi, "Privatus: Wallet-friendly privacy protection for smart meters," in *Proc. European Symp. Re. Comput. Security*, Pisa, Italy, pp. 343–360, Sept. 2012.

[88] J. Chin, G. Giaconi, T. Tinoco De Rubira, D. Gündüz, and G. Hug, "Considering time correlation in the estimation of privacy loss for consumers with smart meters," in *Proc. Power Syst. Computation Conf.*, Dublin, Ireland, pp. 1–7, June 2018.

[89] J. X. Chin, T. T. D. Rubira, and G. Hug, "Privacy-protecting energy management unit through model-distribution predictive control," *IEEE Trans. Smart Grid*, vol. 8, no. 6, pp. 3084–3093, Nov. 2017.

11 Data Quality and Privacy Enhancement

Meng Wang and Joe H. Chow

11.1 Introduction

To enhance the situational awareness of the operators, 2,000 phasor measurement units (PMUs) [1] have been recently installed in North America and are creating huge volumes of data about power system operating conditions, at a rate of 30 or 60 samples per second per channel. The abundance of data enables the development of *data-driven methods* for real-time situational awareness and control, which are of vital importance to building a reliable and efficient power grid and prevent blackouts. However, there are limited number of data-driven methods that both have *analytical performance guarantees* and are *applicable to power systems*. Because reliability is critical in power systems, the lack of theoretical guarantees is the major hurdle in applying many existing machine learning tools in power system operations. Moreover, it is difficult to apply existing data analytics tools directly on low-quality PMU data.

Wide-area control using synchrophasor data outperforms traditional controllers by providing faster control loops and coordinated actions. To enable the reliable implementation of synchrophasor-data–based real-time control, these synchrophasor measurements should always be available and accurate. Due to device malfunction, misconfiguration, communication errors, and possible cyber data attacks, synchrophasor data often contain missing data and bad data [2]. Moreover, data privacy is an increasing concern. One can enhance data privacy by adding noise to the data before sending the data to the operator [3–6]. That leads to a reduction in the data accuracy at the operator. One natural question is how to maintain individual user's data privacy and the operator's information accuracy.

The spatial-temporal blocks of PMU data exhibit intrinsic low-dimensional structures such as the low-rank property despite the high ambient dimension. The low-rank model naturally characterizes the data correlations without modeling the power systems. Reference [7] for the first time connected low-rank matrix theory with PMU data recovery, and a body of work continued exploiting the low-rank property to develop computationally efficient data-driven methods for PMU data recovery and information extraction [7–16].

This chapter discusses a common framework of methods for data recovery, error correction, detection and correction of cyber data attacks, and data privacy enhancement by exploiting the intrinsic low-dimensional structures in the high-dimensional spatial-temporal blocks of PMU data. The developed data-driven approaches are computationally efficient with provable analytical guarantees.

11.2 Low Dimensionality of PMU Data

Collect the phasor measurements of m PMU channels across n time instants into a matrix, denoted by X^*. $X^* \in \mathbb{C}^{m \times n}$ can be approximated by a rank-r matrix with r much smaller than m and n. This is the approximate low-rank property of the PMU data matrix [7, 8].[1] Low-rank matrices exist widely in various applications. For example, the rating matrix of multiple users against multiple movies is approximately low rank despite the huge number of users and movies. Low-rank matrices also characterize images and videos due to correlations in the data points.

A PMU data matrix is more special than just being a low-rank matrix. Let $x(t) \in \mathbb{C}^m$ denote the measurements from m measurement channels at time t. The measurement matrix is $X^* = [x(1), x(2), \ldots, x(n)]$. For an arbitrary $k \geq 1$, the Hankel matrix is defined as

$$\mathcal{H}_k(X^*) = \begin{bmatrix} x(1) & x(2) & \cdots & x(n+1-k) \\ x(2) & x(3) & \cdots & x(n+2-k) \\ \vdots & \vdots & & \vdots \\ x(k) & x(k+1) & \cdots & x(n) \end{bmatrix}. \quad (11.1)$$

In the following contents, k is selected as $\frac{n}{2}$ to achieve the best performance both theoretically and numerically. $\mathcal{H}_k(X^*)$ is proved to be low rank provided that the time series can be viewed as the output of a reduced-order linear dynamical system [17, 18]. The rank r depends on the order of the dynamical system. The low-rank Hankel property holds for time series that can be viewed as a sum of r damped or undamped sinusoids (possibly with transformation) and does not hold for an arbitrary low-rank matrix. To see this, imagine randomly permuting the columns of X^*, the resulting permuted matrix is still low rank because a column permutation does not change the rank of the matrix. However, the Hankel matrix of the permuted matrix is no longer low-rank because the temporal correlation is no longer preserved after a column permutation.

The low-rank Hankel property is visualized in a recorded synchrophasor data set. Figure 11.1 (see figure 1 in [19]) shows the data set of 20 seconds with 11 voltage phasors recorded at 30 samples per second, where a disturbance occurs around 2.3 seconds.

Let $X^* = U \Sigma V^H$ denote the singular value decomposition (SVD) of X^*. One can approximate X^* by the rank-r matrix $U \Sigma^{(r)} V^*$, where $\Sigma^{(r)}$ is the diagonal matrix with only the first r largest singular values in Σ. The rank-r approximation error is defined as

$$\|U \Sigma^{(r)} V^* - X^*\|_F / \|X^*\|_F = \|\Sigma - \Sigma^{(r)}\|_F / \|\Sigma\|_F. \quad (11.2)$$

---

[1] Strictly speaking, X^* is approximately low-rank and can be viewed as the summation of a low-rank matrix plus noise. We assume X^* is strictly low-rank for notational simplicity in this chapter, and the results can be extended to low-rank matrices plus noise with little modifications.

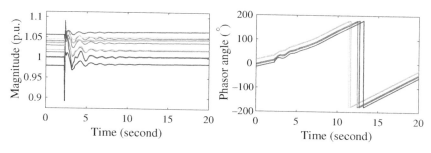

Figure 11.1 The measured voltage phasors.

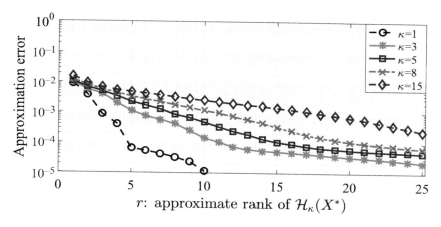

Figure 11.2 The approximation errors of Hankel matrices from X^*.

Figure 11.2 (see figure 2 in [19]) shows the approximation errors to the Hankel matrices using the data in Figure 11.1 with different r and k. All the matrices $\mathcal{H}_k(X^*)$ can be approximated by a rank-2 or rank-3 matrix with a negligible error.

11.3 Missing PMU Data Recovery

PMU data analysis is connected with low-rank matrix theory [20] for the first time in [7, 8] and the missing PMU data recovery problem is formulated as a special case of the general low-rank matrix completion problem [20], which aims to recover a low-rank matrix from partial observations.

One can then employ low-rank matrix completion methods to recover missing data. Low-rank matrix theory solves the data-recovery problems with provable guarantees. As long as the number of observations exceeds a very small lower bound, the data losses/errors can be accurately recovered. These approaches do not require any modeling or prior assumption of power system topology and dynamics.

The existing analysis of low-rank matrix completion methods usually assume that the location of the missing points are randomly and independently selected and characterize the required number of observations to recover the low-rank matrix under this independent erasure model. For a $m \times n$ rank-r matrix (assuming $m \leq n$), the required number of observations at random locations is $O(rn \log^2 n)$.[2] The locations of the missing PMU data, however, are usually not randomly distributed. Because missing data usually result from communication congestions or device malfunction, the locations of missing PMU are often correlated. For instance, some consecutive data points in the same PMU channel might be lost, or measurements across multiple PMUs may be lost at the same time. The existing analysis based on the random erasure assumption for low-rank matrix completion methods does not hold in this case.

Reference [7] provided the first analytical analysis of using low-rank matrix completion methods for recovering correlated data losses during such consecutive time instants or simultaneous data losses across some channels. Theorems 1 and 2 in [7] show that the missing entries can be recovered if $O(n^{2-\frac{1}{r+1}} r^{\frac{1}{r+1}} \log^{\frac{1}{r+1}} n)$ entries at correlated locations are observed in a rank-r $n \times n$ matrix. Note that this number is higher than $O(rn \log^2 n)$ due to the artifacts in the proof techniques. Still, this bound is less than the total number of entries n^2 in the matrix. That means one can recover missing points from partial observations.

Although Reference [7] considers consecutive data losses or simultaneous data losses using low-rank methods, the analysis in Reference [7] only applies to partial data losses across channel or time and does not apply to the extreme case of missing all data points in one PMU channel or missing all the measurements across all channels simultaneously. Mathematically, that corresponds to the cases that a complete row or column is lost in the matrix. In fact, this is the fundamental limit of low-rank matrix completion methods. They cannot recover a completely lost row or column. They can determine the subspace to which that row or column belongs but cannot determine the exact values in that row or column without any other prior information.

Luckily, PMU data have the additional temporal correlations that are not characterized by conventional low-rank matrices. The low-rank Hankel property characterizes the data dynamics without modeling the system. The low-rank Hankel property holds for PMU data series but does not hold for general low-rank matrices. One can exploit the temporal correlations characterized by the low-rank Hankel property to recover simultaneous data losses across all channels, i.e., fully lost columns in the data matrix.

Let Ω denote the set of indices of the entries that are observed. Define an operator $\mathcal{P}_\Omega$ such that $\mathcal{P}_\Omega(M)_{i,j} = M_{i,j}$ if $(i,j) \in \Omega$, and 0 otherwise. Let $M = X^* + N$ denote the noisy measurements, and N is zero if measurements are noiseless. $\mathcal{P}_\Omega(M)$ represents the observed entries. Given partial observations $\mathcal{P}_\Omega(M)$, the data recovery problem is formulated as

$$\min_{X \in \mathbb{C}^{m \times n}} \quad \|\mathcal{P}_\Omega(M - X)\|_F \qquad (11.3)$$
$$\text{s.t.} \quad \text{rank}(\mathcal{H}_k(X)) \leq r,$$

in [17]. The solution to (11.3), denoted by $\hat{X}$, models the estimates of the ground-truth PMU data X^*. The optimization (11.3) is nonconvex due to the rank constraint.

---

[2] We use the notations $g(n) \in O(h(n))$, $g(n) \in \Omega(h(n))$, or $g(n) = \Theta(h(n))$ if as n goes to infinity, $g(n) \leq c \cdot h(n)$, $g(n) \geq c \cdot h(n)$, or $c_1 \cdot h(n) \leq g(n) \leq c_2 \cdot h(n)$ eventually holds for some positive constants c, c_1, and c_2, respectively.

An Accelerated Multi-channel Fast Iterative Hard Thresholding (AM-FIHT) algorithm is developed in [17, 18] to solve (11.3). AM-FIHT is initialized as a low-rank Hankel approximation to $p^{-1}\mathcal{P}_\Omega(M)$. In each iteration, the current estimate $\mathbf{X}_l \in \mathbb{C}^{m \times n}$ first moves along the gradient descent direction $\mathbf{G}_l \in \mathbb{C}^{m \times n}$, with a step size $p^{-1} = \frac{mn}{|\Omega|}$. To improve the convergence rate, the update is further combined with additional heavy-ball term $\beta(W_{l-1} - W_{l-2})$, which represents the update direction in the previous iteration. Next, $\mathcal{H}_k(X_l + p^{-1}G_l) + \beta(W_{l-1} - W_{l-2})$ is projected to a rank-r matrix L_l. W_l and L_l are in $\mathbb{C}^{mk \times (n-k+1)}$. To reduce the computational complexity, it is first projected to the $2r$-dimensional space $\mathcal{S}_l$ and then apply SVD on the rank-$2r$ matrix [21], instead of directly computing its SVD. The rank-r matrix L_{l+1} is obtained in line 7 by thresholding the singular values of the rank-$2r$ matrix W_l. Finally, X_{l+1} is updated by $\mathcal{H}_k^\dagger L_{l+1}$.

The details of AM-FIHT are summarized in Algorithm 11.1. L_l is a rank-r matrix and its SVD is denoted as $L_l = U_l \Sigma_l V_l^*$, where $U_l \in \mathbb{C}^{mk \times r}$, $V_l \in \mathbb{C}^{(n-k+1) \times r}$, and $\Sigma_l \in \mathbb{C}^{r \times r}$. $\mathcal{S}_l$ is the tangent subspace of the rank-r Riemannian manifold at L_l, and for any matrix $Z \in \mathbb{C}^{mk \times (n-k+1)}$, the projection of Z onto $\mathcal{S}_l$ is defined as

$$\mathcal{P}_{\mathcal{S}_l}(Z) = U_l U_l^* Z + Z V_l V_l^* - U_l U_l^* Z V_l V_l^*. \tag{11.4}$$

$\mathcal{Q}_r$ finds the best rank-r approximation as

$$\mathcal{Q}_r(Z) = \sum_{i=1}^{r} \sigma_i u_i v_i^*, \tag{11.5}$$

if $Z = \sum_i \sigma_i u_i v_i^*$ is the SVD of Z with $\sigma_1 \geq \sigma_2 \geq \cdots$. $\mathcal{H}_k^\dagger$ is the Moore–Penrose pseudoinverse of $\mathcal{H}_k$. For any matrix $Z \in \mathbb{C}^{n_c k \times (n-k+1)}$, $(\mathcal{H}_k^\dagger Z) \in \mathbb{C}^{n_c \times n}$ satisfies

$$\langle \mathcal{H}_k^\dagger Z, e_i e_j^T \rangle = \begin{cases} \frac{1}{w_j} \sum_{j_1=1}^{j} Z_{(i-1)n+j_1, j+1-j_1} & \text{if } j \leq k. \\ \frac{1}{w_j} \sum_{j_2=n+1-j}^{n+1-k} Z_{(i-1)n+j+1-j_2, j_2} & \text{if } j \geq k+1. \end{cases} \tag{11.6}$$

Algorithm 11.1 AM-FIHT for Data Recovery from Noiseless Measurements

1: **Input:** the observation matrix $\mathcal{P}_\Omega(M)$, the desired error tolerance parameter ε, the Hankel matrix parameter k, and rank r.
2: Set $W_{-1} = p^{-1} \mathcal{H}_k(\mathcal{P}_\Omega(M))$, $L_0 = \mathcal{Q}_r(W_{-1})$;
3: Initialize $X_0 = \mathcal{H}_k^\dagger L_0$;
4: **for** $l = 0, 1, \ldots, O(\log(1/\varepsilon))$ **do**
5: $\quad G_l = \mathcal{P}_\Omega(M - X_l)$;
6: $\quad W_l = \mathcal{P}_{\mathcal{S}_l}\left(\mathcal{H}(X_l + p^{-1}G_l) + \beta(W_{l-1} - W_{l-2})\right)$;
7: $\quad L_{l+1} = \mathcal{Q}_r(W_l)$;
8: $\quad X_{l+1} = \mathcal{H}_k^\dagger L_{l+1}$;
9: **end for**
10: **return** X_l

where $w_j = \#\{(j_1, j_2) | j_1 + j_2 = j + 1, 1 \le j_1 \le n, 1 \le j_2 \le n - k + 1\}$ as the number of elements in the jth antidiagonal of an $k \times (n - k + 1)$ matrix.

The computational complexity of solving SVD of a matrix in $\mathbb{C}^{mk \times (n-k+1)}$ is generally $O(rmn^2)$. By exploiting the structures, the SVD of $W_l \in \mathbb{C}^{mk \times (n-k+1)}$ can be computed in $O(r^2 mn + r^3)$ [21]. The total per-iteration complexity is $O(r^2 mn + rmn \log n + r^3)$.

Theorem 1 of [17] characterizes the recovery guarantee of AM-FIHT with noiseless measurements, and it is restated as follows.

THEOREM 11.1 *Suppose the number of observations satisfies*
$$|\Omega| \ge O(r^2 \log(n) \log(1/\varepsilon)).$$
Then, with high probability, the iterates X_l's generated by AM-FIHT with $\beta = 0$ satisfy
$$\|X_l - X^*\|_F \le \nu^l \|X_0 - X^*\|_F \tag{11.7}$$
for any arbitrarily small constant $\varepsilon > 0$ and some constant $\nu \in (0, 1)$.

Theorem 11.1 indicates that if the number of noiseless observations is $O(r^2 \log(n) \log(1/\varepsilon))$, then AM-FIHT is guaranteed to recover X^* exactly. Moreover, from (11.7), the iterates generated by AM-FIHT converge linearly to the ground-truth X^*, and the rate of convergence is ν. Because X^* is rank r, if one directly applies a conventional low-rank matrix completion method such as nuclear norm minimization [20, 22, 23], the required number of observations is $O(rn \log^2(n))$. Thus, when n is large, by exploiting the low-rank Hankel structure of correlated time series, the required number of measurements is significantly reduced.

The result can be generalized to noisy measurements. Minor changes are made in AM-FIHT to simplify the theoretical analysis of noisy measurements (see RAM-FIHT in [17]) while AM-FIHT and RAM-FIHT perform almost the same numerically. The right-hand side of (11.7) includes an additional term that is related with noise. See Theorem 4 of [17] for details.

Moreover, the heavy-ball step improves the convergence rate of AM-FIHT. Although it is well known that the heavy-ball method is a standard technique to accelerate graduate descent, the theoretical characterization of the improvement in the convergence rate by the heavy-ball step in solving a nonconvex optimization is novel. The result is stated in the following theorem (see theorem 2 in [17] for details).

THEOREM 11.2 *(Faster convergence with a heavy-ball step) Let X_l's denote the convergent iterates returned by AM-FIHT. There exists an integer s_0 and a constant $q \in (0, 1)$ that depends on β such that*
$$\frac{\|X_{l+1} - X\|_F}{\|X_l - X\|_F} \le q(\beta), \quad \forall l \ge s_0 \tag{11.8}$$
Moreover,
$$q(0) > q(\beta), \tag{11.9}$$
for any small enough β.

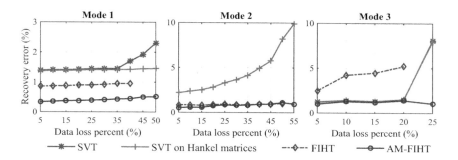

Figure 11.3 Relative recovery error of SVT, FIHT, and AM-FIHT on the PMU data set.

Theorem 11.2 indicates that by adding a heavy-ball term, the iterates still converge linearly to X, and the convergence rate is faster than that without the heavy-ball term.

Note that with measurements in only one channel ($m = 1$), one can construct a low-rank Hankel matrix and apply AM-FIHT. Because the measurements are time series in the same dynamical system and are thus correlated, recovering the missing data simultaneously in a multi-channel Hankel matrix outperforms recovering the missing data separately in each single-channel Hankel matrix. The improvement of recovering missing data in multiple PMU channels simultaneously is analytically characterized in theorem 5 of [17].

Figure 11.3 compares the relative recovery error of AM-FIHT with Singular Value Thresholding (SVT) [24] and Fast Iterative Hard Thresholding (FIHT) [21]. SVT solves the convex relaxation of (11.3). FIHT is a nonconvex algorithm that recovers the missing data of single-channel Hankel matrices, and it is implemented to recover data in each channel separately. The ground-truth data is shown in Figure 11.1, and some data points are erased to test the recovery methods. In Mode 1, data points are erased randomly and independently across channels and time. In Mode 2, data points in all channels are erased simultaneously at some randomly selected time instants. In Mode 3, all data points at some fixed channels are erased simultaneously and consecutively for some time. SVT fails to recover fully lost columns in a low rank and thus, its result is not included in mode 2. AM-FIHT achieves the best performance among all the methods in all three modes.

11.4 Bad Data Correction

Bad data may result from device malfunction, communication error, and cyber data attacks. Conventional bad data detectors usually exploit the correlations in the measurements due to circuit laws and identify the measurements that deviate from the true values significantly. These bad data detections need the system topology and impedance information. Now with the low-rank property, one can locate and correct the bad data using only the measurements without any system topology information.

Let $X^* \in \mathbb{C}^{m \times n}$ denote the low-rank data from m channels in n time instant. Let S^* denote the additive errors in the measurements. The measurements with bad data can be represented by $\mathcal{P}_\Omega(M) = \mathcal{P}_\Omega(X^* + S^*)$. We assume at most s measurements are corrupted, i.e., $\|S^*\|_0 \leq s$, where $\|\cdot\|_0$ measures the number of nonzero entries. The values of the nonzero entries can be arbitrary. The Robust Matrix Completion problem (RMC) aims to recover X^* from $\mathcal{P}_\Omega(M)$. One can formulate it as the following nonconvex optimization problem,

$$\min_{X,S} \quad \|\mathcal{P}_\Omega(M - X - S)\|_F$$
$$\text{s.t.} \quad \text{rank}(X) \leq r \quad \text{and} \quad \|S\|_0 \leq s, \tag{11.10}$$

where the nonconvexity results from the rank and the sparsity constraints. When Ω is respect to the full observations, then the problem is reduced to recovering X^* from $M = X^* + S^*$, which is called Robust Principal Component Analysis (RPCA).

Because nonconvex optimization problems usually suffer from spurious local optimums, several approaches have relaxed the nonconvex rank and ℓ_0-norm terms in (11.10) into the corresponding convex nuclear norm and ℓ_1-norm. Under mild assumptions, X^* and S^* are indeed the solution to the convex relaxation (see, e.g., [25, 26] for RMC and [27–29] for RPCA). However, the convex relaxation is still time-consuming to solve; fast algorithms based on alternating minimization or gradient descent were developed recently to solve the nonconvex problem directly, for example [30, 31] for RMC and [32–34] for RPCA.

If the fraction of nonzeros in each column and row of S^* is at most $\Theta(\frac{1}{r})$, then both the convex method in [25] and the nonconvex method in [35] are proven to be able to recover X^* successfully. If all the observations in a column are corrupted, however, with the prior assumption that the matrix is low-rank, one can locate the corrupted column at best but cannot recover the actual values in either RPCA [36] or RMC [28]. Because every column lies in the r-dimensional column subspace, at least r entries of each column are needed to determine the exact values of that column.

Conventional RPCA and RMC methods cannot recover the fully corrupted columns, which correspond to the case that all the measurements in all PMU channels at the same time instant are all bad data. One can exploit the temporal correlations in the PMU data to correct fully corrupted columns without any model information. As discussed earlier, the Hankel matrix of the PMU data matrix is also low-rank, and the low-rank Hankel property characterizes the temporal correlations in the time series of a dynamical system. Exploiting the low-rank Hankel property, the error correction problem from partial observations can be formulated as

$$\min_{X,S} \quad \|\mathcal{P}_\Omega(M - X - S)\|_F$$
$$\text{s.t.} \quad \text{rank}(\mathcal{H}_k(X)) \leq r \quad \text{and} \quad \|S\|_0 \leq s, \tag{11.11}$$

in [37, 38]. Moreover, an alternating minimization algorithm named structured alternating projections (SAP) with the recovery guarantee is provided in [37, 38].

Algorithm 11.2 Structured Alternating Projections (SAP)

1: **Input** Observations $\mathcal{P}_\Omega(M)$, the error tolerance parameter ε, the rank r, the largest singular value $\sigma_1 = \sigma_1(\mathcal{H}_k(X^*))$, and convergence criterion η.
2: **Initialization** $X_0 = 0$, $\xi_0 = \eta \sigma_1$.
3: **for** t from 1 to r **do**
4:     **for** l from 0 to $L = O(\log(1/\varepsilon))$ **do**
5:         $S_l = \mathcal{T}_{\xi_l}(\mathcal{P}_\Omega(M - X_l))$;
6:         $W_l = \mathcal{H}_k\left(X_l + p^{-1}(\mathcal{P}_\Omega(M - X_l) - S_l)\right)$;
7:         $\xi_{l+1} = \eta\left(\sigma_{t+1}(W_l) + (\frac{1}{2})^l \sigma_t(W_l)\right)$;
8:         $L_{l+1} = \mathcal{Q}_t(W_l)$;
9:         $X_{l+1} = \mathcal{H}_k^\dagger L_{l+1}$;
10:     **end for**
11:     **if** $\eta \sigma_{t+1}(W_L) \leq \frac{\varepsilon}{\sqrt{mn}}$ **then**
12:         Return X_{T+1};
13:     **end if**
14:     $X_0 = X_{T+1}$, $\xi_0 = \xi_{T+1}$;
15: **end for**

SAP contains two stages of iterations. The target rank t increases from 1 gradually in the outer loop. SAP estimates the component of X^* that correspond to the tth largest singular value. The resulting matrix in the tth stage is used as the initializations in the $(t+1)$th stage. In the lth iteration of the inner loop, SAP updates the estimated sparse error matrix S_l and data matrix X_{l+1} based on S_{l-1} and X_l. S_l is obtained by a hard thresholding over the residual error between M and X_l. The updating rule of X_{l+1} is similar to AM-FIHT without the heavy-ball step. SAP first updates X_l by moving along the gradient descent direction. Then, W_l is calculated as the projection of the updated X_l to the Hankel matrix space. Finally, X_{l+1} is obtained by $\mathcal{H}_k^\dagger L_{l+1}$, and L_{l+1} is updated by truncating W_l to a rank-t matrix. The thresholding ξ_l decreases as l increases, and outliers with large magnitudes in S^* are identified and removed gradually.

Details of SAP are summarized in Algorithm 11.2. $\mathcal{T}_\xi$ is the hard thresholding operator,

$$\mathcal{T}_\xi(Z)_{i,j} = Z_{ij} \quad \text{if} \quad |Z_{ij}| \geq \xi, \quad \text{and } 0 \quad \text{otherwise.} \tag{11.12}$$

The operations $\mathcal{Q}_t$ and $\mathcal{H}_k^\dagger$ follow the same definitions in (11.5) and (11.6). Note that $\sigma_1(\mathcal{H}_k(X^*))$ may not be computed directly. In practice, $p^{-1}(\mathcal{H}_k \mathcal{P}_\Omega(M))$ can be used as an estimation for $\sigma_1(\mathcal{H}(X^*))$.

Calculating the best rank-t approximation in line 8 dominates the computation complexity, and it can be solved in $O(tmn \log(n))$ for a Hankel matrix $\mathcal{H}_k(Z) \in \mathbb{C}^{mk \times (n-k+1)}$ (see [18, 21, 37, 38]). Hence, the computational complexity per iteration is $O(rmn \log(n))$, and the total computational complexity is $O(r^2 mn \log(n) \log(1/\varepsilon))$.

The following theorem in [37] characterizes the recovery performance of SAP.

THEOREM 11.3 *Suppose the number of observations satisfies*

$$|\Omega| \geq O\left(r^3 \log^2(n) \log(1/\varepsilon)\right) \quad (11.13)$$

and each row of S^* contains at most $\Theta(\frac{1}{r})$ fraction of outliers. Then, the output X and S returned by SAP satisfy

$$\|X - X^*\|_F \leq \varepsilon, \quad (11.14)$$

$$\|S - \mathcal{P}_\Omega(S^*)\|_F \leq \varepsilon, \quad (11.15)$$

$$\text{and} \quad \text{Supp}(S) \subseteq \text{Supp}(\mathcal{P}_\Omega(S^*)), \quad (11.16)$$

with high probability.

Equation (11.16) means that any identified location in S corresponds to an actual outlier in the data. Theorem 11.3 indicates that the resulting X returned by SAP is ε-close to the ground truth X^* as long as the number of observations exceeds $O(r^3 \log^2(n) \log(1/\varepsilon))$, and each row of S^* has at most $\Theta(\frac{1}{r})$ fraction of outliers, where ε is the desired accuracy. The number of iterations rL depends on $\log(1/\varepsilon)$. Therefore, the algorithm also enjoys a linear convergent rate.

If there is no missing data, i.e., $\Omega = \{(k,t)|1 \leq k \leq m, 1 \leq t \leq n\}$, (11.10) is reduced to the RPCA problem. The existing results in [35] for RMC and [29, 33] for RPCA can tolerate at most $\Theta(\frac{1}{r})$ fraction of outliers per row and per column. The fully corrupted columns can be located but not corrected by existing approaches in either RPCA [36] or RMC [28]. SAP outperforms conventional RMC and RPCA algorithms because SAP can recovery fully corrupted columns, while no existing methods can. In fact, SAP can tolerate at most $\Theta(\frac{1}{r})$ fraction of corrupted columns. One can also apply an existing RPCA algorithm such as AltProj [33] on the structured Hankel matrix $\mathcal{H}_k(M)$. Then Altproj can recover the corrupted columns correctly. However, the computational time per iteration of Altproj is $O(rmn^2)$, which is larger than $O(rmn \log(n))$ by SAP.

Figure 11.4 compares the performance of SAP with R-RMC [35], a nonconvex algorithm to solve the Robust Matrix Completion problem, as well as using Alternating Direction Method of Multipliers (ADMM) to solve the convex relaxation of (11.10). The ground truth PMU data is shown in Figure 11.1 from $t = 2.5$s to 20s. The results

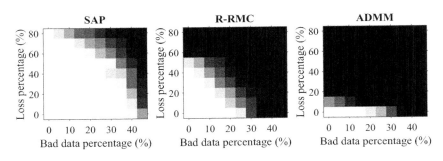

Figure 11.4 Performance of SAP, R-RMC, and ADMM on the PMU data set.

in Figure 11.4 are averaged over 50 independent trials. A white block corresponds to all successes, and a black block corresponds to all failures. The data losses and outliers are at random locations. SAP achieves the best performance among these methods. The convex method performs poorly because the original data matrix is a very flat matrix ($m \ll n$). Additional experiments are tested when full-column data losses and full-column data errors happen at randomly selected columns. SAP shows a similar performance as in Figure 11.4, while the other two methods fail because they cannot handle fully corrupted or lost columns.

11.5 Location of Corrupted Devices

Cyber data attacks (firstly studied in [39]) can be viewed as "the worst interacting bad data injected by an adversary" [40]. Malicious intruders with system configuration information can simultaneously manipulate multiple measurements so that these attacks cannot be detected by any bad data detector. Because the removal of affected measurements would make the system unobservable, these attacks are termed as "unobservable attacks"[3] in [40].

If a cyber intruder injects data attacks to some PMUs constantly, the PMUs under attack do not provide any accurate measurements at any time instant. The identification and location of affected PMUs under attack becomes a challenging task. Moreover, because all the measurements in the affected PMUs are corrupted, one cannot recover the ground-truth data accurately, unless other prior information is leveraged. An attack location approach is developed in [12] to locate the affected PMUs. The intuition is that even though an intruder can constantly inject data attacks that are consistent with each other at each time instant, as long as the intruder does not know the system dynamics, one can identify the attacks by comparing time series of different PMUs and locating the PMUs that exhibit abnormal dynamics.

Let $X^* \in \mathbb{C}^{m \times n}$ denote the ground-truth data in m channels across n time instants. Let S^* denote the additive errors injected by cyber intruders. Note that S^* is column-sparse because intruders might only change the measurements of a few PMU channels due to the resource constraints. That means S^* only has a few nonzero rows, and all the entries in these rows can be nonzero. Let $S^*_{i\cdot} \in \mathbb{C}^{1 \times n}$ denote the ith row of S^*. To locate the corrupted channels, one can solve the following convex optimization problem,

$$\min_{X \in \mathbb{C}^{m \times n}, S \in \mathbb{C}^{m \times n}} \|X\|_* + \lambda \sum_{i=1}^{m} \|S_{\cdot i}\|_2 \quad (11.17)$$

$$\text{s.t. } M = X + S,$$

where the first term in the objective measures the nuclear norm, which is the sum of the singular values. Minimizing the nuclear norm promotes a low-rank solution, and minimizing the sum of the row-norms of S promotes a row-sparse solution. The predetermined weight λ achieves the tradeoff between these two objectives. This convex

---

[3] The term "unobservable" is used in this sense throughout the chapter.

formulation is known to be able to locate fully corrupted rows in a low-rank matrix [36]. If the low-rank matrix has a small incoherence and the number of corrupted rows is below a certain threshold, the solution to (11.17) returns a matrix $\hat{S}$, and the location of the nonzero rows in $\hat{S}$ and S^* are the same.

One can further enhance the attack identification performance by exploiting the structures in these cyber data attacks. Existing bad data detectors (combined with state estimators) can identify and correct some data attacks if the affected measurements are not consistent with other observations. Under the π equivalent model for a transmission line, let Z^{ij} and Y^{ij} denote the impedance and admittance of the transmission line between bus i and bus j. Current I^{ij} from bus i to bus j is related to bus voltage V^i and V^j by

$$I^{ij} = \frac{V^i - V^j}{Z^{ij}} + V^i \frac{Y^{ij}}{2}. \tag{11.18}$$

If PMUs are installed to measure I^{ij}, V^i, and V^j, then the injected cyber data attacks should still satisfy the Equation (11.18) at each time instant, otherwise these attacks can be detected by an exciting bad data detector. Then an intruder with the topology information of the power system should choose the values of the injected errors such that the circuit laws are still satisfied to pass the existing bad data detectors. That is the reason why cyber data attacks are also considered as interacting bad data.

To model the correlations in the injected data attacks, define a matrix $W \in \mathbb{C}^{p \times n}$ as follows. If the kth PMU channel measures the voltage phasor of bus j, $W_{kj} = 1$; if it measures the current phasor from bus i to bus j, then $W_{ki} = 1/Z^{ij} + Y^{ij}/2$, $W_{kj} = -1/Z^{ij}$; $W_{kj} = 0$ otherwise. Let $Z^* \in \mathbb{C}^{p \times n}$ contain the ground-truth bus voltage phasors of all p buses in the system in n time steps. The PMU measurements and the state variables are related by

$$X^* = W Z^*. \tag{11.19}$$

To pass the existing bad data detectors, the injected errors S^* should also satisfy

$$S^* = W D^* \tag{11.20}$$

for some row sparse matrix $D^* \in \mathbb{C}^{p \times n}$. That means the attacks are injected to alter the estimation of bus voltage phasors of buses corresponding to the nonzeros in D^*. Note that m is greater than p when the system contains enough PMUs. Then the obtained measurements under attack can be written as

$$M = X^* + W D^*. \tag{11.21}$$

To locate the affected bus voltage phasors, as well as the corresponding affected PMU channels, reference [12] proposes to solve

$$\min_{X \in \mathbb{C}^{m \times n}, D \in \mathbb{C}^{p \times n}} \|X\|_* + \lambda \sum_{i=1}^{p} \|D_{\cdot i}\|_2 \tag{11.22}$$

$$\text{s.t. } M = X + W D.$$

Let $\hat{X}$ and $\hat{D}$ denote the solution to (11.22). Let $\hat{\mathcal{J}}$ denote the set of row indices in $\hat{D}$ that are nonzero. Let $\mathcal{J}^*$ denote the set of row indices in D^* that are nonzero. The identification guarantee is stated as follows.

THEOREM 11.4 *If the number of nonzero rows in D^* is at most $\tilde{k}$, there exists $\lambda_{\min}$ and $\lambda_{\max}$ that depend on $\tilde{k}$, the incoherence of X^*, and some properties of W such that for any $\lambda \in [\lambda_{\min}, \lambda_{\max}]$, the solution $\hat{X}, \hat{D}$ satisfy*

$$\hat{\mathcal{J}} = \mathcal{J}^*, \qquad (11.23)$$

and for all $i \notin \mathcal{J}^*$,

$$\hat{X}_{\cdot i} = X^*_{\cdot i}. \qquad (11.24)$$

Theorem 11.4 guarantees that the affected bus voltage phasors are correctly identified from (11.23), and then the affected PMUs can be correctly located by the nonzero rows in $\hat{S} = W\hat{D}$. Then, the "clean" PMU measurements that are not affected by cyber data attacks could be identified, as shown in (11.24). Moreover, the recovery is also successful when the number of nonzero rows in D^* is zero. Thus, the false alarm rate is zero.

The range $\lambda \in [\lambda_{\min}, \lambda_{\max}]$ is computed in [12] for special systems, but in general, one cannot compute them directly as the incoherence of X^* is usually unknown. However, as $\lambda_{\min}$ decreases and $\lambda_{\max}$ increases, $\tilde{k}$ decreases. Therefore, the range of a proper λ increases when the number of affected bus voltage phasors decreases. If an intruder only affects a few PMU channels, it is very likely that the approach performs similarly with a wide range of λ.

The approach can be generalized to the case of noisy measurements, i.e.,

$$M = X^* + WD^* + N, \qquad (11.25)$$

where the matrix N contains the noise in the measurements. Equation (11.22) can be revised to handle noise by replacing the equality constraint with

$$\sum_{i,j} |M_{ij} - X_{ij} - (WD)_{ij}|^2 \leq \eta. \qquad (11.26)$$

In this case, the solution to the revised problem, denoted by $\tilde{X}$ and $\tilde{D}$, is close to a pair $\hat{X}$ and $\hat{D}$ that satisfies (11.23) and (11.24), and the difference depends on the noise η. The exact relationship is characterized in theorem 2 of [12].

11.6 Extensions in Data Recovery and Correction

11.6.1 Data Recovery Using Time-Varying Low-Dimensional Structures to Characterize Long-Range Temporal Correlations

When the underlying dynamical system changes slowly, the generated time series have long-range correlations. The time series are truncated into a sequence of matrices, and one can exploit the correlations with past matrices to enhance the data estimation accuracy.

First proposed in [41] to study low-rank matrices and extended [42] to incorporate sparse errors, *a sequence of low-rank matrices with correlated subspaces* models the long-range correlations. Let $M^d \in \mathbb{C}^{m \times n_1}$ denote the measurements at the dth time segment d, with each segment including n_1 time steps. $M^d = X^{d*} + S^{d*}$, where the low-rank (assuming to be rank-r) X^{d*} models the ground-truth data, and the sparse matrix S^{d*} with at most s nonzeros models the sparse errors. At time segment d, the goal is to recover X^{d*}. Using $\{M^t\}_{t=1}^d$ would enhance the recovery performance compared to using M^d only.

One can solve the following nonconvex problem (11.27) and use its solution $\hat{X}$ as the estimate of X^d.

$$\min_{X, S \in \mathbb{R}^{n_1 \times n_2}} \sum_{t=1}^d \omega_t \|\mathcal{P}_{\Omega^t}(X + S) - \mathcal{P}_{\Omega^t}(M^t)\|_2^2$$
$$\text{s.t.} \quad \|X\|_\infty \leq \alpha, \|S\|_\infty \leq \alpha, \quad (11.27)$$
$$\text{rank}(X) \leq r, \sum_{ij} 1_{[S_{ij} \neq 0]} \leq s,$$

The non-negative weights $\{\omega_t\}_{t=1}^d$ are predetermined, and $\sum_{t=1}^d \omega_t = 1$. The idea is to impose the constraint of similarities between the estimated X^d and the past observations through the objective function. The optimization (11.27) is nonconvex due to the constraints. An approximation algorithm is developed in [43] to solve (11.27) through alternating minimization.

11.6.2 Data Recovery under Multiple Disturbances

Disturbances of various magnitude happen in power systems constantly, and the rank of the PMU data matrix increases significantly when the system experiences multiple disturbances. Because the required number of observations to recover the remaining ones depends linearly on the rank, applying low-rank matrix completion methods directly when the system is under multiple disturbances requires a large number of observations.

A novel nonlinear model, termed as "union and sums of subspaces," is developed in [10, 11] to characterize the low-dimensional structure of the PMU data blocks when the system is under multiple disturbances. Under this model, some data points in C^m belong to one of k r-dimensional subspaces, while the rest belong to the sums of a few subspaces. Each data point represents the time series in n time instants of one PMU channel. Data points in the same subspace are affected by the same local event, while data points in the sum of subspaces are influenced by multiple events.

A data recovery method that exploits the low-dimensional property when multiple disturbances happen is developed in [10, 11]. The required number of observations to determine the missing points of an $m \times n$ matrix with rows belonging to the "union and sums" of k r-dimensional subspaces is proved to be in the order of $r^2 n \log^3 m$. This number is much smaller than the number $krn \log^2 m$ by directly using low-rank matrix completion methods directly when k is large.

11.6.3 Online Data Recovery

AM-FIHT and SAP are batch methods that recover the data in a time window simultaneously. Online data recovery methods such as OLAP [7] and OLAP-H [19] can fill in the missing points and correct bad measurements of streaming data. Compared with batch methods, these methods require a smaller memory size and have lower computational complexity. Online low-rank methods exist, e.g., GROUSE [44], PETRELS [45], and ReProCS [46], but they cannot differentiate the onset of a power system disturbance from bad data and cannot handle simultaneous data losses or data errors across all channels.

OLAP-H [19] has the distinctive feature from the existing online low-rank approaches: (1) it differentiates an event from bad data; (2) it handles simultaneous and consecutive data losses/errors; and (3) OLAP-H adapts to the dimensionality change due to the onset and offset of an event in power systems. To identify an event, OLAP-H computes the rank of a Hankel matrix of a short window, permutes the time, and computes the rank of the Hankel matrix of the permuted data. If the ranks of the two Hankel matrices differ much, then OLAP-H declares a system event. Otherwise, OLAP-H treats them as bad data and corrects the values.

11.7 Data Recovery from Privacy-Preserving Measurements

Currently, large amounts of synchrophasor measurements are transmitted to the operator through multiple layers of Phasor Data Concentrators (PDCs). The data transmission consumes significant bandwidth, which leads to congestions and data losses. Moreover, a malicious intruder might eavesdrop the communication or even alter some data to mislead the operator.

A privacy-preserving and communication-reduced data-collection and information recovery framework is proposed in [16, 43]. In the sensors or PDCs, random noises are first added to the sensing data, and then the resulting real value is mapped to one of K discrete values. The highly quantized measurements are transmitted to the operator. The operator collectively processes all the data and recovers the information (see Figure 11.5).

Let $X^* \in \mathbb{R}^{m \times n}$ denote the actual data in m channels across n time instants. Let $S^* \in \mathbb{R}^{m \times n}$ denote the sparse additive errors in the measurements. The rank of X^* is r ($r \ll n$), and S^* has at most s nonzero entries. Assume $\|X^*\|_\infty \leq \alpha$ and $\|S^*\|_\infty \leq \alpha$ for some constant α, where the infinity norm $\|\cdot\|_\infty$ is the largest entry-wise absolute value. $M = X^* + S^*$ denotes the measurements that are partially erroneous. Let $N \in \mathbb{R}^{m \times n}$ denote the noise matrix which has i.i.d. entries with a known cumulative distribution function $\Phi(z)$. The Random Gaussian noise is considered here for the presentation simplicity. Given a positive constant K and the quantization boundaries $\omega_0 < \omega_1 < \cdots < \omega_K$, a K-level quantized noisy measurement Y_{ij} based on M_{ij} satisfies

$$Y_{ij} = \mathcal{Q}(X^*_{ij} + C^*_{ij} + N_{ij}), \quad \forall (i, j), \tag{11.28}$$

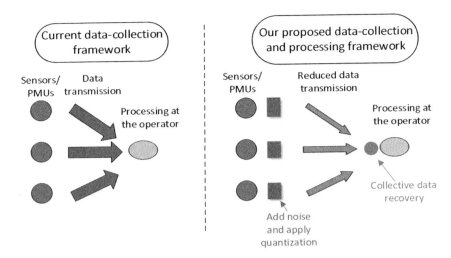

Figure 11.5 Comparison of the current framework and our proposed data-collection and processing framework.

where the operator Q maps a real number to one of the K labels, i.e., $Q(x) = l$ if $\omega_{l-1} < x \leq \omega_l$. Then, we have

$$Y_{ij} = l \text{ with probability } f_l(M_{ij}), \forall (i, j), \tag{11.29}$$

where $\sum_{l=1}^{K} f_l(M_{ij}) = 1$, and

$$f_l(M_{ij}) = P(Y_{ij} = l | M_{ij}) = \Phi(\omega_l - M_{ij}) - \Phi(\omega_{l-1} - M_{ij}). \tag{11.30}$$

Given quantized observations Y, the noise distribution Φ, and quantization boundaries $\{\omega_0, \ldots, \omega_K\}$, the data recovery problem for the operator is to estimate the actual data X^*.

The data recovery problem can be formulated as a nonconvex constrained maximum likelihood estimation problem. One can estimate (X^*, S^*) by $(\hat{X}, \hat{S})$, where

$$(\hat{X}, \hat{S}) = \arg\min_{X,S} - \sum_{i=1}^{m} \sum_{j=1}^{n} \sum_{l=1}^{K} \mathbf{1}_{[Y_{ij}=l]} \log(f_l(X_{ij} + S_{ij})),$$

$$\text{s.t. } X, S \in \mathbb{R}^{m \times n}, \|X\|_\infty \leq \alpha, \|S\|_\infty \leq \alpha, \tag{11.31}$$

$$\text{rank}(X) \leq r, \sum_{i=1}^{m} \sum_{j=1}^{n} \mathbf{1}_{[S_{ij} \neq 0]} \leq s.$$

THEOREM 11.5 *As long as s is at most $\Theta(m)$, i.e., the number of corrupted measurements per row is bounded, then with probability at least $1 - C_1 e^{-C_2 m}$, any global minimizer $(\hat{X}, \hat{S})$ of (11.31) satisfies*

$$\|\hat{X} - X^*\|_F / \sqrt{mn} \leq O(\sqrt{r/\min(m,n)}). \tag{11.32}$$

for some positive constants C_1 and C_2. Moreover, for any algorithm that takes $Y = X^* + C^* + N$ as the input and returns $\hat{L}$ and $\hat{S}$, there always exists a pair X^* and C^* such that the returned estimates $\hat{L}, \hat{S}$ by this algorithm satisfy

$$\frac{\|\hat{X} - X^*\|_F}{\sqrt{mn}} \geq \Theta(\sqrt{r/\min(m,n)}). \tag{11.33}$$

Theorem 11.5 indicates that the recovery error, measured by $\frac{\|\hat{X}-X^*\|_F}{\sqrt{mn}}$, is at most in the order of $\sqrt{\frac{r}{\min(m,n)}}$, as long as the number of corruptions per row is bounded [16, 43]. Thus, the recovery error decreases to zero as the number of channels and the number of data samples increase. An operator with large amounts of data can achieve a very small error. In contrast, a cyber intruder with an access to partial measurements cannot recover the information accurately even using the same data recovery approach. Combining (11.32) and (11.33), one can infer that the recovery error by solving (11.31) is order-wise optimal. Thus, very little information is lost by quantization.

A projected gradient method is developed in [43] to solve the nonconvex quantized matrix recovery problem. The rank constraint on X in (11.31) is converted to a rank-r factorization of X, i.e., $X = UV^T$, where $U \in \mathcal{R}^{m \times r}$, $V \in \mathcal{R}^{n \times r}$. In each iteration, the algorithm updates the estimates of U, V, and S sequentially by applying the alternating gradient descent method. The step sizes can be selected using a backtracking line search using Armijo's rule or by calculating Lipschitz constants. A hard thresholding is applied to S afterward to meet the sparsity constraint. The infinity norm constraints on X and S can be achieved by hard thresholding in each iteration. The per-iteration complexity is $O(mnr)$. Starting from any initial point, every sequence generated by the algorithm converges to a critical point of (11.31) [43].

Figure 11.6 shows the original data, the quantized value with 5% of corruptions, the recovered data, and the recovered data after filtering of two PMU channels. One can see that the details of the time series are masked in the quantized measurements. The recovered data are noisier than the actual data because noise is added to each channel with a noise level comparable to the signal level before quantization. Still, the overall

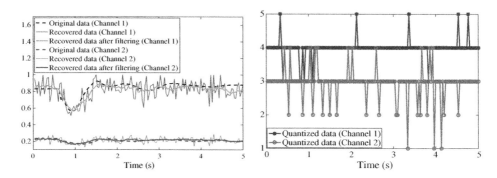

Figure 11.6 Original data, quantized data, recovered data, and recovered data after filtering of two PMU channels ($K = 5$, Gaussian noise $\mathcal{N}(0, 0.18^2)$).

trend of the time series are correctly recovered. If needed, the noise in the recovered data could be removed by applying a low-pass filter.

One can extend this setup to the case that the ground-truth data X^* is not exactly low-rank but contains data points from multiple subspaces. The row vectors of X^* are in $\mathbb{R}^n$, and each vector belongs to one of p different d-dimensional subspaces in $\mathbb{R}^n$ ($d \ll n$). Let r denote the rank of X^*, then $r \leq pd$. r can be large when p is large. The model presents the case that measurements are affected by multiple disturbances, and time series affected by the same event belong to the same subspace.

Then there exists a coefficient matrix $C^* \in \mathbb{R}^{m \times m}$ such that $X^* = C^* X^*$, $C_{ii}^* = 0$ for all $i = 1, \ldots, m$, and C_{ij}^* is zero if the ith row and the j row of X^* do not belong to the same subspace. Following the terminologies in [47], one can say that X^* satisfies the *self-expressive property*, and C^* satisfies the *subspace-preserving property*.

The data recovery and clustering problem can be formulated by adding new constraints $X = CX$, $\|C_i.\|_0 \leq d$, $C_{ii} = 0, i = 1, \ldots, m$ to (11.31) [48]. An algorithm is proposed in [48] to solve the recovery and clustering problem. Factorize X into $X = UV$. In each iteration, the algorithm updates the estimations of U, V, X, S, and C sequentially by first moving along the gradient descent direction and then projecting to the constraint set. The per-iteration complexity of the algorithm is $O(m^2 r)$, and the algorithm converges to a critical point from any initial point.

In summary, the proposed framework has the following advantages. First, data privacy can be enhanced in a simple fashion. Adding noise and applying quantization are known to be able to enhance data privacy [49–51], and both can be readily implemented. Second, the data transmission rate is significantly reduced by sending the highly quantized data instead of the raw data. Third, no complicated agreement is needed between the PMUs, PDCs, and the operator. The operator only knows the noise distribution but does not know the actual noise. The additive noise does not need to be communicated to the operator. Last but not least, a high recovery accuracy is achieved only by the operator with large amounts of measurements, and data privacy is maintained against a cyber intruder.

11.8 Conclusion

This chapter discusses the recent works on PMU data analytics by exploiting low-dimensional structures. Under the same framework, a class of computationally efficient methods with provable analytical guarantees are developed for missing data recovery, bad data correction, and the detection of cyber data attacks. Moreover, a privacy-preserving data-collection and information retrieval method is developed by exploiting the low-dimensional structures.

The authors would like to thank Bruce Fardanesh and George Stefopoulos at New York Power Authority; Evangelos Farantatos and Mahendra Patel at Electric Power Research Institute; Frankie Qiang Zhang, Xiaochuan Luo, and Slava Maslennikov at ISO-NE for providing recorded PMU data sets and helpful discussions. The first author also wants to thank Matthew Berger and Lee M. Seversky for technical discussions during her visit to Air Force Research Lab in the summer of 2015.

The results described in this chapter are obtained through the research support in part by NSF 1508875, EPRI 1007316, ARO W911NF-17-1-0407, and the ERC Program of NSF and DoE under the supplement to NSF Award EEC-1041877 and the CURENT Industry Partnership Program.

References

[1] A. Phadke and J. Thorp, *Synchronized Phasor Measurements and Their Applications*. New York: Springer, 2008.

[2] A. Silverstein and J. E. Dagle, "Successes and challenges for synchrophasor technology: An update from the North American synchrophasor initiative," in *Proc. 45th Hawaii International Conference on System Sciences*, pp. 2091–2095, 2012.

[3] M. Backes and S. Meiser, "Differentially private smart metering with battery recharging," in *Data Privacy Management and Autonomous Spontaneous Security*. Berlin and Heidelberg: Springer, pp. 194–212, 2014.

[4] D. Egarter, C. Prokop, and W. Elmenreich, "Load hiding of household's power demand," in *2014 IEEE International Conference on Smart Grid Communications (SmartGridComm)*, pp. 854–859, 2014.

[5] S. McLaughlin, P. McDaniel, and W. Aiello, "Protecting consumer privacy from electric load monitoring," in *Pro. 18th ACM Conference on Computer and Communications Security*, pp. 87–98, 2011.

[6] M. Savi, C. Rottondi, and G. Verticale, "Evaluation of the precision-privacy tradeoff of data perturbation for smart metering," *IEEE Transactions on Smart Grid*, vol. 6, no. 5, pp. 2409–2416, 2015.

[7] P. Gao, M. Wang, S. G. Ghiocel, J. H. Chow, B. Fardanesh, and G. Stefopoulos, "Missing data recovery by exploiting low-dimensionality in power system synchrophasor measurements," *IEEE Transactions on Power Systems*, vol. 31, no. 2, pp. 1006–1013, 2016.

[8] P. Gao, M. Wang, S. G. Ghiocel, and J. H. Chow, "Modeless reconstruction of missing synchrophasor measurements," in *Proc. IEEE Power and Energy Society (PES) General Meeting*, pp. 1–5, 2014.

[9] M. Wang, J. H. Chow, P. Gao, X. T. Jiang, Y. Xia, S. G. Ghiocel, B. Fardanesh, G. Stefopolous, Y. Kokai, and N. Saito, "A low-rank matrix approach for the analysis of large amounts of power system synchrophasor data," in *Proc. IEEE Hawaii International Conference on System Sciences (HICSS)*, pp. 2637–2644, 2015.

[10] P. Gao, M. Wang, J. H. Chow, M. Berger, and L. M. Seversky, "Matrix completion with columns in union and sums of subspaces," in *Proc. IEEE Global Conference on Signal and Information Processing (GlobalSIP) 2015*, pp. 785–789, 2015.

[11] ——, "Missing data recovery for high-dimensional signals with nonlinear low-dimensional structures," *IEEE Transactions on Signal Processing*, vol. 65, no. 20, pp. 5421–5436, 2017.

[12] P. Gao, M. Wang, J. H. Chow, S. G. Ghiocel, B. Fardanesh, G. Stefopoulos, and M. P. Razanousky, "Identification of successive 'unobservable' cyber data attacks in power systems," *IEEE Transactions on Signal Processing*, vol. 64, no. 21, pp. 5557–5570, 2016.

[13] M. Wang, P. Gao, S. G. Ghiocel, J. H. Chow, B. Fardanesh, G. Stefopoulos, and M. P. Razanousky, "Identification of 'unobservable' cyber data attacks on power grids," in *Proc. IEEE International Conference on Smart Grid Communications (SmartGridComm)*, pp. 830–835, 2014.

[14] W. Li, M. Wang, and J. H. Chow, "Fast event identification through subspace characterization of PMU data in power systems," in *Proc. IEEE Power and Energy Society (PES) General Meeting*, 2017.

[15] ——, "Real-time event identification through low-dimensional subspace characterization of high-dimensional synchrophasor data," *IEEE Transactions on Power Systems*, vol. 33, no. 5, pp. 4937–4947, 2018.

[16] P. Gao, R. Wang, M. Wang, and J. H. Chow, "Low-rank matrix recovery from quantized and erroneous measurements: Accuracy-preserved data privatization in power grids," in *Proc. Asilomar Conference on Signals, Systems, and Computers*, 2016.

[17] S. Zhang, Y. Hao, M. Wang, and J. H. Chow, "Multi-channel missing data recovery by exploiting the low-rank Hankel structures," in *Proc. Int. Workshop Comput. Adv. Multi-Sensor Adaptive Process. (CAMSAP)*, pp. 1–5, 2017.

[18] ——, "Multi-channel Hankel matrix completion through nonconvex optimization," *IEEE J. Sel. Topics Signal Process., Special Issue on Signal and Information Processing for Critical Infrastructures*, vol. 12, no. 4, pp. 617–632, 2018.

[19] Y. Hao, M. Wang, J. H. Chow, E. Farantatos, and M. Patel, "Model-less data quality improvement of streaming synchrophasor measurements by exploiting the low-rank Hankel structure," *IEEE Transactions on Power Systems*, vol. 33, no. 6, pp. 6966–6977, 2018.

[20] E. J. Candès and B. Recht, "Exact matrix completion via convex optimization," *Foundations of Comput. Math.*, vol. 9, no. 6, pp. 717–772, 2009.

[21] J.-F. Cai, T. Wang, and K. Wei, "Fast and provable algorithms for spectrally sparse signal reconstruction via low-rank Hankel matrix completion," *Applied and Computational Harmonic Analysis*, vol. 46, no. 1, pp. 94–121, 2017.

[22] E. Candès and T. Tao, "The power of convex relaxation: Near-optimal matrix completion," *IEEE Transactions on Information Theory*, vol. 56, no. 5, pp. 2053–2080, 2010.

[23] D. Gross, "Recovering low-rank matrices from few coefficients in any basis," *IEEE Transactions on Information Theory*, vol. 57, no. 3, pp. 1548–1566, 2011.

[24] J.-F. Cai, E. J. Candès, and Z. Shen, "A singular value thresholding algorithm for matrix completion," *SIAM Journal on Optimization*, vol. 20, no. 4, pp. 1956–1982, 2010.

[25] Y. Chen, A. Jalali, S. Sanghavi, and C. Caramanis, "Low-rank matrix recovery from errors and erasures," *IEEE Transactions on Information Theory*, vol. 59, no. 7, pp. 4324–4337, 2013.

[26] O. Klopp, K. Lounici, and A. B. Tsybakov, "Robust matrix completion," *Probability Theory and Related Fields*, vol. 169, no. 1, pp. 523–564, Oct. 2017.

[27] E. J. Candès, X. Li, Y. Ma, and J. Wright, "Robust principal component analysis?" *Journal of the ACM (JACM)*, vol. 58, no. 3, p. 11, 2011.

[28] Y. Chen, H. Xu, C. Caramanis, and S. Sanghavi, "Robust matrix completion with corrupted columns," in *Proc. International Conference on Machine Learning*, 2011.

[29] D. Hsu, S. M. Kakade, and T. Zhang, "Robust matrix decomposition with sparse corruptions," *IEEE Transactions on Information Theory*, vol. 57, no. 11, pp. 7221–7234, 2011.

[30] Q. Gu, Z. Wang, and H. Liu, "Low-rank and sparse structure pursuit via alternating minimization," in *Proc. 19th International Conference on Artificial Intelligence and Statistics*, vol. 51, pp. 600–609, 2016.

[31] A. Kyrillidis and V. Cevher, "Matrix alps: Accelerated low rank and sparse matrix reconstruction," in *2012 IEEE Statistical Signal Processing Workshop (SSP)*, pp. 185–188, 2012.

[32] Y. Chen and M. J. Wainwright, "Fast low-rank estimation by projected gradient descent: General statistical and algorithmic guarantees," *arXiv:1509.03025*, 2015.

[33] P. Netrapalli, U. Niranjan, S. Sanghavi, A. Anandkumar, and P. Jain, "Non-convex robust PCA," in *Advances in Neural Information Processing Systems*, pp. 1107–1115, 2014.

[34] X. Yi, D. Park, Y. Chen, and C. Caramanis, "Fast algorithms for robust PCA via gradient descent," in *Advances in Neural Information Processing Systems 29*, pp. 4152–4160, 2016.

[35] Y. Cherapanamjeri, K. Gupta, and P. Jain, "Nearly-optimal robust matrix completion," *arXiv preprint arXiv:1606.07315*, 2016.

[36] H. Xu, C. Caramanis, and S. Sanghavi, "Robust PCA via outlier pursuit," *IEEE Transactions on Information Theory*, vol. 58, no. 5, pp. 3047–3064, May 2012.

[37] S. Zhang and M. Wang, "Correction of simultaneous bad measurements by exploiting the low-rank Hankel structure," in *Proc. IEEE International Symposium on Information Theory*, pp. 646–650, 2018.

[38] ——, "Correction of corrupted columns through fast robust Hankel matrix completion," *IEEE Transactions on Signal Processing*, vol. 67, no. 10, pp. 2580–2594, 2019.

[39] Y. Liu, P. Ning, and M. K. Reiter, "False data injection attacks against state estimation in electric power grids," *ACM Transactions on Information and System Security (TISSEC)*, vol. 14, no. 1, p. 13, 2011.

[40] O. Kosut, L. Jia, R. Thomas, and L. Tong, "Malicious data attacks on smart grid state estimation: Attack strategies and countermeasures," in *Proc. IEEE International Conference on Smart Grid Communications (SmartGridComm)*, pp. 220–225, 2010.

[41] L. Xu and M. Davenport, "Dynamic matrix recovery from incomplete observations under an exact low-rank constraint," in *Advances in Neural Information Processing Systems 29*, pp. 3585–3593, 2016.

[42] P. Gao and M. Wang, "Dynamic matrix recovery from partially observed and erroneous measurements," in *Proc. IEEE International Conference on Acoustics, Speech and Signal Processing (ICASSP)*, 2018.

[43] P. Gao, R. Wang, M. Wang, and J. H. Chow, "Low-rank matrix recovery from noisy, quantized and erroneous measurements," *IEEE Transactions on Signal Processing*, vol. 66, no. 11, pp. 2918–2932, 2018.

[44] L. Balzano, R. Nowak, and B. Recht, "Online identification and tracking of subspaces from highly incomplete information," in *Proc. Allerton Conference Communication, Control, and Computing*, pp. 704–711, 2010.

[45] Y. Chi, Y. C. Eldar, and R. Calderbank, "PETRELS: Parallel subspace estimation and tracking by recursive least squares from partial observations," *IEEE Transactions on Signal Processing*, vol. 61, no. 23, pp. 5947–5959, 2013.

[46] C. Qiu, N. Vaswani, B. Lois, and L. Hogben, "Recursive robust PCA or recursive sparse recovery in large but structured noise," *IEEE Transactions on Information Theory*, vol. 60, no. 8, pp. 5007–5039, 2014.

[47] E. Elhamifar and R. Vidal, "Sparse subspace clustering: Algorithm, theory, and applications," *IEEE Transactions on Pattern Analysis and Machine Intelligence*, vol. 35, no. 11, pp. 2765–2781, 2013.

[48] R. Wang, M. Wang, and J. Xiong, "Data recovery and subspace clustering from quantized and corrupted measurements," *IEEE Journal of Selected Topics in Signal Processing, Special Issue on Robust Subspace Learning and Tracking: Theory, Algorithms, and Applications*, vol. 12, no. 6, pp. 1547–1560, 2018.

[49] G. Dán and H. Sandberg, "Stealth attacks and protection schemes for state estimators in power systems," in *Proc. IEEE Smart Grid International Conference on Communications (SmartGridComm)*, pp. 214–219, 2010.

[50] A. Reinhardt, F. Englert, and D. Christin, "Enhancing user privacy by preprocessing distributed smart meter data," in *Proc. Sustainable Internet and ICT for Sustainability (SustainIT)*, pp. 1–7, 2013.

[51] S. Xiong, A. D. Sarwate, and N. B. Mandayam, "Randomized requantization with local differential privacy," in *Proc. IEEE Int. Conf. Acoust Speech Signal Process.*, pp. 2189–2193, 2016.

Part IV

Signal Processing

12 A Data Analytics Perspective of Fundamental Power Grid Analysis Techniques

Danilo P. Mandic, Sithan Kanna, Yili Xia, and Anthony G. Constantinides

12.1 Introduction

The problem of tracking the power system frequency has been of practical and theoretical interest for decades, as variations in system frequency indicate a mismatch between the supply and demand of electricity [1]. The area is undergoing a resurgence of interest as future low-inertia grids are bound to experience rapid frequency excursions, which calls for faster and more accurate frequency tracking algorithms [2]. Therefore, it comes as no surprise that this task has attracted a wide-range of solutions that include: Fourier transform approaches [3], adaptive notch filters [4], Newton-type algorithms [5], state-space methods [6], and frequency demodulation techniques [7].

System frequency is the most important power quality parameter, and its accurate estimation underpins power system monitoring, analysis, and synchronization.

Data Analytics Solutions: Opportunities. Frequency estimation in power systems is routinely performed from the three phase voltages, based on the Clarke and related transforms; these were designed for stable grids and nominal conditions. However, smart grids exhibit a prominent random component, owing to the intermittency of renewables and smart loads. This introduces dynamically unbalanced conditions which affect frequency estimates due to [8, 9]:

- Inadequacy of the Clarke and related transforms for unbalanced system conditions;
- Interharmonics introduced by some loads (furnaces, cyclo-converters) that are not integer multiples of the fundamental frequency – these cannot be estimated spectrally and tend to drift over time, which affects systems prone to resonance (low damping- or a high Q-factor);
- Need for observations of high temporal granularity to understand issues such as the power draw of short phases, for example, in a synchronization that only takes a fraction of a millisecond.

Issues in modern power system analysis that cannot be adequately addressed using these classic, Circuit Theory–based, approaches include:

- Loss of mains detection from voltage dips and off-nominal frequencies; this is critical for system balance as various imbalances will be much more prominent in low-inertia grids of the future;

- Current system frequency estimators are unreliable in the presence of voltage sags, harmonics, and rapid frequency jump type of imbalances (even though system frequency remains nominal); this causes false alarms and unnecessary corrective actions that may lead to larger-scale failures [10];
- The IEEE 1547 Standard specifies that a distributed generation source (Smart Grid) must disconnect from a locally islanded system within 2 seconds, it also must disconnect for a sagging voltage under high demand (described by IEEE Standard 1159–1995). However, disconnecting a large number of local generators (e.g., solar) can cause the low-voltage condition to accelerate [11]; and
- Current analyses compute features over predefined time intervals, such as 200 ms, 3 s, 1 m, 10 m, and 2 h. These are adequate for standard grids with power reserve and high inertia (International Standard IEC 6100-4-30) but not for low-inertia grids of the future. For example, PQ variations are currently calculated over a 200 ms window – too coarse for real-time monitoring and synchronization in smart grids where the required time scales are in the region of 2 ms and below.

However, owing to the lack of a coherent and unifying analysis framework, the underlying commonalities between various frequency tracking methods are often overlooked. For example, the extensive benchmarking of the various algorithms focuses mainly upon their performance differences and not on their theoretical connections [12–14]. Amongst the different methods to track the system frequency, the rate of change of the phase angle of the voltage phasor has been the most widely used frequency estimation technique in the electricity grid. For example, the important IEEE synchrophasor standards, like the C37.118.1-2011 and C37.118.1a-2014, use the rate of change of the phase angle of the voltage phasor as a benchmark for all other techniques [15, 16].

Our aim is to show that tracking the system frequency based on the rate of change of the voltage phasor angle has deep connections with well-known concepts like frequency demodulation (FM) and maximum likelihood estimation. Specifically, we show that the voltage phasor angle-based frequency tracker can be interpreted as a special case of an FM demodulation scheme whereby the fundamental frequency of the power system (e.g., 50 Hz or 60 Hz) represents the carrier frequency. Secondly, the rate of change of the phase angle is shown to represent the maximum likelihood frequency estimate of a frequency modulated sinusoid. For rigor, the analysis formulates the problem in the complex-domain to cater for both single-phase and three-phase systems. In this way, it has been possible to both cast the classical Clarke transform into the Principal Component Analysis (PCA) framework and to provide in-depth analysis and interpretation of the Clarke transform as a dimensionality reduction scheme. This has also enabled us to analyze the inadequacy of Clarke transform from a data-driven PCA perspective. These connections have enabled us to use the notion of complex noncircularity and widely linear models to add one more degree of freedom into the analysis, which has resulted in so-called Smart Clarke and Smart Park (SCT and SPT) transforms that operate equally well and in real time for both balanced and unbalanced power systems. We finally discuss the various implications of the findings through practical examples and further reinforce connections with other frequency tracking methods such as the phase locked loop (PLL) and extended Kalman filter (EKF) [17].

12.2 Problem Formulation

Consider discrete time observations, x_k, of a complex-valued voltage signal, s_k, embedded in a white noise process η_k, given by

$$x_k = s_k + \eta_k,$$
$$s_k = Ae^{j(\omega_\circ k + \phi_k)} + Be^{-j(\omega_\circ k + \phi_k)}, \quad (12.1)$$

where $\omega_\circ = 2\pi\frac{f_\circ}{f_s}$ is the fundamental frequency, f_s the sampling frequency, and k the discrete time instant. The real-valued amplitudes are A and B, while the time-varying phase is ϕ_k. For generality, the signal model, s_k in (12.1), was chosen so that it can represent both single-phase and three-phase voltages; for $B = A$, the model in (12.1) degenerates into a single-phase voltage, $s_k = 2A\cos(\omega_\circ k + \phi_k)$, while, for $B = 0$, the model represents the $\alpha\beta$ voltage of a balanced system, i.e. $s_k = Ae^{j(\omega_\circ k + \phi_k)}$. Furthermore, the case $B \neq A$ characterizes the Clarke transform of a general unbalanced three-phase system [1].

The instantaneous frequency, ω_k, of the signal s_k in (12.1) is then given by

$$\omega_k = \omega_\circ + \Delta\omega_k = \omega_\circ + (\phi_k - \phi_{k-1}). \quad (12.2)$$

Such a decomposition of the power system frequency into the fundamental frequency component, $\omega_\circ$, and rate of change of phase, $\Delta\omega_k = (\phi_k - \phi_{k-1})$, is convenient as the system frequency, $f_\circ$, is fixed and known to be either 50 Hz or 60 Hz. The frequency estimation task therefore boils down to tracking the instantaneous deviation, $\Delta\omega_k$, from the fundamental frequency.

12.3 Background: Tracking the Instantaneous Frequency

12.3.1 Rate of Change of the Voltage Phase Angle

To track the instantaneous frequency deviation, $\Delta\omega_k$, of the signal x_k defined in (12.1), the standard frequency estimation algorithm in power systems first computes the phasor, X_k, using the fundamental component of the stationary discrete Fourier transform (DFT) given by [3, 16]

$$X_k \stackrel{\text{def}}{=} \frac{\sqrt{2}}{N}\sum_{n=0}^{N-1} x_{k-n} e^{-j\frac{2\pi}{N}(k-n)} \quad (12.3a)$$

where the window length, N, is chosen as a ratio of the sampling frequency and the fundamental frequency, that is

$$N = f_s/f_\circ. \quad (12.3b)$$

The data index in the exponent within the stationary DFT definition in (12.3a) was chosen to be $(k - n)$ instead of the usual $(k + n)$ to indicate that the latest data sample used in the phasor at time k is the sample x_k.

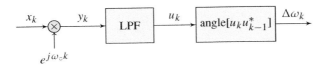

Figure 12.1 Block diagram of a general fixed frequency demodulation scheme.

Next, the frequency deviation of the phasor X_k can be calculated using

$$\Delta\omega_k = \text{angle}[X_k] - \text{angle}[X_{k-1}]$$
$$= \text{angle}[X_k X_{k-1}^*] \quad (12.3c)$$

where the operator $\text{angle}[X_k]$ obtains the angle of the complex variable X_k and $(\cdot)^*$ denotes the complex conjugation operator. Using the frequency decomposition in (12.2), the instantaneous frequency estimate can therefore be obtained as

$$\hat{\omega}_{\text{DFT},k} = \omega_\circ + \text{angle}[X_k X_{k-1}^*]. \quad (12.3d)$$

In practice, to obtain smoother estimates, postprocessing (filtering) is carried out in the form of fitting a linear or quadratic model to the phase angle sequence, $\text{angle}[X_k]$, obtained from (12.3d).

12.3.2 Fixed Frequency Demodulation

Frequency modulation (FM) is a fundamental concept in communications. The standard heterodyne demodulation scheme of the signal x_k in (12.1) is given by [7, 14]

$$y_k = x_k e^{-j\omega_\circ k} \quad (12.4a)$$
$$u_k = \text{LPF}[y_k] \quad (12.4b)$$
$$\hat{\omega}_{\text{FM},k} = \omega_\circ + \text{angle}[u_k u_{k-1}^*]. \quad (12.4c)$$

The step in (12.4a) in our analysis is known as the heterodyne or demixing step, with $\omega_\circ$ being a known carrier frequency that is chosen to be the fundamental power system frequency, while the operator $\text{LPF}[\cdot]$ in (12.4b) denotes a low-pass filtering operation performed by any suitable low-pass filter with a cutoff frequency $\omega_{\text{cut}} < \omega_\circ$. The demodulated instantaneous frequency is then obtained in (12.4c) from the rate of change of the phase angles of the low-pass filtered signal u_k; the block diagram of an FM demodulator is shown in Figure 12.1.

12.4 Frequency Estimation from Voltage Phase Angles versus Frequency Demodulation

The intrinsic relationship between the DFT-based frequency estimation in (12.3a)–(12.3d) and the FM scheme in (12.4a)–(12.4c) can now be derived without any assumption on the measurement x_k. The essential step comes from representing the DFT-based phasor computation in (12.3a) using its recursive form as [3]

$$X_k = X_{k-1} + \frac{\sqrt{2}}{N}(x_k - x_{k-N})e^{-j\frac{2\pi}{N}k}, \qquad (12.5)$$

where the recursive formulation of the phasor computation in (12.5) is also known as the recursive DFT [18]. The derivation of (12.5) from (12.3a) exploits the fact that only the two data points, x_k and x_{k-N}, differ in the phasor estimates X_k and X_{k-1}; this removes the necessity to recompute the sum in (12.3a) at every time instant.

Following that, the relationship between the phasor computation in (12.5) and the FM demixing step in (12.4a) can be further clarified by noticing that the data length, N, in (12.3b) and the demodulating frequency, $\omega_\circ$, can be related through $\omega_\circ = 2\pi f_\circ/f_s = 2\pi/N$. Therefore, the recursive phasor computation in (12.5) can now be expressed as

$$X_k = X_{k-1} + \frac{\sqrt{2}}{N}(x_k - x_{k-N})e^{-j\omega_\circ k}. \qquad (12.6)$$

Next, observe that the terms $x_k e^{-j\omega_\circ k}$ and $x_{k-N}e^{-j\omega_\circ k}$ in (12.6) are related to the demodulated signal y_k in (12.4a) through

$$y_k = x_k e^{-j\omega_\circ k}, \qquad y_{k-N} = x_{k-N}e^{-j\omega_\circ(k-N)} = x_{k-N}e^{-j\omega_\circ k}, \qquad (12.7)$$

which yields the relationship

$$X_k = X_{k-1} + \frac{\sqrt{2}}{N}(y_k - y_{k-N}). \qquad (12.8)$$

The difference equation in (12.8) represents a low-pass filtering operation on the signal y_k, and can be expressed as

$$X_k = H(z^{-1})y_k \qquad (12.9)$$

whereby the infinite impulse response (IIR) filter transfer function, $H(z^{-1})$, is given by [18]

$$H(z^{-1}) = \frac{\sqrt{2}}{N}\frac{1-z^{-N}}{1-z^{-1}} \qquad (12.10)$$

with z^{-1} as the discrete-time delay operator.

Equivalence with Recursive DFT. Steps (12.8)–(12.10) state that the standard DFT-based phasor estimation in (12.3a) implicitly performs the low-pass filtering operation required by the FM demodulation scheme. In other words, the phasor computation based on the DFT in (12.3a) performs both the demixing and low-pass filtering steps of the FM demodulation scheme in (12.4a)–(12.4b). From this interpretation, it follows that the frequency estimation based on the rate of change of the voltage phasor angle in (12.3d) is identical to the frequency demodulation step in (12.4c).

REMARK 12.1 From (12.5)–(12.10), we have shown that the frequency estimation scheme based on the rate of change of the phase angle of the voltage phasor is equivalent to a general FM demodulation scheme. This opens up a number of opportunities for enhanced frequency estimation, outlined in the following text.

REMARK 12.2 The only difference between the FM and the DFT-based schemes is that the DFT-based phasor and frequency estimation algorithm restricts the low pass filter, $LPF[\cdot]$, in step (12.4b), to assume only the transfer function in (12.10). Because the choice of the low-pass filter block in FM demodulation allows for any suitable low-pass structure, the phasor-based frequency estimation can be interpreted as a special case of the fixed frequency demodulation scheme. This is an important insight because the phasor and frequency estimation algorithms based on (12.3a) can now be generalized to a wider range of low-pass structures compared to the restrictive DFT definition.

Illustrative Example. The interpretation in Remark 12.2 gives the designers of phasor measurement unit (PMU) algorithms the freedom to choose low-pass filters with characteristics that are favorable to the application at hand, without having to resort to the strict definition of the synchronized phasor in (12.5).

To illustrate this advantage, consider the magnitude response of the recursive DFT transfer function in (12.10), with $N = f_s/f_\circ = 20$, $f_s = 1$ kHz and $f_\circ = 50$ Hz. Figure 12.4 shows that the filter in (12.10) does not exhibit steep roll-off. In contrast, a more desirable magnitude response of a third-order Butterworth filter with the cutoff frequency, $\omega_{\text{cut}} = 2$Hz, is also shown in Figure 12.2.

The benefits of having a filter with sharp roll-off can be illustrated through the following frequency tracking task. The signal $x_k = \cos(2\pi \frac{f}{f_s}k) + \eta_k$, with a signal-to-noise ratio of 50 dB and frequency $f = 50$ Hz, undergoes a step-change to 50.5 Hz after 0.5 s. For a sampling frequency of $f_s = 1$ kHz, the instantaneous frequency of the signal x_k was estimated using: (i) the recursive DFT frequency estimator based on (12.5)–(12.3d) and (ii) the FM demodulator in (12.4a)–(12.4c) with a Butterworth filter (frequency response shown in Figure 12.2).

Figure 12.3 shows that both methods were able to accurately track the nominal frequency of 50 Hz. However, for off-nominal frequencies, the recursive DFT performed poorly and required further postprocessing (e.g., secondary low-pass filtering). The problem can be avoided by selecting an appropriate low-pass filter, without having to rely on the filter in (12.10).

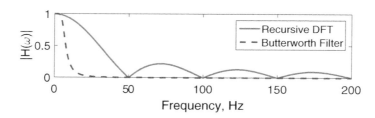

Figure 12.2 Magnitude responses of a third-order Butterworth filter with cutoff frequency $\omega_{\text{cut}} = 2$ Hz, and the Recursive DFT filter in (12.10).

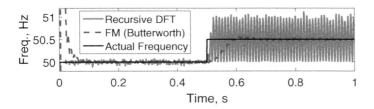

Figure 12.3 Frequency estimates from the recursive DFT and FM demodulator.

12.5 Smart DFT-Based Frequency Estimation

When the off-nominal frequency is taken into considered by setting $\omega_k = \omega_\circ + \Delta\omega_k = 2\pi(f_\circ + \Delta f_k)$ and considering a high level signal-to-noise ratio (SNR), the DFT-based phasor estimate X_k in (12.3a) can be expressed as [19, 20]

$$X_k = \underbrace{\frac{\sqrt{2}A}{N} e^{-j2\pi k(f_\circ+\Delta f_k)/Nf_\circ} e^{j\pi(N-1)\Delta f_k/Nf_\circ} \frac{\sin(\frac{\pi\Delta f_k}{f_\circ})}{\sin(\frac{\pi\Delta f_k}{Nf_\circ})}}_{c_k}$$

$$+ \underbrace{\frac{\sqrt{2}B}{N} e^{j2\pi k(f_\circ+\Delta f_k)/Nf_\circ} e^{-j(N-1)(2f_\circ+\Delta f_k)/Nf_\circ} \frac{\sin(\frac{\pi(2f_\circ+\Delta f_k)}{f_\circ})}{\sin(\frac{\pi(2f_\circ+\Delta f_k)}{Nf_\circ})}}_{d_k} \quad (12.11)$$

Upon defining the exponential kernel r as

$$r = e^{j\frac{\omega_k}{Nf_\circ}} \quad (12.12)$$

from (12.11), it follows that

$$X_k = c_k + d_k = r^{-1}c_{k-1} + rd_{k-1} \quad (12.13)$$
$$X_{k-1} = c_{k-1} + d_{k-1} \quad (12.14)$$
$$X_{k-2} = rc_{k-1} + r^{-1}d_{k-1} \quad (12.15)$$

After some algebraic manipulations, the following time-series relationship among three consecutive DFT fundamental components can be found as

$$hX_{k-1} = X_k + X_{k-2} \quad (12.16)$$

where

$$h = r + r^{-1} = 2\cos(\omega_k/Nf_\circ) \quad (12.17)$$

which can be estimated by

$$\hat{h} = (X_k + X_{k-2})/X_{k-1} \quad (12.18)$$

Therefore, the instantaneous frequency estimate of the smart DFT (SDFT) technique can be obtained as

$$\hat{\omega}_{\text{SDFT},k} = N f_\circ \cos^{-1}(\Re(\hat{h})/2) \qquad (12.19)$$

where $\Re(\cdot)$ denotes the real part of a complex number.

12.6 Maximum Likelihood Frequency Estimation

We shall next show that the rate of change of phase angle, which is present in both the DFT-based scheme in (12.3d), and the FM demodulation in (12.4c), at the same time represents the maximum likelihood instantaneous frequency estimator. Consider a single-tone complex sinusoid embedded in white Gaussian noise, given by

$$v_k = A e^{j\omega k} + \tilde{\eta}_k, \quad \tilde{\eta}_k \sim \mathcal{N}(0, \sigma_{\tilde{\eta}}^2), \qquad (12.20)$$

where the signal model in (12.20) can represent the phasor, $v_k = X_k$ in (12.3a), and the low-pass filtered FM signal in (12.4b), $v_k = u_k$. This is because both the phasor computation in (12.3a) and the FM demodulation steps in (12.4a)–(12.4b) serve to filter out the negative sequence component, $B e^{-j(\omega_\circ k + \phi_k)}$, from the original signal, x_k, to give a signal in the form of (12.20).

The maximum likelihood (ML) estimation of the frequency, ω, in (12.20) is performed by the so-called periodogram maximizer [21]

$$\hat{\omega}_{\text{ML}} = \underset{\omega}{\mathrm{argmax}}\, \frac{1}{N} \left| \sum_{n=0}^{N-1} v_n e^{-j\omega n} \right|^2 \qquad (12.21)$$

where the ML solution in (12.21) assumes that the signal frequency ω is constant within a block of N data samples. However, if the instantaneous frequency is time-varying, using N data samples to obtain a single frequency estimate, as described in (12.21), is invalid. Therefore, $(N-1)$ estimates[1] of the frequency are required, where the unknown frequency parameter vector is given by

$$\hat{\boldsymbol{\omega}}_{\text{ML}} = [\hat{\omega}_{\text{ML},1}, \hat{\omega}_{\text{ML},2}, \ldots, \hat{\omega}_{\text{ML},N-1}]^{\mathrm{T}}. \qquad (12.22)$$

To obtain the instantaneous frequency estimates, the ML scheme in (12.21) can be successively applied on two consecutive data points, where the frequency estimate at a time instant $(k+1)$ is given by

$$\begin{aligned}
\hat{\omega}_{\text{ML},k+1} &= \underset{\omega}{\mathrm{argmax}}\, \frac{1}{2} \left| \sum_{n=0}^{1} v_{k+n} e^{-j\omega n} \right|^2, \quad k = 0, \ldots, N-2 \\
&= \underset{\omega}{\mathrm{argmax}}\, \frac{1}{2} |v_k + v_{k+1} e^{-j\omega}|^2 \\
&= \underset{\omega}{\mathrm{argmax}}\, \frac{1}{2}(|v_k|^2 + |v_{k+1}|^2) + \Re\{v_{k+1} v_k^* e^{-j\omega}\} \qquad (12.23)
\end{aligned}$$

---

[1] The maximum number of instantaneous frequency estimates that can be obtained from a block of N samples is $(N-1)$, because at least two data samples are required for a single frequency estimate.

Because the term $(|v_k|^2 + |v_{k+1}|^2)$ is independent of the frequency estimate, ω, the maximization procedure in (12.23) can be expressed as

$$\hat{\omega}_{\text{ML},k+1} = \arg\max_{\omega} \text{Re}\{v_{k+1} v_k^* e^{-j\omega}\} \quad (12.24)$$

The value of ω which maximizes (12.24) is therefore given by[2]

$$\hat{\omega}_{\text{ML},k+1} = \text{angle}[v_{k+1} v_k^*] \quad (12.25)$$

REMARK 12.3 *The ML frequency estimate in (12.25) is identical to the instantaneous frequency estimate obtained from the rate of change of phase angles in the DFT algorithm in (12.3d) and the FM demodulation scheme in (12.4c).*

REMARK 12.4 *To some readers, Remark 12.3 might seem obvious, that is, with two data samples, the only sensible option to estimate the frequency is to compute the difference between the phase angles of v_k. The point here, however, is to show that the intuition behind the frequency being the derivative of the phase angle is faithfully encoded in the mathematics of estimation theory.*

12.7 Real-World Case Studies

Windowing and Low Pass Filtering

The interpretation in Remark 12.1 gives the designers of PMU algorithms the freedom to choose low-pass filters with characteristics that are favorable to the application at hand, without having to resort to the strict definition of the synchronized phasor in (12.5). It is important to note that phasor computations based on the DFT commonly employ data windows, so that the DFT-based phasor computation in (12.3a) takes the form

$$X_k = \frac{\sqrt{2}}{G} \sum_{n=0}^{N-1} w_n x_{k-n} e^{-j\frac{2\pi}{N}(k-n)} \quad (12.26)$$

where w_n are the coefficients of a suitable window function and $G = \sum_{n=0}^{N-1} w_n$. Examples of window functions include the triangle, Hamming, and Hanning windows, which are used to suppress the level of side lobes and to decrease the width of the main lobe in the DFT spectrum [22]. Given the ideas presented in this work, windowing can now be interpreted as a process of changing the structure of the low-pass filter in the demodulation step.

To examine the impact of the various low-pass filtering schemes on the frequency demodulation (and hence DFT-based schemes), Figure 12.4 illustrates the magnitude and phase responses of the low-pass filter transfer functions of: (i) the implicit rectangular windowing[3] of the DFT in (12.10), with $N = f_s/\hat{f}_\circ = 40$, $f_s = 2$ kHz, and $f_\circ = 50$ Hz; (ii) a third-order IIR Butterworth low-pass filter with the cutoff

---

[2] Proof is provided in the Appendix.
[3] The DFT expression in (12.3a) can be interpreted as a rectangular windowed DFT in (12.26) with $w_n = 1, \forall n$.

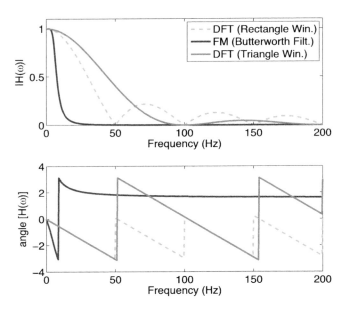

Figure 12.4 Magnitude and phase responses of a rectangle-windowed and triangle-windowed low-pass filter employed by the DFT, respectively in (12.3a) and (12.26), in comparison with a third-order IIR Butterworth filter with a cutoff frequency $\omega_{\text{cut}} = 2$ Hz in (12.4b).

frequency, $\omega_{\text{cut}} = 2$ Hz and (iii) the low-pass filtering operation of the windowed DFT operation in (12.26) with a triangle window of length $N = 40$. Observe that both the triangle-windowed DFT in (12.26) and the Butterworth filter have lower side-lobe levels compared to the original rectangle-windowed DFT.

The benefits of having a filter with low side-lobe levels can be illustrated through the following frequency tracking task. The single-phase voltage

$$x_k = \cos\left(2\pi \frac{f}{f_s} k\right) + \eta_k, \qquad (12.27)$$

with a signal-to-noise ratio of 50 dB and frequency $f = 50$Hz, undergoes a step-change to 49.5Hz after 0.5s. For a sampling frequency of $f_s = 2$ kHz, the instantaneous frequency of the signal x_k was estimated using: (i) the DFT frequency estimator based on (12.3a)–(12.3d); (ii) the FM demodulator in (12.4a)–(12.4c) with a third-order Butterworth filter (frequency response shown in Figure 12.4); and (iii) the triangle windowed DFT in (12.26).

Figure 12.5 shows that both the DFT and FM demodulation schemes were able to accurately track the nominal frequency of 50 Hz. However, for off-nominal frequencies, the rectangle windowed DFT performed poorly and required further postprocessing (e.g., secondary low-pass filtering). To this end, a frequency estimation scheme based on a triangle-windowed DFT was employed to avoid the problem of frequency oscillations at off-nominal frequencies. The performance advantage of using a windowed DFT with a better low-pass structure compared to (12.10) is illustrated in Figure 12.6.

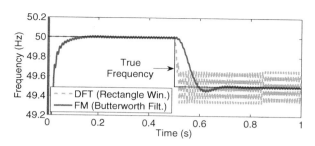

Figure 12.5 Frequency estimates from the DFT-based phase angle difference and the FM demodulator that employs a low-pass IIR Butterworth filter.

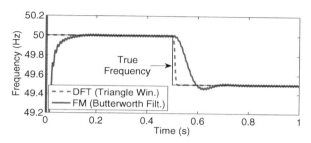

Figure 12.6 Frequency estimates from the windowed DFT and FM demodulator.

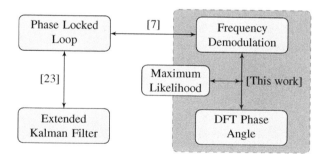

Figure 12.7 Connections between various frequency estimation schemes.

Connections with Other Frequency Tracking Methods

The links presented in this work between the FM- and DFT-based frequency tracking complement the work in [23] that states that for high SNR, the well-known extended Kalman filtering (EKF)–based frequency estimation is equivalent to the phase locked loop (PLL) algorithm. In addition, the PLL, which is essentially a variable frequency demodulation technique, is an extension of the fixed frequency demodulation algorithm in (12.4a)–(12.4c) [7]. As shown in Figure 12.7, by relating the DFT-based frequency tracking scheme to FM demodulation and maximum likelihood estimation, we have now provided a new dimension toward the unification of the most popular power system frequency estimators.

12.8 Meeting the Needs of Unbalanced Smart Grids: Smart Clarke and Park Transforms

It is a prerequisite for future smart grids to move away from the traditional high-inertia load-following operating strategy to a dynamic scenario that involves low inertia, smart loads, and renewable generation, which causes the three-phase systems to operate in a dynamically unbalanced condition. This, in turn, yields unequal phase voltages and nonuniform phase separations that need to be catered for in real time.

Three-Phase Systems and the Clarke and Park Transformations. Frequency estimation in power systems is routinely performed from the three-phase voltages. These were designed for stable grids and nominal conditions, based on a sampled three-phase voltage measurement vector, s_k, which at a discrete time instant k, is given by

$$s_k = \begin{bmatrix} v_{a,k} \\ v_{b,k} \\ v_{c,k} \end{bmatrix} = \begin{bmatrix} V_a \cos(\omega k + \phi_a) \\ V_b \cos(\omega k + \phi_b - \frac{2\pi}{3}) \\ V_c \cos(\omega k + \phi_c + \frac{2\pi}{3}) \end{bmatrix}, \qquad (12.28)$$

where V_a, V_b, V_c are the amplitudes of the phase voltages $v_{a,k}, v_{b,k}, v_{c,k}$, while $\omega = 2\pi f T$ is the fundamental angular frequency, with f the fundamental power system frequency and T the sampling interval. The phase values for phase voltages are denoted by ϕ_a, ϕ_b, and ϕ_c.

The Clarke transform, also known as the $\alpha\beta$ transform, aims to change the basis of the original 3D vector space where the three-phase signal s_k in (12.35) resides, to yield the Clarke-transformed $v_{0,k}, v_{\alpha,k}, v_{\beta,k}$ voltages in the form

$$\begin{bmatrix} v_{0,k} \\ v_{\alpha,k} \\ v_{\beta,k} \end{bmatrix} = \sqrt{\frac{2}{3}} \underbrace{\begin{bmatrix} \frac{\sqrt{2}}{2} & \frac{\sqrt{2}}{2} & \frac{\sqrt{2}}{2} \\ 1 & -\frac{1}{2} & -\frac{1}{2} \\ 0 & \frac{\sqrt{3}}{2} & -\frac{\sqrt{3}}{2} \end{bmatrix}}_{\text{Clarke matrix}} \underbrace{\begin{bmatrix} v_{a,k} \\ v_{b,k} \\ v_{c,k} \end{bmatrix}}_{s_k}, \qquad (12.29)$$

The quantities $v_{\alpha,k}$ and $v_{\beta,k}$ are referred to as the α and β sequences, while the term $v_{0,k}$ is called the zero-sequence, as it is null when the three-phase signal s_k is balanced, to yield the reduced 2D interpretation (due to the orthogonality of v_α and v_β, we usually write $v_{\alpha\beta,k} = v_{\alpha,k} + jv_{\beta,k}$)

$$\begin{bmatrix} v_{\alpha,k} \\ v_{\beta,k} \end{bmatrix} = \sqrt{\frac{2}{3}} \underbrace{\begin{bmatrix} 1 & -\frac{1}{2} & -\frac{1}{2} \\ 0 & \frac{\sqrt{3}}{2} & -\frac{\sqrt{3}}{2} \end{bmatrix}}_{\text{Reduced Clarke matrix: } \mathbf{C}} \begin{bmatrix} v_{a,k} \\ v_{b,k} \\ v_{c,k} \end{bmatrix}. \qquad (12.30)$$

The Park transform (also known as the dq transform) multiples the Clarke's $\alpha\beta$ voltages in (12.30) with a time-varying unit-determinant rotation matrix, $\mathbf{P}_\theta$, which serves as a rotating frame to produce the Park voltages, $v_{d,k}, v_{q,k}$ [24]

$$\begin{bmatrix} v_{d,k} \\ v_{q,k} \end{bmatrix} = \underbrace{\begin{bmatrix} \cos(\theta_k) & \sin(\theta_k) \\ -\sin(\theta_k) & \cos(\theta_k) \end{bmatrix}}_{\text{Park Matrix: } \mathbf{P}_\theta} \begin{bmatrix} v_{\alpha,k} \\ v_{\beta,k} \end{bmatrix}. \qquad (12.31)$$

where $\theta_k = \omega_\circ k$. The orthogonal direct and quadrature components, $v_{d,k}$ and $v_{q,k}$, within the Park transformation can also be conveniently combined into a complex variable $v_{dq,k} = v_{d,k} + j v_{q,k}$.

The Clarke $\alpha\beta$ voltage in (12.30) is a complex variable, $s_k \stackrel{\text{def}}{=} v_{\alpha,k} + j v_{\beta,k}$, or more compactly

$$s_k = \mathbf{c}^H \mathbf{s}_k, \qquad \mathbf{c} \stackrel{\text{def}}{=} \sqrt{\frac{2}{3}} [1, e^{-j\frac{2\pi}{3}}, e^{j\frac{2\pi}{3}}]^T \qquad (12.32)$$

where $\mathbf{c}$ is the complex Clarke transform vector in (12.30) (for a spatial DFT interpretation, see Part 1).

Degrees of Freedom in Clarke Voltage. Upon combining with the phasors in (12.35), the complex Clarke voltage, s_k, admits a physically meaningful representation through the counter-clockwise rotating *positive-sequence voltage*, $\bar{V}_+$, and the clockwise rotating *negative-sequence voltage*, $\bar{V}_-$, both rotating at the system frequency, ω, to take the form

$$s_k = \frac{1}{\sqrt{2}} \left(\bar{V}_+ e^{j\omega k} + \bar{V}_-^* e^{-j\omega k} \right), \qquad (12.33)$$

where [25]

$$\bar{V}_+ = \frac{1}{\sqrt{3}} \left[V_a e^{j\phi_a} + V_b e^{j\phi_b} + V_c e^{j\phi_c} \right] \qquad (12.34)$$

$$\bar{V}_-^* = \frac{1}{\sqrt{3}} \left[V_a e^{-j\phi_a} + V_b e^{-j\left(\phi_b + \frac{2\pi}{3}\right)} + V_c e^{-j\left(\phi_c - \frac{2\pi}{3}\right)} \right].$$

REMARK 12.5 *For balanced three-phase power systems, with $V_a = V_b = V_c$ and $\phi_a = \phi_b = \phi_c$, the negative sequence voltage component within the Clarke voltage, $\bar{V}_-^*$ in (12.33), vanishes and the Clarke voltage has a single degree of freedom. For unbalanced systems, characterized by inequal phase voltage amplitudes and/or phases, and thus $\bar{V}_-^* \neq 0$, the Clarke voltage in (12.33) exhibits* **two degrees of freedom**, *a signature of system imbalance.*

Clarke Transform as a Principal Component Analyzer

Consider again the three phase voltages from (12.1) and (12.35), written as

$$\mathbf{s}_k = \begin{bmatrix} v_{a,k} \\ v_{b,k} \\ v_{c,k} \end{bmatrix} = \begin{bmatrix} V_a \cos(\omega k + \phi_a) \\ V_b \cos(\omega k + \phi_b - \frac{2\pi}{3}) \\ V_c \cos(\omega k + \phi_c + \frac{2\pi}{3}) \end{bmatrix} = \frac{1}{2} \begin{bmatrix} \bar{V}_a \\ \bar{V}_b \\ \bar{V}_c \end{bmatrix} e^{j\omega k} + \frac{1}{2} \begin{bmatrix} \bar{V}_a^* \\ \bar{V}_b^* \\ \bar{V}_c^* \end{bmatrix} e^{-j\omega k}$$

$$(12.35)$$

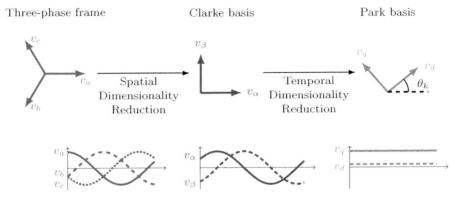

Figure 12.8 Comparison of the reference frames of the Clarke and Park Transforms [26].

where V_a, V_b, V_c are the amplitudes of the phase voltages $v_{a,k}, v_{b,k}, v_{c,k}$, while $\omega = 2\pi f T$ is the fundamental angular frequency, with f the fundamental power system frequency and T the sampling interval. The phase values for phase voltages are denoted by $\phi_a, \phi_b,$ and ϕ_c.

Now, we can write

$$\mathbf{v} \stackrel{\text{def}}{=} \begin{bmatrix} \bar{V}_a, & \bar{V}_b, & \bar{V}_c \end{bmatrix}^T \quad \Rightarrow \quad s_k = \frac{1}{2}\left(\mathbf{v}e^{j\omega k} + \mathbf{v}^* e^{-j\omega k}\right) \tag{12.36}$$

so that

$$s_k s_k^H = \frac{1}{4}\left(\mathbf{v}\mathbf{v}^H + \mathbf{v}^*\mathbf{v}^T + \mathbf{v}\mathbf{v}^T e^{2j\omega k} + \mathbf{v}^*\mathbf{v}^H e^{-2j\omega k}\right) \quad \longrightarrow \quad = \frac{1}{2}\left(\mathbf{v}_r \mathbf{v}_r^T + \mathbf{v}_i \mathbf{v}_i^T\right) \tag{12.37}$$

where $\mathbf{v}_r = \text{Re}\{\mathbf{v}\}$ and $\mathbf{v}_i = \text{Im}\{\mathbf{v}\}$ denote the real and imaginary part of the phasor vector $\mathbf{v}$. This, in turn, means that the covariance matrix of the three phase voltages

$$\mathbf{R}_s = \lim_{N \to \infty} \frac{1}{N} \sum_{k=0}^{N-1} s_k s_k^H \tag{12.38}$$

is of Rank-2, as it is governed by the two Rank-1 outer products of vectors in (12.37). The covariance matrix of three-phase power voltages is therefore rank-deficient.

Clarke Transform as PCA. For simplicity, consider the normalized phasors relative to $\bar{V}_a$. We also define the relative phase voltages, $\delta_i \stackrel{\text{def}}{=} \bar{V}_i / \bar{V}_a$, $i \in \{a,b,c\}$, with $\delta_a = 1$, to yield $\bar{\mathbf{v}} = \begin{bmatrix} 1, & \delta_b, & \delta_c \end{bmatrix}^T$, so that

$$\delta_b = \frac{V_b}{V_a} e^{j(\phi_b - \phi_a - \frac{2\pi}{3})}, \quad \delta_c = \frac{V_c}{V_a} e^{j(\phi_c - \phi_a - \frac{2\pi}{3})}. \tag{12.39}$$

The covariance matrix, $\mathbf{R}_s^u$, of an unbalanced system can now be examined in terms of the voltage imbalances (such as the voltage sags in Figure 12.10), and also phase mismatch. These both reflect the effects of off-nominal amplitude/phase conditions that we model by the complex-valued imbalance ratios, δ_b and δ_c, to give the general covariance matrix (for both balanced and unbalanced systems) in the form

$$\mathbf{R}_s^u = \frac{1}{2} \begin{bmatrix} 1 & |\delta_b|\cos(\angle\delta_b) & |\delta_c|\cos(\angle\delta_c) \\ |\delta_b|\cos(\angle\delta_b) & |\delta_b|^2 & |\delta_b||\delta_c|\cos(\angle\delta_b - \angle\delta_c) \\ |\delta_c|\cos(\angle\delta_c) & |\delta_b||\delta_c|\cos(\angle\delta_b - \angle\delta_c) & |\delta_c|^2 \end{bmatrix}. \tag{12.40}$$

In balanced conditions, with $\delta_a = \delta_b = \delta_c = 1$, the covariance matrix reduces to

$$\mathbf{R}_s = \frac{1}{2}\begin{bmatrix} 1 & -0.5 & -0.5 \\ -0.5 & 1 & -0.5 \\ -0.5 & -0.5 & 1 \end{bmatrix}. \tag{12.41}$$

and its eigen-decomposition, $\mathbf{R}_s = \mathbf{Q\Lambda Q}^T$, yields

$$\mathbf{Q}^T = \sqrt{\frac{2}{3}}\begin{bmatrix} \frac{\sqrt{2}}{2} & \frac{\sqrt{2}}{2} & \frac{\sqrt{2}}{2} \\ 1 & -\frac{1}{2} & -\frac{1}{2} \\ 0 & \frac{\sqrt{3}}{2} & -\frac{\sqrt{3}}{2} \end{bmatrix}, \quad \mathbf{\Lambda} = \frac{1}{4}\begin{bmatrix} 0 & 0 & 0 \\ 0 & 1.5 & 0 \\ 0 & 0 & 1.5 \end{bmatrix}. \tag{12.42}$$

REMARK 12.6 *The eigenmatrix of the phase voltage covariance matrix, $\mathbf{Q}^T$ in 12.42, is identical to the Clarke transformation matrix in (12.29). Therefore, in the case of balanced three-phase power system voltages, their variance can be fully explained by the two eigenvectors associated with the nonzero eigenvalues (principal directions) – again identical to the $\alpha\beta$ Clarke-transform matrix in (12.30). This offers a modern interpretation of the Clarke transform as a PCA that projects the three-dimensional three-phase voltage onto a 2D subspace spanned by the two largest eigenvectors of the data covariance matrix.*

Example: Clarke Transform as a PCA Procedure. From (12.42), we can see that the eigenvectors that are associated with the nonzero eigenvalues of the phase voltage covariance matrix are given by

$$\mathbf{q}_2 = \sqrt{\frac{2}{3}}\left[1, -\frac{1}{2}, -\frac{1}{2}\right]^T \quad \text{and} \quad \mathbf{q}_3 = \sqrt{\frac{2}{3}}\left[0, \frac{\sqrt{3}}{2}, -\frac{\sqrt{3}}{2}\right]^T \tag{12.43}$$

and are identical to the basis vectors of the Clarke transform. Figure 12.8 illustrates the equality between the Clarke transform and PCA for balanced system voltages, and their gradually increasing disparity with an increase in system imbalance. Observe that in all cases, the "static" plane designated by the Clarke transform and the "data-driven" PCA plane share the eigenvector $\mathbf{q}_3$ in (12.43).

Clarke and Symmetrical Transforms as a 3-Point DFT

We now offer an interpretation of the well–known inadequacy of current power system analysis techniques in unbalanced grid scenarios, through a link with the effects of incoherent sampling in spectral analysis.

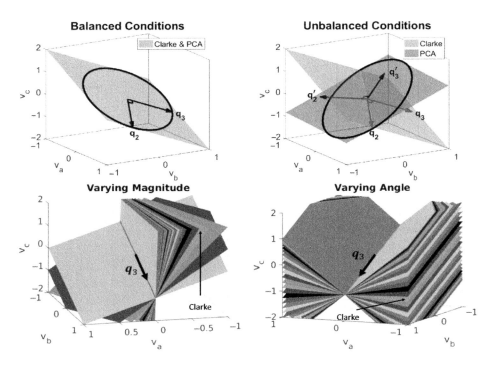

Figure 12.9 Dimensionality reduction that is inherently performed by the Clarke transform and its interpretation as a PCA for balanced power systems. *Top left:* The Clarke and PCA frames coincide for a balanced power system. *Top right:* The Clarke frame is static and it cannot designate the correct directions in data for unbalanced systems, while the PCA always designates the correct directions in data. *Bottom left:* The directions in data when the amplitudes of the phases b and c vary between 0 and ∞ (the static Clarke plane is in the middle of the set of planes). *Bottom right:* The directions in data for when the phase angle between the phases b and c which varies from 0 to π (the static Clarke plane is in the middle of the set of planes).

Symmetrical Transform as a Spatial DFT. The vector of three-phase time-domain voltages, $\mathbf{s}_k$ in (12.35), is typically considered as a collection of three univariate signals. However, observe that the phase voltage samples within $\mathbf{s}_k$ can also be treated as three samples of a monocomponent signal rotating at a *spatial frequency* $\Omega = -\frac{2\pi}{3}$. From this viewpoint, the phasor vector, $\mathbf{v}$ in a balanced system is given by

$$\mathbf{v} = \begin{bmatrix} 1, & e^{j\Omega}, & e^{j2\Omega} \end{bmatrix}^{\mathsf{T}}. \tag{12.44}$$

It is now obvious that $\mathbf{v}$ can be treated as a single sinusoid rotating at a spatial frequency of $\Omega = -\frac{2\pi}{3}$, whereby the elements of $\mathbf{v}$ are the corresponding phase voltages $v_{a,k}, v_{b,k}$, and $v_{c,k}$.

REMARK 12.7 *Under unbalanced conditions, the phasor vector, $\mathbf{v}$, does not represent a single complex-valued spatial sinusoid because it contains the individual phasors with different amplitudes and a nonuniform phase separation, as defined in* (12.35).

Consider now the DFT of the phasor vector, $\mathbf{v} = [v_0, v_1, v_2]^T \in \mathbb{C}^{3\times 1}$, given by

$$X[k] = \frac{1}{\sqrt{3}} \sum_{n=0}^{2} v_n e^{-j\frac{2\pi}{3}nk}, \quad k = 0, 1, 2$$

and its equivalent matrix form

$$\begin{bmatrix} X[0] \\ X[1] \\ X[2] \end{bmatrix} = \frac{1}{\sqrt{3}} \begin{bmatrix} 1 & 1 & 1 \\ 1 & e^{-j\frac{2\pi}{3}} & e^{-j2\times\frac{2\pi}{3}} \\ 1 & e^{-j2\times\frac{2\pi}{3}} & e^{-j\frac{2\pi}{3}} \end{bmatrix} \begin{bmatrix} \bar{V}_a \\ \bar{V}_b \\ \bar{V}_c \end{bmatrix} = \frac{1}{\sqrt{3}} \begin{bmatrix} 1 & 1 & 1 \\ 1 & a & a^2 \\ 1 & a^2 & a \end{bmatrix} \begin{bmatrix} \bar{V}_a \\ \bar{V}_b \\ \bar{V}_c \end{bmatrix} \tag{12.45}$$

where $a = e^{-j\frac{2\pi}{3}}$. The three-point DFT in (12.45) therefore yields a stationary component $X[0]$ and two other components, $X[1]$ and $X[2]$, which rotate at the respective spatial frequencies $\frac{2\pi}{3}$ and $-\frac{2\pi}{3}$.

REMARK 12.8 *The spatial DFT in (12.45) is identical to and represents a mathematical justification for the Symmetrical Component Transform. Observe that the stationary DFT component, $X[0]$, corresponds to the zero-sequence phasor, $\bar{V}_0$, while the fundamental DFT components, $X[1]$ and $X[2]$, represent, respectively, the positive- and negative-sequence phasors. This makes it possible to analyze three-phase component transforms from a Spectral Estimation perspective, and thus offers both enhanced physical insights into the interpretation of imperfections of these transforms and new avenues for the mitigation of these issues.*

Park Transform as an FM Demodulation Scheme. Similar to the complex-valued representation of the Clarke transform, the complex-valued version of the Park transform in (12.31) is given by

$$v_k \stackrel{\text{def}}{=} v_{d,k} + jv_{q,k} \tag{12.46}$$

which, in analogy to (12.32) can also be compactly represented as

$$v_k = e^{-j\omega_\circ k} \mathbf{c}^H \mathbf{s}_k = e^{-j\omega_\circ k} s_k, \tag{12.47}$$

where $s_k = v_{\alpha,k} + jv_{\beta,k}$ is the Clarke voltage. Observe the "circular," time-varying, rotation frame designated by $e^{-j\omega_\circ k}$, which connects the Clarke and Park transforms.

REMARK 12.9 *From a modern perspective viewpoint, the Park transform in (12.47) can be interpreted as a frequency demodulation (FM) scheme [7] of the $\alpha\beta$ voltage, whereby the demodulating frequency is the nominal system frequency $\omega_\circ$, as illustrated in Figure 12.1. The demodulated instantaneous frequency is then obtained from the rate of change of the phase angles of the low-pass filtered signal u_k.*

Therefore, for a balanced three-phase power system operating at the fundamental frequency $\omega_\circ$, the Park transform yields the stationary positive sequence phasor, shown in Figure 12.8 and given by

$$v_k = \bar{V}_+. \tag{12.48}$$

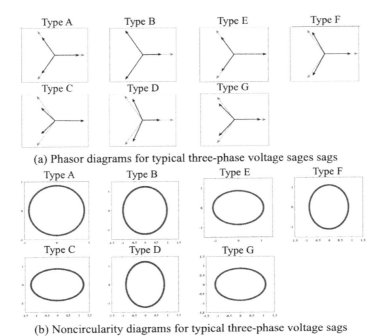

(a) Phasor diagrams for typical three-phase voltage sages sags

(b) Noncircularity diagrams for typical three-phase voltage sags

Figure 12.10 The effects of voltage sags on voltage phasors and complex noncircularity properties. The dashed lines designate a set of balanced three-phase voltage phasors under nominal power system conditions, as in Figure 12.8. Notice the change in magnitudes and phase separations during faults and their unique signatures in the corresponding circularity diagrams.

Sources of Bias in Park Transform When Used in Dynamically Unbalanced Smart Grid. In both current grids that incorporate renewables, and especially in the future Smart Grid, the three-phase voltages will be rarely perfectly balanced and the system frequency will never be operating at exactly the fundamental frequency, ω_0 [27]. From (12.33), the complex-valued dq voltage for a general unbalanced three-phase system, which operates at an **off-nominal** system frequency ω, is given by

$$v_k = \bar{V}_+ e^{j(\omega - \omega_\circ)k} + \bar{V}_-^* e^{-j(\omega + \omega_\circ)k}, \tag{12.49}$$

while the Park transform is designed for the nominal frequency, $\omega_\circ$. Therefore, the imperfect "demodulation" effect of the Park transform at an off-nominal frequency $\omega_\circ \neq \omega_\circ$ explains the spurious frequency estimation by the standard Park transform, as illustrated in Figure 12.12.

However, if an unbalanced system is operating at the **nominal system frequency**, $\omega = \omega_\circ$, but **off-nominal phase voltage/phase values**, the Park dq voltage in (12.49) becomes

$$v_k = \bar{V}_+ + \bar{V}_-^* e^{-j2\omega_\circ k}. \tag{12.50}$$

A Data Analytics Perspective of Fundamental Power Grid Analysis Techniques

Table 12.1 Signal processing interpretations of three-phase transformations.

Transform	Interpretation
Symmetrical [28]	Spatial DFT
Clarke [29]	PCA
Park [24]	FM demodulation

Table 12.2 Output of the Clarke and Park transformations.

Transform	Balanced	Power System Condition Unbalanced
Clarke [29]	$\bar{V}_+ e^{j\omega k}$	$\bar{V}_+ e^{j\omega k} + \bar{V}_- e^{-j\omega k}$
Park [24]	$\bar{V}_+$	$\bar{V}_+ e^{j(\omega-\omega_o)k} + \bar{V}_-^* e^{-j(\omega+\omega_o)k}$

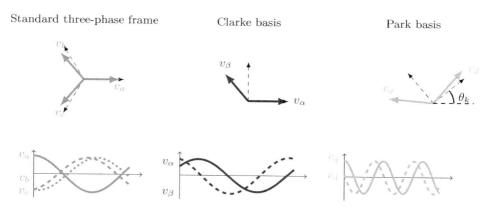

Figure 12.11 Effects of unbalanced three-phase power systems on the accuracy of classical three-phase reference frames [30]. The nominal conditions are designated by broken lines in the top panel. Observe that both the standard Clarke and Park reference frames are unsuitable for unbalanced phase voltages/phases and the operation at off-nominal frequencies, as seen from the oscillatory Park output for a static off-nominal frequency, instead of the two straight lines, as in Figure 12.8.

which again is consistent with the output of an FM demodulator. This paves the way for a novel treatment of power system imbalances from the Communication Theory perspective, as summarized in Table 12.1.

REMARK 12.10 *Figure 12.11 shows that during unbalanced system conditions, the optimal reference frames (basis vectors) for the $\alpha\beta$ and dq voltages are different from the nominal ones defined by the classical Park and Clarke transforms.*

Table 12.2 summarizes the functional expressions for the Clarke and Park transforms under both balanced and unbalanced conditions of the electricity grid. As the Clarke and Park transforms will not yield accurate outputs under unbalanced system conditions (noncircular), their adaptive (referred to as "smart" here) versions are required to enable: (i) accounting for voltage imbalances to provide a dynamically balanced Clarke output (circular) and (ii) tracking instantaneous changes in system frequency through an "adaptive" Park transform.

Teaching Old Power Systems New Tricks: Adaptive Clark and Park Transforms for Unbalanced Power Systems

It has been widely accepted that the minimum mean square error (MMSE) estimator for a complex-valued process, y_k, based on a complex-valued regressor, $\mathbf{x}_k$ is a straightforward extension of the corresponding real-valued one, and assumes the **strictly linear** form

$$\hat{y}_k = E\{y_k|\mathbf{x}_k\} = \mathbf{h}^H \mathbf{x}_k \qquad (12.51)$$

where $\mathbf{h}$ is a set of complex-valued coefficients. However, given that

$$\hat{y}_{r,k} = E\{y_{r,k}|\mathbf{x}_{r,k}, \mathbf{x}_{i,k}\} \quad \hat{y}_{i,k} = E\{y_{i,k}|\mathbf{x}_{r,k}, \mathbf{x}_{i,k}\} \qquad (12.52)$$

and using the well-known identities, $x_r = (x + x^*)/2$ and $x_i = (x - x^*)/2\jmath$, the only adequate estimator for the generality of complex data is **widely linear**, and is given by [31, 32]

$$\hat{y}_k = \mathbf{h}^H \mathbf{x}_k + \mathbf{g}^H \mathbf{x}_k^H. \qquad (12.53)$$

A comparison with the general unbalanced $\alpha\beta$ voltage in (12.33), which is a sum of two complex-valued sinusoids rotating in opposite directions, can be represented by a widely linear autoregressive (WLAR) model, given by [10, 31, 32]

$$s_k = h^* s_{k-1} + g^* s_{k-1}^* \qquad (12.54)$$

REMARK 12.11 *A comparison with (12.33) and (12.49) shows that, unlike the standard "strictly linear" model in (12.51) with only the single coefficient vector h, the two sets of WLAR coefficients $h, g \in \mathbb{C}$ not only contain all the necessary information related to the system frequency, ω, but also this gives the required additional degree of freedom for the analysis of unbalanced power systems.*

The level of voltage imbalance in the power system can be defined through the voltage unbalance factor (VUF), given by [27]

$$\text{VUF:} \qquad \kappa \stackrel{\text{def}}{=} \bar{V}_-/\bar{V}_+. \qquad (12.55)$$

so that, as showed in [10], the system frequency can be expressed through the WLAR coefficients, h and g, and the VUF, κ, to yield

$$e^{j\omega} = h^* + g^*\kappa \quad \text{and} \quad e^{-j\omega} = h^* + \frac{g^*}{\kappa^*} \qquad (12.56)$$

Upon solving for the system frequency, ω, and VUF, κ, this yields

$$e^{j\omega} = \text{Re}\{h\} + j\sqrt{\text{Im}^2\{h\} - |g|^2},\qquad(12.57)$$

$$\kappa = \frac{\bar{V}_-}{\bar{V}_+} = \frac{j}{g^*}\left(\text{Im}\{h\} + \sqrt{\text{Im}^2\{h\} - |g|^2}\right).\qquad(12.58)$$

REMARK 12.12 *From (12.55) and (12.56), we can now not only optimally estimate the behavior of unbalanced smart grids, but also this can serve as a basis for the design of advanced algorithms (referred to as "smart") that can even adequately operate based on standard, strictly linear models.*

Self-Balancing Clarke and Park Transforms

We next show that the knowledge of the VUF, κ in (12.58), and its relation to the degree of complex noncircularity (see also Figure 12.10), makes it possible to eliminate the negative sequence phasor, $\bar{V}_-$, from the expression for $\alpha\beta$ voltage s_k. Consider the expression [30, 33]

$$m_k \stackrel{\text{def}}{=} \sqrt{2}\left(s_k - \kappa^* s_k^*\right)\qquad(12.59)$$

$$= \bar{V}_+ e^{j\omega k} + \bar{V}_-^* e^{-j\omega k} - \frac{\bar{V}_-^*}{\bar{V}_+^*}\left(\bar{V}_+^* e^{-j\omega k} + \bar{V}_- e^{j\omega k}\right)$$

$$= \bar{V}_+\left(1 - |\kappa|^2\right)e^{j\omega k}.\qquad(12.60)$$

Because κ is already available from the WLAR coefficients in (12.58), the effects of voltage imbalance on the Clarke's $\alpha\beta$ voltage can be removed through

$$\bar{m}_k = m_k/(1 - |\kappa|^2) = \bar{V}_+ e^{j\omega k}.\qquad(12.61)$$

REMARK 12.13 *The voltage $\bar{m}_k$ contains only the positive sequence voltage, $\bar{V}_+$, and is thus unaffected by the effects of system imbalance, whereby the degree of system imbalance (cf. degree of noncircularity of Clarke voltage) is reflected in a nonzero negative sequence, $\bar{V}_-$. The procedure in (12.61) can be regarded as an* **adaptive "Smart" Clarke transform***, the output of which is always identical to the correct $\alpha\beta$ voltage of a balanced system (in case the system imbalance arises from voltage sags).*

Finally, from the estimated drifting system frequency (through $e^{j\omega_k}$ in (12.57) and κ_k in (12.58)), the SCT and the SPT can be summarized as [26, 30, 34]

Smart Clarke Transform: $\quad \bar{m}_k = \sqrt{2}(s_k - \kappa_k^* s_k^*)/(1 - |\kappa_k|^2)\qquad(12.62a)$

Smart Park Transform: $\quad \tilde{m}_k = e^{-j\omega_k k}\bar{m}_k.\qquad(12.62b)$

The Smart Clarke and Park transforms can be implemented in realtime using a suitable adaptive learning algorithm (e.g., least mean square [LMS] or Kalman filter) to track the VUF, κ_k, and system frequency, ω_k [26, 35]. For illustration, we present the

Table 12.3 Reference voltage frames for classical (static) three-phase transforms and the introduced adaptive (smart) ones.

	Power System Condition	
Transform	Balanced	Unbalanced
Clarke [29]	$\bar{V}_+ e^{j\omega k}$	$\bar{V}_+ e^{j\omega k} + \bar{V}_- e^{-j\omega k}$
Balancing [33]	$\bar{V}_+ e^{j\omega k}$	$(1 - |\kappa|^2)\bar{V}_+ e^{j\omega k}$
Smart Clarke	$\bar{V}_+ e^{j\omega k}$	$\bar{V}_+ e^{j\omega k}$
Park [24]	$\bar{V}_+$	$\bar{V}_+ e^{j(\omega-\omega_\circ)k} + \bar{V}_-^* e^{-j(\omega+\omega_\circ)k}$
Smart Park	$\bar{V}_+$	$\bar{V}_+$

Algorithm 12.1 Adaptive Clarke/Park Transform

Input: Original three-phase voltages, s_k, learning rate, μ

At each time instant $k > 0$:

1: Obtain the Clarke transform : $s_k = \sqrt{2}\,\mathbf{c}^H \mathbf{s}_k$
2: Update the weights of ACLMS

$$\varepsilon_k = s_k - (h_{k-1}^* s_{k-1} + g_{k-1}^* s_{k-1}^*)$$
$$h_k = h_{k-1} + \mu s_{k-1} \varepsilon_k^*$$
$$g_k = g_{k-1} + \mu s_{k-1}^* \varepsilon_k^*$$

3: Use (12.62a) and (12.62b) to obtain κ_k ad $e^{j\omega k}$
4: Calculate adaptive Clarke transform: $\bar{m}_k = (s_k - \kappa_k^* s_k^*)/(1 - |\kappa_k|^2)$
5: Calculate adaptive Park transform: $\tilde{m}_k = e^{-j\omega_k k}\bar{m}_k$

---

adaptive Clarke/Park transform in Algorithm 12.1, with the augmented complex least mean square (ACLMS) [32, 36] used to estimate the information bearing WLAR coefficients h and g.

REMARK 12.14 *The smart "self-balancing" versions of the Clarke and Park transforms perform, referred to as the Smart Clarke and Smart Clarke transforms, are capable of performing accurately the respective dimensionality reduction and rotation operations, regardless of the drifts in system frequency or level of voltage/phase imbalance. Table 12.3 summarizes the functional expressions for these adaptive transforms, which make it possible to use standard analysis techniques designed for nominal conditions in general unbalanced systems (including the strictly linear ones), to yield a bias-free operation.*

Example: Operation of the Smart Clarke and Smart Park Transforms. Figure 12.12 shows the direct and quadrature Park voltages, $v_{d,k}$ and $v_{q,k}$, obtained from both the original Park transform defined in (12.31), and the adaptive Smart Park transform, $\tilde{m}_k$ in (12.62b) and Algorithm 12.1. Observe the oscillating output of the original Park

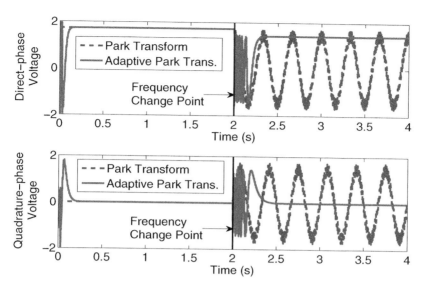

Figure 12.12 Frequency estimation in unbalanced power grids, whereby at $t = 2s$ the system experienced a frequency change. Observe the self-stabilizing nature of the Smart Park transform in the presence of system imbalance, as it settled to a constant value (solid line) and thus a correct frequency reading soon after the frequency drop starting from $t = 2s$. The classical Park transform was not designed for off-nominal conditions and exhibited an oscillatory behaviour under system imbalance (broken line). See also Figures 12.8 and 12.11.

transform (broken line) when the system frequency suddenly changed to a lower, off-nominal, value starting from $t = 2s$. However, the Smart Park transform was able to converge to a stationary (nonoscillatory) correct phasor soon after the occurrence of system imbalance. For more detail and additional insight, we refer to Figures 12.8 and 12.11.

Finally, Figure 12.13 illustrates the operation of the SCT in real time, in the presence of two types of voltage imbalance, Type A and Type D voltage sags, which occurred respectively at samples $k = 100$ and $k = 100$.

REMARK 12.15 *Observe that the use of widely linear models, elaborated upon in the preceding text, enables us to:*

- *Accurately track system frequency in both balanced and unbalanced conditions;*
- *Assess the type of system fault through knowledge of the degree of second-order noncircularity of Clarke's voltage, as e.g., every voltage sag has its own "noncircularity signature" as shown in Figure 12.10 and a recent patent [17];*
- *In this way, prevent operators from sending false alarms through the system; and*
- *Equalize the noncircular system trajectories, to be able to use standard techniques that would otherwise provide biased and inconsistent estimates in unbalanced system conditions.*

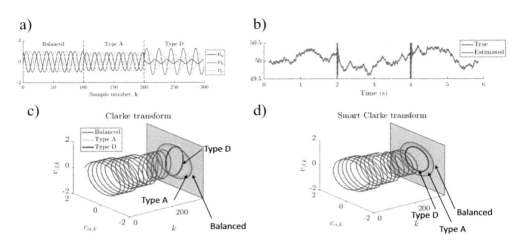

Figure 12.13 Self-balancing nature of the Smart Clarke transform for a power system operating at a nominal system frequency, ω_o, but in the presence of Type A voltage sag followed by Type D voltage sag. a) The original phasor voltage values are balanced up until sample $k = 100$, then they undergo a Type A voltage sag, followed by Type D voltage sag from the sample $= 200$. b) The corresponding frequency estimates using the Smart Clarke Transform in Algorithm (12.1). c) The circularity diagrams for the balanced case, Type-A voltage sag and Type D voltage sag. d) The self-stabilizing effect of the Smart Clarke transform, which exhibits a circular trajectory even for the Type D voltage sag.

12.9 Conclusion

The operation of the future smart grids requires close cooperation between the power systems and data analytics communities, such as those working in signal processing and machine learning. A major prohibitive factor in this endeavor has been a lack of common language; for example, the most fundamental power analysis techniques, such the Clarke and Park transform, introduced, respectively in 1943 and 1929, have been designed from a circuit theory perspective and only for balanced "nominal" system conditions, characterized by high grid inertia. This renders such methodologies both inadequate for the demands of modern, dynamically unbalanced, smart grids and awkward for linking up with data analytics communities. To help bridge this gap, we have provided modern interpretations of the frequency estimation techniques and standard analysis methods, such the Clarke and related transforms through the modern subspace, demodulation, and complex noncircularity concepts.

We have first shown that DFT-phasor–based frequency tracking is a special case of the FM demodulation scheme. By formulating the fixed frequency demodulation scheme in the complex domain, we have shown that frequency estimation based on the phase angle difference of the DFT-phasor is equivalent to demodulating a frequency modulated sinusoid with a fixed frequency oscillator that oscillates at the fundamental power system frequency of 50Hz or 60Hz. Furthermore, for voltage signals corrupted by a white Gaussian noise process, the frequency estimator obtained through the rate of change of the phase angle has been shown to be the maximum likelihood instantaneous frequency estimator. The so achieved unification of the standard frequency estimation schemes promises much enhanced flexibility in the design of PMU algorithms and opens new avenues for further work on the optimal design and analysis of frequency trackers. This includes the interpretation of the Clarke transform as a PCA, and the introduction of the novel Smart Clarke and Smart Park transforms, which are second-order optimal in both balanced and unbalanced system conditions.

These have served as a mathematical lens into the inadequacies of current methodologies under unbalanced power system conditions, and have enabled us to create a framework for the understanding and mitigation of the effects of off-nominal system frequency and dynamically unbalanced phase voltages and phases. All in all, such a conceptual insight permits seamless migration of ideas, in a bidirectional way, between these normally disparate communities and helps demystify power system analysis for data analytics practitioners.

Appendix: Proof of (12.25)

Consider a complex-valued variable $a = |a|e^{j\phi_a}$. The maximization procedure in (12.24) takes the form

$$\begin{aligned}
\hat{\omega}_{ML} &= \underset{\omega}{\operatorname{argmax}} \ \operatorname{Re}\{ae^{-j\omega}\} \\
&= \underset{\omega}{\operatorname{argmax}} \ \operatorname{Re}\{|a|e^{j\phi_a}e^{-j\omega}\} \\
&= \underset{\omega}{\operatorname{argmax}} \ |a|\cos(\phi_a - \omega).
\end{aligned} \quad (12.63)$$

The expression in (12.63) is clearly maximized when $\hat{\omega}_{ML} = \omega = \phi_a$. □

The authors wish to thank Ahmad Moniri, Shengxi Li, and Zeyang Yu, from Imperial College London, for their help in various stages of the preparation of this chapter.

References

[1] S. Kanna, S. Talebi, and D. Mandic, "Diffusion widely linear adaptive estimation of system frequency in distributed power grids," in *Proc. of the IEEE International Energy Conference (ENERGYCON)*, pp. 772–778, May 2014.

[2] P. Tielens and D. Van Hertem, "The relevance of inertia in power systems," *Renewable and Sustainable Energy Reviews*, vol. 55, pp. 999–1009, 2016.

[3] A. G. Phadke, J. S. Thorp, and M. G. Adamiak, "A new measurement technique for tracking voltage phasors, local system frequency, and rate of change of frequency," *IEEE Transactions on Power Apparatus and Systems*, vol. PAS-102, no. 5, pp. 1025–1038, May 1983.

[4] H. So and P. Ching, "Adaptive algorithm for direct frequency estimation," *IEE Proceedings Radar, Sonar and Navigation*, vol. 151, no. 6, pp. 359–364, Dec. 2004.

[5] V. V. Terzija, "Improved recursive Newton-type algorithm for power system relaying and measurement," *IEE Proceedings – Generation, Transmission and Distribution*, vol. 145, no. 1, pp. 15–20, Jan. 1998.

[6] A. A. Girgis and T. L. D. Hwang, "Optimal estimation of voltage phasors and frequency deviation using linear and non-linear Kalman filtering: Theory and limitations," *IEEE Transactions on Power Apparatus and Systems*, vol. PAS-103, no. 10, pp. 2943–2951, Oct. 1984.

[7] M. Akke, "Frequency estimation by demodulation of two complex signals," *IEEE Transactions on Power Delivery*, vol. 12, no. 1, pp. 157–163, Jan. 1997.

[8] M. H. J. Bollen, "Voltage sags in three-phase systems," *IEEE Power Engineering Review*, vol. 21, no. 9, pp. 8–15, 2001.

[9] M. H. J. Bollen, I. Y. H. Gu, S. Santoso, M. F. McGranaghan, P. A. Crossley, M. V. Ribeiro, and P. F. Ribeiro, "Bridging the gap between signal and power," *IEEE Signal Processing Magazine*, vol. 26, no. 4, pp. 11–31, 2009.

[10] Y. Xia, S. Douglas, and D. Mandic, "Adaptive frequency estimation in smart grid applications: Exploiting noncircularity and widely linear adaptive estimators," *IEEE Signal Processing Magazine*, vol. 29, no. 5, pp. 44–54, 2012.

[11] G. W. Arnold, "Challenges and opportunities in smart grid: A position article," *Proceedings of the IEEE*, vol. 99, no. 6, pp. 922–927, 2011.

[12] P. Kootsookos, "A review of the frequency estimation and tracking problems," in *Tech. Rep.* Cooperative Research Centre for Robust and Adaptive Systems, Defence Science and Technology Organisation, Australia, Feb. 1999.

[13] D. W. P. Thomas and M. S. Woolfson, "Evaluation of frequency tracking methods," *IEEE IEEE Transactions on Power Delivery*, vol. 16, no. 3, pp. 367–371, Jul. 2001.

[14] M. M. Begovic, P. M. Djuric, S. Dunlap, and A. G. Phadke, "Frequency tracking in power networks in the presence of harmonics," *IEEE Transactions on Power Delivery*, vol. 8, no. 2, pp. 480–486, Apr. 1993.

[15] "IEEE standard for synchrophasor measurements for power systems – Amendment 1: Modification of selected performance requirements," *IEEE Std C37.118.1a-2014 (Amendment to IEEE Std C37.118.1-2011)*, pp. 1–25, Apr. 2014.

[16] "IEEE standard for synchrophasor measurements for power systems," *IEEE Std C37.118.1-2011 (Revision of IEEE Std C37.118-2005)*, pp. 1–61, Dec. 2011.

[17] D. P. Mandic, Y. Xia, and D. Dini, "Frequency estimation," Patent, US Patent 9995774, June 2018.

[18] E. Jacobsen and R. Lyons, "The sliding DFT," *IEEE Signal Processing Magazine*, vol. 20, no. 2, pp. 74–80, Mar. 2003.

[19] J. Yang and C. Liu, "A precise calculation of power system frequency and phasor," *IEEE Transactions on Power Delivery*, vol. 15, no. 2, pp. 494–499, Apr. 2000.

[20] Y. Xia, Y. He, K. Wang, W. Pei, Z. Blazic, and D. Mandic, "A complex least squares enhanced smart DFT technique for power system frequency estimation," *IEEE Transactions on Power Delivery*, vol. 32, no. 3, pp. 1270–1278, June 2017.

[21] D. C. Rife and R. R. Boorstyn, "Single-tone parameter estimation from discrete-time observations," *IEEE Transactions on Information Theory*, vol. 20, no. 5, pp. 591–598, 1974.

[22] M. H. Hayes, *Statistical Digital Signal Processing and Modeling*. New York: John Wiley & Sons, 1996.

[23] D. L. Snyder, *The State-Variable Approach to Continuous Estimation*. Cambridag, MA: MIT Press, 1969.

[24] R. H. Park, "Two-reaction theory of synchronous machines generalized method of analysis – Part I," *Transactions of the American Institute of Electrical Engineers*, vol. 48, no. 3, pp. 716–727, July 1929.

[25] G. C. Paap, "Symmetrical components in the time domain and their application to power network calculations," *IEEE Transactions on Power Systems*, vol. 15, no. 2, pp. 522–528, May 2000.

[26] D. P. Mandic, S. Kanna, Y. Xia, A. Moniri, A. Junyent-Ferre, and A. G. Constantinides, "A data analytics perspective of power grid analysis. Part 1: The Clarke and related transforms," *IEEE Signal Processing Magazine*, vol. 36, no. 2, pp. 110–116, 2019.

[27] A. von Jouanne and B. Banerjee, "Assessment of voltage unbalance," *IEEE Transactions on Power Delivery*, vol. 16, no. 4, pp. 782–790, Oct. 2001.

[28] C. L. Fortescue, "Method of symmetrical co-ordinates applied to the solution of polyphase networks," *Transactions of the American Institute of Electrical Engineers*, vol. 37, no. 2, pp. 1027–1140, 1918.

[29] E. Clarke, *Circuit Analysis of A.C. Power Systems*. New York: Wiley, 1943.

[30] S. Kanna, A. Moniri, Y. Xia, A. G. Constantinides, and D. P. Mandic, "A data analytics perspective of power grid analysis. Part 2: Teaching old power systems new tricks," *IEEE Signal Processing Magazine*, vol. 36, no. 3, pp. 110–117, 2019.

[31] B. Picinbono, "On Circularity," *IEEE Transactions on Signal Processing*, vol. 42, no. 12, pp. 3473–3482, 1994.

[32] D. P. Mandic and V. S. L. Goh, *Complex Valued Nonlinear Adaptive Filters: Noncircularity, Widely Linear and Neural Models*. New York: Wiley, 2009.

[33] Y. Xia, K. Wang, W. Pei, and D. P. Mandic, "A balancing voltage transformation for robust frequency estimation in unbalanced power systems," in *Proceedings of the Asia Pacific Signal and Information Processing Association Annual Summit and Conference (APSIPA)*, pp. 1–6, Dec. 2014.

[34] S. Kanna and D. Mandic, "Self-stabilising adaptive three-phase transforms via widely linear modelling," *Electronics Letters*, vol. 53, no. 13, pp. 875–877, 2017.

[35] D. P. Mandic, S. Kanna, and A. G. Constantinides, "On the intrinsic relationship between the least mean square and Kalman filters [Lecture Notes]," *IEEE Signal Processing Magazine*, vol. 32, no. 6, pp. 117–122, Nov 2015.

[36] S. Javidi, M. Pedzisz, S. L. Goh, and D. P. Mandic, "The augmented complex least mean square algorithm with application to adaptive prediction problems," *Proceedings of the 1st IARP Workshop on Cognitive Information Processing*, pp. 54–57, 2008.

13 Graph Signal Processing for the Power Grid

Anna Scaglione, Raksha Ramakrishna, and Mahdi Jamei

13.1 Preliminaries

The signals of interest to monitor the state of the grid are the alternating voltages and currents that carry electric power through its lines. Figure 13.1 shows the signals of interest in electric power delivery systems. These signals exhibit typically a slow amplitude and phase modulation around a carrier (60 Hz for the United States). Charles Proteus Steinmetz [1] in 1893 expressed the power flow equations in a way that made apparent the relationship between the amplitude and the relative phases of the voltage carrier and the transfer of power. For decades, power systems have relied on *state estimation* algorithms [2, 3] to unravel what is the instantaneous amplitude and phase of the alternating voltage (AC) at its buses. This is because no tool to measure the signals synchronously was available given the wide area of sensor deployment necessary to reconstruct the state of the system. This had been the status quo until the advent of the synchrophasor technology, or phasor measurement unit (PMU), invented in 1988 by Dr. Arun G. Phadke and Dr. James S. Thorp at Virginia Tech [4]. The invention of PMUs was enabled by the opening to civilian applications in the 1980s, of the far more accurate version of the global positioning system (GPS), allowing sensors to synchronize across the wide area interconnections that exist in most advanced countries.

A synchrophasor measurement device extracts the phasor signal using universal coordinated time (UTC) as a time reference for its samples. Hence, the measurement for all locations can be aligned, offering an overall picture of the power system at a certain instant of time.

Given a signal $x(t)$, the phasor is its complex equivalent representation $\hat{x}(t)$ (conventionally called the *complex envelope*).

$$x(t) = A(t)\cos(\omega t + \theta(t)), \quad \leftrightarrow \quad \hat{x}(t) = A(t)e^{j\theta(t)}. \tag{13.1}$$

As long as the frequency ω is given, the mapping is one to one

$$x(t) = \Re\left\{\sqrt{2}\hat{x}(t)e^{j\omega t}\right\}. \tag{13.2}$$

The most widely accepted standard that describes the architecture for data transfer of PMU devices is the IEEE C37.118.1-2011 - IEEE Standard for Synchrophasor Measurements for Power Systems [5]; the standard is primarily concerned with how to communicate its samples in an interoperable and secure manner but it also indicates,

Table 13.1 Table listing some current and future uses of PMUs.

Current uses	Voltage stability and phase angle monitoring Linear state estimation Event replay
Future uses	Automated, autonomous system protection schemes Advanced PMU deployment and data sharing Machine learning for anomalous event identification

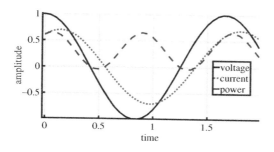

Figure 13.1 The AC signals of interest in electric power delivery systems.

among other things, a reference model for the signal processing associated with the phasor extraction. In practice, the implementations often use proprietary algorithms, and only the communication standard is followed. With reporting rates of 30 to 60 samples per second, PMUs provide information about the power system at a rate and scale that is unprecedented relative to their power flow sensors predecessors. They provide new opportunities to carry out data analytics for power systems and at a greater level of security because the state variable is directly measured, rather than being computed by fusing several data, some of which may be unreliable. Papers [6, 7] present the historical development of PMUs and their commercialization along with their working and applications. Since the invention of PMUs, their deployments have steadily increased in size worldwide and have become an important tool in wide-area measurement for monitoring the electrical grid [8]. As reported by the North American SynchroPhasor Initiative (NASPI),[1] in North America the number of PMUs has grown from 200 research grade units in 2009 to more than 2,500 production-grade units deployed in both the United States and Canada. The claim is that these units streaming data give almost 100% visibility into the bulk power system. Even though they have not yet completely supplanted the legacy technology of sensory control and data acquisition (SCADA), there is a clear trajectory toward that goal. White papers by NASPI routinely discuss new applications and challenges pertaining to PMU data. Table. 13.1 lists few current and future uses of PMUs. More recently, in [9] they highlight data mining methods and provide examples that show how data mining has been applied by the power grid community. PMUs are starting to appear also in distribution systems [10].

---

[1] www.naspi.org/reference-documents

Structure in PMU signals

It has been broadly validated by simulations and experimentally that the PMU signals observed exhibit two behaviors. Normally PMU data tend to be confined to a much smaller dimension compared to the size of the data record in both space and time. However, PMU signals register a sudden change in structure, which occurs in the presence of fast transients and faults. For monitoring purposes, the first kind of data correspond to the *null hypothesis*, and are typical in periods where the grid is sailing through in its normal slowly varying operating conditions. The second kind of data, i.e. the alternative hypothesis, is typically indicative of some trouble and can present various interesting signatures that depend on the nature of the underlying events. From the statistical signal processing perspective it is interesting to notice that the situation for PMU data is the diametrical opposite of what is encountered in radio detection and ranging (RADAR) applications, where the null hypothesis correspond to a high rank covariance (i.e., white noise) while the signal is low rank in the presence of a target.

This clearly means that when PMU data are highly structured they are naturally amenable to dimensionality reduction [11, 12]. The feature parameters describing the low-dimensional space can be used to interpret macroscopically the specific trend of the system. As shown in [11, 13, 14], using data from few PMUs installed in Central New York Power system, the data exhibit a low-rank structure. Faults change this low-rank structure, a fact that several papers [15–17] used to detect the events and segment the data.

Given that the equations that explain the data are well known and accurate, it may seem futile to look at the data through the lens of graph signal processing (GSP) [18, 19]. But GSP results are developed according to a general theory, whose goal is to bring about tools for graph signals analysis that are a direct generalization of digital signal processing (DSP), which could prove very valuable. The goal of this chapter is understanding the graph-spectral properties of the signals, which are typically explained through the linear generative model using graph filters. Are PMU a graph signal that obeys the linear generative model prevalent in the literature? If so, what kind of graph-filter structure and excitation justifies the graph-spectral properties? Can we derive new strategies to sense and process these data based on GSP? By introducing the link between PMU data and GSP on the right footing, one can understand if and to what extent we can use the GSP tools, and how can we use the basic equations to give us theoretical insight that support the observations.

13.2 Graph Signal Processing for the Power Grid

As electrical wiring forms a graph where nodes are referred to as *buses* and edges are referred to as *lines*, it is natural to see the measurement data as graph signals. The goal of this chapter is to analyze electrical grid measurements that we henceforth refer to as *grid graph signals* using tools of GSP. While it is obvious that the PMU samples are a graph signal, what is nontrivial is to show that these signals fit a generative model that includes a graph filter. The goal of this chapter is to show that it is possible by interpreting well-known results in power systems analysis that model PMU data.

GSP leverages topological understanding to extend ideas from discrete signal processing. Pivotal to the development of GSP is the generalization of the notion of shift and of filtering of graph signals [18–20]. The graph shift operator (GSO) is meant to be analogous to a time-delay operator, or spatial shift in images. The goal of GSP is to extend fundamental insights that come from the frequency analysis for time series to the domain of signals indexed by graphs. This objective limits the options on what operators are meaningful GSOs. An example of a GSO that is used often is the graph Laplacian matrix associated with a graph. By identifying the correct GSO, we can extend well-known results in GSP to power system data without losing the associated physical interpretation. For electric power signals, we confirm the intuition that the appropriate shift operator is the system admittance matrix. The latter, is a complex and symmetric graph Laplacian operator where the edge weights are the line admittances. The question we want to address is whether we can consider it as the GSO for the voltage phasor measurements and characterize the spectral properties of the data from first principles, based on the physics of the system.

Much work has been done to understand topological features of the electric grid and that knowledge is used solve problems such as optimal placement of generation and measurement units [21], creating synthetic grids [23] among many others. A comprehensive overview on the graph-theoretical concepts that are used in power systems can be found in a survey paper [24]. As explained previously, a series of papers relied on the observation that voltage and current measurements from the grid lie in a low-dimensional subspace [11, 14, 25]. This fact has been utilized for anomaly detection [15, 17], reconstructing missing measurements [13, 17] and other applications.

By studying the relationship between GSP and voltage phasor measurements we are able to show that the low dimensionality of the voltage measurements can be interpreted as GSP result, by showing that such measurements are the result of a *low-pass graph filter*. This result allows us to apply the theory of sampling and recovery of graph signals [26] to learn the relationship between observability of the grid, sampling graph signals and optimal placement of PMUs. Also, by establishing the connection with GSP, we gain insight on how to recover the community structure of the graph from voltage measurements [27], establishing the relationship between principal subspace geometry and the characteristics of the underlying graph topology.

Because most GSP developments are for real signals, we first provide a brief exposition of GSP that are based on complex Laplacian matrix GSO operators in Section 13.3. In Section 13.4, we use a classical approach from power systems transient analysis to establish that the voltage phasor measurements from the electric grid can be interpreted as the output of a low-pass graph filter with a low-rank input excitation. Importantly the low-pass response is expressed in terms of a GSO that is identical to the system admittance matrix. The excitation signal, instead, comes from the generators that are sparse in the grid, explaining why voltage measurement data lies in a low-dimensional subspace in the spatial domain. In Section 13.5, sampling results are presented that connect this finding with the sampling results in the GSP literature. We also investigate time-varying power measurements using time-vertex graph signal processing. A related result is that GSP also offers methods for PMU data compression thanks to the fact that only few important low-graph frequency components describe with high degree of accuracy the original signal. We connect the GSP concepts studied in the previous sections to look at two inference problems in Section 13.6. One is that of fault localization in undersampled grid data regime in Subsection 13.6.1 and the other being identification of the underlying community structure in the electrical grid in Subsection 13.6.2. Section 13.7 concludes the chapter also outlining future work to be undertaken. We start by establishing that voltage measurements are the result of a low-rank excitation to a low-pass graph filter. This excitation is nothing but the response of the generators in the system. The excitation, i.e. the voltage phasors of the generator buses have an interesting structure that can again be associated with a graph filter albeit with a different GSO.

Notation: Boldfaced lowercase letters are used for vectors, x and uppercase for matrices, $\mathbf{A}$. Transpose of vector or matrix is $x^\top, \mathbf{A}^\top$ and conjugate transpose as $\mathbf{A}^\dagger$. The operation $\Re\{.\}, \Im\{.\}$ denote the real and imaginary parts of the argument. The operation diag(.) implies extraction of the diagonal values of a matrix $\mathbf{A}$ when the input is a matrix and creating a diagonal matrix with elements from a vector if the input is a vector x.

13.3 Complex-Valued Graph Signal Processing

PMU are complex data and to develop a GSP framework to interpret them is necessary to expand GSP models to complex valued GSOs. The goal of this section is to briefly review GSP concepts for complex valued graph signals and their Laplacian matrices with complex-valued entries. Consider an *undirected* graph with nodes $i \in \mathcal{N}$ and edges $(i, j) \in \mathcal{E}$, $\mathcal{G} = (\mathcal{N}, \mathcal{E})$. The edges of the graph are assigned complex-valued weights $w_{ij} \in \mathbb{C}, w_{ij} = 0, (i, j) \notin \mathcal{E}$. Let a graph signal $x \in \mathbb{C}^{|\mathcal{N}|}$ be defined as values $[x]_i$ occurring at each node $i \in \mathcal{N}$. Define the GSO to be the complex-valued graph Laplacian $\mathbf{L} \in \mathbb{C}^{|\mathcal{N}| \times |\mathcal{N}|}$

$$[\mathbf{L}]_{i,j} = \begin{cases} \sum_{k \in \mathcal{N}_i} w_{i,k}, & i = j \\ -w_{i,j}, & i \neq j \end{cases} \tag{13.3}$$

Note that Laplacian matrix is complex symmetric but not necessarily *Hermitian*. For a diagonalizable matrix,[2] the eigenvalue decomposition is given by $\mathbf{L} = \mathbf{S} \Lambda \mathbf{S}^{-1}$ where Λ is the diagonal matrix with eigenvalues $\lambda_0, \lambda_1, \ldots \lambda_{|\mathcal{N}|}$ on the principal diagonal and $\mathbf{S}$ are the eigenvectors. However, the eigenvectors $\mathbf{S}$ can be transformed to give *complex orthogonal* eigenvectors $\mathbf{U}$ where $\mathbf{U} \triangleq \mathbf{SD}$, $\mathbf{D} \in \mathbb{C}^{|\mathcal{N}|}$ and $\mathbf{D}$ is a block diagonal matrix where the size of each block is the algebraic multiplicity of the corresponding eigenvalue (see theorem 4.4.13 in [29]). Thus, for diagonalizable complex symmetric matrices we can write,

$$\mathbf{L} = \mathbf{U} \Lambda \mathbf{U}^T, \qquad \mathbf{U}^T \mathbf{U} = \mathbf{U} \mathbf{U}^T = \mathbb{I} \tag{13.4}$$

Note that there is no conjugation i.e., $\mathbf{U}^\dagger \mathbf{U} \neq \mathbb{I}$. For complex-valued signals occurring on graph $\mathcal{G}$, we define graph Fourier transform (GFT) using the complex orthogonal basis $\mathbf{U}$. The GFT of a graph signal x, $\tilde{x}$ and the inverse GFT are given by $\tilde{x} = \mathbf{U}^T x$ and

---
[2] If the matrix is non diagonalizable, Jordan decomposition and the equivalent matrices can be used [28].

$x = \mathbf{U}, \tilde{x}$ respectively, where $[\tilde{x}]_m$ is the frequency component that corresponds to the mth eigenvalue λ_m. The order of frequencies is based on total variation (TV) [18, 30] of graph signal x, $S_1(x)$ that is defined using discrete p Dirchlet form with $p = 1$ as in [31] as

$$S_1(x) = \|\mathbf{L}x\|_1 \implies S_1(u_i) = \|\mathbf{L}u_i\|_1 = |\lambda_i| \|u_i\|_1 \qquad (13.5)$$

After normalizing the eigenvectors such that $\|u_i\|_1 = 1 \forall i$, it is clear that $S_1(u_i) > S_1(u_j)$ $\implies |\lambda_i| > |\lambda_j|$. Hence, the *ascending* order of eigenvalues corresponds to increase in frequency, $|\lambda_0| = 0 \le |\lambda_1| \le |\lambda_2| \cdots |\lambda_{|\mathcal{N}|}|$. This ordering is not unique because two distinct complex eigenvalues can have the same magnitude. Then, ties can be resolved based on value of real or imaginary values as appropriate [19].

Graph filters: A graph filter $\mathcal{H}(.)$ takes a graph signal x as input to provide output, $v = \mathcal{H}(x), v \in \mathbb{C}^{|\mathcal{N}|}$. A linear shift invariant graph filter is a matrix polynomial in $\mathbf{L}$. Specifically, let $\mathcal{H}(x) = \mathbf{H}x$,

$$\mathbf{H} = \sum_{k=0}^{L-1} h_k \mathbf{L}^k = \mathbf{U}\mathrm{diag}(\tilde{h})\mathbf{U}^T, \quad [\tilde{h}]_i = \sum_{k=0}^{L-1} h_k \lambda_i^k \qquad (13.6)$$

$h_k \in \mathbb{C}$, $i = 1, 2, \ldots |\mathcal{N}|$. The frequency response of the filter is given by elements in $\tilde{h}$. A *low-pass* graph filter retains lower frequency components of a graph input signal until the cutoff frequency λ_m while higher frequency components corresponding to $\lambda_i, i > m$ are zero.

13.3.1 Time-Vertex Frequency Analysis

If graph signal at time t is given as x_t, let $\mathbf{X}$ define a matrix concatenating columns of graph signals at different times:

$$\mathbf{X} = \begin{bmatrix} x_1 & x_2 & \cdots & x_T \end{bmatrix} \qquad (13.7)$$

A joint time-vertex signal is defined as the vectorized form of $\mathbf{X}$, $\mathrm{vec}(\mathbf{X})$. The underlying graph is a Cartesian product of the graph, $\mathcal{G}$ and the discrete-time graph that is usually represented as a cycle graph of length T. The joint time-vertex analysis in GSP [32] studies both the graph-frequency and discrete-time frequency response. The joint time-vertex Fourier transform (JFT) is defined as the application of discrete Fourier transform (DFT) and GFT sequentially (regardless of order),

$$\tilde{\mathbf{X}} = \mathrm{GFT}\left(\mathrm{DFT}[\mathbf{X}]\right) = \mathrm{JFT}(\mathbf{X}) = \mathbf{U}^T \mathbf{X} \mathbf{W} \qquad (13.8)$$

where $\mathbf{W}$ is the normalized DFT matrix whose (t, ℓ) entry is defined as

$$[\mathbf{W}]_{t,\ell} = \frac{1}{\sqrt{T}} e^{-j2\pi(\ell-1)/T}, \ell = 1, 2, \ldots T \qquad (13.9)$$

To find equivalence with GFT, the JFT can also be written in vector form as

$$\mathrm{vec}\left(\tilde{\mathbf{X}}\right) = \mathbf{U}_J^T \mathrm{vec}(\mathbf{X}) \qquad (13.10)$$

where $\mathbf{U}_J = \mathbf{U} \otimes \mathbf{W}$ with $\otimes$ referring to Kronecker product of the two matrices.

13.4 Voltage Measurements as Graph Signals

Consider a network of generators and nongenerator/load buses that can be represented as a graph $\mathcal{G}$ with undirected edge set $\mathcal{E}$ and vertex set is a union between set of generator nodes, G and nongenerator nodes/load, N, $i \in \{G \cup N\} = \mathcal{N}$, $\mathcal{G} = (\mathcal{E}, \mathcal{N})$. PMU installed on node i provides measurements of voltage and current phasors at time t, $V_i(t) = |V_i(t)|e^{j\theta_i(t)}$, $I_i(t) = |I_i(t)|e^{j\phi_i(t)}$, respectively. Let the voltage phasor for a generator at time t be $E_i(t) = |E_i(t)|e^{\delta_i(t)}$, $i \in G$. The branch admittance matrix $\mathbf{Y}$ for the network is also the complex-valued Laplacian matrix and is given by,

$$[\mathbf{Y}]_{m,n} = \begin{cases} \sum_{k \in \mathcal{N}_m} y_{mk}, & m = n \\ -y_{m,n}, & m \neq n \end{cases}, \quad y_{mn} \in \mathbb{C} \tag{13.11}$$

where y_{ij} is the admittance of the branch between nodes i and j if $(i, j) \in \mathcal{E}$. To model PMU data according to grid GSP, we show next that the appropriate graph shift operator is the system admittance matrix $\mathbf{Y}$ with eigenvalue decomposition as in (13.4), $\mathbf{Y} = \mathbf{U}\mathbf{\Lambda}\mathbf{U}^T$. $\mathbf{Y}$ is complex symmetric but need not be Hermitian. Therefore, the eigenvalues may not be *real*.

A typical approximation to capture the signal dynamics models loads buses $\ell \in N$ as constant impedances [24]. Given the apparent power that corresponds to load, the ratio of the apparent power and the voltage amplitude square at a particular load bus gives the value of the equivalent admittance at time t, $\mathbf{y}_\ell(t) \in \mathbb{C}^N$ (we revisit and further clarify this point in Section 13.4.1).

Shunt elements from all buses is denoted by $\mathbf{y}_{sh} \in \mathbb{C}^{|\mathcal{N}|}$ and generator admittances at nodes $g \in G$ as $\mathbf{y}_g \in \mathbb{C}^{|G|}$. Let $\mathbf{Y}_G$ and $\mathbf{Y}_L(t)$ be diagonal matrices that contain internal admittance of generators and loads, respectively:

$$\mathbf{Y}_G \triangleq \operatorname{Diag}(\mathbf{y}_g + \mathbf{y}_{sh}^g), \qquad \mathbf{Y}_L(t) \triangleq \operatorname{Diag}(\mathbf{y}_\ell(t) + \mathbf{y}_{sh}^\ell). \tag{13.12}$$

Partitioning the system between generator and nongenerator nodes makes the following well-known relationship between the complex phasor vectors explicit:

$$\begin{bmatrix} \mathbf{Y}_{gg} + \mathbf{Y}_G & \mathbf{Y}_{g\ell} \\ \mathbf{Y}_{g\ell}^T & \mathbf{Y}_{\ell\ell} + \mathbf{Y}_L(t) \end{bmatrix} \begin{bmatrix} \mathbf{V}_g(t) \\ \mathbf{V}_\ell(t) \end{bmatrix} = \begin{bmatrix} \operatorname{Diag}(\mathbf{y}_g) \mathbf{E}(t) \\ \mathbf{0} \end{bmatrix}, \tag{13.13}$$

where $\mathbf{Y}_{gg}$ is the Laplacian for the part of the network connecting generators, $\mathbf{Y}_{g\ell}$ includes the portion connecting generators and loads and $\mathbf{Y}_{\ell\ell}$ corresponds to the section of the grid connecting the loads buses among themselves. Equation (13.13) is commonly referred to as differential algebraic equations (DAEs) [24] and follows from Kirchoff's current law and Ohm's law and from converting the loads into equivalent impedances. Let

$$\mathbf{v}_t \triangleq \begin{bmatrix} \mathbf{V}_g(t) \\ \mathbf{V}_\ell(t) \end{bmatrix}, \quad \mathbf{\Delta}(t) \triangleq \begin{bmatrix} \mathbf{Y}_G & \mathbf{0} \\ \mathbf{0} & \mathbf{Y}_L(t) \end{bmatrix}, \quad \mathbf{x}_t \triangleq \begin{bmatrix} \operatorname{Diag}(\mathbf{y}_g) \mathbf{E}(t) \\ \mathbf{0} \end{bmatrix} \tag{13.14}$$

From (13.13) voltage phasor measurements, v_t, are written as

$$v_t = (Y + \Delta(t))^{-1} x_t, \qquad (13.15)$$

and we have the following observation from [33],

OBSERVATION 13.1 *In quasi-steady state, voltage phasor measurements, v_t, is approximately the output of a slow-time varying low-pass graph filter, $\mathcal{H}(Y) \triangleq (d(t)\mathbb{I} + Y)^{-1}$, when fed with a low-rank excitation x_t:*

$$v_t \approx \mathcal{H}(Y) x_t + \eta_t, \qquad v_t, x_t, \eta_t \in \mathbb{C}^{|\mathcal{N}|} \qquad (13.16)$$

The scalar $d \in \mathbb{C}$ can be chosen such that $\|\mathcal{H}(Y)(\Delta(t) - d\mathbb{I})\|_2 < 1$.

Observation 13.1 shows that generators are the drivers of the brunt of the variations observed in the PMU data, and that effectively loads can be viewed as inducing relatively slow time variations in the graph filter response. The generators buses, which are relatively few compared to the other buses, end up providing the excitation x_t for the PMU signal observation. It is easy to see that x_t is a low-rank excitation because we can write

$$x_t = P z_t, \quad P^T = \begin{bmatrix} \text{Diag}(y_g) & 0 \end{bmatrix}, z_t = E(t). \qquad (13.17)$$

Also rank$(P) \leq |G| \ll |\mathcal{N}|$, which means that the number of nodes exciting the grid is the number of generators. The low-pass nature of the graph filter, $\mathcal{H}(Y)$ is mainly due to the inversion of system admittance matrix when load and shunt elements are added. Approximation error η_t also captures the slow-time varying nature of the load. Note, however, that if load changes drastically, the corresponding node also becomes part of the excitation vector.

Single-pole IIR graph filter, $\mathcal{H}(Y) = (\mathbb{I} + d^{-1} Y)$ is a $(\tilde{K}, c)$ low-pass graph filter [27] when

$$c = \frac{\max\{|\lambda_1 + d|, |\lambda_2 + d|, \ldots, |\lambda_{\tilde{K}} + d|\}}{\min\{|\lambda_{\tilde{K}+1} + d|, |\lambda_2 + d|, \ldots, |\lambda_{|\mathcal{N}|} + d|\}} < 1 \qquad (13.18)$$

Note that adding d can change the relative ordering of frequencies in general. However, the system admittance matrix Y is quite sparse and has a high condition number, which implies that there many eigenvalues close to zero. Thus, for power systems we can safely assume that adding d will not affect the ordering in smallest $\tilde{K}$ eigenvalues and make it behave as a $(\tilde{K}, c)$ low-pass graph filter. The first implication of observation 13.1 is that it provides an explanation for the low-rank nature of phasor measurement data exploited in several algorithms to denoise and interpolate PMU data [11, 14, 25]. In the subsection that follows, we further clarify properties of the electrical power grid that help to analyze the coefficients of the filter $\mathcal{H}(Y)$.

13.4.1 Coefficients of the Grid Graph Filter

From (13.16) follows that the coefficients or the taps of $\mathcal{H}(Y)$ are mainly decided by the scalar parameter d; this parameter can be derived from generator, shunt, and load

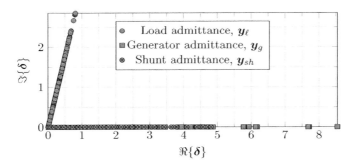

Figure 13.2 Diagonal elements of Δ, in ACTIVSg2000 i.e. δ where symbols indicate different types of admittance values.

admittances values. To demonstrate what is expected in terms of actual values, we take data from ACTIVSg2000 test case that, a synthetic MATPOWER testcase that emulates the properties of the Electric Reliable Council of Texas (ERCOT) system. The PMU data for its 2,000 buses are generated through simulations. In the system there are 432 generator buses and the rest are nongenerator buses [34]. Figure 13.2 shows the values of generator, load, and shunt admittance values on a complex plane. As seen, the values of generator admittance, $\mathbf{y}_g$ are well approximated by purely imaginary values, $[\mathbf{y}_g]_i \approx 1/(jX_{d,i}) = -j\tilde{y}_{g,i}$ where $X_{d,i}$ is the steady-state impedance of a generator i. Both $\mathbf{y}_{sh}^g$ and $\mathbf{y}_{sh}^\ell$ small and purely imaginary shunt values, i.e., $\mathbf{y}_{sh} = \begin{bmatrix} \mathbf{y}_{sh}^g & \mathbf{y}_{sh}^\ell \end{bmatrix}^T = -j\mathbf{b}_{sh}$. The loads are modeled to be constant impedance although in reality impedance varies slowly in time. For the test case considered, they are computed using power at operating point, $[\mathbf{y}_\ell]_i = S_i^*$ in per unit where S_i is the complex power at node i. The values here are complex, $\mathbf{y}_\ell = \mathbf{g}_\ell - j\mathbf{b}_\ell$ as seen in the figure. Also in ACTIVSg2000 test case, system admittance matrix has purely imaginary values, $Y = -jB$. This is indeed more general and not limited to the test case. From now on, we consider purely imaginary system admittance matrix. Hence,

$$\mathcal{H}(Y) \approx j(B + jd\mathbb{I})^{-1} = j\mathbf{U}(\Lambda + jd\mathbb{I})^{-1}\mathbf{U}^T \qquad (13.19)$$

Since B is a real-symmetric Laplacian matrix, $\Lambda \succeq 0$ with the smallest eigenvalue being 0 and $\mathbf{U} \in \mathbb{R}^{|\mathcal{N}| \times |\mathcal{N}|}$ is orthonormal. In [33] it was shown the d such that $\|\mathcal{H}(Y)(\Delta - d\mathbb{I})\|_2 < 1$ holds and that maximally reduces $\|j(\Delta - d\mathbb{I})\|_2$ is:

$$d = (j/2)\max_i \left[\tilde{\mathbf{y}}_g + \mathbf{b}_{g,sh}\right]_i = (j/2)\tilde{d} \qquad (13.20)$$

In other words, it corresponds to half the maximum admittance among generators. From Figure 13.2 real and imaginary parts in $j\Delta$ are positive because the values are admittances. The entries, therefore, lie in the positive quadrant. Let $a_{min}, a_{max} \in \mathbb{C}$ correspond to the points in entries of $j\Delta$ that are closest to and farthest from the origin, respectively. Then jd is chosen to be the *mid-point* on the line joining a_{min}, a_{max} so that $\|j(\Delta - d\mathbb{I})\|_2 = \max\{|a_{min} - jd|, |a_{max} - jd|\}$. In an electric grid, there are many nodes that are intermediate buses, for which the shunt admittance is very low so that $a_{min} \approx 0$.

Generator admittances are usually greater in absolute value than those of loads. Because jg_ℓ is the only imaginary component in $j\Delta$, without loss of generality, we can assume that $a_{max} = \max_i [\tilde{\mathbf{y}}_g + \mathbf{b}_{g,sh}]_i + j0$. As seen, they all lie in the positive quadrant and $a_{min} = 0 + j0$ and $a_{max} = 0 + j\tilde{d}$ (marked). All the values were normalized by the base MVA. From the test case, $\tilde{d} = 8.544$ and $d = j\tilde{d}/2 = j4.2720$. It was found that with this d, $\|\mathcal{H}(Y)(\Delta - d\mathbb{I})\|_2 = 0.9973$.

Thus, the filter $\mathcal{H}(Y)$ is $(\tilde{K}, c)$ low pass with

$$c = \left(\lambda_{\tilde{K}} + (\tilde{d}/2)\right)\left(\lambda_{\tilde{K}+1} + (\tilde{d}/2)\right)^{-1} \tag{13.21}$$

Smaller c alludes to a better low-pass nature of the filter.

13.5 Sampling and Compression for Grid-Graph Signals

Let T samples of voltage measurements from PMU be collected to form a measurement matrix:

$$\mathbf{V}_t = \begin{bmatrix} \mathbf{v}_t & \mathbf{v}_{t+1} & \cdots & \mathbf{v}_{t+T} \end{bmatrix} \in \mathbb{C}^{N \times T} \tag{13.22}$$

Consider the vectorized form of $\mathbf{V}_t$, i.e. $\mathbf{v} \triangleq \text{vec}(\mathbf{V}_t) \in \mathbb{C}^{|N|T}$ as a graph-signal. The underlying graph is a Cartesian product of electrical graph, $\mathcal{G}$ and the discrete-time graph that is usually represented as a cycle graph of length T. The joint time-vertex analysis in GSP [32] is a means to look at spatio-temporal patterns in PMU data and study both the graph-frequency and discrete-time frequency response as summarized in Subsection 13.3.1. Utilizing this idea, we can provide a sampling theory framework for grid-graph signals. Using the JFT as defined in (13.10) with the operator $\mathbf{U}_J = \mathbf{U} \otimes \mathbf{W}$ where $\mathbf{W}$ is the normalized DFT matrix for certain period T we have:

$$\mathbf{v} = \mathbf{U}_J(\tilde{\mathbf{v}}), \qquad \tilde{\mathbf{v}} = \mathbf{U}_J^T \mathbf{v} \tag{13.23}$$

where $\tilde{\mathbf{v}}$ is the JFT of $\mathbf{v}$.

For illustration, two cases are considered. The first one presents the data from the ACTIVSg2000 case already introduced. The data are from a transient simulation of the system in a quasi-steady state regime (i.e., no disturbances, just normal load variations at each bus to which the generators react) [35]. The second case is that of real-world data from the grid managed by the ISO New England [36] with an actual transient although for this case the admittance matrix is not available. For both the cases, the sampling rate was 30 samples per second. Figure 13.3(a) show the magnitude and phase of voltage phasors at five locations in the measurement matrix $\mathbf{V}_t$ for 20 seconds selected from the data of the ACTIVSg2000 case to highlight the low-rank nature of PMU data especially in quasi-steady state. Similarly Figures 13.4(a) and 13.4(b) show the magnitude and phase for a transient case in which some equipment issues in a large generator create oscillations with growing magnitude during an interval lasting 40 seconds. Although the time series are featuring notable oscillations, spatially the effects are observed at multiple locations of the New England power system, with all the streams except one exhibiting consistent trends due to the low-pass graph-filtering effect in space.

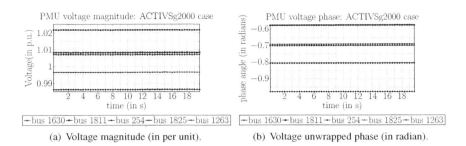

(a) Voltage magnitude (in per unit).   (b) Voltage unwrapped phase (in radian).

Figure 13.3 PMU voltage measurement V_t for 20 seconds. Note the very slow variation both in time (x-axis) and space (y-axis).

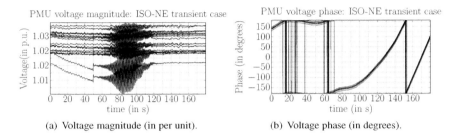

(a) Voltage magnitude (in per unit).   (b) Voltage phase (in degrees).

Figure 13.4 PMU voltage measurement V_t for 180 seconds during a transient in ISO-NE case. (Data taken from source in [36])

13.5.1 Dimensionality Reduction of PMU Data

As seen in Section 13.4 voltage and consequently current graph signals are outputs of a (K, c) low-pass filter that have frequency content limited to few smaller eigenvalues according to (13.16). This characteristic is called band-limited nature of the graph signals, which means that there is a cutoff frequency λ_K such that frequency content corresponding to λ_{K+1} and higher is negligible [37]. From (13.18), we see that c decides the decay of energy into different frequency bins because it is not a perfect low-pass filter. The graph-frequency response of the grid-graph signals at time t, $\tilde{v}_t$ is then,

$$\tilde{v}_t = \mathbf{U}^T \mathcal{H}(Y) x_t = j (\mathbf{\Lambda} + jd\mathbb{I})^{-1} \tilde{x}_t \qquad (13.24)$$

The response is pertinent on both the low-pass filter and the input $\tilde{x}_t = \mathbf{U}^T P z_t$. The combined effect is that the energy is limited to a relatively small set of graph frequency coefficients compared to the size of the system. Additionally, grid-graph signals also vary temporally even in quasi-steady state due to slow-varying loads and generators matching production with demand. This implies that the DFT of the grid-graph signals has only a few nonzero components in the lower frequencies in a case like ACTIVSg2000. Therefore, the JFT of PMU signals, v also has a small support both relative to the size of the system and the number of samples in time that are processed by the JFT. Further analysis, that we omit here for brevity but shown in [38], shows that the generator dynamics induce this effect.

Graph Signal Processing for the Power Grid

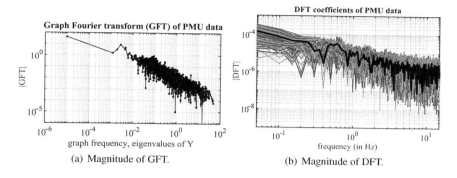

(a) Magnitude of GFT.

(b) Magnitude of DFT.

Figure 13.5 GFT and DFT for PMU voltage measurement $\mathbf{V}_t$ for 20 seconds.

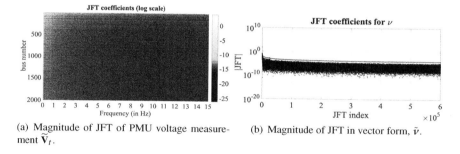

(a) Magnitude of JFT of PMU voltage measurement $\widetilde{\mathbf{V}}_t$.

(b) Magnitude of JFT in vector form, $\tilde{\nu}$.

Figure 13.6 GFT of PMU voltage measurement $\mathbf{V}_t$. Sparsity in both time and graph-frequency domain is observed.

In Figures 13.5(a) and 13.5(b), the magnitude of coefficients for both the GFT and DFT of $\mathbf{V}_t$ is shown, respectively, for the ACTIVSg2000 case. Also, JFT coefficients, $\widetilde{\mathbf{V}}_t$ and $\tilde{\nu}$ for the same case are shown in Figures 13.6(a) and 13.6(b). Note that the coefficients in the joint Fourier domain are quite sparse and there are only a few high-energy coefficients.

Compression

Based on the most significant coefficients, define $\mathbf{U}_K \in \mathbb{C}^{|\mathcal{N}| \times K}$ as the collection of K eigenvectors of the system admittance matrix $\mathbf{Y}$ corresponding to the smallest K eigenvalues because it is a low-pass filter. Similarly, let $\mathbf{W}_L$ be the L lowest frequencies of interest temporally. With this,

$$\hat{\mathbf{U}}_J \triangleq \mathbf{U}_K \otimes \mathbf{W}_L \in \mathbb{C}^{|\mathcal{N}|T \times KL} \tag{13.25}$$

Consequently, a bandlimiting operator can be defined as, $\mathcal{B} = \hat{\mathbf{U}}_J \hat{\mathbf{U}}_J^T, \in \mathbb{C}^{|\mathcal{N}| \times KL}$ projects signal into subspace spanned by columns of $\hat{\mathbf{U}}_J$. As $KL \ll |\mathcal{N}|T$ due to sparsity in the graph and time-frequency domains, it is expected that:

$$\mathcal{B}\nu \approx \nu \tag{13.26}$$

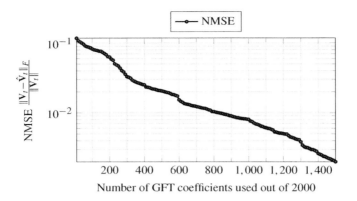

Figure 13.7 Residual distortion versus GFT coefficients used for reconstruction.

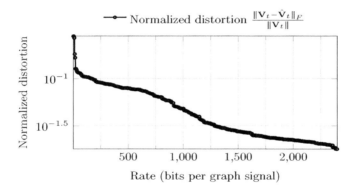

Figure 13.8 Empirical rate distortion curve obtained quantizing the GFT coefficients.

i.e., the signal v is approximately localized over K-graph and L time frequencies. Thus, PMU signals are sparse in JFT domain. This observation explains the low dimensionality of PMU data and confirms that the data set is an excellent candidate for dimensionality reduction and compression. A simple scheme for compression is to retain only the K (KL) graph (joint) Fourier coefficients and have the approximated signals $\hat{v}_t$ and $\hat{v}$, respectively.

$$\hat{v}_t = \mathbf{U}_K \begin{bmatrix} \mathbb{I}_K & \mathbf{0} \end{bmatrix} \tilde{v}, \tag{13.27}$$

$$\hat{v} = \hat{\mathbf{U}}_J \begin{bmatrix} \mathbb{I}_{KL} & \mathbf{0} \end{bmatrix} \tilde{v} \tag{13.28}$$

Figures 13.7 and 13.8 highlight the result of such an approach using the GFT coefficients and shows how the normalized mean squared error (NMSE) defined as $\frac{\|\mathbf{V}_t - \hat{\mathbf{V}}_t\|_F}{\|\mathbf{V}_t\|}$ changes with number of coefficients and bits per graph signal used for the approximation, respectively.

For the compression we sort the JFT coefficients by energy and choose the topmost ones for PMU data compression. The performance of this compression scheme is seen in Figure 13.9 where we define distortion measure as the mean-squared error (MSE),

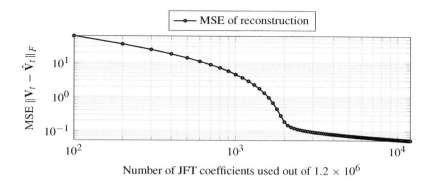

Figure 13.9 Performance of compression algorithm using JFT coefficients.

$\|\mathbf{V}_t - \widehat{\mathbf{V}}_t\|_F$ for reconstruction of PMU data starting with 100 JFT coefficients. One can see a sharp decrease in MSE and good reconstruction. The results, however, need to be taken with a grain of salt, as the ACTIVSg2000 case data are generated for a quasi-steady state condition in which the time variations are minimal. But that is not always the case, as seen in Figures 13.4(a) and 13.4(b). Hence more interesting segments of data naturally lead to a higher rate distortion curve. This means that the JFT is an appropriate feature space to not only compress but also to detect informative segments of the data and classify the events.

13.5.2 Sampling and Recovery of PMU Grid-Graph Signals

There has been significant interest in the problem of optimal PMU placement in the literature from various perspectives. A more recent line of work focused on placing PMUs strategically to dwarf stealth attacks to state estimation (see, e.g., [39, 40]). A related problem is that of recovering missing or incorrect data that occur at random [13, 14, 25, 41]. The fact that PMU signals have a sparse support in the GFT (and JFT) domain provides the theoretical underpinning that support the arguments made in the literature, as discussed next.

The Placement of PMUs as a GSP Sampling Problem

Placing PMU sensors can be thought of as the problem of sampling a graph signal in a fixed spatial pattern. Graphs do not generally have a regular structure, and therefore the placement pattern is not as easy to discern as it is for time series or images, where the optimum sampling pattern is obtained by simple decimation. The pattern is important because a bad one may hamper the interpolation of the voltages at buses without the PMUs, particularly when the band-limited assumption is based on an approximation. Finding a sampling pattern can be expressed as the task of finding a sampling mask $\mathbf{M}_S \in \{0, 1\}^{N \times T}$ such that we observe only $\mathbf{M}_S \odot \mathbf{V}_t$ samples with a total of $|S|$ nonzero entries, where $\odot$ refers to element-wise/Hadamard product. From (13.26) in vector form we have:

$$\mathbf{v}_S = \mathbf{P}_S^T \mathbf{v} \approx \mathbf{P}_S^T \mathcal{B} \mathbf{v} \tag{13.29}$$

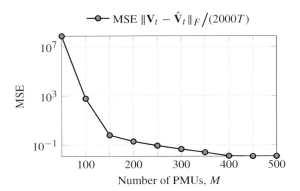

Figure 13.10 Performance of compression after optimal placement of M PMUs. $K = 40$.

where $\boldsymbol{P}_S \in \mathbb{R}^{\mathcal{N}T \times |S|}$ has coordinate vectors in each column choosing an entry from $\boldsymbol{v}$ and let $\mathcal{D}_S = \boldsymbol{P}_S \boldsymbol{P}_S^T = \mathrm{diag}\,[\mathrm{vec}\,(\boldsymbol{M}_S)]$ From theorem 4.1 in [26], condition for perfect recovery of band-limited signals is

$$\mathrm{rank}(\mathcal{D}_S \hat{\mathbf{U}}_J) \geq KL, \implies |S| \geq KL \tag{13.30}$$

As said, the total number of samples needed are at least as many as significant coefficients in the frequency domain. In GSP having $|S| \geq KL$ alone is not, in general, a sufficient condition because the location of the $|S|$ for a given graph structure also dictates the recovery performance. Intuitively, this can happen if the GFT basis has sparse components that are dominant in the graph signal and the sensor's placement does not overlap with such components. Also, there is a residual error in the model and, when interpolating, some sampling patterns may amplify the effects of these errors.

A criterion to design the sampling mask $\boldsymbol{M}_S$ is introduced in [26, 37], where the optimum mask is obtained by maximizing the smallest singular value of $\mathcal{D}_S \hat{\mathbf{U}}_J$ for a given the number of samples $|S| \geq KL$. Note that here we are not looking at random sampling masks in space and time because in the placement problem the position of the PMUs is permanent. Consider then the spatial sampling mask for M PMU measurements, $\mathcal{D}_M = \mathrm{diag}(\mathbf{1}_M)$ that picks M locations. The optimal placement of M PMUs maximizes $\sigma_{min}(\mathcal{D}_M \mathbf{U}_K)$ which, intuitively, amounts to choosing the rows of $\mathbf{U}_K$ with the smallest possible coherence or, in simpler terms, as close as possible to be orthogonal. For the placement of M PMUs, using a convex relaxation of the same problem, the optimal sampling for reducing model-mismatch errors is given by:

$$\min_{f} -\lambda_{min}\left(\mathbf{U}_K^T \mathrm{diag}[f] \mathbf{U}_K\right)$$
$$\text{subject to} \quad \mathbf{1}^T f = M, \mathbf{0} \preceq f \preceq \mathbf{1} \tag{13.31}$$

The location of top M entries of the solution f corresponds to locations of optimally placed PMUs. Figure 13.11 shows the location of placed PMUs in the ACTIVSg case for $M = 50$ PMUs with $K = 40$. We see that the choice of location is closely tied

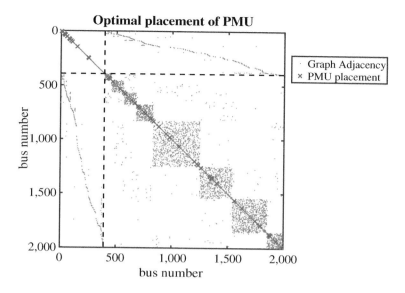

Figure 13.11 Optimal placement of $M = 50$ PMUs. $K = 40$.

with the community structure evident in the ACTIVSg case. This is a topic that we will revisit in the next two sections of this chapter. Figure 13.10 shows the performance of reconstruction of graph signals after optimal placement. With optimally placed PMUs, to recover the matrix $\mathbf{V}_t$, number of measurements (in time) needed are approximately KL/M.

Missing PMU Data Recovery

The underlying generative model responsible for low-rank nature of PMU data that has been established in the previous section helps explain the success that many past works, such as [13, 14, 25, 41], have attained in recovering missing PMU data using matrix completion methods. Even when contiguous data in time has been missing as in [42], the authors used a Hankel matrix and were still able to recover missing data, which is consistent with looking at sparsity in the JFT domain, instead of the GFT domain only. For illustration purposes here we reproduce the data-recovery algorithm formulated as a matrix completion for the Hankel matrix in [43], for denoising PMU data from the ISO-NE transient case 3 where we added corrupted measurements and also for recovering missing data from a block of PMUs. The algorithm solution is an instance of the robust principal component analysis (RPCA). Specifically, let the corrupted measurement be $\hat{\mathbf{V}}$ while the actual data is $\mathbf{V}$, the solution is given by:

$$\min_{\mathbf{X}} \|\mathbb{H}(\mathbf{X})\|_* + \gamma \|\mathbb{P}_\Omega(\mathbb{H}(\mathbf{X}) - \hat{\mathbf{V}})\|_1 \qquad (13.32)$$

where $\mathbb{H}(\mathbf{X})$ refers to Hankel matrix arrangement using columns of $\mathbf{X}$ as defined in [43] and $\mathbb{P}_\Omega$ is an operator that chooses the subset of measurements available. Figures 13.12(a) and 13.12(b) show the recovery performance when sparse noise is added to

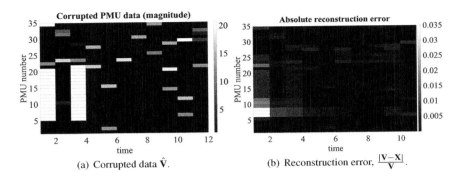

Figure 13.12 Performance of missing PMU data recovery using Hankel matrix structure.

measurements along with missing data. Naturally, the error is slightly higher where the data is missing. Yet the reconstruction error is not too large.

Even in the case of transient, it is possible to recover missing data. This is because even if not low pass, in this case the perturbation is localized spatially but the location of the event is sampled and the signal is no longer low pass, yet it is sparse in the DFT domain. In the next section, we study how placing sparsely PMUs spatially, rather than missing a few samples, can affect ones ability to localize an event. More importantly, we provide insights on how the grid topology affects the localization accuracy and connect this with the problem of placement/sampling of graph signals we just discussed.

Anomaly or Fault Detection
As should be clear at this point, changes in the dominant components of the GFT/JFT spectrum signal anomalies. This supports the evidence that a strategy to detect anomalies exploits the knowledge of low-rank nature of PMU measurements, V_t for the null hypothesis and the deviation from this hypothesis to label the data as an anomaly. For instance [15] proposes subspace tracking combined with change detection to identify the pre- and post-fault data. Similarly in [16], the low-rank nature of PMU data is used for event identification by projecting data to the low-dimensional space estimated from pilot PMU data. In [17], a dictionary of events is built in real time to identify faults or anomalies in real time by looking at residuals with respect to the estimated principal components. Any of these approaches can be undertaken for fault detection. These techniques provide a primitive for segmenting the data into pre- and post-fault portions, which can be helpful in identifying the root cause of an event that contributed to change the network in some fashion, for instance a line or current fault.

13.6 The Role of the Topology on Grid-GSP Data

The ACTIVSg case features a distinctive community structure. That structure is found in real grids and it is naturally linked to the nonuniform distribution of population. The question is how this feature, that is inevitable in physical infrastructure networks, affects the graph signals that result from these systems. The two inference problems we study next provide an insight on the impact of this structural characteristic on the algebraic properties of the grid signals observed. The problems are 1) fault localization using undersampled grid graph signals and 2) identification of community structure of graph signals with PMU data.

13.6.1 Fault Localization Using Undersampled Grid-Graph Signals

Line or current faults, switching events, voltage sags, and faults at generators are associated with events that start with a localized spatio-temporal pattern, eventually smoothened by the low-pass response of the network. It is intuitive that to pinpoint the location of such faults, one needs to be able to reconstruct sharp spatio-temporal derivatives, i.e. signals that have a wide GFT/JFT spectrum. The question is to what extent one is able to localize faults in a system using a small number of PMU measurements, which could be sufficient in a quasi-steady state regime but are not sufficient if the signal has a broad GFT/JFT spectrum. The optimal placement of M PMUs, in this case, would not guarantee a good reconstruction but rather be aimed at detecting, with the maximum possible resolution, the fault in space and time. By *resolution* in localizing an event on a graph we refer to the ability to localize the correct ℓ-hop neighborhood, with ℓ as small as possible. Although there are both parametric and nonparametric approaches to localization such as in [44–46], the connection to topology was first made in [21].

Based on the discussion in Section 13.5.2, we can assume that the time instant at which a detected fault started and ended is known because we are able to determine when high JFT components are first observed and when they end.

Assume that the measurements pre-fault and post-fault currents are given by

$$\text{Pre-fault: } \mathbf{I}_0 = \mathbf{Y}\mathbf{v}_0, \quad \text{Post-fault: } \mathbf{I}_F + \mathbf{I}_E = \mathbf{Y}\mathbf{v}_F \tag{13.33}$$

with $\mathbf{I}_E$ a sparse vector with 1 nonzero entry and equal to injected fault current at faulty bus. Subtracting pre-fault from post-fault,

$$(\mathbf{I}_F - \mathbf{I}_0) + \mathbf{I}_E = \delta\mathbf{I} + \mathbf{I}_E = \mathbf{Y}(\mathbf{v}_F - \mathbf{v}_0) = \mathbf{Y}\delta\mathbf{v} \tag{13.34}$$

The current and voltages can be parsed as available and unavailable PMUs measurements:

$$\begin{pmatrix} \delta\mathbf{I}_a \\ \delta\mathbf{I}_u \end{pmatrix} + \begin{pmatrix} \mathbf{I}_E^a \\ \mathbf{I}_E^u \end{pmatrix} = \begin{pmatrix} \mathbf{Y}_{aa} & \mathbf{Y}_{au} \\ \mathbf{Y}_{au}^T & \mathbf{Y}_{uu} \end{pmatrix} \begin{pmatrix} \delta\mathbf{V}_a \\ \delta\mathbf{V}_u \end{pmatrix}. \tag{13.35}$$

Let us define the vector z obtained from the available measurements as follows:

$$z = \overbrace{(\mathbb{I} \mid -(\mathbf{Y}_{aa} - \mathbf{Y}_{au}\mathbf{Y}_{uu}^{-1}\mathbf{Y}_{au}^T))}^{H} \begin{pmatrix} \delta\mathbf{I}_a \\ \delta\mathbf{V}_a \end{pmatrix} \tag{13.36}$$

Rearranging (13.35), the following measurement model for z can be obtained, tying the observation to the faults:

$$z = (-\mathbb{I} \mid \overbrace{Y_{au}Y_{uu}^{-1}}^{C}) \overbrace{\begin{pmatrix} I_E^a \\ I_E^u \end{pmatrix}}^{s} + \overbrace{Y_{au}Y_{uu}^{-1}}^{C} \overbrace{\delta I_u}^{\epsilon} \tag{13.37}$$

$$z = [-\mathbb{I} \mid C] s + C\epsilon \tag{13.38}$$

where s is sparse with fault current at nonzero entries and ϵ is the noise term with complex normal distribution, $\epsilon \sim \mathcal{CN}(0, \sigma_\epsilon^2 \mathbb{I})$. Let the singular value decomposition (SVD) be $C = U_C \Sigma_C W_C^\dagger$, whitening ϵ, i.e. multiplying $\Sigma_C^{-1} U_C^\dagger$, gives:

$$d = \Sigma_C^{-1} U_C^\dagger z = Fs + \varepsilon \tag{13.39}$$

where $\varepsilon = W_C^\dagger \epsilon \sim \mathcal{CN}(0, \sigma_\epsilon^2 \mathbb{I})$, and $F = [-\Sigma_C^{-1} U_C^\dagger \mid W_C^\dagger]$. Model in (13.38) and (13.39) is a well-known linear model with a sparse input and additive Gaussian noise. However, remember that measurements d are obtained as a linear combination through F of graph signals at different nodes i.e. s. Fault localization is then equivalent to recovering the support of s. It is assumed that s is 1-sparse, i.e. only one location at fault.

The sparsity of s underscores the problem with the undersampled nature of the grid-graph signals d. Its excitation is no longer a bandlimited graph signal; all graph-frequency components are important, which makes the recovery not possible when the number of PMUs is much less than number of nodes in the graph. It is then required to operate under this loss of resolution and place PMUs to obtain maximum resolution possible. Following [21], fault localization can be formulated as a multiple hypothesis testing problem based on (13.39) where the goal is to identify the fault location ℓ. With assumptions about the distribution of s for each hypothesis ℓ, the conditional distribution of d for hypothesis $\mathcal{H}_\ell$ is given by

$$d | \mathcal{H}_\ell \sim \mathcal{CN}(F\mu_\ell, \sigma_\epsilon^2 \mathbb{I} + F\Phi_\ell F^\dagger) \tag{13.40}$$

where $\mu_\ell = \mathbb{E}[s]$, $\Phi_\ell = \mathbb{E}[ss^H | \mathcal{H}_\ell]$. The maximum-likelihood (ML) detector of the fault location is then

$$\ell^* = \underset{\ell}{\operatorname{argmax}}\, \lambda_\ell(d) = \underset{\ell}{\operatorname{argmin}}\, (d - m_\ell)^H \Psi_\ell^{-1} (d - m_\ell) + \ln(\pi^K |\Psi_\ell|). \tag{13.41}$$

where $m_\ell = F\mu_\ell$, $\Psi_\ell = F\Phi_\ell F^H$. If for a given fault at location ℓ^*, the value of log-likelihood $\lambda_{\ell^*}(d)$ is close to $\lambda_k(d)$ ($k \neq \ell^*$), the location of the fault can be misidentified. A cluster then can be defined as a set of nodes for which $\lambda_\ell(d) \approx \lambda_{\ell^*}(d)$ under $y|\mathcal{H}_{\ell^*}$. The distance between conditional distribution of data under different hypotheses is characterized by the Kullback–Leibler (KL) Divergence, $D_{KL}(f(d|\mathcal{H}_\ell) \| f(d|\mathcal{H}_k))$. In this case,

$$D_{KL}(f(d|\mathcal{H}_\ell) \| f(d|\mathcal{H}_k)) \propto (r_{\ell k})^{-1}, \quad r_{\ell k} = \frac{|f_\ell^\dagger f_k|}{\|f_\ell\| \|f_k\|} \tag{13.42}$$

where f_ℓ is the ℓth column of F. Higher the $r_{\ell k}$, more the possibility of misclassification between ℓ and k hypothesis. In other words, k, ℓ will belong to the same cluster. More small size clusters of vectors of F with high mutual correlation imply finer localization resolution. The mutual coherence of matrix F [47] is:

$$\mu(F) = \max_{k,\ell, k \neq \ell} r_{\ell k} \qquad (13.43)$$

From the theory of recovery of sparse signals [47], it is known that when

$$\|s\|_0 \leq \frac{1}{2}(1 + \mu(F)^{-1}) \qquad (13.44)$$

then ℓ_1 minimization can recover s. In general, the lower is the mutual coherence of a matrix, the better the recovery performance. In this problem, $\|s\|_0 = 1$ which implies that it is desired to have $\mu(F) \leq 1$ for recovery. To compute the mutual coherence, consider

$$F^\dagger F = \begin{bmatrix} U_C \Sigma_C^{-2} U_C^\dagger & -U_C \Sigma_C^{-1} W_C^\dagger \\ -W_C \Sigma_C^{-1} U_C^\dagger & \mathbb{I} \end{bmatrix}. \qquad (13.45)$$

The magnitude of each entry of the matrix $F^H F$, with suitable normalization, corresponds to $r_{\ell k}$ except for the diagonal entries. Naturally, the magnitude depends on the spectrum Σ of the matrix $C = Y_{au} Y_{uu}^{-1}$, which must be as *flat* as possible so as to lower the mutual coherence of the matrix F. This suggests to minimize the so-called Shatten infinity norm of the matrix $Y_{au} Y_{uu}^{-1}$, which is the infinity norm $\|\sigma\|_\infty$ of the vector σ containing the nonzero singular values of $Y_{au} Y_{uu}^{-1}$. The design strategy is aimed precisely at providing a flat spectrum for C. As an illustration, we consider a modified IEEE-34 bus case as in [48]. Figure 13.13(a) shows this case while marking four optimally placed PMUs. In Figure 13.13(b), we see the proximity of log-likelihood values for a fault at bus 820 to those of its neighboring buses and the resulting loss of resolution while localizing the fault. Figure 13.14 shows the correlation coefficient $r_{\ell k}$ between two columns of F for an IEEE-34 bus case related to phase-A for optimal and nonoptimal placement. Gray areas are clusters and amongst them is hard to discriminated. Smaller the clusters, better the discrimination.

Note the resemblance with optimal placement problem for recovery of grid-graph sampled signals in Subsection 13.5.2. In both of the problems, the optimal placement corresponds to choosing uncorrelated rows of the matrix U_K and $C = Y_{au} Y_{uu}^{-1}$, respectively. With U_K, it is choosing a representative row per "cluster" in a graph. Similarly, in C, choosing the rows of $Y_{au} Y_{uu}^{-1}$ to be uncorrelated is attained by having nonoverlapping support among of the rows in $Y_{au} Y_{uu}^{-1}$ and having Y_{uu} as close as possible to a block diagonal matrix, where the diagonal blocks as matched to the nonzero portions of rows Y_{au}. The question that we explore next through a simple example is the relationship between orthogonality of the rows of this matrix and topology.

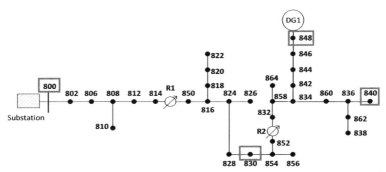

(a) Reduced IEEE-34 test case with added generator. Boxes show optimal placement of PMUs.

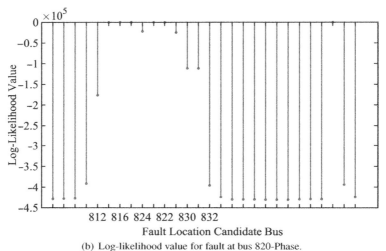

(b) Log-likelihood value for fault at bus 820-Phase.

Figure 13.13 Fault localization for IEEE-34 bus case.

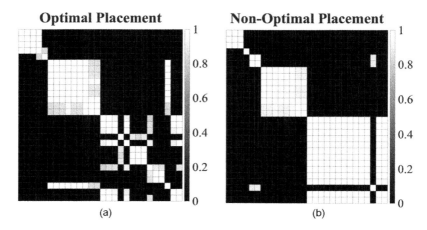

Figure 13.14 Correlation of columns of F for IEEE-34 bus case related to phase-A for a) optimal placement b) nonoptimal placement. Values of correlation less than the threshold of $\tau = 0.814$ are truncated to zero.

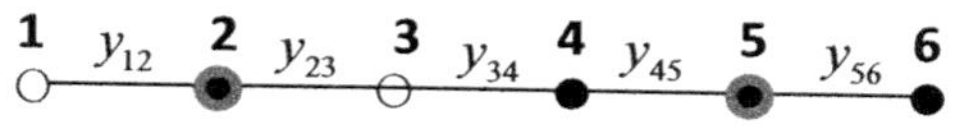

Figure 13.15 Network line diagram with PMUs at buses 2 and 5. (From [21])

Sampling, Resolution and Community Structure
Consider, for example, the system in Figure 13.15, if the PMUs are placed at bus 2 and 5, the admittance matrix is partitioned in the desirable way as:

$$Y = \begin{pmatrix} y_1 & 0 & -y_{12} & -y_{23} & 0 & 0 \\ 0 & y_5 & 0 & 0 & -y_{45} & -y_{56} \\ -y_{12} & 0 & y_2 & 0 & 0 & 0 \\ -y_{23} & 0 & 0 & y_3 & -y_{34} & 0 \\ 0 & -y_{45} & 0 & -y_{34} & y_4 & 0 \\ 0 & -y_{56} & 0 & 0 & 0 & y_6 \end{pmatrix}. \quad (13.46)$$

This design is clearly distancing the PMUs in the graph and dividing the network in neighborhoods, each associated with one of the adjacent sensors and two hop neighbors where measurements are unavailable. This means that the ambiguity that remains is confined to a connected set of buses that are topologically close to the PMU sensors. Hence, the optimum resolution for a certain grid is tied to the same intrinsic topological properties that are studied in graph clustering. There is a basic intuition regarding this result. The nonzero elements of the admittance matrix tend to have very similar values and signs. We can clearly obtain a Y_{aa} that is diagonal by not having an edge between two sensors. More importantly, that small coherence among the rows in Y_{au} basically requires that there is little overlap among the neighborhoods that are connected to each sensor. This also implies that Y_{uu} will reflect the community structure of the system. In turn, this means that for a GSO with similar weights, the best sampling patterns tend to place sensors in separate communities.

13.6.2 Identification of Community Structure in the Electrical Grid

In [27] it was established that the eigenvectors of the data covariance of graph signals can be used to unveil the community structure of the underlying graph, when signals are the output of a low-pass graph filter. However, the result does not hold if a GSO has edge weights with positive and negative, or complex-valued weights. In this case, the ratio-cut minimization formulation cannot rely on such Laplacian operators and the algorithm proposed based on spectral clustering returns meaningless results. But when the GSO has weights with a common phase, the technique of community detection makes sense. This is the case with nearly purely imaginary valued system admittance matrix. Hence, the voltage phasor measurements are good candidates to reveal the community structure of the power grid. The electric grid has an inherent community structure that is driven by the fact that population density is highly uneven, and its spatial distribution determines the graph density [35]. Let us assume that there are K communities. The grid is also designed such that each community is served by a portfolio of generators, which means

Algorithm 13.1 Blind community detection for grid-graph signals

Require: Voltage phasors $\{v_t\}_{t=1}^T$ and desired number of communities K.
1: Compute sample covariance matrix $C_v = T^{-1}\sum_{i=1}^{T} v_t v_t^\dagger$.
2: Find the K eigenvectors corresponding to top K eigenvalues. Denote them by $\hat{U}_K \in \mathbb{C}^{|\mathcal{N}| \times K}$.
3: Define a higher dimensional matrix $\hat{U}_{2K} \in \mathbb{R}^{|\mathcal{N}| \times 2K}$ as $\hat{U}_{2K} = \begin{bmatrix} \Re\{\hat{U}_K\} & \Im\{\hat{U}_K\} \end{bmatrix}$.
4: Apply k-means clustering method that minimizes the objective

$$F = \min_{\mathcal{A}_1,\dots,\mathcal{A}_K} \sum_{k=1}^{K} \sum_{i \in \mathcal{A}_k} \|\hat{u}_i^{\text{row}} - \frac{1}{|\mathcal{A}_k|} \sum_{m \in \mathcal{A}_k} \hat{u}_m^{\text{row}}\| \quad (13.47)$$

where $\hat{u}_i^{\text{row}}$ is the ith row of $\hat{U}_{2K}$.
Ensure: K communities $\hat{\mathcal{A}}_1, \dots, \hat{\mathcal{A}}_K$

---

that there are more generators than communities, $K < |\mathcal{G}|$. From proposition 13.1, PMU voltage phasor data v_t is the output of a low-pass graph filter $\mathcal{H}(Y)$. Assuming that system admittance matrix is purely imaginary, the graph filter has the form as in (13.19). The eigenvectors U are real valued and orthonormal and the connection between spectral clustering [49] and the blind community detection as in [27] is seamless. The community structure can be recovered using eigenvectors of [27], $C_v = T^{-1}\sum_{i=1}^{T} v_t v_t^\dagger$ where † is the complex conjugate operator. Each outer product yields (from (13.19))

$$v_t v_t^\dagger = \mathcal{H}(Y) x_t x_t^\dagger (\mathcal{H}(Y))^\dagger + C_\eta \quad (13.48)$$
$$= U(\Lambda + jd\mathbb{I})^{-1} U^T x_t x_t^\dagger U(\Lambda - jd\mathbb{I})^{-1} U^T + C_\eta \quad (13.49)$$

The rank of the excitation covariance matrix C_x,

$$C_x = T^{-1}\sum_{i=1}^{T} x_t x_t^\dagger = P\left(T^{-1}\sum_{i=1}^{T} z_t z_t^\dagger\right) P^\dagger \quad (13.50)$$

is roughly the order of number of generators $|\mathcal{G}|$ and is usually less than number of communities to be detected, $K < \text{rank}(P)$. Because $\mathcal{H}(Y)P$ can be considered the sketch of the graph filter that is low pass with $(\tilde{K}, c)$ [27], the rank of C_v will be approximately K and, therefore, the top K eigenvectors of covariance matrix C_v can be used instead of the lowest K eigenvectors of graph Laplacian Y. Therefore, as in spectral clustering [50], a k-means clustering on the top K eigenvectors of C_v correspond to the smallest eigenvectors of Y and can be used for community detection [27]. Clustering using k-means is performed in a higher dimension due to complex valued eigenvectors. Algorithm 13.1 specifies the steps undertaken for community detection. Also, when the DC approximation holds [33], k-means clustering on the covariance matrix of phase angle measurements, $C_\theta = T^{-1}\sum_{i=1}^{T} \theta_t \theta_t^T$, which has real-valued eigenvectors and nonnegative eigenvalues, can also recover the community structure [33]. Figure 13.16(a) shows the ordered system admittance matrix Y and has purely imaginary values. The

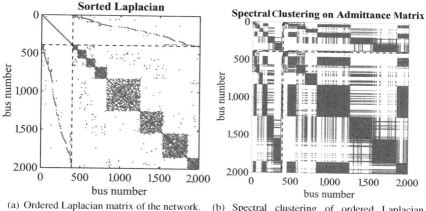

(a) Ordered Laplacian matrix of the network.

(b) Spectral clustering of ordered Laplacian matrix of the network. The k-means optimal objective value $F^* = 1.8921$.

Figure 13.16 Spectral clustering with PMU data.

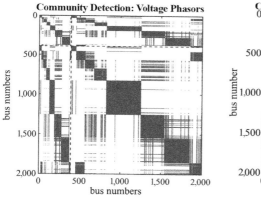
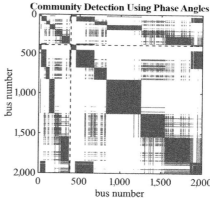

(a) Community detection using complex voltage phasors v_t of the network. The k-means optimal objective value is $\tilde{F}_y = 2.3369$.

(b) Community detection using phase angles, θ of the network. The k-means optimal objective value is $\tilde{F}_\theta = 2.9233$.

Figure 13.17 Blind community detection using PMU data.

top left corner represents Y_{gg} and the bottom right $Y_{\ell\ell}$. The community structure results in a block-diagonal Y. Each community signifies a population cluster [35].

As ground truth, we performed spectral clustering on the eigenvectors **U** of Laplacian Y wherein the goal was to recover K communities. The support of recovered adjacency matrix for the clusters is shown in Figure 13.16(b). Then, we utilized algorithm 13.1 and clustered the eigenvectors of the data covariance matrix C_v where voltage phasor measurements were used. The corresponding support of adjacency matrix is given in Figure 13.17(a). Also, to test the efficacy of DC approximation, eigenvectors from the covariance matrix formed by phase angles θ_t, C_θ were clustered and the results are shown in Figure 13.17(b). For both of the preceding techniques for community detection, the k-means optimal objective values were $\tilde{F}_y = 2.3369$ and $\tilde{F}_\theta = 2.9233$ when using complex voltage phasors v_t and phase angle θ, respectively. The respective difference from optimal objective value for spectral clustering and and the community detection techniques were found to be $\tilde{F}_y - F^* = 0.4448$ and $\tilde{F}_\theta - F^* = 1.0312$. This is indicative of the better performance using voltage phasor measurements. However, the DC approximation roughly holds for this data set and, therefore, the performance is not bad either.

13.7 Conclusion

The basic tenet of this chapter is that PMU signals can be viewed as the output of a low-pass graph filter with a low-rank excitation with a GSO that is the admittance matrix of the system. This fact serves as the principle to derive a number of results that link the data to the underlying grid characteristics in a mathematically rigorous manner, explaining empirical findings that confirm the theory discussed.

References

[1] C. P. Steinmetz, "Complex quantities and their use in electrical engineering," in *Proceedings of the International Electrical Congress*, pp. 33–74, 1893.

[2] A. Monticelli, "Electric power system state estimation," *Proceedings of the IEEE*, vol. 88, no. 2, pp. 262–282, 2000.

[3] A. Gomez-Exposito and A. Abur, *Power System State Estimation: Theory and Implementation*. New York and Basel: CRC Press, 2004.

[4] A. G. Phadke, J. S. Thorp, and M. G. Adamiak, "A new measurement technique for tracking voltage phasors, local system frequency, and rate of change of frequency," *IEEE Transactions on Power Apparatus and Systems*, no. 5, pp. 1025–1038, 1983.

[5] K. Martin, G. Brunello, M. Adamiak, G. Antonova, M. Begovic, G. Benmouyal, P. Bui, H. Falk, V. Gharpure, A. Goldstein et al., "An overview of the IEEE standard c37. 118.2 – Synchrophasor data transfer for power systems," *IEEE Transactions on Smart Grid*, vol. 5, no. 4, pp. 1980–1984, 2014.

[6] A. G. Phadke, "Synchronized phasor measurements – A historical overview," in *IEEE/PES Transmission and Distribution Conference and Exhibition*, vol. 1. IEEE, pp. 476–479, 2002.

[7] A. Phadke and J. Thorp, "History and applications of phasor measurements," in *2006 IEEE PES Power Systems Conference and Exposition*. IEEE, pp. 331–335, 2006.

[8] A. Monti, C. Muscas, and F. Ponci, *Phasor Measurement Units and Wide Area Monitoring Systems*. Cambridge, MA: Academic Press, 2016.

[9] E. Farantatos and B. Amidan, "Data mining techniques and tools for synchrophasor data," NASPI Engineering Analysis Task Team (EATT), Tech. Rep., 2019.

[10] J. Liu, J. Tang, F. Ponci, A. Monti, C. Muscas, and P. A. Pegoraro, "Trade-offs in PMU deployment for state estimation in active distribution grids," *IEEE Transactions on Smart Grid*, vol. 3, no. 2, pp. 915–924, 2012.

[11] P. Gao, M. Wang, S. G. Ghiocel, J. H. Chow, B. Fardanesh, and G. Stefopoulos, "Missing data recovery by exploiting low-dimensionality in power system synchrophasor measurements," *IEEE Transactions on Power Systems*, vol. 31, no. 2, pp. 1006–1013, 2016.

[12] J. E. Tate, "Preprocessing and Golomb–Rice encoding for lossless compression of phasor angle data," *IEEE Transactions on Smart Grid*, vol. 7, no. 2, 2016.

[13] M. Wang, "Data quality management of synchrophasor data in power systems by exploiting low-dimensional models," in *2017 51st Annual Conference on Information Sciences and Systems (CISS)*, 2017.

[14] P. Gao, M. Wang, S. G. Ghiocel, and J. H. Chow, "Modeless reconstruction of missing synchrophasor measurements," in *2014 IEEE PES General Meeting – Conference & Exposition*, 2014.

[15] M. Jamei, A. Scaglione, C. Roberts, E. Stewart, S. Peisert, C. McParland, and A. McEachern, "Anomaly detection using optimally-placed PMU sensors in distribution grids," *IEEE Transactions on Power Systems*, vol. 33, no. 4, pp. 3611–3622, July 2018.

[16] L. Xie, Y. Chen, and P. R. Kumar, "Dimensionality reduction of synchrophasor data for early event detection: Linearized analysis," *IEEE Transactions on Power Systems*, vol. 29, no. 6, pp. 2784–2794, 2014.

[17] W. Li, M. Wang, and J. H. Chow, "Real-time event identification through low-dimensional subspace characterization of high-dimensional synchrophasor data," *IEEE Transactions on Power Systems*, vol. 33, no. 5, Sept. 2018.

[18] A. Sandryhaila and J. M. F. Moura, "Discrete signal processing on graphs: Frequency analysis," *IEEE Transactions on Signal Processing*, vol. 62, no. 12, June 2014.

[19] N. Tremblay, P. Gonçalves, and P. Borgnat, "Design of graph filters and filterbanks," in *Cooperative and Graph Signal Processing Principles and Applications*. Cambridge, MA: Elsevier, pp. 299–324, 2018.

[20] A. Sandryhaila and J. M. Moura, "Discrete signal processing on graphs," *IEEE Transactions on Signal Processing*, vol. 61, no. 7, Apr. 2013.

[21] M. Jamei, A. Scaglione, and S. Peisert, "Low-resolution fault localization using phasor measurement units with community detection," in *2018 IEEE International Conference on Smart Grid Communications*, Oct. 2018.

[22] M. Jamei, R. Ramakrishna, T. Tesfay, R. Gentz, C. Roberts, A. Scaglione and S. Peisert, "Phasor Measurement Units Optimal Placement and Performance Limits for Fault Localization," *IEEE Journal on Selected Areas in Communications*, vol. 38, no. 1, pp 180–192, January 2020.

[23] Z. Wang, A. Scaglione, and R. Thomas, "Generating statistically correct random topologies for testing smart grid communication and control networks," *IEEE Transactions on Power Systems*, vol. 1, no. 1, pp. 28–39, 2010.

[24] T. Ishizaki, A. Chakrabortty, and J.-I. Imura, "Graph-theoretic analysis of power systems," *Proceedings of the IEEE*, vol. 106, no. 5, pp. 931–952, May 2018.

[25] N. Dahal, R. L. King, and V. Madani, "Online dimension reduction of synchrophasor data," in *IEEE PES Transmission and Distribution Conference and Exposition (T & D)*, 2012.

[26] M. Tsitsvero, S. Barbarossa, and P. Di Lorenzo, "Signals on graphs: Uncertainty principle and sampling," *IEEE Transactions on Signal Processing*, vol. 64, no. 18, pp. 4845–4860, 2016.

[27] H.-T. Wai, S. Segarra, A. E. Ozdaglar, A. Scaglione, and A. Jadbabaie, "Blind community detection from low-rank excitations of a graph filter," *IEEE Transactions on Signal Processing*, vol. 68, pp. 436–451, 2020.

[28] G. H. Golub and C. F. V. Loan, *Matrix Computations*, 3rd ed. Baltimore, MD: Johns Hopkins University Press, 1996.

[29] R. A. Horn and C. R. Johnson, *Matrix Analysis*. Cambridge: Cambridge University Press, 1990.

[30] S. Mallat, *A wavelet Tour of Signal Processing: The Sparse Way*. Burlington, MA: Academic Press, 2008.

[31] R. Singh, A. Chakraborty, and B. Manoj, "Graph fourier transform based on direcetd Laplacian," in *International Conference on Signal Processing and Communications (SPCOM)*, 2016.

[32] F. Grassi, A. Loukas, N. Perraudin, and B. Ricaud, "A time-vertex signal processing framework: Scalable processing and meaningful representations for time-series on graphs," *IEEE Transactions on Signal Processing*, vol. 66, no. 3, pp. 817–829, Feb. 2018.

[33] R. Ramakrishna and A. Scaglione, "On modeling voltage phasor measurements as graph signals," in *IEEE Data Science Workshop (DSW) 2019*, June 2019.

[34] A. B. Birchfield, T. Xu, K. M. Gegner, K. S. Shetye, and T. J. Overbye, "Grid structural characteristics as validation criteria for synthetic networks," *IEEE Transactions on Power Systems*, vol. 32, no. 4, pp. 3258–3265, July 2017.

[35] A. B. Birchfield, T. Xu, and T. J. Overbye, "Power flow convergence and reactive power planning in the creation of large synthetic grids," *IEEE Transactions on Power Systems*, vol. 33, no. 6, 2018.

[36] S. Maslennikov, B. Wang, Q. Zhang, F. Ma, X. Luo, K. Sun, and E. Litvinov, "A test cases library for methods locating the sources of sustained oscillations," in *IEEE PES General Meeting*, Boston, MA, 2016.

[37] P. D. Lorenzo, S. Barbarossa, and P. Banelli, *Cooperative and Graph Signal Processing*. Cambridge, MA: Elsevier, ch. 9, June 2018.

[38] R. Ramakrishna and A. Scaglione, "Grid-graph signal processing (Grid-GSP): A graph signal processing framework for the power grid" (in review). *IEEE Transactions on Signal Processing*, 2020.

[39] S. Bi and Y. J. Zhang, "Graphical methods for defense against false-data injection attacks on power system state estimation," *IEEE Transactions on Smart Grid*, vol. 5, no. 3, May 2014.

[40] J. Hao, R. J. Piechocki, D. Kaleshi, W. H. Chin, and Z. Fan, "Sparse malicious false data injection attacks and defense mechanisms in smart grids," *IEEE Transactions on Industrial Informatics*, vol. 11, no. 5, Oct. 2015.

[41] M. Liao, D. Shi, Z. Yu, Z. Yi, Z. Wang, and Y. Xiang, "An alternating direction method of multipliers based approach for PMU data recovery," *IEEE Transactions on Smart Grid*, 2018.

[42] M. Wang and et al., "A low-rank matrix approach for the analysis of large amounts of power system synchrophasor data," in *2015 48th Hawaii International Conference on System Sciences*, 2015.

[43] S. Zhang, Y. Hao, M. Wang, and J. H. Chow, "Multi-channel Hankel matrix completion through nonconvex optimization," *IEEE Journal of Selected Topics in Signal Processing*, vol. 12, no. 4, pp. 617–632, 2018.

[44] S. M. Brahma, "Fault location in power distribution system with penetration of distributed generation," *IEEE Transactions on Power Delivery*, vol. 26, no. 3, pp. 1545–1553, 2011.

[45] I. Dzafic, R. A. Jabr, S. Henselmeyer, and T. Donlagic, "Fault location in distribution networks through graph marking," *IEEE Transactions on Smart Grid*, vol. 9, no. 2, pp. 1345–1353, 2018.

[46] H. Jiang, J. J. Zhang, W. Gao, and Z. Wu, "Fault detection, identification, and location in smart grid based on data-driven computational methods," *IEEE Transactions on Smart Grid*, vol. 5, no. 6, pp. 2947–2956, 2014.

[47] D. Donoho and M. Elad, "Optimally sparse representation in general (nonorthogonal) dictionaries via ℓ_1 minimization," *Proceedings of the National Academy of Sciences*, 2003.

[48] W. H. Kersting, "Radial distribution test feeders," in *Power Engineering Society Winter Meeting, 2001. IEEE*, vol. 2. pp. 908–912, 2001.

[49] U. Von Luxburg, "A tutorial on spectral clustering," *Statistics and Computing, Springer*, vol. 17, no. 4, pp. 395–416, 2007.

[50] A. Y. Ng, M. I. Jordan, and Y. Weiss, "On spectral clustering: Analysis and an algorithm," in *Advances in Neural Information Processing Systems*, Cambridge, MA: MIT Press, pp. 849–856, 2002.

14 A Sparse Representation Approach for Anomaly Identification

Hao Zhu and Chen Chen

14.1 Introduction

The rise of synchrophasor technology in the last decades offers unprecedented opportunities for enhanced situational awareness in power systems [1]. Nowadays, thousands of phasor measurement units (PMUs) have been deployed to the wide-area transmission grids worldwide. The PMUs can complement the legacy supervisory control and data acquisition (SCADA) systems at power system control centers, with synchronized samples of the voltage and current phasor data at the subsecond rates. Thanks to the high spatial and temporal resolution, synchrophaosr measurements have enabled fast and accurate identification of anomaly disturbance events in today's power grids.

This chapter aspires to provide an overview of tackling the power system anomaly identification problem through the introduction of a sparse representation model. Though not exhaustive by any means, we consider here three types of anomaly events, namely power line outages, fault events, and phase imbalance. Interestingly for all three types of anomalies, the synchrophasor measurement data can be related to all candidate events using a general sparse representation model. The latter have allowed for the development of efficient solution techniques using compressed sensing algorithms, such as the orthogonal matching pursuit (OMP) and Lasso regression. Both methods enjoy tractable computational complexity and are very suitable for real-time implementations. Under this setting, the impact of phasor measurement unit (PMU) meter placement and synchrophasor data uncertainty is further investigated, focusing on the line outage identification problem.

Notation: Lower- (upper-)case boldface letters denote column vectors (matrices), and calligraphic letters stand for sets. The conjugate of a complex-valued object (scalar, vector, or matrix) x is denoted by x^*; $\Re\{x\}$ and $\Im\{x\}$ are its real and imaginary parts, and $j := \sqrt{-1}$. Superscripts T and H stand for transpose and conjugate-transpose, respectively. A diagonal matrix having vector $\mathbf{x}$ on its main diagonal is denoted by $dg(\mathbf{x})$; whereas, the vector of diagonal entries of $\mathbf{X}$ is $dg(\mathbf{X})$. The notation $\mathcal{N}(\boldsymbol{\mu}, \boldsymbol{\Sigma})$ represents the Gaussian distribution with mean $\boldsymbol{\mu}$ and covariance matrix $\boldsymbol{\Sigma}$.

14.2 Power Grid Modeling

This section introduces notation and briefly reviews the power flow equations; for detailed exposition see, e.g., [2, 3], and references therein. A power system can be represented by the graph $\mathcal{G} = (\mathcal{B}, \mathcal{L})$, where the node set $\mathcal{B}$ comprises its N_b buses, and the edge set $\mathcal{L}$ its N_l transmission lines. Here, we will introduce both the network Ohm's law for ac systems and the linearized "dc" power flow model.

A transmission line $(n, k) \in \mathcal{L}$ running across buses $n, k \in \mathcal{B}$ is modeled by its total series admittance $y_{nk} = g_{nk} + jb_{nk}$. If $\mathcal{V}_n$ is the complex voltage at bus n, the current $\mathcal{I}_{nk}$ flowing from bus m to bus n over line (m, n) is

$$\mathcal{I}_{nk} = y_{nk}(\mathcal{V}_n - \mathcal{V}_k). \tag{14.1}$$

The current $\mathcal{I}_{kn}$ coming from the other end of the line can be expressed symmetrically. For simplicity, the line shunt susceptance and the tap ratio for transformers are omitted here. Kirchhoff's current law dictates that the current injected into bus n is $\mathcal{I}_n = \sum_{n \in \mathcal{B}_n} \mathcal{I}_{nk}$, where $\mathcal{B}_n$ denotes the set of buses directly connected to bus n. If vector $\mathbf{i} \in \mathbb{C}^{N_b}$ collects all nodal currents, and $\mathbf{v} \in \mathbb{C}^{N_b}$ all nodal voltages, the two vectors are linearly related through the bus admittance matrix, or *Ybus* matrix $\mathbf{Y}$, as

$$\mathbf{i} = \mathbf{Y}\mathbf{v}. \tag{14.2}$$

Similar to (14.2), line currents can be stacked in the $(2N_l)$-length vector $\mathbf{i}_f$, and expressed as a linear function of nodal voltages

$$\mathbf{i}_f = \mathbf{Y}_f \mathbf{v} \tag{14.3}$$

for some properly defined $2N_l \times N_b$ complex matrix $\mathbf{Y}_f$ [cf. (14.1)].

The complex power injected into bus n will be denoted by $\mathcal{S}_n := P_n + jQ_n$. Because by definition $\mathcal{S}_n = \mathcal{V}_n \mathcal{I}_n^*$, the vector of complex power injections $\mathbf{s} = \mathbf{p} + j\mathbf{q}$ can be expressed as

$$\mathbf{s} = dg(\mathbf{v})\mathbf{i}^* = dg(\mathbf{v})\mathbf{Y}^*\mathbf{v}^*. \tag{14.4}$$

The power flowing from bus n to bus k over line (n, k) is $\mathcal{S}_{nk} = \mathcal{V}_n \mathcal{I}_{nk}^*$. Clearly, complex power injections/flows are quadratic functions of the voltage phasor $\mathbf{v}$. Expressing (14.4) explicitly using the polar form of $\mathcal{V}_n = V_n e^{j\theta_n}$ leads to the well-known *nonlinear* "ac" power flow models.

The linearized approximation between real power $\mathbf{p}$ and all voltage phase angles in vector $\theta \in \mathbb{R}^{N_b}$ has been widely adopted, known as the *linear* "dc" power flow model. Roughly speaking, if the system is lossless and voltage regulation is perfect everywhere, one can express the admittance $y_{nk} = jb_{nk}$ by setting the series resistance to 0 while keep $V_n = 1$ as constant. This way, the real power flowing from bus n to bus k over line (n, k) is approximated by

$$P_{nk} = -b_{nk}(\theta_n - \theta_k). \tag{14.5}$$

Similar to the currents, the power injected to bus n balances all the line flows originating from it; that is, $P_n = \sum_{k \in \mathcal{B}_n} P_{nk}$. Hence, the linear dc power flow asserts that

$$\mathbf{p} = \mathbf{B}\boldsymbol{\theta} \qquad (14.6)$$

where $\mathbf{B}$ is the so-called *Bbus* matrix.

Note that matrices $\mathbf{Y}$ and $\mathbf{B}$ share very similar structure, as they are uniquely determined by the line series parameters and the graph topology information in $\mathcal{L}$. To formulate this topological effect concretely, the two matrices can be viewed as the weighted Laplacian matrix of the graph $\mathcal{G}$. To this end, consider the $N_b \times N_l$ bus-line incidence matrix $\mathbf{M}$, see e.g., [4, pg. 56], formed by column vectors $\{\mathbf{m}_\ell\}_{\ell=1}^{N_l}$ of length N_b. With subscript ℓ corresponding to the line (n,k), the column $\mathbf{m}_\ell$ has all its entries equal to 0 except the nth and kth, which take the value 1 and -1, respectively. Clearly, the rank-one matrix $\mathbf{m}_\ell \mathbf{m}_\ell^T$ has nonzero entries on its diagonal positions (n,n) and (k,k), which equal 1, and also on its (n,k) and (k,n) positions, which equal -1. Thus, when summing such rank-one matrices for distinct lines, those having common m or n are superimposed on the diagonal, while those off the diagonal still have values equal to -1. This property establishes readily the following representation of the network topology matrix

$$\mathbf{B} = \sum_{\ell=1}^{N_l} \left(-b_\ell \mathbf{m}_\ell \mathbf{m}_\ell^T\right) = \mathbf{M}\mathbf{D}_b\mathbf{M}^T \qquad (14.7)$$

where the diagonal matrix $\mathbf{D}_b$ has its ℓth diagonal entry $-b_\ell$ equal to b_{nk}, if ℓ corresponds to line (n,k). One can similarly express Ybus matrix $\mathbf{Y}$ as in (14.7) by replacing the weights using y_ℓ's.

As graph Laplacian, matrix $\mathbf{B}$ is symmetric, and has rank $(N-1)$ if the power network is connected because its null space is only spanned by $\mathbf{1}$; see e.g., [4, pg. 469]. However, this is typically not an issue for $\mathbf{Y}$ as the line shunt susceptance omitted earlier on would induce additional diagonal terms and thus makes the Ybus matrix of full rank. Rank deficiency of $\mathbf{B}$ gives rise to multiple solutions for $\boldsymbol{\theta}$ in (14.6). This ambiguity issue can be fixed by choosing one generator bus as the reference bus with its phase angle set to zero. Accordingly, all other θ_n's denote their differences relative to the reference bus; see e.g., [3, pg. 76]. Clearly, with the reference bus convention, the $(N_b - 1) \times (N_b - 1)$ matrix $\mathbf{B}$ has full rank, and can be formed by the reduced $\mathbf{M}$ matrix after removing the corresponding row of the incidence matrix.

14.3 Sparse Representations of Anomaly Events

Anomaly events arise in large-scale power systems due to changes of network topology or connectivity, grounding fault events, or imperfect phase balancing. For each type of anomaly events, this section develops a sparse overcomplete representation that can capture all possible scenarios of simultaneous anomaly events, with the nonzero entries of unknown vector indicating occurred events. This overarching sparse model will allow for efficient identification of anomaly events from real-time synchrophasor data.

To formalize the measurement model, for given locations of PMUs we denote $\mathcal{B}_I \in \mathcal{B}$ as the subset of buses with voltage phasor data, and $\mathcal{L}_J \in \mathcal{L}$ as that of lines with current phasor data.

14.3.1 Power Line Outages

Identifying outages and generally changes of lines is particularly critical for constructing the correct network models. Suppose that due to changes in the grid, e.g., cascading failures in an early stage, several outages occur on lines, collected in the subset $\tilde{\mathcal{L}} \subseteq \mathcal{L}$. Line outages on the transmission network yield the post-event graph $(\mathcal{B}, \mathcal{L}')$, with $\mathcal{L}' := \mathcal{L} \setminus \tilde{\mathcal{L}}$.

To analyze the post-event grid, it is typically assumed that quasi-steady state is achieved quickly following the line outages and that no islanding is formed in $\mathcal{L}'$; see e.g., [5–7]. Under these assumptions, the linear dc power model for the post-event system is given by $\mathbf{p}' = \mathbf{p} + \eta = \mathbf{B}'\theta'$, where $\mathbf{B}'$ is the Bbus matrix of $(\mathcal{B}, \mathcal{L}')$, and η accounts for the small perturbations due to variations in loads and generation. The latter can usually be modeled as zero-mean (possibly Gaussian) vector with covariance matrix $\sigma_\eta^2 \mathbf{I}$; see e.g., [8]. Note that the difference $\tilde{\mathbf{B}} := \mathbf{B} - \mathbf{B}'$ can be expressed as the weighted Laplacian for the outage lines in $\tilde{\mathcal{L}}$ [cf. (14.7)]

$$\tilde{\mathbf{B}} = \sum_{\ell \in \tilde{\mathcal{L}}} \left(-b_\ell \mathbf{m}_\ell \mathbf{m}_\ell^T \right). \tag{14.8}$$

By defining the phase angle difference $\tilde{\theta} := \theta' - \theta$, one can write

$$\mathbf{B}\tilde{\theta} = (\tilde{\mathbf{B}} + \mathbf{B}')\theta' - \mathbf{B}\theta = \tilde{\mathbf{B}}\theta' + \mathbf{p}' - \mathbf{p}$$
$$= \sum_{\ell \in \tilde{\mathcal{L}}} s_\ell \mathbf{m}_\ell + \eta \tag{14.9}$$

where $s_\ell := -b_\ell \mathbf{m}_\ell^T \theta'$, $\forall \ell \in \tilde{\mathcal{L}}$. Note that instantaneous voltage phase angles can be measured in real time at the subset of buses $\mathcal{B}_I \in \mathcal{B}$. Solving (14.9) with respect to (wrt) $\tilde{\theta}$, and extracting the rows corresponding to the buses in $\mathcal{B}_I$ leads to the following expression

$$\tilde{\theta}_I = [\mathbf{B}^{-1}]_I \left(\sum_{\ell \in \tilde{\mathcal{L}}} s_\ell \mathbf{m}_\ell \right) + [\mathbf{B}^{-1}]_I \eta \tag{14.10}$$

where $[\mathbf{B}^{-1}]_I$ is constructed from the rows corresponding to buses in $\mathcal{B}_I$, of the inverse matrix $\mathbf{B}^{-1}$.

We present an overcomplete representation capturing all possible line outages to tackle the combinatorial complexity in the model (14.10). First, notice that the inversion in (14.10) introduces colored perturbation. To account for this, consider the compact singular value decomposition (SVD) of the fat matrix $[\mathbf{B}^{-1}]_I = \mathbf{U}_I \mathbf{\Sigma}_I \mathbf{V}_I^T$, where the square diagonal matrix $\mathbf{\Sigma}_I$ comprises all its nonzero singular values. Upon defining $\mathbf{y} := \mathbf{\Sigma}_I^{-1} \mathbf{U}_I^T \tilde{\theta}_I$, the data model in (14.10) reduces to a sparse linear regression one with unknown regression coefficients contained in the $L \times 1$ vector $\mathbf{s}$, whose ℓth entry equals s_ℓ, if $\ell \in \tilde{\mathcal{L}}$, and 0 otherwise. Thus, the additive model in (14.10) can be written as

$$\mathbf{y} = \mathbf{V}_I^T \left(\sum_{\ell \in \tilde{\mathcal{L}}} s_\ell \, \mathbf{m}_\ell \right) + \mathbf{V}_I^T \boldsymbol{\eta} = \mathbf{V}_I^T \mathbf{M} \mathbf{s} + \mathbf{V}_I^T \boldsymbol{\eta}$$
$$= \mathbf{A}\mathbf{s} + \mathbf{V}_I^T \boldsymbol{\eta} \qquad (14.11)$$

where the transformed incidence matrix $\mathbf{A} := \mathbf{V}_I^T \mathbf{M}$, now viewed as a regression matrix, captures all transmission lines in $\mathcal{L}$.

14.3.2 Fault Events

Accurately and timely locating fault events from real time can accelerate the restoration process of faulted lines without costly manual line examinations. Common line faults in power systems include line-to-ground or line-to-line short circuits [9, 10]. Each of these fault events can be modeled by additional current sources connected to the buses adjacent to the faulted line. To this end, consider a line-to-ground fault on line $(n, k) \in \mathcal{L}$ that distance r away from bus n. Hence, the total series impedance z_{nk} (as inverse of y_{nk}) can be separated into the two parts. The fault can be modeled by a current source of $\mathcal{I}^{lg}$ to the ground, and thus the voltage difference follows from the Ohm's law

$$\mathcal{V}_n - \mathcal{V}_k = r z_{nk} \mathcal{I}_{nk} + (1-r) z_{nk} (-\mathcal{I}_{kn}) = r z_{nk} \mathcal{I}_{nk} + (1-r) z_{nk} (\mathcal{I}_{nk} - \mathcal{I}^{lg}) \quad (14.12)$$

where the last equality is due to Kirchhoff's current law. Accordingly, the line current from both ends are updated as [cf. (14.1)]

$$\mathcal{I}_{nk} = y_{nk}(\mathcal{V}_n - \mathcal{V}_k) + (1-r)\mathcal{I}^{lg} \qquad (14.13)$$
$$\mathcal{I}_{kn} = y_{nk}(\mathcal{V}_k - \mathcal{V}_n) + r\mathcal{I}^{lg} \qquad (14.14)$$

Therefore, a line-to-ground fault on line (n, k) can be represented by two fictitious current sources connected to both end buses, respectively. Line-to-line faults can be modeled similarly by using two current sources. They are omitted here for simplicity; please refer to [10] for the details.

Suppose line-to-ground faults occur on several lines, collected in the subset $\tilde{\mathcal{L}}$, each represented by the pair $(\mathcal{I}_\ell^{lg}, r_\ell)$, $\forall \ell \in \tilde{\mathcal{L}}$. Hence, the post-event network Ohm's law gives to

$$\mathbf{i}' = \mathbf{Y}\mathbf{v}' = \mathbf{Y}\mathbf{v} + \sum_{\ell \in \tilde{\mathcal{L}}} \mathbf{r}_\ell \mathcal{I}_\ell^{lg}. \qquad (14.15)$$

where vector $\mathbf{r}_\ell$ has similar sparse pattern to $\mathbf{m}_\ell$, with two nonzero entries equal to $(1 - r_\ell)$ and r_ℓ, respectively. Note that the last summand term in (14.15) represents the fictitious current sources due to the faults. Similar to (14.11), one can construct an $L \times 1$ vector $\mathbf{i}^{lg}$, whose ℓth entry equals $\mathcal{I}_\ell^{lg}$, if $\ell \in \tilde{\mathcal{L}}$, and 0 otherwise. Thus, the voltage difference $\tilde{\mathbf{v}} := \mathbf{v}' - \mathbf{v}$ has the following sparse linear representation

$$\tilde{\mathbf{v}} = \mathbf{Y}^{-1} \left(\sum_{\ell \in \tilde{\mathcal{L}}} \mathbf{r}_\ell \mathcal{I}_\ell^{lg} \right) = \mathbf{Y}^{-1} \mathbf{R} \, \mathbf{i}^{lg} \qquad (14.16)$$

where matrix $\mathbf{R}$ collects all $\mathbf{r}_\ell$'s as column vectors. Note that the difference of all line current phasors $\tilde{\mathbf{i}}_f$ is linearly related to $\tilde{\mathbf{v}}$ due to (14.3). Hence, the linear measurement model for selected entries in $[\tilde{\mathbf{v}}; \tilde{\mathbf{i}}_f]$ corresponding to $\mathcal{B}_I$ and $\mathcal{L}_J$, as given by

$$\mathbf{y} = \mathbf{AR}\,\mathbf{i}^{lg} + \eta. \tag{14.17}$$

where the perturbation η represents the modeling mismatch and measurement error. Slightly different from (14.11), the input sparse faults to model (14.17) have certain structure constraints, reflected by the columns of matrix $\mathbf{R}$ that are sparse with unknown parameter r_ℓ. Hence, the solution method will be slightly more complicated to meet such structural constraints, as detailed in Section 14.4.

14.3.3 Phase Imbalance

Phase imbalance in three-phase power systems may cause serious contingencies such as load shedding, insulation degradation, and generator outages; see e.g., [11, 12]. During normal operations, the three-phase voltage waveform per bus is assumed purely sinusoidal of the same frequency ($\omega_o + \Delta$) where ω_o is the nominal grid frequency and Δ stands for the frequency deviation. Unbalanced system settings due to load currents can cause the per-phase voltage magnitude to differ and voltage phase angle to be imperfectly separated. Under this setting, all three-phase voltages at bus n can be represented by the positive and negative sequence phasors, C_n^+ and C_n^-, respectively. For a balanced system, C_n^+ is exactly the complex voltage phasor $\mathcal{V}_n$ that we have defined so far, and $C_n^- = 0$. Due to phase imbalance, the negative sequence phasor C_n^+ becomes nonzero.

For reduced communication complexity, the PMU at bus n can report the so-termed positive sequence voltage signal through symmetrical component transformation, given by

$$v_n^+[t] = e^{j\gamma \frac{\omega_o + \Delta}{\omega_o} t} C_n^+ + e^{-j\gamma \frac{\omega_o + \Delta}{\omega_o} t} C_n^- \tag{14.18}$$

where ω_o and Δ are the respective nominal and deviation values for frequency, while $\gamma = 2\pi/T$ with the sampling rate at T times per cycle. Similarly, the three-phase current signals are also transformed into the positive sequence representation $\{i_{nk}^+[t]\}$ before being sent to the control center. Again, recalling the linearity in (14.3), the measurement model for selected entries in $[\tilde{\mathbf{v}}^+[t]; \tilde{\mathbf{i}}_f^+[t]]$ corresponding to $\mathcal{B}_I$ and $\mathcal{L}_J$ is given by

$$\mathbf{y}[t] = e^{j\gamma \frac{\omega_o + \Delta}{\omega_o} t} \mathbf{Ac}^+ + e^{-j\gamma \frac{\omega_o + \Delta}{\omega_o} t} \mathbf{Ac}^- + \eta[t], \tag{14.19}$$

while $\mathbf{c}^+$ ($\mathbf{c}^-$) collects all the positive (conjugate negative) sequence phasors per bus, and vector $\eta[t]$ is stands for noise perturbation. Assuming the majority of buses are perfectly balanced, we have vector $\mathbf{c}^-$ to be naturally sparse as its elements related to the balanced buses are equal to zero. Hence, the problem of determining the imbalance condition of each of the N_B buses again becomes a sparse regression one to identifying the nonzero entries of $\mathbf{c}^-$.

The model (14.19) has both the full vector $\mathbf{c}^+$ and sparse one $\mathbf{c}^-$ as unknown coefficients. Interestingly, one can separate the problem of estimating both vectors by transforming the measurement vector; see details in [12]. With frequency deviation Δ known or accurately estimated from, e.g., phase-lock loop (PLL), one can determine the positive and negative sequence components of $\{\bar{y}[t]\}$, respectively

$$\mathbf{y}^+ := \frac{1}{T}\sum_t e^{-j\gamma\frac{\omega_0+\Delta}{\omega_0}t}\mathbf{y}[t], \quad \mathbf{y}^- := \frac{1}{T}\sum_t e^{j\gamma\frac{\omega_0+\Delta}{\omega_0}t}\mathbf{y}[t]. \qquad (14.20)$$

Moreover, by defining the complex number $f := \frac{1}{T}\sum_t e^{-j2\gamma\frac{\omega_0+\Delta}{\omega_0}t}$, the measurement model (14.19) can be projected to form

$$\tilde{\mathbf{y}} := \frac{\mathbf{y}^- - f^*\mathbf{A}\mathbf{A}^\dagger\mathbf{y}^+}{1-|f|^2} = \mathbf{A}\mathbf{c}^- + \eta, \qquad (14.21)$$

which is a sparse linear model with only $\mathbf{c}^-$ as the unknown. Hence, we can develop efficient solvers for (14.21) similarly to (14.11).

To sum up, all three types of anomalies events can be captured by a sparse linear model that encompasses simultaneous occurrence of multiple events. To make the presentation more concise, the rest of chapter will focus on the sparse line outage model (14.11), while introducing relevant modifications for the structured fault model in (14.17). Note that if the location for nonzero entries in $\mathbf{s}$, or $\mathcal{L}_o$, is already known, then finding $\mathbf{s}$ can be easily solved by the following least-squares (LS) problem

$$\min_{\{s_\ell\}_{\ell\in\mathcal{L}_o}} \left\|\mathbf{y} - \left(\sum_{\ell\in\mathcal{L}_o}\mathbf{a}_\ell s_\ell\right)\right\|_2^2. \qquad (14.22)$$

However, the main challenge is to efficiently identify the nonzero entries in the sparse vector, which is the subject of the ensuing section.

14.4 Efficient Solvers Using Compressed Sensing

This section presents the solution techniques that can efficiently identifying the anomaly events based on the class of sparse overcomplete representation developed in Section 14.3. In essence, the overcomplete representation in these linear models allows one to cast the problem as one of estimating the *sparse* vector therein. The key *premise* is that the number of anomaly events is a small fraction of the total number of possibilities. This holds even for multiple simultaneous anomalies such as line outages occurring during cascading failures, at least in the early stage when only a small number of lines start to fail. The same premise holds for the case of faulted lines or phase imbalanced buses. Under this sparsity constraint, the unknown vector in each model has only a few nonzero entries. In turn, estimating the unknown by exploiting its sparsity falls under the realm of sparse signal reconstruction approaches.

Reconstruction of sparse signals has become very popular after the emergence of compressive sampling (CS) theory; see e.g., [13–15] and references therein. Efficient approaches of sparse signal recovery can be broadly grouped into two categories: those relying on greedy approximation schemes, and those minimizing the ℓ_1-norm of the sparse vector. The first category is rooted on the matching pursuit (MP) algorithm for signal approximation [16]. MP has undergone improvements that led to the orthogonal matching pursuit (OMP) algorithm [13], that draws its popularity from its computational simplicity and guaranteed performance. Those greedy algorithms are especially easy to implement, mostly looking "heuristic" in selecting the nonzero entries. Albeit heuristic, greedy schemes such as the OMP have been shown capable of recovering the optimal representation of an exactly sparse signal provided that the regression matrix (a.k.a. dictionary) satisfies the so-termed "coherence" conditions [13]. The second category relies on the ℓ_1-norm of the wanted vector offering a convex relaxation of the ℓ_0-norm based minimization problems, which are known to be NP-hard but can yield the most sparse solution to an underdetermined linear system of equations; see e.g., [14, 15]. Popular solvers of the resultant optimization problems are variants of the basis pursuit (BP) algorithm [15], or the Lasso [17]. Albeit a relaxation, this approach has been shown to exhibit guaranteed performance of recovering sparse signal vectors; see e.g., [14] and references therein. As a convex quadratic program (QP), the global minimizer of the ℓ_1-norm regularized Lasso problem can be efficiently obtained under various sparsity levels using coordinate descent (CD) iterations [18].

14.4.1 Greedy Anomaly Identification Using OMP

Given $\mathbf{y}$ and $\mathbf{A}$, the greedy OMP algorithm finds sparse solutions to (14.11) with respect to a fixed number of nonzero entries (a.k.a. sparsity level) κ of the sought solution $\hat{\mathbf{s}}^\kappa$ to (14.11). The number κ can be directly related to the maximum number of anamolies, and its choice can be determined for each type of events.

Greedy schemes generally aim to approximate the vector $\mathbf{y}$ by successively selecting columns $\{\mathbf{a}_\ell\}_{\ell=1}^L$ of $\mathbf{A}$ to superimpose with weights given by the wanted nonzero entries of $\mathbf{s}$. The selection criteria may be different but all share the common idea of choosing the next column as the one that correlates "best" with the current approximation error. Updating the sparsity level κ one by one, the approximation error vector per step is given by

$$\mathbf{r}^\kappa := \mathbf{y} - \mathbf{A}\hat{\mathbf{s}}^\kappa, \quad \kappa \geq 1; \quad \mathbf{r}^0 = \mathbf{y}. \tag{14.23}$$

Furthermore, with $\mathcal{L}^\kappa$ denoting the subset of indices corresponding to nonzero entries in $\hat{\mathbf{s}}^\kappa$, once a new column ℓ^κ is chosen at step κ for inclusion in $\mathcal{L}^\kappa$, the ℓ^κth entry of $\hat{\mathbf{s}}^\kappa$ will be nonzero compared to $\hat{\mathbf{s}}^{\kappa-1}$. The orthogonality in OMP is manifested by updating the estimate $\hat{\mathbf{s}}^\kappa$ through LS fitting of $\mathbf{y}$ using all the columns in $\mathcal{L}^\kappa$, such that $\mathbf{r}^\kappa$ in (14.23) becomes orthogonal to all the chosen columns in $\mathcal{L}^\kappa$. The OMP-based solution to (14.11) with maximum sparsity level $\kappa_{\max}$ is tabulated as Algorithm 14.1.

Because Algorithm 14.1 chooses the column by comparing its correlation with the residual error vector, it is useful to normalize all columns first, so that comparison is in essence performed based on the correlation coefficient. Also, due to its greedy update, the sequence of solutions $\{\hat{\mathbf{s}}^\kappa\}_{\kappa=1}^{\kappa_{\max}}$ is nested in terms of their nonzero entries, but not necessarily in the exact entry values after the orthogonal LS fitting step. In addition to computational efficiency, numerical tests have demonstrated that Algorithm 14.1

Algorithm 14.1 (OMP): Input $\mathbf{y}$, $\mathbf{A}$, and $\kappa_{\max}$. Output $\{\hat{\mathbf{s}}^\kappa\}_{\kappa=1}^{\kappa_{\max}}$.

Initialize $\mathbf{r}^0 = \mathbf{y}$, and the column subset $\mathcal{L}^0 = \emptyset$.
for $\kappa = 1, \ldots, \kappa_{\max}$ **do**
  Choose $\ell^\kappa \in \arg\max_\ell |\mathbf{a}_\ell^T \mathbf{r}^{\kappa-1}|$ (arbitrarily breaking ties), and update $\mathcal{L}^\kappa := \mathcal{L}^{\kappa-1} \bigcup \{\ell^\kappa\}$.
  Estimate $\hat{\mathbf{s}}^\kappa := \arg\min_\mathbf{s} \|\mathbf{y} - \mathbf{As}\|_2^2$, with its ℓth entry $s_\ell = 0$ fixed for all $\ell \notin \mathcal{L}^\kappa$.
  Update $\mathbf{r}^\kappa := \mathbf{y} - \mathbf{A}\hat{\mathbf{s}}^\kappa$.
end for

---

exhibits very competitive reconstruction performance [7]. Analytically, the latter depends on the so-termed coherence of matrix $\mathbf{A}$ that is defined as

$$\mu := \max_{\ell \neq \ell'} |\mathbf{a}_\ell^T \mathbf{a}_{\ell'}|. \tag{14.24}$$

By adapting [13, thm. B], the following proposition can be established for the present context of identifying anomalies.

PROPOSITION 14.1   *(Exact recovery of OMP) Consider the noise-free case, i.e., $\eta = \mathbf{0}$ in (14.11), and suppose that the actual vector $\mathbf{s}_o$ has its number of nonzero entries L_o satisfying*

$$L_o < (\mu^{-1} + 1)/2. \tag{14.25}$$

Given the vector $\mathbf{y} = \mathbf{As}_o$ and matrix $\mathbf{A}$, the OMP in Algorithm 14.1 is capable of recovering the unknown $\mathbf{s}_o$ exactly, if the greedy steps stop once $\|\mathbf{r}^\kappa\|_2 = 0$.

The greedy OMP procedure in Algorithm 12.1 is directly applicable to the phase imbalance problem based on the projected sparse linear model in (14.22). This was termed as the projected OMP (POMP) method in [12]. Nonetheless, the structured sparse model for fault events in (14.17) has additional constraints, requiring some modifications of the basic OMP procedure. Note that the unknown vector $\mathbf{Ri}^{lg}$ to (14.17) has two nonzero entries for a single line-to-ground short at line (n, k), corresponding to both buses n and k. To account for such structured sparse constraints, a modified greedy method termed as StructMP was proposed in [10] for fault location. Specifically, to compare the fitness of each line ℓ, one needs to compute the mismatch error by searching for the best fault parameters $(r, \mathcal{I}^{lg})$, as given by

$$e^\ell := \min_{0 \leq r \leq 1, \mathcal{I}^{lg} \in \mathbb{C}} \|\mathbf{y} - (1-r)\mathcal{I}^{lg}\mathbf{a}_n - r\mathcal{I}^{lg}\mathbf{a}_k\|_2^2. \tag{14.26}$$

Accordingly, at each step κ, the line ℓ^κ is chosen as the one with the smallest mismatch error as determined by (14.26). After the set of chosen lines $\mathcal{L}^\kappa$ is updated, both parameters in the pair $(r, \mathcal{I}^{lg})$ need to be reestimated for all chosen lines. Note that the minimization problem (14.26) is nonconvex as it involves the product of the two groups of fault parameters. To tackle this, the StructMP algorithm uses an alternating minimization scheme that cyclically updates each group of parameters with the other group fixed with guaranteed convergence to a local minimum.

14.4.2 Lassoing Anomalies Using CD

Lasso promotes sparse solutions by regularizing the LS error criterion with the ℓ_1-norm of $\mathbf{s}$. Specifically, the coefficient vector whose nonzero entries identify line outages is found as

$$\hat{\mathbf{s}}^\lambda := \arg\min_{\mathbf{s}} \|\mathbf{y} - \mathbf{As}\|_2^2 + \lambda\|\mathbf{s}\|_1 \tag{14.27}$$

where λ is a tuning parameter to be determined. Thanks to the overcomplete representation (14.11) and the ℓ_1-norm relaxation, the line outage identification is reduced to the convex QP in (14.27). Hence, its global minimizer can be efficiently obtained using general-purpose convex solvers, such as interior-point algorithms [19, ch. 11]. Instead of these off-the-shelf solvers, the CD iterative solver [18] will be adapted because it has been shown to exploit the specific problem structure and can yield the so-termed regularization path of Lasso-based solutions as a function of λ [20]. This approach is advocated especially for sparse unknown vectors of high dimension; that is, for a large number of lines N_L.

To allow for a sequence of λ values, consider first (14.27) with a fixed λ. CD optimizes the regularized LS cost (14.27) by cyclically minimizing over the coordinates, namely, the scalar entries of $\mathbf{s}$. It yields successive estimates of each coordinate, while keeping the rest fixed. Suppose that the ℓth entry $s_\ell(i)$ is to be found. Precursor entries $\{s_1(i), \ldots, s_{\ell-1}(i)\}$ have been already obtained in the ith iteration along with postcursor entries $\{s_{\ell+1}(i-1), \ldots, s_L(i-1)\}$ are also available from the previous $(i-1)$-st iteration. Thus, the effect of these given entries can be removed from $\mathbf{y}$ by forming

$$\mathbf{e}_\ell(i) := \mathbf{y} - \sum_{j=1}^{\ell-1} \mathbf{a}_j s_j(i) - \sum_{j=\ell+1}^{L} \mathbf{a}_j s_j(i-1) \tag{14.28}$$

where $\mathbf{a}_j$ denotes the jth column of matrix $\mathbf{A}$. Using (14.28), the vector optimization problem in (14.27) reduces to the following scalar one with $s_\ell(i)$ as unknown: $s_\ell(i) = \arg\min_{s_\ell}[\|\mathbf{e}_\ell(i) - \mathbf{a}_\ell s_\ell\|_2^2 + \lambda|s_\ell|]$. This is the popular *scalar* Lasso problem, which admits a closed-form solution expressed using a soft thresholding operator as (see e.g., [20])

$$s_\ell(i) = \text{sign}\left(\mathbf{a}_\ell^T \mathbf{e}_\ell(i)\right) \left[\frac{|\mathbf{a}_\ell^T \mathbf{e}_\ell(i)|}{\|\mathbf{a}_\ell\|_2^2} - \frac{\lambda}{2\|\mathbf{a}_\ell\|_2^2}\right]_+, \quad \forall \ell \tag{14.29}$$

where $[\chi]_+ := \chi$, if $\chi > 0$, and zero otherwise.

Cycling through the closed forms in (14.28) and (14.29) explains why CD here is faster than general-purpose convex solvers. The residual update in (14.28) is feasible iteratively, while the soft thresholding operation in (14.29) is also very fast. Following the basic convergence results in [18], the CD iteration is provably convergent to the global optimum $\hat{\mathbf{s}}^\lambda$ of (14.27), as asserted in the following proposition.

PROPOSITION 14.2 (*Convergence of CD*) *Given the parameter λ and arbitrary initialization, the iterates $\{\mathbf{s}(i)\}$ given by (14.29) converge monotonically to the global optimum $\hat{\mathbf{s}}^\lambda$ of the anomaly identification problem in (14.27).*

The identification path is further obtained by applying the CD method for a decreasing sequence of λ values. Larger λ's in (14.29) force more entries of $\mathbf{s}(i)$ to be nulled. Hence, if a large enough parameter λ is picked, the corresponding $\hat{\mathbf{s}}^\lambda$ will eventually become $\mathbf{0}$. With a decreasing sequence of λ's, $\hat{\mathbf{s}}^\lambda$ for a large λ can be used as a *warm start* for solving (14.27) with the second-largest λ. This way, the CD-based anomaly solution path of (14.27), as tabulated in Algorithm 14.2, exploits both the efficient scalar solution in (14.29) as well as warm starts, to ensure reduced complexity and algorithmic stability. It is possible to modify the CD-based algorithm to account for the structured constraints of fault event model (14.17). Specifically, the scalar Lasso problem per entry of $\mathbf{s}$ can be updated to the one that solves for both $(r, \mathcal{I}^{lg})$, similar to (14.26).

Compared to enumerating all possible anomaly combinations, the two proposed algorithms incur linear complexity on the total number of anomalies. The recursive OMP Algorithm 14.1 has complexity $\mathcal{O}(\kappa_{\max} L)$. For Algorithm 14.2, each CD iteration entails only L scalar operations, while further acceleration is possible by exploiting sparsity and warm starts across CD iterations. Table 14.1 lists the runtime comparisons of the proposed algorithms on the line outage identification problem and the exhaustive

---

Algorithm 14.2 (CD): Input $\mathbf{y}$, $\mathbf{A}$, and a decreasing sequence of λ values. Output $\{\hat{\mathbf{s}}^\lambda\}$ in (14.27) for each λ.

Initialize with $i = -1$ and $\mathbf{s}(-1) = \mathbf{0}$.
for each λ value from the decreasing sequence **do**
    repeat
        Set $i := i + 1$.
        for $\ell = 1, \ldots, L$ **do**
            Compute the residual $\mathbf{e}_\ell(i)$ as in (14.28).
            Update the scalar $s_\ell(i)$ via (14.29).
        end for
    until CD convergence is achieved.
    Save $\hat{\mathbf{s}}^\lambda = \mathbf{s}(i)$ for the current λ value.
    Initialize with $i = -1$ and $\mathbf{s}(-1) = \mathbf{s}(i)$ as the warm start for the next λ.
end for

Table 14.1 Average runtime (in seconds) for the line outage identification problem using three test cases.

	ES	OMP	CD
Single, 118-bus	5.9e−2	1.3e−4	4.9e−2
Double, 118-bus	4.1	2.7e−4	9.7e−2
Single, 300-bus	0.15	4.0e−4	0.39
Double, 300-bus	22.5	1.1e−3	0.97
Single, 2383-bus	1.2	1.4e−2	2.6
Double, 2383-bus	–	2.9e−2	4.2

search (ES) using three test cases of different size; see more details in [7]. As the number of outaged lines goes from single to double, both propose algorithms have just doubled the computational time, while the ES method increases the time by the order of the network size (i.e., number of buses). Hence, linear complexity in the number of outaged lines by the OMP and CD algorithms can greatly reduce the computational time for large-scale power grids.

14.5 Meter Placement

This section focuses on the problem of placing PMUs to maximize the performance for identifying anomalies. Given the similarity of the three problems, again we focus on the line outage identification problem here. Given the budget of N_P PMU devices, let $\mathcal{B}_P \subseteq \mathcal{B}$ denote the subset of N_P locations chosen for PMU installation. Accordingly, let $\tilde{\boldsymbol{\theta}}_P$ be the observable sub-vector of $\tilde{\boldsymbol{\theta}}$, collecting phasor angles at buses in $\mathcal{B}_P$; and similarly for other subvectors to be defined soon. Extracting only the rows in (14.11) corresponding to $\mathcal{B}_P$, one can obtain

$$\tilde{\boldsymbol{\theta}}_P = \sum_{\ell \in \tilde{\mathcal{L}}} s_\ell \, \mathbf{a}_{\ell, P} \qquad (14.30)$$

where $\mathbf{a}_{\ell, P}$ is also constructed by selecting the entries from $\mathcal{B}_P$.

Assuming all signature vectors $\{\mathbf{a}_\ell\}$ have been normalized to be of unit Euclidean norm, the OMP method would identify the outage line as

$$\hat{\ell} := \arg\max_{k \in \mathcal{L}} |\mathbf{y}_{\ell, P}^T \mathbf{a}_{k, P}|. \qquad (14.31)$$

Without loss of generality (Wlog), the sign of $\mathbf{y}_\ell$ can be flipped to make sure that $\mathbf{y}_{\ell, P}^T \mathbf{a}_{\ell, P} \geq 0$ always holds. Hence, the outaged line ℓ is correctly identified if and only if (iff)

$$\mathbf{y}_{\ell, P}^T \mathbf{a}_{\ell, P} \geq |\mathbf{y}_{\ell, P}^T \mathbf{a}_{k, P}|, \quad \forall k \in \mathcal{L}. \qquad (14.32)$$

A weighting coefficient $\alpha_\ell \in [0, 1]$ is introduced here that associates with any line ℓ. The value α_ℓ is chosen depending on the impact if the system operator can correctly detect the outage event occurring at line ℓ. One simple choice is the uniform weighting rule where $\alpha_\ell = (1/L)$, $\forall \ell \in \mathcal{L}$. However, in practical power systems, some transmission lines such as those high-voltage ones could carry much larger power flows, and thus they are of higher priority for the control center to monitor. To account for this, it is possible to choose α_ℓ to be proportional to the pre-event real power flow on line ℓ. With $\{\alpha_\ell\}$ given, the weighted success rate of identifying all single line-outage events is given for any subset $\mathcal{B}_P$, as

$$f(\mathcal{B}_P) := \sum_{\ell \in \mathcal{L}} \alpha_\ell \mathbb{1}\{\mathbf{y}_{\ell, P}^T \mathbf{a}_{\ell, P} \geq |\mathbf{y}_{\ell, P}^T \mathbf{a}_{k, P}|, \, \forall k \in \mathcal{L}\} \qquad (14.33)$$

where the indication function $\mathbb{1}(\cdot) = 1$ if the statement is true, otherwise 0. Given the budget of N_P PMU devices, the problem of interest becomes how to select the locations to maximize the weighted success rate, as given by

$$f^\star := \max_{\mathcal{B}_P \subseteq \mathcal{B}} f(\mathcal{B}_P)$$
$$\text{s. t. } |\mathcal{B}_P| \leq N_P. \quad (14.34)$$

The PMU placement problem (14.34) is complicated by two issues. First, the optimization variable $\mathcal{B}_P$ is a discrete subset, which implies a combinatorial problem complexity in the number of candidate bus locations. Second, the indicator function is noncontinuous, making it difficult to analyze the objective cost. As shown in [21], the problem (14.34) can be reformulated as a nonconvex mixed-integer nonlinear program (MINLP), and thus is generally NP-hard; see e.g., [22]. To tackle the second issue on indicator function, it is possible to approximate it using a tractable linear objective function. Nonetheless, it has been shown in [21] that the resultant linear programming (LP) formulation would still exhibit a large optimality gap as it relaxes the discrete constraint set. To directly account for such constraint, one can adopt the branch-and-bound (BB) algorithm [23] by successively estimating the lower and upper bounds for the optimal solution, until the two approach to each other. Hence, if convergent the BB algorithm can achieve the global optimum of (14.34). Nonetheless, the main concern for the BB algorithm is its computational complexity, which in worst case may be equivalent to the enumeration solutions.

Interestingly, it turns out that a simple greedy search approach works well for the PMU placement problem (14.34). Its complexity grows linearly in the total number of PMUs, namely N_P. In addition, it can gracefully handle a variety of objective functions. For example, for the line outage identification problem a max-min error norm criterion was proposed in [24], while a log-likelihood objective related to multinomial logistic regression was considered [25]. As for the fault location problem, the coherence of sensing matrix $\mathbf{A}$ as defined in (14.24) was used in [10]. The greedy search method has been shown effective for all these objective functions. Hence, we will focus on introducing the greedy method as a general solution for PMU placement.

14.5.1 A Greedy Search Method

The greedy method is tabulated in Algorithm 14.3. This heuristic approach is initialized with a preselected subset of buses $\mathcal{B}_P^o$, such as the empty set ∅ if no bus has been selected so far. It proceeds by iteratively choosing the next PMU bus location that achieves the largest marginal cost in terms of (14.33), one location at a time until the solution $\mathcal{B}_P^G$ reaches the cardinality budget N_P.

Clearly, the complexity of such iterative scheme grows only linearly with the number of available PMU devices, which makes it extremely suitable for implementation in large-scale systems. In addition to the low complexity, the buses selected at an early stage are always maintained even if the budget N_P increases later on, which is termed as the *consistency property* for the solutions of the greedy method. This nested property

Algorithm 14.3 (Greedy Search): Input N_P, $\{\mathbf{y}_\ell\}$, $\{\mathbf{a}_\ell\}$, and the initial set $\mathcal{B}_P^o$. Output $\mathcal{B}_P^G = \mathcal{B}_P^o \cup \{n_\rho\}_{\rho=1}^{N_P - |\mathcal{B}_P^o|}$.

Initialize $\mathcal{B}_P^G \leftarrow \mathcal{B}_P^o$, and compute all the dot-product vectors $\mathbf{c}_{\ell k} = \mathbf{y}_\ell^T \mathbf{a}_k$, for any $k, \ell \in \mathcal{L}$.

for $\rho = 1, \ldots, N_P - |\mathcal{B}_P^o|$ **do**
  Find $n_\rho = \arg\max_{n \notin \mathcal{B}_P^G} f(\mathcal{B}_P^G \cup \{n\})$;
  Update $\mathcal{B}_P^G \leftarrow \mathcal{B}_P^G \cup \{n_\rho\}$.
end for

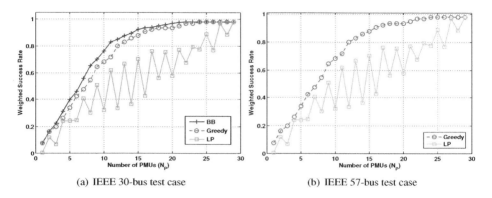

(a) IEEE 30-bus test case   (b) IEEE 57-bus test case

Figure 14.1 The achievable detection success rate versus the number of PMU devices N_P using the LP approximation, the greedy search, and the BB algorithms.

also makes it a very attractive scheme for PMU placement in general because it can easily accommodate any future system expansion plans if more PMU devices become available. For this reason, the greedy method has been very popular in several other PMU placement work for, e.g., improving the power system state estimation in [26, 27].

REMARK 14.1 *(Greedy's performance.) Although the greedy method seems to be simply heuristic, it is provable to achieve an approximation ratio of $(1 - 1/e)$ to the optimum for some specific discrete set optimization problems, as in [26]. This performance guarantee can be established if the problem objective function is shown to satisfy the so-termed submodularity property; see e.g., [28]. In general, the objective function does not need to be strictly submodular as long as it can be approximated as a submodular one under some conditions; see e.g., [29]. Based on the numerical results in [21], it is very likely that the detection success rate function (14.33) is approximately submodular. This can help explain why the greedy solution achieves the near-optimal performance in practical systems.*

The greedy search method is compared to the aforementioned LP approximation and BB algorithms on two IEEE test cases, as shown in Figure 14.1; see more comparisons in [21]. With the high computational order of BB algorithm, its performance for the larger 57-bus test case has not been provided. The success rate attained by the LP approximation could be very low for some cases, especially if N_P is around half of the total number of buses. This is because it fails to account for the discrete constraint set of the placement problem (14.34). Interestingly, the greedy heuristic is extremely competitive as compared to the global optimum attained by the BB algorithm, as shown in Figure 14.1(a). As the greedy algorithm is very attractive in terms of computational time, this result is very encouraging. The success rate function attained by the greedy search exhibits the property of diminishing marginal return, thus it is very likely that the objective function here can be approximated to a submodular function. This could explain the near-optimal performance of the greedy heuristic method.

14.6 Uncertainty Quantification

In this section, we consider the effects of measurement uncertainty on anomaly identification using multihypothesis test with Bayesian risk analysis. The objective is to find the anomaly event in an optimal way, i.e., minimizing the Bayesian risk, to integrate the prior probabilities of anomaly events and the costs of wrong detection. Based on that, the impact of PMU uncertainty is analyzed in this general paradigm, under which an anomaly identification procedure is proposed accordingly. Again, for simplicity the line outage problem is considered here.

Let $\mathcal{K} := \{1, 2 \ldots, K\}$ denote the index set of the outage event set $\mathcal{L}$, and define $\bar{\mathcal{K}} := \mathcal{K} \bigcup \{0\}$ to include normal state index by 0. Let H_k denote the hypothesis test corresponding to the event k. According to classical detection theory [30], the decision rules of the multihypothesis test are to find the regions in $\mathbb{R}^M$ corresponding to each hypothesis in an optimal manner, i.e., if the observation y is within the region k, the hypothesis H_k will be chosen. Let Γ denote the observation space, and the subset $\Gamma_k \subset \Gamma$, $k \in \bar{\mathcal{K}}$ denote the region of choosing the hypothesis H_k. We can see all the decision regions constitute the observation space, and any two decision regions do not overlap, i.e., $\bigcup_{k \in \bar{\mathcal{K}}} \Gamma_k = \Gamma$, $\Gamma_i \bigcap \Gamma_j = \emptyset$, $\forall i, j \in \bar{\mathcal{K}}$.

Provided that the hypothesis H_k is true, the conditional probability density function (PDF) of the observation of phasor angle difference y is denoted by $f(y|H_k)$. Based on the assumption of a normal distribution of noise $\mathcal{N}(\mathbf{0}, \mathbf{\Sigma})$, $f(y|H_k)$ has a normal distribution as $\mathcal{N}(\boldsymbol{\mu}_k, \mathbf{\Sigma})$, where $\boldsymbol{\mu}_k$ denote the phasor angle difference for event k. Thus, the probability of choosing H_j, given H_k is true, can be computed as the integral of $f(y|H_k)$ over the decision region Γ_j, i.e., $\Pr(H_j|H_k) = \int_{\Gamma_j} f(y|H_k) \mathrm{d}y$.

Under the Bayesian criterion [30], a cost will be assigned to each decision. In particular, we assume a nonnegative number C_{jk} for $j, k \in \bar{\mathcal{K}}$ as the cost incurred by choosing the hypothesis H_j when the hypothesis H_k is true. This assumption in our problem comes from the fact that the detection results may have different impacts on the power system, e.g., the followed protections and remedy procedures may incur distinct operation costs, and especially the wrong detection results will trigger false reactions that may introduce enormous costs. If the hypothesis H_k is true, the cost of the line outage detection, denoted by R_k, can be expressed as: $R_k = \sum_{j \in \bar{\mathcal{K}}} C_{jk} \Pr(H_j | H_k) = \sum_{j \in \bar{\mathcal{K}}} C_{jk} \int_{\Gamma_j} f(y|H_k) \mathrm{d}y$.

We assume that the prior probability information for each hypothesis is given for the detection problem. This probability is due to distinct features of outage line events. For example, the line that always operates near its power flow capacity may have a high probability of an outage, or the line that is in a severe environment may be prone to having an outage. Let $q_0, q_1, \ldots, q_K$ denote the prior probabilities of occurrences of hypotheses $H_0, H_1, \ldots, H_K$, respectively, and let them satisfy $q_0 + q_1 + \cdots + q_K = 1$. The Bayesian risk, denoted by R, being the overall expected cost incurred by the decision rules, can thus be expressed as:

$$R = \sum_{k \in \bar{\mathcal{K}}} q_k R_k = \sum_{k \in \bar{\mathcal{K}}} \sum_{j \in \bar{\mathcal{K}}} q_k C_{jk} \int_{\Gamma_j} f(y|H_k) \mathrm{d}y. \tag{14.35}$$

The Bayesian criterion is to minimize Bayesian risk R by determining the decision regions (Γ_k) of these hypotheses, i.e., $\min_{\Gamma_j, j \in \bar{\mathcal{K}}} \sum_{k \in \bar{\mathcal{K}}} \sum_{j \in \bar{\mathcal{K}}} q_k C_{jk} \int_{\Gamma_j} f(y|H_k) \mathrm{d}y$.

14.6.1 Optimal Decision Rules

The decision regions for the Bayesian criterion can be expressed as a set of decision rules in the multihypotheses test [30], in which each decision rule excludes a hypothesis. Specifically, let's denote $\mathcal{H}_{k\bar{j}}$, the set of hypotheses with regard to H_j and H_k, as $\mathcal{H}_{k\bar{j}} = \{H_k\} \cup \{H_s | s \in \bar{\mathcal{K}} \setminus \{k, j\}\}$. The optimal decision rule regarding $\mathcal{H}_{k\bar{j}}$ and $\mathcal{H}_{j\bar{k}}$ can be derived as [31]:

$$q_j(C_{kj} - C_{jj}) f(y|H_j) \underset{\mathcal{H}_{k\bar{j}}}{\overset{\mathcal{H}_{j\bar{k}}}{\gtrless}} q_k(C_{jk} - C_{kk}) f(y|H_k) + \sum_{s \in \bar{\mathcal{K}} \setminus \{k, j\}} q_s(C_{js} - C_{ks}) f(y|H_s). \tag{14.36}$$

To choose H_k, K sets of hypotheses have to be chosen, i.e., $\bigcap_{j \in \bar{\mathcal{K}} \setminus \{k\}} \mathcal{H}_{k\bar{j}} = H_k$. The likelihood-ratio-based decision rules can then be expressed as:

$$q_j(C_{kj} - C_{jj}) \Lambda_{jk}(y) \underset{\mathcal{H}_{k\bar{j}}}{\overset{\mathcal{H}_{j\bar{k}}}{\gtrless}} q_k(C_{jk} - C_{kk}) + \sum_{s \in \bar{\mathcal{K}} \setminus \{k, j\}} q_s(C_{js} - C_{ks}) \Lambda_{sk}(y), \tag{14.37}$$

where the likelihood ratio $\Lambda_{jk}(y)$ is defined as $\Lambda_{jk}(y) = \frac{f(y|H_j)}{f(y|H_k)}$.

For normal distribution of y as $\mathcal{N}(\mu_k, \Sigma)$, given that the hypothesis H_k is true, the likelihood ratio can be written as $\Lambda_{jk}(y) = \exp(\beta_{jk}^T \cdot y + \gamma_{jk})$, where the coefficient vector β_{jk} and the scalar γ_{jk} are:

$$\beta_{jk} = [(\mu_j^T - \mu_k^T) \Sigma^{-1}]^T, \quad \gamma_{jk} = \frac{1}{2}(\mu_k^T \Sigma^{-1} \mu_k - \mu_j^T \Sigma^{-1} \mu_j). \tag{14.38}$$

So the decision rules in this scenario can be expressed as:

$$q_j(C_{kj} - C_{jj})\exp(\boldsymbol{\beta}_{jk}^{\mathrm{T}} \cdot \mathbf{y} + \gamma_{jk}) \underset{\mathcal{H}_{kj}}{\overset{\mathcal{H}_{j\bar{k}}}{\gtrless}} q_k(C_{jk} - C_{kk})$$
$$+ \sum_{s \in \bar{\mathcal{K}} \setminus \{k,j\}} q_s(C_{js} - C_{ks})\exp(\boldsymbol{\beta}_{sk}^{\mathrm{T}} \cdot \mathbf{y} + \gamma_{sk}).$$
(14.39)

To detect and identify the outage line(s) event, K decision rules of (14.39) need to be carried out, and each decision excludes one hypothesis. Similar to the bubble sort algorithm, the procedure of identifying the final hypothesis H_k is summarized in Algorithm 14.4.

Algorithm 14.4 Line outage identification under the Bayesian criterion
---
1: Initialization: $\mathcal{S} = \mathcal{K}$, $k = 0$, randomly choose $j \in \mathcal{S}$.
2: **repeat**
3:    Compute coefficient vectors $\boldsymbol{\beta}_{jk}$, $\boldsymbol{\beta}_{sk}$, $\forall s \in \mathcal{K} \setminus \{j,k\}$, and scalars γ_{jk}, γ_{sk}, $\forall s \in \mathcal{K} \setminus \{j,k\}$.
4:    Compute the following two quantities given the observation vector $\mathbf{y}$:

$$U_j = q_j(C_{kj} - C_{jj})\exp(\boldsymbol{\beta}_{jk}^{\mathrm{T}}\mathbf{y} + \gamma_{jk})$$
$$U_k = q_k(C_{jk} - C_{kk}) + \sum_{s \in \bar{\mathcal{K}} \setminus \{j,k\}} q_s(C_{js} - C_{ks})\exp(\boldsymbol{\beta}_{sk}^{\mathrm{T}}\mathbf{y} + \gamma_{sk})$$

5:    Update $\mathcal{S} \leftarrow \mathcal{S} \setminus \{j\}$.
6:    **if** $U_j > U_k$ **then**
7:        Update $k \leftarrow j$
8:    **end if**
9:    **if** $\mathcal{S} \neq \emptyset$ **then**
10:       Randomly choose $j \in \mathcal{S}$.
11:   **end if**
12: **until** $\mathcal{S} = \emptyset$
13: Decide that the hypothesis H_k is true.

14.6.2 Effects of PMU Uncertainty on Bayesian Risk

We can see that the number of decision rules (14.37) for all hypotheses is $\binom{K+1}{2}$. If we pick a hypothesis H_w and compute the likelihood ratios of all other hypotheses to hypothesis H_w, all the $\binom{K+1}{2}$ decision rules can be expressed as linear combinations of K likelihood ratios $\Lambda_{jw}(\mathbf{y})$, $\forall j \in \bar{\mathcal{K}} \setminus \{w\}$. In this K-dimensional space spanned by these likelihood ratios, it is obvious that each decision rule is a hyperplane to partition the space. These hyperplanes are uniquely determined by the prior probabilities of outage line events and the detection costs. Thus the decision regions Γ_k of all hypotheses H_k in this space can be obtained. The PMU uncertainty in terms of the covariance matrix Σ

impacts the Bayesian risk, and the Monte Carlo simulations with the optimal decision rules in Algorithm 14.4 can be applied to quantify the effects.

Here we use a simple example based on the IEEE 9-bus system with three generators and three loads (shown in Figure 14.2(a)) to illustrate the effects; see more details in [31]. The MATPOWER [32] toolbox is utilized to generate the phasor angle measurements as well as the pertinent power flows. All the parameters of the 9-bus test system can be referred in the file "case9.m" in the MATPOWER software. We consider single-line outage events here. The set of possible outage lines is assumed to be $\mathcal{E} = \{3, 5, 8\}$, with the corresponding index set as $\mathcal{K} = \{1, 2, 3\}$. The prior probabilities ($q_k, k \in \mathcal{K}$) of these three events are 0.3, 0.2, and 0.5, respectively. Bus 5 and bus 8 are assumed to be equipped with PMUs to provide phasor angle measurement vector y, i.e., $\mathcal{N}_O = \{5, 8\}$. The precomputed phasor angle difference due to each line outage event in $\mathcal{K}$ can be obtained using (14.6) as:

$$\mu_1 = \Delta\theta_O^{(1)} = \begin{bmatrix} -3.217° \\ 8.602° \end{bmatrix}, \mu_2 = \Delta\theta_O^{(2)} = \begin{bmatrix} 1.263° \\ -3.378° \end{bmatrix},$$

$$\mu_3 = \Delta\theta_O^{(3)} = \begin{bmatrix} 4.584° \\ 21.665° \end{bmatrix}.$$

Under the Bayesian criterion, we assume the costs of correct detection to be zero and the costs of wrong detection to be random positive numbers. The detection costs matrix $\mathbf{C}$ used in the test case is shown in the following text, with (j, k)th entry as the cost C_{jk}, $\forall j, k \in \mathcal{K}$.

$$\mathbf{C} = \begin{bmatrix} 0 & 1.901 & 2.319 \\ 9.210 & 0 & 1.562 \\ 7.008 & 4.185 & 0 \end{bmatrix}.$$

We investigate how the PMU uncertainty Σ affects the Bayesian risk using Monte Carlo simulation, as shown in Figure 14.2(b). We use scalar ϕ to indicate the PMU uncertainty level. Specifically, both the case of the homogeneous PMU uncertainty

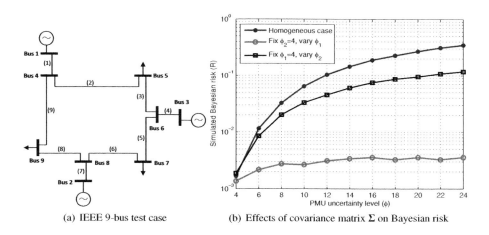

(a) IEEE 9-bus test case (b) Effects of covariance matrix Σ on Bayesian risk

Figure 14.2 An illustration of effect of PMU uncertainty on line outage detection.

($\boldsymbol{\Sigma} = \phi \mathbf{I}$, $\phi > 0$) and the case of nonhomogeneous PMU uncertainty in which we vary one PMU's uncertainty each time are illustrated. We can see that the simulated Bayesian risk increases with the PMU uncertainty level. Specifically, the second PMU (at bus 8) has a larger influence on the Bayesian risk.

14.7 Conclusion

This chapter has reviewed some of the recent results on power system anomaly identification by utilizing a sparse overcomplete representation. The modeling of three types of anomaly events have been considered using PMU phasor data, namely line outages, fault events, and phase imbalance. Based on a general sparse representation model that can capture all these three types of anomalies, two efficient solvers using compressed sensing techniques were presented. The computational order for the OMP and CD algorithms grows linearly with the number of anomaly events, making the two methods very suitable for real-time implementations. Furthermore, the PMU meter placement problem was discussed for improving the anomaly identification performance, along with a greedy search method at potentially near-optimal performance. Finally, the effects of uncertain PMU data were investigated under the Bayesian risk analysis framework, providing a fresh perspective to the performance guarantees for the anomaly identification problem.

Nonetheless, several technically challenging and practically pertinent issues remain to be addressed for the anomaly identification problem. The classical bus-branch representation as adopted here has been increasingly recognized to be insufficient to provide the grid model in the operation stage. Instead, the more complete yet complicated *node-breaker representation* is necessary to enable the visibility at the level of basic switching devices. Moreover, moving the power grid monitoring tasks toward a *model-free* paradigm has been a major trend to cope up with the modeling uncertainty and malicious intrusions. Advances in machine learning and statistical signal processing, such as sparse and low-rank models, missing and incomplete data, online change detection, deep learning, and (multi)kernel-based learning to name a few, are currently advocated for grid monitoring tasks toward the vision of smarter energy systems.

Acknowledgements

The authors wish to thank the following colleagues and co-authors who contributed to their joint publications that the material of this chapter was extracted from: Drs. G. B. Giannakis, Y. Shi, and J. Wang. H. Zhu was supported by NSF grants 1802319 and 1807097. C. Chen was supported by the U.S. Department of Energy, Office of Electricity, under contract DE-AC02-06CH11357.

References

[1] A. G. Phadke and J. S. Thorp, *Synchronized Phasor Measurements and Their Applications*. New York: Springer, 2008.

[2] G. B. Giannakis, V. Kekatos, N. Gatsis, S.-J. Kim, H. Zhu, and B. Wollenberg, "Monitoring and optimization for power grids: A signal processing perspective," vol. 30, no. 5, pp. 107–128, Sept. 2013.

[3] A. J. Wood and B. F. Wollenberg, *Power Generation, Operation, and Control*, 2nd ed. New York: Wiley & Sons, 1996.

[4] D. B. West et al., *Introduction to Graph Theory*. Upper Saddle River, NJ: Prentice Hall, vol. 2, 1996.

[5] J. E. Tate and T. J. Overbye, "Line outage detection using phasor angle measurements," vol. 23, no. 4, pp. 1644–1652, 2008.

[6] ——, "Double line outage detection using phasor angle measurements," in *Proc. IEEE Power & Energy Society General Meeting*, pp. 1–5, 2009.

[7] H. Zhu and G. B. Giannakis, "Sparse overcomplete representations for efficient identification of power line outages," vol. 27, no. 4, pp. 2215–2224, 2012.

[8] A. Schellenberg, W. Rosehart, and J. Aguado, "Cumulant-based probabilistic optimal power flow (P-OPF) with Gaussian and Gamma distributions," vol. 20, no. 2, pp. 773–781, 2005.

[9] G. Feng and A. Abur, "Fault location using wide-area measurements and sparse estimation," vol. 31, no. 4, pp. 2938–2945, 2016.

[10] I. Rozenberg, Y. Beck, Y. C. Eldar, and Y. Levron, "Sparse estimation of faults by compressed sensing with structural constraints," vol. 33, no. 6, pp. 5935–5944, 2018.

[11] N. C. Woolley and J. V. Milanovic, "Statistical estimation of the source and level of voltage unbalance in distribution networks," *IEEE Transactions on Power Delivery*, vol. 27, no. 3, pp. 1450–1460, 2012.

[12] T. Routtenberg and Y. C. Eldar, "Centralized identification of imbalances in power networks with synchrophasor data," vol. 33, no. 2, pp. 1981–1992, 2018.

[13] J. A. Tropp, "Greed is good: Algorithmic results for sparse approximation," *IEEE Transactions on Information theory*, vol. 50, no. 10, pp. 2231–2242, 2004.

[14] E. Candes and T. Tao, "Near-optimal signal recovery from random projections: Universal encoding strategies?" *IEEE Transactions on Information Theory*, vol. 12, no. 52, pp. 5406–5425, 2006.

[15] S. S. Chen, D. L. Donoho, and M. A. Saunders, "Atomic decomposition by basis pursuit," *SIAM review*, vol. 43, no. 1, pp. 129–159, 2001.

[16] S. G. Mallat and Z. Zhang, "Matching pursuits with time-frequency dictionaries," *IEEE Transactions on Signal Processing*, vol. 41, no. 12, pp. 3397–3415, 1993.

[17] R. Tibshirani, "Regression shrinkage and selection via the Lasso," *Journal of the Royal Statistical Society: Series B (Methodological)*, vol. 58, no. 1, pp. 267–288, 1996.

[18] P. Tseng, "Convergence of a block coordinate descent method for nondifferentiable minimization," *Journal of Optimization Theory and Applications*, vol. 109, no. 3, pp. 475–494, 2001.

[19] S. Boyd and L. Vandenberghe, *Convex Optimization*. New York: Cambridge University Press, 2004.

[20] J. Friedman, T. Hastie, and R. Tibshirani, "Regularization paths for generalized linear models via coordinate descent," *Journal of Statistical Software*, vol. 33, no. 1, p. 1, 2010.

[21] H. Zhu and Y. Shi, "Phasor measurement unit placement for identifying power line outages in wide-area transmission system monitoring," in *Proc. 47th Hawaii Intl. Conf. System Sciences (HICSS)*, pp. 2483–2492, 2014.

[22] S. Burer and A. N. Letchford, "Non-convex mixed-integer nonlinear programming: A survey," *Surveys in Operations Research and Management Science*, vol. 17, no. 2, pp. 97–106, 2012.

[23] E. L. Lawler and D. E. Wood, "Branch-and-bound methods: A survey," *Operations Research*, vol. 14, no. 4, pp. 699–719, 1966.

[24] Y. Zhao, A. Goldsmith, and H. V. Poor, "On PMU location selection for line outage detection in wide-area transmission networks," in *2012 IEEE Power and Energy Society General Meeting*. IEEE, pp. 1–8, 2012.

[25] T. Kim and S. J. Wright, "PMU placement for line outage identification via multinomial logistic regression," *IEEE Transactions on Smart Grid*, vol. 9, no. 1, pp. 122–131, 2018.

[26] Q. Li, T. Cui, Y. Weng, R. Negi, F. Franchetti, and M. D. Ilic, "An information-theoretic approach to PMU placement in electric power systems," *IEEE Transactions on Smart Grid*, vol. 4, no. 1, pp. 446–456, 2013.

[27] V. Kekatos, G. B. Giannakis, and B. Wollenberg, "Optimal placement of phasor measurement units via convex relaxation," *IEEE Transactions on Power Systems*, vol. 27, no. 3, pp. 1521–1530, 2012.

[28] G. L. Nemhauser, L. A. Wolsey, and M. L. Fisher, "An analysis of approximations for maximizing submodular set functions – I," *Mathematical Programming*, vol. 14, no. 1, pp. 265–294, 1978.

[29] M. X. Goemans, N. J. Harvey, S. Iwata, and V. Mirrokni, "Approximating submodular functions everywhere," in *Proceedings of the Twentieth Annual ACM-SIAM Symposium on Discrete Algorithms*, pp. 535–544, 2009.

[30] H. L. Van Trees, *Detection, Estimation, and Modulation Theory*. New York: John Wiley and Sons, 2001.

[31] C. Chen, J. Wang, and H. Zhu, "Effects of phasor measurement uncertainty on power line outage detection," *IEEE Journal of Selected Topics in Signal Processing*, vol. 8, no. 6, pp. 1127–1139, 2014.

[32] R. D. Zimmerman, C. E. Murillo-Sanchez, and R. J. Thomas, "MATPOWER: Steady-state operations, planning and analysis tools for power systems research and education," *IEEE Transactions on Power Systems*, vol. 26, no. 1, pp. 12–19, Feb. 2011.

Part V

Large-Scale Optimization

15 Uncertainty-Aware Power Systems Operation

Daniel Bienstock

15.1 Introduction: Power Systems Operations in Flux

Modern society expects cheap and reliable power delivery – it is a resource that is needed for industrial development and more broadly to make modern life feasible. As previously underdeveloped regions of the world aspire to join the industrial world, the need for effective power transmission will grow.

In parallel, the power industry and the very nature of "power" are changing, for a variety of reasons. First, there is well-known interest in developing alternative sources for power, in particular, power based on renewables such as solar and wind power. Putting aside the appeal for renewables based on environmental issues, the question as to the long-term economic justification for reliance on renewables appears to have been settled – there is clear justification. However, renewable sources of power come with an inherent and unavoidable hindrance, which is real-time variability, which as we will discuss in the following text, must be handled in a satisfactory manner. Thus, to put it in simple language, power systems operations have to be rethought so as to accommodate significant real-time variability or, more broadly, "noise."

Second, modern power systems operations are more participatory. Technologies such as distributed energy resources (e.g., rooftop solar panels), "smart" loads that algorithmically respond to market signals, electrical vehicles that can provide grid support, and others, contribute a different type of uncertainty that is more structural and possibly more difficult to predict. An extreme form of this type of challenge is the (up to now) hypothetical possibility of an attack on a power grid.

Third, the power engineering community is now embarking on an (explicit) algorithimic modernization. Whereas work on fundamental problems such as the ACOPF problem [1–4] effectively languished for several decades, in recent years we have seen substantial new activity in this domain that will surely have an impact on industrial practice. The ongoing competition on ACOPF problems sponsored by ARPA-E [5] is a witness to this trend. Additionally, the overwhelming force of modern data science and machine learning is having an impact on grid operation as well, aided in a timely manner by the advent of very rapid and accurate grid sensors (PMUs) that are now being deployed, and that report data 30 times per second or faster.

The combination of all these factors is causing the power community to focus attention on short-term grid uncertainty or, more broadly, short-term variability. Here, "short-term" refers to spans of time lasting minutes or much less. There are important operational reasons why this particular time frame is of interest. For example, power lines may become thermally overloaded (e.g., as a result of sudden increase in renewable output) but can operate in this fashion so long as the overloads are corrected within reasonable time, and these margins are measured in minutes. Larger generators may also require on the order of minutes to adjust their output. At the opposite end of the operational spectrum, there are risky events, such as frequency excursions, that must be dealt with swiftly.

Both very short term-term actions and the slower-moving events must be handled in joint fashion, because, for example, frequency corrections requires generation changes. In this chapter we will examine various computational problems and algorithmic and modeling tools that have been recently developed, so as to address issues arising in this time domain. "Uncertainty" and "volatility" and such terms will also concern stochastic effects restricted to the same duration. We will also discuss likely future developments.

Grid planners also face other computational issues that are affected by uncertainty. For example, decisions on which generators to deploy during the next (say) 12-hour period must also deal with stochastics. In this chapter, we do not consider that problem, although there is certainly an overlap because a poor choice of which generators to commit to operation over the next day may result in lack of real-time flexibility. Another important topic is that of *markets* and *pricing*. We will briefly touch on these issues in the following text, because they do have a short-term impact.

This chapter is organized as follows. Sections 15.1.1–15.1.3 provide an introduction to power systems operations and control. Section 15.1.3 introduce power flow problems. Section 15.1.4 describe how power systems react to real-time changes in demands. Sections 15.2 and 15.2.1 discuss reformulation of power-flow problems that are uncertainty aware, with Sections 15.2.2–15.2.7 devoted to technical issues arising in modeling and solution of such reformulations. Section 15.3 discusses appropriate data-driven estimation models, and Section 15.4 concerns the adaptation of financial risk measures to power-grid operations.

15.1.1 Power Engineering in Steady State

Modern power grids rely on a robust three-tier control mechanism that encompasses time frames lasting fractions of a second to tens of minutes. Stability is a keyword that captures two concepts: steady power delivery and equipment protection. In addition stability must be attained in an efficient, i.e. economically justifiable manner. The three-tier control structure binds decisions made by solving optimization problems to near-instantaneous behavior arising from physics and equipment. It would thus be artificial to simply focus on the optimization procedures. An important ingredient that we will touch on is that some of the constraints encountered in the optimization are "soft" while others are very hard.

To understand the issues outlined in the preceding text and examined in the text that follows, a basic understanding of power engineering is needed. Here we will provide a brief overview. The discussion we provide next will serve to motivate new optimization

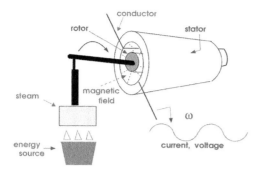

Figure 15.1 A simplified view of traditional power generation.

methodologies that have been proposed in recent years, and new directions of research that are likely to prove influential. We strongly recommend to the interested reader the well-written and comprehensive textbooks [1–3]. Also see [4].

We begin with AC (alternating current) power generation (see Figure 15.1). The elements shown in this simplified picture include a (typically) fossil fuel energy source, which is consumed so as to produce high-pressure steam, which is in turn used to rotate the "rotor," essentially a large and powerful electromagnet. The rotor turns inside a hollow cylinder known as the "stator," which has bundles of conducting wire looped around it.

The action of time-varying magnetic field induces a current on these wires; current technically is measured as rate of movement of charge across a unit area cross-section (of the conductors) per unit time. The current is also characterized by a voltage, i.e. potential electrical energy. The product of voltage times current has the units of energy per unit time, that is to say, power. Thus, from a simplified perspective, a generator generates electrical current at a certain voltage, rather than power. Additionally, the rotor is turning at a notionally constant frequency ω (60 Hz in the United States and 50 Hz in most other countries). The picture we have painted is here is overly simplified – we recommend to the interested reader further reading in the previously cited textbooks.

Suppose we have a transmission system, that is to say, a network where the nodes (*buses* in power engineering language) represent locations that may include generators and/or demand (or *load*) connected by power lines, or *branches*. A system of nonlinear equations is used to describe the steady-state behavior of the system. To this effect, consider a generator located at a bus k, and suppose that there is a line between k and another bus, km. Based on the preceding discussion the (real-time) voltage at k will be of the form

$$V_k(t) = \tilde{V}_k \cos(\omega t + \theta_k). \tag{15.1}$$

while the current injected into line km at k will be of the form

$$I_{km}(t) = \tilde{I}_{km} \cos(\omega t + \phi_{km}). \tag{15.2}$$

An analysis based on laws of physics will yield that the voltage at *any* bus k will be of the general form (15.1) while the current on any line km will be like (15.2). Hence, the (real-time) power injected into a line km at k

$$S_{km}(t) = V_k(t) I_{km}(t) \tag{15.3}$$

is also time periodic. While these considerations yield nominally exact representations of physical quantities, they are inconvenient when considering a system in steady state, which at this point means a system where all loads and generation are constant. Note that $S_{km}(t)$ can be positive or negative, however its average value (over time) is what matters because this average value multiplied by time indicates total energy being transported. Yet simply relying on this average overly simplifies the picture because it provides an inadequate understanding of power losses.

Rather than using (15.1)–(15.3) directly, it is customary to use a "phasor" representation of average voltage and current in steady state, i.e. representations as complex numbers of the form

$$V_k = |V_k| e^{j\theta_k}, \quad I_{km} = |I_{km}| e^{j\phi_{km}} \tag{15.4}$$

where $j \doteq \sqrt{-1}$. Furthermore, laws of physics yield that

$$I_{km} = y_{km}(V_k - V_m) \tag{15.5}$$

where the complex number y_{km} is the *admittance* of line km, and that the power injected into km at k is of the form

$$S_{km} = V_k I_{km}^*, \tag{15.6}$$

Here, representation (15.5) assumes a standard power line; when transformers are present a further elaboration is needed. We stress that (15.6) does not fully represent "power," rather, this is an average time *summary* or approximate representation of power. As such, both real and imaginary parts of S_{km} (resp., active and reactive power) have a meaningful interpretation. This point is made more readily understandable when we consider loads, or demands. A load can be constant; consider the example of a traditional lightbulb. But, in particular, the load arising from an electrical motor is time dependent (and periodic) and is also therefore represented by a phasor, i.e. a complex number, and likewise with the output of a generator.

A direct interpretation of active and reactive power, and their relationship, is not simple and is beyond the scope of this chapter. A qualitative understanding is in fact best obtained by considering a number of numerical studies. For example, a well-known fact is that "reactive power does not travel far" (i.e., reactive power losses, on lines, are nontrivial) and that a shortage of reactive power in a sector of a grid will result in declining voltage magnitudes. In turn, declining voltage magnitudes cause the grid to respond by increasing current magnitudes so as to deliver desired amounts of active power to loads, and that the combination of these two effects tend to cause larger losses of active power on lines (which are thermal losses). A collection of such insights provides, possibly, the best intuitive understanding of the nontrivial relationship between grid structure, voltage, and active and reactive power.

Equation (15.6) can be used as the basis for network equations. For a given bus k let us denote by $\delta(k)$ the set of all lines of the form km. Then we can write the **complex-valued** power balance equation

$$\text{(generation at } k) - \text{(load at } k) = \sum_{km \in \delta(k)} S_{km}. \tag{15.7}$$

The collection of all such equations describes power balance across the system. Note that if we use (15.5) and (15.6) to represent each term S_{km} on the right-hand side of (15.7) is a quadratic function of the voltages V_k and V_m. In fact,

$$\text{(generation at } k) - \text{(load at } k) = \sum_{km \in \delta(k)} V_k y^*_{km}(V_k^* - V_m^*). \tag{15.8}$$

15.1.2 Power-Flow Problems

Suppose that we are given a transmission system described by a network with known admittances y_{km} and (at least for the time being) the generation and load at every bus k are known. We would like to ascertain what the resulting physics will be: What are the voltage levels, currents, and power injections? We can obtain this information by solving the system of (quadratic and complex-valued) equations (15.8) on the voltages V_k. If n is the total number of buses this is a system on n complex variables and n complex quadratic constraints. The problem of computing the unknowns so as to satisfy the system of all equations (15.8) is known as the *power flow problem*, or PF.

However, motivated by the way the grid operates, it is best to think of this computational task as follows. First, we write

$$\mathcal{G} = \text{set of generator buses}, \mathcal{D} = \text{set of demand buses}. \tag{15.9}$$

For convenience, in the following text, we may allow any bus to provide generation (but at value zero) or demand (also at zero). We also assume that a generator bus has just one generator. Further, for any bus k

$$P_k^g + jQ_k^g \doteq \text{complex power generation at } k \tag{15.10a}$$
$$P_k^d + jQ_k^d \doteq \text{complex load at } k, \tag{15.10b}$$

so that (15.7) becomes

$$(P_k^g - P_k^d) + j(Q_k^g - Q_k^d) = \sum_{km \in \delta(k)} V_k y^*_{km}(V_k^* - V_m^*), \ \forall \ k. \tag{15.11}$$

In practice, when solving the power-flow problem we assume that

$$\forall k \in \mathcal{G}, |V_k| \text{ and } P_k^g \text{ are given, and} \tag{15.12a}$$
$$\forall k \in \mathcal{D}, P_k^d \text{ and } Q_k^d \text{ are given.} \tag{15.12b}$$

Additionally, we usually have bounds

$$\text{for all buses } k, \quad V_k^{\min} \leq |V_k| \leq V_k^{\max}. \tag{15.13}$$

The power-flow problem is to solve the system of equations (15.11) subject to (15.12), again a system of $2n$ real, nonlinear equations with the $2n$ variables $(\theta_k, Q_k^g)_{k \in \mathcal{G}}$, $(|V_k|, \theta_k)_{k \in \mathcal{D}}$) including, usually (15.13). Solving such a system is tantamount to discovering the physics consistent with a given level of (complex) demand and (active) generation at given voltage levels. This view is simplistic, however, for a number of important reasons.

First, the solution to the system *may not be unique!* Though well known in the community, this is an extremely thorny issue that has not been adequately resolved. Second, loads are inherently noisy quantities, and the geometry of the problem is extremely complex; potentially, small changes in load could result in large and nontrivial changes to the solution [6]. Finally, system (15.11)–(15.12) may be *infeasible*. In part this is due to the fact that it is difficult to formulate a precise relationship between total generation and total load, due to *transmission losses*. A feasible system may become infeasible after a small change in loads, admittance values, or voltage magnitude requirements. The first two types of quantities are difficult to know, precisely, in real time. The voltage requirements are set by the grid operator with equipment safety in mind, yet also allow for some flexibility.

15.1.3 Three-Level Grid Control

Power-grid control mechanisms have evolved over the last 100+ years. In current practice there is a hierarchical system that operates at three levels. All three are of relevance to this chapter because the increasing level of uncertainty in operations impacts all of them and is adding pressure to, effectively, merge the three levels into a continuous-time algorithmic control scheme. First, we will deal with the top level of this three-tier structure, the optimal power-flow computation, introduced next, and further developed in Section 15.1.3. Sections 15.2 and 15.2.1 will describe uncertainty-aware formulations of the optimal power-flow problem (OPF). Much of the material in the text that follows draws from work in [7–11] and other citations we will provide.

The Optimal Power-Flow Problem (OPF)

The optimal power-flow problem seeks to compute a minimum-cost generation mix so as to meet known levels of demands over a period spanning several minutes. Mathematically it can be written in the following form, with details provided as follows:

$$\text{Min} \sum_k c_k(P_k^g) \tag{15.14a}$$

Subject to: $\tag{15.14b}$

$$\sum_{km \in \delta(k)} V_k y_{km}^*(V_k^* - V_m^*) = (P_k^g - P_k^d) + j(Q_k^g - Q_k^d), \quad \forall k \tag{15.14c}$$

$$P_k^{\min} \le P_k^g \le P_k^{\max}, \ Q_k^{\min} \le Q_k^g \le Q_k^{\max}, \ V_k^{\min} \le |V_k| \le V_k^{\max} \quad \forall k \quad (15.14\text{d})$$
$$|\theta_k - \theta_m| \le \theta_{km}^{\max} \quad \forall km \quad (15.14\text{e})$$
$$|V_k y_{km}^*(V_k^* - V_m^*)| \le L_{km} \quad \forall km. \quad (15.14\text{f})$$

Here, as before, we write a complex voltage in the form (15.4). We have seen constraints (15.14c) and (15.14d). Constraint (15.14e) limits the phase angle change across a line (a stability constraint), and (15.14f) is a limit on power (sometimes, a limit on current may be used) and typically represents a thermal limitation. Finally, the functions c_k measure generation cost and, typically, are convex increasing.

Problem (15.14) is the archetypal ACOPF problem that has recently received a great deal of attention. It was first conceived by Carpentier [12]. See [13] for a history of the problem and for a more recent survey, see [14]. The problem is both nonlinear and nonconvex. From a technical standpoint, deciding feasibility is, already, a strongly NP-hard problem [15–17]. Substantial experimental success has been attained by general-purpose nonlinear solvers, such as KNITRO [18] and IPOPT [19], which can find excellent quality solutions in tens of seconds, even on systems with thousands of buses. Yet solvers of this type can fail, may even misdiagnose a problem as infeasible, and may prove too slow. Perhaps of more importance, it may be the case that formulation (15.14) is, unto itself, problematic. We can outline a number of justifications for this statement:

(a) Problem (15.14) needs to be solved often and reliably. In the current practice a problem of this type should be solved every five minutes. This is a substantial decrease over previous practice, and under future conditions we can expect a shorter interval still, aiming for near real-time intelligent adjustment of generation in response to changing conditions. Even though solvers like IPOPT and KNITRO are, empirically, remarkably efficient we may wish for stronger guarantees of successful solvability.

(b) The numerical parameters in the formulation, such as the admittances y_{km} and the line limits L_{km} are not precisely known. In particular, admittances represent electrical properties of conductors that are exposed to ambient conditions such as weather, and can experience substantial changes that may be difficult to precisely diagnose – an important point here is that transmission lines may be hundreds of miles long. The precise determination of line temperature (a principal factor to take into account when setting line limits) and line temperature *trend* under changing environmental conditions, is a nontrivial task. See, e.g. [20]. A complex standard, IEEE Standard 738 [21] is used in practice to estimate line temperature. In summary, (15.14) is a nonlinear, nonconvex problem with uncertain parameters, an arguably dangerous combination.

(c) A special uncertainty category applies to demands. The solution of (15.14) provides generator output over the (say) next five minutes, but what demand values are we using in the computation? It turns out that demands are noisy (in real time) and may furthermore exhibit trends, as in, for example, the periods during the day when demands are ramping up toward a peak. Further, many generators (especially, large generators) cannot alter their output instantaneously – they require time to ramp up (or even, down). Thus the output of a generator during a given time window is not going to be exactly as expected by formulation (15.14). Putting aside the issue of how the industry does resolve these types of uncertainty, it seems almost a mistake to solve an extremely complex problem when we do not even know, precisely, what the crucial data elements are. Should we not focus on a simpler problem than (15.14) that is more malleable?

The DC Approximation

The DC approximation to (15.14) is a widely used, linearly constrained, convex objective simplification that aims to address some of the previously mentioned concerns. To arrive at this approximation we note that, used verbatim in computation, formulation (15.14) may run into numerical scaling issues because the quantities under consideration (voltages, currents, loads, and admittances) have vastly different scales. A well-known industry practice, the *per-unit system* addresses this point. In particular, under the per-unit system, all voltage magnitudes $|V_k|$ will equal approximately 1.0, which is enforced by the bounds $V_k^{\min}$ and $V_k^{\max}$. To arrive at the DC approximation we indeed assume that $|V_k|=1$ for all k. We also ignore reactive power (!) and finally assume that the admittances are such that no line has thermal (active power) losses. Under these assumptions (see any of the textbooks cited previously for a full justification) the active power on line km is of the form

$$p_{km} = b_{km}(\theta_k - \theta_m), \qquad (15.15)$$

for an appropriate nonzero constant b_{km} (the *susceptance* of km).[1] Under these simplifications, the ACOPF problem is replaced by DCOPF:

$$\text{Min} \sum_k c_k(P_k^g) \qquad (15.16a)$$

Subject to: $\qquad (15.16b)$

$$\sum_{km \in \delta(k)} b_{km}(\theta_k - \theta_m) = P_k^g - P_k^d \quad \forall k \qquad (15.16c)$$

$$|b_{km}(\theta_k - \theta_m)| \leq L_{km} \quad \forall km \qquad (15.16d)$$

$$P_k^{\min} \leq P_k^g \leq P_k^{\max} \quad \forall k. \qquad (15.16e)$$

This is a convex, linearly constrained problem. As noted previously, the c_k functions are convex quadratics; problem (15.16) is quickly solved by many solvers. Additionally, optimal dual variables can be readily used for pricing considerations, a topic outside the scope of this chapter. The DC approximation is very common in the practice. It does not, however, bypass issue (c) noted in the preceding text, i.e. data uncertainty.

---

[1] Technically, b_{km} is defined as x_{km}^{-1} where x_{km} is the reactance, typically a positive number. In any case, $b_{km} \neq 0$.

15.1.4 Responding to Changing Conditions

As we discussed previously a potential weakness in solving (15.14) or (15.16) is that the demands cannot be known, precisely, in advance. In fact demands are inherently noisy, either through what could be termed stochastic variability, or noise, or ramping up or down, and generator outputs are also, frequently, adjusting in response to an OPF-commanded directive. How, exactly, does the grid behave because loads and the nominal generator outputs are not quite correct much of the time? It is important to notice that the sum of left-hand sides of (15.16c) is zero, meaning that total generation equals total demand, a condition that is almost never true in real time.

This issue is best viewed from the perspective of an extreme event, such as the sudden loss of a generator, or a sudden drop or increase in demands. Such events, if rare, do take place, and grids must be ready to respond. For example, there is a well-known "N-1" requirement that a sudden generator failure must be gracefully handled. In any of these events, we find ourselves in a situation in which total generator output is at a substantially wrong level.

The industry currently handles all sudden generation-load mismatches through a two-tier control mechanism that supports the OPF computation. In the context of this chapter, this particular mechanism is of utmost importance because the mechanism could plausibly also handle uncertainty in power sources such as wind and solar.

To fix ideas, it is best to imagine a scenario in which, all of a sudden, a net change in loads takes place. Suppose for example that a generator that is providing positive output were to suddenly fail (or "trip"). The precise behavior is rather complex but can be outlined as follows (see [1–4]):

1. System frequency, the parameter ω given in (15.1) and (15.2), will drop. To understand this phenomenon consider again Figure 15.1. Keeping the rotor (a massive object) spinning at 50 or 60Hz requires a large amount of energy. By slowing down, each rotor effectively donates a fraction of that kinetic energy toward meeting the electrical demands. The precise numerical frequency change, at each rotor, will be inversely proportional to its inertia.

    More precisely, in a purely dynamical setting (time measured: less than a second because the load change) we should denote by $w_k(t)$ the frequency at bus k at time t; $w_k = \frac{\partial \theta_k}{\partial t}$. Let k be a generator bus. Then we have a relationship given by the swing equation (here in simplified form)

$$A_k \frac{\partial w_k}{\partial t} = A_k \frac{\partial^2 \theta_k}{\partial t^2} = P_k^m - P_k^e, \tag{15.17}$$

   where $A_k > 0$ is a constant related to inertia, P_k^m is *mechanical* power input by the generator into the rotor, and P_k^e is the total *electrical* power generated at k, i.e. the sum of load at k plus power flows injected into lines at k.[2] At the time

---

[2] A more complete version would include a dampening term.

of the load change, in principle we know both terms in the right-hand side of (15.17) for each generator k. One can numerically estimate network-wide evolution of electrical properties (voltages and power flows) through a discrete-time procedure that alternates between updates to the solution to the system (15.17) with solution of a (steady state!) AC power-flow system. This last computation is used to estimate voltage angles at nongenerator buses. Such estimates (through numerical differentiation) yield estimates of frequency at nongenerator buses. Power flows (also obtained in the AC power-flow computation) are used to approximate the P_k^e terms in (15.17). We have presented the process in highly simplified form; it is a numerically challenging computation and the state-of-the-art implementations run far slower than the actual physical process.

In any case, following a period of "dynamics," frequency will stabilize at a new systemwide value, lower than before the load change. To first order, the frequency drop will be proportional to the net load change (which is an increase in net loads in our discussion). This first phase of network control, which is primarily control through physics, is known as primary response. One final point is that if the load change is large enough, the frequency change will be likewise large, and generators will "trip" (turn off) for safety reasons.

2. Following primary response the system is left operating at a lower frequency, an undesirable condition for stability reasons. Unless the load change were to correct itself, rapidly, after a few seconds the frequency drop will be noted at control centers and the net load increase will be estimated (e.g., using the frequency change). This load increase is now corrected by commanding suitable generators to output more power precisely so as to correct the imbalance. In detail, the transmission system will include a set $\mathcal{R}$ of *participating* or *responding* generators, and for each $k \in \mathcal{R}$ a value $\alpha_k \geq 0$ such that $\sum_{k \in \mathcal{R}} \alpha_k = 1$. Suppose that the net load increase is estimated by Δ. Then

$$\text{for each } k \in \mathcal{R}, \text{ generator } k \text{ alters its output by } \alpha_k \Delta. \tag{15.18}$$

This mechanism is a simplified view of what is usually known as AGC (area generation control). Expression (15.18) also applies when $\Delta < 0$, i.e., we have a net load decrease, and clearly attains balancing. The typical application of AGC addresses normal imbalances in demand versus supply arising from typical demand variation, or stochasticity, rather than an extreme event as in the example in the preceding text. Such stochasticity is "well-behaved": analysis of high-volume, actual sensor data [22] reveals that the variations are usually sub-Gaussian rather than heavy-tailed, with small deviations.

Note that the load estimates in OPF can benefit from recently observed AGC applications. Furthermore, OPF reschedules ("redispatches") generators so as to relieve the generators in $\mathcal{R}$, i.e. make them available for future applications of AGC.

We have now described a three-level control hierarchy of primary response, AGC, and OPF. Of course, this is a highly simplified summary; however, the indicated three-tier control responsibilities over different time scales present an accurate picture of current grid operation.

AGC and Stochastic Sources

Suppose that a given transmission system includes stochastic power sources, e.g., renewables. A substantial amount of research has been invested on the economics of such a proposition, especially the long-term economics, which would concern the statistics of the long-term variability of output. However, it turns out that the more challenging stochastics involves short-term, or near-instantaneous changes. In principle, AGC provides a mechanism for addressing the resulting power mismatches, assuming of course that a sufficient amount of generation is available for balancing purposes. However, this philosophy has encountered a number of obstacles. In increasing order of importance, these difficulties can be outlined as follows.

First, renewable sources are often located in geographical areas that are not near major loads and not near traditional generation. As a result, the power flows arising from stochastics in renewable output and resulting applications of AGC could affect large geographical areas and also could be "complex" due to power-flow physics, that is to say, the precise trajectory of renewable power flows and AGC-response power flows may be complex.

Second, it has been observed that renewables such as wind power may experience a high degree of instantaneous variability, say, on the order of 30%. A more precise way to put this is that the short-term standard deviation of output is of the same order of magnitude as the mean. If the transmission system has attained a high degree of penetration by renewables then we will potentially see a large amount of instantaneous generation shortfall or excess, and thus, large concomitant AGC-induced power flows.

Finally, the actual stochastics of renewable output seem quite different than demand-induced stochastics. Significant intermittency results in real-time load-generation mismatch characteristics that prove challenging for traditional generation-based AGC to follow. Additionally, the distribution of real-time output could possibly display strong timewise correlation, which would require persistence of AGC response. Generally speaking, renewable output peaks can produce strong power flows, in unusual patterns, and such power flows can overload equipment. Variability of output has been correlated with turbulence arising from storm "cut-out." See [23–25] for background in these topics.

In summary, the magnitude and phenomenology of real-time variability in renewable output challenges current grid balancing mechanisms, in particular AGC and OPF. A large body of recent research has been devoted to algorithmically recasting these two mechanisms so as to make them better aware of (a) each other and (b) complex, possibly heavy-tailed stochastics. The emphasis is on economically maintaining grid safety goals, by which we mean avoiding equipment overload and stress; these two terms need to be made more precise.

15.2 Safe DC-OPF

Let us consider the research agenda just outlined. The goal is to suggest new uncertainty-aware grid operation paradigms, however in such a way that they are implementable without wholesale changes to the practice. Hence the AGC mechanism (15.18) should not be replaced, but rather the participation factors α_k should be set with some attention to ambient conditions and in particular to anticipated near-real-time behavior of stochastic power sources. Here we will focus the discussion to modifications on DC-based OPF.

It is natural to embed a computation of the α within OPF itself, since the power flows resulting from the application of AGC will mix with those from nonresponding generators and from stochastic sources. Moreover, AGC carries an economic cost (namely, in the operation of the responding generators). Equipment safety already appears in standard OPF computations, namely in constraints (15.16e) and (15.16d) and such constraints will have to be updated. In addition, *variability* of power flows and control output should be taken into account in the computation because such variability could prove stressful on equipment. As an example, variability of power flows will include (in the AC setting) voltage variability, resulting in frequent transformer tap operation (a topic not dealt with in this chapter). See e.g., [26].

To address these goals we will provide a generic template, which will be made specific later later. Some notation is needed to explain the template; we present it below for future reference; the notation will be explained in the following text.

Terminology and conventions

$\mathcal{N}$ = set of buses, $n = \mathcal{N}$
$\mathcal{L}$ = set of line, $m = \mathcal{L}$
$\bar{w}_k$ = expected value of stochastic output at bus k
$\mathbf{w}_k$ = variable component of stochastic output at bus k, with zero mean
$\mathcal{W}$ = support for distribution of $\mathbf{w}$
$\mathcal{W}$ = distribution of stochastic outputs $\mathbf{w}$; $\Omega \doteq \mathbf{Var}(\mathbf{w})$
$\bar{P}_k^g$ = expected output of nonstochastic generator at bus k
α = matrix of participation factors α_{kj}; $\mathcal{A}$ = set of allowable matrices
b_{km} = susceptance of line km (nonzero).
$\check{B}$ = pseudo-inverse of bus admittance matrix B.

As the list of terminology and conventions suggests, generation at a bus k will in general consist of output from a "traditional" generator, which may or may not participate in AGC, plus output from stochastic sources at k (which might be none). The stochastic output at k equals $\bar{w}_k + \mathbf{w}_k$ where $\bar{w}_k$ is constant and $\mathbf{w}_k$ is uncertain, with zero mean (here, and in the following text, we use boldface to denote uncertain quantities and stochastic operators). In other words $\bar{w}_k$ can be interpreted as the expectation of the stochastic output in the OPF time window of interest. The term distribution used to describe $\mathcal{W}$ is meant to be sufficiently generic to include specific probability distributions as well as coarser, data-driven models. Also, our notion of a stochastic source can include, for example, "smart loads" that respond to an unknown set of incentives. It may even be the case that for certain buses k, $\bar{w}_k = 0$ but $\mathbf{w}_k$ is nontrivial (nonzero).

The quantity $\bar{P}_k^g$ is the expected output of a nonstochastic generator at k. Here we note that "nonstochastic" would include traditional generators and also nontraditional controllable sources that may become available in the near future, such as batteries. If the generator does not participate in AGC, this quantity will represent the constant output of the generator during the OPF window.[3] Hence, in general, the (real-time) output of (traditional) generator bus k can be written in the following form

$$P_k^g \doteq \bar{P}_k^g - \sum_{j \in \mathcal{N}} \alpha_{kj} \mathbf{w}_j. \tag{15.19}$$

In this equation we have introduced a new concept. Rather than having a participation factor of the form α_k, we indicate by α_{kj} the (per unit deviation) response of generator k to a deviation at bus j. This is a more flexible version of AGC; see e.g., [8]. We recover standard AGC by imposing that for a given k all α_{kj} be equal to a common value α_k, so that (15.19) reads $P_k^g = \bar{P}_k^g - \alpha_k \sum_{j \in \mathcal{N}} \mathbf{w}_j$ (this is termed the *global* case in [8]). Further, the values α_{kj} may be additionally constrained. For example $\alpha_{kj} = 0$ if bus k does not participate in AGC. The set $\mathcal{A}$ describes such constraints (and we will assume that, e.g., it is convex and otherwise simple enough).

Using these conventions, we have that the total real-time output at bus k, in all cases, equals

$$\bar{P}_k^g + \bar{w}_k - \sum_{j \in \mathcal{N}} \alpha_{kj} \mathbf{w}_j + \mathbf{w}_k. \tag{15.20}$$

So far we have assumed that loads (demands) are certain; however, uncertainty in demand stochastic output, with no change in the notation or in the algorithmics given in the following text. We will thus view P_k^d as the deterministic demand at k and assume that all other uncertainty is covered by the random variable $\mathbf{w}$. Under the DC model, then, we must have that

$$\sum_{k \in \mathcal{N}} \left(\bar{P}_k^g + \bar{w}_k - \sum_{j \in \mathcal{N}} \alpha_{kj} \mathbf{w}_j + \mathbf{w}_k \right) = \sum_{k \in \mathcal{N}} P_k^d \quad \forall \mathbf{w} \in \mathcal{W}. \tag{15.21}$$

We need to point out that this equation, while guaranteeing balancing, is an idealized statement. First, the AC power-flow model is a steady-state representation of the underlying physics, and the DC model is an approximation to the AC model. Hence, in principle, an action that results in a small violation of the DC power-flow equations may not prove significant, especially if brief.

In fact, in an OPF computation we are imputing constant generator output, while in reality a generator redispatch (relative to the previous OPF computation) will cause generation to ramp-up or ramp-down over a period of time. During this period we may see small discrepancies between generation and demand. Hence, in principle, rather than insisting on the "for all" in (15.21) we might only ask that the statement in (15.21)

---

[3] Ramping characteristics of generators are ignored.

holds with sufficiently high probability, and/or that the discrepancy between right-hand and left-hand sides be small, with high probability. To the best of our knowledge this approach has not been taken in the literature, and we will not take it here, but it should be noted.

Continuing with (15.21), recall that we are assuming that the random variables $\mathbf{w}_j$ have zero expectation. Thus under (15.20) the expected total output at bus k equals $\bar{P}_k^g + \bar{w}_k$. It follows from (15.20) that to attain balancing, on average, we must have that

$$\sum_{k \in \mathcal{N}} (\bar{P}_k^g + \bar{w}_k) = \sum_{k \in \mathcal{N}} P_k^d \qquad (15.22)$$

which is the same as saying that

$$\sum_{k \in \mathcal{N}} \left(\mathbf{w}_k - \sum_{j \in \mathcal{N}} \alpha_{kj} \mathbf{w}_j \right) = 0, \quad \forall \mathbf{w} \in \mathcal{W}. \qquad (15.23)$$

Note that (15.22), although completely reasonable, at this point appears to be an operative assumption. In fact it is required by the DC model, under reasonable assumptions on the distribution of $\mathbf{w}$, if we insist on constant balancing (as we said we will).

To understand this point, note that a sufficient condition for (15.23) is that (see [8])

$$1 = \sum_{k \in \mathcal{N}} \alpha_{kj} \quad \forall j \in \mathcal{N}. \qquad (15.24)$$

But (15.23) describes a hyperplane in $\mathbf{w}$-space. If the distribution $\mathcal{W}$ is full dimensional then the linear form in the right-hand side of (15.23) must be identically zero, which means, as pointed out in [27] that (15.24) is necessary for constant balancing.

Thus, we will consider (15.24) as one of the requirements defining the set $\mathcal{A}$ of allowable participation factor matrices. Once again we see an opportunity for a different model: if we ask that, e.g., (15.23) hold with sufficiently high probability then (15.24) is no longer necessary. Of course (as in preceding text) we will again have a model that does not guarantee constant balancing (even on average). In any case in what follows we will impose (15.22).

With these ideas in mind, we can now set up a generic template for a stochastics-aware variant of OPF that also sets participation factors.

- First, rather than the generation cost minimized in (15.16), we will seek to minimize a function of the form $\mathcal{C}(\bar{P}^g, \alpha, \mathcal{W})$ where the function $\mathcal{F}$ should be interpreted as a metric that takes into account the (static) generation mix given by the vector $\bar{P}^g$ and the participation matrix α; however, this metric may of course also depend on the distribution $\mathcal{W}$. As an example, we could have

$$\mathcal{C}(\bar{P}^g, \alpha, \mathcal{W}) = \sum_{k \in \mathcal{N}} \mathbf{E}(c_k(P_k^g)), \qquad (15.25)$$

i.e. the expected cost of controllable generation ($\mathbf{E}$ = expectation under $\mathcal{W}$). This type of objective, and close variants, has been the primary focus in the literature. See [7–9, 28]. However in the context of stochastic variability, direct economic cost may not be the only target of concern. One could in addition incorporate cost estimates of variability of equipment operation (such as frequency of ramping) and, generally, the cost function could be set up to discourage grid operating modes with high variability. Such modifications would amount to incorporating additional terms in the expression for $\mathcal{C}(\bar{P}^g, \alpha, \mathcal{W})$ given in the preceding text. We will return to these points later.

- In the context of renewable incorporation, one concern outlined in the preceding text and in the literature is the potential for "congestion," by which the community means high usage and overloads of power lines. It is important to understand how such overloads are managed in current operations. Consider inequality (15.16d) that constrains the flow magnitude on line km. Constraints of this type are "soft" – in normal operation this constraint may be violated, with the proviso that the infeasibility (in optimization terms) is corrected within an appropriate time span.

  The constraint is of a *thermal* nature. If a line experiences flows exceeding its limit, the line temperature will start to increase and if this increase reaches a critical point the line will trip (turn itself off) for protective reasons, a highly undesirable outcome that may compromise the entire transmission system. The exact process by which temperature rises is complex and well understood (see the preceding discussion on IEEE Standard 738). In fact, rather than a single limit L_{km} there are as many as three limits, each with a maximum sojourn time, that is to say, a stipulation that states the maximum amount of time that the line can spend above that limit, with grid operators expected to correct an overload before the time limit is reached.

  The question is then how to frame this very complex set of rules into a simple yet effective constraint to incorporate into an optimization problem; preferably, a convex constraint. For a given line km we will use a constraint of the form

  $$\mathcal{P}_{km}(\mathbf{p}_{km}, L_{km}) \leq \epsilon,$$

  where $0 < \epsilon < 1$ is a desired tolerance, and $\mathbf{p}_{km}$ indicates the stochastic flow on line km as per the DC powerflow model (see Equation (15.15)). The left-hand side of this inequality is again a function that quantifies the stochastic risk that $\mathbf{p}_{km}$ exceeds the limit L_{km}. For example, under an appropriate model for $\mathcal{W}$, this function could measure the probability that the line flow exceeds L_{km}, though this is by no means the only choice.

- A final ingredient is a stochastic constraint, superficially like that just described, but that is now used to protect generation equipment, broadly defined. As per Equation (15.20) the (controllable) real-time output at generator k equals

  $$P_k^g \doteq \bar{P}_k^g - \sum_{j \in \mathcal{N}} \alpha_{kj} \mathbf{w}_j. \qquad (15.26)$$

  This quantity must fall in the range $[P_k^{\min}, P_k^{\max}]$. Unlike the discussion on power line limits, this is a hard constraint. Enforcing this constraint is tricky, not the least because the expression in (15.26) will be unbounded for the distributions $\mathcal{W}$ that have been considered and that will be considered in the following text. Even if the quantity were not unbounded, guaranteeing that the constraint is satisfied with 100% probability would likely yield a very conservative constraint. With that idea in mind, instead we will impose a constraint that, again, is superficially similar to that just used for lines, i.e.

$$\mathcal{P}_k(P_k^g, P_k^{\min}, P_k^{\max}) \leq \epsilon',$$

where $\epsilon' \ll \epsilon$. For example, we might be saying that the probability that the output of generator k falls outside the desired range is at most ϵ'. Again this is not the only reasonable choice, and one could instead rely on concepts such as *value-at-risk* and *conditional-value-at-risk* [29, 30], which may be more appropriate.

We are now ready to present an initial, and very general, formulation for a "safe" or uncertainty-aware DC-OPF formulation:

$$\text{Min } C(\bar{P}^g, \alpha, \mathcal{W}) \tag{15.27a}$$

Subject to: (15.27b)

$$\alpha \in \mathcal{A} \tag{15.27c}$$

$$\sum_{km \in \delta(k)} \mathbf{p}_{km} = \bar{P}_k^g - P_k^d + \bar{w}_k - \sum_{j \in \mathcal{N}} \alpha_{kj} \mathbf{w}_j + \mathbf{w}_k, \quad \forall k \in \mathcal{N}, w \in \mathcal{W} \tag{15.27d}$$

$$\mathcal{P}_{km}(\mathbf{p}_{km}, L_{km}) \leq \epsilon \quad \forall km \in \mathcal{L}. \tag{15.27e}$$

$$\mathcal{P}_k(P_k^g, P_k^{\min}, P_k^{\max}) \leq \epsilon' \quad \forall k \in \mathcal{N} \tag{15.27f}$$

We have discussed constraints (15.27c), (15.27f), and (15.27e); constraint (15.27d) is just the stochastic variant of the DC balance equation (15.16c). Of course, (15.27) is not a usable formulation because of the explicit random variables therein.

15.2.1 Specific Formulations

To render practicable versions of the preceding problem we will first replace (15.27d) with a simpler expression. Recall that under the DC model (15.15), $\mathbf{p}_{km} = b_{km}(\boldsymbol{\theta}_k - \boldsymbol{\theta}_m)$ where for a bus k, $\boldsymbol{\theta}_k$ is the stochastic phase angle under the DC model. So we can rewrite (15.27d) in matrix form as

$$B\boldsymbol{\theta} = \bar{P}^g - P^d + \bar{w} - \alpha \mathbf{w} + \mathbf{w}, \quad \forall w \in \mathcal{W} \tag{15.28}$$

Here, B is the (DC version of) the *bus admittance matrix*, and is defined by

$$B_{km} = \begin{cases} -b_{km}, & \forall km \in \mathcal{L} \\ \sum_{km \in \mathcal{L}} b_{kj}, & m = k \\ 0, & \text{otherwise} \end{cases} \tag{15.29}$$

As such, B is a *generalized Laplacian* matrix, an object with a long history in algebraic graph theory. See, e.g., [31–33]. System (15.28) has many important properties.

An important fact fact is the following: suppose we add the rows of system (15.28). The left-hand side of the sum will be zero, by definition of the matrix B (or by inspection from (15.27d) because under the DC model $\mathbf{p}_{km} = -\mathbf{p}_{mk}$). Hence

$$0 = \sum_{k \in \mathcal{N}} \left(\bar{P}_k^g - P_k^d + \bar{w}_k \right) + \sum_{k \in \mathcal{N}} \mathbf{w}_k - \sum_{j \in \mathcal{N}} \left(\sum_{k \in \mathcal{N}} \alpha_{kj} \right) \mathbf{w}_j \tag{15.30}$$

$$= \sum_{k \in \mathcal{N}} \left(\bar{P}_k^g - P_k^d + \bar{w}_k \right) \tag{15.31}$$

by (15.24). In other words, we recover (the intuitive) observation that total controllable generation plus total stochastic generation must equal total load, which we had already pointed out in (15.22). The derivation that produced (15.31) used (15.24), which we justified by the assumption that the distribution of $\mathbf{w}$ should be "full dimensional." Without that assumption, (15.31) may not hold, and the assumption that total generator output, on average, equals total demand (i.e. (15.22) or (15.31)) does not necessarily hold.

As pointed out in the preceding text, a, system operator might, in principle (but likely not in practice), choose to violate the balancing condition $\sum_{k \in \mathcal{N}} (\bar{P}_k^g + \bar{w}_k) = \sum_{k \in \mathcal{N}} P_k^d$ by a small amount and for a short period if doing so confers some economic advantage without causing risk (and assuming that repeated frequency excursions and AGC applications are tolerable). However, we will insist on the balancing condition, in what follows. In this regard, note that if as in the case of OPF-type problems the generation quantities $\bar{P}_k^g$ are variables, then (15.31) constitutes a constraint to be satisfied.

Proceeding in this form, in summary, system (15.28) has (at least) one degree of freedom and one redundant constraint. We should obtain an equivalent system by fixing a variable θ_k to zero, and dropping a constraint. In fact,

LEMMA 15.1 *Suppose the network is connected and that the b_{km} values are all nonzero. Then the $(n-1) \times (n-1)$ submatrix obtained by removing from B any row and any column is invertible.*

This fact is essentially folklore; a proof can be read in the preceding citations.

We can now put these two observations together. Let us pick an arbitrary bus r (sometimes called the *reference* bus). We will compute solutions to (15.28) by fixing $\theta_r = 0$, thus removing the one degree of freedom (in the solutions) and by ignoring constraint r, which is redundant. To that effect, let $\hat{B}_r$ be the submatrix of by removing from B the row and column corresponding to r. To fix ideas assume $r = n$. Then write:

$$\check{B}_n = \begin{pmatrix} \hat{B}_n^{-1} & 0 \\ 0 & 0 \end{pmatrix}. \tag{15.32}$$

When $r \neq n$ we appropriately adjust the construction. The matrix $\check{B}_r$ may be termed, with some abuse of language, a *pseudo-inverse* of B (more properly, a Moore–Penrose pseudo-inverse). To unclutter notation we will simply write $\check{B}$ in what follows.

Thus the *unique* solution to (15.28) with $\theta_r = 0$ is given by

$$\theta = \check{B} \left(\bar{P}^g - P^d + \bar{w} - \alpha \mathbf{w} + \mathbf{w} \right). \tag{15.33}$$

The central point here is that

$$\sum_{k \in \mathcal{N}} \left(\bar{P}_k^g - P_k^d + \bar{w}_k \right) = 0 \tag{15.34}$$

(i.e. (15.31)), together with $\alpha \in \mathcal{A}$ is equivalent to (15.28). Thus, while (15.33) is still stochastic, we will employ it to simplify the uncertainty-aware DC-OPF formulation (15.27): first, we impose (15.34) as a constraint. Further, we use (15.33) and the DC flow equation $b_{km}(\theta_k - \theta_m)$ to substitute $\mathbf{p}_{km}$ in (15.27e) with an appropiate expression, namely

$$\mathbf{p}_{km} = b_{km} \pi_{km}^\top (\bar{P}^g - P^d + \bar{w} - \alpha \mathbf{w} + \mathbf{w}), \quad (15.35)$$

where for each line $km \in \mathcal{L}$, the (row) vector $\pi_{km}^\top \doteq \check{B}_k - \check{B}_m$.

Effectively we will be eliminating flows and phase angles from the formulation. Note that, crucially, (15.33) and (15.35) are linear in the optimization variables, $\bar{P}^g$ and α.

In preparation for simplifying (15.27e) we can make some observations regarding the random variables $\mathbf{p}_{km}$. Using (15.35), for any line km

$$\bar{p}_{km} \doteq \mathbf{E}(\mathbf{p}_{km}) = b_{km} \pi_{km}^\top (\bar{P}^g - P^d + \bar{w}) \quad (15.36)$$

regardless of the stochastic distribution. Up to now, of course, we have been vague about the meaning of the term "distribution." We will pretend, for the time being, that a precise distribution is known, and later relax the assumption.

Hence for any line km,

$$\mathbf{p}_{km} = \bar{p}_{km} + b_{km} \pi_{km}^\top (I - \alpha) \mathbf{w} \quad (15.37)$$

and so, writing $\Omega \doteq \mathbf{Var}(\mathbf{w})$ we have that

$$\mathbf{Var}(\mathbf{p}_{km}) = b_{km}^2 \pi_{km}^\top (I - \alpha) \Omega (I - \alpha^\top) \pi_{km}. \quad (15.38)$$

We stress that this expression holds without distribution or independence assumptions. We finally have enough machinery in place that we can restate constraint (15.27e) of our safe DC-OPF formulation, at least in the case of some distributions. To do so, we can first write a constraint that would be easy to justify to a nonmathematician.

Consider a requirement on each line km of the form

$$|\mathbf{E}(\mathbf{p}_{km})| + \eta \sqrt{\mathbf{Var}(\mathbf{p}_{km})} \leq L_{km}, \quad (15.39\mathrm{a})$$

i.e.,

$$|b_{km} \pi_{km}^\top (\bar{P}^g - P^d + \bar{w})| + \eta \sqrt{b_{km}^2 \pi_{km}^\top (I - \alpha) \Omega (I - \alpha^\top) \pi_{km}} \leq L_{km}, \quad (15.39\mathrm{b})$$

where $\eta > 0$ is a "safety" parameter chosen by the network operator. In the completely general setting that we have used so far, this constraint, although intuitively appealing, does not provide any guarantees whatsoever. However, it might well be possible to choose ν based on empirical evidence. The constraint, as written, can be seen as a stochastics-aware line limit constraint. Moreover, by (15.36), $\mathbf{E}(\mathbf{p}_{km})$ is linear in the optimization variables (the $\bar{P}^g$) while $\mathbf{Var}(\mathbf{p}_{km})$ is positive-semidefinite quadratic in the variables (α) and so (15.39b) can be restated as a (convex) cone constraint, of a type that can routine be handled by any modern optimization suite.

Having dealt, if incompletely, with constraint (15.27e) we can now turn to (15.27f), which can receive, superficially at least, a similar treatment. The goal of this constraint is to maintain the random variable P_k^g safely within the desired range $[P_k^{\min}, P_k^{\max}]$. In terms of the upper bound, we can thus write another safety constraint,

$$E(P_k^g) + \eta' \sqrt{\mathrm{Var}(P_k^g)} \leq P_k^{\max}. \qquad (15.40)$$

and similarly with the lower bound.

15.2.2 The Gaussian Case

Here we will first explain the modeling approach that has been favored in the literature, and which yields, from (15.27e), a precise chance constraint. This approach makes a number of assumptions about the distribution of **w** which may be questionable, namely that the $\mathbf{w}_k$ are (a) Gaussian and (b) pairwise independent. Under the Gaussian assumption, (15.37) yields that $\mathbf{p}_{km}$ is normally distributed with mean $\bar{p}_{km}$ and variance

$$\mathrm{Var}(\mathbf{p}_{km}) = b_{km}^2 \pi_{km}^\top (I - \alpha) D (I - \alpha^\top) \pi_{km}. \qquad (15.41)$$

with $D = \mathrm{diag}(\sigma_1^2, \ldots, \sigma_n^2)$ where $\sigma_k^2 \doteq \mathrm{Var}(\mathbf{w}_k)$, $1 \leq k \leq n$. Further, let us assume that the specific version of constraint (15.27e) that we pursue is that which enforces that the *probability* that line km is overloaded (i.e. that $|\mathbf{p}_{km}| > L_{km}$) is less than ϵ. Because $\mathbf{p}_{km}$ is Gaussian,

$$\mathrm{Prob}(\mathbf{p}_{km} > L_{km}) < \epsilon \quad \text{iff} \quad E(\mathbf{p}_{km}) + \eta(\epsilon)\sqrt{\mathrm{Var}(\mathbf{p}_{km})} \leq L_{km} \qquad (15.42a)$$

and

$$\mathrm{Prob}(\mathbf{p}_{km} < -L_{km}) < \epsilon \quad \text{iff} \quad E(\mathbf{p}_{km}) - \eta(\epsilon)\sqrt{\mathrm{Var}(\mathbf{p}_{km})} \geq -L_{km} \qquad (15.42b)$$

where $\eta(\epsilon) = \Phi^{-1}(1 - \epsilon)$, and Φ^{-1} is the inverse function to the Gaussian CDF, i.e. for any real t, $\Phi(t) = \int_{-\infty}^{t} e^{-x^2/2} dx$. Probabilistic requirements of this type are known as *chance* constraints; the two expressions to the right of the "iffs" can be abbreviated as

$$|E(\mathbf{p}_{km})| + \eta(\epsilon)\sqrt{\mathrm{Var}(\mathbf{p}_{km})} \leq L_{km}.$$

We have thus recovered (15.39) albeit with a specific choice of the safety parameter η. The alert reader may notice that imposing, separately, that $\mathrm{Prob}(\mathbf{p}_{km} > L_{km}) < \epsilon$ and that $\mathrm{Prob}(\mathbf{p}_{km} < -L_{km}) < \epsilon$ is not equivalent to imposing that

$$\mathrm{Prob}(\mathbf{p}_{km} > L_{km} \quad \text{or} \quad \mathbf{p}_{km} < -L_{km}) < \epsilon,$$

a point that we will return to later. In carrying out these derivations, we have been replicating work by Nemirovsky and Shapiro [34].

Now we turn to constraint (15.27f), which can receive a similar treatment. Note that by (15.26), $\mathrm{Var}(P_k^g) = \sigma_k^2 + \sum_{j \in \mathcal{N}} \alpha_{kj}^2 \sigma_j^2$. We obtain, again, a constraint of the form (15.40) with a specific interpretation for η'. It is explicitly given by

$$\bar{P}_k^g + \Phi^{-1}(1 - \epsilon') \sqrt{\sigma_k^2 + \sum_{j \in \mathcal{N}} \sigma_j^2 \alpha_{kj}^2} \leq P_k^{\max}, \qquad (15.43)$$

with a similar chance constraint used to handle the lower generation limit.

> **Comment**
>
> From a modeling perspective, there is a fundamental difference between the chance constraint $\mathrm{Prob}(|\mathbf{p}_{km}| > L_{km}) < \epsilon$ used to protect a line km, and a chance constraint $\mathrm{Prob}(P_k^g > P_k^{\max}) < \epsilon'$. Imposing either constraint does not make the event that we are protecting from impossible, rather it makes its probability small. In the case of lines, an overload can be acceptable if, indeed, it is rarely observed. Hence a chance constraint seems quite appropriate. In the case of generators, however, their upper or lower limit should not be violated (at all). We can capture this fact, up to a point, by insisting that $\epsilon' \ll \epsilon$, or, in the language of the safety constraints, that η' be significantly larger than η (e.g., three standard deviations rather than two).
>
> Yet even this approach seems somewhat inadequate. First, making ϵ' too small (i.e. η' too large), for many generators, may yield an infeasible optimization problem or a very expensive solution. Second, the event where a generator's output, according to the model, does exceed, e.g its upper limit, might still take place even if, according to the model, that is unlikely. Conditional on the event taking place the excess in principle could be very large. Then we would see a severe AGC shortfall. One can conceive of solutions to deploy in this regard, i.e. keeping a special "standby" generator or generators that are able to supply the needed power.
>
> A different approach would be to rely on the allied concepts of value-at-risk (VaR), and, especially, conditional-value-at-risk (CVaR) [29, 30], instead of chance constraints.
>
> For example, in the case of a generator k its CVaR at level $1 - \epsilon$ would be the expected value of the excess generation over the maximum, i.e. $\max\{P_k^g - P_g^{\max}, 0\}$, *conditioned* over events with probability at least $1 - \epsilon$. Placing an upper bound on CVaR, as a constraint in an OPF computation, would likely provide a more effective way to plan for backup generation than using a chance constraint.

The final ingredient in obtaining a concrete formulation for the uncertainty-aware DC-OPF problem (15.27) in the Gaussian, independent case is a development of the cost function $\mathcal{C}(\bar{P}^g, \alpha, \mathcal{W})$. The prior literature has relied on the expectation of total cost, i.e. $\sum_k \mathbf{E}(c_k(P_k^g))$, though this is not the only possibility, and we will suggest a variance-aware version of this objective later.

In the industry, the typical generation cost functions c_k range from linear to convex piecewise-quadratic. To simplify matters, let us assume that $c_k(x) = c_{k2}x^2 + c_{k1}x + c_{k0}$ for each k, where $c_{k2} \geq 0$. A routine computation yields the expression for $\mathbf{E}(c_k(P_k^g))$ that is used in (15.44).

We are now able to write a precise formulation for problem (15.27) in the Gaussian, independent case, with inputs ϵ, ϵ':

$$\mathrm{Min} \sum_{k \in \mathcal{N}} c_{k0}(\bar{P}_k^g + \alpha^T D\alpha) + c_{k1}\bar{P}_k^g + c_{k0} \qquad (15.44\mathrm{a})$$

Subject to: (15.44b)

$$\alpha \in \mathcal{A} \quad (15.44c)$$

$$\sum_{k \in \mathcal{N}} (\bar{P}_k^g + \bar{w}_k) = \sum_{k \in \mathcal{N}} P_k^d \quad (15.44d)$$

$$|b_{km} \pi_{km}^T (\bar{P}^g - P^d + \bar{w})| +$$
$$\Phi^{-1}(1-\epsilon)\sqrt{b_{km}^2 \pi_{km}^T (I-\alpha)D(I-\alpha^T)\pi_{km}} \leq L_{km} \quad \forall km \in \mathcal{L}, \quad (15.44e)$$

$$\bar{P}_k^g + \Phi^{-1}(1-\epsilon')\sqrt{\sigma_k^2 + \sum_{j \in \mathcal{N}} \alpha_{kj}^2 \sigma_j^2} \leq P_k^{\max}, \quad \forall k \in \mathcal{N} \quad (15.44f)$$

$$\bar{P}_k^g - \Phi^{-1}(1-\epsilon')\sqrt{\sigma_k^2 + \sum_{j \in \mathcal{N}} \alpha_{kj}^2 \sigma_j^2} \geq P_k^{\min}, \quad \forall k \in \mathcal{N}. \quad (15.44g)$$

This is a second-order cone program [35] that can be tackled with a variety of commercial and open-source optimization packages. As we discuss, the preceding formulation is not always necessarily practicable and a solution strategy may be needed.

15.2.3 Numerical Algorithmics

Second-order cone programs of the form (15.44) have been studied by several authors, see [7–10]. Also see [11, 36]. Typically, these studies have focused on the "global" case in which for any k, all participation factors α_{kj} are equal to a constant α_k, and $\mathcal{A}$ is described by $\sum_k \alpha_k = 1$, $\alpha_k \geq 0\,\forall k$. On a system with n buses and m lines, the total number of variables and constraints[4] will be on the order of $n + m$.

Surprisingly, on systems with thousands of buses (and lines), direct solution of (15.44) with an SOCP solver may not yield good results. It is worth noting that the MATPOWER [37] library now ships with realistic cases with tens of thousands of buses. In such cases formulation (15.44) may have on the order of a million constraints. The difficulties (observed in the previously mentioned literature) include slow running times and numerical issues, such as failure to converge. When we consider problems of the more general form (15.27) also formulated as SOCPs, the difficulties increase. The difficulties are significant because speed is essential, and more advanced planning models may require the solution of multiple problems instances of type (15.27).

An important point, however, is that there is an *opportunistic* algorithm that has achieved good results, and that this algorithm points out important structure about realistic instances of problem (15.44) and more generally (15.27). Here we will outline the core aspects of this algorithm as well as another, theoretically stronger algorithm, and briefly compare the two.

The difficulties we have just alluded to concern, in particular, the conic inequalities (15.44e). Every inequality of this type can be substituted with an equivalent system of the general form

---

[4] A secondary point is that the preceding formulation will not be used verbatim; a number of auxiliary variables and constraints will be introduced, but nevertheless the size of the system will remain $O(n+m)$.

$$|x_{km}| + y_{km} \leq L_{km} \qquad (15.45a)$$

$$\sqrt{\sum_{j \in \mathcal{N}} z_{km,j}^2} \leq y_{km} \qquad (15.45b)$$

where x_{km}, y_{km} and the $z_{km,j}$ are new variables; plus the linear constraints:

$$x_{km} = b_{km}\pi_{km}^\top(\bar{P}^g - P^d + \bar{w}) \quad \text{and} \quad z_{km} = \sqrt{\Phi^{-1}(1-\epsilon)b_{km}(I-\alpha)\pi_{km}\sqrt{D}}. \qquad (15.46)$$

Suppose we were to perform this reformulation of the conics (15.44e) in problem (15.44), and then simply ignore (i.e., remove) all constraints (15.45b) (but not (15.45a)). The resulting problem will be far easier than the original formulation (15.44) – the absence of the conics is the explanation. We will have a *linearly* constrained problem, modulo (15.44f) and (15.44g), which are conics but very simple structure.

Of course, the resulting formulation is a weakening of the original formulation. In particular, if we were to solve the new formulation, and $(\hat{z}_{km}, \hat{y}_{km})$ is the appropriate subvector of the optimal solution corresponding to a line km, we might have that $\sqrt{\sum_{j \in \mathcal{N}} \hat{z}_{km,j}^2} > \hat{y}_{km}$, i.e., (15.45b) is violated. What can be done in that case is to add the linear constraint

$$y_{km} \geq \sqrt{\sum_{j \in \mathcal{N}} \hat{z}_{km,j}^2} + \frac{1}{\sqrt{\sum_{j \in \mathcal{N}} \hat{z}_{km,j}^2}} \sum_j (z_{km,j} - \hat{z}_{km,j})\hat{z}_{km,j}. \qquad (15.47)$$

The linear function on the right-hand side is a first-order estimate of the left-hand side of (15.45b). Hence any (z_{km}, y_{km}) satisfying (15.45b) satisfies (15.47) and we are correct in imposing (15.47), which, incidentally, cuts off the point $(\hat{z}_{km}, \hat{y}_{km})$.

We thus obtain a (very simple) algorithm:

Algorithm 15.1 Cutting-plane algorithm for problem (15.44)

c.1. We begin with the formulation of (15.44) with the conics (15.44e) omitted.
c.2. At each iteration of the algorithm we solve a relaxed problem.
c.3. At a given iteration if the current solution a corresponding conic (15.44e) by a sufficiently large enough amount, then add corresponding the cut (15.47).
c.4. Otherwise, we stop.

In developing this idea, we have rediscovered a very old algorithm due to Kelley [38]. What we are doing is to approximate a conic with a polyhedral outer approximation constructed using gradients. This idea can be applied to any optimization problem with convex constraints; the problem is that a straightforward application of the idea will in general render a very slow algorithm that may run into numerical difficulties and as a result may halt after many (increasingly slow) iterations without a valid solution. Despite this difficulty, since its inception the idea has been repeatedly applied, and in a huge variety of contexts.

Algorithm 15.1 was implemented in [7], which describes experiments using several cases in the MATPOWER library corresponding to systems based on the Polish grid,

Table 15.1 Performance of Algorithm 15.1 on Polish grid examples.

Case	Buses	Generators	Lines	Time (s)	Iterations	Barrier Iterations
Polish1	2,383	327	2,896	13.64	13	535
Polish2	2,746	388	3,514	30.16	25	1,431
Polish3	3,120	349	3,693	25.45	23	508

with 2,000–3,000 buses, approximately 5,000 lines, and hundreds of generators. The experiments used the global control case ($\alpha_{kj} = \alpha_k$ for all k) with some 50 sources of stochastic injection, modeling different levels of overall average stochastic generation as a fraction of the total and different levels of stochastic variance. The computation used values for ϵ and ϵ' corresponding to two or three standard deviations. Step c.3 was implemented by only introducing a cut for the particular line km that attained the largest violation, rather than adding a cut whenever a violation occurred.

Table 15.1, copied from [7], describes experiments using several versions of the Polish system in MATPOWER. "Iterations" shows the number of steps (c.2–c.3) executed by Algorithm 15.1, while "Barrier iterations" shows the total number of steps required by the algorithm that was used to solve relaxed problems in (c.2) (implemented using CPLEX [39]). Finally, "time" is the total time required by the procedure. This performace vastly outstripped that of CPLEX applied to the full SOCP.

The reason for this success, *a posteriori*, is that out of thousands of lines, only a handful would give rise to infeasibilities discovered in iterations c3. And, in fact, it was usually the *same* handful of lines, not just during one particular run of the algorithm, but during different runs using different parameter choices (e.g., amount of stochastic generation). One way to interpret this empirical observation is to say that a realistic grid is not a random object – power flows will follow nontrivial trajectories, and the introduction of stochastics exposes lines to risk in a nonuniform way. From this perspective, a very useful function of the cutting-plane algorithm is that it quickly discovers this important set of risky lines.

Cutting-Plane Solution of Standard DC-OPF

In fact, it turns out that the idea embodied in Algorithm 15.1, or rather an idea that is closely related, is already known in the industry [40, 41] in the context of the DC-OPF problem (15.16). Using $\mathbf{p}_{km} = b_{km} \pi_{km}^T (\bar{P}^g - P^d)$ (as in (15.35)) we can remove (15.16c) and rewrite (15.16d), obtaining the equivalent reformulation of DC-OPF:

$$\text{Min} \sum_k c_k(P_k^g) \tag{15.48a}$$

Subject to: (15.48b)

$$|b_{km} \pi_{km}^T (\bar{P}^g - P^d)| \leq L_{km} \quad \forall km \tag{15.48c}$$

$$P_k^{\min} \leq P_k^g \leq P_k^{\max} \quad \forall k. \tag{15.48d}$$

The parameters π_{km} are known under various names, such as *shift factors*, or *power transfer distribution factors* but, as we have said, are computed easily from an appropriate pseudo-inverse of the B matrix. While (15.48) is equivalent to (15.16) it suffers from the disadvantage that constraints (15.48c) are very dense (under adequate assumptions they are, provably, completely dense). Moreover, the vectors π_{km} require many digits for accurate representation.

Superficially, then, this reformulation of DC-OPF does not appear promising. However, as observed in the practice, only a few of the inequalities (15.48c) will be binding. And an experienced operator will likely know, in advance, which lines are likely to give rise to binding constraints. One can thus implement an iterative algorithm to solve the problem (15.48) that discovers, opportunistically, those constraints (15.48c) that are needed. The discovery of the violated inequalities (like our step c.3) would in an initial phase only examine those lines that are, *a priori*, expected to be binding, rather than *all* lines. The restricted implementation of step c.3 will significantly speed up computation in the initial stage. Following the initial stage one switches to a full search over all lines, and ideally the full search will be performed a small number of times. In a complete implementation we would deploy some cut management mechanism (i.e., a mechanism that removes some of the cuts (15.48c) if they are deemed not useful after some step). However, even without such elaborations, this opportunistic algorithm, properly instrumented, ends running far faster than a straight-solve of (15.16). The algorithm is only superficially like our method 15.1 mentioned previously, but the central point is that both algorithms leverage the empirically observed detail that only a few line limit constraints are "important."

A Formal Study

Recent work [42] studies the issue just discussed. They find that, indeed, across many large cases in the MATPOWER library, the number of binding constraints (15.16d) (or (15.48c)) is quite small in the solution to DC-OPF (and other problems, including DC-Unit commitment) – typically a few percent and frequently below 1%. In fact, critically, the set of binding constraints remains unchanged even under large changes of the loads.

This observation motivates the development a formal procedure for *a priori* constraint screening. Recall that the OPF computation will be periodically repeated (once every five minutes, e.g.). The procedure suggested by [42] would, rather less frequently, invest some computational effort in identifying a set of line limit constraints that will remain binding even under significant changes of the loads. The operator could afford to spend a nontrivial amount of computational effort on such a procedure (see the following text for an example) if it is sporadically run – indeed the computation would be separate from the DC-OPF computation and could be performed in parallel. In the case of the chance-constrained DC-OPF problem, the screening process could also be used to explicitly impose those conic constraints arising from identified lines, for example as the starting point for the cutting-plane procedure. Also see [8].

As an example of the procedure, let us consider a specific line ij and consider a problem of the form

$$\text{Max } b_{ij}(\theta_i - \theta_j) \tag{15.49a}$$

Subject to: (15.49b)

$$\sum_{km \in \delta(k)} b_{km}(\theta_k - \theta_m) = P_k^g - P_k^d \quad \forall k \tag{15.49c}$$

$$|b_{km}(\theta_k - \theta_m)| \le L_{km} \quad \forall km \tag{15.49d}$$

$$P_k^{\min} \le P_k^g \le P_k^{\max} \quad \forall k, \tag{15.49e}$$

$$P^d \in \mathbf{D}. \tag{15.49f}$$

This problem maximizes the DC flow on line ij. Constraints (15.49c)–(15.16e) are the standard DC power-flow constraints. However, in this formulation the loads P^d are variables; $\mathbf{D}$ is a set that includes the current loads but allows for possibly large changes. Clearly, if the value of the objective is smaller than L_{ij} then the upper limit constraint for line ij is redundant even if the loads were to change to any value in the set $\mathbf{D}$. A similar formulation is used to minimize the flow on ij.

Performing this computation on each line will identify unneeded line limit constraints (upper or lower limits). The computation could be expensive because in principle it needs to be performed for every line. However (a) the computation, as we have said, can be performed separately from the operational DC-OPF computation and (b) the computation can be parallelized.

In computational experiments presented in [42] it is shown that even under very large load changes (on the order of 50%, e.g.) modeled using the set $\mathbf{D}$ previously mentioned, the proportion of active line limit constraints grows to fewer than 10%. These observations provide strong justification for elaborate constraint screening mechanisms such as those outlined previously and, more generally, for cut management procedures.

A Provably Strong Reformulation of Conic Constraints

Kelley's algorithm, as we have described already, has been deployed in many problem contexts. In a general implementation one needs a well thought-out criterion for deciding which cuts to introduce in (the general version of) step c.3 of Algorithm 15.1, from possibly many cuts that are violated by the current solution (depending on the specific problem context, there could be infinitely many valid cuts). The main theoretical difficulty is that we have not presented a convergence analysis of the algorithm. Indeed, practical implementations may suffer from the well-known "tailing-off" effect where the iterate vectors are cut off by very small amounts, and we produce a sequence of near-parallel cuts, without clear convergence. Is there a theoretically stronger alternative?

Here, we are concerned with the procedure as applied already, i.e. used to produce linear cuts that act as surrogates for convex conic constraints. Nemirovsky and Ben–Tal [43] consider a generic conic-quadratic program, i.e. a problem of the form

$$\text{Min}_x \left\{ f^T x : A_0 x \ge b_0, \sqrt{A_h x - b_h} \le c_h^T x + d_h, \ h = 1, \ldots, m \right\}. \tag{15.50}$$

Problem (15.44) can be reformulated in this way. We could approach problem (15.50) using Kelley's algorithm as already mentioned. In contrast, [44] proposes a linear reformulation, or more precisely, a linear *lifted relaxation* with the following attributes. Let N be the dimension of x and let M be the total number of rows in matrices A_h, $0 \leq h \leq m$, and let $0 < \epsilon < 1$ be a given tolerance.

(a) The reformulation, $\mathbf{LP}(\epsilon)$, is a linear program of the form

$$\text{Min}_{x,u} \left\{ f^T x : A_0 x \geq b_0, \ Px + Qu \geq g \right\}. \tag{15.51}$$

The number of variables and constraints is $O((N+M)\log(1/\epsilon))$.

(b) Any feasible solution x to (15.50) can be lifted to a feasible solution (x,u) to $\mathbf{LP}(\epsilon)$.

(c) Given a feasible solution (x,u) to $\mathbf{LP}(\epsilon)$ we have that every $1 \leq h \leq m$

$$\sqrt{A_h x - b_h} \leq (1+\epsilon)(c_h^T x + d_h).$$

In brief by (b) we have that the projection of $\mathbf{LP}(\epsilon)$ to x-space will be a polyhedron that contains the feasible set for (15.50), and thus solving $\mathbf{LP}(\epsilon)$ will provide an underestimate to the value of (15.50). Moreover, (c) shows that simply solving the linear program $\mathbf{LP}(\epsilon)$ will provide a provably good surrogate for (15.50). Finally, (a) shows that this is a polynomial-time transformation (including an appropriate dependence on ϵ).

In summary, this approach is a provably accurate and provably efficient replacement for Kelley's algorithm. On the negative side, however, this replacement is not opportunistic – in particular it will not leverage the type of empirical observation that was discussed in the preceding text. The lifted formulation, despite its polynomial size, can be quite large. Given the experiments discussed previously we would estimate that a reliance on this procedure will possibly not pay off. One may still ask the question as to whether Kelley's algorithm or the Nemirovsky–Ben–Tal method is superior for broad problem classes of the form (15.50). In experiments using Gurobi [45], neither method dominates [46].

15.2.4 Two-Tailed versus Single-Tailed Models

We have detailed how, in the Gaussian independent case, for a line km we obtain convex representations for each of the two chance constraints

$$\text{Prob}(\mathbf{p}_{km} > L_{km}) < \epsilon \quad \text{and} \quad \text{Prob}(\mathbf{p}_{km} < -L_{km}) < \epsilon. \tag{15.52}$$

But as we noted, this requirement is not equivalent to the ideal chance constraint, i.e.

$$\text{Prob}(|\mathbf{p}_{km}| > L_{km}) < \epsilon. \tag{15.53}$$

Indeed, imposing (15.52) may only yield $\text{Prob}(|\mathbf{p}_{km}| > L_{km}) < 2\epsilon$. This raises the question: Does (15.53) admit a convex, compact representation?

This issue has been taken up in [47], which shows that, remarkably, under fairly general assumptions (including the Gaussian case) constraint (15.53) has a compact, convex representation. In the case of the general stochastic flow on a line km given by (15.37), i.e.

$$\mathbf{p}_{km} = \bar{p}_{km} + b_{km}\pi_{km}^\top(I - \alpha)\mathbf{w}$$

the representation is as follows:

$$b_{km}^2 \pi_{km}^\top (I - \alpha)\Omega(I - \alpha^\top) + s^2 \leq \epsilon(L_{km} - \pi) \tag{15.54}$$
$$|\bar{p}_{km}| \leq s + \pi \tag{15.55}$$
$$L_{km} \geq \pi \geq 0, \ s \geq 0. \tag{15.56}$$

Here, π and s are two additional variables. It should prove straightforward to incorporate this formulation into a cutting-plane procedure such as that described previously without increased computational cost. An interesting question is whether this sharper formulation gives rise to significant savings. See [48] for related investigations in the AC case.

15.2.5 Gaussian Data or Not?

Formulation (15.44) was developed under the assumption that load uncertainty is Gaussian and that the different stochastic sources are independent. Are those valid assumptions? This is an important topic that is still being debated. Some of the initial discussions on this topic were to some extent complicated by the fact that we are interested in real-time or near-real time time frame, or more precisely the time frame in which AGC operates, rather than long-term distributions – days or months, and this distinction was not always clear. An initial study [24], performed in the appropriate time domain, does seem to argue for independence.

The justification for the Gaussian and independent assumptions given in [7] goes as follows: real-time variability is due to turbulence, e.g., storm cut off as pointed out in [25]. Turbulence is characterized by vortices of wind activity, with the typical vortex having a diameter of less than $100m$, which exceeds the typical separation of wind turbines in a wind farm. Thus, the turbulence-driven output from a wind farm is the sum of independent but similar random variables. Such an argument does provide a valid claim for Gaussianity, but how about independence (between different wind farms). The aforementioned storm cut-off effect might argue for geographical correlation, at any given point in time, between wind farms that are experiencing the weather front change.

And how about data-driven models? Recently, utilities have begun to deploy fast, accurate sensors known as PMUs (phasor measurement units). Such sensors report on the order of 30–100 times per second, perform filtering (i.e., anti-aliasing) and provide accurate readings as per an error standard [49, 50]. PMUs report (complex) voltage and current at their location, from which one can deduce power flows. Reference [22] reports on output from some 240 PMUs over a two-year period (amounting to $28Tb$ of data) from an industrial partner [22] from a region with high renewable penetration (also see Section 15.3). Statistical analyses show that voltages are frequently not Gaussian – though they fail a Gaussianity test due to being too *light-tailed*, or sub-Gaussian, rather than otherwise. Note that the voltage data being analyzed shows the aggregated system behavior, i.e. the sum of stochastic injection due to wind variability, AGC actions, and other ambient stochastics and also control measures.

This fact suggests that a generic line safety constraint (15.39b) would be justified. With $\eta = \Phi^{-1}(1-\epsilon)$ it would amount as a perhaps overly conservative chance-constraint – recall that the square root term in (15.39b) indicates the standard deviation of flow on line km, under all distributions; but additionally a quantity that we expect to have a sub-Gaussian distribution. For a different approach to handling non-Gaussian distributions, see [51] (and the following text). In Section 15.3 we will return to the study in [22].

15.2.6 Robust Optimization

The field of robust optimization concerns solution of problems where some of the data parameters are not precisely known, and one desires to compute a solution that does not commit to any one realization of the data at the expense of becoming a very bad solution should the data vary way from this realization. See [44, 52] for general background.

Ideas from robust optimization have been applied to chance-constrained models for DC-OPF. To fix ideas, consider the generic safety constraint (15.39b) for a line km, rewritten here for convenience

$$|\bar{p}_{km}| + \eta\sqrt{b_{km}^2 \pi_{km}^\top (I-\alpha)\Omega(I-\alpha^\top)\pi_{km}} \leq L_{km}$$

where we write $\bar{p}_{km} = b_{km}\pi_{km}^\top(P^g - P^d + \bar{w})$. In this expression Ω, which is the covariance matrix for the stochastic injection vector $\mathbf{w}$, may not be precisely known, even under the assumption that the shape of the distribution for $\mathbf{w}$ is known. In general, however, Ω will be estimated from data and it is of interest to rephrase constraint (15.39b) to take possible estimation errors into account.

One way to do this is to adopt ideas from robust optimization. Thus, let $\bar{\Omega}$ denote a particular *estimate* for Ω, and let A denote a set (in matrix space) that contains $\bar{\Omega}$. The reader may think of A as a ball with center $\bar{\Omega}$. Then rather than simply setting $\Omega = \bar{\Omega}$ in (15.39b), we would like to extend (15.39b) to all $\Omega \in A$, i.e. to impose (in our uncertainty aware DC-OPF formulation) the constraint:

$$|\bar{p}_{km}| + \eta \max_{\Omega \in A}\left\{\sqrt{b_{km}^2 \pi_{km}^\top(I-\alpha)\Omega(I-\alpha^\top)\pi_{km}}\right\} \leq L_{km}. \tag{15.57}$$

This expression captures the essence of robust optimization. In the Gaussian, independent case in which Ω is the diagonal of variances of stochastic injections (what we called D in (15.41)) this is an ambiguous chance-constraint [34, 53–55]. Note that by varying the size of the set A (e.g., by varying the radius of a ball around a point estimate $\bar{\Omega}$) we control the degree of risk aversion in assuming misestimation.

Even though (15.57) can amount to an infinite set of conic constraints, it can be handled using a cutting-plane procedure for a variety of "ambiguity" sets A [7]. In brief, for a line km and given $\bar{p}_{km}$ and participation matrix $\bar{\alpha}$, we need to check whether

$$|\bar{p}_{km}| + \eta \max_{\Omega \in A} \left\{ \sqrt{b_{km}^2 \pi_{km}^\top (I - \bar{\alpha}) \Omega (I - \bar{\alpha}^\top) \pi_{km}} \right\} > L_{km}, \tag{15.58}$$

and if not, to construct a linear inequality that cuts off $(\bar{p}_{km}, \bar{\alpha})$. This amounts to checking whether (for an appropriate constant L'_{km})

$$\max_{\Omega \in A} \left\{ b_{km}^2 \pi_{km}^\top (I - \bar{\alpha}) \Omega (I - \bar{\alpha}^\top) \pi_{km} \right\} > L'_{km}. \tag{15.59}$$

This optimization problem, in Ω, can be efficiently solved if, e.g., A is an ellipsoid or is polyhedral. If the optimizer Ω^* does satisfy (15.59) then we proceed as in (15.47) to obtain a cut in $(\bar{p}_{km}, \alpha)$-space that cuts off $(\bar{p}_{km}, \bar{\alpha})$ by linearizing the expression $\sqrt{b_{km}^2 \pi_{km}^\top (I - \alpha) \Omega^* (I - \alpha^\top) \pi_{km}}$. See [7] for details.

Recent work has focused on various robust versions of DC-OPF (see [11, 28, 47]).

15.2.7 The AC Case

The preceding discussions on uncertainty-aware formulations of OPF focused on the DC case. AC models are, of course, more challenging due to their nonconvexity. Some recent work [48, 56, 57] has suggested interesting work-arounds that enable modeling of stochastics.

A comprehensive model is presented in [56], which considers grid support mechanisms that go beyond the simple AGC control discussed previously, including reactive power generation (at renewables), voltage control, and reserve generation capacity. Such mechanisms are described using realistic response policies. The nonlinear AC constraints are linearized through their first-order Taylor expansion; chance-constraints are modeled as was done previously, assuming Gaussian distributions.

The approach in [57] (extending work in [51]) relies on a technique known as *polynomial chaos*. Polynomial chaos expansions of random variables has been described as "Fourier series for random variables." When performing a polynomial chaos expansion (PCE) of a random variable **w** one chooses a "basis" of polynomials $\psi_k(\omega)$, $k = 1, 2, \ldots,$. This basis has been chosen so as to be orthogonal under the probability measure defining **w**. Having chosen the basis, we can expand $\mathbf{w} = \sum_{k=0}^{+\infty} x_k \psi_k$ where the x_k are reals computed by taking an appropriate inner product of **w** and ψ_k. The expansion can be truncated to a finite number of terms. With the expansion in place, a stochastic flow quantity (such as the expression in (15.35), in the DC case) is rewritten as a sum of polynomials in **w**; crucially, *moments* of flow quantities can be analytically computed in terms of the coefficients x_k given previously. This feature permits the efficient representation of chance constraints (see [57] for details).

To conclude this section we point out an approach that appears not to have been considered, in the AC case. Let us assume that we consider a case of chance constraints (or safety constraints) for which an explicit algebraic representation is available (such as an SOCP constraint, or even a nonconvex constraint). Of course, the overall problem we write will be nonconvex because of the AC power-flow equation. However, one could consider solution (or local solution) of the problem using a nonlinear solver such as KNITRO [18] or IPOPT [19]. This approach is suggested by the remarkable success that these solvers have attained on standard AC-OPF problems, where, as certified by convex relaxations, they often compute solutions of excellent quality, frequently faster (sometimes much faster) than the relaxations.

15.3 Data-Driven Models and Streaming Covariance Estimation

Consider again constraint (15.39b) Some of the preceding discussions focused on whether or not the underlying stochastics are Gaussian. However, a practical approach would accept the use of this constraint as a straightforward strengthening of the standard line limit constraint, and instead would focus on appropriate estimation of the matrix Ω from data. We remind the reader that the quantity inside the square root is the variance of power flow on line km, under the DC model.

In this context, an additional and very important fact uncovered by the investigation in [22] described previously is that PMU output seems to have low rank, or more precisely, that covariance matrices of PMU output have low rank, from a numerical perspective. Table 15.2 shows typical behavior. Here we consider the output of 50 PMUs over 1 minute; eigenvalues have been scaled so that the largest takes value 1.0. In addition to the rapidly declining eigenvalues we also see large spectral gaps in particular among the top four or five eigenvalues. Similar behavior can be observed over multiple time scales, and can also be observed in voltage magnitudes.

Given that the utility that produced the data in [22] does employ a large amount of wind power, we would expect that the covariance matrix of stochastic injections, i.e. the matrix Ω, also has low rank. Thus, the goal of development of rapid, memory-efficient data-driven methods for computing accurate estimates of Ω, in the low rank regime, becomes pertinent. A data-driven estimate of Ω would be used in the safety constraints (15.39b) as part of a variance-aware DC-OPF computation.

We note that a large amount of research has been devoted to covariance estimation from sample data; however, a point of interest is to carry out the estimation using data in *streaming* mode: this means that at any point the estimation procedure should rely

Table 15.2 Top scaled eigenvalues for covariance matrix of phase angles.

#	Scaled Eigenvalue
1	1.000
2	0.078
3	0.012
4	0.009
5	0.007
6	0.004
7	0.003
8	0.002
9	0.001

on recent observations, with old enough observations discarded. In the grid data context, this requirement would be motivated by a desire to protect the computation from features appearing in old (stale) data. An additional goal would be to save memory – this goal might be less compelling in today's transmission systems (which might only be endowed with a few hundred sensors) but may become important in future power grids if the number of sensors becomes very large.

An important development in this domain is the analysis of the *noisy power method* used in [58]. This is an adaptation to the popular power method for approximate principal component analysis (PCA). In short, the method in [58]

- Concerns a time series of d-dimensional data;
- Assumes that the data is drawn from a Gaussian *spiked covariance* model (see below) or, more generally, from a sufficiently light-tailed distribution; and
- Aims to compute a close estimate of the subspace spanned by the top k eigenvectors of the covariance matrix of the data (i.e., the eigenvectors corresponding to the top k eigenvalues) where $k \ll d$.

Under appropriate assumptions, the method in [58] requires a number of samples that grows linearly in kd to obtain an appropriate estimation within error ϵ (in a matrix norm sense) from the space spanned by the true subspace being estimated. Note that $O(kd)$ is a lower bound because this is the amount of space needed to write k vectors of dimension d. The number of samples that is derived defines the streaming aspect of the model – observations older than this number can be discarded, and if fewer observations can be kept, the error bound will correspondingly suffer (see [58] for details). The "appropriate assumptions" we noted in the preceding text are nontrivial and include the assumption of a large gap between the kth and $(k+1)^{st}$ eigenvalues. While this assumption appears justified, e.g., by the Table 15.2, it may be difficult to verify on an online fashion.

The spiked covariance model [59, 60] mentioned is of relevance here, in particular in light of Table 15.2 (i.e., the implicit low-rank structure evinced by this table). The model posits a zero-mean, d-dimensional time series x_t generated by a process of the form

$$\mathbf{x} = A\mathbf{z} + \mathbf{v}, \tag{15.60}$$

where A is an unknown $d \times k$ matrix with orthogonal columns, $\mathbf{z}$ is k-dimensional, and $\mathbf{v}$ is d-dimensional. Neither $\mathbf{z}$ nor $\mathbf{v}$ is observed directly – only the x_t are observed. Further, the entries in $\mathbf{z}$ are drawn from a standard Gaussian distribution (and are pairwise independent), while the entries $\mathbf{v}$ represent noise and are also Gaussian, pairwise independent, and independent of $\mathbf{z}$, and with covariance $\sigma^2 I_d$. It is implicitly assumed that the quantity σ^2 is small. In any case, it follows that $\mathbf{x}$ is Gaussian, with covariance $AA^\top + \sigma^2 I_d$. Ignoring the noise term we have a low-rank (when $k \ll d$) covariance matrix.

Note that if, for $1 \leq j \leq k$, we write λ_j for the norm of the jth column of A, process (15.60) can be rewritten as $x_t = U\Lambda z_t + v_t$, where the columns of U are orthonormal and $\Lambda = \text{diag}(\lambda_1, \ldots, \lambda_k)$. Under these conditions, the covariance of x_t takes the form

$U\Lambda^2 U^\top + \sigma^2 I_d$. It is the subspace spanned by the columns of U that is estimated by the method in [58]. An important point is that this is not the same as actual covariance matrix estimation (i.e., estimation of $U\Lambda^2 U^\top$) because there remains to estimate the actual eigenvalues.

A relevant extension of the spiked covariance model is that where the underlying stochastics is nonstationary, i.e. (15.60) is replaced with a model of the type $x_t = A_t z_t + v_t$. This is an important extension, in the streaming case, because it is of interest to detect the change as soon as possible, whereas a computation assuming a static model such as (15.60) may provide an incorrect estimate. In [61] the authors show how to optimally adapt the noisy power method to the nonstationary case.

To complete this section we note that spiked covariance models seem closely related to the independently developed concept of *factor models*, arising in *capital asset pricing models* or CAPM, developed in the 1960s; this work merited a Nobel Prize in Economics. See [62] for an overview of related topics. Here we are given a d-dimensional time series x_t generated by a process of the form

$$\mathbf{x} = V\mathbf{z} + \boldsymbol{\epsilon}.$$

The matrix V is $d \times k$ with $k \ll d$. Moreover the entries in $\mathbf{z}$ and $\boldsymbol{\epsilon}$ are Gaussian, and the entries in $\boldsymbol{\epsilon}$ are pairwise independent and independent of $\mathbf{z}$. This is superficially similar to a spiked covariance model, except that the columns in V are general (not orthogonal) and $\mathbf{z}$ has general covariance matrix F (i.e., the entries in z are not necessarily independent). It follows that the covariance matrix of $\mathbf{x}$ has the structure

$$VFV^\top + \operatorname{diag}(s_1^2, \ldots, s_d^2)$$

where s_j^2 is the variance of ϵ_j. Thus, it appears that factor models are better-suited for modeling uncertainty in, e.g., renewables in that they allow correlation among the "factors" $\mathbf{z}$, unlike spiked covariance models. As mentioned previously, the topic of interest would be the rapid, accurate, and space efficient estimation of $VFV^\top$ from data.

Factor models for covariance are used in financial portfolio management; in this context the goal of providing *robust* covariance estimators has been tackled in [63]. We will return to this particular topic, and related estimations of $VFV^\top$, in Section 15.4.

15.4 Tools from Financial Analytics: Variance-Aware Models, Sharpe Ratios, Value-at-Risk

Here we return to some of the issues outlined in at the start of this chaptaer. As the concept of power engineering operations continues to evolve toward shorter time frames of control, and as the sources of variability become more pronounced, grids are becoming more exposed to *risk*. Another, possibly related form of risk will be the exposure to *markets*. Power systems rely on markets to deliver power on a reliable basis, the probability of a default in generator by a market participant becomes more pressing when there is increased exogenous variability.

OPF computations can be made aware of risk, not simply as a quantity to be kept small or otherwise controlled, but as a parameter to be hedged against. Indeed, different operators may display different proclivity toward risk – and be prepared to pay, i.e. deploy adequate resources, when a planned control action fails. Our discussion on the use of conditional value-at-risk to control generator output (Section 15.2.2) is a simple example of this paradigm.

The power engineering community has initiated some work in this direction. See e.g., [64–67]. In [68] the authors consider a modification of safety-constrained DC-OPF where the objective includes an explicit term to account for "system variance." That is to say, instead of solving an optimization problem that seeks to minimized expected cost of controllable generation, i.e. $\sum_{k \in \mathcal{N}} \mathbf{E}(c_k(P_k^g))$ in (15.25), we seek to minimize an expression of the form

$$\sum_{k \in \mathcal{N}} \mathbf{E}(c_k(P_k^g)) + \lambda \mathbf{V} \qquad (15.61)$$

where $\mathbf{V}$ is the variance of a quantity of interest, and $\lambda > 0$ is a risk-aversion parameter. As examples of this problem we could have

- $\mathbf{V}$ = the variance of cost of controllable generation, in which case (15.61) is the well-known mean-variance optimization problem; and
- $\mathbf{V}$ = variance of power flow on some subset of the equipment, e.g., a specific subset of transformers.

In either case λ enforces a tradeoff between cost and risk. Setting $\lambda = 0$ recovers the standard safety-constrained optimization problem. Experiments in [68] show, however, that $\lambda = 0$ may incur a high variance measure, whereas setting $\lambda > 0$ (even if small) will often greatly decroease variance, without increasing generation cost; this phenomenon is due, again, to the fact that system risk seems concentrated on a few "congested" lines. Reference [68] also considers the use of Sharpe-ratios [69] to exercise the cost-risk tradeoff.

We expect that in the coming future, explicit risk quantification and associated hedging will see increased use in power engineering operations.

15.5 Conclusion

As power grids evolve toward algorithmic control in shorter time frames, and increased use of data, the use of data-driven stochastic risk abatement will become increasingly necessary. In this chapter we reviewed several central topics and recent results. We expect that the near future will see rapid growth in this direction.

References

[1] G. Andersson, *Modelling and Analysis of Electric Power Systems*. Zurich: Power Systems Laboratory, ETH Zürich, 2004.

[2] A. Bergen and V. Vittal, *Power Systems Analysis*. Toronto: Prentice-Hall, 1999.

[3] J. D. Glover, M. S. Sarma, and T. J. Overbye, *Power System Analysis and Design*. Toronto: CENGAGE Learning, 2012.

[4] D. Bienstock, *Electrical Transmission System Cascades and Vulnerability: An Operations Research Viewpoint*. Philadelphia: Society for Industrial and Applied Mathematics, 2015.

[5] ARPA-E, "Grid Optimization Competition," https://gocompetition.energy.gov/

[6] I. Hiskens and R. Davy, "Exploring the power flow solution space boundary," *IEEE Trans. Power Systems*, vol. 16, pp. 389–95, 2001.

[7] D. Bienstock, M. Chertkov, and S. Harnett, "Chance-constrained DC-OPF," *SIAM Review*, vol. 56, pp. 461–495, 2014.

[8] L. Roald, S. Misra, T. Krause, and G. Andersson, "Corrective control to handle forecast uncertainty: A chance constrained optimal power flow," *IEEE Transactions on Power Systems*, vol. 32, no. 2, pp. 1626–1637, Mar. 2017.

[9] L. Roald and G. Andersson, "Chance-constrained AC optimal power flow: Reformulations and efficient algorithms," *IEEE Transactions on Power Systems* 33, pp 2906–2918, 2018.

[10] L. Roald, T. Krause, and G. Andersson, "Integrated balancing and congestion management under forecast uncertainty," in *2016 IEEE International Energy Conference (ENERGYCON)*, pp. 1–6, Apr. 2016.

[11] M. Vrakopoulou and I. A. Hiskens, "Optimal policy-based control of generation and hvdc lines in power systems under uncertainty," in *2017 IEEE Manchester PowerTech*, pp. 1–6, June 2017.

[12] J. Carpentier, "Contribution á l'étude du dispatching économique," *Bulletin de la Société Française des Électriciens*, vol. 8, pp. 431–447, 1962.

[13] M. Cain, R. ONeill, and A. Castillo, "History of optimal power flow and formulations," fERC staff technical paper, 2012.

[14] D. K. Molzahn and I. A. Hiskens, "A survey of relaxations and approximations of the power flow equations," *Foundations and Trends in Electric Energy Systems*, vol. 4, p. 1–221, 2019.

[15] M. Garey and D. Johnson, *Computers and Intractability: A Guide to the Theory of NP-Completeness*. New York: W. H. Freeman, 1979.

[16] D. Bienstock and A. Verma, "Strong NP-hardness of AC power flows feasibility," *arXiv preprint arXiv:15.07315*, 2015.

[17] A. Verma, "Power grid security analysis: An optimization approach," Ph diss., Columbia University, 2009.

[18] R. Byrd, J. Nocedal, and R. Waltz, "KNITRO: An integrated package for nonlinear optimization," in *Large-Scale Nonlinear Optimization*, G. di Pillo and M. Roma, eds. Springer, pp. 35–59, 2006.

[19] A. Wächter and L. T. Biegler, "On the implementation of a primal-dual interior point filter line search algorithm for large-scale nonlinear programming," *Mathematical Programming*, vol. 106, pp. 25–57, 2006.

[20] D. Bienstock, J. Blanchet, and J. Li, "Stochastic models and control for electrical power line temperature," *Energy Systems*, vol. 7, pp. 173–192, 2016.

[21] IEEE, "IEEE standard for calculating the current-temperature of bare overhead conductors," *Std 738-2006 (Revision of IEEE Std 738-1993)*, pp. 1–59, Jan. 2007.

[22] D. Bienstock, M. Chertkov, and M. Escobar, "Learning from power system data stream: phasor-detective approach," 2018.
[23] "CIGRE working group considers solutions to risks from high penetration of intermittent renewables," www.cigreaustralia.org.au/news/in-the-loop-news/cigre-working-group-matches-solutions-to-identified-risks-from-large-penetration-of\penalty\z@-intermittent-renewables/
[24] "Examining the variability of wind power output in the regulation time frame," *11th International Workshop on Large-Scale Integration of Wind Power into Power Systems as Well as on Transmission Networks for Offshore Wind Power Plants*, 2012.
[25] CIGRE Working Group 08 of Study Committee C6, "Technical brochure on grid integration of wind generation," *International Conference on Large High Voltage Electric Systems*, 2009.
[26] S. Baghsorkhi and I. Hiskens, "Impact of wind power variability on sub-transmission networks," in *Power and Energy Society General Meeting*. IEEE, pp. 1–7, 2012.
[27] D. Bienstock, G. Muñoz, C. Matke, and S. Yang, "Robust linear control of storage in transmission systems, and extensions to robust network control problems," *CDC 2017*, 2017.
[28] M. Lubin, Y. Dvorkin, and S. Backhaus, "A robust approach to chance constrained optimal power flow with renewable generation," vol. 31, no. 5, pp. 3840–3849, Sept. 2016.
[29] R. T. Rockafellar and S. Uryasev, "Optimization of conditional value-at-risk," *Journal of Risk*, vol. 2, pp. 21–41, 2000.
[30] P. Krokhmal, J. Palmquist, and S. Uryasev, "Portfolio optimization with conditional value-at-risk objective and constraints," *Journal of Risk*, vol. 4, pp. 11–27, 2002.
[31] F. Chung, *Spectral Graph Theory*. CMBS Lecture Notes 92, American Mathematical Society, Providence, RI, 1997.
[32] S. Chaiken and D. Kleitman, "Matrix tree theorems," *Journal of Combinatorial Theory, Series A*, vol. 24, no. 3, pp. 377–381, 1978.
[33] R. Ahuja, T. Magnanti, and J. Orlin, *Network Flows: Theory, Algorithms, and Applications*. New York: Prentice Hall, 1993.
[34] A. Nemirovski and A. Shapiro, "Convex approximations of chance constrained programs," *SIAM Journal on Optimization*, vol. 17, no. 4, pp. 969–996, 2006. http://link.aip.org/link/?SJE/17/969/1
[35] F. Alizadeh and D. Goldfarb, "Second-order cone programming," *Mathematical Programming*, vol. 95, pp. 3–51, 2001.
[36] L. Roald, F. Oldewurtel, T. Krause, and G. Andersson, "Analytical reformulation of security constrained optimal power flow with probabilistic constraints," in *Proceedings of the Grenoble PowerTech, Grenoble, France*, 2013.
[37] R. D. Zimmerman, C. E. Murillo-Sánchez, and D. Gan, "MATPOWER, a MATLAB power system simulation package," *IEEE Transactions on Power Systems*, vol. 26, no. 1, pp. 12–19, 2011.
[38] J. E. Kelley, "The cutting-plane method for solving convex programs," *Journal of the Society for Industrial and Applied Mathematics*, vol. 8, pp. 703–712, 1960.
[39] IBM, "ILOG CPLEX Optimizer," www-01.ibm.com/software/integration/optimization/cplex-optimizer/
[40] D. Molzahn, personal communication, 2016.
[41] K. Hedman, personal communication, 2017.
[42] L. Roald and D. Molzahn, "Implied constraint satisfaction in power systems optimization: The impacts of load variations," *arXiv:1903.11337v2*, 2019.

[43] A. Ben-Tal and A. Nemirovski, "On polyhedral approximations of the second-order cone," *Math. Oper. Res.*, vol. 26, no. 2, pp. 193–205, May 2001. http://dx.doi.org/10.1287/moor.26.2.193.10561

[44] R. Ben-Tal and A. Nemirovski, "Robust solutions of linear programming problems contaminated with uncertain data," *Mathematical Programming*, vol. 88, pp. 411–424, 2000.

[45] GUROBI, "Gurobi Optimizer," www.gurobi.com/

[46] E. Rothberg, personal communication, 2019.

[47] W. Xie and S. Ahmed, "Distributionally robust chance constrained optimal power flow with renewables: A conic reformulation," *IEEE Transactions on Power Systems*, vol. 33, no. 2, pp. 1860–1867, 2018.

[48] L. Roald and G. Andersson, "Chance-constrained ac optimal power flow: Reformulations and efficient algorithms," *IEEE Transactions on Power Systems*, vol. 33, no. 3, pp. 2906–2918, May 2018.

[49] K. Narendra, D. Rangana, and A. Rajapakse, "Dynamic performance evaluation and testing of phasor measurement unit (PMU) as per IEEE C37.118.1 Standard," 2018.

[50] Y. Tang, G. N. Stenbakken, and A. Goldstein, "Calibration of phasor measurement unit at NIST," *IEEE Transactions on Instrumentation and Measurement*, vol. 62, pp. 1417–1422, June 2013.

[51] T. Mülpfordt, T. Faulwasser, L. Roald, and V. Hagenmeyer, "Solving optimal power flow with non-Gaussian uncertainties via polynomial chaos expansion," in *CDC, 2017*. IEEE, 2017.

[52] D. Bertsimas and M. Sim, "Price of robustness," *Operations Research*, vol. 52, pp. 35–53, 2004.

[53] A. Nemirovski and A. Shapiro, "Scenario approximations of chance constraints," in: "Probabilistic and randomized methods for design under uncertainty," Calafiore and Campi (Eds.), pp. 3–48, 2005.

[54] E. Erdogan and G. Iyengar, "Ambiguous chance constrained problems and robust optimization," *Mathematical Programming*, vol. 107, pp. 37–61, 2007.

[55] G. C. Calafiore and L. El Ghaoui, "On distributionally robust chance-constrained linear programs," *Journal of Optimization Theory and Applications*, vol. 130, no. 1, pp. 1–22, 2006.

[56] M. Lubin, Y. Dvorkin, and L. Roald, "Chance constraints for improving the security of ac optimal power flow," *IEEE Transactions on Power Systems*, vol. 34, no. 3, pp. 1908–1917, May 2019.

[57] T. Mülpfordt, L. Roald, V. Hagenmeyer, T. Faulwasser, and S. Misra, "Chance-constrained ac optimal power flow: A polynomial chaos approach," *arXiv:1903.11337v2*, 2019.

[58] M. Hardt and E. Price, "The noisy power method: A meta algorithm with applications," in *NIPS*, pp. 2861–2869, 2014.

[59] I. M. Johnstone and D. Paul, "PCA in high dimensions: An orientation," *Proceedings of the IEEE*, vol. 106, no. 8, pp. 1277–1292, Aug. 2018.

[60] D. Paul, "Asymptotics of empirical eigenstructure for high dimensional spiked covariance model," *Statistica Sinica*, vol. 17, pp. 1617–1642, 2007.

[61] A. Shukla, D. Bienstock, and S. Yun, "Non-stationary streaming PCA," 2019.

[62] D. Luenberger, *Investment Science*. Oxford: Oxford University Press, 1997.

[63] D. Goldfarb and G. Iyengar, "Robust portfolio selection problems," *Mathematical Programming*, vol. 28, pp. 1–38, 2001.

[64] A. Schellenberg, W. Rosehart, and J. Aguado, "Cumulant based stochastic optimal power flow (s-opf) for variance optimization," in *IEEE Power Engineering Society General Meeting, 2005*, Vol. 1 473–478, June 2005.

[65] M. Davari, F. Toorani, H. Nafisi, M. Abedi, and G. B. Gharehpetian, "Determination of mean and variance of LMP using probabilistic DCOPF and T-PEM," in *2008 IEEE 2nd International Power and Energy Conference*, pp. 1280–1283, Dec. 2008.

[66] R. Jabr, "A conic optimal power flow with uncertain prices," in *Power Systems Computation Conference (PSCC), 2005*, pp. 1–7, 2005.

[67] H. Y. Yamin and S. M. Shahidehpour, "Risk and profit in self-scheduling for GenCos," *IEEE Transactions on Power Systems*, vol. 19, no. 4, pp. 2104–2106, Nov. 2004.

[68] D. Bienstock and A. Shukla, "Variance-aware optimal power flow," in *Power Systems Computation Conference (PSCC)*, pp. 1–8, 2018 (also see arXiv:1711.00906).

[69] W. F. Sharpe, "The sharpe ratio," *The Journal of Portfolio Management*, vol. 21, p. 49–58, 1994.

16 Distributed Optimization for Power and Energy Systems

Emiliano Dall'Anese and Nikolaos Gatsis

16.1 Introduction

This chapter provides an overview of distributed optimization methodologies, utilized to decompose the execution of decision-making and monitoring tasks across devices and modules (with sensing and computing capabilities) that communicate over an information layer. Applications in power systems and the accompanying architectural frameworks are outlined. The literature on distributed methods for solving convex and nonconvex optimization problems is vast, and theoretical contributions emerge at a fast pace as a result of efforts across different research areas. This chapter contains a sample of representative frameworks; links to other important topics are, however, suggested.

The chapter starts by stating a unifying optimization problem setup and related assumptions (Section 16.2). Motivating applications associated with tasks as diverse as linearized and relaxed AC optimal power flow (OPF) (Section 16.2.2), DC optimal power OPF (Section 16.2.3), demand-side management (Section 16.2.4), state estimation (Section 16.2.5), and optimization of wind farm power output (Section 16.2.6) are presented. The first class of algorithms detailed in this chapter are dual (sub)gradient methods and variants (Section 16.3). Mathematical structures leading to distributed methods that are supported by a star topology are identified; this involves a two-way communication between grid operators and devices in the field. The chapter then covers the emerging concept of measurement-based gradient methods (Section 16.4), where distributed algorithmic solutions are designed to replace synthetic models of power and energy grids with measurements of pertinent electrical quantities. The supporting information layer, which involves a message-passing between sensors, grid operators, and controllable devices is discussed. The next class of algorithms to be presented are consensus-based methods (Section 16.5). In particular, the chapter outlines a general strategy to derive distributed algorithms with general communication topologies by leveraging appropriate problem reformulations that are suitable for gradient-based methods and the alternating direction method of multipliers. This setting includes multi-area control paradigms in power systems.

The focus of Sections 16.3–16.5 is on distributed algorithms for convex programs. However, references to pertinent works in the domain of distributed algorithms for nonconvex problems are provided (Section 16.6.1), alongside pointers to additional distributed optimization paradigms such as Lyapunov optimization (Section 16.6.2) and online measurement-based algorithms (Section 16.6.3).

For optimization methods and the applications in power systems, the chapter provides only some representative references. Due to strict page limits, it is not possible to provide an extensive list of links to prior works in context. As a prerequisite, a good familiarity with basic concepts in optimization theory is expected. Optimization textbooks such as [1] and [2] can be consulted for basic definitions and concepts.

16.2 General Problem Setup and Motivating Applications

To outline a number of decision-making and monitoring tasks associated with power system operations in a unified manner, consider a generic model where a power grid features N controllable devices or assets with control inputs collected in the vectors $\mathbf{x}_i \in \mathbb{R}^{C_i}$, $i = 1, \ldots, N$. The control actions of the N devices are mapped into pertinent electrical states $\mathbf{y} \in \mathbb{R}^S$ through the mapping $\mathbf{y} = \mathcal{M}(\mathbf{x}, \mathbf{w})$, with $\mathcal{M} : \mathbb{R}^C \times \mathbb{R}^W \to \mathbb{R}^S$ a network map, $C := \sum_{i=1}^{N} C_i$, $\mathbf{x} = [\mathbf{x}_1^\mathsf{T}, \ldots, \mathbf{x}_N^\mathsf{T}]^\mathsf{T}$, and $\mathbf{w} \in \mathbb{R}^W$ a vector of W uncontrollable inputs or disturbances. Hereafter, the dependency of $\mathbf{y}$ on $\mathbf{x}$ and $\mathbf{w}$ is also denoted using the shorthand notation $\mathbf{y}(\mathbf{x}, \mathbf{w})$.

Throughout this chapter, tools and methodologies are outlined based on the following generic problem formulation:

$$\min_{\{\mathbf{x}_i \in \mathcal{X}_i\}_{i=1}^{N}} f(\mathbf{x}) \tag{16.1a}$$

$$\text{subject to: } \mathbf{g}(\mathbf{y}(\mathbf{x}, \mathbf{w})) \leq \mathbf{0} \tag{16.1b}$$

where $f : \mathbb{R}^C \to \mathbb{R}$ is a function capturing given performance metrics or costs, possibly associated with both the control inputs $\{\mathbf{x}_i\}_{i=1}^{N}$ and the states $\mathbf{y}$; $\mathcal{X}_i \subset \mathbb{R}^{C_i}$ is a set capturing constraints on $\mathbf{x}_i$; and, $\mathbf{g} : \mathbb{R}^C \to \mathbb{R}^M$ is a vector-valued function modeling M constraints associated with $\mathbf{y}$ (the inequality is meant entry-wise). For subsequent developments, let $\mathcal{X} := \mathcal{X}_1 \times \mathcal{X}_2 \times \ldots \times \mathcal{X}_N$. When the problem (16.1) is convex, the notation $\{\mathbf{x}_i^\star\}_{i=1}^{N}$ is hereafter used for a globally optimal solution. However, when the problem is nonconvex, $\{\mathbf{x}_i^\star\}_{i=1}^{N}$ denotes a point satisfying the Karush–Kuhn–Tucker (KKT) conditions [1].

Examples of tasks in power systems that are well-modeled by problem (16.1) are provided shortly in Sections 16.2.2–16.2.6; the chapter expands on algorithmic frameworks associated with some of those problems whenever appropriate problem reformulations and optimization tools are introduced. The chapter considers applications that involve convex problems in the form of (16.1); this is the case, for example, when the network map $\mathbf{y} = \mathcal{M}(\mathbf{x}, \mathbf{w})$ is linear or affine, and the functions f and $\mathbf{g}$ are convex, closed, and proper.

Problem (16.1) could be solved in a *centralized* fashion, upon collecting at a central processor or computing module (available at, e.g., the grid operator) the feasible sets $\{\mathcal{X}_i\}_{i=1}^{N}$ associated with the N controllable devices or assets, the individual cost functions, and measurements of the exogenous inputs $\mathbf{w}$. Once the problem is solved, optimal solutions $\{\mathbf{x}_i^\star\}_{i=1}^{N}$ could then be sent to the N controllable devices; see Figure 16.1 for a qualitative illustration. This centralized architecture might be impractical in

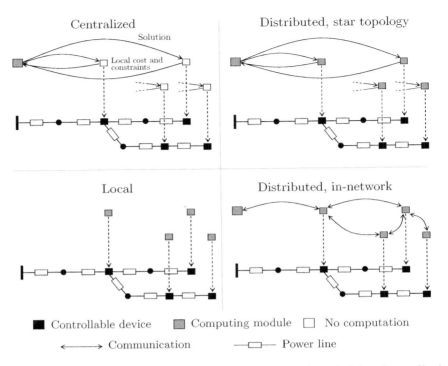

Figure 16.1 Qualitative illustration of differences in the operating principles of centralized, distributed, and local optimization paradigms in power systems; the example is illustrated for distribution systems. Continuous lines and arrows signify communications; dotted lines imply commands to the devices.

realistic power systems applications; in fact, among possible downsides, one has that the central processor represents an isolated point of failure. Furthermore, this strategy requires the central processor to have access to information regarding hardware constraints and cost functions of all the devices, with the latter possibly specifying private preferences of users/customers in a power grid. The main objective of this chapter is to overview methodologies to solve (16.1)—and problems that are subsumed by (16.1)—in a distributed manner; the term *distributed* here refers to a setting where the solution of an optimization problem is decoupled across processors, devices, or users, coordinating over a communication network. With reference to the illustration in Figure 16.1, distributed algorithms involve the sequential execution of algorithmic updates where the variables $\{x_i\}_{i=1}^N$ are updated locally at each device or processors, followed by the exchange of pertinent variables; the algorithmic steps are designed so that the users partaking into the distributed optimization process can collectively solve problem (16.1), while retaining controllability of their devices and without sharing their cost functions. Two architectures will be considered:

- A star topology, also referred to as gather-and-broadcast in some applications; this involves a message passing between a central node (i.e., the grid operator) and the controllable devices.

- A strategy with peer-to-peer communication, also referred to as in-network optimization in some communities, where nodes communicate through a connected communication network.

Of course, designing distributed algorithms for power systems is not straightforward; proper problem reformulations might be necessary so that optimization toolboxes can lend themselves to a distributed strategy with a desired communication topology. The chapter stresses this point by presenting equivalent reformulations of (16.1) leading to different distributed strategies.

It is also worth pointing out that distributed algorithms are not to be confused with *local* strategies, where each device computes its decision x_i based on a local cost function and measurement, and without communicating with other devices; see Figure 16.1. In some communities, this setting is also referred to as decentralized. These methods are not within the scope of this chapter.

16.2.1 Notation and Assumptions

In this chapter, upper-case (lower-case) boldface letters will be used for matrices (column vectors), and $(\cdot)^\mathsf{T}$ denotes transposition. $\mathbb{R}^N_+$ denotes the spacve of N-dimensional vectors with nonnegative entries. For a given $N \times 1$ vector $\mathbf{x} \in \mathbb{R}^N$, the Euclidean norm is denoted by $\|\mathbf{x}\|_2 := \sqrt{\mathbf{x}^\mathsf{T}\mathbf{x}}$, and x_i denotes the ith entry of $\mathbf{x}$. For a function $f: \mathbb{R}^N \to \mathbb{R}$, $\nabla_\mathbf{x} f(\mathbf{x})$ returns the gradient vector of $f(\mathbf{x})$ with respect to $\mathbf{x} \in \mathbb{R}^N$. Moreover, $\text{proj}_{\mathcal{X}}\{\mathbf{x}\}$ denotes a closest point to $\mathbf{x}$ in the convex set $\mathcal{X}$, namely $\text{proj}_{\mathcal{X}}\{\mathbf{x}\} = \arg\min_{\mathbf{y} \in \mathcal{X}} \|\mathbf{x} - \mathbf{y}\|_2$ [2]. Finally, $\mathbf{0}$ and $\mathbf{1}$ are vectors of all zeros and ones of appropriate dimensions.

This chapter considers applications where (16.1) is a convex problem. The following assumptions are considered throughout.

ASSUMPTION 16.1 *Functions f and $\mathbf{g}$ are continuous, convex, closed, and proper over $\mathcal{X}$.*

ASSUMPTION 16.2 *The sets $\{\mathcal{X}_i\}_{i=1}^N$ are convex and compact.*

ASSUMPTION 16.3 (Slater's constraint qualification) *There exists a Slater vector $\bar{\mathbf{x}} \in \mathcal{X}$ such that $g_j(\mathbf{y}(\bar{\mathbf{x}}, \mathbf{w})) < 0$ for all $j = 1, \ldots, M$.*

Assumptions 16.1 and 16.3 are supposed to hold for any vector $\mathbf{w}$ for which we wish to solve (16.1).

The compactness in Assumption 16.2 and continuity in Assumption 16.1 imply by Weierstrass's theorem that an optimal solution exists. Let f^* denote the optimal value for problem (16.1). It also follows from Assumptions 16.1 and 16.2 that a bound G on the function $\mathbf{g}(\mathbf{x})$ exists:

$$\|\mathbf{g}(\mathbf{x})\| \leq G, \text{ for all } \mathbf{x} \in \mathcal{X}. \tag{16.2}$$

Additional assumptions will be introduced in subsequent sections, based on the specific optimization methods and applications in power systems.

Let $\lambda \in \mathbb{R}_+^M$ denote the vector of dual variables associated with (16.1b); then, the Lagrangian function associated with (16.1) is given by

$$\mathcal{L}(\mathbf{x}, \lambda) := f(\mathbf{x}) + \lambda^\mathsf{T} \mathbf{g}(\mathbf{y}(\mathbf{x}, \mathbf{w})). \tag{16.3}$$

The dual function is defined as the minimization of the Lagrangian function with respect to the primal variables; that is,

$$v(\lambda) = \min_{\{\mathbf{x}_i \in \mathcal{X}_i\}_{i=1}^N} \{f(\mathbf{x}) + \lambda^\mathsf{T} \mathbf{g}(\mathbf{y}(\mathbf{x}, \mathbf{w}))\} \tag{16.4}$$

The dual problem is defined as follows:

$$v^* = \max_{\lambda \in \mathbb{R}_+^M} v(\lambda) \tag{16.5}$$

Convexity (Assumption 16.1) and Slater's constraint qualification (Assumption 16.3) imply that there is no duality gap; that is, $f^* = v^*$, and that the set of optimal solutions to the dual problem is nonempty and bounded [3]. See [1] and [2] for a comprehensive review of duality theory.

Representative applications from power systems that fit the general formulation (16.1) and typically satisfy the previously mentioned assumptions are presented next.

16.2.2 Optimization of Power Distribution Systems

Consider a distribution grid featuring a variety of controllable devices such as photovoltaic (PV) systems, energy storage systems, controllable loads, and electric vehicles—hereafter referred to as distributed energy resources (DERs). Let $S_i := P_i + jQ_i$ the complex power injection of a device i, and define $\mathbf{x}_i := [P_i^\mathsf{T}, Q_i^\mathsf{T}]^\mathsf{T}$. In this case, $\mathbf{y} = \mathcal{M}(\mathbf{x})$ maps powers injections into, e.g., voltages, voltage magnitudes, currents, and power flows, depending on the states included in $\mathbf{y}$ and the constraints included in the problem [4, 5].

For example, the map can be constructed using the admittance matrix $\mathbf{Y}$ of the network [6]. Letting $\mathbf{v}$ be the vector of complex voltages in the nodes, the map $\mathcal{M}(\mathbf{x})$ can be constructed based on manipulations of the AC power-flow equation $\mathbf{s}_{inj} = \mathbf{v} \circ \mathbf{Y}^* \mathbf{v}^*$, where $\circ$ denotes entrywise product, $*$ denotes the conjugate of a complex number, and $\mathbf{s}_{inj}$ is the vector of net injected complex powers. The resulting mapping $\mathbf{y} = \mathcal{M}(\mathbf{x})$ is in general nonlinear and nonconvex.

An AC optimal power flow (OPF) problem for distribution systems can therefore be put in the form of (16.1), with the following features:

- For a device i, the set $\mathcal{X}_i$ models hardware constraints. For example, for a PV system, the set $\mathcal{X}_i$ is given by $\mathcal{X}_i = \{\mathbf{x}_i : 0 \leq P_i \leq P_i^{av}, P_i^2 + Q_i^2 \leq (S_i^{inv})^2\}$, where P_i^{av} is the maximum power available (based on given irradiance conditions) and S_i^{inv} is the inverter capacity. As another example, for an electric vehicle charger, one can set $\mathcal{X}_i = \{\mathbf{x}_i : -P_i^{max} \leq P_i \leq 0, Q_i = 0\}$, where P_i^{max} is the maximum charging rate.

- The function $f_i(\mathbf{x}_i)$ models costs associated with the device i. For example, the cost associated with a PV inverter can be $f_i(\mathbf{x}_i) = a_i(P_i^{av} - P_i)^2 + b_i(P_i^{av} - P_i)$, with $a_i \geq 0, b_i \geq 0$, to minimize the power curtailment. One can also add a cost associated with $\mathbf{y}$; for example, when $\mathbf{y}$ collects the voltage magnitudes, one may want to minimize $f_0(\mathbf{y}) = \sum_m (y_m - |V^{ref}|)^2$, with V^{ref} a given voltage reference. The overall objective in this case is $f_0(\mathbf{y}) + \sum_i f_i(\mathbf{x}_i)$.
- The constraint $\mathbf{g}(\mathbf{y}) \leq \mathbf{0}$ models operational requirements. For example, if $\mathbf{y}$ collects the voltage magnitudes, then one can set the constraints $V^{min} \leq y_m \leq V^{max}$ for all nodes, where V^{min} and V^{max} are minimum and maximum, respectively, voltage magnitudes. Similar arguments apply to currents and power flows, or even net power at the substation.

See, for example, the representative references [7–10] for a complete formulation of the AC OPF problem for distribution systems.

The AC OPF is nonconvex and NP-hard. Methods to obtain convex surrogates of the AC power-flow equations, and consequently of the map $\mathbf{y} = \mathcal{M}(\mathbf{x})$, include convex relaxations and linear approximations. For convex relaxations, see the representative works [11–13]. Linear approximations of the AC power-flow equations are developed for example in [14–17], with the latter dealing with multiphase systems with both wye and delta connections. Using these linear models, it turns out that voltage magnitudes, magnitudes of the line currents, and power at the substation, which can all be included in $\mathbf{y}$, can be expressed as $\mathbf{y} = \mathbf{Ax} + \mathbf{Bw} + \mathbf{b}$, where $\mathbf{A}, \mathbf{B}$, and $\mathbf{b}$ are constructed as shown in the references, and $\mathbf{w}$ is a vector collecting the uncontrollable powers in the network. Using this linear model, and using a convex function $\mathbf{g}$, yield a convex surrogate of the AC OPF in the form of (16.1). See also [9], [18], and [15].

For example, suppose that PV systems are to be controlled to ensure that voltage limits are satisfied. Then, using a linear model for the voltage magnitudes $[|V_1|, \ldots, |V_{N_d}|]^T \approx \mathbf{Ax} + \mathbf{Bw} + \mathbf{b}$, where $\mathbf{x}$ stacks the active and reactive powers of the PV systems and N_d is the number of nodal voltages to be controlled, one can formulate the following problem:

$$\min_{\{\mathbf{x}_i=[P_i,Q_i]^T \in \mathcal{X}_i\}_{i=1}^N} \sum_{i=1}^N f_i(\mathbf{x}_i) = a_i(P_i^{av} - P_i)^2 + b_i(P_i^{av} - P_i) + c_i Q_i^2 \quad (16.6a)$$

$$\text{subject to: } \mathbf{Ax} + \mathbf{Bw} + \mathbf{b} - \mathbf{1}V^{max} \leq \mathbf{0} \quad (16.6b)$$

$$\mathbf{1}V^{min} - \mathbf{Ax} - \mathbf{Bw} - \mathbf{b} \leq \mathbf{0}. \quad (16.6c)$$

Additional constraints on currents can be added, upon deriving approximate linear relationships. The assumptions of Section 16.2.1 are typically satisfied for this rendition of the AC OPF problem; for example, Assumption 16.3 is satisfied by simply considering the no-load voltage profile [5].

It is worth emphasizing at this point that when linear approximations are used, the electrical states $\mathbf{y}$ can be eliminated from the optimization problem using the map $\mathbf{y} = \mathbf{Ax} + \mathbf{Bw} + \mathbf{b}$. This was the case in the AC OPF of (16.6). In general, the elimination or inclusion of $\mathbf{y}$ as an optimization variable can give rise to different distributed solvers for the same problem.

16.2.3 Optimization of Power Transmission Systems

Transmission grids are typically modeled by the DC power-flow equations, which lend computational tractability to generator scheduling in the context of electricity markets [19], [4]. Consider a transmission grid with N buses, where for simplicity it is assumed that each bus has a generating plant with power output P_{G_i} and given load P_{L_i}, $i = 1, \ldots, N$. In this case, we have that $\mathbf{x}_i = [P_i]$. The demands are organized in the vector $\mathbf{l}$. The set $\mathcal{X}_i$ is an interval with lower and upper bounds on the plant's power output. The objective function takes the form $f(\mathbf{x}) = \sum_{i=1}^{N} f_i(\mathbf{x}_i)$, where $f_i(\mathbf{x}_i) = C_i(P_{G_i})$ is the cost of operating plant i.

The electrical states $\mathbf{y}$ of interest here are the resulting power flows on the lines. These are given by

$$\mathbf{y} = \mathbf{D}\mathbf{A}^\top \mathbf{B}^\dagger (\mathbf{x} - \mathbf{l}) \tag{16.7}$$

where $\mathbf{D}$ is a diagonal matrix with the inverse line reactances on the diagonal, $\mathbf{A}$ is the branch-to-node incidence matrix, $\mathbf{B}$ is the weighted Laplacian of the graph with the inverse line reactances as weights, and $\dagger$ denotes the pseudoinverse. The constraint $\mathbf{g}(\mathbf{y}) \leq \mathbf{0}$ represents the limits on power flows stated as $|\mathbf{y}| \leq \mathbf{y}^{\max}$, where the absolute value is taken entrywise. Introducing (16.7) into $\mathbf{g}(\mathbf{y})$ results in the constraint (16.1b) written as $\sum_{i=1}^{N} \mathbf{g}_i(\mathbf{x}_i) \leq \mathbf{0}$.

This formulation can be extended to multiperiod setups that include generator ramp constraints [19]. In addition, if the demands $\mathbf{l}$ are elastic, then the vector $\mathbf{x}$ can be extended to include the elastic demands, and the objective function (16.1a) will include the negative of the demand utility functions [19]. Related problems include postcontingency decisions [20]; coordinated scheduling of transmission and distribution grids [21]; and scheduling of demand response aggregators [22].

16.2.4 Demand-Side Management

Consider a load-serving entity (LSE) providing power to an aggregation of M users with controllable resources. The objective is to formulate a demand-side management problem that allows scheduling of end-user loads by maximizing the social welfare of the system [23–25]. For a time horizon $\{1, \ldots, T\}$, vector $\mathbf{p}_m \in \mathbb{R}^T$ includes the total power consumption of end-user m at the T periods; and let vector $\mathbf{s} \in \mathbb{R}^T$ denote the power provided by the utility at the T time slots. Here, we have $N = M + 1$, $\mathbf{x}_i = \mathbf{p}_i$ for $i = 1, \ldots, M$, and $\mathbf{x}_N = \mathbf{s}$. The objective function takes the form

$$f(\mathbf{x}) = \sum_{m=1}^{M} U_m(\mathbf{p}_m) + C(\mathbf{s}) \tag{16.8}$$

where $U_m(\mathbf{p}_m)$ is the utility function, or benefit, of end-user m from consuming power $\mathbf{p}_m$.

The electrical states $\mathbf{y}$ of interest here are the demand-supply deficiencies in each period, that is,

$$y = \sum_{m=1}^{M} \mathbf{p}_m - \mathbf{s}. \tag{16.9}$$

The constraint function is simply $g(\mathbf{y}) = \mathbf{y}$, and thus constraint (16.1b) enforces that adequate supply is provided by the LSE.

Capacity constraints can be included for $\mathbf{s}$ in the form of $\mathbf{s} \in \mathcal{S} = \{\mathbf{s} \in \mathbb{R}^T \mid \mathbf{0} \leq \mathbf{s} \leq \mathbf{s}^{\max}\}$. The constraint set for vector $\mathbf{p}_m$ is derived from the fact that $\mathbf{p}_m$ is the Minkowski sum of the individual appliance consumption vectors, that is,

$$\mathbf{p}_m \in \mathcal{P}_m = \left\{ \mathbf{p}_m \in \mathbb{R}^m \,\middle|\, \mathbf{p}_m = \sum_{a \in \mathcal{A}_m} \mathbf{p}_{m,a}, \, \mathbf{p}_{m,a} \in \mathcal{P}_{m,a}, \, a \in \mathcal{A}_m \right\} \tag{16.10}$$

where $\mathcal{A}_m$ is the set of appliances of end-user m, $\mathbf{p}_{m,a} \in \mathbb{R}^T$ contains the power consumptions of appliance $a \in \mathcal{A}_m$ over the T periods, and $\mathcal{P}_{m,a}$ defines the appliance power and scheduling constraints. The term "appliance" is used here generically. The following list provides pertinent examples of appliances.

- The first class includes load without intertemporal constraints, whose utility is derived from operating the device at desired setpoints. These may be varying throughout the period $\{1, \ldots, T\}$.
- The second class includes loads that must consume a specified amount of energy over a period, such as charging an electric vehicle.
- Another class pertains to loads with thermodynamical states, e.g., temperatures. Examples are heating or cooling.
- Finally, a battery energy storage system can be included, that can be charged or discharged during the scheduling horizon.

Depending on the level of modeling details, the previous examples can be described using convex constraints [24, 25]. In this case, the resulting problem (16.1) is convex, and the methods described in the ensuing sections are applicable. Typical modeling approaches that yield nonconvex sets $\mathcal{P}_m$ are described in [26] and [27]. Pointers to the literature on distributed optimization approaches for nonconvex optimization problems, including the demand-side management problem with appliances subject to nonconvex constraints, are provided in Section 16.6.

Finally, it is worth noting that the problem of scheduling the charging of a fleet of electric vehicles has similar structure to the demand-side management problem described in this section; see, e.g., [28] and references therein.

16.2.5 State Estimation

The power system state estimation (PSSE) problem amounts to determining the electrical states $\mathbf{y}$ of a transmission or distribution grid given a set of measurements of certain electrical quantities [29, 30]. The states are typically the nodal voltages expressed in either polar coordinates, when supervisory control and data acquisition (SCADA) systems are used for measurements, or in rectangular coordinates, when phasor measurement units (PMUs) are used. The following measurements are typically obtained depending on the instrumentation technology and are collected in the vector $\hat{\mathbf{z}}$:

- SCADA systems: real and reactive power flows on lines and injections; voltage magnitudes; current magnitudes
- PMUs: real and imaginary parts of voltage and line currents (phasors)

A linear measurement model $\hat{\mathbf{z}} = \mathbf{Hy} + \mathbf{n}$ is adopted for simplicity of exposition, where $\mathbf{n}$ is Gaussian noise. This model is appropriate in case of PMUs; or when linear approximations of the AC power-flow equations are employed and most notably under the Gauss–Newton iterative procedure. Then, the PSSE amounts to the following least-squares problem:

$$\min_{\mathbf{y}} \|\hat{\mathbf{z}} - \mathbf{Hy}\|_2^2 \qquad (16.11)$$

which can be seen as a special case of (16.1) when no constraints are present. Robust counterparts of (16.11) are discussed in, e.g., [31].

16.2.6 Optimization of Wind Farms

Consider a wind farm with N wind turbines, and consider the problem of maximizing the aggregate power output of the wind farm. The generation of the wind turbines is affected by the wake effects, where the wind speed downstream a turbine is reduced. Wake effects in wind farms are challenging to model due to complex turbulence effects and wake-terrain interactions, but models are available [32]. Wakes can be steered away from downstream turbines by controlling the yaw angles of the upstream turbine. Consider then an optimization problem of the form (16.1) where

- Vector $\mathbf{x}$ collects the yaw angles of the N turbines.
- The state $\mathbf{y}(\mathbf{x}, \mathbf{w})$ represents the active power outputs of the N turbines, as a function of the yaw angles and disturbances $\mathbf{w}$.
- The function $\mathcal{M}$ (or pertinent linearizations of this function) captures models that give the output powers of the turbines as a function of the yaw angles.
- The cost can be set to $f(\mathbf{x}) = -\sum_{i=1}^{N} y_i(\mathbf{x}, \mathbf{w})$ to maximize the power produced by the wind farm.
- Sets $\mathcal{X}_i$ represent constraints on the yaw angles, and constraint $\mathbf{g}(\mathbf{y}) \leq \mathbf{0}$ is not present.

See, for example, [33–35].

16.3 Dual Methods for Constrained Convex Optimization

The theme of this section is to demonstrate the application of dual methods toward the development of distributed solvers for problem (16.1). The focus is on applications where the mapping $\mathcal{M}$ is affine, and the objective and constraint functions $f(\mathbf{x})$ and

$g(x)$ are, respectively, written as $f(\mathbf{x}) = \sum_{i=1}^{N} f_i(\mathbf{x}_i)$ and $\mathbf{g}(\mathbf{y}(\mathbf{x})) = \sum_{i=1}^{N} \mathbf{g}_i(\mathbf{x}_i)$. The resulting optimization problem is called *separable*. This was the case in the motivating examples of Sections 16.2.2–16.2.4.

The chief instrument of this method is the dualization of the constraint (16.1b) using a Lagrange multiplier vector. The resulting algorithm entails iteratively updating the Lagrange multiplier vector (dual variable) and the decisions $\mathbf{x}_i$, $i = 1, \ldots, N$ (primal variables). As will be explained, the star communication topology is the most pertinent for this scheme, where the central node performs dual update, and each of the remaining nodes perform the primal update.

Dual methods have been very popular schemes for developing distributed resource management algorithms in various applications beyond power systems; see, e.g., [36] for communication networks.

16.3.1 Dual Method

The basic dual method is described first. The separability of problem (16.1) enables writing the Lagrangian (16.3) as a sum of N terms, with each term pertaining to variable $\mathbf{x}_i$:

$$\mathcal{L}(\mathbf{x}, \lambda) = \sum_{i=1}^{N} [f_i(\mathbf{x}_i) + \lambda^\top \mathbf{g}_i(\mathbf{x}_i)]. \tag{16.12}$$

The corresponding dual function can be written as a sum of N terms as follows:

$$v(\lambda) = \sum_{i=1}^{N} \min_{\mathbf{x}_i \in \mathcal{X}_i} \{f_i(\mathbf{x}_i) + \lambda^\top \mathbf{g}_i(\mathbf{x}_i)\}. \tag{16.13}$$

Let $k = 1, 2, 3, \ldots$ denote the iteration index of the dual method. The dual method iterates over two steps, namely, computing primal variables and updating the Lagrange multipliers. The method is initialized with arbitrary Lagrange multipliers $\lambda(1) \in \mathbb{R}_+^M$. At iteration index $k = 1, 2, 3, \ldots$, the primal variables are updated by minimizing the Lagrangian function in (16.12), yielding $\mathbf{x}_i(k+1)$. Then, the Lagrange multipliers are updated by taking a step in the direction $\sum_{i=1}^{N} \mathbf{g}_i(\mathbf{x}_i(k+1))$. These steps are expressed mathematically as follows:

$$\mathbf{x}_i(k+1) \in \arg\min_{\mathbf{x}_i \in \mathcal{X}_i} \{f_i(\mathbf{x}_i) + \lambda^\top(k)\mathbf{g}_i(\mathbf{x}_i)\}, \ i = 1, \ldots, N \tag{16.14a}$$

$$\lambda(k+1) = \text{proj}_{\mathbb{R}_+^M} \left\{ \lambda(k) + \alpha(k) \sum_{i=1}^{N} \mathbf{g}_i(\mathbf{x}_i(k+1)) \right\} \tag{16.14b}$$

where in (16.14b), $\alpha(k)$ is a stepsize at iteration k. The vector $\sum_{i=1}^{N} \mathbf{g}_i(\mathbf{x}_i(k+1))$ is a *subgradient* of the dual function (16.13) at the value $\lambda(k)$ [37]. Two stepsize rules will be considered:

(R1) Constant stepsize: $\alpha(k) = \alpha > 0$ for all $k = 1, 2, 3, \ldots$
(R2) Harmonic series: $\alpha(k) = \alpha/(k+c)$, for $k = 1, 2, \ldots$, where α and c are positive constants

When the objective function $f(\mathbf{x})$ is not strongly convex, convergence is established in terms of the running primal and dual averages, defined as

$$\bar{\mathbf{x}}_i(k) = \frac{1}{k} \sum_{l=1}^{k} \mathbf{x}_i(l), \ i = 1, \ldots, N; \ k = 1, 2, \ldots \quad (16.15a)$$

$$\bar{\lambda}(k) = \frac{1}{k} \sum_{l=1}^{k} \lambda(l), \ k = 1, 2, \ldots \quad (16.15b)$$

It is worth emphasizing that the running averages can be computed recursively, which facilitates the practical implementation:

$$\bar{\mathbf{x}}(k) = \frac{1}{k}\mathbf{x}(k) + \frac{k-1}{k}\bar{\mathbf{x}}(k-1), \ k = 2, 3, \ldots \quad (16.16)$$

and similarly for $\bar{\lambda}(k)$.

Convergence results for the two stepsizes are presented next, followed by a discussion on the distributed implementation of (16.14). Proofs of the following propositions can be found for a more general algorithm than (16.14) in [25]; see also [38, 39]. The first set of results pertains to the convergence of the primal and dual iterates with a constant stepsize.

PROPOSITION 16.1 Under Assumptions 16.1, 16.2, and 16.3, and with stepsize (R1), the sequence $\bar{\lambda}(k)$ satisfies

$$\liminf_{k \to \infty} v(\bar{\lambda}(k)) \geq v^* - \frac{1}{2}\alpha G^2. \quad (16.17)$$

The previous equation asserts that the dual function evaluated at $\lambda(k)$ may be suboptimal by an amount $\frac{1}{2}\alpha G^2$, where G was introduced in Section 16.2.1. This suboptimality can be controlled by the stepsize α. A tradeoff emerges here, as smaller stepsize implies smaller suboptimality, but potentially slower convergence.

PROPOSITION 16.2 Under Assumptions 16.1, 16.2, and 16.3, and with stepsize (R1), the sequence $\bar{\mathbf{x}}(k)$ satisfies the following:

1. $\lim_{k \to \infty} \left\| proj_{\mathbb{R}_+^M} \left[\sum_{i=1}^{N} \mathbf{g}_i(\bar{\mathbf{x}}_i(k)) \right] \right\|_2 = 0$
2. $\limsup_{k \to \infty} f(\bar{\mathbf{x}}(k)) \leq f^* + \frac{1}{2}\alpha G^2.$
3. $\liminf_{k \to \infty} f(\bar{\mathbf{x}}(k)) \geq f^*.$

The first part of the proposition asserts that running average $\bar{\mathbf{x}}(k)$ becomes asymptotically feasible because the projection of $\mathbf{g}(\bar{\mathbf{x}}(k))$ to the positive orthant tends to zero. The second and third part of the proposition affirms that the sequence $\bar{\mathbf{x}}(k)$ incurs a suboptimality of no more than $\frac{1}{2}\alpha G^2$. The tradeoff that was observed after Proposition 16.1 is also at play here.

The second set of results deals with primal and dual convergence under the stepsize given by the harmonic series. The main difference is that the suboptimality evidenced in Propositions 16.1 and 16.2 now shrinks to zero.

PROPOSITION 16.3 *Under Assumptions 16.1, 16.2, and 16.3, and with stepsize (R2), the sequence $\lambda(k)$ converges to an optimal dual solution.*

The previous proposition claims that under the harmonic series stepsize, one does not need to take running averages; the sequence of multipliers converges.

PROPOSITION 16.4 *Under Assumptions 16.1, 16.2, and 16.3, and with stepsize (R2), the sequence $\bar{\mathbf{x}}(k)$ satisfies the following:*

1. $\lim_{k\to\infty} \left\| proj_{\mathbb{R}_+^M} \left[\sum_{i=1}^N \mathbf{g}_i(\bar{\mathbf{x}}_i(k)) \right] \right\|_2 = 0$
2. $\lim_{k\to\infty} f(\bar{\mathbf{x}}(k)) = f^*$.

Proposition 16.4 asserts that the sequence $\bar{\mathbf{x}}(k)$ becomes asymptotically feasible, and asymptotically achieves the optimal value f^* of the problem; it is thus asymptotically optimal.

Propositions 16.3 and 16.4 reveal that no suboptimality is incurred when a diminishing stepsize is used, such as (S2). The drawback, however, is that convergence may become slower, as the value of $a(k)$ becomes smaller.

Propositions 16.1–16.4 characterize the primal and dual convergence of the method given by 16.14 and (16.15) and applied to the convex problem (16.1). Under both stepsize rules, primal averaging is needed to obtain (near-)optimal control decisions $\mathbf{x}_i$. It is worth mentioning that if the objective function (16.1a) is strictly convex, then primal averaging is not necessary. Specifically, suppose that each of the functions $f(\mathbf{x}_i)$ is strictly convex. In this case, the following hold [40, Sec. 6.3]:

- The dual function (16.4) is continuously differentiable;
- The vector $\sum_{i=1}^N \mathbf{g}_i(\mathbf{x}_i(k+1))$ in (16.14b) becomes a *gradient* of the dual function at $\lambda(k)$; and
- The minimizer in (16.14a) is unique.

In other words, the iteration (16.14b) becomes a gradient method for solving the dual problem (16.5). Convergence of the iterates in (16.14b) to a dual optimal solution follows from standard theory for gradient methods [2]; and correspondingly, the iterates in (16.14a) converge to a primal optimal solution. In regard to the constant stepsize, the right value may not be straightforward to compute; see, e.g., [40, Thm. 6.1]. In practice, a sufficiently small value is chosen.

In terms of *distributed* implementation, the iterations 16.14 can be carried out by a set of $1 + N$ computing nodes connected using a star communication topology, as depicted in Figure 16.2. Specifically, the primal update (16.14a) is performed at a computing module corresponding to variable $\mathbf{x}_i$, while another computing node corresponding to the center of the star performs the dual update (16.14b).

At the beginning of every iteration, the following steps take place:

S1. The central node transmits $\lambda(k)$ to the computing modules of devices $1, \ldots, N$;
S2. Computing module $i = 1, \ldots, N$ updates $\mathbf{x}_i(k+1)$ using (16.14a) and $\bar{\mathbf{x}}(k+1)$ using (16.16);
S3. Computing module $i = 1, \ldots, N$ transmits $\mathbf{x}_i(k+1)$ back to the central node; and

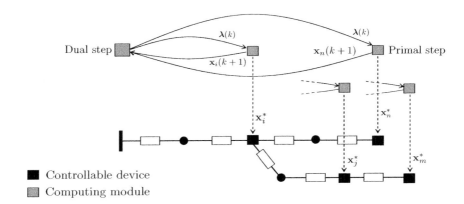

Figure 16.2 Illustrative example of the distributed implentation of the dual method (16.14), for an application in power distribution systems. The grid operator performs the dual update step and transmits $\lambda(k)$ to the computing modules of devices $i = 1, \ldots, N$. These update and transmit $\mathbf{x}_i(k+1)$ it back to the grid operator.

S4. The subgradient $\sum_{i=1}^{N} \mathbf{g}_i(\mathbf{x}_i(k+1))$ is formulated at the central node, and the Lagrange multiplier is updated according to (16.14b).

Step S3 assumes that the constraint functions $\mathbf{g}_i(\mathbf{x}_i)$ are available at the central node. If the constraint functions $\mathbf{g}_i(\mathbf{x}_i)$ are only known locally at the respective computing module $i = 1, \ldots, N$, then the vector $\mathbf{g}_i(\mathbf{x}_i(k+1))$ is transmitted back to the central node, instead of $\mathbf{x}_i(k+1)$. The latter scenario does not necessitate availability of $f(\mathbf{x}_i)$, $\mathbf{g}(\mathbf{x}_i)$, or $\mathcal{X}_i$ at the central node, and may be motivated by privacy considerations. The subgradient computation proceeds with Step S4.

As the message exchanges in the previously mentioned dual algorithm take place over communication links that may be susceptible to errors or noise, it is possible that messages may be lost. The work in [25] analyzes an *asynchronous* version of the algorithm (16.14) suitable for such setups.

16.3.2 Further Reading: Cutting Plane and Bundle Methods

The subgradient-based dual method described in the previous section can exhibit slow convergence. This issue could be remedied by changing the Lagrange multiplier update rule (16.14b) to one that is leveraging more than one previous subgradients. The cutting plane and bundle methods are viable approaches to this end and are briefly described next.

The main idea of the cutting-plane method is to solve (16.5) sequentially by replacing the dual function (16.4) with a piecewise linear overestimator. The method is called *disaggregated* when a different overestimator is used for each of the N summands in (16.13). The overestimators are computed using the previous subgradients $\mathbf{g}_i(\mathbf{x}_i(k))$ provided by (16.14a). Bundle methods add a quadratic regularizer, called *proximal term*, to the objective function, which can enhance the convergence over the cutting-plane method. Details are provided in [40, Ch. 7], [2, Ch. 7], and [41]. It turns out that for the Lagrange multiplier update step, an optimization problem must be solved, which is in contrast to the simple subgradient step of (16.14b).

The advantages of the disaggregated cutting-plane method over its standard version for transmission network optimization with demand response aggregators have been reported in [22], and further improvements are delivered by the disaggregated bundle method [42]. Related techniques for demand-side management with the objective of peak load minimization have been reported in [43]. Applications to nonconvex problems in power networks are described in Section 16.6.

16.4 Model-Based and Measurement-Based Primal-Dual Methods

This section overviews primal-dual gradient methods for solving convex optimization problems of the form of (16.1). We will consider two settings:

- *Model-based algorithms*: This refers to the case in which the network map $\mathbf{y} = \mathcal{M}(\mathbf{x}, \mathbf{w})$ is utilized in the design of the algorithm;
- *Measurement-based algorithms*: The guiding principles behind the design of measurement-based algorithmic solutions is that measurements of the network output $\mathbf{y}$ replace the mathematical expression of the map $\mathcal{M}(\mathbf{x}, \mathbf{w})$ in the updates of the algorithm.

Measurement-based methods, therefore, avoid a complete knowledge of the function $\mathcal{M}(\mathbf{x}, \mathbf{w})$, relax requirements on sensing capabilities, and facilitate the development of distributed architectures, as explained shortly. Notice that the dual methods explained in Section 16.3 were model based; although this section focuses on primal-dual gradient methods, the same design principles can be applied to developing measurement-based dual methods; references will be provided as further readings.

The section focuses on primal-dual methods with constant step sizes; this is motivated by existing works on quantifying the progress of the algorithm in finite number of iterations [3], and by the real-time implementation of measurement-based methods [44]. The literature on primal-dual gradient methods is extensive; see, as representative contributions, the works [45] and [3] dealing with primal-dual subgradient methods. However, when dealing with subgradients, convergence results can be established in terms of approximate primal and dual solutions using running averages of the primal and dual sequences. Additional convergence results based on Fenchel-type operators can be found in, e.g., [46]. This subsection focuses on regularized primal-dual methods [47] applied to convex and strongly smooth functions because (1) convergence results for the iterates can be established (as opposed to running averages of the sequences or convergence in terms of objective function); (2) linear convergence results can be obtained; and (3) this setting facilitates the design of measurement-based algorithms.

Recall that the network map is of the form $\mathbf{y}(\mathbf{x}, \mathbf{w}) = \mathbf{A}\mathbf{x} + \mathbf{B}\mathbf{w} + \mathbf{b}$, and the functions f and $\mathbf{g}$ are convex. It is also assumed that $f(\mathbf{x}) = \sum_{i=1}^{N} f_i(\mathbf{x}_i)$. Differently from

Section 16.3, the vector-valued function is not necessarily assumed to decompose as $\sum_{i=1}^{N} \mathbf{g}_i(\mathbf{x}_i(k))$.

16.4.1 Model-Based Regularized Gradient Methods

Consider problem (16.1), and assume that that functions g_j, $j = 1, \ldots M_I$ are nonlinear and convex, whereas g_j, for $j = M_I + 1, \ldots M$, are linear or affine. Instead of Assumption 16.1, the following stronger version is adopted in this section.

ASSUMPTION 16.4 *The function $f(\mathbf{x})$ is r-strongly convex and continuously differentiable. The gradient map $\nabla_{\mathbf{x}} f(\mathbf{x})$ is Lipschitz continuous with constant $L_f > 0$ over $\mathcal{X}$. For each $j = 1, \ldots, M_I$, $g_j(\mathbf{y})$ is convex and continuously differentiable, with a Lipschitz continuous gradient with constant $L_{g_m} > 0$.*

If the function f is not strongly convex, one can add to the cost in (16.1) a Tikhonov regularization $\frac{r}{2}\|\mathbf{x}\|_2^2$ with $r > 0$, which would render Assumption 16.4 valid. Furthermore, Section 16.2.1 asserted that under Assumption 16.3 the set of dual optimal solutions Λ^* to (16.5) is nonempty and bounded. The following estimation of Λ^* will be useful for the convergence results of this section [3]:

$$\Lambda^* \subset \mathcal{D} := \left\{ \lambda \in \mathbb{R}_+^M : \sum_{i=1}^{M} \lambda_i \leq \frac{f(\bar{\mathbf{x}}) - v(\bar{\lambda})}{\min_{i=1,\ldots,M}\{-g_i(\bar{\mathbf{x}})\}} \right\} \quad (16.18)$$

where $v(\cdot)$ is the dual function (16.4) and $\bar{\lambda}$ is any dual feasible variable (i.e., for any $\bar{\lambda} \geq \mathbf{0}$, $v(\bar{\lambda})$ represents a lower bound on the optimal dual function). It is also worth noticing that, from the continuity of the gradients of $\mathbf{g}$ and the compactness of $\mathcal{X}$, there exists a scalar $M_g < +\infty$ such that $\|\nabla_{\mathbf{x}} \mathbf{g}(\mathbf{x})\|_2 \leq M_g$, with $\nabla_{\mathbf{x}} \mathbf{g} := [\nabla_{\mathbf{x}} g_1, \ldots, \nabla_{\mathbf{x}} g_M]^\mathsf{T}$.

Define the *regularized* Lagrangian function as:

$$\mathcal{L}_r(\mathbf{x}, \lambda) := f(\mathbf{x}) + \lambda^\mathsf{T} \mathbf{g}(\mathbf{y}(\mathbf{x}, \mathbf{w})) - \frac{d}{2}\|\lambda\|_2^2. \quad (16.19)$$

where $d > 0$, and notice that $\mathcal{L}_r(\mathbf{x}, \lambda)$ is strongly convex in $\mathbf{x}$ and strongly concave in λ. As explained in [47], considering a regularized Lagrangian function is critical for the derivation of the ensuing linear convergence results; if one uses the standard Lagrangian function, then results similar to [46] on the optimal objective function can be derived. Let $\hat{\mathbf{z}}^* := [(\hat{\mathbf{x}}^*)^\mathsf{T}, (\hat{\lambda}^*)^\mathsf{T}]^\mathsf{T}$ be the unique saddle point of the problem:

$$\min_{\mathbf{x} \in \mathcal{X}} \max_{\lambda \in \mathcal{D}} \mathcal{L}_r(\mathbf{x}, \lambda). \quad (16.20)$$

Because of the regularization, $\hat{\mathbf{z}}^*$ is not an optimal primal-dual solution associated with problem (16.1). The error, though, that one incurs can be bounded as follows:

$$\|\hat{\mathbf{x}}^* - \mathbf{x}^*\|_2^2 \leq \frac{d}{2r} \max_{\lambda^* \in \mathcal{D}} \|\lambda^*\|_2; \quad (16.21)$$

see [47] for a more in-depth discussion of (16.21).

With these assumptions in place, the (regularized) primal-dual algorithm for solving (16.20) amounts to the sequential execution of the following steps:

$$\mathbf{x}(k+1) = \text{proj}_{\mathcal{X}} \left\{ \nabla_{\mathbf{x}} f(\mathbf{x}(k)) + \sum_{j=1}^{M} \lambda_j(k) \mathbf{A}^{\mathsf{T}} \nabla_{\mathbf{x}} g_j(\underbrace{\mathbf{A}\mathbf{x}(k) + \mathbf{B}\mathbf{w} + \mathbf{b}}_{=\mathbf{y}(k)}) \right\} \quad (16.22a)$$

$$\lambda(k+1) = \text{proj}_{\mathcal{D}}\{\lambda(k) + \alpha \mathbf{g}(\underbrace{\mathbf{A}\mathbf{x}(k) + \mathbf{B}\mathbf{w} + \mathbf{b}}_{=\mathbf{y}(k)})\} \quad (16.22b)$$

where k is the iteration index and $\alpha > 0$ is the stepsize. In (16.22), we stress the role of the network map $\mathbf{y}(\mathbf{x}, \mathbf{w}) = \mathbf{A}\mathbf{x} + \mathbf{B}\mathbf{w} + \mathbf{b}$ in the algorithmic steps. The main convergence results for (16.22) are stated next, where $\mathbf{z}(k) := [\mathbf{x}^{\mathsf{T}}(k), \lambda^{\mathsf{T}}(k)]^{\mathsf{T}}$.

THEOREM 16.1 *Consider the sequence $\{\mathbf{z}(k)\}$ generated by the algorithm (16.22). Under Assumptions 16.2, 16.3, and 16.4, the distance between $\mathbf{z}(k)$ and the optimizer $\hat{\mathbf{z}}^*$ of (16.20) at iteration k can be bounded as:*

$$\|\mathbf{z}(k) - \hat{\mathbf{z}}^*\|_2 \leq c(\alpha)^k \|\mathbf{z}(0) - \hat{\mathbf{z}}^*\|_2, \quad c(\alpha) := [1 - 2\alpha\eta_\phi + \alpha^2 L_\phi^2]^{\frac{1}{2}} \quad (16.23)$$

where $L_\phi := \sqrt{(L_f + r + M_d + \xi_\lambda L_g)^2 + (M_d + d)^2}$, $\eta_\phi := \min\{r, d\}$, $L_g := \sqrt{\sum_{m=1}^{M_I} L_{g_m}^2}$, *and* $\xi_\lambda := \max_{\lambda \in \mathcal{D}} \|\lambda\|_2$.

Proofs of this theorem can be found in [47] and [44].

A direct consequence of Theorem 16.1 is that, if $\alpha < 2\eta_\phi/L_\phi^2$, then $c(\alpha) < 1$; therefore, $\mathbf{z}(k)$ converges *linearly* to $\hat{\mathbf{z}}^*$. In view of (16.21), it further follows that the algorithms converges linearly to a neighborhood of the solution of (16.1).

Regarding *distributed implementations* of the algorithm (16.22), it is important to notice that the update (16.22a) can be decomposed across devices $i = 1, \ldots, N$ only in some specific cases. For example, suppose that the constraint in (16.1) is of the forms $g_j(\mathbf{x}) = \mathbf{C}\mathbf{x} - \mathbf{c}$ (i.e., in case of linear map and linear constraint); then, (16.22a) can be decomposed across devices $i = 1, \ldots, N$, and the algorithm can be implemented in a distributed manner using the star communication topology in Figure 16.2. By contrast, when one has a constraint function of the form $g_j(\mathbf{x}) = \|\mathbf{C}\mathbf{x} - \mathbf{c}\|^2 - a$, then (16.22a) is not amenable to distributed solution. The measurement-based method explained next will enable one to develop a distributed algorithm even when (16.22a) does not decompose across devices.

16.4.2 Measurement-Based Gradient Methods

Measurement-based methods can address the following potential drawbacks of the model-based algorithm (16.22) when applied to the power grid problems explained in Section 16.2:

- Algorithmic updates require grid operators to measure the exogenous inputs $\mathbf{w}$; this may be impractical because it requires pervasive metering throughout the network (some entries of $\mathbf{w}$ might not be even observable).

- Imperfect knowledge of the network map $\mathbf{y}(\mathbf{x},\mathbf{w})$ might drive the network operation to points that might not be implementable.
- The mathematical structure of the map $\mathbf{y}(\mathbf{x},\mathbf{w})$ and the constraints may prevent a distributed implementation of the update (16.22).

The idea behind measurement-based methods is to replace synthetic models $\mathbf{y}(\mathbf{x},\mathbf{w})$ or even functional evaluations of the constraints $\mathbf{g}(\mathbf{y})$ with *measurements* or estimates. Accordingly, let $\hat{\mathbf{y}}(k)$ denote a measurement of the network state $\mathbf{y}$ at iteration k of the algorithm, and consider the following modification of (16.22):

$$\mathbf{x}_i(k+1) = \mathrm{proj}_{\mathcal{X}_i}\left\{\nabla_{\mathbf{x}_i} f_i(\mathbf{x}_i) + \sum_{j=1}^{M} \lambda_j(k)\mathbf{A}^\mathsf{T}\nabla_{\mathbf{x}}g_j(\hat{\mathbf{y}}(k))\right\}, i=1,\ldots,N \quad (16.24\mathrm{a})$$

$$\lambda(k+1) = \mathrm{proj}_{\mathcal{D}}\{\lambda(k) + \alpha \mathbf{g}(\hat{\mathbf{y}}(k))\} \quad (16.24\mathrm{b})$$

It is clear that the algorithm (16.24) affords a distributed implementation, even when the gradient of g_j does not decompose across $i=1,\ldots,N$. In particular, at each iteration:

S1. Sensors collect measurements $\hat{\mathbf{y}}(k)$ and transmit them to the grid operator;
S2. Grid operator computes (16.24b) and $\xi(k) := \sum_{j=1}^{M}\lambda_j(k)\mathbf{A}^\mathsf{T}\nabla_{\mathbf{x}}g_j(\hat{\mathbf{y}}(k))$;
S3. Grid operator transmits $\lambda(k)$ and $\xi(k)$ to devices $i=1,\ldots,N$; and
S4. Each individual device performs (16.24a).

This message-passing strategy is illustrated in Figure 16.3, and it is typically referred to as "gather-and-broadcast" [48]; this is because the measurements are gathered at the grid operator, and the dual variables are broadcasted to the devices. It is also worth pointing out the following features:

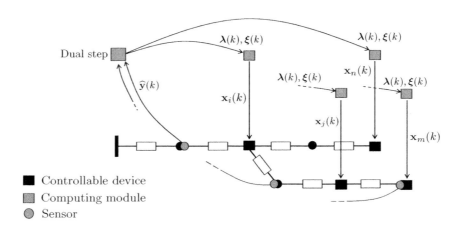

Figure 16.3 Illustrative example of the distributed implementation of the measurement-based primal-dual method (16.2), for an application in power distribution systems. The method is inherently implemented in real time, where commands $\{\mathbf{x}_i\}_{i=1}^{N}$ are sent to the devices at each iteration.

- Algorithm (16.24) does not require one to estimate the vector **w**; this is important especially in problems in transmission and distribution systems, where **w** is a possibly very large vector collecting the noncontrollable active and reactive powers across the network.
- This algorithm is inherently implemented in real time: in fact, the commands $\{\mathbf{x}_i(k)\}_{i=1}^{N}$ are sent to the devices at each iteration k (with k now denoting, in fact, a temporal index).

Example For the linearized AC OPF problem (16.6), one has the following message-passing strategy assuming that voltage measurements are gathered by meters, DERs, or distribution-PMUs:

S1. Collect measurements of voltages and transmit them to the grid operator;
S2. Grid operator computes (16.24b) and $\xi(k)$ based on the voltage measurements;
S3. Grid operator transmits $\lambda(k)$ and $\xi(k)$ to PV systems $i = 1, \ldots, N$; and
S4. Each PV system performs (16.24a) to compute output powers.

Similar strategies can be obtained for other DERs. □

Example For the problem outlined in Section 16.2.3, the following distributed algorithm can be implemented

S1. Grid operator collects measurements of line flows $\mathbf{y} = \mathbf{D}\mathbf{A}^\mathsf{T}\mathbf{B}^\dagger(\mathbf{x} - \mathbf{l})$;
S2. Grid operator computes (16.24b) and $\xi(k)$ based on the measurements of the line flows;
S3. Grid operator transmits $\lambda(k)$ and $\xi(k)$ to generators; and
S4. Each generator performs (16.24a) to compute the output power.

Measurements of **x** can also be introduced in the algorithm. □

Convergence of (16.24) is investigated next. To this end, the following assumption is introduced.

ASSUMPTION 16.5 *There exists a scalar $e_y < +\infty$ such that the measurement error can be bounded as $\|\hat{\mathbf{y}} - \mathbf{y}(\mathbf{x}, \mathbf{w})\|_2 \leq e_y$ for all $\mathbf{x}$, and for any $\mathbf{w}$ for which problem (16.1) is solved.*

Based on this assumption, the following convergence result holds.

THEOREM 16.2 *Consider the sequence $\{\mathbf{z}(k)\}$ generated by the algorithm (16.24). Under Assumptions 16.2, 16.3, 16.4, and 16.5, the distance between $\mathbf{z}(k)$ and the optimizer $\hat{\mathbf{z}}^*$ of (16.20) at iteration k can be bounded as:*

$$\|\mathbf{z}(k) - \hat{\mathbf{z}}^*\|_2 \leq c(\alpha)^k \|\mathbf{z}(0) - \hat{\mathbf{z}}^*\|_2 + \sum_{\ell=0}^{k-1} c(\alpha)^\ell \alpha e \qquad (16.25)$$

where

$$e = \left((M_\lambda M_I \max_{j=1,\dots,M_I} \{L_{g_j}\})^2 \|\mathbf{A}\|_2^2 e_y^2 + M_g^2 e_y^2 \right)^{\frac{1}{2}}, \quad (16.26)$$

with $M_\lambda := \sup_{k \geq 1} \max_{\lambda \in \mathcal{D}^{(k)}} \|\lambda\|_1$.

Based on this assumption, the following convergence result holds.

COROLLARY 16.1 *Suppose that $c(\alpha) < 1$. Then, the sequence $\{\mathbf{z}^{(k)}\}$ generated by algorithm (16.24) converges linearly to $\{\hat{\mathbf{z}}^{(*,k)}\}$ up to an asymptotic error bound given by:*

$$\limsup_{k \to \infty} \|\mathbf{z}(k) - \hat{\mathbf{z}}^*\|_2 \leq \frac{\alpha e}{1 - c(\alpha)}. \quad (16.27)$$

A proof of these results can be found in [49] and [44]. Clearly, if $e = 0$, then $\mathbf{z}(k)$ converges asymptotically to $\hat{\mathbf{z}}^*$.

16.4.3 Further Reading

Although this section focused on primal-dual gradient methods, measurement-based counterparts of the dual methods explained in Section 16.3 can be found in, e.g., [50]. The main architecture is similar to the one illustrated in Figure 16.3.

Convergence results for the case of measurements-based primal-dual methods in case of Lagrangian functions that are not strongly convex and strongly concave rely on regret analysis; this can be found in [46] and [44]. Additional works on measurements-based methods can be found in [18] where voltage measurements are leveraged in distribution systems; [51] where algorithmic frameworks are based on a manifold formalism; [52], [48] for frequency control in transmission systems; and, [53] for transmission systems.

16.5 Consensus-Based Distributed Optimization

Sections 16.3 and 16.4 described model-based algorithmic solutions that lead to a star communication topology; However, measurement-based methods using gradients lead to a gather-and-broadcast strategy. The present section focuses on the design of distributed in-network algorithms that can be implemented using a more generic message-passing strategy; see Figure 16.1. In particular, with reference to Figure 16.4, consider a case in which a power grid is partitioned into A control areas; each area is supposed to be equipped with an area controller, monitoring and optimizing the operation of the subgrid and devices within its control area. In some applications, one area can boil down to a single node.

In transmission systems, a control area can be an ISO region or a balancing authority [4]. In distribution systems, an area can refer to a microgrid (thus advocating the concept of interconnected microgrids) or simply a soft microgrid, a community choice aggregation, or a lateral or a feeder. In terms of monitoring and control tasks, this setting is relevant for many applications, including the following:

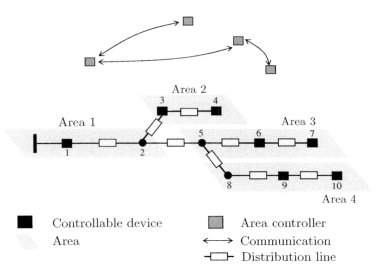

Figure 16.4 Illustrative example of a partitioning of a distribution system in areas. Subvectors $\mathbf{x}^{(i)}$ and $\mathbf{y}^{(i)}$ pertain to area i and may contain electrical quantities pertaining to a link to a neighboring areas or a node of neighboring areas.

- Multi-area power system state estimation [29, 31].
- Multi-area AC OPF [54–56].
- Distributed inference methods and sparsity-regularized rank minimization for power systems applications [57, 58].

Focusing on multi-area state estimation and optimization applications, let $\mathcal{A} := \{1, \ldots, A\}$ denote the set of areas, and let $\mathcal{A}^{(j)} \subset \mathcal{A}$ be set of areas physically connected to area j through a line. In the algorithms presented here, the communication strategy will be tied to the physical coupling of the areas. For example, with reference to Figure 16.4, one has that $\mathcal{A}^{(1)} = \{2, 3\}$, $\mathcal{A}^{(2)} = \{1\}$, $\mathcal{A}^{(3)} = \{1, 4\}$, and $\mathcal{A}^{(4)} = \{3\}$. As a remark, in the case of distributed inference problems, the partitioning in areas and the consequent communication strategy does not need to match the physical topology of the power network.

Assume that $\bar{\mathbf{x}}^{(j)} \in \mathbb{R}^{N^{(j)}}$ is a vector pertaining to area j; similarly, $\bar{\mathbf{y}}^{(j)} \in \mathbb{R}^{S^{(j)}}$ is a vector of states pertaining to area j. Given the electrical coupling across areas, an area j might include in its local vectors $\bar{\mathbf{x}}^{(j)}$ and $\bar{\mathbf{y}}^{(j)}$ electrical quantities that pertain to a either a line connecting two areas or a neighboring node; therefore, for two neighboring areas i and j, the vector $\bar{\mathbf{x}}^{(i)}$ may partially overlap with $\bar{\mathbf{x}}^{(j)}$, or the vector $\mathbf{y}^{(i)}$ may partially overlap with $\mathbf{y}^{(j)}$. For example, for a state estimation problem in the system in Figure 16.4, $\bar{\mathbf{y}}^{(2)}$ contains the voltages of nodes 3 and 4 (inside area 2) and of node 2 (of the neighboring area 1). To motivate why this is necessary, suppose that node 3 measures the power flow on the line (2, 3). This power flow on the line that connects nodes 2 and 3 is a function of the voltages at nodes 2 and 3, and therefore, area 2 must include the voltage at node 2 in its state vector. Likewise, $\bar{\mathbf{y}}^{(1)}$ contains the voltages of nodes 1, 2, 3, and 5.

Example For the state estimation problem (cf. Section 16.2.5), the vector $\{\bar{\mathbf{y}}^{(j)}\}_{j \in \mathcal{A}}$ collects voltages (in either polar or rectangular form, depending on whether SCADA or PMUs are used). The measurement vector $\hat{\mathbf{z}}^{(j)}$ of area j includes measurements within the area, and also the power flows on the tie lines connecting the areas. Given $\{\hat{\mathbf{z}}^{(j)}\}_{j \in \mathcal{A}}$, the task is to solve the following problem:

$$\min_{\{\bar{\mathbf{y}}^{(j)}\}_{j=1}^{A}} \sum_{j=1}^{A} \|\hat{\mathbf{z}}^{(j)} - \mathbf{H}^{(j)}\bar{\mathbf{y}}^{(j)}\|_2 \qquad (16.28)$$

where $\mathbf{H}^{(j)}$ is the observation matrix for the area j; see [29] and [31]. □

The ensuing discussion focuses on optimization problems stated in terms of the variables $\{\bar{\mathbf{x}}^{(j)}\}_{j \in \mathcal{A}}$ and without coupling constraints of the form $\mathbf{g}(\mathbf{y}(\mathbf{x}, \mathbf{w})) \leq \mathbf{0}$. The very same solution approach is applicable to a formulation that includes the vectors $\{\bar{\mathbf{y}}^{(j)}\}_{j \in \mathcal{A}}$, such as a multi-area state estimation setup. Generalization to problems with coupling constraints will be provided afterward.

Let $f^{(j)} : \mathbb{R}^{N^{(j)}} \to \mathbb{R}$ be a function associated with the area j, and consider the following optimization problem:

$$\min_{\{\bar{\mathbf{x}}^{(j)} \in \mathcal{X}^{(j)}\}_{j=1}^{A}} \sum_{j=1}^{A} f^{(j)}(\bar{\mathbf{x}}^{(j)}) \qquad (16.29)$$

where $\mathcal{X}^{(j)}$ is a convex set. For example, the objective in (16.29) is a least squares cost for the state estimation problem (16.28); or, (16.29) can represent a low-rank matrix completion problem as in [57].

Recalling that the vector $\bar{\mathbf{x}}^{(j)}$ collects variables of area j and of some nodes of neighboring areas, it turns out that (16.29) is not decomposable across areas; therefore, (16.29) would need to be solved centrally. To facilitate the development of a distributed algorithm, one standard trick is to introduce auxiliary variables that can allow one to decouple the cost functions in (16.29) on a per-area basis; see, for example, [59, 60]. To this end, for every two neighboring areas k and i, identify their common variables; then, let $\mathbf{x}_i^{(j)}$ be a vector representing a *copy* available at area j of the elements that vectors $\bar{\mathbf{x}}^{(j)}$ and $\bar{\mathbf{x}}^{(i)}$ share. Likewise, $\mathbf{x}_j^{(i)}$ is a *copy* available at area i of the elements that vectors $\bar{\mathbf{x}}^{(j)}$ and $\bar{\mathbf{x}}^{(i)}$ share. For example, in Figure 16.4, $\mathbf{x}_2^{(1)}$ and $\mathbf{x}_1^{(2)}$ contain a copy of the variables pertaining to nodes 2 and 3; as a another example, $\mathbf{x}_4^{(3)}$ and $\mathbf{x}_3^{(4)}$ contain a copy of the variables pertaining to nodes 5 and 8.

With these local copies of overlapping variables, let the vector $\mathbf{x}^{(j)}$ stack the variables pertaining to the nodes of area j as well as the copies $\{\mathbf{x}_i^{(j)}\}_{i \in \mathcal{A}^{(j)}}$. Therefore, problem (16.29) can be equivalently reformulated in the following way:

$$\min_{\{\mathbf{x}^{(j)} \in \mathcal{X}^{(j)}\}_{j=1}^{A}} \sum_{j=1}^{A} f^{(j)}(\mathbf{x}^{(j)}) \qquad (16.30\text{a})$$

$$\text{subject to: } \mathbf{x}_i^{(j)} = \mathbf{x}_j^{(i)}, \forall i \in \mathcal{A}^{(j)}, \forall j \qquad (16.30\text{b})$$

where the cost function now decouples across areas, and the consensus constraints (16.30b) enforce neighboring areas to agree on their shared variables.

Problem (16.30) is now in a form where a dual gradient method can be applied to decompose the solution of (16.30) across areas. However, to cover an additional optimization method that is widely employed to develop distributed algorithms, the alternating direction method of multipliers (ADMM) is the subject of the ensuing section; see, e.g., [61], and the tutorials [58] and [60].

16.5.1 ADMM-Based Distributed Algorithms

Consider the following reformulation of (16.30):

$$\min_{\{\mathbf{x}^{(j)} \in \mathcal{X}^{(j)}\}_{j=1}^{A}, \{\mathbf{z}_{ij}\}} \sum_{j=1}^{A} f^{(j)}(\mathbf{x}^{(j)}) \quad (16.31a)$$

$$\text{subject to: } x_i^{(j)} = z_{ij}, \forall i \in \mathcal{A}^{(j)}, \forall j = 1, \ldots, A \quad (16.31b)$$

where additional auxiliary optimization variables z_{ij} have been introduced. Let $\{\nu_{i,j}\}$ be the dual variables associated with (16.31b); in particular, $\nu_{i,j}$ is associated with the constraint $x_i^{(j)} = z_{ij}$, whereas $\nu_{j,i}$ is associated with the constraint $x_j^{(i)} = z_{ij}$. Consider then the following augmented Lagrangian function:

$$\mathcal{L}_a(\{\mathbf{x}\}, \{\mathbf{z}_{ij}\}, \{\nu_{i,j}\}) := \sum_{j=1}^{A} f^{(j)}(\mathbf{x}^{(j)})$$

$$+ \sum_{j=1}^{A} \left[\sum_{i \in \mathcal{A}^{(j)}} \left(\nu_{i,j}^{\mathsf{T}}(x_i^{(j)} - z_{ij}) + \frac{r}{2} \|x_i^{(j)} - z_{ij}\|_2^2 \right) \right] \quad (16.32)$$

where $r > 0$ is a parameter; see discussion in e.g., [60, 61]. Based on (16.32), the ADMM amounts to the sequential execution of the following three steps (k is the iteration index):

$$\{\mathbf{x}^{(j)}(k+1)\} = \arg \min_{\mathbf{x}^{(j)} \in \mathcal{X}^{(j)}} \mathcal{L}_a(\{\mathbf{x}^{(j)}\}, \{\mathbf{z}_{ij}(k)\}, \{\nu_{i,j}(k)\}) \quad (16.33a)$$

$$\{\mathbf{z}_{ij}(k+1)\} = \arg \min_{\mathbf{z}_{ij}} \mathcal{L}_a(\{\mathbf{x}^{(j)}(k+1)\}, \{\mathbf{z}_{ij}\}, \{\nu_{i,j}(k)\}), \forall i, j \quad (16.33b)$$

$$\nu_{i,j}(k+1) = \nu_{i,j}(k) + r(x_i^{(j)}(k+1) - z_{ij}(k+1)), \forall i, j. \quad (16.33c)$$

Algorithm (16.33) affords a distributed solution, where the updates of the variables are decoupled across area controllers. However, steps (16.33) can be simplified to better highlight the message-passing structure of this ADMM-based algorithm; see for, example, [31, 56, 58, 60].

With a slight abuse of notation, denote as $x_i^{(j)}$ the ith entry of vector $\mathbf{x}^{(j)}$. Let $\mathcal{A}_i^{(j)}$ be the set of neighboring areas sharing the variable $x_i^{(j)}$ with area j, and let $\mathcal{I}^{(j)} := \{i = 1, \ldots, N^{(j)} : \mathcal{A}_i^{(j)} \neq \emptyset\}$; that is, $\mathcal{I}^{(j)}$ is the set of indices for the entries of $\mathbf{x}^{(j)}$ that are shared with neighboring areas. With this notation in place, the steps (16.33) can be rewritten as explained next.

PROPOSITION 16.5 *When the dual variables are initialized to zero, the ADMM algorithm in* (16.33) *produces the same sequence* $\{\mathbf{x}^{(j)}(k)\}_{k\in\mathbb{N}}$ *as the following steps, performed at each area* $j \in \mathcal{A}$:

$$\mathbf{x}^{(j)}(k+1) = \arg\min_{\mathbf{x}^{(j)}\in\mathcal{X}} f^{(j)}(\mathbf{x}^{(j)}) + \frac{r}{2}\sum_{i\in\mathcal{I}^{(j)}} |\mathcal{A}_i^{(j)}|(x_i^{(j)} - \rho_i^{(j)}(k))^2 \quad (16.34a)$$

$$\varepsilon_i^{(j)}(k+1) = \frac{1}{|\mathcal{A}_i^{(j)}|}\sum_{\ell \in \mathcal{A}_i^{(j)}} x_i^{(\ell)}(k+1), \forall i \in \mathcal{I}^{(j)} \quad (16.34b)$$

$$\rho_i^{(j)}(k+1) = \rho_i^{(j)}(k) + \varepsilon_i^{(j)}(k+1) - \frac{x_i^{(j)}(k) - \varepsilon_i^{(j)}(k)}{2}, \forall i \in \mathcal{I}^{(j)}. \quad (16.34c)$$

The variables $\{\mathbf{x}^{(j)}(0)\}$ *are initialized to arbitrary values,* $\rho_i^{(j)}(0)$ *is set to* $\rho_i^{(j)}(0) = (x_i^{(j)}(0) - \varepsilon_i^{(j)}(0))/2$, *and* $\varepsilon_i^{(j)}(0)$ *is set as in* (16.34b).

The proof of this proposition can be found in [31] and [60]. Thus, with appropriate initialization of the dual variables and manipulation of the iterations in (16.33), the variables $\mathbf{z}_{ij}$ and $\mathbf{v}_{i,j}$ can be eliminated, and the intermediate variables $\varepsilon_i^{(j)}$ and $\rho_i^{(j)}$ are introduced.

The steps (16.34) can be implemented using the communication structure in Figure 16.4 by executing the following steps at each area controller:

S1. Update $\mathbf{x}^{(j)}(k)$, and transmit $x_i^{(j)}(k+1), i \in \mathcal{I}^{(j)}$, to neighboring areas;
S2. Receive $x_j^{(\ell)}(k+1)$ from neighboring areas $\mathcal{A}_i^{(j)}, i \in \mathcal{I}^{(j)}$; and
S3. Update $\varepsilon_i^{(j)}$ and $\rho_i^{(j)}, i \in \mathcal{I}^{(j)}$. Go to S1.

The ADMM algorithm (16.33) – and, therefore, the algorithm (16.34) – is known to converge under fairly mild conditions; for rigorous convergence proofs, see for example [62] and the tutorial [60]. In a nutshell, the proof rely on a reformulation of the problem in the form

$$\min_{\mathbf{x}\in\mathcal{X}} \sum_j f^{(j)}(\mathbf{x}^{(j)}), \text{ subject to } \mathbf{A}\mathbf{x} + \mathbf{B}\mathbf{z} = 0 \quad (16.35)$$

where $\mathbf{x}$ and $\mathbf{z}$ stack the variables $\{\mathbf{x}^{(j)}\}$ and $\{\mathbf{z}_{ij}\}$, and the matrices $\mathbf{A}$ and $\mathbf{B}$ can be constructed based on the constraints in (16.31). Then, the following results are in order:

- Suppose that $f^{(j)}(\mathbf{x}^{(j)}), j = 1, \ldots, A$ are closed, proper, and convex, and have Lipschitz gradients. Then, (16.33) converge to a pair of optimal primal and dual solutions of (16.31). See [60].
- Suppose that $f^{(j)}(\mathbf{x}^{(j)}), j = 1, \ldots, A$ are strongly convex and the matrix $\mathbf{A}$ is full rank; then, (16.33) converge linearly to the unique primal-dual solution of (16.31); see [62] and [63].

16.5.2 A More General Formulation

The algorithm (16.34) is suitable for problems in power systems monitoring such as multi-area power system state estimation as well as inference and sparsity-regularized rank minimization methods. The common feature of these formulations is that the have no coupling constraints in the form $g(y(x, w)) \leq 0$; the constraints in (16.31b) are introduced to facilitate the implementation of ADMM. This section presents a more general formulation where the ADMM is applied to a constrained problem. This formulation is suitable, for example, for a multi-area AC OPF when linear approximations or convex relaxations of the AC power-flow equations are employed.

Consider the area-based model of the previous section, and suppose now that y_{ji} is a vector of states that area j has in common with area $i \in \mathcal{A}_j$. Consider, for example, Figure 16.4: if y represents voltages, then $y_{12} = [V_2, V_3]^T$ (with V_i the voltage at node i), $y_{13} = [V_2, V_5]^T$, and so on [56]. A similar procedure is applied when y_{ji} contains line currents or power flows. Similarly to the previous section, let $y_{ji}^{(j)}$ be a *copy* of y_{ji} available at area j, and let $y_{ji}^{(i)}$ be a *copy* of y_{ji} available at area i. Therefore, a constraint on the states of area j can be expressed as $g^{(j)}(y^{(j)}) \leq 0$, with $y^{(j)}$ a vector collecting the states inside area j and the shared states $\{y_{ji}^{(j)}\}$.

Suppose now that area j has a linear map for $y_{ji}^{(j)}$, based on approximations of the power-flow equations or linear equations emerging from a semidefinite relaxation approach; that is, for an OPF problem, a generic linear model for the approximations of the power-flow equations inside area j is of the form $y_{ij}^{(j)} = C_{ji} x^{(j)} + c_{ij} + \sum_{\ell \in \mathcal{A}^{(j)} \setminus \{j\}} D y_{\ell j}^{(j)}$, where $y_{ij}^{(j)}$ are linearly related to the powers $x^{(j)}$ and the states $y_{\ell j}^{(j)}$ shared with other areas. The term $D y_{\ell j}^{(j)}$ is relevant when $y_{ij}^{(j)}$ represent power flows. Then, the following problem can be formulated:

$$\min_{\substack{\{x^{(j)} \in \mathcal{X}_i\}_{i=1}^N \\ \{y_{ij}^{(j)}\}}} \sum_{j=1}^A f_0^{(j)}(y^{(j)}) + f^{(j)}(x^{(j)}) \tag{16.36a}$$

subject to: $g^{(j)}(y^{(j)}) \leq 0, \forall j$ (16.36b)

$$y_{ij}^{(j)} = C_{ji} x^{(j)} + c_{ij} + \sum_{\ell \in \mathcal{A}^{(j)} \setminus \{j\}} D y_{\ell j}^{(j)}, \forall i \in \mathcal{A}_j, \forall j \tag{16.36c}$$

$$y_{ij}^{(j)} = y_{ij}^{(i)}, \forall i \in \mathcal{A}^{(j)}, j = 1, \ldots, A. \tag{16.36d}$$

Upon defining the set $\mathcal{Y}^{(j)} := \{x^{(j)}, y^{(j)} : g^{(j)}(y^{(j)}) \leq 0, y_{ij}^{(j)} = C_{ji} x^{(j)} + c_{ij} + \sum_{\ell \in \mathcal{A}^{(j)} \setminus \{j\}} D y_{\ell j}^{(\ell)}, \forall i \in \mathcal{A}_j\}$, one has that (16.36) can be put in the form:

$$\min_{\substack{\{x^{(j)} \in \mathcal{X}_i\}_{i=1}^N \\ \{x^{(j)}, y^{(j)} \in \mathcal{Y}^{(j)}\}}} \sum_{j=1}^A f_0^{(j)}(y^{(j)}) + f^{(j)}(x^{(j)}) \tag{16.37a}$$

subject to: $y_{ij}^{(j)} = y_{ij}^{(i)}, \forall i \in \mathcal{A}^{(j)}, j = 1, \ldots, A.$ (16.37b)

Problem (16.37) has the same mathematical structure of (16.30). Therefore, the ADMM method can be applied to solve (16.37) in a distributed setting, with the communication topology in Figure 16.4.

Example Consider a multi-area AC OPF where a linear model is utilized. In this case, if $\mathbf{y}_{ij}^{(j)}$ denotes the active and reactive power flow entering area j from area i, the equation $\mathbf{y}_{ij}^{(j)} = \mathbf{C}_{ji}\mathbf{x}^{(j)} + \mathbf{c}_{ij} + \sum_{\ell \in \mathcal{A}^{(j)} \setminus \{j\}} \mathbf{D}\mathbf{y}_{\ell j}^{(\ell)}$ simply corresponds to the power balance equation for the area j. Thus, applying the ADMM algorithm to (16.37) yields in this case a message massing of the following form at each iteration:

S1. Each area j updates locally the vectors $\mathbf{y}^{(j)}$ and $\mathbf{x}^{(j)}$.
S2. Area j transmits $\mathbf{y}_{ij}^{(j)}$ to the neighboring areas $i \in \mathcal{A}^{(j)}$.
S3. Area j receives $\mathbf{y}_{ij}^{(i)}$ from the neighboring areas $i \in \mathcal{A}^{(j)}$.
S4. Area j updates $\varepsilon^{(j)}$ and $\rho^{(j)}$. Go to S1. □

16.5.3 Further Reading

Distributed algorithms based on ADMM with node-to-node message passing are proposed for the AC OPF with semidefinite relaxation in [64] and with second-order cone relaxation in [65]. The work in [66] develops an ADMM-based distributed algorithm for a stochastic version of the AC OPF, where the power flow equations are simultaneously relaxed for different values of renewable energy injections, but are coupled through demand response decisions. In [67], a two-stage architecture is presented where local controllers coordinate with other controllers at neighboring buses through a distributed algorithm that relies on a local exchange of information.

16.6 Further Reading

16.6.1 Distributed Methods for Nonconvex Optimization

Nonconvexity may emerge from a nonconvex cost and/or nonconvex constraints, or because of discrete variables. Results of the references below pertain to either locally optimal solutions (for unconstrained problems), or KKT points (for constrained problems).

Unconstrained nonconvex problems. Suggested references start with the case of unconstrained problems with a nonconvex cost function. Assuming that the cost function is smooth and has Lipschitz continuous gradient, consensus-based algorithms are developed in [68] based on a proximal primal-dual method. In particular, [68] enables distributed nonconvex optimization with global sublinear rate guarantees. This algorithm is applicable to nonconvex formulations for the PSSE (see Section 16.2.5) and for optimization of wind farms (see Section 16.2.6); for the latter, see [35].

Constrained nonconvex problems. For problems with nonconvex cost and constraints such as the AC OPF, the recent work in [69] proposes a two-level distributed algorithm with peer-to-peer message passing; the general strategy involves a reformulation for constrained nonconvex programs that enables the design of a two-level algorithm, which embeds a specially structured three-block ADMM at the inner level in an augmented Lagrangian method; convergenge results are provided. Additional distributed algorithms for the nonconvex AC OPF can be found in [70] and [55]. Regarding measurement-based algorithms for nonconvex problem, see the recent works in [71] and [72]; these works, however, deal with centralized algorithms.

Discrete variables. Among other applications, discrete variables show up in the definition of sets $\mathcal{X}_i$ that describe load constraints in demand-side management setups. Representative references to dual methods that deal with such problems include [73, 74], and [50].

16.6.2 Lyapunov Optimization

Lyapunov optimization is pertinent when the network operator wishes to minimize a *long-term average* cost given by $\lim_{T \to \infty} \frac{1}{T} \sum_{t=1}^{T} f(\mathbf{x}(t))$, where t indexes time. The decisions $\mathbf{x}(t)$ are coupled across time because networked storage devices are typically considered in this setup; see, e.g., [75–79] and references therein. The algorithms developed in this context entail observing at time t the current state of the network $\mathbf{w}(t)$, which is assumed stochastic, and solving an optimization problem for $\mathbf{x}(t)$. Details as well as strategies for distributed schemes are provided in the previously mentioned references.

16.6.3 Online Measurement-Based Optimization of Power Systems

Distributed algorithms typically require a number of steps and communication rounds to converge to a solution. One question pertains to the convergence of algorithms (centralized or distributed) when the problem inputs, cost, and constraints change during the execution of the algorithmic steps (i.e., these quantities vary at each iteration or every few iterations); this setting leads to a problem of tracking a solution that changes in time.

Recently, online measurement-based primal-dual methods for convex problems were proposed in [49] and [44], and convergence results were established; the framework is applicable to linearized AC OPF problems as well as demand-side management, and it affords a star communication topology. Centralized algorithms were developed in [80, 81]. Finally, see [82] for a measurement-based online ADMM method.

Acknowledgment

The work of N. Gatsis is supported by the National Science Foundation under Grant No. 1847125.

References

[1] S. Boyd and L. Vandenberghe, *Convex Optimization*. Cambridge: Cambridge University Press, 2004.

[2] D. P. Bertsekas, *Nonlinear Programming*. 3rd ed. Belmont, MA: Athena Scientific, 2016.

[3] A. Nedić and A. Ozdaglar, "Subgradient methods for saddle-point problems," *Journal of Optimization Theory and Applications*, vol. 142, no. 1, pp. 205–228, 2009.

[4] A. J. Wood and B. F. Wollenberg, *Power Generation, Operation, and Control*. New York: John Wiley & Sons, 1984.

[5] W. H. Kersting, *Distribution System Modeling and Analysis*. 2nd ed, Boca Raton, FL: CRC Press, 2007.

[6] M. Bazrafshan and N. Gatsis, "Comprehensive modeling of three-phase distribution systems via the bus admittance matrix," *IEEE Transactions on Power Systems*, vol. 33, no. 2, pp. 2015–2029, Mar. 2018.

[7] M. Farivar, R. Neal, C. Clarke, and S. Low, "Optimal inverter VAR control in distribution systems with high PV penetration," in *IEEE Power and Energy Society General Meeting*. pp. 1–7, 2012.

[8] S. Paudyal, C. A. Cañizares, and K. Bhattacharya, "Optimal operation of distribution feeders in smart grids," *IEEE Transactions on Industrial Electronics*, vol. 58, no. 10, pp. 4495–4503, 2011.

[9] S. S. Guggilam, E. Dall'Anese, Y. C. Chen, S. V. Dhople, and G. B. Giannakis, "Scalable optimization methods for distribution networks with high PV integration," *IEEE Transactions on Smart Grid*, vol. 7, no. 4, pp. 2061–2070, 2016.

[10] K. Christakou, D.-C. Tomozei, J.-Y. Le Boudec, and M. Paolone, "AC OPF in radial distribution networks – Part I: On the limits of the branch flow convexification and the alternating direction method of multipliers," *Electric Power Systems Research*, vol. 143, pp. 438–450, 2017.

[11] J. Lavaei and S. H. Low, "Zero duality gap in optimal power flow problem," *IEEE Transactions on Power Systems*, vol. 27, no. 1, pp. 92–107, 2012.

[12] S. H. Low, "Convex relaxation of optimal power flowpart I: Formulations and equivalence," *IEEE Transactions on Control of Network Systems*, vol. 1, no. 1, pp. 15–27, June 2014.

[13] J. A. Taylor, *Convex Optimization of Power Systems*. Cambridge: Cambridge University Press, 2015.

[14] M. Baran and F. F. Wu, "Optimal sizing of capacitors placed on a radial distribution system," *IEEE Transactions on Power Delivery*, vol. 4, no. 1, pp. 735–743, 1989.

[15] K. Turitsyn, P. Sulc, S. Backhaus, and M. Chertkov, "Options for control of reactive power by distributed photovoltaic generators," *Proceedings of the IEEE*, vol. 99, no. 6, pp. 1063–1073, 2011.

[16] S. Bolognani and S. Zampieri, "On the existence and linear approximation of the power flow solution in power distribution networks," *IEEE Transactions on Power Systems*, vol. 31, no. 1, pp. 163–172, 2016.

[17] A. Bernstein, C. Wang, E. Dall'Anese, J.-Y. Le Boudec, and C. Zhao, "Load flow in multiphase distribution networks: Existence, uniqueness, non-singularity and linear models," *IEEE Transactions on Power Systems*, vol. 33, no. 6, pp. 5832–5843, 2018.

[18] S. Bolognani, R. Carli, G. Cavraro, and S. Zampieri, "Distributed reactive power feedback control for voltage regulation and loss minimization," *IEEE Transactions on Automatic Control*, vol. 60, no. 4, pp. 966–981, Apr. 2015.

[19] A. Gómez-Expósito, A. Conejo, and C. A. Cañizares, eds., *Electric Energy Systems: Analysis and Operation.*, 2nd ed. Boca Raton, FL: CRC Press, 2018.

[20] F. D. Galiana, F. Bouffard, J. M. Arroyo, and J. F. Restrepo, "Scheduling and pricing of coupled energy and primary, secondary, and tertiary reserves," *Proceedings of the IEEE*, vol. 93, no. 11, pp. 1970–1983, Nov. 2005.

[21] Z. Li, Q. Guo, H. Sun, and J. Wang, "Coordinated economic dispatch of coupled transmission and distribution systems using heterogeneous decomposition," *IEEE Transactions on Power Systems*, vol. 31, no. 6, pp. 4817–4830, Nov. 2016.

[22] N. Gatsis and G. B. Giannakis, "Decomposition algorithms for market clearing with large-scale demand response," *IEEE Transactions on Smart Grid*, vol. 4, no. 4, pp. 1976–1987, Dec. 2013.

[23] P. Samadi, A. Mohsenian-Rad, R. Schober, V. W. S. Wong, and J. Jatskevich, "Optimal real-time pricing algorithm based on utility maximization for smart grid," in *IEEE International Conference on Smart Grid Communications*, pp. 415–420, Oct. 2010.

[24] L. Chen, N. Li, L. Jiang, and S. H. Low, "Optimal demand response: Problem formulation and deterministic case," in *Control and Optimization Methods for Electric Smart Grids*, A. Chakrabortty and M. D. Ilić, eds. New York: Springer pp. 63–85, 2012.

[25] N. Gatsis and G. B. Giannakis, "Residential load control: Distributed scheduling and convergence with lost AMI messages," *IEEE Transactions on Smart Grid*, vol. 3, no. 2, pp. 770–786, June 2012.

[26] Z. Chen, L. Wu, and Y. Fu, "Real-time price-based demand response management for residential appliances via stochastic optimization and robust optimization," *IEEE Transactions on Smart Grid*, vol. 3, no. 4, pp. 1822–1831, Dec. 2012.

[27] C. Chen, J. Wang, Y. Heo, and S. Kishore, "MPC-based appliance scheduling for residential building energy management controller," *IEEE Transactions on Smart Grid*, vol. 4, no. 3, pp. 1401–1410, Sept. 2013.

[28] L. Zhang, V. Kekatos, and G. B. Giannakis, "Scalable electric vehicle charging protocols," *IEEE Transactions on Power Systems*, vol. 32, no. 2, pp. 1451–1462, Mar. 2017.

[29] A. Gómez-Expósito and A. Abur, *Power System State Estimation: Theory and Implementation*. Boca Raton: FL: CRC Press, 2004.

[30] M. Kezunovic, S. Meliopoulos, V. Venkatasubramanian, and V. Vittal, *Application of Time-Synchronized Measurements in Power System Transmission Networks*. Cham, Switzerland: Springer, 2014.

[31] V. Kekatos and G. B. Giannakis, "Distributed robust power system state estimation," *IEEE Transactions on Power Systems*, vol. 28, no. 2, pp. 1617–1626, 2013.

[32] T. Burton, D. Sharpe, N. Jenkins, and E. Bossanyi, *Wind Energy Handbook*. Chichester, UK: John Wiley & Sons, 2001.

[33] P. A. Fleming, A. Ning, P. M. Gebraad, and K. Dykes, "Wind plant system engineering through optimization of layout and yaw control," *Wind Energy*, vol. 19, no. 2, pp. 329–344, 2016.

[34] Z. Dar, K. Kar, O. Sahni, and J. H. Chow, "Wind farm power optimization using yaw angle control," *IEEE Transactions on Sustainable Energy*, vol. 8, no. 1, pp. 104–116, Jan 2017.

[35] J. King, E. Dall'Anese, M. Hong, and C. J. Bay, "Optimization of wind farms using proximal primal-dual algorithms," in *American Control Conference*, June 2019.

[36] M. Chiang, S. H. Low, A. R. Calderbank, and J. C. Doyle, "Layering as optimization decomposition: A mathematical theory of network architectures," *Proceedings of the IEEE*, vol. 95, no. 1, pp. 255–312, Jan. 2007.

[37] D. P. Bertsekas, A. Nedić, and A. E. Ozdaglar, *Convex Analysis and Optimization*. Belmont, MA: Athena Scientific, 2003.

[38] T. Larsson, M. Patriksson, and A.-B. Strömberg, "Ergodic, primal convergence in dual subgradient schemes for convex programming," *Mathematical Programming*, vol. 86, no. 2, pp. 283–312, 1999.

[39] A. Nedić and A. Ozdaglar, "Approximate primal solutions and rate analysis for dual subgradient methods," *SIAM Journal on Optimization*, vol. 19, no. 4, pp. 1757–1780, 2009.

[40] A. Ruszczyński, *Nonlinear Optimization*. Princeton, NJ: Princeton University Press, 2006.

[41] S. Feltenmark and K. Kiwiel, "Dual applications of proximal bundle methods, including Lagrangian relaxation of nonconvex problems," *SIAM Journal on Optimization*, vol. 10, no. 3, pp. 697–721, 2000.

[42] Y. Zhang, N. Gatsis, and G. B. Giannakis, "Disaggregated bundle methods for distributed market clearing in power networks," in *Proceedings of the IEEE Global Conference on Signal and Information Processing*, pp. 835–838, Dec. 2013.

[43] P. McNamara and S. McLoone, "Hierarchical demand response for peak minimization using Dantzig-Wolfe decomposition," *IEEE Transactions on Smart Grid*, vol. 6, no. 6, pp. 2807–2815, Nov. 2015.

[44] A. Bernstein, E. Dall'Anese, and A. Simonetto, "Online primal-dual methods with measurement feedback for time-varying convex optimization," *IEEE Transactions on Signal Processing*, 2019, preprint https://arxiv.org/abs/1804.05159

[45] Y. Nesterov, "Primal-dual subgradient methods for convex problems," *Mathematical Programming*, vol. 120, no. 1, pp. 221–259, 2009.

[46] A. Yurtsever, Q. T. Dinh, and V. Cevher, "A universal primal-dual convex optimization framework," in *Advances in Neural Information Processing Systems*, pp. 3150–3158, 2015.

[47] J. Koshal, A. Nedić, and U. Y. Shanbhag, "Multiuser optimization: Distributed algorithms and error analysis," *SIAM Journal on Optimization*, vol. 21, no. 3, pp. 1046–1081, 2011.

[48] F. Dörfler and S. Grammatico, "Gather-and-broadcast frequency control in power systems," *Automatica*, vol. 79, pp. 296–305, 2017.

[49] E. Dall'Anese and A. Simonetto, "Optimal power flow pursuit," *IEEE Transactions on Smart Grid*, vol. 9, no. 2, pp. 942–952, Mar. 2018.

[50] X. Zhou, E. Dall'Anese, and L. Chen, "Online stochastic optimization of networked distributed energy resources," *preprint arXiv:1711.09953*, 2017.

[51] A. Hauswirth, S. Bolognani, G. Hug, and F. Dorfler, "Projected gradient descent on Riemannian manifolds with applications to online power system optimization," in *54th Annual Allerton Conference on Communication, Control, and Computing*, pp. 225–232, Sept. 2016.

[52] N. Li, L. Chen, C. Zhao, and S. H. Low, "Connecting automatic generation control and economic dispatch from an optimization view," in *American Control Conference*, Portland, OR, June 2014.

[53] A. Jokić, M. Lazar, and P. Van den Bosch, "Real-time control of power systems using nodal prices," *International Journal of Electrical Power & Energy Systems*, vol. 31, no. 9, pp. 522–530, 2009.

[54] F. J. Nogales, F. J. Prieto, and A. J. Conejo, "A decomposition methodology applied to the multi-area optimal power flow problem," *Annals of Operations Research*, vol. 120, no. 1-4, pp. 99–116, 2003.

[55] B. H. Kim and R. Baldick, "Coarse-grained distributed optimal power flow," *IEEE Transactions on Power Systems*, vol. 12, no. 2, pp. 932–939, 1997.

[56] E. Dall'Anese, H. Zhu, and G. B. Giannakis, "Distributed optimal power flow for smart microgrids," *IEEE Transactions on Smart Grid*, vol. 4, no. 3, pp. 1464–1475, 2013.

[57] P. Gao, M. Wang, S. G. Ghiocel, J. H. Chow, B. Fardanesh, and G. Stefopoulos, "Missing data recovery by exploiting low-dimensionality in power system synchrophasor measurements," *IEEE Transactions on Power Systems*, vol. 31, no. 2, pp. 1006–1013, 2016.

[58] G. B. Giannakis, Q. Ling, G. Mateos, I. D. Schizas, and H. Zhu, "Decentralized learning for wireless communications and networking," in *Splitting Methods in Communication, Imaging, Science, and Engineering*. Cham, Switzerland: Springer, pp. 461–497, 2016.

[59] I. D. Schizas, A. Ribeiro, and G. B. Giannakis, "Consensus in ad hoc WSNs with noisy links–part I: Distributed estimation of deterministic signals," *IEEE Transactions on Signal Processing*, vol. 56, no. 1, pp. 350–364, 2008.

[60] S. Boyd, N. Parikh, E. Chu, B. Peleato, and J. Eckstein, "Distributed optimization and statistical learning via the alternating direction method of multipliers," *Foundations and Trends in Machine Learning*, vol. 3, no. 1, pp. 1–122, 2011.

[61] D. P. Bertsekas and J. N. Tsitsiklis, *Parallel and Distributed Computation: Numerical Methods*. Englewood Cliffs, NJ: Prentice Hall, 1989.

[62] W. Deng and W. Yin, "On the global and linear convergence of the generalized alternating direction method of multipliers," *Journal of Scientific Computing*, vol. 66, no. 3, pp. 889–916, 2016.

[63] M. Hong and Z.-Q. Luo, "On the linear convergence of the alternating direction method of multipliers," *Mathematical Programming*, vol. 162, no. 1–2, pp. 165–199, 2017.

[64] B. Zhang, A. Y. Lam, A. D. Domínguez-García, and D. Tse, "An optimal and distributed method for voltage regulation in power distribution systems," *IEEE Transactions on Power Systems*, vol. 30, no. 4, pp. 1714–1726, 2015.

[65] Q. Peng and S. H. Low, "Distributed optimal power flow algorithm for radial networks, I: Balanced single phase case," *IEEE Transactions on Smart Grid*, vol. 9, no. 1, pp. 111–121, Jan. 2018.

[66] M. Bazrafshan and N. Gatsis, "Decentralized stochastic optimal power flow in radial networks with distributed generation," *IEEE Transactions on Smart Grid*, vol. 8, no. 2, pp. 787–801, Mar. 2017.

[67] B. A. Robbins, C. N. Hadjicostis, and A. D. Domínguez-García, "A two-stage distributed architecture for voltage control in power distribution systems," *IEEE Transactions on Power Systems*, vol. 28, no. 2, pp. 1470–1482, May 2013.

[68] M. Hong, D. Hajinezhad, and M.-M. Zhao, "Prox-PDA: The proximal primal-dual algorithm for fast distributed nonconvex optimization and learning over networks," in *International Conference on Machine Learning*, pp. 1529–1538, 2017.

[69] K. Sun and A. Sun, "A two-level distributed algorithm for general constrained non-convex optimization with global convergence," *Preprint arXiv:1902.07654*, 2019.

[70] T. Erseghe, "Distributed optimal power flow using ADMM," *IEEE Transactions on Power Systems*, vol. 29, no. 5, pp. 2370–2380, Sept. 2014.

[71] Y. Tang, K. Dvijotham, and S. Low, "Real-time optimal power flow," *IEEE Transactions on Smart Grid*, vol. 8, no. 6, pp. 2963–2973, 2017.

[72] Y. Tang, E. Dall'Anese, A. Bernstein, and S. H. Low, "A feedback-based regularized primal-dual gradient method for time-varying nonconvex optimization," in *IEEE Conference on Decision and Control*, Dec. 2018.

[73] S. Kim and G. B. Giannakis, "Scalable and robust demand response with mixed-integer constraints," *IEEE Transactions on Smart Grid*, vol. 4, no. 4, pp. 2089–2099, Dec. 2013.

[74] S. Mhanna, A. C. Chapman, and G. Verbič, "A fast distributed algorithm for large-scale demand response aggregation," *IEEE Transactions on Smart Grid*, vol. 7, no. 4, pp. 2094–2107, July 2016.

[75] B. Li, T. Chen, X. Wang, and G. B. Giannakis, "Real-time energy management in microgrids with reduced battery capacity requirements," *IEEE Transactions on Smart Grid*, vol. 10, no. 2, Mar. 2019.

[76] S. Sun, M. Dong, and B. Liang, "Distributed real-time power balancing in renewable-integrated power grids with storage and flexible loads," *IEEE Transactions on Smart Grid*, vol. 7, no. 5, pp. 2337–2349, Sept. 2016.

[77] J. Qin, Y. Chow, J. Yang, and R. Rajagopal, "Distributed online modified greedy algorithm for networked storage operation under uncertainty," *IEEE Transactions on Smart Grid*, vol. 7, no. 2, pp. 1106–1118, Mar. 2016.

[78] N. Gatsis and A. G. Marques, "A stochastic approximation approach to load shedding in power networks," in *IEEE International Conference on Acoustics, Speech and Signal Processing*, Florence, Italy, pp. 6464–6468, May 2014.

[79] Y. Guo, M. Pan, Y. Fang, and P. P. Khargonekar, "Decentralized coordination of energy utilization for residential households in the smart grid," *IEEE Transactions on smart grid*, vol. 4, no. 3, pp. 1341–1350, Sept. 2013.

[80] A. Bernstein, L. Reyes Chamorro, J.-Y. Le Boudec, and M. Paolone, "A composable method for real-time control of active distribution networks with explicit power set points. Part I: Framework," *Electric Power Systems Research*, vol. 125, pp. 254–264, Aug. 2015.

[81] A. Hauswirth, I. Subotić, S. Bolognani, G. Hug, and F. Dörfler, "Time-varying projected dynamical systems with applications to feedback optimization of power systems," in *IEEE Conference on Decision and Control (CDC)*. pp. 3258–3263, 2018.

[82] Y. Zhang, E. Dall'Anese, and M. Hong, "Dynamic ADMM for real-time optimal power flow," in *IEEE Global Conference on Signal and Information Processing*, Nov. 2017.

17 Distributed Load Management

Changhong Zhao, Vijay Gupta, and Ufuk Topcu

The essence of power system operation is to maintain system security at minimal cost. The most important security requirements include maintaining instantaneous balance between generation and load power and keeping frequency close to its nominal value, i.e., 60 Hz in the United States. These security requirements must be satisfied at minimal operational cost, whether it is the cost for generator fuel consumption and emissions, or the cost to compensate for user discomfort when their electric loads are controlled by the system operator.

Traditional power system operations to fulfill these requirements primarily rely on controlling large generators. The underlying philosophy is to schedule the generators to match forecasted demand. However, this approach will not work in the future as the deepening penetration of intermittent renewable generation, such as wind and solar, introduces increasingly large and fast variations in power supply. In this case, the large generators may not be able to provide adequate control capacity or respond sufficiently fast to balance the variations, which motivates us to control alternative resources, especially electric loads.

The idea of load management dates back to the late 1970s. Schweppe et al. advocated in a 1980 paper its deployment to "assist or even replace turbine-governed systems and spinning reserve." They also proposed to use spot prices to incentivize the users to adapt their consumption to the true cost of generation at the time of consumption. Remarkably it was emphasized back then that such adaptive loads would "allow the system to accept more readily a stochastically fluctuating energy source, such as wind or solar generation" [1]. This point is echoed over the last decade, e.g., in [2–7], which argue for "grid-friendly" appliances, such as refrigerators, water or space heaters, ventilation systems, and air conditioners, as well as plug-in electric vehicles to help manage energy imbalance in power systems.

To exploit the full potential of load management, a set of important issues need to be addressed, including: (1) scalability and flexibility of the control system to support autonomous and plug-and-play operations of controllable loads; (2) coordination between controllable loads, as well as coordination between loads and generators, to ensure a predictable and stable system behavior; and (3) optimization of comfort levels of controllable load users measured by utility functions. Addressing these issues calls for the transformation of power systems from a centralized, hierarchical control architecture to a *distributed* architecture.

To facilitate the historic transformation discussed in the preceding text, this chapter introduces our work on distributed load management. Specifically, Section 17.1 introduces a set of distributed protocols for electric vehicle (EV) charging, which provide analytical convergence guarantees of individual EV charging profiles without requiring the customers to share their charging constraints with the distribution utility company (or system operator).

These distributed protocols require messages to be exchanged among the distribution utility company and the customers regarding possible price profiles and desired charging profiles in response. This exchange requires secure and low-latency two-way communication between the distribution utility company and the EV customers, which may be costly to achieve. In Section 17.2, we propose an online learning version of the distributed charging algorithm. This version requires only one-way communication from the distribution utility company to the customers about the pricing profiles of *previous* days.

The preceding EV charging schemes are based on a simplified system model where the power network connecting the distribution utility company and EVs is not explicitly modeled. In Section 17.3, we extend these schemes in two aspects. First, we consider general controllable loads that not only include EVs but also refrigerators, air conditioners, etc.; second, we consider the dynamic model of power network that connects the controllable loads, and focus on design of distributed feedback control and stability of dynamic power network under such control.

17.1 Distributed Charging Protocols for Electric Vehicles

Consider a scenario, also studied in [8–11], where an electric utility company negotiates with N electric vehicles (EVs) over T time slots of length ΔT on their charging profiles. The utility is assumed to know (precisely predict) the inelastic base demand profile (aggregate non-EV demand) and aims to shape the aggregate charging profile of EVs to flatten the total demand (base demand plus EV demand) profile. Each EV can charge after it plugs in and needs to be charged a specified amount of electricity by its deadline. In each time slot, the charging rate of an EV is a constant. Let $D(t)$ denote the base load in slot t, $r_n(t)$ denote the charging rate of EV n in slot t, $r_n := (r_n(1), \ldots, r_n(T))$ denote the charging profile of EV n, for $n \in \mathcal{N} := \{1, \ldots, N\}$ and $t \in \mathcal{T} := \{1, \ldots, T\}$. Roughly speaking, this optimal control problem formalizes the intent of flattening the total demand profile, which is captured by the objective function

$$L(r) = L(r_1, \ldots, r_N) := \sum_{t=1}^{T} U\left(D(t) + \sum_{n=1}^{N} r_n(t)\right). \tag{17.1}$$

In (17.1) and hereafter, $r := (r_1, \ldots, r_N)$ denotes a charging profile of all EVs. The map $U : \mathbb{R} \to \mathbb{R}$ is strictly convex.

The charging profile r_n of EV n is considered to take values in the interval $[0, \bar{r}_n]$ for some given $\bar{r}_n \succeq 0$. To impose arrival time and deadline constraints, $\bar{r}_n$ is considered to be time-dependent with $\bar{r}_n(t) = 0$ for slots t before the arrival time and after the deadline of EV n. Hence

$$0 \leq r_n(t) \leq \bar{r}_n(t), \; n \in \mathcal{N}, \; t \in \mathcal{T}. \tag{17.2}$$

For EV $n \in \mathcal{N}$, let B_n, $s_n(0)$, $s_n(T)$, and η_n denote its battery capacity, initial state of charge, final state of charge, and charging efficiency. The constraint that EV n needs to reach $s_n(T)$ state of charge by its deadline is captured by the total energy stored over time horizon

$$\eta_n \sum_{t \in \mathcal{T}} r_n(t) \Delta T = B_n(s_n(T) - s_n(0)), \; n \in \mathcal{N}. \tag{17.3}$$

Define the charging rate sum

$$R_n := B_n(s_n(T) - s_n(0))/(\eta_n \Delta T)$$

for $n \in \mathcal{N}$. Then, the constraint in (17.3) can be written as

$$\sum_{t=1}^{T} r_n(t) = R_n, \; n \in \mathcal{N}. \tag{17.4}$$

DEFINITION 17.1 *Let $U : \mathbb{R} \to \mathbb{R}$ be strictly convex. A charging profile $r = (r_1, \ldots, r_N)$ is*

1) *feasible, if it satisfies the constraints (17.2) and (17.4);*
2) *optimal, if it solves the optimal charging (OC) problem*

$$\text{OC} \begin{cases} \min_{r_1, \ldots, r_N} & \sum_{t=1}^{T} U\left(D(t) + \sum_{n=1}^{N} r_n(t)\right) \\ \text{s.t.} & 0 \leq r_n(t) \leq \bar{r}_n(t), \; t \in \mathcal{T}, n \in \mathcal{N}; \\ & \sum_{t=1}^{T} r_n(t) = R_n, \; n \in \mathcal{N}; \end{cases} \tag{17.5}$$

3) *valley-filling, if there exists $A \in \mathbb{R}$ such that*

$$\sum_{n \in \mathcal{N}} r_n(t) = [A - D(t)]^+, \; t \in \mathcal{T}.$$

If the objective is to track a given demand profile G rather than to flatten the total demand, the objective function can be modified as

$$\sum_{t=1}^{T} U\left(D(t) + \sum_{n=1}^{N} r_n(t) - G(t)\right). \tag{17.6}$$

As a means toward developing a decentralized charging protocol, [8, 9] established a series of properties of optimal charging profiles. We now review some of these properties without their proofs. For notational simplicity, for a given charging profile $r = (r_1, \ldots, r_N)$, let

$$R_r := \sum_{n \in \mathcal{N}} r_n$$

denote its corresponding aggregate charging profile.

PROPERTY 17.1 *If a feasible charging profile r is valley-filling, then it is optimal.*

Define $\mathcal{F}_n := \{r_n \mid 0 \leq r_n \leq \bar{r}_n, \sum_{t \in \mathcal{T}} r_n(t) = R_n\}$ as the set of feasible charging profiles for EV n. Then,

$$\mathcal{F} := \mathcal{F}_1 \times \cdots \times \mathcal{F}_N$$

is the set of feasible charging profiles $r = (r_1, \ldots, r_N)$.

PROPERTY 17.2 *If $\mathcal{F}$ is nonempty, optimal charging profiles exist.*

Valley-filling is our intuitive notion of optimality. However, it may not be always achievable. For example, the "valley" in inelastic base demand may be so deep that even if all EVs charge at their maximum rate, it is still not completely filled, e.g., at 4:00 in Figure 17.1 (bottom). Besides, EVs may have stringent deadlines such that the potential for shifting the load over time to yield valley-filling is limited. The notion of optimality in Definition 17.1 relaxes these restrictions as a result of Property 17.2. Moreover, it agrees with the intuitive notion of optimality when valley-filling is achievable as a result of Property 17.1, illustrated in Figure 17.1 (top).

DEFINITION 17.2 *Two feasible charging profiles r and r' are equivalent, provided that $R_r = R_{r'}$, i.e., r and r' have the same aggregate charging profile. We denote this relation by $r \sim r'$.*

It is easy to check that the relation $\sim$ is an equivalence relation. Define equivalence classes $\{r' \in \mathcal{F} \mid r' \sim r\}$ with representatives $r \in \mathcal{F}$, and

$$\mathcal{O} := \{r \in \mathcal{F} \mid r \text{ optimal}\}$$

as the set of optimal charging profiles.

THEOREM 17.1 *If $\mathcal{F}$ is nonempty, then $\mathcal{O}$ is nonempty, compact, convex, and an equivalence class of the relation $\sim$.*

COROLLARY 17.1 *Optimal charging profile may not be unique.*

THEOREM 17.2 *The set $\mathcal{O}$ of optimal charging profiles does not depend on the choice of U. That is, if r^* is optimal with respect to a strictly convex utility function, then r^* is also optimal with respect to any other strictly convex utility function.*

The optimal solution to problem OC provides a uniform means for defining optimality even when valley-filling is not achievable, and Theorem 17.2 implies that this optimality notion is intrinsic, independent of the choice of U.

Having established key properties of the optimal solutions for problem OC, we now follow the presentation of an algorithm from [8, 9] to solve the problem in a decentralized manner. While [8, 9] developed both synchronous and asynchronous algorithms, we focus on the former (see Figure 17.2).

Figure 17.3 shows the information exchange between the utility company and the EVs for the implementation of this algorithm. Given the "price" profile broadcast by the utility, each EV chooses its charging profile independently, and reports back to the

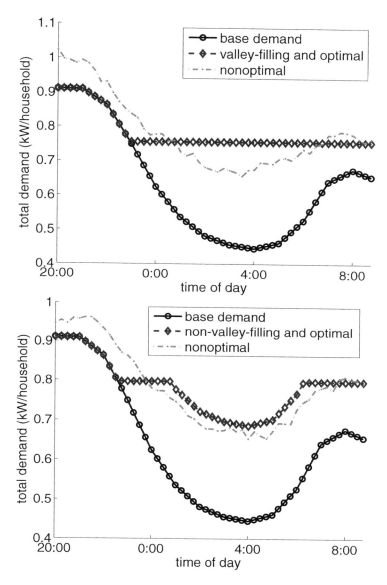

Figure 17.1 Base demand profile is the average residential load in the service area of Southern California Edison (SCE) from 20:00 on 02/13/2011 to 9:00 on 02/14/2011 [9]. Optimal total demand profile curve corresponds to the outcome of DCA (decentralized charging algorithm; an algorithm introduced later in this chapter) with $U(x) = x^2$. With different specifications for EVs (e.g., maximum charging rate r_n^{max}), optimal charging profile can be valley-filling (top figure) or non-valley-filling (bottom figure). A hypothetical nonoptimal curve is shown with dash-dot line.

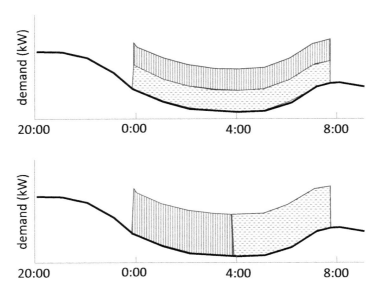

Figure 17.2 An example of equivalent charging profiles. In both top and bottom figures, the two regions correspond to the charging profiles of two different EVs. Because aggregate charging profiles in both figures are equal, the charging profile r in the top figure is equivalent to the charging profile r' in the bottom figure.

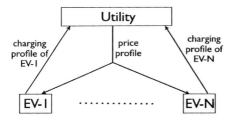

Figure 17.3 Schematic view of the information flow between the utility and the EVs. Given the "price" profile, the EVs choose their charging profiles independently. The utility guides their decisions by altering the "price" profile based on total demand profile.

utility. The utility guides their behavior by altering the "price" profile. We assume U' is Lipschitz with the Lipschitz constant $\beta > 0$, i.e.,

$$|U'(x) - U'(y)| \leq \beta |x - y|$$

for all x, y.

Decentralized Charging Algorithm

Given the scheduling horizon $\mathcal{T}$, the maximum number K of iterations, the error tolerance $\epsilon > 0$, the base load profile D, the number N of EVs, the charging rate sum R_n, and the charging rate upper bound $\bar{r}_n$ for EV $n \in \mathcal{N}$, pick a step size γ satisfying

$$0 < \gamma < \frac{1}{N\beta}.$$

1. Initialize the "price" profile and the charging profile as

$$p^0(t) := U'(D(t)), \quad r_n^0(t) := 0$$

   for $t \in \mathcal{T}$ and $n \in \mathcal{N}$, $k \leftarrow 0$.
2. The utility broadcasts γp^k to all EVs.
3. Each EV $n \in \mathcal{N}$ calculates a new charging profile r_n^{k+1} as the solution to the following optimization problem

$$\min_{r_n} \sum_{t \in \mathcal{T}} \gamma p^k(t) r_n(t) + \frac{1}{2}\left(r_n(t) - r_n^k(t)\right)^2 \quad (17.7)$$

$$\text{s.t. } 0 \leq r_n(t) \leq \bar{r}_n(t), \ t \in \mathcal{T};$$

$$\sum_{t \in \mathcal{T}} r_n(t) = R_n,$$

   and reports r_n^{k+1} to the utility.
4. The utility collects charging profiles r_n^{k+1} from the EVs, and updates the "price" as

$$p^{k+1}(t) := U'\left(D(t) + \sum_{n=1}^{N} r_n^{k+1}(t)\right) \quad (17.8)$$

   for $t \in \mathcal{T}$.
   If $\|p^{k+1} - p^k\| \leq \epsilon$, return p^{k+1}, r_n^{k+1} for all n.
5. If $k < K$, $k \leftarrow k + 1$, and go to step (2).
   Else, return p^K, r_n^K for all n.

The "price" signal p is a control signal used by the utility company to guide the EVs in choosing their charging profiles, and is not necessarily the real electricity price.

In each iteration, the algorithm can be split into two parts. In the first part, EV n updates its charging profile to minimize its objective function as (17.7). There are two terms in the objective: the first term is the "cost" and the second term penalizes deviations from the profile computed in the previous iteration. The extra penalty term ensures convergence of DCA, and vanishes as $k \to \infty$ (see Theorem 17.4). Hence, the objective function of each EV reduces to its "cost" as $k \to \infty$. Intuitively, the smaller γ is, the more significant the penalty term becomes, the less r_n^{k+1} is going to deviate from r_n^k, and the more likely DCA will converge. When $\gamma < \frac{1}{N\beta}$, DCA converges to optimal charging profiles (see Theorem 17.3). The upper bound $\bar{\gamma} := \frac{1}{N\beta}$ of γ is inversely proportional to the number N of EVs and the Lipschitz constant β of U'. For large N, each EV has to update its charging profile more slowly because the aggregate profile update is roughly N times amplified. Hence, the penalty term should be larger, and $\bar{\gamma}$ should be smaller. When the Lipschitz constant β is larger, the same difference in total demand is going to cause a larger difference in U', or p according to (17.8). Hence, for the two terms in (17.7) to remain the same scale, $\bar{\gamma}$ should decrease. In conclusion, the upper bound $\bar{\gamma} = \frac{1}{N\beta}$ agrees with our intuition. In the second part of the iteration, the utility updates the "price" profile according to (17.8). It sets higher prices for slots with higher total demand, to give EVs the incentive to shift their energy consumption to slots with lower total demand.

Let the superscript k for each variable denote its respective value in iteration k. For example, r_n^k denotes the charging profile of EV n in iteration k. Similarly,

$$R^k := \sum_{n=1}^{N} r_n^k$$

denotes the aggregate charging profile in iteration k.

LEMMA 17.1 *If the set $\mathcal{F}$ of feasible charging profiles is nonempty, then the inequality*

$$\langle \gamma p^k, r_n^{k+1} - r_n^k \rangle \leq - \left\| r_n^{k+1} - r_n^k \right\|^2 \tag{17.9}$$

holds for $n \in \mathcal{N}$ and $k \geq 1$.

LEMMA 17.2 *If $\mathcal{F}$ is nonempty, then for $n \in \mathcal{N}$ and $k \geq 1$, $r_n^{k+1} = r_n^k$ if and only if, for all $r_n \in \mathcal{F}_n$,*

$$\langle p^k, r_n - r_n^k \rangle \geq 0. \tag{17.10}$$

Recall that $\mathcal{O}$ denotes the set of optimal charging profiles.

THEOREM 17.3 *If $\mathcal{F}$ is nonempty, then $r^k \to \mathcal{O}$ as $k \to \infty$.*

COROLLARY 17.2 *A charging profile r is stationary for DCA, i.e., if $r^{\bar{k}} = r$ for some $\bar{k} \geq 0$ then $r^k = r$ for all $k \geq \bar{k}$, if and only if $r \in \mathcal{O}$.*

THEOREM 17.4 *Let r^* be an optimal charging profile. If $\mathcal{F}$ is nonempty, then*

- *The aggregate charging profile converges to that of r^*, i.e.,*

$$\lim_{k \to \infty} R^k = R_{r^*};$$

- *The price profile converges to that corresponding to r^*, i.e.,*

$$\lim_{k \to \infty} p^k = U'(D + R_{r^*});$$

- *For each EV n, the difference between two consecutive charging profiles vanishes, i.e.,*

$$\lim_{k \to \infty} \left\| r_n^{k+1} - r_n^k \right\| = 0.$$

Theorem 17.3 shows the convergence of r^k to the optimal set $\mathcal{O}$ while Theorem 17.4 focuses on the convergence of R^k and p^k to the optimal value R_{r^*} and $U'(D + R_{r^*})$.

We conclude this section with an example from [9] which compared DCA to another algorithm from [12] (referred to as "DAP" for "deviation from average penalty").

We choose the average residential load profile in the service area of South California Edison from 20:00 on 02/13/2011 to 9:00 on 02/14/2011 as the average base demand profile per household. According to the typical charging characteristics of EVs in [13], we set $r_n^{max} = 3.3$kW. We assume that the charging rate $r_n(t)$ of EV n at slot t can take continuous values in $[0, r_n^{max}]$ after EV n plugs in for charging and before its deadline. We consider the penetration level of $N = 20$ EVs in 100 households. The planning horizon is from 20:00 to 9:00 the next day, and divided into 52 slots of 15 minutes, during which the charging rate of an EV is a constant. The map $U : \mathbb{R} \to \mathbb{R}$ is taken to be $U(x) = \frac{1}{2}x^2$. As in [12], we choose the price function and parameters for Algorithm DAP as $p(x) = 0.15x^2$, $c = 1$, and $\delta = 0.15$.

Homogeneous Case

Although DCA obtains optimal charging profiles, whatever the parameters $\bar{r}_n(t)$ and R_n are, we simulate the homogeneous case to compare with Algorithm DAP because Algorithm DAP. In this example, all the EVs have the same deadline at 9:00 the next morning. Figure 17.4 shows the average total demand profile in each iteration of DCA and DAP in a homogeneous case. Both algorithms converge to a flat charging profile. Moreover, DCA converges with a single iteration, while DAP takes several iterations to converge.

Nonhomogeneous Case

Figure 17.5 shows the average total demand profiles at convergence of DCA and DAP in a nonhomogeneous case in which EVs have different charging capacities. DCA still converges to a flat charging profile in a few iterations while DAP no longer converges to a flat charging profile.

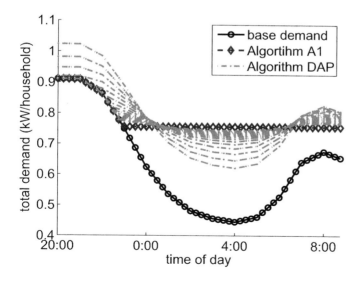

Figure 17.4 All EVs plug in at 20:00 with deadline at 9:00 on the next day, and have 10kWh charging capacity. Multiple dash-dot curves correspond to total demand in different iterations of Algorithm DAP.

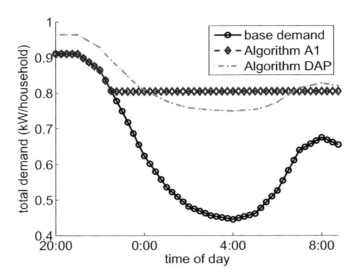

Figure 17.5 All EVs plug in at 20:00 with deadline at 9:00 on the next day, but have different charging capacities uniformly distributed in $[0, 20]$kWh.

17.2 An Online-Learning–Based Implementation

The algorithm considered (and its various extensions) has analytical convergence guarantees and does not require the customers to share their charging constraints with the utility company. However, they require a series of messages to be exchanged among the utility company and the customers regarding possible price profiles and desired charging profiles in response. As the available power supply and the customer requirements for charging their EVs change from day to day, these messages need to be exchanged daily to calculate the charging profiles. This requires secure and low-latency two-way communication between the utility company and the EV customers, which may be costly to achieve. Instead, we propose an online learning version of the algorithm that requires only one-way communication from the utility company to the customers about the pricing profiles of *previous* days. The price that we pay for this one-way communication is a slower convergence of the charging profiles to the optimal ones, and a relaxation of the sense in which the charging profiles are optimal.

Some relevant references that apply regret minimization to demand response (DR) are [14–16]. In [14], real-time electricity pricing strategies for DR are designed using regret minimization. However, the focus of that work is on optimizing the utility function for the utility company, and the customer behavior is assumed to be such that the load change is linear in the price variation. The objective of [15] is to design pricing policies for the customers having price responsive loads. The exact demand function of the customer is assumed to be unknown to the pricing policy maker. In [15], the utility company is the only decision maker, whereas in the discussion that follows, the utility company and the EV customers are all decision makers. In [16], regret minimization is used to learn the charging behavior of the EV customers. The price responsiveness for a community of customers is captured through a conditional random field model. The regret minimization algorithm is adopted to learn the parameters of the model.

We consider the same problem formulation as mentioned previously. The specific difference is in the information flow. The utility company monitors the total load and publishes the price profile for the previous day as realized according to a fixed and known pricing policy. The customers decide on the charging schedules for the next day with access to these pricing profiles for the previous days. No other communication occurs between the utility company and the customers, or among the customers.

Online Learning Framework

We now adopt a regret minimization framework to solve both the optimization problem of the company and the customer. Let L be a λ-strongly convex function with respect to a given norm $\|\cdot\|$. Let $D_L(\cdot,\cdot)$ denote the Bregman divergence [17] with respect to L. Let $\|\cdot\|_*$ denote the norm that is dual to $\|\cdot\|$. Let ∇L denote the gradient of L and ∇L^{-1} denote the inverse mapping of ∇L.

For the ith EV customer, the decision variable is her charging profile on the kth day and a charging profile is feasible if its charging rate at every time satisfies the specified lower and upper bounds, and the customer has the required amount S_i of charge at the end. We define r_i^* as the optimal feasible charging profile that minimizes the price p_i^k paid by the customer. Note that this profile does not change from one day to the next.

In the regret minimization framework, the notion of regret is used to measure the performance of an online algorithm [18]. For customer $i \in \mathcal{N}$, the customer regret after K days R_i is defined as the difference between the cumulative cost function value of the charging profiles r_i^k, $k = 1,\ldots,K$ generated by an online algorithm and the one generated by x_i^*, i.e.,

$$R_i(K, r_i^k) := \sum_{k=1}^{K} p_i^k(r_i^k) - \min_{x_i \in \mathcal{F}_i} \sum_{k=1}^{K} p_i^k(r_i). \tag{17.11}$$

Notice that the suboptimal charging profile x_i^* can only be calculated in hindsight after K days have elapsed.

We adopt the optimistic mirror descent (OMD) algorithm [17] to generate the charging profile update which minimizes the regret (17.11). On each day, the regret minimization algorithm generates the charging profile update without knowing the current objective function (and its gradient). Specifically, the OMD algorithm iteratively applies the updates

$$\begin{aligned} h_i^{k+1} &= \nabla L_i^{-1}\left(\nabla L_i(h_i^k) - \eta_i \nabla p_i^k(r_i^k)\right) \\ r_i^{k+1} &= \operatorname*{argmin}_{r_i \text{ is feasible}} \eta_i r_i^T M_i^{k+1} + D_{L_i}(r_i, h_i^{k+1}), \end{aligned} \tag{17.12}$$

where $\eta_i \in \mathbb{R}$ is an algorithm parameter, h_i^k is an intermediate update of the charging profile. For easy of presentation, for the vector $h_i^k \in \mathbb{R}^T$, $L_i(h_i^k)$ is set to $L_i(h_i^k) = \frac{\|h_i^k\|^2}{2}$. The ith customer may have a prediction M_i^k for the gradient of the cost function p_i^k. Intuitively, the iteration (17.12) updates the charging profile toward the negative gradient direction and projects it onto the set of feasible charging profiles.

The regret minimization framework for the company can be similarly defined. We skip the details for conciseness.

Convergence Results

The following result summarizes the convergence of the charging profile updates generated by the OMD algorithm.

PROPOSITION 17.1 *(Convergence of regret):* For every r_i^* that is feasible, the iteration (17.12) converges in the sense that

$$R_i(K, r_i^k) \le \frac{1}{\eta_i} P_i + \frac{\eta_i}{2} \sum_{k=1}^{K} \|\nabla p_i^k(r_i^k) - M_i^k\|_*^2, \quad (17.13)$$

where

$$R_i(K, r_i^k) := \sum_{k=1}^{K} p_i^k(r_i^k) - \sum_{k=1}^{K} p_i^k(r_i^*),$$
$$P_i := \max_{x_i \text{ is feasible}} L_i(r_i) - \min_{x_i \in \mathcal{F}_i} L_i(r_i). \quad (17.14)$$

In particular, if η_i is chosen as $O(1/\sqrt{K})$, then the average regret, i.e., $R_i(K)/K$, converges to zero as $K \to \infty$. A similar convergence result holds for the OMD algorithms followed by the utility company as well.

Thus, as the number of days increases, the average performance of the charging profiles generated by the OMD algorithm approaches the performance that is obtained by the charging profiles r^* and r_i^*, $i \in \mathcal{N}$, respectively.

Design of the Pricing Function

There are no guarantees that the solutions r_i^*, $i \in \mathcal{N}$ obtained by the customers will sum up to the optimal solution obtained by the company. In fact, unless the pricing function

p_i^k is carefully designed, these solutions will not be the same because the objectives of the utility company and the EV customers are different. In fact, the natural choice of p_i^k as

$$p_i^k(r_i^k) = \left(\sum_{j=1}^{N} r_j^k + D^k\right)^T r_i^k \qquad (17.15)$$

does not lead to the charging profiles $(r_1^*, \ldots, r_N^*)$ that reduce the regret of the utility company to zero. We now propose a choice of p_i^k to ensure that when each customer minimizes her regret, the aggregated charging profile minimizes the utility company's regret.

PROPOSITION 17.2 *If p_i^k is chosen as*

$$p_i^k(r_i^k) = \left(\frac{1}{2}r_i^k + \sum_{j \neq i}^{N} r_j^k + D^k\right)^T r_i^k, \quad i \in \mathcal{N}, \qquad (17.16)$$

the customers adopt the iteration (17.12), and $\eta_u = \frac{1}{2}\eta_i$, then the average regret of the utility company converges to zero as the total number of days goes to infinity.

To update the charging profile on day k, the ith customer needs to know $2r_i^{k-1} + \sum_{j \neq i} r_j^{k-1} + D^{k-1}$ or $\sum_j r_j^{k-1} + D^{k-1}$ depending on which pricing function is adopted. The utility company can simply publish the total load information for the previous day. The customers do not need to have full knowledge about how their consumption will map to a corresponding expenditure.

Extensions

The basic framework presented in the preceding text can be extended in various directions.

Regret with Respect to the Optimal Charging Profiles

The regrets defined previously measure the difference between the performance of the charging profiles generated by our algorithm and the performance that is obtained by the charging profiles r^* and r_i^*, $i \in \mathcal{N}$ that are the optimal profiles that do not vary from one day to the next. In that sense, these are static regrets. Instead, we can define tracking regret that characterizes the difference between the cumulative cost of the charging profiles generated by our algorithm and the cumulative cost of executing the optimal charging profiles r^{k*} that vary from one day to the next (but can be calculated only in hindsight).

THEOREM 17.5 *If $\eta_u = O(1/\sqrt{K})$, the OMD algorithm yields that the tracking regret is of the order $O(\sqrt{K}[1 + \sum_{k=1}^{K} \|r^{k*} - r^{k+1*}\|])$.*

Note that this regret bound increases as the variation of the optimal sequence of decisions $\sum_{k=1}^{K} \|r^{k*} - r^{k+1*}\|$ increases. If the optimal solution remains the same from one day to the next, then the tracking regret is of the order $O(\sqrt{K})$. However, if the optimal solution varies significantly from one day to the next, then the average tracking regret will not necessarily converge to zero. Note also that if the utility company has perfect prediction of the gradient of the cost function, then the utility company can set $\eta \to \infty$ to ensure that the regret bound is zero.

Presence of Inelastic Customers

The discussion so far assumed that all customers were rational in the sense that they wanted to choose their charging profile to minimize their cost. Furthermore, they were elastic in scheduling their charging (within the prespecified constraints). We now assume that some customers are either irrational or inelastic. Suppose that N_I out of N customers are inelastic. Denote the set of inelastic customers by $\mathcal{N}_I$. For every inelastic customer $i \in \mathcal{N}_I$, we assume that her charging profile remains the same from day to day and is not updated. We also set all customers' predictions to zeros, i.e., $M_i^k = 0$, $i \in \mathcal{N}$, $k \in \mathbb{N}_{>0}$. The update of the charging profile for inelastic customer $i \in \mathcal{N}_I$ can thus be written as

$$h_i^{k+1} = \nabla L_i^{-1}\left(\nabla L_i(h_i^k) - \eta_i \left(\nabla c_i^k(r^k) + \epsilon_i^k\right)\right),$$
$$r_i^{k+1} = \underset{x_i \in \mathcal{F}_i}{\operatorname{argmin}}\, D_{L_i}(r_i, h_i^{k+1}), \qquad (17.17)$$

where ϵ_i^k is an error term that quantifies the inconsistency between the updates as desired by the utility company for each customer to execute and the inelastic customer's behavior. Due to the presence of the inelastic customers, the ability of the aggregated solution to be valley-filling and hence to minimize the cost function in problem is decreased. We can quantify the performance loss. In particular, we find that the average regret converges to a constant whose size depends on the error terms ϵ_i, $i \in \mathcal{N}_I$ and the charging constraints of the inelastic customers.

Numerical Examples

We conclude with the following example from [19]. Assume that there are 20 customers. A time slot representing an interval of 30 minutes is used. There are $T = 24$ time slots. The starting time is set to 8:00 pm. For simplicity, we consider that all EV customers charge their EVs from the 9th to the 16th time slots. On the first day, the initial charging profiles are assumed to be uniformly distributed over the time slots. The maximum charging rate is set to $r_i^{\text{up}}(t) = 2$ kW, $i \in \mathcal{N}$ and the desired sum $S_i = 10$ kW, $i \in \mathcal{N}$. The simulation is carried out for total $K = 200$ days. We set the parameters $\eta_i = 0.05/\sqrt{K}$, $i \in \mathcal{N}$. We first examine the convergence of the static regret. The base load profile is given in Figure 17.6. The prediction M_i^k, $i \in \mathcal{N}$, $k \in \mathbb{N}_{>0}$ is set to $M_i^k(r_i^k) = \frac{1}{k-1}\sum_{\bar{k}=1,\dots,k-1} \nabla p_i^{\bar{k}}(r_i^{\bar{k}})$, $i \in \mathcal{N}$. Figure 17.7 shows that the average regrets converge to zero and the average regret with the prediction converges faster than the one without the prediction. Figure 17.8 shows the static regrets with and without the prediction. Figure 17.8 shows that the regrets are sublinear functions of the number of days.

Distributed Load Management

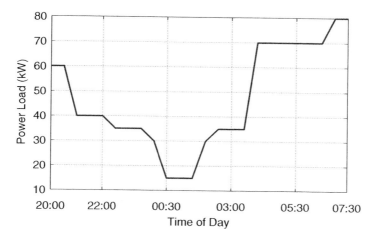

Figure 17.6 Base load profile from 8:00 pm to 7:30 am (the next day).

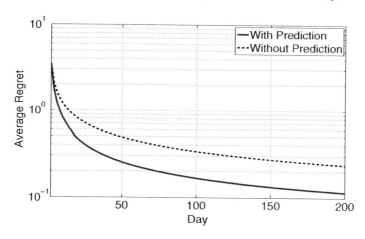

Figure 17.7 Average regrets generated by OMD with and without prediction.

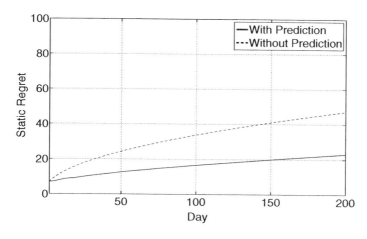

Figure 17.8 Static regrets generated by OMD with and without prediction.

Similar results for the case of a base load profile that does not remain the same from day to day and when inelastic customers are present are shown in [19].

Thus, we have designed a framework for distributed charging control of EVs using online learning and online convex optimization. The proposed algorithm can be implemented without low-latency two-way communication between the utility company and the EV customers, which fits in with the current communication infrastructure and protocols in the smart grid.

17.3 Distributed Feedback Control of Networked Loads

This section extends the distributed EV charging schemes mentioned previously in two aspects. First, we consider general controllable loads that not only include EVs but also refrigerators, air conditioners, dryers, etc., which can temporarily adjust their power consumption to provide power balance, frequency regulation, and voltage regulation services to the grid. Second, we consider the dynamic model of power network that connects the controllable loads, and focus on design of distributed feedback control and stability of dynamic power network under such control.

In this section, we will introduce our work on load-side primary frequency control [20]. To the best of our knowledge, this work has the first network model and stability analysis of distributed feedback control of networked loads. References are also provided for other load-side control functions, such as secondary frequency control, that are developed by us.

The power network under consideration is modeled as a directed and connected graph $(\mathcal{N}, \mathcal{E})$ where $\mathcal{N}$ is the set of buses (nodes) and $\mathcal{E}$ is the set of lines connecting the buses. Directions of lines provide reference directions but not indicate actual directions of power flow, so the assignment of directions can be arbitrary. We use $(i, j) \in \mathcal{E}$ and $i \to j$ interchangeably to denote a line directed from bus i to bus j, and use "$i : i \to j$" and "$k : j \to k$" to denote the sets of buses that are predecessors and successors of bus j, respectively.

The network has two types of buses: generator buses and load buses. A generator bus has an AC generator that converts mechanic power into electric power. A load bus has only electric loads but no generator. We denote the set of generator buses by $\mathcal{G}$ and the set of load buses by $\mathcal{L}$. The dynamic model of power network is described by the following differential-algebraic equations:

$$\dot{\omega}_j = -\frac{1}{M_j} \left(D_j \omega_j + d_j - P_j^m + \sum_{k:j \to k} P_{jk} - \sum_{i:i \to j} P_{ij} \right), \quad \forall j \in \mathcal{G} \quad (17.18)$$

$$0 = D_j \omega_j + d_j - P_j^m + \sum_{k:j \to k} P_{jk} - \sum_{i:i \to j} P_{ij}, \quad \forall j \in \mathcal{L} \quad (17.19)$$

$$\dot{P}_{ij} = B_{ij} (\omega_i - \omega_j), \quad \forall (i, j) \in \mathcal{E} \quad (17.20)$$

where, for every bus $j \in \mathcal{N}$, ω_j is the local frequency deviation from its nominal value (60 Hz in the United States), d_j is the deviation of aggregate controllable load power from its nominal value or user-preferred value, P_j^m models uncontrolled changes in power injection, such as load startup, generator loss, or changes in nondispatchable renewable generation, P_{ij} is the deviation of active power flow on line (i, j) from its nominal value. Positive constant M_j measures inertia of generator bus j, and the term $D_j \omega_j$, where D_j is a positive constant, models the damping effect of frequency sensitive loads, such as induction motors, at bus j. Constant B_{ij} is given, and in practice it can be calculated from the nominal voltage magnitudes at buses i and j, the phase angle difference between buses i and j, and the inductance of line (i, j).

An equilibrium point of the dynamical system (17.18)–(17.20) is a state (ω, P) where $\dot{\omega}_j = 0$ for $j \in \mathcal{G}$ and $\dot{P}_{ij} = 0$ for $(i, j) \in \mathcal{E}$, i.e., where frequency at all the buses and power flow on all the lines are constant over time.

Given uncontrolled changes $P^m = (P_j^m, j \in \mathcal{N})$ of power injections. How should we adjust the controllable loads d_j to balance power generation and load in a way that minimizes the aggregate disutility for changing these loads? In general we can design state feedback controllers of the form $d_j(t) := d_j(\omega(t), P(t))$, prove the feedback system is globally asymptotically stable, and evaluate the aggregate disutility at the equilibrium point. Here we take an alternative approach by directly formulating our goal as an optimal load control (OLC) problem and derive feedback control as a distributed algorithm to solve OLC. Let $\hat{d}_j := D_j \omega_j$ be the aggregate change of frequency-sensitive uncontrollable load at bus j. Then OLC minimizes the total cost over d and $\hat{d}$ while balancing generation and load across the network:

$$\text{OLC:} \quad \min_{\underline{d} \leq d \leq \overline{d}, \hat{d}} \quad \sum_{j \in \mathcal{N}} \left(c_j(d_j) + \frac{1}{2D_j} \hat{d}_j^2 \right) \quad (17.21)$$

$$\text{subject to} \quad \sum_{j \in \mathcal{N}} (d_j + \hat{d}_j) = \sum_{j \in \mathcal{N}} P_j^m \quad (17.22)$$

where $\underline{d}_j$ and $\overline{d}_j$ are given constants that bound the controllable load change d_j. We assume the following condition:

CONDITION 17.1 *OLC is feasible. The cost functions c_j are strictly convex and twice continuously differentiable on $[\underline{d}_j, \overline{d}_j]$.*

The choice of cost functions is based on physical characteristics of electric loads and how user comfort levels change with load power. Examples of cost functions can be found for EVs in [8, 9, 21] and air conditioners in [22]; see, e.g., [23, 24] for other cost functions that satisfy Condition 17.1.

The objective function of the dual problem of OLC is

$$\sum_{j \in \mathcal{N}} \Phi_j(v) := \sum_{j \in \mathcal{N}} \min_{\underline{d}_j \leq d_j \leq \overline{d}_j, \hat{d}_j} \left(c_j(d_j) - v d_j + \frac{1}{2D_j} \hat{d}_j^2 - v \hat{d}_j + v P_j^m \right)$$

where the minimization can be solved explicitly as

$$\Phi_j(v) := c_j(d_j(v)) - vd_j(v) - \frac{1}{2}D_j v^2 + vP_j^m \qquad (17.23)$$

with

$$d_j(v) := \min\left\{\max\left\{c_j'^{-1}(v), \underline{d}_j\right\}, \overline{d}_j\right\}. \qquad (17.24)$$

This objective function has a scalar variable v and is not separable across buses $j \in \mathcal{N}$. Its direct solution hence requires coordination across buses. We propose the following **distributed** version of the dual problem over the vector $v := (v_j, j \in \mathcal{N})$, where each bus j optimizes over its own variable v_j which are constrained to be equal at optimality:

DOLC: $\quad \max_v \; \Phi(v) := \sum_{j \in \mathcal{N}} \Phi_j(v_j)$

subject to $\quad v_i = v_j, \quad \forall (i,j) \in \mathcal{E}.$

We have the following results that are proved in [20]. Instead of solving OLC directly, these results suggest solving DOLC and recovering the unique optimal point $(d^*, \hat{d}^*)$ of OLC from the unique dual optimal v^*.

LEMMA 17.3 *The objective function Φ of DOLC is strictly concave over $\mathbb{R}^{|\mathcal{N}|}$.*

LEMMA 17.4 1. *DOLC has a unique optimal point v^* with $v_i^* = v_j^* = v^*$ for all $i, j \in \mathcal{N}$.*
2. *OLC has a unique optimal point $(d^*, \hat{d}^*)$ where $d_j^* = d_j(v^*)$ and $\hat{d}_j^* = D_j v^*$ for all $j \in \mathcal{N}$.*

To derive a distributed solution for DOLC consider its Lagrangian

$$L(v, \pi) := \sum_{j \in \mathcal{N}} \Phi_j(v_j) - \sum_{(i,j) \in \mathcal{E}} \pi_{ij}(v_i - v_j) \qquad (17.25)$$

where $v \in \mathbb{R}^{|\mathcal{N}|}$ is the (vector) variable for DOLC and $\pi \in \mathbb{R}^{|\mathcal{E}|}$ is the associated dual variable for the dual of DOLC. Hence π_{ij}, for all $(i, j) \in \mathcal{E}$, measure the cost of not synchronizing the variables v_i and v_j across buses i and j. Using (17.23)–(17.25), a partial primal-dual algorithm for DOLC takes the form

$$\dot{v}_j = \gamma_j \frac{\partial L}{\partial v_j}(v, \pi) = -\gamma_j\left(d_j(v_j) + D_j v_j - P_j^m + \pi_j^{\text{out}} - \pi_j^{\text{in}}\right), \forall j \in \mathcal{G} \qquad (17.26)$$

$$0 = \frac{\partial L}{\partial v_j}(v, \pi) = -\left(d_j(v_j) + D_j v_j - P_j^m + \pi_j^{\text{out}} - \pi_j^{\text{in}}\right), \quad \forall j \in \mathcal{L} \qquad (17.27)$$

$$\dot{\pi}_{ij} = -\xi_{ij}\frac{\partial L}{\partial \pi_{ij}}(v, \pi) = \xi_{ij}(v_i - v_j), \qquad \forall (i,j) \in \mathcal{E} \qquad (17.28)$$

where γ_j, ξ_{ij} are positive stepsizes and $\pi_j^{\text{out}} := \sum_{k: j \to k} \pi_{jk}$, $\pi_j^{\text{in}} := \sum_{i: i \to j} \pi_{ij}$. We interpret (17.26)–(17.28) as an algorithm iterating on the primal variables v and dual variables π over time $t \geq 0$. Set the stepsizes to be:

$$\gamma_j = M_j^{-1}, \qquad \xi_{ij} = B_{ij}.$$

Then (17.26)–(17.28) become identical to (17.18)–(17.20) if we identify v with ω and π with P, and use $d_j(\omega_j)$ defined by (17.24) for d_j in (17.18) and (17.19). This means that the frequency deviations ω and the branch flows P are, respectively, the primal and dual variables of DOLC, and the network dynamics together with frequency-based load control execute a primal-dual algorithm for DOLC.

For convenience, we collect system dynamics and load control equations:

$$\dot{\omega}_j = -\frac{1}{M_j}(d_j + \hat{d}_j - P_j^m + P_j^{\text{out}} - P_j^{\text{in}}), \quad \forall j \in \mathcal{G} \quad (17.29)$$

$$0 = d_j + \hat{d}_j - P_j^m + P_j^{\text{out}} - P_j^{\text{in}}, \quad \forall j \in \mathcal{L} \quad (17.30)$$

$$\dot{P}_{ij} = B_{ij}(\omega_i - \omega_j), \quad \forall (i,j) \in \mathcal{E} \quad (17.31)$$

$$\hat{d}_j = D_j \omega_j, \quad \forall j \in \mathcal{N} \quad (17.32)$$

$$d_j = \min\{\max\{c_j'^{-1}(\omega_j), \underline{d}_j\}, \overline{d}_j\}. \quad \forall j \in \mathcal{N}. \quad (17.33)$$

The dynamics (17.29)–(17.32) are automatically carried out by the power system while the active control (17.33) needs to be implemented at controllable loads. We have the following result regarding the load-controlled system (17.29)–(17.33).

THEOREM 17.6 *Starting from any initial point, $(d(t), \hat{d}(t), \omega(t), P(t))$ generated by (17.29)–(17.33) converges to a limit $(d^*, \hat{d}^*, \omega^*, P^*)$ as $t \to \infty$ such that*

1. *$(d^*, \hat{d}^*)$ is the unique vector of optimal load control for OLC;*
2. *ω^* is the unique vector of optimal frequency deviations for DOLC;*
3. *P^* is a vector of optimal line power flows for the dual of DOLC.*

Detailed proof of Theorem 17.6 can be referred to [20], and here we only provide a sketch. First, we establish equivalence between the set of optimal points (ω^*, P^*) of DOLC and its dual and the set of equilibrium points of (17.29)–(17.33). Denote both sets by Z^*. Then, we show that if $(\mathcal{N}, \mathcal{E})$ is a tree network, Z^* contains a unique equilibrium point (ω^*, P^*); otherwise (if $(\mathcal{N}, \mathcal{E})$ is a mesh network) Z^* has an uncountably infinite number (a subspace) of equilibria with the same ω^* but different P^*. Next, we use a Lyapunov argument to prove that every trajectory $(\omega(t), P(t))$ generated by (17.29)–(17.33) approaches a nonempty, compact subset Z^+ of Z^* as $t \to \infty$. Hence, if $(\mathcal{N}, \mathcal{E})$ is a tree network, any trajectory $(\omega(t), P(t))$ converges to the unique optimal point (ω^*, P^*); if $(\mathcal{N}, \mathcal{E})$ is a mesh network, we show with a more careful argument that $(\omega(t), P(t))$ still converges to a point in Z^+, as opposed to oscillating around Z^+. Theorem 17.6 then follows from Lemma 17.4. The Lyapunov function we find is:

$$U(\omega, P) = \frac{1}{2}(\omega_\mathcal{G} - \omega_\mathcal{G}^*)^T M_\mathcal{G}(\omega_\mathcal{G} - \omega_\mathcal{G}^*) + \frac{1}{2}(P - P^*)^T B^{-1}(P - P^*). \quad (17.34)$$

where $(\omega^*, P^*) = (\omega_\mathcal{G}^*, \omega_\mathcal{L}^*, P^*) \in Z^*$ is an arbitrary equilibrium point, $M_\mathcal{G} := \text{diag}(M_j, j \in \mathcal{G})$, and $B := \text{diag}(B_{ij}, (i,j) \in \mathcal{E})$.

This result confirms that, with the proposed algorithm, frequency adaptive loads can balance power and synchronize frequency after a disturbance in the power network, just as the droop control of the generators currently does. Moreover, our design ensures minimal aggregate disutility at equilibrium caused by changes in loads from their nominal power usage. Our result has four important implications. First, the local frequency deviation on each bus conveys exactly the right information about the global power imbalance for the loads to make local decisions that turn out to be globally optimal. This allows a completely decentralized solution without explicit communication to or among the loads. Second, the global asymptotic stability of the primal-dual algorithm of DOLC suggests that ubiquitous continuous decentralized load participation in primary frequency control is stable. Third, we present a "forward engineering" perspective where we start with the basic goal of load control and derive the frequency-based controller and the swing dynamics as a distributed primal-dual algorithm to solve the dual of OLC. In this perspective, the controller design mainly boils down to specifying an appropriate optimization problem (OLC). Fourth, the opposite perspective of "reverse engineering" is useful as well where, given an appropriate frequency-based controller design, the network dynamics will converge to a unique equilibrium that inevitably solves OLC with an objective function that depends on the controller design.

We illustrate the performance of OLC through the simulation of the 68-bus New England/New York interconnection test system [25]. The single line diagram of the 68-bus system is shown in Figure 17.9. We run the simulation on Power System Toolbox. The simulation model is more detailed and realistic than the analytic model used in the preceding text. The detail of the simulation model including parameter values can be found in the data files of the toolbox.

In the test system there are 35 load buses with a total load power of 18.23 GW. We add three loads to buses 1, 7, and 27, each making a step increase of 1 pu (based on 100 MVA). We also select 30 load buses to perform OLC. In the simulation we use the same bounds $[\underline{d}, \overline{d}]$ with $\underline{d} = -\overline{d}$ for each of the 30 controllable loads, and call the value of

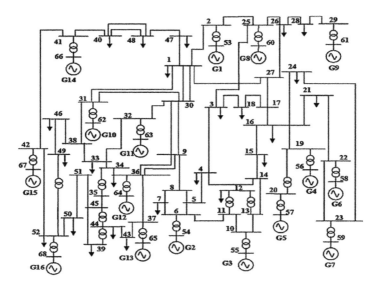

Figure 17.9 Single line diagram of the 68-bus test system.

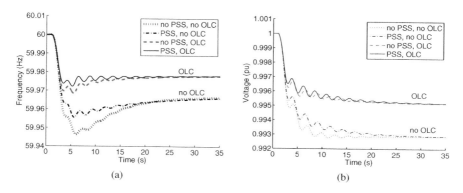

Figure 17.10 The (a) frequency and (b) voltage on bus 66 for cases (1) no PSS, no OLC; (2) with PSS, no OLC; (3) no PSS, with OLC; and (4) with PSS and OLC.

$30 \times \bar{d}$ the total size of controllable loads. We present simulation results in the following text with different sizes of controllable loads. The disutility function of controllable load d_j is $c_j(d_j) = d_j^2/(2\alpha)$, with identical $\alpha = 100$ pu for all the loads.

We look at the impact of OLC on both the steady state and the transient response of the system, in terms of both frequency and voltage. We present the results with a widely used generation-side stabilizing mechanism known as power system stabilizer (PSS) either enabled or disabled. Figures 17.10(a) and 17.10(b), respectively, show the frequency and voltage on bus 66, under four cases: (1) no PSS, no OLC; (2) with PSS, no OLC; (3) no PSS, with OLC; and (4) with PSS and OLC. In both cases, (3) and (4), the total size of controllable loads is 1.5 pu. We observe in Figure 17.10(a) that whether PSS is used or not, adding OLC always improves the transient response of frequency, in the sense that both the overshoot and the settling time (the time after which the difference between the actual frequency and its new steady-state value never goes beyond 5% of the difference between its old and new steady-state values) are decreased. Using OLC also results in a smaller steady-state frequency error. Cases (2) and (3) suggest that using OLC solely without PSS produces a much better performance than using PSS solely without OLC. The impact of OLC on voltage, with and without PSS, is qualitatively demonstrated in Figure 17.10(b). Similar to its impact on frequency, OLC improves significantly both the transient and steady state of voltage with or without PSS. For instance the steady-state voltage is within 4.5% of the nominal value with OLC and 7% without OLC.

To better quantify the performance improvement due to OLC we plot in Figures 17.11(a)–17.11(c) the new steady-state frequency, the lowest frequency (which indicates overshoot) and the settling time of frequency on bus 66, against the total size of controllable loads. PSS is always enabled. We observe that using OLC always leads to a higher new steady-state frequency (a smaller steady-state error), a higher lowest frequency (a smaller overshoot), and a shorter settling time, regardless of the total size of controllable loads. As the total size of controllable loads increases, the steady-state error and overshoot decrease almost linearly until a saturation around 1.5 pu. There is a

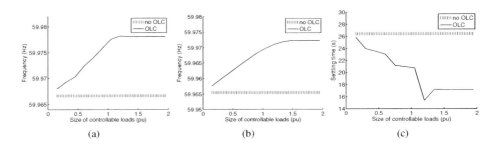

Figure 17.11 The (a) new steady-state frequency, (b) lowest frequency, and (c) settling time of frequency on bus 66, against the total size of controllable loads.

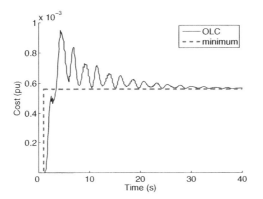

Figure 17.12 The cost trajectory of OLC compared to the minimum cost.

similar trend for the settling time. In summary, OLC improves both the steady-state and transient performance of frequency, and in general deploying more controllable loads leads to bigger improvement.

To verify the theoretical result that OLC minimizes the aggregate cost (disutility) of load control, Figure 17.12 shows the cost of OLC over time, obtained by evaluating the quantity defined in (17.21) using the trajectory of controllable and frequency-sensitive loads from the simulation. We see that the cost indeed converges to the minimum.

Our extended work on OLC includes the design of distributed load-side secondary frequency control [26], which not only stabilizes frequency after a disturbance but also restores frequency to the nominal value and satisfies other power network operational constraints such as interarea power flow equal to scheduled value and thermal limits on line power flow. Stability of this distributed load control scheme is also analyzed with a more realistic power network model, which includes nonlinear AC power flow and higher-order generator dynamic models [27].

In summary, the distributed load management schemes discussed in this chapter serve as a good complement to the existing generator control schemes in balancing the power system and maintaining appropriate frequency. They, with provable convergence and optimality and significant performance improvement demonstrated by numerical simulations, address a set of important issues in load management, including scalability and flexibility of the control system to support autonomous and plug-and-play operations of controllable loads; coordination between controllable loads, as well as coordination between loads and generators, to ensure a predictable and stable system behavior; and optimization of comfort levels of controllable load users or minimization of aggregate disutility due to the change of load power.

References

[1] F. C. Schweppe, R. D. Tabors, J. L. Kirtley, H. R. Outhred, F. H. Pickel, and A. J. Cox, "Homeostatic utility control," *IEEE Transactions on Power Apparatus and Systems*, vol. PAS-99, no. 3, pp. 1151–1163, May 1980.

[2] A. Brooks, E. Lu, D. Reicher, C. Spirakis, and B. Weihl, "Demand dispatch," *IEEE Power and Energy Magazine*, vol. 8, no. 3, pp. 20–29, May-June 2010.

[3] A. Molina-Garcia, F. Bouffard, and D. S. Kirschen, "Decentralized demand-side contribution to primary frequency control," *IEEE Transactions on Power Systems*, vol. 26, no. 1, pp. 411–419, Feb. 2011.

[4] D. S. Callaway and I. A. Hiskens, "Achieving controllability of electric loads," *Proceedings of the IEEE*, vol. 99, no. 1, pp. 184–199, Jan. 2011.

[5] N. Lu et al., "Design considerations for frequency responsive grid friendly appliances," in *IEEE/PES Transmission and Distribution Conference and Exhibition*, Dallas, TX, 2006, pp. 647–652, May 2006.

[6] J. A. Short, D. G. Infield, and L. L. Freris, "Stabilization of grid frequency through dynamic demand control," *IEEE Transactions on Power Systems*, vol. 22, no. 3, pp. 1284–1293, Aug. 2007.

[7] N. Li, L. Chen, and S. H. Low, "Optimal demand response based on utility maximization in power networks," in *IEEE Power and Energy Society General Meeting*, Detroit, MI, July 2011.

[8] L. Gan, U. Topcu, and S. Low, "Optimal decentralized protocol for electric vehicle charging," in *IEEE Conference on Decision and Control/European Control Conference*, Orlando, FL, pp. 5798–5804, Dec. 2011.

[9] L. Gan, U. Topcu, and S. H. Low, "Optimal decentralized protocol for electric vehicle charging," *IEEE Transactions on Power Systems*, vol. 28, no. 2, pp. 940–951, May 2013.

[10] L. Gan, A. Wierman, U. Topcu, N. Chen, and S. H. Low, "Real-time deferrable load control: Handling the uncertainties of renewable generation," in *Proceedings of International Conference on Future Energy Systems (ACM e-Energy)*, Berkeley, CA, pp. 113–124, 2013.

[11] L. Gan, U. Topcu, and S. H. Low, "Stochastic distributed protocol for electric vehicle charging with discrete charging rate," in *IEEE Power and Energy Society General Meeting*, San Diego, CA, July 2012.

[12] Z. Ma, D. S. Callaway, and I. A. Hiskens, "Decentralized charging control of large populations of plug-in electric vehicles," *IEEE Transactions on Control Systems Technology*, vol. 21, no. 1, pp. 67–78, Jan. 2013.

[13] A. Ipakchi and F. Albuyeh, "Grid of the future," *IEEE Power and Energy Magazine*, vol. 7, no. 2, pp. 52–62, Mar.-Apr. 2009.

[14] S.-J. Kim and G. B. Giannakis, "Real-time electricity pricing for demand response using online convex optimization," in *IEEE Conference on Innovative Smart Grid Technologies*, Washington DC, Feb. 2014.

[15] L. Jia, Q. Zhao, and L. Tong, "Retail pricing for stochastic demand with unknown parameters: An online machine learning approach," in *Allerton Conference on Communication, Control, and Computing*, Monticello, IL, pp. 1353–1358, Oct. 2013.

[16] N. Y. Soltani, S.-J. Kim, and G. B. Giannakis, "Online learning of load elasticity for electric vehicle charging," in *IEEE International Workshop on Computational Advances in Multi-Sensor Adaptive Processing*, Saint Martin, France, pp. 436–439, Dec. 2013.

[17] A. Rakhlin and K. Sridharan, "Online learning with predictable sequences," in *Conference on Learning Theory (COLT)*, Princeton, NJ, pp. 993–1019, June 2013.

[18] M. Zinkevich, "Online convex programming and generalized infinitesimal gradient ascent," in *International Conference on Machine Learning (ICML)*, Washington DC, pp. 928–936, Aug. 2003.

[19] W.-J. Ma, V. Gupta, and U. Topcu, "Distributed charging control of electric vehicles using online learning," *IEEE Transactions on Automatic Control*, vol. 62, no. 10, pp. 5289–5295, Oct. 2017.

[20] C. Zhao, U. Topcu, N. Li, and S. H. Low, "Design and stability of load-side primary frequency control in power systems," *IEEE Transactions on Automatic Control*, vol. 59, no. 5, pp. 1177–1189, May 2014.

[21] Z. Ma, D. S. Callaway, and I. A. Hiskens, "Decentralized charging control of large populations of plug-in electric vehicles," *IEEE Transactions on Control Systems Technology*, vol. 21, no. 1, pp. 67–78, Jan. 2013.

[22] B. Ramanathan and V. Vittal, "A framework for evaluation of advanced direct load control with minimum disruption," *IEEE Transactions on Power Systems*, vol. 23, no. 4, pp. 1681–1688, Nov. 2008.

[23] M. Fahrioglu and F. L. Alvarado, "Designing incentive compatible contracts for effective demand management," *IEEE Transactions on Power Systems*, vol. 15, no. 4, pp. 1255–1260, Nov. 2000.

[24] P. Samadi, A.-H. Mohsenian-Rad, R. Schober, V. W. Wong, and J. Jatskevich, "Optimal real-time pricing algorithm based on utility maximization for smart grid," in *IEEE Conference on Smart Grid Communications*, Gaithersburg, MD, pp. 415–420, Oct. 2010.

[25] G. Rogers, *Power System Oscillations*. Berlin: Springer Science & Business Media, 2012.

[26] E. Mallada, C. Zhao, and S. Low, "Optimal load-side control for frequency regulation in smart grids," *IEEE Transactions on Automatic Control*, vol. 62, no. 12, pp. 6294–6309, Dec. 2017.

[27] C. Zhao, E. Mallada, S. H. Low, and J. Bialek, "Distributed plug-and-play optimal generator and load control for power system frequency regulation," *International Journal of Electrical Power & Energy Systems*, vol. 101, pp. 1–12, Oct. 2018.

18 Analytical Models for Emerging Energy Storage Applications

I. Safak Bayram and Michael Devetsikiotis

To sustain economic prosperity, global competitiveness, and quality of life, secure, reliable, and affordable energy resources are needed. Electrical energy systems are uniquely critical as they play a key facilitating role across various important industries such as water desalination and distribution, communication, Internet, and health care, and very few sectors would survive without a continuous supply of electricity [1]. In the electric power industry, a new set of challenges have arisen due to the proliferation of disruptive technologies such as renewable energy systems and electric vehicles, and more complex market operations leveraged by consumer participation through smart meters and information technologies. Moreover, growing environmental concerns are forcing utilities to lower peak diesel-fired electricity production. To that end, there is a pressing need to improve grid flexibility to ease system operations and decarbonize electricity sector [2]. Energy storage systems (ESS) have emerged as viable solutions to address the aforementioned issues by providing a "buffer" zone between supply and demand.

Electricity is a commodity that is delivered through power grids that connect bulk electricity suppliers to geographically dispersed customers, and several grid-related services must be simultaneously carried out by a wide range of technologies. In legacy power grids, supply and demand must match in real time, with tight control on the supply side, and traditionally in a demand-following fashion. Moreover, operational rules are set by off-line analyses and manual interventions are made after contingency events based on incomplete information. By contrast, smart power grids are envisioned to offer finer-scale pricing and scheduling, improved utilization of system assets, and maximized system reliability. To design, analyze, and control such a complex system, necessary mathematical capabilities need to be developed to ensure optimal system operations. For instance, due to power grid's large interconnected nature, the behavior of the system as a whole is often not easily understood from local events and small disturbances in the grid affect other parts of the network. Because experimentation of such capital-intensive networks is not viable, mathematical models are needed to understand underlying governing dynamics and provide insights to system planners and decision makers [3].

This is also the case for energy storage technologies as the future cost of energy storage technologies has uncertainties and the values that energy storage can bring to the power systems are multidimensional and includes economic, social, and environmental benefits. From economic sustainability, computational and inference foundations are needed to address intermittence of supply, the variability of demand per meteorological parameters and electricity tariffs, and operational limits of electrical networks. Environmental sustainability, however, is linked to carbon savings and climate change mitigation, while social benefits include empowerment of sustainable development by increasing the security and quality of supply. At the present moment, energy storage problems are based on fixed-term cost-benefit analyses with very rough assumptions excluding details mentioned in the preceding text.

Moreover, with the shift toward smart power grids, new system architectures are demanded to address problems in emerging applications such as community energy storage systems and electric vehicle charging stations. Such applications are outside the traditional energy storage system applications, hence require special attention. In this chapter, we propose a stochastic modeling approach to size and operate energy storage units in emerging applications and provide insights into complex system operations. Before proceeding to the developed models, we present a thorough analysis of an energy storage systems landscape by classifying technologies according to the form of storage, grid applications, and size-discharge durations.

18.1 Energy Storage Systems Landscape

Energy storage systems are considered to be the next disruptive technology that could transform the power landscape [4]. Unit energy storage prices have been rapidly decreasing due to increasing research and development efforts and growing demand for electric vehicles and consumer electronics. Thus especially, major storage manufacturers have been scaling up lithium-based battery manufacturing, and storage has already become economical for many grid applications. As of June 2016, the energy storage capability in the United States was around 2% of the installed generation, while it was 10% in Europe and 15% in Japan. However, there has been an intensified push to expand energy storage capabilities of grids through new legislation to promote public and private partnership projects [5]. For instance, the number of US energy storage projects increased by 174% from 2013 to 2018, while there were more than 1,300 operational projects around the globe [6]. Moreover, the emergence of state-level energy storage demonstration and deployment policies have driven energy storage applications. California bill AC 2514 enacted in 2010 requires investor-owned utilities to adopt 1.3 GW energy storage by 2020. Similarly, Massachusetts has set a target of procuring 200 MWh energy storage by 2020, while the State of New York's goal is to reach 15 GW installed storage capacity by 2025 [4].

18.1.1 Energy Storage Classification

Energy can be stored in a device in a number of forms and can be converted into electrical form, later on, to be used in power system operations. The most common classification methodologies for energy storage systems are by the form of storage and

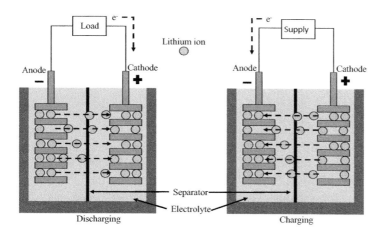

Figure 18.1 Lithium-ion battery charge and discharge schemes.

application type. Energy storage systems can be classified into six categories according to the form of energy and details are listed in the following text [7].

Electrochemical Storage

Energy is stored using electrochemical reactions. Batteries that are nonrechargeable and disposable are called primary batteries. These type of batteries are commonly used in portable electronic devices. However, secondary batteries are rechargeable and the most famous example is lithium-ion battery, which is also widely used in smart grid applications. Therefore, we present a brief overview of the working principles of this battery type as follows. Li-ion battery consist of an anode made of a graphitic carbon (C_6), a cathode made of a lithium metal oxide ($LiCoO_2$, $LiNiO_2$, $LiMO_2$, etc.), and an electrolyte that is made of lithium salt ($LiClO_4$ or $LiPF_6$) dissolved in organic carbonates [7]. During battery charging, lithium atoms become ions and migrate toward anode through the electrolyte and combine with the electrons. During discharge, however, lithium ions migrate toward cathode whereas the electrons move through the external circuit toward the anode. An overview of charge and discharge levels is presented in Figure 18.1. Other important electrochemical technologies are sodium sulfur, lead-acid, and flow batteries.

Mechanical Storage

Energy is stored in mechanical forms and the most famous examples include flywheels, compressed air energy storage (CAES), and pumped hydro storage units. Pumped hydro storage (PHS) has a long history, has technology maturity, and represents the vast majority of worldwide bulk storage. In PHS applications, during off-peak hours, water is pumped into a high-level reservoir and it is released during peak hours to power magnetic tribunes to generate electricity. The typical rating of a PHS is more than 1MW, thereby mainly used for bulk energy supply. In CAES, energy is stored in the form of high-pressure air in an underground cavern or over ground tanks. Flywheels, however, are becoming more popular as they have a low maintenance cost, long life, and negligible environmental impact. In flywheels, energy is stored in the form of rotating kinetic energy using a rotating mass in a frictionless environment.

Chemical Storage

Energy is stored in chemical fuels that can be used for power generation and for transport. Fuel cells and solar fuels are two of the most famous examples. Hydrogen fuel cell vehicles are also the topic of various active research field. For instance, hydrogen energy storage has two processes to store and produce electricity. Hydrogen is typically produced by water electrolysis and stored in a high-pressure container. However, stored hydrogen is used in a fuel cell to produce electricity. The advantages of hydrogen cells include high-energy density and less noise and pollution during the production phase. However, disadvantages include high cost and low round-trip efficiency.

Electric Field

Capacitors can store electrical energy without a need for any conversion. Supercapacitors and ultracapacitors are widely used in power systems to store a high amount of power in short time durations. Such storage devices have low-energy density and high-energy dissipation due to high rates of self-discharge [8]. Therefore, capacitors are typically used in applications like smoothing the output of power supplies, power quality correction, and energy recovery in mass transit systems.

Magnetic Field

Similar to capacitors, magnetic coils can store electrical energy in the magnetic field which can be charged and discharged quickly and very high power can be provided for a brief period of time. A superconducting magnetic field energy storage (SMES) system has three main components: a superconducting coil unit, a power conditioning subsystem, and a refrigeration and vacuum subsystem [8]. The main disadvantages of SMES are high capital cost and high sensitivity to small temperature variations which cause energy loss. Therefore, they have low market penetration.

Thermal Energy Storage

Thermal energy is stored at very low or high temperatures in insulated repositories to be used in peak hours. Thermal energy storage (TES) is typically divided into two groups: (1) low-temperature TES, which contains aquiferous low-temperature TES and cryogenic energy storage; and (2) high-temperature TES such as molten salt and sensible heat storage types. Low-temperature TES are commonly used for cooling air inside buildings: during off-peak hours ice is made and stored inside an ice-bank and the stored ice is used to cool the building during the day. In Table 18.1, market share for energy storage technologies by type is presented. It can be seen that there is a growing interest in electrochemical batteries, which are mostly dominated by lithium-ion technology.

Table 18.1 Worldwide energy storage technologies by type [6]

Technology Type	No. of Projects	Rated Power (MW)
Thermal	220	3,275
Electro-chemical	994	3,301
Pumped Hydro	351	183,007
Hydrogen Storage	13	20

18.1.2 Energy Storage Applications

Energy storage systems can be employed at various grid-related services throughout the generation, transmission, distribution, and end-user chain. The value of the storage is unique for each application and economic viability changes significantly across applications. For instance, 82% of the annual energy storage capacity in the United States in 2015 was in two markets: PJM for frequency regulation (front meter) and California for demand charge management (behind the meter). Next, we provide an overview of energy storage applications and services [9].

End-User Applications

End users can adopt energy storage technologies for (1) *time of use energy management*, (2) *demand charge management*, (3) *back-up power*, and (4) *power quality management*. In the first two applications, the overall aim is to store cheap electricity either from the grid or from distributed renewables and use it during peak hours. Demand charge management is mostly applied to commercial and industrial customers while time of use tariffs is valid for all customer types including residential ones. Back-up power is one obvious motivation for customers who require a continuous power supply, i.e., hospitals and data centers. Finally, power quality management is particularly desired by industrial customers as they require highly conditioned power. It is noteworthy that end-user applications are behind the meter applications that are owned and operated by the customers.

Distribution Network

A distribution circuit connects a high-voltage transmission system with low-voltage customers. Overall, energy storage systems can be used for (1) *voltage support*, (2) *reliability enhancement*, and (3) *capacity deferral and peak shaving*. Energy storage at distribution networks has become more important with the penetration of renewables and electric vehicle charging at customer premises. Depending on the network loading levels and the location of distributed solar generation sources, bidirectional electricity injected back to the grid can lead to overvoltage condition. Similarly, EV charging, due to high amounts of charging power, causes under voltage conditions particularly during peak hours. Storage units can smoothen voltage profile and reduce wear and tear on other network elements such as tap changers, capacitors, etc.

At a microgrid scale, storage units can increase reliability to support microgrid operations during islanding mode either as the main supply of energy or support renewable operations. As the demand for electricity increases, distribution network components get closer to their operating limits and require capital-intensive system upgrades. Employing storage units can potentially defer system upgrades by reducing peak demand.

Transmission Network
Energy storage located at transmission networks can *defer required investments* by providing extra capacity in specific locations that would otherwise necessitate additional transmission capacity. The use of storage unit further reduces stress on transmission network elements and increase their lifetimes. Storage units can also *alleviate the congestion* during peak hours and hence lower unit electricity cost by allowing more economic generators to produce more electricity. This is particularly important because there are growing constraints on the deployment of new transmission lines [10].

Generation and Resource Adequacy
Energy storage systems at generation side can be used for (1) *peak capacity deferral*, (2) *bulk energy time shifting*, (3) *frequency regulations*, and (4) *spinning/non-spinning reserves*. First two applications are similar to ones in the distribution network while the last two applications are related to ancillary services market. Storage units with quick

Table 18.2 An overview of energy storage types and applications in North Carolina [9].

	Electro-chemical				Mechanical			Electrical		Chemical		Thermal		
	Lithium-ion	Sodium Sulfur	Flow	Lead-Acid	Flywheel	Pumped Hydro	Compressed Air	Super Capacitor	SMES	H2 Electrolysis	Synthetic Methane	Chilled Water	Ice	Water Heater
End-User Services														
TOU/Energy Man.	✓	✓	✓	✓	✓							✓	✓	✓
Demand Charge Man.	✓	✓	✓	✓	✓							✓	✓	✓
Distributed Energy Man.	✓	✓	✓	✓								✓	✓	✓
Power Quality Man.	✓	✓	✓	✓				✓	✓			✓	✓	✓
Distribution Network														
Voltage Sup. and Con.	✓	✓	✓	✓								✓	✓	✓
Microgrid Rel.	✓	✓	✓	✓								✓	✓	✓
Capacity Def.	✓		✓	✓								✓	✓	✓
Transmission Services	✓	✓	✓	✓		✓	✓							
Trans. Investment Def.	✓	✓	✓	✓		✓	✓							
Trans. Congestion Rel.														
Generation Res. Adeq.														
Bulk Energy	✓	✓	✓	✓		✓	✓			✓	✓			
Peak Capacity Def.	✓	✓	✓	✓		✓	✓			✓	✓			
Spin./Non-Spin.	✓	✓	✓	✓						✓	✓			
Frequency Reg.	✓	✓	✓	✓		✓	✓	✓	✓					

ramping capabilities are used to follow real-time market signals for primary frequency regulation or longer timescales (minutes) of secondary frequency regulation. This way system operators are able to keep system frequency at 60 or 50Hz during contingency events. Federal Energy Regulatory Commission in the United States has passed a rule to remove barriers to allow electrical energy storage units to participate in energy markets [11]. The final draft of the rule shows that maximum capacity requirement for market participation will be less than 100kW, which would allow for high penetration of storage units. An overview of energy storage applications and typical storage types are presented in Table 18.2.

18.2 Comparing Energy Storage Solutions

Previous sections provided a detailed overview of the energy storage landscape. It has been shown that a wide range of storage technologies exists as a candidate solution for various grid applications, while no energy storage option is perfectly suited for every grid applications. A natural question arises on how to compare and choose the most appropriate storage technology [12]. There are a number of important considerations for technology selection, but the most important ones are defined as follows. The first key characteristic are energy that represents the total stored energy (kWh) and the second key characteristic is the instantaneous power at which storage can be charged and discharged. A comparison of energy storage technologies according to energy and power is presented in Figure 18.2. It can be seen that bulk energy storage applications require long durations of discharge and high energy capacity. Therefore, only pumped hydro and compressed air energy storage technologies fit for it. However, an application like frequency regulation and power quality improvement requires a

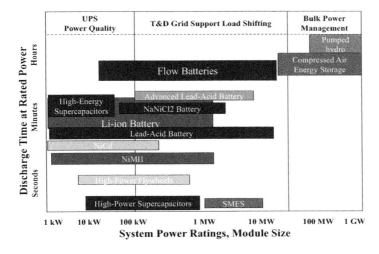

Figure 18.2 Energy storage comparison by power and duration of support.

significant amount of power in short durations. Therefore, supercapacitors or superconducting magnetic energy storage units are the most suitable technologies. The second important characteristics are reliability and durability. Mechanical storage technologies can operate indefinitely with routine maintenance, while electrochemical batteries such as redox and lithium-ion are depleted over time due to material degradation. In addition, electrochemical battery explosions are growing a major concern and safety is becoming an important criterion. The last characteristics are the levelized cost of storage (LCOS), which calculates the lifetime cost of the storage project. It is important to note that LCOS calculations are project-specific and rely on simplifying assumptions. For instance, LCOS estimation for electrochemical batteries rely on complex factors such as ambient temperature and depth of discharge, while most calculations only consider the best-operating conditions [12]. Also, note that most of the current storage assessment studies are based on LCOS calculations.

18.2.1 Technology Lock-In

Of all energy storage technologies discussed in the previous section, lithium batteries are uniquely positioned and increasing their market share due to relative maturity in technology, flexibility, and suitability for a wide range of applications. In the 1990s, growing consumer electronics market has given lithium-ion batteries a first mover advantage, while the push toward electric vehicles has intensified its lead in the energy market. In fact, research on the innovation of technology suggests that once a technology type manages to bring the most capacity to the market, it is very likely for that particular technology to be the most cost-competitive. Reference [13] shows that storage costs decrease proportional with the installed capacity. If current trends continue, Li-ion batteries will entrench its lead in the storage market.

However, if Li-ion storage becomes entirely ready for mass adoption, its "dominant design" is likely to drive out competing technologies. Innovation studies define this situation as technology "lock-in" [4]. For a technology to be dominant, it should be complemented by economic and social processes as well. For instance, once users gain confidence in storage technology, positive network externalities will increase the value and utility of the product. Hence, it is very likely for an energy storage system to have a turning point in the near future and penetration rates can be boosted. In most recent literature, lithium-based batteries are considered, especially for studies considered distribution network, microgrids, and renewable integration [14].

18.3 Analytical Problems in Energy Storage Systems

As energy storage technologies are increasingly becoming available for network operators, a number of challenges arise to enable efficient system operation and planning. This is particularly important for distribution side as they are being transformed into active networks that handle responsive customers, intermittent renewables, and large loads represented by electric vehicles [15]. One of the most important challenges is related to the optimal ESS location to maximize the benefits of storage units. The optimal location problem is usually project-dependent and takes into consideration various parameters including network structure, the location of distributed generators, spatiotemporal characteristics of loads, and target applications. In the current literature, most of the siting problems involve power-flow studies and are solved by computer simulation tools such as OpenDSS, GridLAB-D, Matlab, PSCAD, and DIgSILENT [14]. Analytical models for distributed storage units also exist, for instance in [16, 17] AC power flow is converted into a linearized model and demand is assumed to be deterministic. Then, the cost of energy storage is calculated at every bus (e.g., IEEE 14-bus system) and minimum cost bus is chosen. Such approximation is critical because power networks impose nonconvex power-flow constraints that make siting problems NP-hard.

Other group of analytical studies focuses on energy storage sizing. Due to their high acquisition cost, optimal sizing of storage units is essential for the correct operation of the grids. As mentioned in the earlier sections, traditional methods for optimal sizing is related to cost-benefit analyses with rough assumptions. However, recent research presents new analytical models using optimization methods such as genetic algorithms, particle swarm optimization, and mixed integer programming. Probabilistic modeling such as Markov chains and Monte Carlo methods are also used to size storage units. Some of the details of the related literature are presented as follows. Reference [18] provides a sizing approach for single industrial customers for peak saving applications. The sizing problem is solved by maximizing the net benefit, which is the aggregate sum of savings minus the operation costs, and one-time acquisition cost. Similarly, reference [19] provides a sizing framework using similar cost models for a microgrid, but it also considers savings stemming from the storage of energy generated from renewable resources. The work in [20] develops a sizing approach based on stochastic network calculus to size the storage units coupled with solar power generation. It employs *outage probability* as the main performance metric to provision the resources. Another probabilistic approach is presented in [21], which couples forecasting errors in wind generation with the storage unit and sizes the storage unit according to a desired level of accuracy. Moreover, [22] solves the sizing problem by using stochastic dynamic programming to minimize the operation cost and employs the storage device for balancing the load in a wind farm application.

In addition to sizing, the optimal storage operation is also an important research problem addressed by several analytical studies. Optimal operational strategy considers a variety of parameters such as system demand, the forecast of renewables, electricity prices, network conditions, and storage degradation [23]. The overall aim is to maximize the benefits offered by storage units by optimally charging and discharging it. Considering the fact that cost variables are time-dependent (prices, renewable generation, etc.), dynamic or stochastic programming techniques are widely used for online decision making. If system parameters are given as *a priori*, nonlinear optimization can be used to determine optimal decision making at each time step.

18.4 Case Study: Sizing On-Site Storage Units in a Fast Plug-In Electric Vehicle Charging Station

Over the last decade, there has been a renewed push toward the adoption of plug-in electric vehicles (PEV). In Norway, more than half of the new car registration were PEVs in summer 2018, in the United States, cumulative PEV sales have reached one million, and global sales have surpassed four million [24]. To fuel such a growing demand, a network of public fast-charging stations are needed. Such stations draw significant amounts of power from the grid, varies between 25kW and 125kW per charger, but 50kW is typical and threatens grid reliability. Moreover, operators of fast-charging stations are typically billed on a utility's service tariff with a three-part rate: fixed customer charge, energy charge (per-kWh), and demand charge (per-kW). For instance, electricity tariffs proposed by Tacoma Public Utilities [25] are as follows: $80 fixed charge per month, 4.9 cents/kWh for energy charges, and 8.51 USD/kW for demand charges. It is important to note that with current tariffs, demand charges are multiplied with the number concurrent PEV chargings. In the United States, demand charges may go up to 90 USD/kW depending on the region and season. For instance, in a station with two 50kW fast chargers, the cost of demand charges took up to 90% of the total service charge [26] in the summer 2017 in San Diego. If the station operator reflects this cost to customers, PEVs will lose its competitiveness in fuel cost. In addition, such charges significantly reduce the profit margin of fast-charging station operators and PEV adoption rates.

ESS have been employed by commercial and industrial customers to cut demand charges. A similar concept can be applied to fast-charging stations: an on-site energy storage system can be charged during low demand durations and discharged during peak hours to keep the power drawn from the grid below a demand-charge threshold. In this section, we introduce a PEV fast-charging station architecture furnished with an ESS to charge PEVs. Moreover, we propose a stochastic model to capture the charging station's operational characteristics, which inherently connects the capacity planning problem with the target customer service rates. We define the notions of outage probability or blocking probability as a stochastic measure for evaluating the performance of a station. The main goal is to keep peak demand below a demand charge level and to maintain an acceptable level of outage probability. To that end, the proposed model has the following operating principles:

- Similar to commercial customers, fast-charging stations are assumed to have long-term contracts with the utilities, where predetermined tariffs include demand charges. Therefore, stations draw constant power from the grid.
- An ESS is used to keep stochastic customer load. During peak demand hours, energy storage can be used to charge more vehicles, while during off-peak hours the storage unit can be charged with the grid power.
- PEV service times depend on the charging power rate, supporting power electronics, and battery technology. Yet, the average charging takes about half an hour for most

PEV models. Therefore, it is rational to assume that arriving PEVs prefer not to join the system, therefore, we model the station as a "bufferless" systems. In such systems, outage or blocking probability arise as natural performance metrics.

18.4.1 Stochastic Model

Based on the system description given in the preceding proposed station model has the following operational settings. The power drawn from the grid is discretized to C slots, hence C PEVs can be charged at the same time. Similarly, the energy storage unit can charge up to B PEVs when it is in a fully charged state. If all station resources are committed to other customers, an arriving customer will be blocked. It is noteworthy that station power C is determined by the agreement with the station operator, while the size of B is used to minimize demand charges and provide a good level of station performance.

In the proposed model, customer arrivals are modeled by a Poisson process with parameter λ. This assumption is used because in the Poisson process customers' arrivals behave statistically independently of each other. Moreover, there are a variety of different PEV models with different battery packs and charging durations. Therefore, the service time of customers is assumed to be exponentially distributed with a rate of μ. Similarly, the charging duration of one slot in the energy storage device is also exponentially distributed with parameter ν. It is noteworthy that this rate depends on the employed storage technology parameters such as power rating and charging efficiency. In this model, customers are first charged by the grid power, and if the total demand exceeds C kW, on-site ESS starts to serve other customers. It is important to note that the facility can serve up to $C + B$ customers simultaneously. To that end, the primary aim is to provide a good quality of service to the customers.

Based on the given system settings, a fast-charging station can be modeled by a 2-dimensional continuous-time Markov chain. As shown in Figure 18.3, each state is described by the number of PEVs in the station and the battery charge level. Moreover, the horizontal state transitions represent customer arrivals and departures, while the vertical ones correspond to ESS charge and discharge events. For instance, at state (1,0) three transitions can happen: (1) upon an arrival of a new PEV, system moves to (2,0) state with rate λ; (2) if the existing PEV completes service, system moves to (0,0) state with rate μ; and (3) if the ESS is charged, system moves to (1,1) state with rage ν. Notice that transition rates written in bold represent PEV charging from the ESS and the dotted states represent the states peak demand duration in which ESS is being used to meet customer demand.

The total number of system states is

$$T = (C + 1)(B + 1) + \sum_{i=1}^{B} i.$$

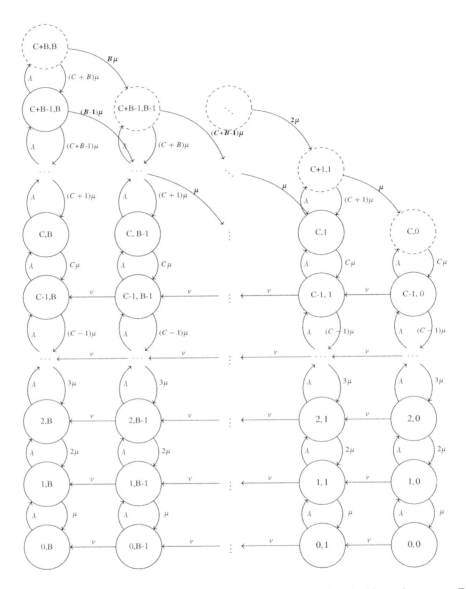

Figure 18.3 Markov chain model of the fast-charging station equipped with on-site storage. Each state is represented by the number of customers and ESS charge level.

It can be seen that to see that a unique steady-state distribution represented by vector $\pi = (\pi_1, \pi_2, \ldots, \pi_T)$ exists and can be calculated by solving

$$\pi Q = 0 \quad \text{and} \quad \sum_{i=1}^{T} \pi_i = 1. \tag{18.1}$$

In the preceding equation, the $T \times T$ matrix Q contains state transition rates and is called the infinitesimal generator matrix. Note that the elements of Q satisfy

$$q_{ij} \geq 0, \quad \forall i \neq j \quad \text{and} \quad q_{ii} = -\sum_{j \neq i} q_{ij}, \quad \forall i. \tag{18.2}$$

It is important to note that the rightmost states (depicted with dotted borders), namely (C,0), (C+1,1),...,(C+B,B) are outage states. To calculate steady-state distribution vector π, we sort the states from 1 to T starting from (0,0) to $(C+B, B)$. Then, the blocking probability can be calculated according to

$$\sum_{i=1}^{C} \pi \left(\frac{i(i+2C+1)}{2} \right).$$

Moreover, the infinitesimal generator matrix given in Equation (18.1) can be constructed as

$$Q = \begin{pmatrix} -(\lambda + \nu) & \lambda & \cdots & 0 \\ \mu & -(\lambda + \nu + \mu) & \cdots & 0 \\ \vdots & \vdots & \ddots & \vdots \\ 0 & 0 & \cdots & -(C+B)\mu \end{pmatrix}. \tag{18.3}$$

18.4.2 Numerical Examples

We evaluate the system performance for grid power $C = 4$ and $C = 8$, and for arrival rate ranges of $0.5 \leq \lambda \leq 10$ with increments of 0.5. Furthermore, energy storage size B is increased from 1 to 10. It is noteworthy that mean PEV charging rate (μ) and mean ESS charging rate (ν) are a function of power rating and efficiency of the storage unit. Hence, these parameters are considered as technological constraints and assumed to be constant through the evaluations.

The results depicted in Figures 18.4 and 18.5 provide insights about the system dynamics. For a given customer load and a target outage performance, the system operator can choose the right values for grid power and the energy storage size. Moreover, grid power C has a strong weight on affecting station performance as shown in the differences between the two results. One important observation is that under heavy traffic regime, choosing larger storage size does not improve the system performance as the grid power is utilized by PEV demand and storage is not charged.

It is noteworthy that the proposed model discretizes resource levels, this way steady-state probabilities can be calculated in an efficient manner. However, for a large number of end users, there will be more frequent charge and discharge events. Hence, the system state will follow a "fluid" model and a case study is presented next.

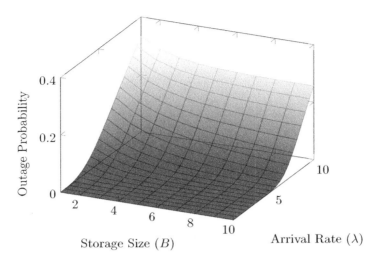

Figure 18.4 Performance evaluation of the proposed station model with $C = 4$.

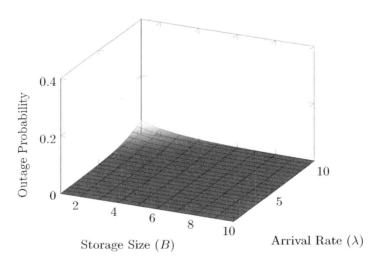

Figure 18.5 Performance evaluation of the proposed station model with $C = 8$.

18.5 Case Study: Community Energy Storage System Sizing

One of the most promising energy storage business cases is to deploy of storage units at residential neighborhoods to reduce peak demand, provide standby power, and enhance power quality especially for distribution feeders with sizable PV integration. Such sharing-based applications typically have higher storage utilization than the case in which each customer acquires a storage unit. In this case study, we propose a community energy storage architecture. Electrical loads fed by the storage is modeled stochastically and similar to the previous case study, total demand is met by power drawn from the grid and the storage unit.

System details can be described as follows. Electrical demands of N users are met both by the power grid and a shared ESS, in which serving capacity fluctuates over time. The grid power at time t is denoted by $C(t)$ and ESS depletion level is represented by $S(t)$ where time index is a positive number. Furthermore, the energy storage unit has the following parameters:

1. Energy rating or the size of the storage (kWh) is denoted by B.
2. Power rating of the ESS is the rate at which storage can be charged or discharged. The charging rate is assumed to be $P_c \leq C_t, \forall t$, and discharge rate P_d is proportional to the energy rating B as $B = P_d \times$ (desired support duration). The length of the support duration is typically equivalent to the length of the peak hour.
3. The efficiency of a charge-discharge cycle is denoted by $\eta \in [0, 1]$. For instance, typical efficiencies of lithium-ion batteries are above 90% that is $\eta \geq 0.9$.
4. Dissipation losses occur due to leakage and are ignored to simplify problem formulation.

Moreover, customer demands are grouped into K classes according to the magnitude of the loads, and denoted by $\{R_k\}$, where R_k stands for the demand per unit time for customer type k. Let N_k denote the number of customers of type k and the 1 by K vector N represents the number of customers of each type, that is $N = (N_1, \ldots, N_K)$. As established in [27–29] the electrical loads of each user can be well-represented by a two-state "On/Off" process. Let us define the binary variable s_t^{ik} to denote the state of user i of type k at time t such that

$$s_t^{ik} = \begin{cases} 1 & \text{consumer } i \text{ is On} \\ 0 & \text{consumer } i \text{ is Off} \end{cases}. \tag{18.4}$$

When a customer is in the "On" state, power is drawn from the grid and/or storage unit. The On duration is represented by a stochastic parameter to capture the various types of consumers' demands. Specifically, the On duration of type k customer's demand follows an exponential distribution with parameter μ_k. Moreover, transitions from "Off" to "On," are generated randomly and according to a Poisson process with parameter λ_k. Therefore, for each customer i of type k at any time t we have

$$\mathbb{P}(s_t^{ik} = 0) = \frac{\mu_k}{\lambda_k + \mu_k}, \tag{18.5}$$

and

$$\mathbb{P}(s_t^{ik} = 1) = \frac{\lambda_k}{\lambda_k + \mu_k}. \tag{18.6}$$

To formulate the operation dynamics of the storage unit, we define $L_{ik}(t)$ as the service request of consumer i of type k. Then, the following cases define ESS state of charge level: (1) the storage can be fully charged, and the total demand is less than grid power C_t; (2) the storage can be in completely discharged state, and the total customer demand is more than the grid power C_t; and (3) the storage can be in any partially charged state with any customer demand. Therefore, the rate of change in the storage state of charge level $S(t)$ follows

$$\frac{dS(t)}{dt} = \begin{cases} 0, & \text{if } S(t) = B \text{ and } \sum_{k=1}^{K}\sum_{i=1}^{N_k} \eta L_{ik}(t) < C(t) \\ 0, & \text{if } S(t) = 0 \text{ and } \sum_{k=1}^{K}\sum_{i=1}^{N_k} \eta L_{ik}(t) > C(t) \\ \eta(C(t) - \sum_k\sum_i L_{ik}(t)), & \text{otherwise} \end{cases} \quad (18.7)$$

Due to probabilistic nature of the station load, only stochastic guarantees can be provided and storage sizing problem can be solved by analyzing system outage probabilities, which similar to the previous case study, represents the probability of events when the all system resources are in use and new demand is rejected. As previously discussed, $S(t)$ denotes the state of charge level, hence, an outage event occurs when the load exerted on the storage unit exceeds its capacity B. Therefore, let us define ε-outage storage capacity, denoted by $B(\varepsilon)$, as the minimum choice of storage size B such that the probability of outage does not exceed ε that is defined in $(0, 1)$, i.e.,

$$B(\varepsilon) = \begin{cases} \min & B \\ \text{s.t.} & \mathbb{P}(S(t) \geq B) \leq \varepsilon, \forall t \end{cases} \quad (18.8)$$

The goal is to determine $B(\varepsilon)$ based on grid capacity $C(t)$, the number of users N, and their associated consumption patterns. For the sake of simplicity, we scale the storage parameters and B/η is redefined as B which is the maximum amount of energy that can be stored, while P_d and P_c represent the actual power ratings.

18.5.1 Energy Storage Operational Dynamics

We start our analyses by assuming that energy storage serves a single type of customer ($K = 1$), meaning that each customer has the same level of load and total demand can be modeled by N independent On-Off processes. Such a system can be composed into a continuous time birth-death process with system states represented by the number of active customers n that varies from 0 to N. When the system is at state n, nR_p units of power is drawn from the storage where R_p is the electrical load level of a single customer. The composite model is depicted in Figure 18.6 and, similar to the previous

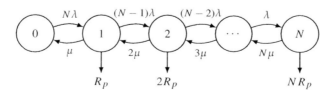

Figure 18.6 Markov model for N users and single user type (K=1). Each consumer becomes active ("On") at rate λ and becomes inactive ("Off") at rate μ.

case study (see the (18.3) for matrix form), associated infinitesimal generator matrix Q, in which the row elements sum to zero, for $i, n \in \{0, \ldots, N\}$ we have

$$Q[i,n] = \begin{cases} -((N-i)\lambda + i\mu) & n = i \\ i\mu & n = i-1 \ \& \ i > 0 \\ (N-i)\lambda & n = i+1 \ \& \ i < N \\ 0 & \text{otherwise} \end{cases}. \quad (18.9)$$

Moreover, steady-state probability distributions for each state can be calculated by (18.1). We define $F_i(t, x)$ as the cumulative distribution function (cdf) of the ESS state of charge level when $i \in \{0, \ldots, N\}$ consumers are actively using the storage unit at time t, that is,

$$F_i(t, x) = \mathbb{P}\big(S(t) \leq x\big). \quad (18.10)$$

To that end, a vector of cdfs can be defined as

$$\boldsymbol{F}(t,x) \triangleq [F_0(t,x), F_1(t,x), \ldots, F_N(t,x)]. \quad (18.11)$$

Based on the preceding definition, we present a lemma that delineates a differential equation that admits the cdf vector as its solution and is instrumental for analyzing the probability of outage events, that is, $\mathbb{P}\big(\sum_{i=1}^{N} L_i(t) > C(t) + B\big)$.

LEMMA 18.1 *The cdf vector $\boldsymbol{F}(t, x)$ satisfies*

$$\frac{d\boldsymbol{F}(t,x)}{dx} \cdot D = \boldsymbol{F}(t,x) \cdot Q, \quad (18.12)$$

where D is a diagonal matrix defined as

$$D \triangleq \text{diag}\big[-C(t)\mu, (1 - C(t))\mu, \ldots, (N - C(t))\mu\big], \quad (18.13)$$

and matrix Q is defined in (18.9).

Proof To compute the probability density functions, we find the expansion of $F_i(t, x)$ for an incremental change Δt in t, that is, $F_i(t+\Delta t, x)$. Notice that during an incremental time Δt, three elementary events can occur:

1. One idle customer may become active, that is, active customer index i increases to $i + 1$;
2. One active consumer may become idle, that is, active customer index i reduces to $i - 1$; or
3. The number of active consumers remains unchanged.

Because the durations of arrival and departure statistics of customers are exponentially distributed as cdf $F_i(t,x)$ can be expanded

$$F_i(t+\Delta t, x) = \underbrace{[N-(i-1)] \cdot (\lambda \Delta t) \cdot F_{i-1}(t,x)}_{\text{one customer added}} + \underbrace{[i+1] \cdot (\mu \Delta t) \cdot F_{i+1}(t,x)}_{\text{one customer removed}} \tag{18.14}$$

$$+ \underbrace{[1-((N-i)\lambda + i\mu)\Delta t] \cdot F_i(t, x-(i-C(t))\cdot \mu \Delta t)}_{\text{no change}} + o(\Delta t^2), \tag{18.15}$$

where $o(\Delta t^2)$ represents the probabilities of the compound events and tends to zero more rapidly than Δt^2 (and Δt) as $\Delta t \to 0$. Next, by passing the limit

$$\lim_{\Delta t \to 0} \frac{F_i(t+\Delta t, x)}{\Delta t}$$

it can be readily verified that (18.14) can be rewritten as

$$\frac{\partial F_i(x,t)}{\partial t} = [N-(i-1)] \cdot (\lambda) \cdot F_{i-1}(t,x) + [i+1] \cdot (\mu) \cdot F_{i+1}(t,x) \tag{18.16}$$

$$- [(N-i)\lambda + i\mu] \cdot F_i(t,x) - (i-C(t)) \cdot (\mu) \cdot \frac{\partial F_i(t,x)}{\partial x}, \tag{18.17}$$

where we have defined $F_{-1}(t,x) = F_{N+1}(t,x) = 0$. Recall that the design is intended for a long-term steady-state operation we have $\partial F_i(x,t)/\partial t = 0$. Hence, (18.16) can be rewritten as

$$(i-C(t))\cdot(\mu)\cdot\frac{\partial F_i(t,x)}{\partial x} = [N-(i-1)]\cdot(\lambda)\cdot F_{i-1}(t,x) + [i+1]\cdot(\mu)\cdot F_{i+1}(t,x)$$

$$- [(N-i)\lambda + i\mu] \cdot F_i(t,x). \tag{18.18}$$

By concatenating all the equations (18.18) for all $i \in \{0,\ldots,N\}$ we obtain the compact form

$$\frac{d\mathbf{F}(t,x)}{dx} \cdot \mathbf{D} = \mathbf{F}(t,x) \cdot \mathbf{Q}. \tag{18.19}$$

$\square$

We express the first-order differential equation given in (18.19) as a sum of exponential terms. Hence, for a general solution we need to calculate $(N+1)$ eigenvalues of the matrix QD^{-1} and the general solution can be expressed as [30]:

$$\mathbf{F}(t,x) = \sum_{i=0}^{N} \alpha_i \, \boldsymbol{\phi}_i \, \exp(z_i x), \tag{18.20}$$

where z_i is the ith eigenvalue of QD^{-1} with the associated eigenvector $\boldsymbol{\phi}_i$, which satisfy $z_i \boldsymbol{\phi}_i D = \boldsymbol{\phi}_i Q$. The coefficients $\{\alpha_0, \ldots, \alpha_N\}$ are determined by the boundary conditions, which are $F_i(t, 0) = 0$ and $F_i(t, \infty) = 1$.

To calculate the probability distribution in (18.20), we need to determine the coefficients $\{\alpha_i\}$, the eigenvalues of QD^{-1}, and the eigenvectors $\{\boldsymbol{\phi}_i\}$. Notice that, because

$x \geq 0$ and $F_j(t,x)$ are upper bounded by 1, all the positive eigenvalues and the corresponding α_i must be set to zero, which reduces the computational burden and (18.20) simplifies to

$$F(t,x) = \sum_{i:Re[z_i] \leq 0} \alpha_i \, \phi_i \, \exp(z_i x). \tag{18.21}$$

It can be further seen that because $z_i \phi_i D = \phi_i Q$, one of the eigenvalues must be zero. Then by setting $z_0 = 0$, the corresponding eigenvector can be computed from $\phi_0 Q = \mathbf{0}$. Earlier, we showed that the steady-state probability distribution π of the $N+1$ state Markov chain can be computed from the same equation, that is $\pi Q = \mathbf{0}$. Because, the eigenvector ϕ_0 is known and one of the eigenvalues is $z_0 = 0$, we can write $\phi_0 = \pi$. Therefore, (18.21) further simplifies to [30]:

$$F(t,x) = \pi + \sum_{i:Re[z_i]<0} \alpha_i \, \phi_i \, \exp(z_i x). \tag{18.22}$$

18.5.2 Storage Sizing for Single Customer ($K=1, N=1$)

We start the computation of probability distributions given in (18.22) for a simple case with a single user and, hence, single customer type. Then, we will expand the findings for a general case with multiple users and customer types. For the single user case, the infinitesimal generator matrix Q defined in (18.9) is

$$Q = \begin{bmatrix} -\lambda & \lambda \\ \mu & -\mu \end{bmatrix}. \tag{18.23}$$

To derive an expression for $F(t,x)$, we need to find the eigenvalues of QD^{-1}, that is, z_0 and z_1, where D is defined in (18.13). Based on (18.13) we find that

$$QD^{-1} = \begin{bmatrix} \frac{1}{C(t)} \cdot \frac{\lambda}{\mu} & -\frac{1}{1-C(t)} \cdot \frac{\lambda}{\mu} \\ -\frac{1}{C(t)} & -\frac{1}{1-C(t)} \end{bmatrix}. \tag{18.24}$$

Hence, the eigenvalues are

$$z_0 = 0 \quad \text{and} \quad z_1 = \frac{\chi}{C(t)} - \frac{1}{1-C(t)}, \tag{18.25}$$

where we have defined $\chi \triangleq \frac{\lambda}{\mu}$. It can be verified that the eigenvector associated with z_1 is $\phi_1 = [1-C(t), C(t)]$. Therefore, according to (18.22) we have

$$F(t,x) = \pi + \alpha_1 \, \phi_1 \, \exp(z_1 x). \tag{18.26}$$

Finally, by computing the coefficient α_1, CDF vector $F(t,x)$ can be completely characterized by leveraging the boundary condition $F_1(t,0) = 0$, which leads to

$$F_1(t,0) = \pi_1 + \alpha_1 \, C(t) = 0, \tag{18.27}$$

where we have that $\pi_1 = \frac{\lambda}{\lambda+\mu}$. Therefore

$$\alpha_1 = -\frac{\chi}{C(t)(1+\chi)}, \quad (18.28)$$

which characterizes both cdfs $F_0(t,x)$ and $F_1(t,x)$ according to

$$F_0(t,x) = \pi_0 + \alpha_1(1 - C(t))\exp(z_1 x) \text{ and } F_1(t,x) = \pi_1 + \alpha_1 C(t)\exp(z_1 x).$$

Consequently, the probability that the storage charge level $S(t)$ falls below a target level x is given by

$$\mathbb{P}(S(t) \le x) = F_0(x) + F_1(x) = 1 + \alpha_1 \exp(z_1 x). \quad (18.29)$$

Given this closed-form characterization for the distribution of $S(t)$, we can now evaluate the probability term $\mathbb{P}(S_t > B)$, which is the constraint in the sizing problem formalized in (18.8). Specifically, for any instantaneous realization of $C(t)$ denoted by c we have

$$\mathbb{P}(S_t > B) = \int_{C(t)} \mathbb{P}(S(t) > B \mid C(t) = c) \, f_{C(t)}(c) \, dc \quad (18.30)$$

$$= -\int_{C(t)} \alpha_1 \exp(z_1 B) \, f_{C(t)}(c) \, dc$$

$$= \int_{C(t)} \frac{\chi}{c(1+\chi)} \exp\left(\frac{B\chi}{c} - \frac{B}{1-c}\right) f_{C(t)}(c) \, dc.$$

To that end, by noting that parameter $z_1 = \frac{\chi}{c} - \frac{1}{1-c}$ is negative, the probability term $\mathbb{P}(S(t) > B)$ becomes strictly decreasing in B. Therefore, the minimum storage size B that satisfies the probabilistic guarantee $\mathbb{P}(S(t) > B) \le \varepsilon$ has a unique solution corresponding to which this constraint holds an equality. For the setting in which grid capacity $C(t)$ is constant c we have

$$B(\varepsilon) = \frac{c(1-c)}{\chi - \chi c - c} \cdot \log \frac{\varepsilon c(1+\chi)}{\chi}. \quad (18.31)$$

18.5.3 Storage Sizing for Multiuser Case ($K = 1, N > 1$)

Next, we present a closed-form for the probability term $\mathbb{P}(S_t \le x)$ for arbitrary positive values of N, which is denoted by $F_N(x)$. Computing all $F_N(x)$ for large N becomes computationally cumbersome and possibly prohibitive as it involves computing the eigenvalues and eigenvectors of QD^{-1}. By emphasizing an observation that for a large number of customers $N \gg 1$, the largest eigenvalues are the main contributors to the probability distribution, and [31] shows that, the asymptotic expression for $F_N(x)$ is given by

$$F_N(x) = \frac{1}{2}\sqrt{\frac{u}{\pi f(\varsigma)(\varsigma + \lambda(1-\varsigma))N}} \times \exp(-N\varphi(\varsigma) - g(\varsigma)x) \quad (18.32)$$

$$\times \exp(-2\sqrt{\{f(\varsigma)(\varsigma + \lambda(1-\varsigma))Nx\}}),$$

where,

$$f(\varsigma) \triangleq \log\left(\frac{\varsigma}{\lambda(1-\varsigma)}\right) - 2\frac{\varsigma(1+\lambda) - \lambda}{\varsigma + \lambda(1-\varsigma)},$$

$$u \triangleq \frac{\varsigma(1+\lambda) - \lambda}{\varsigma(1-\lambda)},$$

$$\varphi(\varsigma) \triangleq \varsigma\log(\varsigma) + (1-\varsigma)\log(1-\varsigma) - \varsigma\log(\varsigma) + \log(1+\lambda),$$

$$g(\varsigma) \triangleq z + 0.5\,(\varsigma + \lambda(1-\varsigma))\frac{\psi(1-\varsigma)}{f(\varsigma)},$$

$$z \triangleq (1-\lambda) + \frac{\lambda(1-2\varsigma)}{(\varsigma + \lambda(1-\varsigma))},$$

$$\text{and } \psi \triangleq \frac{(2\varsigma - 1)(\varsigma(1+\lambda) - \lambda)^3}{\varsigma(1-\varsigma)^2(\varsigma + \lambda(1-\varsigma))^3}.$$

In the set of equations given in the preceding text, the unit time is a single average "On" time $(1/\mu)$. Moreover, κ and ς are defined as the energy storage per customer (B/N) and the grid power allocated per one source, respectively. Furthermore, we denote the variable υ as the power above the mean demand allocated per user as $\upsilon = \varsigma - \frac{\lambda}{1+\lambda}$.

18.5.4 Storage Sizing for the General Case ($K > 1$, $N > 1$)

We proceed to analyze storage dynamics for the general which contains an arbitrary number of customers and customer types. It is noteworthy that the system becomes a K dimensional Markov process and the total number of customers of each class can be represented by vector $\mathbf{N} = (N_1, \ldots, N_K)$. Moreover, the number customers of type k that are active at time t is denoted by vector $\mathbf{n}(t) = [n_1(t) \ldots n_k(t)]$. Moreover, state space of the Markov process is denoted by $\mathbf{n}(t)$ and transitions between states can occur between neighboring states. To that end, we present the transition matrix $\bar{Q} = \{\bar{Q}(i,n)\}$ as

$$\bar{Q}(\mathbf{n}, \Delta_k^+(\mathbf{n})) = (N_k - n_k)\lambda_k, \quad (18.33)$$

$$\bar{Q}(\mathbf{n}, \Delta_k^-(\mathbf{n})) = n_k \mu_k, \quad (18.34)$$

$$\bar{Q}(\mathbf{n}, \mathbf{n}') = 0, \text{otherwise}. \quad (18.35)$$

where

$$\Delta_k^+(n_1, \ldots, n_k, \ldots, n_K) = (n_1, \ldots, n_k + 1, \ldots, n_K)$$
$$\Delta_k^-(n_1, \ldots, n_k, \ldots, n_K) = (n_1, \ldots, n_k - 1, \ldots, n_K).$$

It is noteworthy that, in the single-class ESS model, the transitions can only occur between two adjacent states. In the multiclass case, however, the transitions can occur with $2 \times k$ different neighbors.

So far, our analyses show that for $K > 1$, it is infeasible to calculate eigenvalues, therefore, we adopt the method proposed in [32] to efficiently evaluate the eigenvalues

z_i and the coefficients α_i. Our main goal is to determine ω_k, which is a deterministic quantity acting as a surrogate for the actual aggregate stochastic demand [33]. This way, allocating ω_k amount of resources to each customer class will statistically satisfy blocking probability targets.

Let parameter $\xi = 1 - \varepsilon^{1/B} \in [0, 1]$ represent the susceptibility of the storage unit to load patterns. Assume that we are able to choose $\mathbf{N} = (N_1, \ldots, N_K)$ in a way that ensures $\mathbb{P}(S(t) > B) \leq \varepsilon$ for small ε and the eigenvalues are relabeled such that $z_0 \geq z_1 \geq \cdots \geq 0$, then the following theorem holds.

THEOREM 18.1 *Let $\mathcal{B}(B, \varepsilon) = \{\mathbf{N} : \mathbb{P}(S_t \geq B) \leq \varepsilon\}$. For large energy storage size B and for small blocking probability ε, we have,*

$$\lim_{B=\infty, \varepsilon \to 0} \frac{\log \varepsilon}{B} \to \zeta \in [-\infty, 0]$$

Furthermore, let

$$\tilde{\mathcal{B}} = \left\{ \mathbf{N} : \sum_k \omega_k(\xi) N_k < C \right\} \text{ and } \bar{\mathcal{B}} = \left\{ \mathbf{N} : \sum_k \omega_k(\zeta) N_k \leq C \right\},$$

where effective (customer) demand can be computed by [32],

$$\omega_k(\zeta) = \frac{\zeta R_k + \mu_k + \lambda_k - \sqrt{(\zeta R_k + \mu_k - \lambda_k)^2 + 4\lambda_k \mu_k}}{2\zeta}. \tag{18.36}$$

Then, $\tilde{\mathcal{B}} \subseteq \mathcal{B}(B, \varepsilon) \subseteq \bar{\mathcal{B}}$.

Proof For a given set of users, $\mathbf{N}$, the following holds.

$$\frac{\mathbb{P}(S_t \geq B)}{\varepsilon} = \frac{\sum_k \alpha_k (\mathbf{1}^T \boldsymbol{\phi}_k) \exp(z_k B)}{\varepsilon},$$

where z_k's are the nonpositive eigenvalues of $\bar{Q} D^{-1}$, $\boldsymbol{\phi}_k$ are the corresponding eigenvectors, and ones vector is denoted by $\mathbf{1}^T = [1, 1, \ldots, 1]$. Recall that, the system coefficients α_k can be computed by solving the following set of equations.

$$\begin{cases} \sum_k \alpha_k \phi_k = 0, & \text{if } D_k > 0, \\ \alpha_k = 0, & \text{if } z_k > 0, \\ \alpha_0 \mathbf{1}^T \phi_0 = 1, & \text{otherwise.} \end{cases}$$

In [34], it is shown that $z_0 > z_k$, for $\forall k \geq 1$, and z_0 satisfies $C = \sum_k \alpha_k(z_0) N_k$. Then the following holds,

$$\frac{\mathbb{P}(S_t \geq B)}{\varepsilon} = \alpha_0 \mathbf{1}^T \boldsymbol{\phi}_0 \exp^{(z_0 - \zeta) B} (1 + o(1)), \tag{18.37}$$

as storage size becomes unlimited, that is $B \to \infty$. Differentiation shows that, effective demand $\omega_k(y)$ decreases as y increases. Hence, for $\mathbf{N} \in \tilde{\mathcal{B}}$, $z_0 < \zeta$ and $\frac{\mathbb{P}(S_t \geq B)}{\varepsilon} \to 0$ as $B \to \infty$, which shows that $\mathbf{N} \in \mathcal{B}(B, \varepsilon)$. In a similar manner, when $\mathbf{N} \notin \bar{\mathcal{B}}$, then $z_0 > \zeta$ and $\frac{\mathbb{P}(S_t \geq B)}{\varepsilon} \to \infty$, and hence $\mathbf{N} \notin \mathcal{B}(B, \varepsilon)$. This completes the proof. $\square$

The practical meaning of Theorem 18.1 is that (18.8) holds if and only if $\sum_k \omega_k(\zeta) N_k \leq C$. Furthermore, the following two remarks hold.

REMARK 18.1 *The corresponding effective demand for $\zeta = 0$ is $\omega_k(\zeta) = \frac{\lambda_k R_k}{\lambda_k + \mu_k}$, which is the average user demand for each class. Another implication is that, before a blocking event occurs, each customer turns "On" and "Off" their appliances many times that the total load becomes the mean customer demand. This result serves as a lower bound for the effective demand.*

REMARK 18.2 *At the upper boundary, the corresponding effective demand for $\zeta = z$ is its peak demand R_k. In this case, each customer is active at all times, hence peak demand must be allocated.*

18.5.5 Numerical Results

We present several computations to showcase studies for the proposed model. It is worth mentioning that unit time is measured in μ^{-1} and unit load is measured in peak demand (R_p). In the first evaluation, it is assumed that charge request rate per single user λ is set to 0.4 and the average capacity above the mean demand per user is set to $v = 0.03$. Then, the total system capacity becomes $C = 0.3157N$ where N is the total number of users. Next, we compute the storage size for a varying number of users from $N = 100$ to 500. Assuming that a single household can have 15–20 appliances, this range is chosen to mimic a multidwelling community. Results presented in Figure 18.7(a) can be used by system operators to choose storage size based on target outage probability and user population.

We proceed to investigate the case when energy storage is already acquired and the goal is to calculate grid power required to meet a predefined outage probability. This case is evaluated in in Figure 18.7(b) for $N = 200$ appliances. For instance, suppose ESS size $B = 10$kWh is already acquired and the goal is to meet 95% of the demand, and the system operator should draw $0.31 \times 200 = 62$ kW from the grid. Results show

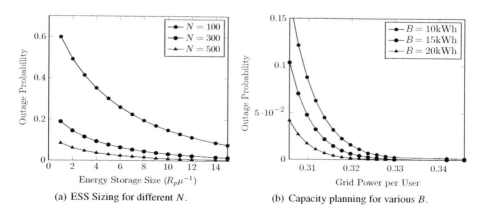

(a) ESS Sizing for different N.

(b) Capacity planning for various B.

Figure 18.7 Outage probability calculations for various system parameters.

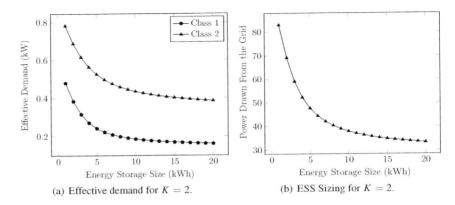

Figure 18.8 ESS sizing for multi-class case.

that as the storage size increases, stress on the grid is reduced as the system requires less amount of grid resources.

Next, we proceed to a multiclass case in which customers own more than one type of appliance. Recall that the main goal for the multiclass model is to compute ω_k (*effective demand*) parameter for each class such that ω_k replaces R_k and the outage probability targets are met. We present the following toy example for further clarification. Let us assume that there are four different levels of demand, that is $R^k = 0.2, 0.4, 0.6, 0.8$ (in kW) and their request statistics are represented by the arrival patterns $\lambda = \{0.3, 0.5, 0.7, 0.9\}$ and the duration of demand $\mu = \{1, 1, 1, 1\}$. Further assume that target outage probability is set to $\varepsilon = 10^{-4}$ and outage storage size $B = 15$kWh. Then, the corresponding *effective demand* becomes $\omega = \{0.0202, 0.0797, 0.1734, 0.2937\}$ (kW).

As a second evaluation, we calculate the *effective demand* for varying storage size. This time the target underflow probability is set as $\varepsilon = 0.0005$ and the customer population is chosen as $N_1 = 100$ and $N_2 = 45$. The results depicted in Figure 18.8(a) shows that instead of provisioning the system according to peak demand ($R_1 = 0.6$ kW, $R_2 = 0.9$ kW), the use of *effective demands* reduces the provisioning of resources tremendously. Our second examination is on the ESS sizing for multiclasses. For the same set of parameters, we evaluate the storage size with respect to power drawn from the grid and results are presented in Figure 18.8(b). Similar to the single class case, this result can be used to size the storage unit for a given grid power, or it can be used to compute the required grid resources for a given storage size.

18.6 Conclusion

In the first part of this chapter, we presented the motivation to develop analytical models to address emerging issues in smart power grids. Next, the landscape of energy storage systems was analyzed and we showed that lithium-based batteries are gaining more market share due to their first-mover advantage and most of the research papers consider lithium-based batteries. In the second half of the chapter, we presented analytical works on storage systems, namely sizing, placement, and operation storage units in power grids. In the last two sections, we presented two case studies for emerging energy storage applications. In the first case study, we considered the sizing problem of on-site storage in a fast-charging station using stochastic modeling. As a second case study, we considered a community energy storage system that is shared by a number of users and accommodates the need of different customer classes. The sizing problem is solved by developing a framework to compute the cumulative distribution function of the energy storage charge and discharge levels. The results provide useful insights for decision makers in such applications.

References

[1] R. Schmalensee, *The Future of Solar Energy: An Interdisciplinary MIT Study*. Energy Initiative, Massachusetts Institute of Technology, 2015.

[2] M. Schwalbe, *Mathematical Sciences Research Challenges for the Next Generation Electric Grid: Summary of a Workshop*. Washington, DC: National Academies Press, 2015.

[3] E. National Academies of Sciences, Medicine et al., *Analytic Research Foundations for the Next-Generation Electric Grid*. Washington, DC: National Academies Press, 2016.

[4] B. W. Hart, D. M. Hart, and N. Austin, "Energy storage for the grid: Policy options for sustaining innovation," *MIT Energy Initiative Working Paper*, 2018.

[5] H.R.5610 – BEST Act of 2018, www.congress.gov/bill/115th-congress/house-bill/5610/text

[6] G. Huff, "DOE global energy storage database," Sandia National Lab. (SNL-NM), Albuquerque, NM, Tech. Rep., 2015.

[7] F. Mumtaz, I. S. Bayram, and A. Elrayyah, "Importance of energy storage system in the smart grid," *Communication, Control and Security Challenges for the Smart Grid*, vol. 2, no. 12, p. 247, 2017.

[8] X. Luo, J. Wang, M. Dooner, and J. Clarke, "Overview of current development in electrical energy storage technologies and the application potential in power system operation," *Applied Energy*, vol. 137, pp. 511–536, 2015.

[9] N. S. E. S. Team, "Energy storage options for north carolina," North Carolina State University, Tech. Rep., 2018.

[10] "Beyond renewable integration: The energy storage value proposition," American Council on Renewable Energy, Tech. Rep., 2016.

[11] FERC Allows Energy Storage to Play in Nationwide Wholesale Markets, www.greentechmedia.com/articles/read/ferc-energy-storage-wholesale-markets#gs.LRR4heEu

[12] R. Kempener and E. Borden, "Battery storage for renewables: Market status and technology outlook," *International Renewable Energy Agency, Abu Dhabi*, p. 32, 2015.

[13] O. Schmidt, A. Hawkes, A. Gambhir, and I. Staffell, "The future cost of electrical energy storage based on experience rates," *Nature Energy*, vol. 2, no. 8, p. 17110, 2017.

[14] C. K. Das, O. Bass, G. Kothapalli, T. S. Mahmoud, and D. Habibi, "Overview of energy storage systems in distribution networks: Placement, sizing, operation, and power quality," *Renewable and Sustainable Energy Reviews*, vol. 91, pp. 1205–1230, 2018.

[15] I. S. Bayram, "Demand-side management for pv grid integration," in *Solar Resources Mapping*. Basel, Switzerland: Springer, pp. 313–325, 2019.

[16] C. Thrampoulidis, S. Bose, and B. Hassibi, "Optimal placement of distributed energy storage in power networks," *IEEE Transactions on Automatic Control*, vol. 61, no. 2, pp. 416–429, 2016.

[17] S. Bose, D. F. Gayme, U. Topcu, and K. M. Chandy, "Optimal placement of energy storage in the grid," in *Decision and Control (CDC), 2012 IEEE 51st Annual Conference Decision and Control, Hawaii*, IEEE, pp. 5605–5612, 2012.

[18] A. Oudalov, R. Cherkaoui, and A. Beguin, "Sizing and optimal operation of battery energy storage system for peak shaving application," in *Proceedings of IEEE Lausanne Power Tech*, Lausanne, Switzerland, pp. 621–625, July 2007.

[19] S. Chen, H. Gooi, and M. Q. Wang, "Sizing of energy storage for microgrids," *IEEE Transactions on Smart Grid*, vol. 3, no. 1, pp. 142–151, Mar. 2012.

[20] Y. Ghiassi-Farrokhfal, S. Keshav, C. Rosenberg, and F. Ciucu, "Solar power shaping: An analytical approach," *IEEE Transactions on Sustainable Energy*, vol. 6, no. 1, pp. 162–170, Jan. 2015.

[21] H. Bludszuweit and J. Dominguez-Navarro, "A probabilistic method for energy storage sizing based on wind power forecast uncertainty," *IEEE Transactions on Power Syst.*, vol. 26, no. 3, pp. 1651–1658, Aug. 2011.

[22] P. Harsha and M. Dahleh, "Optimal management and sizing of energy storage under dynamic pricing for the efficient integration of renewable energy," *IEEE Transactions on Power Systems*, vol. 30, no. 3, pp. 1164–1181, May 2015.

[23] K. Abdulla, J. De Hoog, V. Muenzel, F. Suits, K. Steer, A. Wirth, and S. Halgamuge, "Optimal operation of energy storage systems considering forecasts and battery degradation," *IEEE Transactions on Smart Grid*, vol. 9, no. 3, pp. 2086–2096, 2018.

[24] Cumulative Global EV Sales Hit 4 Million, https://about.bnef.com/blog/cumulative-global-ev-sales-hit-4-million

[25] Tacoma Public Utilities, www.mytpu.org/

[26] G. Fitzgerald and C. Nelder, "Evgo fleet and tariff analysis," Tech. Rep., 2017.

[27] I. S. Bayram, G. Michailidis, M. Devetsikiotis, and F. Granelli, "Electric power allocation in a network of fast charging stations," *IEEE Journal on Selected Areas in Communications*, vol. 31, no. 7, pp. 1235–1246, 2013.

[28] I. S. Bayram, A. Tajer, M. Abdallah, and K. Qaraqe, "Capacity planning frameworks for electric vehicle charging stations with multiclass customers," *IEEE Transactions on Smart Grid*, vol. 6, no. 4, pp. 1934–1943, 2015.

[29] O. Ardakanian, S. Keshav, and C. Rosenberg, "Markovian models for home electricity consumption," in *Proceedings of ACM SIGCOMM Workshop on Green Networking*, Toronto, Canada, pp. 31–36, Aug. 2011.

[30] I. S. Bayram, M. Abdallah, A. Tajer, and K. A. Qaraqe, "A stochastic sizing approach for sharing-based energy storage applications," *IEEE Transactions on Smart Grid*, vol. 8, no. 3, pp. 1075–1084, 2017.

[31] J. A. Morrison, "Asymptotic analysis of a data-handling system with many sources," *SIAM Journal on Appl. Mathematics*, vol. 49, no. 2, pp. 617–637, 1989.

[32] R. J. Gibbens and P. Hunt, "Effective bandwidths for the multi-type UAS channel," *Queueing Systems*, vol. 9, no. 1-2, pp. 17–27, 1991.

[33] F. P. Kelly, "Effective bandwidths at multi-class queues," *Queueing Systems*, vol. 9, no. 1-2, pp. 5–15, 1991.

[34] L. Kosten, "Stochastic theory of data-handling systems with groups of multiple sources," *Performance of Computer-Communication Systems*, vol. 321, p. 331, 1984.

Part VI

Game Theory

19 Distributed Power Consumption Scheduling

Samson Lasaulce, Olivier Beaude, and Mauricio González

19.1 Introduction

In this chapter we consider a scenario in which several consumption entities (households, electric appliances, vehicles, factories, buyers, etc.) generically called "consumers" have a certain energy need and want to have this need to be fulfilled before a set deadline. A simple instance of such a scenario is the case of a pool of electric vehicles (EV) which have to recharge their battery to a given state of charge (SoC) within a given time window set by the EV owner. Therefore, each consumer has to choose at any time the consumption power so that the accumulated energy reaches a desired level. This is the problem of power consumption scheduling. We assume the scheduling operation to be distributed. First, it has to be distributed decision-wise that is, each consumer is free to make its own decision in terms of choosing its consumption power. Second, it has to be distributed information-wise that is, the scheduling algorithm (when it is implemented by a machine, which is the most common scenario) or procedure only relies on local or scalable information. To schedule the consumption of what we call the flexible power (that is, the part of the consumption that leaves degrees of freedom on how to consume), the consumer will need a certain knowledge about the nonflexible part of the consumption. This part will be called the *exogenous* or noncontrollable consumption. The typical scenario considered in this chapter is that a day-head decision has to be made and some knowledge (perfect/imperfect forecast or statistics on the exogenous part) is available.

To design appropriate distributed power consumption scheduling policies, we mainly resort to tools from game theory, optimization, and optimal control. In this chapter, the use of static or one-shot game models is sufficient. Indeed, game-theoretic notions such as the Nash equilibrium (NE) and ordinal potentiality are used to analyze the convergence points of the described distributed algorithms. To design the latter the key idea is to use the sequential best-response dynamics (BRD).

The chapter is structured as follows. In Section 19.2, the consumption profiles are imposed to be noninterruptible and correspond to rectangular windows. Therefore, the power scheduling problem boils down to choosing the time instant at which the consumption operation should start. We describe a distributed algorithm that is built from the sequential BRD. To prove convergence of this algorithm we conduct the equilibrium analysis of the associated game. Each consumer (or consumption device) is assumed to have a given performance criterion, called *utility function*. The algorithm uses a given forecast of the nonflexible part of the consumption. The latter can be imperfect but these imperfections are not accounted for in the design. In Section 19.3, we exploit the possible dynamical structure of the problem. The utility function of a consumer may depend on state variables (e.g., the price of electricity or the distribution transformer hot-spot temperature) and the deterministic state evolution law is exploited. Here again, the possible presence of forecasting errors is not accounted for *a priori*. In Section 19.4, the idea is to take into account of forecasting errors and, for this, Markov decision processes are used to model the problem under consideration. Both in Sections 19.3 and 19.4, a common utility function is assumed for all consumers. This situation corresponds to a team game, and does not require a particular equilibrium analysis.

19.2 When Consumption Profiles Have to Be Rectangular and Forecasting Errors Are Ignored

19.2.1 System Model

We consider a system with $K \geq 1$ consumers and that the consumption of each consumer has two components: one that is flexible (e.g., an EV) and controllable and one that is fixed. Time is assumed to be slotted and indexed by $t \in \mathcal{T} = \{1, \ldots, T\}$. For example, if the whole time window under consideration is from 5 pm (day number j) to 8 am (day number $j + 1$), there are 30 time-slots ($T = 30$), whose duration is 30 min, on which a consumer may be active or not. The extent to which Consumer i, $i \in \mathcal{K} = \{1, \ldots, K\}$, is active on time-slot t is measured by the load it generates, which is denoted by $v_{i,t}$. The total or *network load* at time t is then expressed as:

$$u_t = \sum_{i=1}^{K} \ell_{i,t} + v_{i,t} \tag{19.1}$$

where $\ell_{i,t}$ (resp. $v_{i,t}$) represents the noncontrollable (resp. controllable) part of the consumption of Consumer i. Figure 19.1 illustrates a typical scenario that is encompassed by the considered model: the figure represents a set of K consumers, each consumer being composed of a household and an EV, which are connected to a distribution transformer.

In Section 19.3, the control or consumed power $v_{i,t}$ can take continuous values. But here, we not only assume that it can only take two values 0 or $P_{\max}$ but also that when consuming, the consumption has to be noninterrupted, which leads to rectangular consumption profiles. This is what Figure 19.1 shows. Although this assumption seems to be restrictive, it can be strongly motivated (see, e.g., [1]). Here we provide a couple of arguments of using such consumption profiles. An important argument is that rectangular profiles are currently being used by existing EVs. Second, for a given consumption start time, consuming at full power without interruption minimizes the delay to acquire the desired quantity of energy. Third, from an optimal control theory perspective, rectangular profiles may be optimal. This happens for instance when the state (e.g., the transformer hot-spot temperature) is monotonically increasing with the control

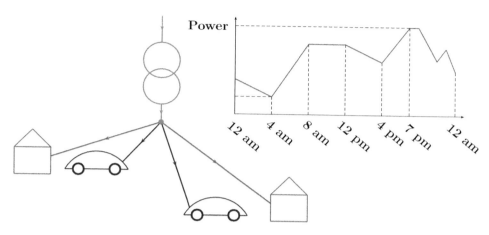

Figure 19.1 A typical scenario that is captured by the model analyzed in this chapter. In this example, each electric vehicle controls its charging power profile to maximize a certain utility function and reach a needed state of charge – for its next trip. For this, each vehicle is assumed to have some knowledge about the day-ahead household consumption.

(i.e., the consumption power). Fourth, profiles without interruption are even required in some important practical scenarios encountered with home energy management. Last but not least, imposing the consumption profile to be rectangular makes the power consumption scheduling policy robust to forecasting errors [1]; this is illustrated at the end of this part.

To complete the description of the model considered in this part, we need to define the performance criterion or metric of a consumer, which is referred to as the utility function of the consumer. A quite general form for the instantaneous utility function for Consumer i at time t is given by functions under the form $f_i(v_{1,t}, \ldots, v_{K,t}, x_t)$ where x_t represents a common state upon which depend all the utility functions. It may represent the (hot-spot) temperature of a distribution transformer, an energy level, the price of electricity, etc. The average or long-term utility is obtained as follows:

$$F_i(\sigma_1, \ldots, \sigma_K) = \frac{1}{T} \sum_{t=1}^{T} f_i(v_{1,t}, \ldots, v_{K,t}, x_t), \quad (19.2)$$

where σ_i represents the power consumption scheduling policy for Consumer i; the policy is typically a sequence of functions that map the available knowledge into actions. Because we consider here rectangular consumption profiles, the power consumption scheduling policy boils down to a simple action or decision namely, the time at which the consumption effectively starts. We denote the start time for Consumer i by $s_i \in \{1, \ldots, T\}$. By assuming that the state at time t that is, x_t only depends on the previous states and the actions, the function F_i will only depend on the action profile $s = (s_1, \ldots, s_K)$. With a small abuse of notations, we will denote the corresponding function by $F_i(s_1, \ldots, s_K)$. To be concrete, let us consider a couple of practical examples.

Example 19.1 If one wants to minimize the average power losses at the distribution network "entrance gate" (e.g., a residential transformer) that supplies the K consumers, one can consider a utility function under the form $\sum_{t=1}^{T} u_t^2$ where u_t would represent here the total current delivered by the distribution node. If one denotes by $I_{\max}$ the maximal current corresponding to the maximal power $P_{\max}$, the long-term utility function merely writes as:

$$F_i^{\text{losses}}(s_1, \ldots, s_K) = -\sum_{t=1}^{T} \left(\sum_{i=1}^{K} \ell_{i,t} + n_t(s_1, \ldots, s_K) I_{\max} \right)^2, \quad (19.3)$$

where $n_t(s_1, \ldots, s_K)$ is the number of consumers who are active at time t; it is given by:

$$n_t(s_1, \ldots, s_K) = \sum_{i=1}^{K} \sum_{t'=1}^{N_i} \mathbb{1}_{[s_i = t - N_i + t']}, \quad (19.4)$$

N_i representing the number of time-slots at full power to meet the energy need of Consumer i. This energy need can be expressed as $E_i = N_i P_{\max} \Delta t$ where Δt is the duration of a time-slot. See, for instance, Figure 19.2.

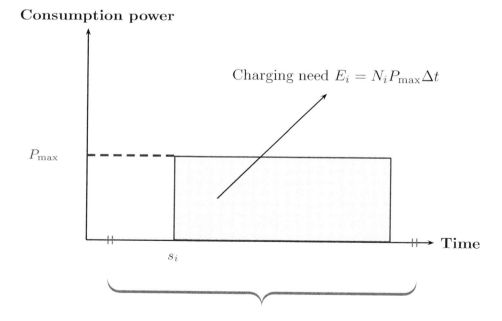

Figure 19.2 Assumed consumption profile.

Example 19.2 Let us now consider an example where the instantaneous utility function depends on a state namely, the hot-spot temperature of a distribution transformer. The system state is assumed to obey the following dynamics [1–4]:

$$x_t = \alpha x_{t-1} + \beta \left(\sum_{i=1}^{K} \ell_{i,t} + n_t(s_1, \ldots, s_K) P_{\max} \right)^2 \\ + \gamma \left(\sum_{i=1}^{K} \ell_{i,t-1} + n_{t-1}(s_1, \ldots, s_K) P_{\max} \right)^2 + c_t, \quad (19.5)$$

where α, β, γ are parameters of the dynamics [1], and c_t is a known deterministic function (typically representing the ambient temperature in Celsius degrees, making $x_t \geq 0$). To maximize the transformer lifetime, a reasonable long-term utility function is as follows:

$$F_i^{\text{lifetime}}(s_1, \ldots, s_K) = -\sum_{t=1}^{T} e^{ax_t}, \quad (19.6)$$

where x_t is given by a certain function of the initial state x_0 and $s_1, \ldots, s_K$.

19.2.2 Power Consumption Scheduling Game: Formulation and Analysis

One of the powerful links between distributed optimization and game theory is that scenarios involving several individual optimizers or decision makers that update their decisions over time may converge to a NE of a certain game. This is one of the reasons why we now define a game of interest, that we will refer to as the power consumption scheduling (PCS) game. Then, we define an NE and conduct the corresponding equilibrium analysis. Among three of the most used mathematical models for a game, we have the strategic or normal form, the extensive form, and the coalitional form. Here, the game is described under strategic form that is, under the form a triplet formed by: the set of players, the strategy sets, and the utility functions. The set of players is merely given by the set of consumers $\mathcal{K}$. The strategy set for a consumer is merely the action set $\mathcal{S}_i = \{1, \ldots, T\}$. The set of action profiles is denoted by $\mathcal{S} = \mathcal{S}_1 \times \ldots \times \mathcal{S}_K$. The utility functions are the functions $F_i(s_1, \ldots, s_K)$. Therefore, the strategic form of the PCS game is given by

$$\mathcal{G} = (\mathcal{K}, (\mathcal{S}_i)_{i \in \mathcal{K}}, (F_i)_{i \in \mathcal{K}}). \quad (19.7)$$

Now, let us define an NE and more precisely a pure NE. For this, we denote by s_{-i} the reduced action profile, that is, the vector of all the actions except action s_i.

DEFINITION 19.1 (Pure NE) *The action profile* $s^\star = (s_1^\star, \ldots, s_K^\star) \in \mathcal{S}$ *is a pure NE if* $\forall i, \forall s_i, \ u_i(s_i, s_{-i}^\star) \leq u_i(s^\star).$

Does the PCS game have a pure NE? First, notice that the PCS game is a discrete game because both the number of players and the number of possible actions (that are consumption starting instants) are discrete and finite. Therefore, it possesses at least one mixed NE that is, a pure NE for which expected utilities and independent probability distributions are considered (see, e.g., [5]). But, for the problem of interest a more suited solution concept is a pure NE. For completely arbitrary utilities, the existence of a pure NE cannot be guaranteed. However, for some typical classes of utility functions, the existence can be proven in a quite simple manner. Indeed, in this chapter we consider scenarios in which the utility function of a consumer depends on the consumption policies of the others through the sum-load or network load u_t. This type of games is often referred to as aggregative games. Under this assumption, the instantaneous utility $f_i(v_{1,t}, \ldots, v_{K,t}, x_t)$ writes as $f_i(u_t, u_{t-1}, \ldots, u_1)$. Using this key observation [1], it is possible to show that the PCS game is an ordinal potential game (OPG), which allows the existence of an NE to be guaranteed. An OPG is defined as follows [6].

DEFINITION 19.2 (Ordinal Potential Game) *A game whose utility functions are $(u_i)_{i \in \mathcal{K}}$ is an OPG if there exists a function Φ such that $\forall i \in \mathcal{K}$, $\forall s = (s_i, s_{-i})$, $\forall s_i' \in \mathcal{S}_i$,*

$$u_i(s_i', s_{-i}) \geq u_i(s_i, s_{-i}) \Leftrightarrow \Phi(s_i', s_{-i}) \geq \Phi(s_i, s_{-i}).$$

Because the function Φ does not depend on the player index, the game analysis amounts, to a large extent, to analyzing an optimization problem. Remarkably, the PCS game under consideration can be shown to be an OPG. Thanks to this property, not only the existence of a pure NE is guaranteed but also we have that the convergence of some distributed PCS algorithms can be guaranteed. This is, for instance, the case for the algorithm described in the next section. The technical reason is that the function Φ can be shown to be a Lyapunov function for the corresponding dynamics. Before describing this algorithm, let us define an important quantity that is, the price of decentralization (PoD). Indeed, an NE is generally not efficient whether its efficiency is measured in terms of sum-utility $\sum_i F_i$ or in terms of Pareto-optimality. One of the most usual ways of assessing the impact of decentralization in a game has been formalized in [7] by defining the notion of price of anarchy. In [1], the authors have slightly modified the latter notion as the price of decentralization (PoD), which we define in the following text. The merit of this definition is just that the price is effectively zero when a distributed algorithm or procedure leads to an equilibrium point that performs as well as the centralized solution in terms of sum-utility.

DEFINITION 19.3 (PoD) *The PoD of $\mathcal{G}$ is defined by*

$$\mathrm{PoD} = 1 - \frac{\max_{s \in \mathcal{S}} w(s)}{\min_{s \in \mathcal{S}^{NE}} w(s)} \tag{19.8}$$

where $\mathcal{S}^{NE}$ is the set of NE of the game.

It can be seen that $0 \leq \mathrm{PoD} \leq 1$ and the larger the PoD, the larger the loss due to decentralization. It is generally difficult to express the preceding quantity as a function of the game parameters [5]. This explains why this quantity is often more relevant from the numerical point of view. Nonetheless, it is possible to characterize it in some special cases. One of the cases in which PoD can be characterized is the limit case of a large number of consumers, that is $K \to \infty$, and when the dynamical system has no inertia or memory. In this asymptotic regime, $\frac{n_t}{K} \to v_t \in \mathbb{R}$ represents the proportion of consumers being active at time t and the analysis of the game $\mathcal{G}$ amounts to analyzing the so-called *nonatomic* counterpart $\mathcal{G}^{\text{NA}}$ of $\mathcal{G}$. In the latter game, the set of players is continuous and given by $\mathcal{K}^{\text{NA}} = [0, 1]$. In the regime of large numbers of consumers, the network load becomes $u_t(v_t) = \ell_t + p_{\max} v_t$; the parameter $p_{\max}$ is introduced for the exogenous demand to scale with K. Indeed, when $K \to +\infty$, if kept fixed, the exogenous demand ℓ_t tends to vanish in comparison to the load induced by the consumers. This is the reason why we introduce the parameter $p_{\max}$ (instead of $P_{\max}$). The obtained nonatomic charging game can be proved to be an OPG and the following result concerning efficiency can be obtained.

PROPOSITION 19.1 (PoD in the nonatomic case ($I \to \infty$)) *Assume that ℓ_t is a nonincreasing function of t. Then we have that PoD $= 0$.*

This result has the merit to exhibit a (particular) scenario in which decentralizing the charging decisions induces no cost in terms of global optimality for the sum-utility. Note that, in particular, if the exogenous demand is either constant or negligible w.r.t. the demand associated with the set of consumers, the preceding assumption holds and there is therefore no efficiency loss due to decentralization.

19.2.3 A Distributed Power Consumption Scheduling Algorithm

The algorithm we describe here has been proposed in [1]. To determine the vector of consumption start times $(s_1, \ldots, s_K)$, the authors proposed an iterative algorithm which is inspired from the sequential BRD. The algorithm is performed offline, which means that the decisions that intervene in the algorithm are intentions but not decisions that have effectively been taken; only the decisions obtained after convergence will be effective and implemented online. Once the consumption start instants are computed, the consumer can effectively consume according to the schedule determined. In its most used form, the BRD operates sequentially such that consumer update their strategies in a round-robin manner. Within round m (with $m \geq 1$) the action chosen by Consumer i (which can be virtual or physical depending on whether one or K entities implement the procedure) is computed as (19.9). The proposed procedure is translated in pseudo-code through Algorithm 19.1.

Comments on Algorithm 19.1.

- In (19.9), when the argmax set is not a singleton, $s_i^{(m)}$ is randomly drawn among the maximum points.
- The quantity $\delta \geq 0$ in Algorithm 19.1 corresponds to the accuracy level wanted for the stopping criteria in terms of convergence.
- To update the consumption power levels m times, $m \times I$ iterations are required.
- The order in which consumers update their actions does not matter to obtain convergence (see, e.g., [8]). However, simulations that are not provided here indicate

---
Algorithm 19.1 The proposed distributed PCS algorithm.

Initialize the round index as $m = 0$. Initialize the vector of consumption start times as $s^{(0)}$.

while $\|s^{(m)} - s^{(m-1)}\| > \delta$ and $m \leq M$ **do**

    Outer loop. Iterate on the round-robin phase index: $m = m + 1$. Set $i = 0$.

    Inner loop. Iterate on the consumer index: $i = i + 1$. Do:

$$s_i^{(m)} \in \arg\max_{s_i \in \mathcal{S}_i} u_i(s_1^{(m)}, s_2^{(m)}, \ldots, s_i, s_{i+1}^{(m-1)}, \ldots, s_I^{(m-1)}) \quad (19.9)$$

    where $s_i^{(m)}$ stands for action of Consumer i in the round robin phase m. Stop when $i = K$ and go to **Outer loop**.

end
---

that some gain in terms of convergence time can be obtained by choosing the order properly. A good rule seems to be to update the consumers' decisions at iteration m in an increasing order in terms of start times as obtained per iteration $m - 1$, which makes the order iteration-dependent.

- The knowledge required to implement Algorithm 19.1 is scenario-dependent. In scenarios in which each decision is computed by a single entity (an aggregator typically), the vector of effective consumption start instants can be computed from its initial value $s^{(0)}$, the (forecasted) sequence of exogenous loads $(\ell_1, \ldots, \ell_T)$, and the parameters that intervene in the utility functions. In scenarios d in which the consumers update their decision, messages have necessarily to be exchanged between the aggregator and the consumers.

- A variation of Algorithm 19.1 can be obtained by updating the consumption policies simultaneously. The main reason why we have not considered the parallel version is that it is known that there is no general analytical result for guaranteeing convergence [5]. When converging, the parallel implementation is faster but because start times are computed offline, convergence time may be seen as a secondary feature.

19.2.4 Numerical Illustration

The purpose of this section is to assess the performance of the distributed PCS algorithm, which is inspired from game-theoretic concepts to existing solutions [1]. More specifically, to assess the impact of not being able to forecast the exogenous demand perfectly we assume that the available exogenous demand profile is given by $\widetilde{\ell}_t = \ell_t + Z$, where $Z \sim \mathcal{N}(0, \sigma_{\text{day}}^2)$. We define the forecasting signal-to-noise ratio (FSNR) by

$$\text{FSNR} = 10 \log_{10} \left(\frac{1}{\sigma_{\text{day}}^2} \times \frac{1}{T_{\text{day}}} \sum_{t=1}^{T_{\text{day}}} (\ell_t)^2 \right) \text{ (dB)} \quad (19.10)$$

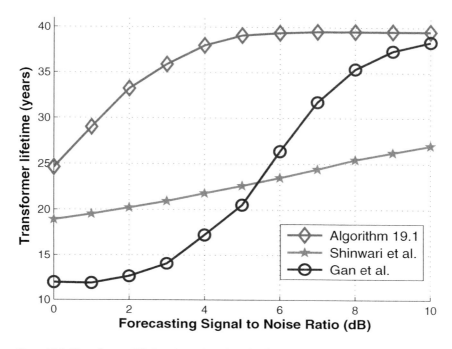

Figure 19.3 Transformer lifetime (years) against the forecasting signal-to-noise ratio (dB) for $I = 10$ electric vehicles; mobility data of Scenario (t), namely those from [1]. This confirms the benefits from using Algorithm 19.1 in presence of forecasting errors.

where $T_{day} = 48$. Figure 19.3 represents the transformer lifetime against FSNR for three consumption schemes [1]. Rectangular profiles have the advantage to be less sensitive to amplitude errors than the valley-filling solutions because the sole parameter to be tuned is the consumption start time. This provides a very strong argument in favor of using rectangular profiles. A parallel with the problem of noise robustness for high-order modulations in digital communications can be drawn and an optimal power level might be identified.

19.3 When Consumption Profiles Are Arbitrary and Forecasting Errors Are Ignored

In contrast with the previous section, the consumption power does not need to be binary anymore but can be take continuous value. More precisely, the consumption profile is no longer rectangular and can be arbitrary. Therefore, the PCS policy for a consumer does not boil down to a single scalar anymore (namely, the consumption start instant). The motivation for this is to have a better performance for the consumer but also to be able to control the system state. In the previous section, the dynamical system was controlled in a one-shot manner and in an open-loop manner. Here, the state evolution law is taken into account explicitly and the state can be controlled. For instance, it is possible to guarantee that $x \leq x_{\max}$, which was not possible with the previous approach. See Figure 19.3.

19.3.1 Introduction to a Dynamic Framework

In this section, we introduce a modeling framework that allows us to integrate dynamical aspects. This analysis is conducted under the assumption of perfect nonflexible load forecasts (deterministic setting), but the considered PCS algorithms can also be fed with imperfect forecasts.

The state is here assumed to be the hot-spot temperature and to be given by the model presented through Example 2, which is here just reformulated for a general network total load u_t (while Section 19.2 focused on rectangular charging profiles):

$$\forall t \in \mathcal{T}, \ x_t = \alpha x_{t-1} + \beta u_t^2 + \gamma u_{t-1}^2 + c_t, \tag{19.11}$$

with α, β, γ defined identically to Example 2, and c_t is a known deterministic function (it typically represents the ambient temperature in Celsius degrees, making $x_t \geq 0$). The couple of initial state and control (x_0, u_0) is assumed to be given.

The metric used to assess the impact of flexible load on the distribution network is assumed to be a linear combination of the transformer ageing (see Example 2) and the power losses (see Example 1). Note that transformer aging being a dynamic metric while power losses are static (only dependent on u_t, and not on past values), this constitutes an interesting combination from the dynamical analysis conducted here.

19.3.2 Distributed Strategies in This Dynamical Setting

In the distributed framework presented here, variables are controlled separately by I decision makers: $i \in \mathcal{I}$ chooses the sequence $\underline{v}_i \triangleq (v_{i,1}, \ldots, v_{i,T})$.

Three distributed charging policies are proposed, all based on the sequential best-response dynamics [5]. Note a key assumption in the dynamical setting of this section: the BRD algorithm is implemented *offline* based on the knowledge of the sequence of nonflexible load levels $(\ell_1, \ldots, \ell_T)$ with $\ell_t = \sum_{i=1}^{K} \ell_{i,t}$. Once the control policies are determined, they can be effectively run *online*. Within round $n + 1$ (with $n \geq 1$) of sequential BRD the action chosen by decision-maker i is computed as:[1]

$$\underline{v}_i^{(n+1)} \in \arg\min_{\underline{v}_i \in \mathcal{V}_i} C\left(\underline{v}_1^{(n+1)}, \ldots, \underline{v}_{i-1}^{(n+1)}, \underline{v}_i, \underline{v}_{i+1}^{(n)}, \ldots, \underline{v}_I^{(n)}\right). \tag{19.12}$$

The three distributed policies are now differentiated, starting with the BRD on arbitrary profiles.

BRD for distributed dynamic charging (DDC) policies. At round $n+1$ decision-maker i solves (19.12) with:

$$\mathcal{V}_i = \{\underline{v}_i = (v_{i,1}, \ldots, v_{i,T}) : \forall t \in \mathcal{T}, 0 \leq v_{i,t} \leq V_{\max} \text{ and } \sum_{t=1}^{T} v_{i,t} \geq S_i\}, \tag{19.13}$$

---
[1] If there are more than one best action, then one of them is chosen randomly.

where the bounds on $v_{i,t}$ account for the (individual) limitations on the consumption power, and the sum-constraint represents the energy need.

Relying on the convexity of the utility function w.r.t. $\underline{v}_i$ given that $ab_1 + b_2 \geq 0$ [9], an element of the argmin set in (19.12) can be obtained by solving the corresponding convex optimization problem, e.g., by using known numerical techniques (e.g., using MATLAB function fmincon). Compared to a centralized approach that would be run by a central operator (e.g., in case of a "Direct-Load-Control" mechanism), the complexity of this distributed strategy is reduced: applying the sequential BRD is linear in the number of rounds needed for convergence (say N, which typically equals 3 or 4) and the number of EVs I; therefore for a numerical routine whose complexity is cubic in the problem dimension, the complexity for the centralized implementation is of the order of $I^3 T^3$ whereas it is of the order of NIT^3 with the distributed implementation. Observe also that, in terms of information, all the model parameters (a, b_1, b_2, q, r, etc.) need to be known by each consumer for this distributed implementation to be run. If this turns out to be a critical aspect in terms of identification in practice (e.g., if the DNO does not want to reveal physical parameters about its transformer), other techniques that only exploit directly measurable quantities such as the sum-load have to be used. This is one of the purposes of the second scheme proposed next.

Iterative valley-filling algorithm (IVFA). Again following the iterative procedure of sequential BRD, this algorithm replaces the exact general minimization algorithm by the valley-filling or water-filling charging algorithm. This algorithm is a quite well-known technique (see, e.g., [10]) to allocate a given additional energy need (which corresponds here to the one induced by the EVs) over time given a primary demand profile (which corresponds here to the non-EV demand). The idea is to charge the EVs when the non-EV demand is sufficiently low. Note that this is optimal in particular when power losses are considered (static metric), i.e., $\alpha = 0$. Here, the novelty relies on the fact that the proposed implementation is an iterative version of the valley-filling algorithm. Indeed, in [10] for instance, valley-filling is used to design a scheduling algorithm but the iterative implementation is not explored. In [11], a distributed algorithm that relies on a parallel implementation (the I charging vectors are updated simultaneously over the algorithm iterations) is proposed. Convergence to the valley-filling solution is obtained by adding a *penalty* (or stabilizing) term to the cost. Note that one of the drawbacks of the latter approach is that the weight assigned to the added term has to be tuned properly. Here, the proposed sequential version does not have this drawback and can be seen as a power system counterpart of the iterative water-filling algorithm used in communications problems [12]. Convergence is ensured thanks to the exact potential property of the associated PCS game for more details on the definition of this game), which is commented more at the end of the present section. Formally, at round $n+1$, the charging power of EV i at time t is updated as

$$v_{i,t}^{(n+1)} = \left[\lambda_i - \ell_t - \sum_{j \in \mathcal{I}, j \neq i} v_{j,t}^{(n)} \right]^+, \qquad (19.14)$$

where $[s]_0^{V_{\max}} = \min(V_{\max}, \max(s,0))$ is the projection operator on the interval of available charging powers and λ_i is a threshold to be chosen. The value of this threshold is obtained by setting $S_i - \sum_{t=1}^{T} v_{i,t}^{(n+1)}$ to zero,[2] because it is easy to see that the sum-load constraint will be active at optimum. Compared to the DDC scheme, an important practical advantage of IVFA is that it relies only on the measure of the total load ℓ_t (it is an "open-loop" scheme). However, both solutions are based on continuous charging power levels ($v_{i,t} \in \mathbf{R}$). This assumption may not be met in some real applications.

The following result guarantees the convergence of the two described distributed charging policies.

PROPOSITION 19.2 *[Convergence] The DDC algorithm, IVFA, and rectangular profiles-based BRD charging algorithm always converge.*

This result can be proved by identifying each of the described distributed policies as the sequential BRD of a certain "auxiliary" strategic-form game. The key observation to be made is that because a common cost function (namely, C) is considered for the I decision makers and the individual control policies are vectors of $\mathbb{R}^T$ instead of general maps from the system state space to the charging power space, the corresponding problem can be formulated as an exact potential strategic-form game [5]. The important consequence of this is that the convergence of dynamics such as the sequential BRD is guaranteed due to the "finite improvement path" property [6]. Note that although Proposition 19.2 provides the convergence for the described policies, the efficiency of the point(s) of convergence is not ensured *a priori*. Typically, this efficiency can be measured relatively to the one obtained with a central operator that would minimize the total cost controlling $(\underline{v}_1, \ldots, \underline{v}_K)$. In the game-theory literature, this is called the "Price of Anarchy"; finding some bounds providing the maximal loss ("anarchy") induced by a distributed implementation is not an easy task in general (see [5]). In the case of EV charging, [1] presents some special cases for which explicit bounds are available (even with a zero loss in the asymptotic case of an infinite number of EVs). In the setting of this section, this question will be addressed numerically in the following part.

19.3.3 Numerical Illustration

Here, we report some numerical results that have obtained for the PCS algorithms we have been describing so far. The retained performance criterion is the distribution transformer lifetime.

Figure 19.4 represents the transformer's lifetime (in years) versus the number of EVs for six scenarios: without EV (dotted horizontal line); the proposed distributed dynamic charging (DDC) policy; the proposed distributed iterative valley-filling algorithm (IVFA); the proposed distributed charging policy using rectangular profiles (the power is either 0 or 3kW); the plug-and-charge policy where the arrival times at home are rounded values of $\epsilon \sim \mathcal{N}(4, 1.5)$ truncated on $[1, T]$. For $K = 20$, advanced charging policies are seen to be mandatory to avoid an immediate replacement of the transformer. The DDC allows the distribution transformer to gain a couple of years in terms of lifetime w.r.t. the IVFA. This is not negligible. The performance degradation is induced

---

[2] $v_{i,t}^{(n+1)}$ can be explicitly obtained in a few simple cases, see, e.g., [13] in the case $V_{\max} = \infty$.

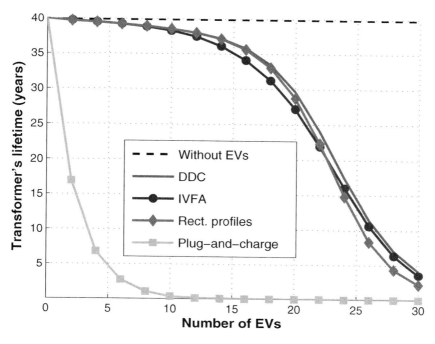

Figure 19.4 Transformer's lifetime against the number of EVs. *The three distributed charging policies perform comparably.*

by implementing the distributed policies instead of the centralized ones. Remarkably, simulations also showed (even this is not plotted here) that the loss induced by the decentralization (in comparison to the centralized algorithms) is very small for the three proposed policies. Distributed algorithms are therefore relevant: for: complexity issues; in the presence of several decision makers; for maintaining a very good performance level.

Furthermore, it has to be reminded that the DDC is the only proposed policy that allows the transformer not to exceed the maximal temperature ($x_{\max} = 150°$ C). Exceeding it will cause a shut down. It turns out that the IVFA and the policy using rectangular profiles cannot guarantee this requirement *a priori*. However, in the simulations conducted here, no shutdown has been observed with these two latter charging policies. Intuitively, this must rely on the trend of the non-EV demand with the evening peak followed by the night "valley."

Figure 19.5 allows one to assess the impact of not being able to forecast the non-EV demand perfectly. We have assumed that the optimization problems were fed with $\tilde{\ell}_t = \ell_t + z$ where z is a zero-mean additive white Gaussian noise with variance σ_{day}^2. It is seen that the distributed policy using the rectangular profile is the most robust one in terms of relative performance loss against forecasting errors. This can be explained by the fact that the charging starting instant is less sensitive to errors in amplitude than the IVFA and DDC. DDC is in the middle because it both aims at considering the memoryless part and the part with memory of the global cost. To complement this analysis on

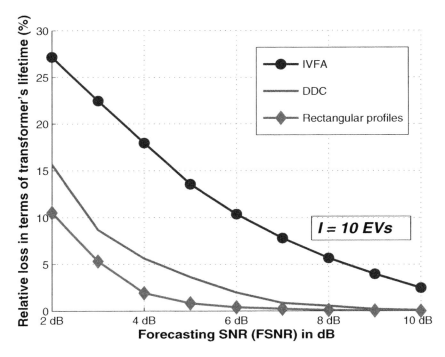

Figure 19.5 Relative performance loss against forecasting SNR (FSNR). *The policy using the rectangular profiles is the more robust.*

the impact of forecasting errors on the performance of the proposed charging policies, Figure 19.4 is the counterpart of Figure 19.3 in the case of a nonperfectly known non-EV demand. In this context, the BRD with rectangular profiles is the more efficient. This also highlights the fact that, for a given FSNR, the loss of performance varies with the number of EVs. On the one hand, Figure 19.4 shows that the transformer's lifetime is really similar in both the DDC and the case with rectangular profiles for $K = 1$–15 EVs. On the other hand, Figure 19.5 provides an almost zero loss in these two cases for a FSNR of 10dB and $K = 10$ EVs. This gives the very similar performance of both policies for $K = 10$ EVs. In contrast, for $K = 15$ EVs, we observe a significant difference.

19.4 When Forecasting Errors Are Accounted

In the previous sections, the effect of the forecasting noise on the noncontrollable load ℓ_t has been undergone. Indeed, the PCS algorithms have been designed by assuming the forecast to be perfect. If the algorithm is robust against noise, this approach is suitable. This is in fact the case for the distributed algorithm of Section 19.2. However, the algorithms of Section 19.3 may not be robust, as shown in [9, 14]. This motivates a modeling that accounts for the effect of the forecasting noise *a priori*. This is the motivation for describing the approach of this section, which is based on using Markov decision processes.

In what follows, a deterministic forecast on the noncontrollable part of the total consumption is not more assumed. Here, the knowledge to schedule the controllable loads consumption is based on a stochastic forecast (statistics) on the aggregated noncontrollable loads consumption. The idea is to take into account forecasting errors that have not been considered in the precedent sections, wherein a perfect/imperfect forecast (i.e., a single noiseless/noisy vector of) the noncontrollable loads is available to scheduling. The principal motivation of using a stochastic forecast is to provide a way to model the exogenous consumption by generating several scenarios, so that a deterministic forecast can be seen as one of these. This sort of forecast model can then be understood as several deterministic forecasts with associated probabilities of occurrence for which large databases exist to extract precise statistics. Moreover, the resulting power consumption scheduling policy can therefore adapt to different noncontrollable consumption events. Because the distribution network experiences an increased amount of variable loads consumption depending on the controllable part, introducing this kind of (adaptable) scheduling policy allows consumers to reduce their impacts on the distribution network. Otherwise, it could produce new load variations and possibly causing transformer overloading, power losses, or moreover increasing the transformer ageing [1, 9, 14].

To be computationally tractable the power scheduling problem under stochastic forecasting, a discretization over is considered. This problem can thus be modeled using finite Markov Decision Processes (MDP) [15] when (suitable) statistics are available. In addition, we can express several requirements on such modeling [16], for example an upper bound on the peak load consumption as a safety condition (encoded as a reachability condition on states in the MDP), a constraint on satisfying all energy need of the consumers as a quantitative reachability objective (e.g., recharging all electric vehicles in the whole time window under consideration), and various optimization criteria as optimization of costs (long-term utility functions) while actions (controllable loads) are selected (e.g., minimizing the ageing of the transformer, the electrical consumption payment).

19.4.1 Markov Decision Process Modeling

Let us consider that the system state x_t is discrete and evolves with the following general dynamics:[3]

$$x_t = g_t(x_{t-1}, v_t; \ell_t) \tag{19.15}$$

where v_t is the controllable load of consumers $v_{1,t}, \ldots, v_{K,t}$ and ℓ_t is the aggregated noncontrollable load consumption, i.e., $\ell_t = \sum_{i=1}^{K} \ell_{i,t}$. Such situations can be conveniently specified in terms of a probability of transition between states of the system. Suppose

---

[3] Note that adding a single period time-lag in the controllable and noncontrollable loads, we can obtain a general representation of the state evolution law: $x_t = \alpha x_{t-1} + \beta(\cdot)^2 + \gamma(\cdot)^2 + c_t$ of the *Example 2*.

that a (discrete) stochastic forecast of ℓ_t is available and denote it by $\widetilde{\ell}_t$. Let $P[\widetilde{\ell}_t]$ the probability distribution for which $\widetilde{\ell}_t$ is characterized (assumed here not dependent on precedent values). The latter together with (19.15) can provide a transition probability description between states as follows:

$$p(x_{t-1}, v_t)(x_t) = P[\{\widetilde{\ell}_t \mid x_t = g_t(x_{t-1}, v_t; \widetilde{\ell}_t)\} \mid x_{t-1}, v_t], \tag{19.16}$$

which is read as the probability that the next state will be x_t, given that the current state is x_{t-1}, and the controllable load selected by consumers is v_t. Note that the deterministic scheduling problem of Section 19.3 is a particular case of the present one. Indeed, when v_t is chosen on x_{t-1}, the next state x_t is full determined by making $p(x_{t-1}, v_t)(x_t) = 1$ and 0 for all other (candidate) next states. In this way, the power scheduling problem modelling by MDP comes naturally as is shown next.

DEFINITION 19.4   An MDP is a 5-tuple $\mathcal{M} = (\mathcal{X}, x_0, \mathcal{V}, p, f)$, where $\mathcal{X}$ is a finite set of (the system) states, $x_0 \in \mathcal{X}$ is a given initial state, $\mathcal{V}$ is a finite action space, and for each time t, $p(x_{t-1}, v_t)(x_t)$ represents the transition probability from state x_{t-1} to state x_t when $v_t \in \mathcal{V}$ is selected and $f(v_t, x_t)$ is the (instantaneous) utility function incurred.

As we said at the beginning, a common utility function is assumed for all consumers in this section. It can be represented by the function f in our MDP model. However, there is still one more important point to be discussed: the way in which the action v_t is chosen. Moreover, how the scheduling policy of consumers is found and constructed. Consider for instance that v_t is effectively the controllable loads of consumers $(v_{1,t}, \ldots, v_{K,t})$. If it is chosen by an only entity (agent), the scheduling problem becomes centralized and no individual decisions are taken into account. In this chapter, each consumer is free to make its own decision in terms of choosing its power consumption policy. Some models have been proposed when these are chosen separately. Multi-agent MDP (MMDP) is a first instance, which generalize an MDP describing sequential decisions in which multiple agents (the consumers) choose an individual action at every time t that jointly optimize a common utility function. However, the optimal cooperative coordination between agents is to communicate at every time step, making the complexity increase exponentially in the number of agents [17]. Moreover, it is generally necessary to consider a joint information space with a centralized approach, rendering this method impractical in most cases and outside of what we are looking for (decentralization). In absence of a central controller choosing the joint scheduling policy, we require some coordination mechanism between agents. Observability is also a subject to take into account. If agents choose their decisions based on incomplete observation on the system states we get into partially observable MDP class (POMDP), where the complexity of a POMDP with one agent is PSPACE[4] even if the state dynamic is stationary and the time-horizon T is smaller than the size of the system state space [18]. When multi-agents are considered in POMDP, decentralized executions raise severe difficulties during coordination even for the joint full observability particular case, namely decentralized MDP, which is based on the assumption that the system state is completely

---
[4] That can be solved in polynomial space.

determined from the current (joint) observations of the agents. However, this type of models lies to the NEXT-complexity[5] even for $K \geq 2$ gents [19]. Using these models to construct a scheduling policy for each consumer is a challenging problem. Indeed, the design of such a policy is difficult (even impossible) in the presence of a large number of states and when coordination has to be implemented at every step of the updating process. Moreover, our power consumption scheduling problem has an energy need constraint of consumers, which is not easy to implement in such models. Here, an approximate BRD between the consumers will be performed using iterative MDPs.

19.4.2 Relaxed Version of MDP Modeling

The main idea is to find the power consumption scheduling policy that optimizes the (joint) expected long-term utility function for one consumer at each iteration, keeping the scheduling policies of all the other consumers fixed. For that, we denote by $\mathcal{M}_i$ the MDP of consumer $i = 1, \ldots, K$ (that will be detailed a little further). Because each one needs to have a markovian structure in its $\mathcal{M}_i$, all the sufficient information to find its optimal scheduling policy must be contained in each state. Thus, we let $\mathcal{E}_i$ the accumulated energy set made by controllable power loads (actions) of the consumer i. Specifically, this set contains the elements

$$e_{i,t} = e_{i,t-1} + \Delta_t v_{i,t}$$

with $e_{i,0} = 0$. If we consider now the states as $\mathcal{X} \times \mathcal{E}_i$, finding an optimal scheduling policy for consumer i is reduced to the well-known Stochastic Shortest Path Problem (SSPP), so that under *a priori* defined target set of states (and finite space of states and actions) $\mathcal{M}_i$ admits an optimal (Markov) scheduling policy. Wherein one can use well-known techniques such as value iteration, linear programming, etc., to find such policy in POLYNOMIAL-time [15]. The target set in $\mathcal{M}_i$ can be naturally defined as the states such that $e_{i,t} \geq S_i$ at final time $t = T$, i.e., satisfying the energy need constraint. In this way, we can express the MDP for consumer i as $\mathcal{M}_i = (\mathcal{X} \times \mathcal{E}_i \times \mathcal{U}, x_0, \mathcal{V}_i, p_i, f)$, where with some abuse of notation, the initial state is $x_0 = (x_0, 0, 0)$, $\mathcal{V}_i$ is the set of actions (controllable loads) for i, $\mathcal{U}$ is the set of total loads and the transition probability p_i between states is described as in (19.16) but for one consumer i:

$$p_i((x_{t-1}, e_{i,t-1}, u_{t-1}), v_{i,t})(x_t, e_{i,t}, u_t) = P_i(\widetilde{\ell}_{-i,t} \in \mathcal{L}_t \mid (x_{t-1}, e_{i,t-1}, u_{t-1}), v_{i,t})$$

where

$$\mathcal{L}_t = \{\widetilde{\ell}_{-i,t} \mid (x_t, e_{i,t}, u_t) = (g_t(x_{t-1}, v_{i,t}; \widetilde{\ell}_{-i,t}), e_{i,t-1} + \Delta_t v_{i,t}, \widetilde{\ell}_{-i,t} + v_{i,t})\}$$

is the set of aggregated loads consumption $\widetilde{\ell}_{-i,t}$ due to the stochastic forecast and the controllable loads of all consumers except i, which evolves in each iteration of the BRD between the consumers (its underlying probability P_i is detailed in the next section). Firstly, the scheduling strategies are initialized sequentially for all consumers in the first iteration $m = 1$ and after, fixing the policies for all consumers except i, the optimal

---
[5] Nondeterministic exponential time.

policy for i is constructed and it is repeated until no policies change. Note that it can be understood as a multiple sequential learning method, where each consumer learns the environment in each MDP to determine its best-response considering other consumers are part of such environment, i.e., part of the stochastic forecast of the noncontrollable loads, allowing often results in local optima [20]. We detail this dynamics in the next section.

19.4.3 Iterative MDP Solution Methodology

As we said earlier, as the decentralized models MMDP, POMDP, etc. sharing explicit state information at each time step between the consumers are computationally costly, we present here an approximated version of BRD between the MDPs of consumers. For this, we update at each i and each round m the amount $\widetilde{\ell}_{-i,t}$, which can be seen as an updating of a matrix representing total power loads from the stochastic forecast and so, adding the power load of the consumers. To develop the dynamic, we define first a correspondence between two MDPs already solved, i.e., an optimal scheduling policy has been constructed in each one. We denote by $\mathcal{M}_i^{(m)}$ the MDP of consumer $i = 1, \ldots, K$ when the round is $m \geq 1$.

DEFINITION 19.5 *Let $\mathcal{M}_i^{(m)}$ and $\mathcal{M}_j^{(n)}$ two MDPs already solved in the rounds $m, n \geq 1$ resp. for the consumers i and j. Suppose that we obtain the values of the total load $u_t^{(n)}$ from the states in $\mathcal{M}_j^{(n)}$ for each t. Fixing a $u_t^{(n)}$, we define by going through the states from the initial state in $\mathcal{M}_i^{(m)}$, the probability that $u_t^{(m)}$ is $u_t^{(n)}$ by:*

$$P_i^{(m)}[u_t^{(m)} = u_t^{(n)}] = \sum_{\substack{(x_\tau, e_{i,\tau}, u_\tau)_{\tau=1}^t \\ \text{s.t. } u_t \in [u_t^{(n)}]_m}} \prod_{\tau=1}^t p_i^{(m)}((x_{\tau-1}, e_{i,\tau-1}, u_{\tau-1}), v_{i,\tau}^*)(x_\tau, e_{i,\tau}, u_\tau)$$

where the summation is over all tuples $(x_\tau, e_{i,\tau}, u_\tau)_{\tau=1}^t$ in $\mathcal{M}_i^{(m)}$ such that $u_t \in [u_t^{(n)}]_m$, with $[u_t^{(n)}]_m$ the equivalence class of elements of $\arg\min_{u_t \in \mathcal{U}^{(m)}} |u_t^{(n)} - u_t|$, and $v_{i,\tau}^*$ is the optimal controllable load selected on the respective state, i.e., $\sigma_i^{(m)}(x_{\tau-1}, e_{i,\tau-1}, u_{\tau-1}) = v_{i,\tau}^*$, for each $\tau = 1, \ldots, t$.

Supposing now that the MDPs are solved sequentially for each consumer, we define in next the profile $\widetilde{\ell}_{-i}^{(m)} = (\widetilde{\ell}_{-i,t}^{(m)})_{t=1}^T$ of the aggregated total load consumption except i at round $m \geq 1$, which will plays the role of the stochastic environment in the MDP $\mathcal{M}_i^{(m)}$.

Round $m = 1, i = 1$: In this case, each value $\widetilde{\ell}_{-1,t}^{(1)}$ is exactly the same as the stochastic forecast $\widetilde{\ell}_t$ of the noncontrollable load consumption because there is no other power load on the distribution network. Thus, the transition probability between states in the (not yet solved) MDP $\mathcal{M}_1^{(1)}$ is given by the known probability distribution of such forecast. Solving $\mathcal{M}_1^{(1)}$, the MDP for the next scheduler can be constructed.

Round $m = 1$, $i > 1$: Suppose that the MDP $\mathcal{M}_{i-1}^{(1)}$ has already been solved. To construct $\mathcal{M}_i^{(1)}$, we need the information of the aggregated load consumption due to the noncontrollable loads and the power loads of the consumers $1, \ldots, i-1$. However, such load is entirely determined by the previously solved $\mathcal{M}_{i-1}^{(1)}$ by going through the states from the initial state, identifying the values of the total loads $u_t^{(1)}$ and so construing equivalence classes to define the respective probability distribution. Thus, each $\tilde{\ell}_{-i,t}^{(1)}$ is the same as $u_t^{(1)}$ from the (solved) $\mathcal{M}_{i-1}^{(1)}$. The probability distribution for $\tilde{\ell}_{-i,t}^{(1)}$ is given by the correspondence of $\mathcal{M}_{i-1}^{(1)}$, see Definition 19.5. This defines the stochastic environment for the transition probability $p_i^{(1)}$ between the states for the (not yet solved) MDP $\mathcal{M}_i^{(1)}$.

Round $m > 1$, $i \geq 1$: As we stated already, all the information to construct $\mathcal{M}_i^{(m)}$ is contained in the previous solved[6] $\mathcal{M}_{i-1}^{(m)}$. However, each total load consumption $u_t^{(m)}$ in $\mathcal{M}_{i-1}^{(m)}$ contains the load of the consumer i made in the previous round $m-1$. For this reason, we define an expected controllable load from $\mathcal{M}_i^{(m-1)}$ to subtract the respective load (action) to each value $u_t^{(m)}$ for each t. Fixing one $u_t^{(m)}$, and considering the optimal power scheduling policy $\sigma_i^{(m-1)}$ of the consumer i constructed in $\mathcal{M}_i^{(m-1)}$, we denote by $\mathcal{S}_{t-1}^{(m-1)}([u_t^{(m)}]_{m-1})$ the set of states at $t-1$ of each state $(x_t, e_{i,t}, u_t)$ in $\mathcal{M}_i^{(m-1)}$ such that $u_t \in [u_t^{(m)}]_{m-1}$ by applying the optimal policy $\sigma_i^{(m-1)}$. For simplicity we write $\mathcal{S}_{t-1}$ for such set. We compute the expected controllable load selected on $\mathcal{S}_{t-1}$ by:

$$\mathbb{E}[\sigma_i^{(m-1)}(\mathcal{S}_{t-1})] = \frac{\sum_{\substack{(x_\tau, e_{i,\tau}, u_\tau)_{\tau=1}^t \\ \text{such that} \\ (x_{t-1}, e_{i,t-1}, u_{t-1}) \in \mathcal{S}_{t-1}, \\ u_t \in [u_t^{(m)}]_{m-1}}} v_{i,t-1}^* \prod_{\tau=1}^t p_i^{(m)}((x_{\tau-1}, e_{i,\tau-1}, u_{\tau-1}), v_{i,\tau}^*)(x_\tau, e_{i,\tau}, u_\tau)}{p_i^{(m-1)}[u_t^{(m-1)} = u_t^{(m)}]}$$

where as before, $v_{i,\tau}^*$ is the controllable load selected on the respective state by the optimal scheduling strategy $\sigma_i^{(m)}$ for each $\tau = 1, \ldots, t$; i.e., $\sigma_i^{(m)}(x_{\tau-1}, e_{i,\tau-1}, u_{\tau-1}) = v_{i,\tau}^*$. Note that the denominator acts a normalization. Thus, the aggregated total load consumption at t except i to construct $\mathcal{M}_i^{(m)}$ is therefore defined by $u_t^{(m)}$ from $\mathcal{M}_{i-1}^{(m)}$ as:

$$\tilde{\ell}_{-1,t}^{(m)} = \max\left\{u_t^{(m)} - \mathbb{E}\left[\sigma_i^{(m-1)}(\mathcal{S}_{t-1}^{(m-1)}([u_t^{(m)}]_{m-1}))\right], \min \tilde{\ell}_t\right\}.$$

The performance of this algorithm is illustrated by Figure 19.6.

---

[6] Well understood that if $i = 1$ and $m > 1$, the information is in $\mathcal{M}_K^{(m)}$.

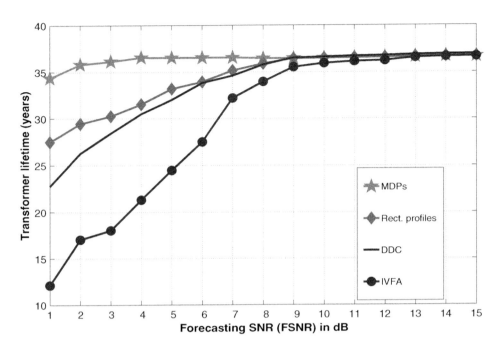

Figure 19.6 Transformer lifetime (years) against the forecasting signal-to-noise ratio (dB) for $I = 15$ electric vehicles. This confirms the benefits from using iterative MDPs in presence of forecasting errors.

References

[1] O. Beaude, S. Lasaulce, M. Hennebel, and I. Mohand-Kaci, "Reducing the impact of distributed EV charging on distribution network operating costs," *IEEE Transactions on Smart Grid*, 2016.
[2] *Guide for Loading Mineral-Oil-Immersed Transformers*. IEEE Std. C57.91-1995, 1995.
[3] IEC, *Loading Guide for Oil-Immersed Power Transformers*, No.- 60354.
[4] L. Rivera and D. Tylavsky, "Acceptability of four transformer top-oil thermal models: Pt. 1: Defining metrics," *Power Delivery, IEEE Trans. on*, vol. 23, no. 2, pp. 860–865, 2008.
[5] S. Lasaulce and H. Tembine, *Game Theory and Learning for Wireless Networks: Fundamentals and Applications*. Cambridge, MA: Academic Press, 2011.
[6] D. Monderer and L. S. Shapley, "Potential games," *Games and Economic Behavior*, vol. 14, no. 1, 1996, pp. 124–143.
[7] C. H. Papadimitriou, "Algorithms, games, and the Internet," *28th International Colloquium on Automata, Languages and Programming*, 2001.
[8] D. P. Bertsekas, "Nonlinear programming," *Athena Scientific*, 1995.
[9] O. Beaude, S. Lasaulce, M. Hennebel, and J. Daafouz, "Minimizing the impact of EV charging on the electricity distribution network," in *2015 European Control Conference (ECC)*. IEEE, 2015, pp. 648–653.
[10] M. Shinwari, A. Youssef, and W. Hamouda, "A water-filling based scheduling algorithm for the smart grid," *Smart Grid, IEEE Transactions on*, vol. 3, no. 2, pp. 710–719, 2012.

[11] L. Gan, U. Topcu, and S. H. Low, "Optimal decentralized protocol for electric vehicle charging," *Power Systems, IEEE Transactions on*, vol. 28, no. 2, pp. 940–951, 2013.

[12] W. Yu, G. Ginis, and J. M. Cioffi, "Distributed multiuser power control for digital subscriber lines," *Selected Areas in Communications, IEEE Journal on*, vol. 20, no. 5, pp. 1105–1115, 2002.

[13] B. Sohet, O. Beaude, and Y. Hayel, "Routing game with nonseparable costs for EV driving and charging incentive design," in *NETwork Games, Control and Optimization, 2018 International Conference*, 2018.

[14] M. González, O. Beaude, P. Bouyer, S. Lasaulce, and N. Markey, "Stratégies d'ordonnancement de consommation d'énergie en présence d'information imparfaite de prévision," in Colloque francophone de traitement du signal et des images (GRETSI), 2017.

[15] M. L. Puterman, *Markov Decision Processes: Discrete Stochastic Dynamic Programming*. New York: John Wiley & Sons, 2014.

[16] P. Bouyer, M. González, N. Markey, and M. Randour, "Multi-weighted Markov decision processes with reachability objectives," *arXiv preprint arXiv:1809.03107*, 2018.

[17] D. V. Pynadath and M. Tambe, "Multiagent teamwork: Analyzing the optimality and complexity of key theories and models," in *Proceedings of the First International Joint Conference on Autonomous Agents and Multiagent Systems: Part 2*. ACM, 2002, pp. 873–880.

[18] C. H. Papadimitriou and J. N. Tsitsiklis, "The complexity of Markov decision processes," *Mathematics of Operations Research*, vol. 12, no. 3, pp. 441–450, 1987.

[19] D. S. Bernstein, R. Givan, N. Immerman, and S. Zilberstein, "The complexity of decentralized control of Markov decision processes," *Mathematics of Operations Research*, vol. 27, no. 4, pp. 819–840, 2002.

[20] G. W. Brown, "Iterative solution of games by fictitious play," *Activity Analysis of Production and Allocation*, vol. 13, no. 1, pp. 374–376, 1951.

20 Electric Vehicles and Mean-Field

Dario Bauso and Toru Namerikawa

20.1 Introduction

By "demand response" one usually denotes a number of measures that allow to decentralize load control in smart grids [1–4]. The ultimate goal is the alteration of the timing and electricity consumption patterns by end-use customers from their normal consumption patterns. This is obtained through changes in the price of electricity or using incentive payments to consumers who provide lower electricity use at off-peak times.

Demand response requires the adoption of smart communication about the past, current, and future state of the grid. The state of the grid is measured by the potential mismatch between supply and demand or by the grid frequency. Appliances are then equipped with smart computation infrastructure and are capable of forecasting the state of the grid and consequently to change their consumption requests. The general setup of this chapter implies fully responsive load control on the part of a population of electric vehicles (EVs). In other words, the EVs operate in a no-prescriptive environment, by this meaning that the EVs are not preprogrammed to operate according to specific charging policies. On the contrary, EVs are designed as smart appliances capable of obtaining autonomously their optimal charging policies as best-responses to the population behavior. The population behavior is sensed by the individual appliances through the mains frequency state [5–8].

The underlying idea in this chapter is the adoption of stochastic charging strategies in the same spirit as in [5, 9, 10]. This means that the EVs select the probability with which to switch to state *on* (charging) or state *off* (discharging). As an example, if an EV sets the probability to $\frac{1}{2}$, this would correspond to the toss of a coin and the switch to charging or discharging depending whether the outcome is head or tail. In a population of EVs, if all appliances adopt the same rule, then we will have 50% of EVs in state *on* and 50% in state *off*. The adoption of stochastic strategies has been studied in [5, 9] and it has been proven to be a sensible alternative to deterministic strategies when it comes to attenuating the *mains frequency* oscillations. Oscillations are usually a direct consequence of the mismatch between supply and demand (see, e.g., [11]). For a correct operation of the grid, the frequency needs to be stabilized around a nominal value (50 Hz in Europe). When demand exceeds supply one observes a negative deviation of the frequency from the nominal value, and vice versa.

In this chapter, we develop a model in which each EV is viewed as a strategic player. The EV is characterized by two state variables, the level of charge and the charging mode, which can be on or off. The dynamical evolution of the state of an EV, also referred to as *microscopic* dynamics, models the change of its level of charge and charging mode. Such dynamics is obtained as a solution of a linear ordinary differential equation. By aggregating dynamics of the whole population of EVs, one obtains the *macroscopic* dynamics of the whole system. EVs are programmed to solve online infinite-horizon receding horizon problems. By doing this, the EVs obtain the best response charging policy that optimizes multiple objectives including energy consumption, deviation of frequency from nominal values, and deviation of the EVs' level of charge from a reference value. One may notice that the objective functional combines individual tasks such as the energy consumption when charging and the deviation from a reference level of charge, and collective tasks such as the individual contribution to the deviation of the mains frequency from the nominal one. Thus the model maximizes the intelligence of the grid. The collective task requires the use of a mean-field term that couples the individual and the population and incentivizes the EV to set to *off* when the frequency is below the nominal value and to set to *on* when the frequency is above the nominal value. Note that the mean-field term represents a cross-coupling term and can be used to capture incentive payments, benefits, or smart pricing policies. A correct design and implementation of the cross-coupling term will result into a shift of the demand from high-peak to off-peak periods.

20.1.1 Highlights of Contributions

In the spirit of prescriptive game theory and mechanism design [12], the main contribution of this chapter is the design of a mean-field game for a population of EVs. We introduce the mean-field equilibrium and investigate ways in which we can obtain an approximation of such equilibrium point using simple calculations. The main idea is to turn the game into a sequence of infinite-horizon receding horizon optimization problems that each EV solves online. The online computation is driven by data made available at regular sampling intervals. We provide analysis of the asymptotic stability of the microscopic and macroscopic dynamics. The physical interpretation of the results corresponds to a level of charge and charging mode for each EV which converge to predefined reference values. We show through simulations that such convergence properties are guaranteed also when the EVs dynamics are affected by an additional stochastic disturbance, which we model using a Brownian motion. Finally, we show the impact of measurement noise on the overall dynamics. Such noise is added to the estimate of the grid frequency and could be also viewed as the consequence of attacks on the part of hackers aiming at compromising the data. An expository work on stochastic analysis and stability is [13].

20.1.2 Literature Overview

In the following, we organize the literature review in two sections. The first section is about dynamic response management. The second section pertains mean-field games.

Demand Response

Dynamic demand management is discussed in [6, 7, 14, 15]. In [14], the authors provide an overview on decentralized strategies to redistribute the load away from peak hours. This idea is further developed in [15].

The use of game theoretic methods is illustrated in [6], where a large population game is developed in which the players are plug-in EVs. Charging policies that redistribute the load away from peaks are now assimilated to Nash-equilibrium strategies (see [16]). In technical jargon, one calls such strategies *valley-filling strategies*. In the present study, we share the same perspective as we show that to stabilize the grid's frequency we can design incentives to the agents to allow them to adjust their strategies and converge to a mean-field equilibrium. This is possible if a game designer can influence the individual objective functions by providing penalties to those players that are charging in peak hours. Valley-filling or more in general "coordination strategies" have been investigated also for thermostatically controlled loads such as refrigerators, air conditioners, and electric water heaters [7].

Mean-Field Games

Mean-field games were formulated by Lasry and Lions in [17] and independently by M. Y. Huang, P. E. Caines, and R. Malhamé in [18, 19]. The mean-field theory of dynamical games is a modeling framework that captures the interaction between a population strategic players and its members. The mean-field approach deals with a system of two partial differential equations (PDEs). The first PDE is a Hamilton–Jacobi–Bellman (HJB) equation that models the ways in which individuals adapt when they observe the population. The second PDE is a Fokker–Planck–Kolmogorov (FPK) equation that describes the density of the strategies of the players. Explicit solutions in terms of mean-field equilibria are available for linear-quadratic mean-field games [20]. Explicit solutions have been recently extended to more general cases in [21].

The idea of extending the state space, which originates in optimal control [22, 23], has been also used to approximate mean-field equilibria in [24].

To account for model misspecifications, robustness and risk-sensitivity have been investigated in [25, 26]. There the first PDE is a Hamilton–Jacobi–Isaacs (HJI) equation in that the game involves additional adversarial disturbances. For a survey on mean-field games and applications we refer the reader to [27]. A first attempt to apply mean-field games to demand response is in [9]. Application of mean-field-based control to power systems is provided in [28] and [29]. Computational algorithms are discussed in [30].

The chapter is organized as follows. In Section 20.2, we state the problem and introduce the model. In Section 20.3, we state and discuss the main results. In Section 20.4, we carry out some numerical studies. In Section 20.5, we provide some discussion. Finally, in Section 20.6, we provide some conclusions.

20.1.3 Notation

The symbol $\mathbb{E}$ indicates the expectation operator. We use ∂_x to denote the first partial derivative with respect to x, respectively. Given a vector $x \in \mathbb{R}^n$ and a matrix $a \in \mathbb{R}^{n \times n}$

we denote by $\|x\|_a^2$ the weighted two-norm $x^T a x$. We denote by $\Pi_{\mathcal{M}}(X)$ the projection of X onto set $\mathcal{M}$. We denote by $]\xi, \zeta[$ the open interval for any pair of real numbers $\xi \leq \zeta$.

20.2 Modeling EVs in Large Numbers

In this section, we assimilate EVs to capacitors and consider their charging and discharging policies, as displayed in Figure 20.1. In the spirit of prescriptive game theory [12], the idea is to design a mean-field game to incentive cooperative behaviors among the EVs.

In a population of EVs, each EV is modeled by a continuous state, which represents the level of charge $x(t)$, and by a binary state $\pi(t) \in \{0, 1\}$, which indicates whether the EV is charging (*on* state, $\pi(t) = 1$) or discharging (*off* state, $\pi(t) = 0$) at time $t \in [0, T]$. Here $[0, T]$ is the time horizon window.

By setting to *on* state, the level of charge increases exponentially up to a fixed maximum level of charge x_{on} whereas in the *off* position the level of charge decreases exponentially up to a minimum level of charge x_{off}. Then, the level of charge of each EV is given by the following differential equations in $[0, T]$:

$$\dot{x}(t) = \begin{cases} -\alpha(x(t) - x_{on}) & \text{if } \pi(t) = 1 \\ -\beta(x(t) - x_{off}) & \text{if } \pi(t) = 0 \end{cases}, \quad (20.1)$$

where $x(0) = x$ represents the boundary condition at initial time, and where the α, β are the rates of charging and discharging, respectively. Both are given positive scalars.

In the spirit of stochastic models as provided in [5, 9], we assume that each EV can be in states *on* or *off* in accordance to probability $\pi \in [0, 1]$. The control input corresponds to the transition rate u_{on} from state *off* to state *on* and the transition rate u_{off} from state *on* to state *off*.

Let us denote $y(t) = \pi(t)$. For any x, y in the

"set of feasible states" $\mathcal{S} :=]x_{off}, x_{on}[\times]0, 1[,$

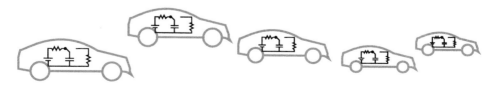

Figure 20.1 Population of EVs assimilated to capacitors.

we obtain the following dynamical system

$$\dot{x}(t) = y(t)\Big[-\alpha(x(t) - x_{on})\Big] + (1 - y(t))\Big[-\beta(x(t) - x_{off})\Big]$$
$$=: f(x(t), y(t)), \ t \in [0,T], \quad x(0) = x,$$
(20.2)

$$\dot{y}(t) = \Big(u_{on}(t) - u_{off}(t)\Big)$$
$$=: g(u(t)), \ t \in [0,T], \quad y(0) = y.$$

A macroscopic description of the model can be obtained by introducing the probability density function $m : [x_{on}, x_{off}] \times [0,1] \times [t,T] \to [0, +\infty[, (x,y,t) \mapsto m(x,y,t)$, for which it holds $\int_{x_{on}}^{x_{off}} \int_{[0,1]} m(x,y,t) dx dy = 1$ for every t in $[0,T)$. Furthermore, let $m_{on}(t) := \int_{x_{on}}^{x_{off}} \int_{[0,1]} y m(x,y,t) dx dy$. Analogously, let $m_{off}(t) := 1 - m_{on}(t)$.

At every time t the grid frequency depends linearly on the discrepancy between the percentage of EVs in state *on* and a nominal value. We refer to such deviation to as *error* and denote as $e(t) = m_{on}(t) - \overline{m}_{on}$. Here $\overline{m}_{on}$ is the nominal value. Note that the more EVs are in state *on* if compared with the nominal value, the more the grid frequency presents negative deviation from the nominal value. In other words, the grid frequency depends on the mismatch between the power supplied and consumed. For sake of simplicity, henceforth we assume that the power supply is constant and equal to the nominal power consumption all the time.

20.2.1 Forecasting Based on Holt's Model

The information on the error $e(t)$ is relevant to let the EVs adjust their best response charging policies. This information is available at discrete times t_k, and after receiving a new data the players first forecast the next value of the error and based on that they compute their best-response strategies over an infinite planning horizon, namely $T \to \infty$.

Once the forecasted error is obtained, the players assume that this value remains fixed throughout the planning horizon. Note that this assumption is mitigated if the interval between consecutive samples is sufficiently small. Once the sequence of optimal controls is obtained, the players implement their first controls until a new sample becomes available. In other words, the players implement a *receding horizon* technique that consists of a multistep ahead *action* horizon. In the following, we denote the length of the interval between consecutive samples as $\delta = t_{k+1} - t_k$. In the following, we can take $\delta = 1$ without loss of generality.

Forecasting is based on the Holt's model, which assumes that the underlying model for the error is linear with respect to time, and has permanent component

$$\mu(t_k) = a(t_k) + b(t_k) t_k.$$

For the error we then have

$$e(t_k) = \mu(t_k) + \epsilon(t_k) = a(t_k) + b(t_k) t_k + \epsilon(t_k)$$

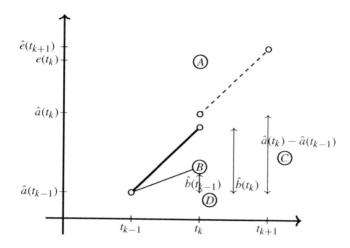

Figure 20.2 One iteration of the forecasting method based on the Holt's model.

When the new observation $e(t_k)$ becomes availabe, the estimate of the constant is updated as follows by

$$\hat{a}(t_k) = \hat{\alpha} e(t_k) + (1 - \hat{\alpha})\left(\hat{a}(t_{k-1}) + \hat{b}(t_{k-1})\right).$$

Similarly, the estimate for the slope is obtained as

$$\hat{b}(t_k) = \hat{\beta}(\hat{a}(t_k) - \hat{a}(t_{k-1})) + (1 - \hat{\beta})\hat{b}(t_{k-1}).$$

Finally, the forecast of the error at time t_{k+1} obtained at time t_k is given by

$$\hat{e}(t_{k+1}) = \hat{a}(t_k) + \hat{b}(t_k).$$

The forecasting iteration at time t_{k-1} using the Holt's model is illustrated in Figure 20.2. The new level $\hat{a}(t_k)$ is in the convex hull between the last obtained sample $e(t_k)$ (indicated by circle A) and the forecasted value from previous iteration $\hat{a}(t_{k-1}) + \hat{b}(t_{k-1})$ (indicated by circle B). Likewise, the new slope $\hat{b}(t_k)$ is in the convex hull between the last observed slope $\hat{a}(t_k) - \hat{a}(t_{k-1})$ (indicated by circle C) and the previous forecast $\hat{b}(t_{k-1})$ (indicated by circle D).

20.2.2 Receding Horizon

The running cost for each player introduces the dependence on the distribution $m(x, y, t)$ through the error $\hat{e}(t_{k+1})$ and is given here:

$$c(\hat{x}(\tau, t_k), \hat{y}(\tau, t_k), \hat{u}(\tau, t_k), \hat{e}(t_{k+1})) = \frac{1}{2}\left(q\hat{x}(\tau, t_k)^2 + r_{on}\hat{u}_{on}(\tau, t_k)^2 + r_{off}\hat{u}_{off}(\tau, t_k)^2\right)$$

$$+ \hat{y}(\tau, t_k)(S\hat{e}(t_{k+1}) + W), \qquad (20.3)$$

where q, r_{on}, r_{off}, and S are opportune positive scalars.

In the running cost (20.3) we have four terms. One term penalizes the deviation of the EVs' levels of charge from the nominal value, which we set to zero. Setting the nominal level of charge to a nonzero value would simply imply a translation of the origin of the axes. This penalty term is then given by $\frac{1}{2}qx(\tau,t_k)^2$.

A second term penalizes fast switching and is given by $\frac{1}{2}r_{on}u_{on}(\tau,t_k)^2$; this cost is zero when either $u_{on}(\tau,t_k)=0$ (no switching) and is maximal when $u_{on}(\tau,t_k)=1$ (probability 1 of switching). The same reasoning applies to the term $\frac{1}{2}r_{off}u_{off}(\tau,t_k)^2$. A positive error forecast $e(t_{k+1}) > 0$, means that demand exceeds supply. Therefore, with the term $y(\tau,t_k)Se(\tau,t_k)$ we penalize the EVs that are in state *on* when power consumption exceeds the power supply ($e(\tau,t_k) > 0$). On the contrary, when the power supply exceeds power consumption, the error is negative, i.e., $e(\tau,t_k) > 0$, and the related term $y(\tau,t_k)Se(\tau,t_k)$ penalizes the EVs that are in state *off*. The last term is $y(\tau,t_k)W$ and is a cost on the power consumption; when the EV is in state *on* the consumption is W. We also consider a terminal cost $g: \mathbb{R} \to [0,+\infty[, x \mapsto g(x)$ to be yet designed.

Let the following update times be given, $t_k = t_0 + \delta k$, where $k = 0, 1, \ldots$. Let $\hat{x}(\tau,t_k)$, $\hat{y}(\tau,t_k)$ and $\hat{e}(t_{k+1})$, $\tau \geq t_k$ be the predicted state of player i and of the error $e(t)$ for $t \geq t_k$, respectively. The problem we wish to solve is the following one.

For all players and times $t_k, k = 0, 1, \ldots$, given the initial state $x(t_k)$, and $y(t_k)$ and $\hat{e}(t_{k+1})$ find

$$\hat{u}^\star(\tau,t_k) = \arg\min \mathcal{J}(\hat{x}(t_k), \hat{y}(t_k), \hat{e}(t_{k+1}), \hat{u}(\tau,t_k)),$$

where

$$\mathcal{J}(\hat{x}(t_k), \hat{y}(t_k), \hat{e}(t_{k+1}), \hat{u}(\tau,t_k)) = \lim_{T \to \infty} \int_{t_k}^{T} c(\hat{x}(\tau,t_k), \hat{y}(\tau,t_k), \hat{u}(\tau,t_k), \hat{e}(t_{k+1})) d\tau \quad (20.4)$$

subject to the following constraints:

$$\dot{\hat{x}}(\tau,t_k) = \hat{y}(\tau,t_k)\Big[-\alpha(\hat{x}(\tau,t_k) - x_{on})\Big] + (1 - \hat{y}(\tau,t_k))\Big[-\beta(\hat{x}(\tau,t_k) - x_{off})\Big]$$
$$=: f(\hat{x}(\tau,t_k), \hat{y}(\tau,t_k)), \quad \tau \in [t_k, T],$$
$$\dot{\hat{y}}(\tau,t_k) = \Big(\hat{u}_{on}(\tau,t_k) - \hat{u}_{off}(\tau,t_k)\Big)$$
$$=: g(\hat{u}(\tau,t_k)), \quad \tau \in [t_k, T], \quad (20.5)$$

The preceding set of constraints involves the predicted state dynamics of the individual player and of the rest of the population through $\hat{e}(t_{k+1})$. The constraints also involve the boundary conditions at the initial time t_k. Note that a player restrains the error to be constant over the planning horizon.

At t_{k+1} a new sample data $e(t_{k+1})$ becomes available. Then the players update their best-response strategies, which we refer to as *receding horizon control strategies*. Consequently, for the individual player, we obtain the *closed-loop* system

$$\dot{\hat{x}}(\tau,t_k) = \hat{y}(\tau,t_k)\Big[-\alpha(\hat{x}(\tau,t_k) - x_{on})\Big] + (1 - \hat{y}(\tau,t_k))\Big[-\beta(\hat{x}(\tau,t_k) - x_{off})\Big]$$
$$=: f(\hat{x}(\tau,t_k), \hat{y}(\tau,t_k)), \quad \tau \in [t_k, T),$$

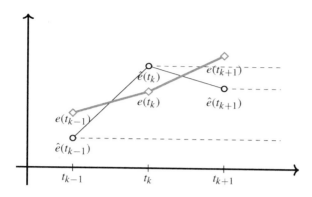

Figure 20.3 One iteration of the forecasting method based on the Holt's model.

$$\dot{y}(\tau, t_k) = \left(u_{on}^{RH}(\tau) - u_{off}^{RH}(\tau) \right)$$
$$=: g(u^{RH}(\tau)), \quad \tau \in [t_k, T),$$
(20.6)

where the receding horizon control law $u^{RH}(\tau)$ satisfies

$$u^{RH}(\tau) = \hat{u}^*(\tau, t_k), \quad \tau \in [t_k, t_{k+1}).$$

Figure 20.3 provides a graphical illustration of the receding horizon model. Given three consecutive samples $e(t_{k-1})$, $e(t_k)$, and $e(t_{k+1})$ (diamonds), the forecasts obtained from the Holt's method are $\hat{e}(t_{k-1})$, $\hat{e}(t_k)$, and $\hat{e}(t_{k+1})$ (circles). The dashed lines indicate that each value is kept fixed throughout the horizon when running the receding horizon optimization.

20.2.3 From Mean-Field Coupling to Linear Quadratic Problem

The receding horizon problem solved at time t_k can be written in terms of the state, and control vectors

$$X(t) = \begin{bmatrix} \hat{x}(t, t_k) \\ \hat{y}(t, t_k) \end{bmatrix}, \quad u(t) = \begin{bmatrix} \hat{u}_{on}(t, t_k) \\ \hat{u}_{off}(t, t_k) \end{bmatrix},$$

and yields the following linear quadratic problem:

$$\inf_{u(\cdot)} \mathbb{E} \int_{t_k}^{T} \left[\frac{1}{2} \left(\|X(t)\|_Q^2 + \|u(t)\|_R^2 \right) + L^T X(t) \right] dt + g(X(T)),$$
$$\dot{X}(t) = A(x)X(t) + Bu(t) + C, \quad \text{in } \mathcal{S}$$
(20.7)

where

$$Q = \begin{bmatrix} q & 0 \\ 0 & 0 \end{bmatrix}, \quad R = r = \begin{bmatrix} r_{on} & 0 \\ 0 & r_{off} \end{bmatrix}, \quad L(\hat{e}(t_{k+1})) = \begin{bmatrix} 0 \\ S\hat{e}(t_{k+1}) + W \end{bmatrix},$$
$$A(x) = \begin{bmatrix} -\beta & k(x) \\ 0 & 0 \end{bmatrix}, \quad B = \begin{bmatrix} 0 & 0 \\ 1 & -1 \end{bmatrix}, \quad C = \begin{bmatrix} \beta x_{off} \\ 0 \end{bmatrix}.$$
(20.8)

and where

$$k(x(t)) = x(t)(\beta - \alpha) + (\alpha x_{on} - \beta x_{off}).$$

Note that the receding horizon problems constitute a sequence of consecutive approximations of the mean-field game given in the following text

$$\begin{cases} \partial_t v(X,t) + \inf_u \left\{ \partial_X v(X,t)^T (AX + Bu + C) + \frac{1}{2} \left(\|X\|_Q^2 \right. \right. \\ \left. \left. + \|u\|_R^2 \right) + L^T X \right\} = 0, \text{ in } \mathcal{S} \times [0,T[, \\ v(X,T) = g(x), \text{ in } \mathcal{S}, \\ u^*(x,t) = \operatorname{argmin}_{u \in \mathbb{R}} \left\{ \partial_X v(X,t)^T (AX + Bu + C) + \frac{1}{2} \|u(t)\|_R^2 \right\}, \quad (b) \end{cases} \quad (20.9)$$

and

$$\begin{cases} \partial_t m(x,y,t) + div[(AX + Bu + C) m(x,y,t)] = 0 \text{ in } \mathcal{S} \times]0,T[, \\ m(x_{on}, y, t) = m(x_{off}, y, t) = 0 \, \forall \, y \in [0,1], \, t \in [0,T], \\ m(x, y, 0) = m_0(x, y) \, \forall \, x \in [x_{on}, x_{off}], \, y \in [0,1], \\ \int_{x_{on}}^{x_{off}} m(x,t) dx = 1 \, \forall \, t \in [0,T]. \end{cases} \quad (20.10)$$

Essentially, the first PDE (20.9) (a) is the HJB equation and can be solved backward with boundary conditions at final time T, see the last line in 20.9 (a). The second equation 20.9 (b) is the optimal closed-loop control $u^*(x,t)$ as minimizers of the Hamiltonian function in the right-hand side. The second PDE (20.10) is the transport equation of the measure m immersed in a vector field $AX+Bu+C$. This equation yields the distribution $m(x,y,t)$ for fixed $u^*(x,t)$ and, consequently, the vector field $AX + Bu^* + C$. Such a PDE has to be solved forward with boundary condition at the initial time (see the fourth line of (20.10)). Finally, once given $m(x, y, t)$ and substituted into the running cost $c(x, y, m, u)$ in (a), we obtain the error

$$\begin{cases} m_{on}(t) := \int_{x_{on}}^{x_{off}} \int_{[0,1]} ym(x, y, t) dx dy \\ \forall t \in [0,T], \\ e(t) = m_{on}(t) - \bar{m}_{on}. \end{cases} \quad (20.11)$$

From the preceding we also obtain

$$\bar{X}(t) = \begin{bmatrix} \bar{x}(t) \\ \bar{y}(t) \end{bmatrix} = \begin{bmatrix} \bar{x}(t) \\ m_{on} \end{bmatrix} = \begin{bmatrix} \int_{x_{on}}^{x_{off}} \int_{[0,1]} xm(x,y,t)dxdy \\ \int_{x_{on}}^{x_{off}} \int_{[0,1]} ym(x,y,t)dxdy \end{bmatrix},$$

and therefore, henceforth we can refer to as mean-field equilibrium solutions any pair $(v(X,t), \bar{X}(t))$ which is solution of (20.9)–(20.10).

20.3 Main Results

In this section, we establish an explicit solution for the receding horizon problem and link it to the mean-field equilibrium. We also prove that under certain conditions the microscopic dynamic is asymptotically stable.

20.3.1 Receding Horizon and Stability

In the previous section we showed that the receding horizon problem is given in the linear quadratic problem that follows:

$$\inf_{u(\cdot)} \int_0^T \left[\frac{1}{2}\left(X(t)^T Q X(t) + u(t)^T R u(t)\right) + L^T X(t)\right] dt + g(X(T)), \quad (20.12)$$

$$\dot{X}(t) = AX(t) + Bu(t) + C \text{ in } S.$$

We are in the position to show, in the next result, that an explicit solution can be obtained by solving three matrix equations. We refer to such solution as a receding horizon equilibrium. This equilibrium is an approximation of the mean-field equilibrium for (20.9)–(20.10).

THEOREM 20.1 *(Receding horizon equilibrium) A receding horizon equilibrium for (20.12) is given by*

$$\begin{cases} v(X,t) = \frac{1}{2} X^T P(t) X + \Psi(t)^T X + \chi(t), \\ \dot{\bar{X}}(t) = [A(x) - BR^{-1}B^T P]\bar{X}(t) \\ \quad - BR^{-1}B^T \bar{\Psi}(t) + C, \end{cases} \quad (20.13)$$

where

$$\begin{cases} \dot{P} + PA(x) + A(x)^T P - PBR^{-1}B^T P + Q = 0 \\ \quad \text{in } [0,T[, \ P(T) = \phi, \\ \dot{\Psi} + A(x)^T \Psi + PC - PBR^{-1}B^T \Psi + L = 0 \\ \quad \text{in } [0,T[, \ \Psi(T) = 0, \\ \dot{\chi} + \Psi^T C - \frac{1}{2}\Psi^T BR^{-1}B^T \Psi = 0 \text{ in } [0,T[, \\ \chi(T) = 0, \end{cases} \quad (20.14)$$

and $\bar{\Psi}(t) = \int_{x_{on}}^{x_{off}} \int_{[0,1]} \Psi(t) m(x,y,t) dx dy$ and where the boundary conditions are obtained by imposing that

$$v(X,T) = \frac{1}{2} X^T P(T) X + \Psi(T) X + \chi(T) = \frac{1}{2} X^T \phi X =: g(X).$$

Furthermore, the receding horizon equilibrium strategy is

$$u^*(X,t) = -R^{-1}B^T[PX + \Psi]. \quad (20.15)$$

Proof Given in the appendix. □

To obtain the closed-loop microscopic dynamics we have to substitute the receding horizon equilibrium strategies $u^* = -R^{-1}B^T[PX + \Psi]$ given in (20.15) in the open-loop microscopic dynamics $\dot X(t) = AX(t) + Bu(t) + C$ as defined in (20.12). Then, the resulting closed-loop microscopic dynamics is

$$\dot X(t) = [A(x) - BR^{-1}B^T P]X(t) - BR^{-1}B^T \Psi(x,e,t) + C. \tag{20.16}$$

In the preceding, and occasionally in the following (see the proof of Theorem 20.1), the dependence of Ψ on x, e, and t is emphasized explicitly. With respect to the closed-loop dynamics (20.16), consider the set of equilibrium points and denote it by $\mathcal{X}$, namely, the set of X such that

$$\mathcal{X} = \{(X,e) \in \mathbb{R}^2 \times \mathbb{R} | [A(x) - BR^{-1}B^T P]X(t) - BR^{-1}B^T \Psi(x,e,t) + C = 0\},$$

and let $V(X,t) = dist(X,\mathcal{X})$. In the next result, we establish a condition under which asymptotic convergence to the set of equilibrium points is guaranteed.

THEOREM 20.2 *(Asymptotic stability)* If it holds

$$\partial_X V(X,t)^T \left([A - BR^{-1}B^T P]X(t) - BR^{-1}B^T \Psi + C\right) < -\|X(t) - \Pi_{\mathcal{X}}(X(t))\|^2 \tag{20.17}$$

then dynamics (20.16) is asymptotically stable, namely, $\lim_{t \to \infty} V(X(t)) = 0$.

Proof Given in the appendix. □

The resulting closed-loop dynamics are then obtained as a collection of the closed-loop dynamics obtained at different intervals. Let the following switching signal be given, $\sigma(t) = k$ for all $t \in [t_k, t_{k+1})$. Then the resulting dynamics can be written as

$$\dot X_{\sigma(t)}(t) = [A(x) - BR^{-1}B^T P]X_{\sigma(t)}(t) - BR^{-1}B^T \Psi(x,e,t) + C. \tag{20.18}$$

In the next section, we provide simulation of the preceding dynamics.

20.4 Numerical Studies

Consider a population of 100 EVs, and set $n = 100$. Simulations are carried out with MATLAB on an Intel®Core™2 Duo, CPU P8400 at 2.27GHz and a 3GB of RAM. Simulations involve a number of iterations $T = 1,000$. Consider a discrete time version of (20.12)

$$X(t + dt) = X(t) + (A(x(t))X(t) + Bu(t) + C)dt. \tag{20.19}$$

Parameters are set as displayed in Table 20.1: the step size $dt = 0.1$, the charging and discharging rates are $\alpha = \beta = 1$, the lowest and highest level of charge are $x_{on} = 1$ and $x_{off} = -1$, respectively; the penalty coefficients are $r_{on} = r_{off} = 20$ and $q = 10$, and the initial distribution is normal with zero-mean and standard deviation $std(m(0)) = 0.3$ for x and mean equal to 0.5 and standard deviation $std(m(0)) = 0.1$ for y.

Table 20.1 Simulation parameters.

α	β	x_{on}	x_{on}	r_{on}, r_{off}	q	$std(m_0(x,.))$	$\bar{m}_0(x,.)$	$std(m_0(y,.))$	$\bar{m}_0(y,.)$
1	1	1	-1	20	10	.3	0	.1	0.5

Table 20.2 Simulation algorithm.

Input: Set of parameters as in Table 20.1.
Output: EVs' states $X(t)$
  1 : **Initialize.** Generate $X(0)$ given $\bar{m}_0$ and $std(m_0)$
  2 : **for** time $iter = 0, 1, \ldots, T-1$ **do**
  3 :   **if** $iter > 0$, **then** compute m_t, $\bar{m}_t$, and $std(m_t)$
          if $iter = t_k$, obtain $e(t_k)$ and forecast $\hat{e}(t_{k+1})$,
          end if
  4 :   **end if**
  5 :   **for** player $i = 1, \ldots, n$ **do**
  6 :   Set $t = iter \cdot dt$ and
          compute control $\tilde{u}(t)$ using current forecast $\hat{e}(t_{k+1})$
  7 :   compute $X(t+dt)$ from (20.19)
  8 :   **end for**
  9 : **end for**
10 : **STOP**

The numerical results are obtained using the algorithm in Table 20.2 for a discretized set of states.

For the optimal control we get

$$u^* = -R^{-1}B^T[PX + \Psi],$$

where P is calculated using the MATLAB in-built function `[P]=care(A,B,Q,R)`. This function takes the matrices as input and returns the solution P to the algebraic Riccati equation. Assuming $BR^{-1}B^T\Psi \approx C$ we get the closed-loop dynamics

$$X(t+dt) = X(t) + [A - BR^{-1}B^TP]X(t)dt.$$

Figure 20.4 displays the time plot of the state of each EV, namely level of charge $x(t)$ (top row) and charging mode $y(t)$ (bottom row) when no noise affects the measure of $e(t_k)$, namely $\epsilon(t_k) = 0$. The dynamic is affected by some Gaussian noise in $\mathcal{N}(0,0.1)$. The simulation is carried out assuming that any 10 iterations a new sample is obtained and any 100 iterations initial states are re-initialized randomly. Figure 20.5 plots the time evolution of the Lyapunov function under additional Gaussian noise. Analogously Figure 20.6 displays the time plot of the state of each EV when Gaussian noise affects the measurement of the error. Figure 20.7 depicts the corresponding Lyapunov function. Finally, Figure 20.8 plots the observed and forecasted data.

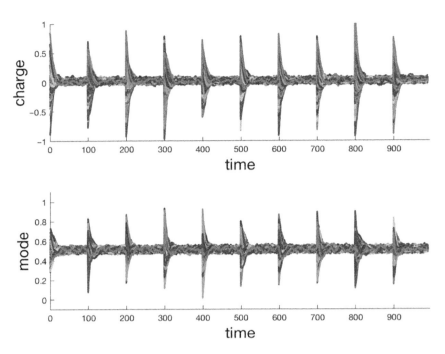

Figure 20.4 Time plot of the state of each EV, namely level of charge $x(t)$ (top row) and charging mode $y(t)$ (bottom row) when no noise affects the measure of $e(t_k)$, namely $\epsilon(t_k) = 0$. Some Gaussian noise in $\mathcal{N}(0, 0.1)$ is added to the dynamics.

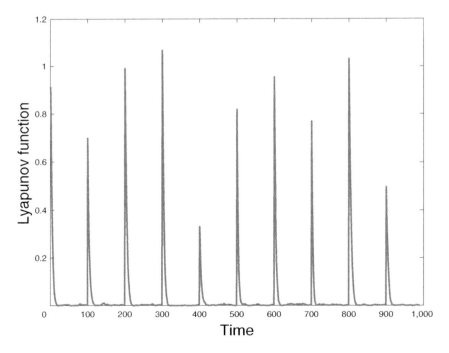

Figure 20.5 Time plot of the Lyapunov function. Some Gaussian noise in $\mathcal{N}(0, 0.1)$ is added to the dynamics.

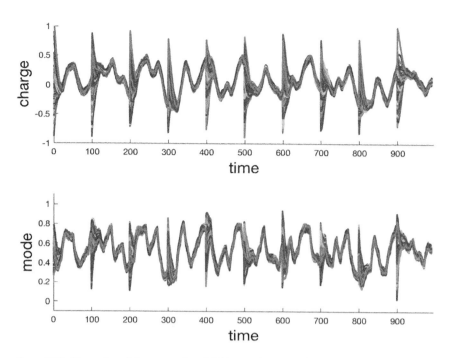

Figure 20.6 Time plot of the state of each EV, namely level of charge $x(t)$ (top row) and charging mode $y(t)$ (bottom row) when Gaussian noise affects the measure of $e(t_k)$, namely $\epsilon(t_k) \in \mathcal{N}(0, 0.5)$. Some Gaussian noise in $\mathcal{N}(0, 0.1)$ is added to the dynamics.

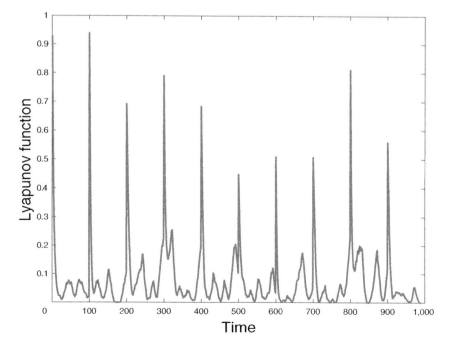

Figure 20.7 Time plot of the Lyapunov function when Gaussian noise affects the measure of $e(t_k)$, namely $\epsilon(t_k) \in \mathcal{N}(0, 0.5)$. Some Gaussian noise in $\mathcal{N}(0, 0.1)$ is added to the dynamics.

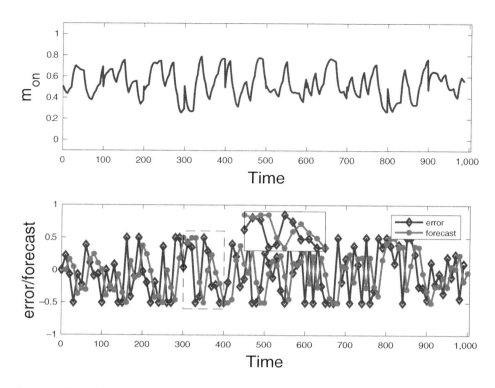

Figure 20.8 Time plot of the observed data $e(t_k)$ and forecasts $\hat{e}(t_{k+1})$.

20.5 Discussion

This chapter contributes to the broader area of dynamic response [5–9, 11, 14, 15], and in turn also to research programs in *smart cities*. Motivations derive from the additional challenges related to increasing size of the population of appliances, which implies a related increase in the computational burdens of centralized solutions.

Relevance of the presented results can be summarized as follows. First, we present a game-theoretic model of a distributed and large-scale system for which no centralized solution can be used due to the massive amount of information data that need to be processed. Our approach suggests to assign part of the regulation burden to the end users by using responsive EVs. Each EV adjusts its charging policy online based on the mismatch demand currently detected and made available to the appliances at specific sampling times. The provided model maximizes the intelligence of the grid by providing the EVs of strategic reasoning. The model is useful when we are dealing with a large number of indistinguishable EVs. By indistinguishable we mean that EVs have the same characteristics and are model using the same parameters. In case of different EVs, the model can be extended to multipopulation games but the approach presented here has still value.

The general understanding we get from the results is that it is possible to obtain solutions that approximate mean-field equilibrium solutions with a reasonably limited amount of calculations. At an equilibrium, each EV is using the best-response charging policy that optimizes, at the same time, multiple individual criteria (level of charge) and collective criteria (balance between supply and demand in the whole grid). The nature of the strategy that we consider is stochastic, in the sense that the EV charges or discharges with a given probability, and this probability is constantly adjusted. Such strategies take the form of closed-loop strategies on the current state, i.e., level of charge and current charging mode. The strategies are optimized over an infinite receding horizon every time new sample data are made available about the current state of the grid (balance between supply and demand). A well-known forecasting method, the Holt's model, is proposed to forecast the future mismatch between supply and demand. The mean-field equilibrium strategies represent the asymptotic limit of Nash equilibrium strategies, and as such they are the best-response strategies of each player, for fixed behavior of the other players. We show that the dynamics of each EV is stable, in terms of level of charge and switching mode. We have also shown that the strategies are robust as convergence is guaranteed also under the assumption that a Gaussian disturbances interferes with the EV's dynamics.

20.6 Conclusion

The model proposed in this chapter advances knowledge on large differential games and provides applications to a population of EVs. The model encorporates measurement disturbances. We have studied robust equilibria and designed stabilizing stochastic charging strategies.

In the future, we will extend our study to dynamic pricing, coalitional production aggregation, and the design of incentives to stabilize aggregation of producers.

Appendix

Proof of Theorem 20.1

Let us start by isolating the HJI part of (20.9). For fixed m_t and for $t \in [0, T]$, we have

$$\begin{cases} -\partial_t v(x,y,t) - \left\{ y\left[-\alpha(x - x_{on}) \right] \right. \\ \left. + (1-y)\left[-\beta(x - x_{off}) \right] \right\} \partial_x v(x,y,t) \\ + \sup_{u \in \mathbb{R}} \left\{ -Bu\, \partial_y v(x,y,t) - \frac{1}{2}qx^2 \right. \\ \left. + \frac{1}{2}u^T r u + y(Se + W) \right\} = 0 \\ \text{in } \mathcal{S} \times]0, T], \\ v(x,y,T) = g(x) \text{ in } \mathcal{S}, \\ u^*(x,t) = -r^{-1} B^T \partial_y v(x,y,t), \end{cases} \quad (20.20)$$

which in a more compact form can be rewritten as

$$\begin{cases} -\partial_t v(X,t) - \sup_u \Big\{ \partial_X v(X,t)^T (AX + Bu + C) \\ \quad + \frac{1}{2}\Big(X^T Q X + u^T R u^T\Big) + L^T X \Big\} = 0, \text{ in } \mathcal{S} \times [0, T[, \\ v(X,T) = g(X) \text{ in } \mathcal{S}, \\ u^*(x,t) = -r^{-1} B^T \partial_y v(X,t). \end{cases}$$

Let us consider the following value function

$$v(X,t) = \frac{1}{2} X^T P(t) X + \Psi(t)^T X + \chi(t),$$

and the corresponding optimal closed-loop state feedback strategy

$$u^* = -R^{-1} B^T [PX + \Psi].$$

Then (20.20) can be rewritten as

$$\begin{cases} \frac{1}{2} X^T \dot{P}(t) X + \dot{\Psi}(t) X + \dot{\chi}(t) \\ + (P(t)X + \Psi(t))^T \Big[-BR^{-1} B^T \Big] \\ \quad \cdot (P(t)X + \Psi(t)) \\ + (P(t)X + \Psi(t))^T (AX + C) \\ + \frac{1}{2}\Big(X(t)^T Q X(t) + u(t)^T R u(t)^T\Big) \\ + L^T X(t) = 0 \text{ in } \mathcal{S} \times [0, T[, \\ P(T) = \phi, \quad \Psi(T) = 0, \quad \chi(T) = 0. \end{cases} \quad (20.21)$$

The boundary conditions are obtained by imposing that

$$v(X,T) = \frac{1}{2} X^T P(T) X + \Psi(T) X + \chi(T) = \frac{1}{2} X^T \phi X =: g(X).$$

Since (20.21) is an identity in X, it reduces to three equations:

$$\begin{cases} \dot{P} + PA(x) + A(x)^T P - PBR^{-1} B^T P \\ \quad + Q = 0 \text{ in } [0, T[, \ P(T) = \phi, \\ \dot{\Psi} + A(x)^T \Psi + PC - PBR^{-1} B^T \Psi + L = 0 \\ \quad \text{in } [0, T[, \ \Psi(T) = 0, \\ \dot{\chi} + \Psi^T C - \frac{1}{2} \Psi^T BR^{-1} B^T \Psi = 0 \\ \quad \text{in } [0, T[, \ \chi(T) = 0. \end{cases} \quad (20.22)$$

To understand the influence of the congestion term on the value function, let us develop the expression for Ψ and obtain

$$\begin{bmatrix} \dot\Psi_1 \\ \dot\Psi_2 \end{bmatrix} + \begin{bmatrix} -\beta & 0 \\ k(x(t)) & 0 \end{bmatrix} \begin{bmatrix} \Psi_1 \\ \Psi_2 \end{bmatrix}$$
$$+ \begin{bmatrix} P_{11} & P_{12} \\ P_{21} & P_{22} \end{bmatrix} \begin{bmatrix} \beta x_{off} \\ 0 \end{bmatrix} \qquad (20.23)$$
$$- \begin{bmatrix} P_{12}(r_{on}^{-1} + r_{off}^{-1})\Psi_2 \\ P_{22}(r_{on}^{-1} + r_{off}^{-1})\Psi_2 \end{bmatrix} + \begin{bmatrix} 0 \\ Se + W \end{bmatrix}.$$

The expression of Ψ then can be rewritten as

$$\begin{cases} \dot\Psi_1 - \beta\Psi_1 + P_{11}\beta x_{off} - P_{12}(r_{on}^{-1} + r_{off}^{-1})\Psi_2 = 0, \\ \dot\Psi_2 + k(x(t))\Psi_1 + P_{21}\beta x_{off} - P_{22}(r_{on}^{-1} + r_{off}^{-1})\Psi_2 \\ + (Se + W) = 0, \end{cases} \qquad (20.24)$$

which is of the form

$$\begin{cases} \dot\Psi_1 + a\Psi_1 + b\Psi_2 + c = 0, \\ \dot\Psi_2 + a'\Psi_1 + b'\Psi_2 + c' = 0. \end{cases} \qquad (20.25)$$

From the preceding set of inequalities, we obtain the solution $\Psi(x(t), e(t), t)$. Note that the term a' depends on x and c' depends on $e(t)$.

Substituting the expression of the mean-field equilibrium strategies $u^* = -R^{-1}B^T[PX + \Psi]$ as in (20.15) in the open-loop microscopic dynamics $\dot X(t) = AX(t) + Bu(t) + C$ introduced in (20.12), and averaging both left-hand side and right-hand side we obtain the following closed-loop macroscopic dynamics

$$\dot{\bar X}(t) = [A(x) - BR^{-1}B^T P]\bar X(t) - BR^{-1}B^T \bar\Psi(t) + C,$$

where $\bar\Psi(t) = \int_{x_{on}}^{x_{off}} \int_{[0,1]} \Psi(x,e,t) m(x,y,t) dx dy$ and this concludes our proof.

Proof of Theorem 20.2

Let $X(t)$ be a solution of dynamics (20.16) with initial value $X(0) \notin \mathcal{X}$. Set $t = \{\inf t > 0| X(t) \in \mathcal{X}\} \le \infty$. For all $t \in [0,t]$

$$V(X(t+dt)) - V(X(t)) = \tfrac{1}{\|X(t)+dX(t)-\Pi_\mathcal{X}(X(t))\|} \|X(t) + dX(t) - \Pi_\mathcal{X}(X(t))\|^2$$
$$- \tfrac{1}{\|X(t)-\Pi_\mathcal{X}(X(t))\|} \|X(t) - \Pi_\mathcal{X}(X(t))\|^2.$$

Taking the limit of the difference above we obtain

$$\dot V(X(t)) = \lim_{dt \to 0} \tfrac{V(X(t+dt))-V(X(t))}{dt}$$
$$= \lim_{dt \to 0} \tfrac{1}{dt} \Big[\tfrac{1}{\|X(t)+dX(t)-\Pi_\mathcal{X}(X(t))\|} \|X(t)+dX(t)$$
$$-\Pi_\mathcal{X}(X(t))\|^2 - \tfrac{1}{\|X(t)-\Pi_\mathcal{X}(X(t))\|} \|X(t) - \Pi_\mathcal{X}(X(t))\|^2 \Big]$$
$$\le \tfrac{1}{\|X(t)-\Pi_\mathcal{X}(X(t))\|} \Big[\partial_X V(X,t)^T \Big([A - BR^{-1}B^T P]X(t)$$
$$- BR^{-1}B^T\Psi + C \Big) + \|X(t) - \Pi_\mathcal{X}(X(t))\|^2 \Big] < 0,$$

which implies $\dot V(X(t)) < 0$, for all $X(t) \notin \mathcal{X}$ and this concludes our proof.

References

[1] M. H. Albadi and E. F. El-Saadany, "Demand response in electricity markets: An overview," *IEEE*, 2007.

[2] J. H. Eto, J. Nelson-Hoffman, C. Torres, S. Hirth, B. Yinger, J. Kueck, B. Kirby, C. Bernier, R.Wright, A. Barat, and D. S.Watson, "Demand response spinning reserve demonstration," *Ernest Orlando Lawrence Berkeley Nat. Lab., Berkeley, CA, LBNL-62761*, 2007.

[3] C. Gellings and J. Chamberlin, *Demand-Side Management: Concepts and Methods*. Lilburn, GA: The Fairmont Press, 1988.

[4] "US Department of Energy, benefits of demand response in electricity markets and recommendations for achieving them," *Report to the United States Congress*, February 2006, http://eetd.lbl.gov.

[5] D. Angeli and P.-A. Kountouriotis, "A stochastic approach to dynamic-demand refrigerator control," *IEEE Transactions on Control Systems Technology*, vol. 20, no. 3, pp. 581–592, 2012.

[6] Z. Ma, D. S. Callaway, and I. A. Hiskens, "Decentralized charging control of large populations of plug-in electric vehicles," *IEEE Transactions on Control System Technology*, vol. 21, no. 1, pp. 67–78, 2013.

[7] J. L. Mathieu, S. Koch, and D. S. Callaway, "State estimation and control of electric loads to manage real-time energy imbalance," *IEEE Transactions on Power Systems*, vol. 28, no. 1, pp. 430–440, 2013.

[8] S. E. Z. Soudjani, S. Gerwinn, C. Ellen, M. Fraenzle, and A. Abate, "Formal synthesis and validation of inhomogeneous thermostatically controlled loads," *Quantitative Evaluation of Systems*, pp. 74–89, 2014.

[9] F. Bagagiolo and D. Bauso, "Mean-field games and dynamic demand management in power grids," *Dynamic Games and Applications*, vol. 4, no. 2, pp. 155–176, 2014.

[10] D. Bauso, "Dynamic demand and mean-field games," *IEEE Transactions on Automatic Controls*, vol. 62, no. 12, pp. 6310–6323, 2017.

[11] M. Roozbehani, M. A. Dahleh, and S. K. Mitter, "Volatility of power grids under real-time pricing," *IEEE Transactions on Power Systems*, vol. 27, no. 4, 2012.

[12] F. Bagagiolo and D. Bauso, "Objective function design for robust optimality of linear control under state-constraints and uncertainty," *ESAIM: Control, Optimisation and Calculus of Variations*, vol. 17, pp. 155–177, 2011.

[13] K. A. Loparo and X. Feng, "Stability of stochastic systems," *The Control Handbook*, pp. 1105–1126, Boca Raton, FL: CRC Press, 1996.

[14] D. S. Callaway and I. A. Hiskens, "Achieving controllability of electric loads," *Proceedings of the IEEE*, vol. 99, no. 1, pp. 184–199, 2011.

[15] R. Couillet, S. Perlaza, H. Tembine, and M. Debbah, "Electrical vehicles in the smart grid: A mean field game analysis," *IEEE Journal on Selected Areas in Communications*, vol. 30, no. 6, pp. 1086–1096, 2012.

[16] T. Başar and G. J. Olsder, *Dynamic Noncooperative Game Theory*. SIAM Series in Classics in Applied Mathematics. Philadelphia: SIAM, 1999.

[17] J.-M. Lasry and P.-L. Lions, "Mean field games," *Japanese Journal of Mathematics*, vol. 2, pp. 229–260, 2007.

[18] M. Huang, P. Caines, and R. Malhamé, "Large population stochastic dynamic games: Closed loop Kean–Vlasov systems and the Nash certainty equivalence principle," *Communications in Information and Systems*, vol. 6, no. 3, pp. 221–252, 2006.

[19] ——, "Large population cost-coupled LQG problems with non-uniform agents: Individual-mass behaviour and decentralized ϵ-Nash equilibria," *IEEE Transactions on Automatic Control*, vol. 52, no. 9, pp. 1560–1571, 2007.

[20] M. Bardi, "Explicit solutions of some linear-quadratic mean field games," *Network and Heterogeneous Media*, vol. 7, pp. 243–261, 2012.

[21] D. Gomes and J. Saúde, "Mean field games models – a brief survey," *Dynamic Games and Applications*, vol. 4, no. 2, pp. 110–154, 2014.

[22] M. Sassano and A. Astolfi, "Dynamic approximate solutions of the HJ inequality and of the HJB equation for input-affine nonlinear systems," *IEEE Transactions on Automatic Control*, vol. 57, no. 10, pp. 2490–2503, 2012.

[23] ——, "Approximate finite-horizon optimal control without PDEs," *Systems & Control Letters*, vol. 62, pp. 97–103, 2013.

[24] D. Bauso, T. Mylvaganam, and A. Astolfi, "Crowd-averse robust mean-field games: approximation via state space extension," *IEEE Transaction on Automatic Control*, vol. 61, no. 7, pp. 1882–1894, 2016.

[25] D. Bauso, H. Tembine, and T. Başar, "Robust mean field games," *Dynamic Games and Applications*, vol. 6, no. 3, pp. 277–303, 2016.

[26] H. Tembine, Q. Zhu, and T. Başar, "Risk-sensitive mean-field games," *IEEE Transactions on Automatic Control*, vol. 59, no. 4, pp. 835–850, 2014.

[27] O. Gueant, J. M. Lasry, and P. L. Lions, "Mean-field games and applications," *Paris-Princeton Lectures*, Springer, pp. 1–66, 2010.

[28] A. C. Kizilkale and R. P. Malhame, "Mean field based control of power system dispersed energy storage devices for peak load relief," in *IEEE Conference on Decision and Control*, pp. 4971–4976, 2013.

[29] ——, "Collective target tracking mean field control for Markovian jump-driven models of electric water heating loads," in *Proc. of the 19th IFAC World Congress, Cape Town, South Africa*, pp. 1867–1872, 2014.

[30] M. Aziz and P. E. Caines, "A mean field game computational methodology for decentralized cellular network optimization," *IEEE Transactions on Control Systems Technology*, vol. 25, no. 2, pp. 563–576, 2017.

21 Prosumer Behavior
Decision Making with Bounded Horizon

Mohsen Rajabpour, Arnold Glass, Robert Mulligan, and Narayan B. Mandayam

21.1 Introduction

Most studies of prosumer decision making in the smart grid have focused on single, temporally discrete decisions within the framework of expected utility theory (EUT) and behavioral theories such as prospect theory (PT). In this work, we study prosumer decision making in a more natural, ongoing market situation in which a prosumer has to decide every day whether to sell any surplus energy units generated by the solar panels on her roof or hold (store) the energy units in anticipation of a future sale at a better price. Within this context, we propose a new behavioral model that extends EUT to take into account the notion of a bounded temporal horizon over which various decision parameters are considered. Specifically, we introduce the notion of a bounded time window (the number of upcoming days over which a prosumer evaluates the probability that each possible price will be the highest) that prosumers implicitly impose on their decision making in arriving at "hold" or "sell" decisions. The new behavioral model assumes that humans make decisions that will affect their lives within a bounded time window regardless of how far into the future their units may be sold. Modeling the utility of the prosumer using parameters such as the offered price on a day, the number of energy units the prosumer has available for sale on a day, and the probabilities of the forecast prices, we fit both traditional EUT and the proposed behavioral model with bounded time windows to data collected from 57 homeowners over 68 days in a simulated energy market. Each prosumer generated surplus units of solar power and had the opportunity to sell those units to the local utility at the price set that day by the utility or hold the units for sale in the future. For most participants, a bounded horizon in the range of 4–5 days provided a much better fit to their responses than was found for the traditional (unbounded) EUT model, thus validating the need to model bounded horizons imposed in prosumer decision making.

The smart grid is a twenty-first-century evolution of the traditional electric power grid resulting from the convergence of several factors including growing demand for electricity and a variety of technological developments, environmental imperatives, and security concerns. Unprecedented changes over the past few years in the smart grid have enabled power utility companies to construct a two-way communications infrastructure to monitor and manage devices across the power system. Smart meters that communicate usage information directly to the power utility and smart appliances that allow consumers to easily schedule their usage at times of low demand are two such elements.

Some utility companies have already implemented incentive programs to influence the timing of consumers' usage to better balance demand. These and other elements of the smart grid infrastructure such as renewable energy generation and storage technologies (e.g., solar panels, wind towers, batteries) have been developed and are being deployed across the United States and internationally.

Two other important characteristics of the smart grid are more distributed power generation capability (versus the traditional centralized power station model) and a larger role for renewable energy resources like wind and solar (and an eventual diminishing reliance on traditional fossil fuel–fired power plants). These two characteristics merge in the proliferation of customer-owned rooftop solar panels (and less frequently, wind turbines). Together with a smart metering technology called "net metering," which allows the flow of electricity in both directions across the meter (from power grid to home and from home to power grid), a new widely distributed power generation infrastructure is emerging in which an individual, so equipped, can be both a consumer and a producer of electric power. These individuals have been labeled "prosumers." They not only consume energy from the power utility but they can also produce energy to power their own homes and can often generate surplus electricity that they can feed back into the grid for an energy credit or sell to the utility at the market price. When also equipped with a storage battery, and multiplied by thousands or tens of thousands, these prosumers become major players in the energy market and a significant consideration in the design and management of the smart grid.

How prosumers behave in this emerging energy management scenario – their perceptions, motivations and decisions – will influence the success of the smart grid [1–4]. For example, how prosumers make decisions about whether and when to sell the surplus energy generated by their solar panels or wind turbines will inform the design of optimal energy management systems. Prosumers can optimize their profits by deciding on the amount of energy to buy or to sell [1–3]. Thus, important technical and economic challenges emerge as prosumers influence demand-side management.

In response to this realization, a substantial literature has emerged with the objective of modeling the behaviors of electricity prosumers in smart grid settings. One early example called out the need to consider the behavior of large numbers of small scale prosumers in smart grid planning efforts [5]. The authors outlined a framework for modeling this behavior, but did not develop such a model. Several other researchers have taken a more quantitative approach, using game-theory to analyze demand-side management and to model prosumer decision behavior in smart grid scenarios. A limitation of these papers is that they consider the prosumers' behavior to be consistent with EUT [6–9]. EUT provides a framework for optimizing the amount of money earned within the narrow context-free constraints of that theory. However, EUT has often been found to not accurately account for how people respond in real-life situations.

To more accurately describe prosumer behavior in realistic smart grid scenarios, a few recent studies have utilized PT for analyzing demand-side management in the smart grid. PT is a descriptive model of prosumer behavior that provides an empirical and mathematical framework that represents subjective utility functions instead of the objective utility functions of EUT [10, 11]. These more recent investigations [1–4], [12–14] have achieved some success in modeling prosumer behavior by assuming that their decision making follows the percepts of PT. To our knowledge however, no studies have observed the decision behavior of actual prosumers in the field, or that of subjects emulating prosumer behavior in a laboratory simulation of a smart grid scenario.

The approach taken here is complementary to the insights provided by PT and extends the previous works in two regards: (1) from a methodological perspective, the current study measures the behaviors of human participants (prosumers) in a realistic smart grid scenario, and (2) from a modeling perspective, it considers the effect of the subjective time frame within which a prosumer operates when making decisions about whether and when to sell their surplus energy in an ongoing market.

Analyses of decision making under uncertainty have typically investigated scenarios in which a single decision is required among two or more alternatives at a fixed point in time. In the smart grid scenario considered here, however, prosumers are engaged in an energy market in which they must make a series of nonindependent decisions over an extended period. For example, if they are generating and storing surplus electric power, and the price at which they can sell that power to the utility company can vary significantly over time due to several factors, they might have to decide on a weekly, daily, or even on an hourly basis, when to sell to maximize their profit.

Within the framework of EUT, if someone has something to sell, there is no parameter corresponding to a finite horizon within which a sale must be made. Furthermore, there is no concept of a time window within that horizon over which a seller chooses to compute the utility of a sale. Consequently, the concept of a time window has only rarely been mentioned or tested. One notable exception is when the seller is selling their own labor. The concept of a time horizon has been discussed by economists in the context of research on optimal job search strategies, looking in particular at the question of whether a job seeker should accept a job at an offered salary or wait for a job at a higher salary. One laboratory study [15, 16] imposed a finite horizon on subjects search decisions, but unlike our 10-week study, sessions were only an hour in length. Also, unlike our realistic smart grid simulation, the authors used an abstract economic decision task rather than a realistic job search scenario.

Another area of research in decision making that shares some characteristics with the prosumer scenario is investing. Like prosumers, individual investors have an opportunity to make a series of decisions over an extended time horizon and to selling none, some, or all their resources. Looking at investment decisions under risk, one topic that has received considerable study has been the so-called disposition effect – the tendency of investors to sell too soon stocks that have increased in value relative to their purchase price and to hold on too long to those that have declined in value [17]. A similar effect is seen in buying and selling real estate [18].

Another investing phenomenon in which a decision horizon can come into play is what Benartzi and Thaler have dubbed myopic loss aversion (MLA) [19]. These authors proposed MLA, a combination of loss aversion and another concept known as mental accounting [20, 21], as an explanation for the "equity premium puzzle" – the fact that in investment markets, equities are priced at a much higher premium (relative to bonds and other more stable investments) than would be predicted by standard economic models.

To the extent that this myopia is temporal in nature – i.e., due to adopting a short-term horizon when a longer-term horizon would be more appropriate, this work could prove helpful in explaining prosumer behavior in an ongoing energy market.

In the present study, we examine prosumer behavior extending over 10 weeks, making it the first meaningful test of the role of time windows in the behavior of sellers. The results reveal that an elaborated EUT model that computes the probability of selling at a higher price within a fixed time window predicts human decision making in the context of energy management better than conventional EUT, providing crucial understanding of prosumer decision making in the smart grid. The elaborated EUT model provides a framework and mathematical method to model how people make decisions within a bounded time window that extends classic EUT, which assumes an unbounded time window.

The rest of this chapter is organized as follows. Section 21.2 describes the study methodology. Section 21.3 presents the mathematical representation of EUT and EUT with a time window, and a method for determining the minimum price at which a prosumer should sell. In Section 21.4 we present the data fitting results and draw some conclusions. Finally, Section 21.5 provides some discussion of the implications of these findings.

21.2 Experimental Design: Simulation of an Energy Market

The decision behavior of 57 household decision makers was studied in a simulation of a 10-week, collapsing horizon, smart energy grid scenario. Homeowners were recruited through an online research participant recruiting service (www.findparticipants.com). Based upon demographic data gathered by the recruiting service and on participants' responses to a prestudy screening survey, 60 participants were recruited, all of whom met the following criteria:

- Homeowner;
- Decision maker in the household regarding energy utility services;
- PC, tablet, or smartphone user with access to the Internet; and
- Committed to participating in the study by playing a short "energy grid game" every day (except holidays) for 10 weeks.

Three subjects dropped out of the study in the first week. The remaining 57 all completed the 10-week study.

Recruitment and Compensation
Subjects were compensated for study participation in a manner that incentivized them to attend carefully and perform well. All subjects earned a fixed amount ($100) for completing the study provided they responded on at least 65 of the days. In addition to this fixed amount, subjects kept the "profit" they earned by selling their surplus energy during the course of the game. Note that all subjects accumulated the same amount of surplus energy on the same schedule over the 70 days. However, the amount of profit

earned could range from about $40 to $120 depending upon how well they played the game, i.e., how adept they were at selling when energy prices were highest. Finally, to further incentivize subjects to attend carefully to their task, the top three performers were awarded bonuses of $25, $50, and $100.

The "Grid Game" Scenario

The framework for the study was a simulated smart grid scenario in which participants acted as prosumers. They were asked to imagine that they had solar panels installed on their property, a battery system to store surplus energy generated by their solar panels, and a means of selling their surplus energy back to the power company for a profit.

In the scenario, the subjects' solar panels generated a small amount of surplus electricity (i.e., over and above household power requirements) on most days of the study. For purposes of this simulation, we avoided references to megawatt or kilowatt hours and simply referred to the units of electricity as units. In an effort to simulate variability introduced by factors like weather and household demand for electricity, the number of surplus units generated on each day was sampled from a distribution in which one unit of surplus power was generated on 50 % of the 70 days; two units on 35 % of the days, and zero units on the remaining 15 % of days.

Surplus energy was stored in the subjects' hypothetical storage batteries. A total of 89 units were generated over the course of the 10-week study. Each day, subjects were given the opportunity to sell some or all the surplus energy they had accumulated in their storage system. The subjects' task was to maximize their profits by selling their surplus electricity when prices were highest.

Energy Price Data

The price paid by the fictional power company for prosumer-generated electricity varied from day to day in a manner that loosely mimicked the variation in wholesale electricity prices. The prices were designed to approximate the range and variability in the wholesale energy market in the United States. The data source for modeling prices was monthly reports of wholesale energy prices published by the U.S. Energy Industry Association (e.g., www.eia.gov/electricity/monthly/update/wholesale_markets.cfm) referencing data from March and April 2016.

A distribution of daily price per unit was created, comprising 15 discrete prices ranging from $0.10 to $1.50 in 10-cent increments. The shape of the distribution was positively skewed, emulating real-world prices, with a mean of $0.60 and a modal price of $0.50. The probability of each price occurring on any given day of the study is shown in Table 21.1. We refer to this set of 15 probabilities as the single-day probabilities (p_i) for the 15 possible prices. The daily price offered to participants for their surplus energy was determined by sampling 70 values from this distribution. After the initial set of 70 daily prices was generated, an offset was applied to weekday (Mon–Fri) and weekend (Sat–Sun) prices such that the average weekday prices ($0.70) were 40 % higher than average weekend prices ($0.41). This offset did not change the characteristics of the overall distribution of prices.

Table 21.1 Single-day probabilities of each of the 15 possible wholesale electricity prices.

Price	0.1	0.2	0.3	0.4	0.5	0.6	0.7	0.8	0.9	1.0	1.1	1.2	1.3	1.4	1.5
p_i	0.03	0.06	0.09	0.12	0.14	0.11	0.09	0.08	0.07	0.06	0.05	0.04	0.03	0.02	0.01

Instructions to Participants

At the beginning of the study, participants were informed about the amount of energy they would generate and the distribution of prices they could expect. They received the following instructions:

> *Here is some important information to help you to maximize your profits from selling your units:*
>
> 1. For the 10-week period of our study, the price per unit paid by our imaginary power company can range from $0.10 (10 cents/unit) to $1.50/unit, in 10-cent increments. So, there are 15 possible prices – $0.10, $0.20, $0.30, etc., up to $1.50 per unit.
> 2. The average price across the 10 weeks will be $0.60/unit (60 cents per unit).
> 3. On average, the price paid on weekdays (Mon–Fri) tends to be somewhat higher than on weekends because demand is greater during the work week.
> 4. Over the 70 days, your system will generate a total of 90 units of surplus electric power. For purposes of our game, you can assume that your battery system can easily store this much power and there is no appreciable decay over time.
>
> Here is how the game will work. Every day before noon Eastern Time you will get a Grid Game e-mail providing you with the following information:
>
> - The number of units of surplus energy generated by your solar panels yesterday. This will always be 0, 1, or 2 units.
> - The total number of units stored in your storage battery. You'll start the game with 5 units.
> - The amount of profit you have earned so far by selling your surplus energy.
> - The price the power company is paying for any units you sell today.
> - A table showing the probability that each of the 15 prices will occur at least once between today and the end of the study. For example, you will see:
>   - $1.50 per Unit – 30% chance
>   - $1.40 per Unit – 45% chance
>   - $1.30 per Unit – 78% chance
>   - ...

- $0.20 per Unit – 59% chance
- $0.10 per Unit – 35% chance

This table of probabilities is also provided to help you maximize your profits. The probabilities are based on actual variability in energy prices in the real energy grid. *They will change gradually as we move through the study, so be sure to check them every day.*

Your task each day will be to respond to the e-mail by answering the following question: Do you want to sell any units today and, if so, how many. In your brief e-mail reply, you will simply answer "Yes" or "No" and, if "Yes" then provide a number of units you want to sell. This number can range from 1 to the total number stored. Be sure to take note of the total units you have stored so that you don't try to sell more than you have. *You must respond no later than 8:00 p.m. Eastern Time every day.*

Before deriving the models, it should first be noted that daily energy prices, which were presented to participants as ranging from $0.10 to $1.50 per unit, were transformed to the values 1 through 15 for model development and data analysis.

21.3 Modeling Approaches

21.3.1 Modeling Prosumer Behavior Using Expected Utility Theory

Prosumer participants in our power grid scenario have to make a new decision every day regarding whether to sell some, all, or none of their surplus energy units to a central utility at the price offered by the utility on that day. To make this decision they must assess the probability that they would receive a higher payoff in the future for those units if they choose not to sell them today.

To model this situation, we assume a range of possible prices that may be offered to the prosumer each day. Further, we assume that the prosumer is provided with a forecast of these prices, i.e., the probabilities of each of these prices occurring on a single day (i.e., the "single day prices" given in Table 21.1). On each day d, a participant has one or more units of energy to sell. The price on offer that day for a unit of energy is denoted by i_d. Let i_d^* denote the unit cutoff price, which is the lowest price on day d for which the gain from a sale on that day is greater than the expected gain from a hold on that day.

The participant decides whether to sell or hold some or all the units available for sale each day. Expected utilities are computed for sell and hold decisions, the components of which are the gains realized from selling and the potential (future) gains from not selling. Hence, the equation for computing the expected utility of selling a unit of energy on day d at price i_d has two components. The first component is the gain realized by selling the unit at price i_d on day d. This is simply i_d. The second component is the possible gain that may be realized by holding the unit on day d and selling on a subsequent day. The possible gain from holding is determined by the probability of selling at a higher price on a subsequent day. The probability of selling at a higher price on a subsequent day is determined by the cutoff price on each subsequent day and the probability of it being met or exceeded by the price on offer on that day. So, the expected utility of selling n units on day d is defined as,

$$\mathbb{E}[U_d(n)] = n i_d + (N_d - n) \left(\sum_{j=i^*_{d+1}}^{I} p_j j + f_{h,d+1} \sum_{j=1}^{i^*_{d+1}-1} p_j \right), \quad (21.1)$$

where N_d is the number of units available for sale on day d, i_d is the offered price on day d, i^*_{d+1} is the cutoff price on day $d+1$, I is the highest price ever offered, p_j is the probability that j would be the unit price on offer on any day, and $f_{h,d+1}$ is the expected gain for a hold on day $d+1$ and is defined recursively as:

$$f_{h,d} = \begin{cases} \sum_{j=i^*_{d+1}}^{I} p_j j + f_{h,d+1} \sum_{j=1}^{i^*_{d+1}-1} p_j & d = 1, \ldots, D-2 \\ \sum_{j=i^*_D}^{I} p_j j & d = D-1 \\ 0 & d = D, \end{cases} \quad (21.2)$$

where D is the last day, hence the highest numbered day in the study. If EUT is considered as a model of human decision making, then the best strategy of selling can be stated as,

$$n^*_{d_{EUT}} = \arg \max_n \mathbb{E}[U_d(n)], \quad (21.3)$$

where $n^*_{d_{EUT}}$ is the value of n that maximizes $\mathbb{E}[U_d(n)]$.

THEOREM 21.1 *EUT model always predicts either selling all available units or nothing. In other words, $n^*_{d_{EUT}}$ is either equal to N_d or zero.*

Proof If we rearrange $\mathbb{E}[U_d(n)]$ as,

$$\mathbb{E}[U_d(n)] = n \left(i_d - \left(\sum_{j=i^*_{d+1}}^{I} p_j j + f_{h,d+1} \sum_{j=1}^{i^*_{d+1}-1} p_j \right) \right)$$
$$+ N_d \left(\sum_{j=i^*_{d+1}}^{I} p_j j + f_{h,d+1} \sum_{j=1}^{i^*_{d+1}-1} p_j \right), \quad (21.4)$$

and

$$s_d := i_d - \left(\sum_{j=i^*_{d+1}}^{I} p_j j + f_{h,d+1} \sum_{j=1}^{i^*_{d+1}-1} p_j \right), \quad (21.5)$$

$$s_{d_1} := \sum_{j=i^*_{d+1}}^{I} p_j j + f_{h,d+1} \sum_{j=1}^{i^*_{d+1}-1} p_j. \qquad (21.6)$$

The values of i^*_d and p_j are always positive. Hence, s_{d_1} is always positive for any values of i^*_d and p_j. Therefore,

$$\mathbb{E}[U_d(n)] = n\left(i_d - s_{d_1}\right) + N_d\left(s_{d_1}\right). \qquad (21.7)$$

As N_d is always positive, the second term in (21.7) is always positive. If $i_d < s_{d_1}$, the first term in (21.7) becomes negative. Hence, to maximize (21.7) the first term should be minimized. As,

$$n = 0, 1, \ldots, N_d, \qquad (21.8)$$

$n = 0$ minimizes the first term or equivalently maximizes (21.7). In addition, if $i_d > s_{d_1}$, both terms in (21.7) are positive. Hence, to maximize (21.7), the first term should be maximized. Considering (21.8), $n = N_d$ maximizes (21.7). It follows that,

$$n^*_{d_{EUT}} = \begin{cases} N_d & s_d > 0 \\ 0 & s_d < 0. \end{cases} \qquad (21.9)$$

$\square$

COROLLARY 21.1 *The sell all or nothing result is independent of cutoff prices, i^*_d.*

COROLLARY 21.2 *The sell all or nothing result is independent of probabilities of prices, p_i.*

While the decision resulting from the EUT-based model is independent of prices, probabilities, and the cutoff price, numerical evaluation of the expected utility function defined in (21.1) requires all the previously mentioned parameters. While the prices and their probabilities are assumed to be known in the model, we next show how the cutoff prices involved in the model can be numerically derived.

Evaluating the Cutoff Price

On day D, the last day of the study, the probability of selling any units at any price on a subsequent day is zero, hence the expected gain from holding on day D is zero. Therefore, regardless of the price on offer on day D, including the lowest possible price, which is 1, a sell decision will be made for all available units on day D, i.e.,

$$i^*_D = 1. \qquad (21.10)$$

The expected utility of selling n units at price i_{D-1} can be obtained from (21.1), (21.2), and (21.10). As (21.2) states, when $d = D - 1$, $f_{h,D} = 0$. As (21.10) states, $i^*_D = 1$. When these values are inserted in (21.1), it readily follows that (21.1) reduces to:

$$\mathbb{E}[U_d(n)] = n i_d + (N_d - n) \left(\sum_{j=i^*_{d+1}}^{I} p_j j \right), \qquad (21.11)$$

and when $N_d = 1$, it reduces to,

$$E[U_d(n)] = \begin{cases} i_d & n = 1 \quad (21.12a) \\ \sum_{j=i_{d+1}^*}^{I} p_j j & n = 0. \quad (21.12b) \end{cases}$$

We set i_d equal to ascending values beginning with $i_D^* = 1$, until we find the lowest value of i_d for which $i_d > \sum_{j=i_{d+1}^*}^{I} p_j j$. We find i_{D-1}^* by,

$$i_{D-1}^* = \lceil i_d \rceil, \quad (21.13)$$

where $\lceil . \rceil$ denotes ceiling function. For $d < D - 1$, the expected gain from holding for a single unit can be given by the recursive Equation (21.2). When $N_d = 1$, the expected utility reduces to,

$$E[U_d(n)] = \begin{cases} i_d & n = 1 \quad (21.14a) \\ \sum_{j=i_{d+1}^*}^{I} p_j j + \left(\sum_{j=1}^{i_{d+1}^*-1} p_j \right) f_{h,d+1} & n = 0, \quad (21.14b) \end{cases}$$

where (21.14a) is the gain for sell when $n = 1$ and (21.14b) is the expected gain for hold when $n = 0$. To compute i_d^*, we find the smallest value of i_d such that,

$$i_d > \sum_{j=i_{d+1}^*}^{I} p_j j + \left(\sum_{j=1}^{i_{d+1}^*-1} p_j \right) f_{h,d+1}, \quad (21.15)$$

i.e., i_d^* is

$$i_d^* = \lceil i_d \rceil, \quad (21.16)$$

for which (21.15) is satisfied. To use (21.14b) to compute the value of i_d^* we need the values of i_{d+1}^* through i_D^*. Hence, to compute the value of each i_d^* we begin with i_D^* then compute the value of the cutoff price for each successively earlier day in the study.

21.3.2 Bounded Horizon Model of Prosumer Behavior

In the bounded horizon model (denoted as EUT$_{TW}$), a prosumer may compute the probability of selling at a higher price over a fixed number of days called a time window until near the end of the study when the remaining days are fewer than the number of days in the prosumer's time window. Let $\tilde{d} = D - d$, be the number of remaining days. So i_w^* is the cutoff price that is the lowest price for which the gain from a sale is greater than the expected gain from a hold for a time window of w days, where $w = \min(\tilde{d}, t)$, and t, the time window, is the number of days over which expected gain for a hold is computed by a participant. The expected utility of selling n units on day d with a time window of w is defined as,

$$E[U_{d,w}(n)] = ni_d + (N_d - n) \left(\sum_{j=i_{w-1}^*}^{I} p_j j + f_{h,w-1} \sum_{j=1}^{i_{w-1}^*-1} p_j \right), \qquad (21.17)$$

where $f_{h,w}$ is the expected gain for hold on day d with a time window of w and is defined as,

$$f_{h,w} = \begin{cases} \sum_{j=i_{w-1}^*}^{I} p_j j + f_{h,w-1} \sum_{j=1}^{i_{w-1}^*-1} p_j & w > 1 \\ \sum_{j=i_{w-1}^*}^{I} p_j j & w = 1 \\ 0 & w = 0. \end{cases} \qquad (21.18)$$

If EUT_{TW} is considered as a model of human decision making, then the best strategy of selling can be stated as,

$$n_{d,w_{EUT}}^* = \arg\max_n E[U_{d,w}(n)]. \qquad (21.19)$$

THEOREM 21.2 *EUT model with a time window always predicts either selling all available units or nothing. In other words, $n_{d,w_{EUT}}^*$ is either equal to N_d or zero.*

Proof If we rearrange $E[U_{d,w}(n)]$ as,

$$E[U_{d,w}(n)] = n \left(i_d - \left(\sum_{j=i_{w-1}^*}^{I} p_j j + f_{h,w-1} \sum_{j=1}^{i_{w-1}^*-1} p_j \right) \right)$$
$$+ N_d \left(\sum_{j=i_{w-1}^*}^{I} p_j j + f_{h,w-1} \sum_{j=1}^{i_{w-1}^*-1} p_j \right), \qquad (21.20)$$

$$s_w := i_d - \left(\sum_{j=i_{w-1}^*}^{I} p_j j + f_{h,w-1} \sum_{j=1}^{i_{w-1}^*-1} p_j \right), \qquad (21.21)$$

and

$$s_{w_1} := \sum_{j=i_{w-1}^*}^{I} p_j j + f_{h,w-1} \sum_{j=1}^{i_{w-1}^*-1} p_j. \qquad (21.22)$$

As i_w^* and p_j are always positive, for any values of i_w^* and p_j, s_{w_1} is always positive. Therefore,

$$E[U_{d,w}(n)] = n(i_d - s_{w_1}) + N_d(s_{w_1}). \qquad (21.23)$$

As N_d is always positive, the second term is always positive. If $i_d < s_{w_1}$, the first term in (21.23) is negative. Therefore, to maximize (21.23), the first term should be minimized. Considering (21.8), $n = 0$ makes the first term minimized or equivalently makes (21.23) maximized. In addition, If $i_d > s_{w_1}$, both terms in (21.23) are positive. Hence, to maximize (21.23), the first term should be maximized. It is clear that $n = N_d$ makes the first term maximized or equivalently makes (21.23) maximized. It follows that,

$$n^*_{d, w_{EUT}} = \begin{cases} N_d & s_w > 0 \\ 0 & s_w < 0. \end{cases} \qquad (21.24)$$

□

COROLLARY 21.3 *The sell all or nothing result is independent of cutoff prices, i^*_w.*

COROLLARY 21.4 *The sell all or nothing result is independent of probabilities of prices, p_i.*

Evaluating the Cutoff Prices

Following the approach outlined in the previous section, the cutoff prices in the EUT$_{TW}$ model can be obtained as follows. When Equations (21.14a) and (21.14b) are used to compute i^*_d, this is also the cutoff price for $t = D - d$. The $D - d$ days remaining in the study are a time window of $D - d$ days. Hence, when $\tilde{d} > t$, t is used. However, when $\tilde{d} < t$ it makes no sense to use t because the probability of a sale beyond $\tilde{d}$ days is zero. When $\tilde{d} \leq t$ the EUT$_{TW}$ model becomes the EUT model. So $\tilde{d}$ is used to compute the cutoff price. So $w = \min(\tilde{d}, t)$ and is used instead of t. Therefore, when $N_d = 1$ the expected utility reduces to:

$$\mathbb{E}[U_{d,w}(n)] = \begin{cases} i_d & n = 1 \quad (21.25a) \\ \sum_{j=i^*_{w-1}}^{I} p_j j + \left(\sum_{j=1}^{i^*_{w-1}-1} p_j \right) f_{h, w-1} & n = 0, \quad (21.25b) \end{cases}$$

where (21.25a) is the gain from a sale ($n = 1$) and (21.25b) is the expected gain from a hold ($n = 0$). Notice that (21.25a) and (21.25b) are identical to (21.14a) and (21.14b) except for the substitution of w, which is decremented, for d, which is incremented. Hence, when $w = \tilde{d}$ then (21.25a) and (21.25b) are identical to (21.14a) and (21.14b) except for notation. $\tilde{d} = D - d$ and so i^*_d and $i^*_{w=\tilde{d}}$ are different ways of denoting the cutoff value for the same day. However, when $w = t$ then (21.25a) and (21.25b) computes the cutoff price for a smaller number of days than is used by (21.14a) and (21.14b). When $w = 0$, whatever units are available for sale will be sold at whatever price is offered.

THEOREM 21.3 *Frequency of selling in EUT with windowing is greater than EUT.*

Proof $f_{h,w}$ in (21.18) is an increasing function of w; that is, the gain from holding increases as window size increases. Reconsidering (21.23), the second term is always positive, and because s_{w_1} is an increasing function of w, therefore, the first term is a decreasing function of w. Consequently, for a constant N_d and a specific i_d, when time window is being increased, the first term becomes smaller. Thus, it reduces the frequency of selling all or it increases the frequency of selling nothing. □

Table 21.2 Number of remaining days in the study for which each cutoff price applies.

Cutoff price, i_d^*	Days remaining, $\tilde{d}$
14	≥31
13	16–30
12	9–15
11	6–8
10	4–5
9	3
8	2
7	1
1	0

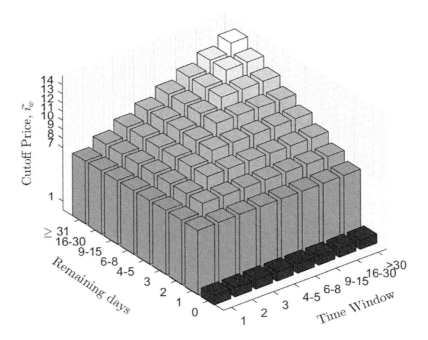

Figure 21.1 Initial cutoff price as a function of time window (TW) and number of days remaining in simulation.

21.4 Data Fitting, Results, and Analysis

Table 21.2 shows the cutoff price as a function of the number of days remaining in the study, computed from Equations (21.15) and (21.16). As shown in the table, the cutoff price necessarily declines as the study approaches its end and the number of days on which a higher price might be offered is reduced. Figure 21.1 shows the cutoff price as a function of both time window and number of days remaining in the simulation.

Note that while the study took place over a 10-week period, no data was collected on two of those days (one holiday and one day on which the data was corrupted). Thus, the following analyses are based on 68 days of data collection.

21.4.1 Data Fitting with Mean Deviation

For each participant, for each day of the study, let N_d be the number of units available for sale, n_d^* be the number of units that a model predicted should be sold, and n_d be the number of units sold. Obviously, the deviation between the proportions of predicted and observed units sold is only meaningful when $N_d > 0$. When $N_d > 0$, the deviation between the proportions of predicted and observed units sold is $|\frac{n_d^*}{N_d} - \frac{n_d}{N_d}|$. The mean deviation for all the days of the study on which $N_d > 0$ is the fit of the model. So, the time window that generates the smallest mean deviation for a participant is the best-fitting time window for that participant. The mean deviation for EUT can be stated as,

$$MD = \frac{\sum_{d \in D'} |\frac{n_d^*}{N_d} - \frac{n_d}{N_d}|}{|D'|}, \tag{21.26}$$

and for EUT$_{TW}$ is,

$$MD = \frac{\sum_{d \in D'} |\frac{n_{d,w}^*(w)}{N_d} - \frac{n_d}{N_d}|}{|D'|}, \tag{21.27}$$

where the set $D' = \{d \mid N_d > 0\}$, and $|D'|$ is the number of days on which $N_d > 0$. Consequently, the optimal time window is fit to the data using

$$w^* = \arg\min_w MD. \tag{21.28}$$

Figure 21.2 shows the histogram frequency distribution over participants for the size of the best-fitting time window.

EUT does not include the concept of a time window. For EUT on each day of the study the probability that each price would be the highest would be computed for the remaining days of the study. For example, because of the distribution of daily price probabilities in this study, until there were only 30 days remaining in the study (a range of from 67 to 31 days remaining) a cutoff price of 14 was predicted.

EUT predicted a sell decision on only four days of the study, including a day when the price offered was a maximum of 15 and the last day of the study, when selling is trivially predicted by every model. Hence, EUT made only two nontrivial sell predictions. In fact, 56 of the 57 participants sold more than four days as shown in Figure 21.3. The mean number of sell decisions for all participants was 20 with a standard deviation of 11.5, and a range from 3 to 56.

The frequency with which our participants made sell decisions demonstrated that most of them adopted time windows over which the probability of a gain was computed rather than using the remaining days in the study to compute the gain from a hold. As

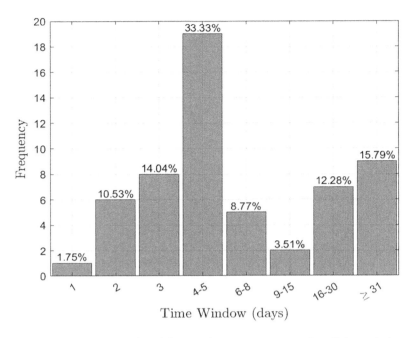

Figure 21.2 The percent of participants whose responses were best fit by each time window.

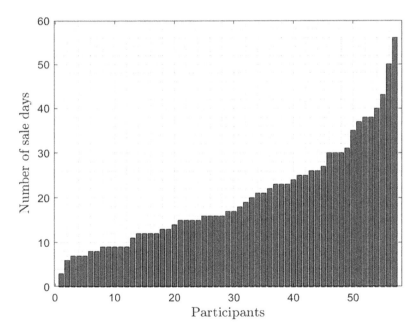

Figure 21.3 The number of sell decisions for each participant in 10-week study.

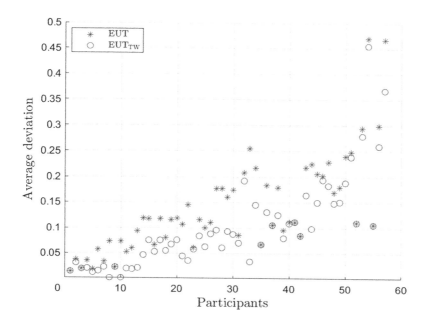

Figure 21.4 Average deviation for EUT and average deviation for EUT with time window.

can be seen from Figure 21.2, the distribution of time windows was bimodal with peaks at 4–5 days and at 31–68 days. However, the peaks are not of equal size. Only 16% of the participants were best fit by a time window ranging from 31–68 days, which is the range consistent with EUT. Data from 84% of the participants were best fit using time windows shorter than 31 days; the great majority of those appearing to use time windows of less than 9 days in determining whether to sell or hold.

Figure 21.4 shows, for each model, the mean deviation between the number of units predicted to be sold and the number of units actually sold by a participant. Each of the 57 participants is represented in the figure by a pair of points – one for the average deviation of their data from the EUT prediction and the other for average deviation from the EUT$_{TW}$ prediction with the best fit of model parameter w. Participants are ordered on the x axis according to the number of days on which they sold at least one unit, from the lowest number of sale days to highest. As can be seen clearly from the plot, EUT$_{TW}$ provides a better fit than EUT.

21.4.2 Data Fitting with Proportional Deviation

A more detailed analysis of the result can be made by computing the proportional deviation between the number of units predicted to be sold by a model and the number of units actually sold by a participant. For a participant, for each day of the study, the absolute difference between the number of units predicted to be sold and the number of units actually sold is computed. Then, the sum of the deviations is divided by the sum of

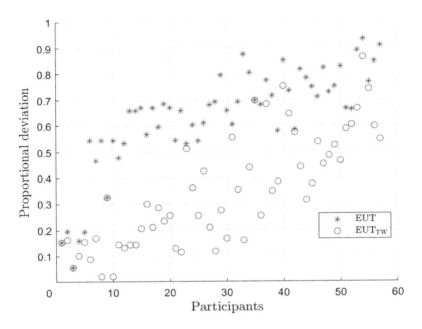

Figure 21.5 Proportional deviation for EUT and proportional deviation for EUT with time window.

the maximum of either the number of units predicted to be sold or the number of units sold. Hence, the proportional deviation for EUT can be stated as,

$$PD = \frac{\sum_{d \in D'} |n_d^* - n_d|}{\sum_{d \in D'} \max(n_d^*, n_d)}, \tag{21.29}$$

and for EUT_{TW} is,

$$PD = \frac{\sum_{d \in D'} |n_{d,w}^*(w) - n_d|}{\sum_{d \in D'} \max(n_{d,w}^*(w), n_d)}. \tag{21.30}$$

Therefore, the optimal time window is fit to the data using

$$w^* = \arg\min_w PD. \tag{21.31}$$

The proportional deviation is a number between 0 and 1, where 0 is a perfect fit of the model predictions to the results. Figure 21.5 shows the proportional deviation for EUT and proportional deviation for EUT_{TW} (with the best fit of model parameter w) for each participant in the study, ordered from those participants who made the fewest number of sales in the grid game to those who made the most. These results demonstrate even more clearly than those plotted in Figure 21.4 that the concepts of horizon and time window could be essential to the understanding of ongoing markets, including the energy market studied here.

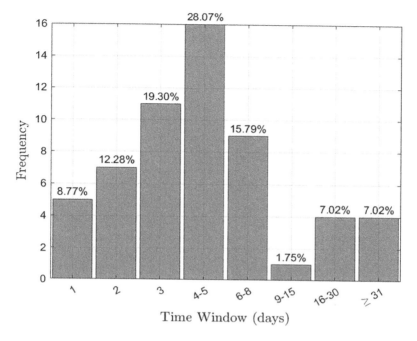

Figure 21.6 The percent of participants whose responses were best fit by each time window based on the proportional deviation method.

Figure 21.6 shows the histogram frequency distribution over participants for the size of the best-fitting time window. This plot is obtained based on fitting procedure for a time window through proportional deviation.

When a large payout has a low daily probability but necessarily has a high probability of occurring over the long term, human participants do not wait around for it to occur; almost never participating in the market at all, as EUT prescribes. Rather, for sell decisions, they make use of time windows that are undoubtedly related in some meaningful way to their lives. As can be seen, EUT_{TW} necessarily provides a better fit than EUT. Because EUT is a special case of EUT_{TW}, at worst, the fits of the two models would be identical. However, whenever a participant in the grid game sold more than the four times predicted by EUT, the fit for EUT_{TW} was better than for EUT.

21.5 Conclusion

Given the parameters defining this study's smart grid scenario, EUT predicts that rational participants would sell their surplus energy units on only four days in the study (the three days when the price exceeded the cutoff price and on the final day). Furthermore, EUT predicts that our prosumers should always sell either none of their units or all of them. Even a cursory analysis of the data revealed that our participants' sell/hold decisions were not aligned with EUT. Many participants sold much more frequently than EUT predicted and often sold only a portion of their energy units.

The addition of a time window parameter to EUT allows for a better fit to our participants' selling behavior. The bimodal nature of the distribution of best-fitting time windows in Figure 21.2 and, to a lesser extent, in Figure 21.6 obviates the individual differences among our participants. A few of those near the far right of the distributions (with time windows exceeding 31 days) are fit just as well by traditional EUT as by EUT_{TW}. Based upon the demographic information gathered prior to the study and on a poststudy questionnaire, these few individuals had slightly more years of education and tended to have taken more math and science courses in school, although neither of these trends were statistically significant. The relationship of decision behavior to numeracy and other cognitive measures would be interesting to explore further in future research.

Parallels between our participants' decision behavior and that of investors in some situations are worth exploring further. The underlying mechanism for the "sell too soon" side of the disposition effect may be similar to that which led many of our smart grid participants to sell energy units much more frequently (and less optimally) than would be predicted by EUT. The finding that most of our study participants often made their sell/hold decisions with a relatively short, self-imposed time window is also conceptually similar to the previously mentioned MLA phenomenon. To the extent that this myopia is temporal in nature – i.e., due to adopting a short-term horizon when longer term would be more appropriate – it seems similar to the bounded time window many of our grid study participants adopted for their sell/hold decision.

Another related concept from investing and behavioral economics is "narrow framing" [20, 21]. Narrow framing holds that investors (or others making decisions under uncertainty) evaluate a given sell/hold decision or other gamble in isolation rather than evaluating it in the context of their entire portfolio and/or a longer-term horizon. That is, they fail to appreciate "the big picture." It may be useful to think of MLA and the bounded time window findings of the present study as instances of narrow framing.

Two other research paradigms where temporal horizons are an important factor in decision making are the previously mentioned job search studies and the temporal discounting literature. In their experiments loosely simulating a job search scenario, Cox and Oaxaca [15, 16] found that study participants terminated their searches sooner than predicted by a risk-neutral linear search model. The data were fit significantly better by a risk-averse model. Their subjects' tendency to decide to terminate their searches and accept suboptimal job offers somewhat parallels our participants' tendency to choose suboptimal, shorter time windows for selling their energy units. However, there are too many differences in the experimental procedures and analytical approaches across the studies to conclude that the underlying behavioral processes are the same.

The temporal (or delay) discounting literature offers another research framework in which to consider findings of the present study. In the typical temporal discounting example, an individual given a choice between a small reward delivered now or soon, and a relatively larger award delivered at a later time, will tend to choose the smaller, sooner reward. Of course, many factors can influence the shape of the discounting function, and can even reverse the preference for smaller–sooner over larger–later. There are excellent reviews of this literature from an economic perspective [22] and from a more psychological perspective [23]. It is tempting to postulate that the choice of relatively short time windows and relatively smaller rewards (unit prices) by many of our grid study participants' shares the same underlying decision-making mechanism with that which leads to temporal discounting. Once again though, large differences in experimental paradigms and modeling approaches make it difficult to directly compare results. Green and Myerson [23] cite extensive evidence for the similarities between temporal discounting and the discounting of probabilistic rewards. Although they tentatively conclude that different underlying mechanisms are at work, their attempt to study both phenomena within a common research framework is laudable.

Acknowledgment

This work is supported in part by the U.S. National Science Foundation (NSF) under the grants 1541069 and 1745829.

References

[1] W. Saad, A. L. Glass, N. B. Mandayam, and H. V. Poor, "Toward a consumer-centric grid: A behavioral perspective," *Proceedings of the IEEE*, vol. 104, no. 4, pp. 865–882, Apr. 2016.

[2] G. E. Rahi, W. Saad, A. Glass, N. B. Mandayam, and H. V. Poor, "Prospect theory for prosumer-centric energy trading in the smart grid," in *2016 IEEE Power Energy Society Innovative Smart Grid Technologies Conference (ISGT)*, pp. 1–5, Sept. 2016.

[3] Y. Wang, W. Saad, N. B. Mandayam, and H. V. Poor, "Integrating energy storage into the smart grid: A prospect theoretic approach," in *2014 IEEE International Conference on Acoustics, Speech and Signal Processing (ICASSP)*, pp. 7779–7783, May 2014.

[4] G. E. Rahi, S. R. Etesami, W. Saad, N. B. Mandayam, and H. V. Poor, "Managing price uncertainty in prosumer-centric energy trading: A prospect-theoretic stackelberg game approach," *IEEE Transactions on Smart Grid*, vol. 10, no. 1, pp. 702–713, Jan. 2019.

[5] I. Lampropoulos, G. M. A. Vanalme, and W. L. Kling, "A methodology for modeling the behavior of electricity prosumers within the smart grid," in *2010 IEEE PES Innovative Smart Grid Technologies Conference Europe (ISGT Europe)*, pp. 1–8, Oct. 2010.

[6] A. Mohsenian-Rad, V. W. S. Wong, J. Jatskevich, R. Schober, and A. Leon-Garcia, "Autonomous demand-side management based on game-theoretic energy consumption scheduling for the future smart grid," *IEEE Transactions on Smart Grid*, vol. 1, no. 3, pp. 320–331, Dec. 2010.

[7] Z. M. Fadlullah, D. M. Quan, N. Kato, and I. Stojmenovic, "GTES: An optimized game-theoretic demand-side management scheme for smart grid," *IEEE Systems Journal*, vol. 8, no. 2, pp. 588–597, June 2014.

[8] W. Saad, Z. Han, H. V. Poor, and T. Basar, "Game-theoretic methods for the smart grid: An overview of microgrid systems, demand-side management, and smart grid communications," *IEEE Signal Processing Magazine*, vol. 29, no. 5, pp. 86–105, Sept. 2012.

[9] N. Hajj and M. Awad, "A game theory approach to demand side management in smart grids," *Springer, Intelligent Systems'2014*, pp. 807–819, 2015.

[10] D. Kahneman and A. Tversky, "Prospect theory: An analysis of decision under risk," *Econometrica*, vol. 47, no. 2, pp. 263–291, 1979.

[11] A. Tversky and D. Kahneman, "Advances in prospect theory: Cumulative representation of uncertainty," *Journal of Risk and Uncertainty*, vol. 5, no. 4, pp. 297–323, 1992.

[12] G. E. Rahi, A. Sanjab, W. Saad, N. B. Mandayam, and H. V. Poor, "Prospect theory for enhanced smart grid resilience using distributed energy storage," in *2016 54th Annual Allerton Conference on Communication, Control, and Computing (Allerton)*, pp. 248–255, Sept. 2016.

[13] L. Xiao, N. B. Mandayam, and H. V. Poor, "Prospect theoretic analysis of energy exchange among microgrids," *IEEE Transactions on Smart Grid*, vol. 6, no. 1, pp. 63–72, Jan. 2015.

[14] Y. Wang, W. Saad, N. B. Mandayam, and H. V. Poor, "Load shifting in the smart grid: To participate or not?" *IEEE Transactions on Smart Grid*, vol. 7, no. 6, pp. 2604–2614, Nov. 2016.

[15] J. C. Cox and R. L. Oaxaca, "Direct tests of the reservation wage property," *The Economic Journal*, vol. 102, no. 415, pp. 1423–1432, 1992.

[16] ——, "Laboratory experiments with a finite-horizon job-search model," *Journal of Risk and Uncertainty*, vol. 2, no. 3, pp. 301–329, 1989.

[17] H. Shefrin and M. Statman, "The disposition to sell winners too early and ride losers too long: Theory and evidence," *The Journal of Finance*, vol. 40, no. 3, pp. 777–790, 1985.

[18] D. Genesove and C. Mayer, "Loss aversion and seller behavior: Evidence from the housing market," *The Quarterly Journal of Economics*, vol. 116, no. 4, pp. 1233–1260, 2001.

[19] S. Benartzi and R. H. Thaler, "Myopic loss aversion and the equity premium puzzle," *Quarterly Journal of Economics*, vol. 110, no. 1, pp. 73–92, 1995.

[20] R. H. Thaler, "Mental accounting matters," *Journal of Behavioral Decision Making*, vol. 12, no. 3, pp. 183–206, 1999.

[21] ——, "Mental accounting and consumer choice," *Marketing Science*, vol. 4, no. 3, pp. 199–214, 1985.

[22] S. Frederick, G. Loewenstein, and T. O'Donoghue, "Time discounting and time preference: A critical review," *Journal of Economic Literature*, vol. 40, no. 2, pp. 351–401, June 2002.

[23] L. Green and J. Myerson, "A discounting framework for choice with delayed and probabilistic rewards." *Psychological Bulletin*, vol. 130, no. 5, pp. 769–792, 2004.

22 Storage Allocation for Price Volatility Management in Electricity Markets

Amin Masoumzadeh, Ehsan Nekouei, and Tansu Alpcan

Nomenclature

Indices:

m, n	Intermittent and classical generators.
i, j	Node.
t	load time (hr).
w	Scenario.

Parameters:

$\alpha_{itw}, \beta_{itw}$	Intercept and slope of the inverse demand function.
$\gamma_{mi}^{ig}, \gamma_{ni}^{sg}, \gamma_{i}^{st}, \gamma_{ij}^{tr}$	Binary parameter to distinguish if the intermittent generator, classical generator, storage firm, and the transmission firm are strategic/regulated, respectively.
c_{ni}^{sg}	Marginal operation and fuel cost of the classical generator.
R_{ni}^{up}, R_{ni}^{dn}	Ramping up and down coefficients of the synchronous generator.
$\eta_{bi}^{ch}, \eta_{bi}^{dis}$	Charge and discharge efficiencies of the storage.
$\zeta_{i}^{ch}, \zeta_{i}^{dis}$	Charge and discharge power coefficient of the storage capacity.
$Q^{ig}, Q^{sg}, Q^{st}, Q^{tr}$	Intermittent generation, classical generation, storage, and transmission capacities.
Ψ_w	Probability of scenario w.

Variables:

q_{mitw}^{ig}	Generation of the intermittent generator.
q_{nitw}^{sg}	Generation of the classical generator.
q_{ijtw}^{tr}	Electricity flow from node j to node i.
$q_{bitw}^{ch}, q_{bitw}^{dis}, q_{bitw}^{st}$	Charge, discharge, and flow of the storage.
μ	Dual variable for its corresponding constraint.

Functions:

$P_{itw}(.)$	Wholesale price.

22.1 Introduction

Recent studies show that the fast growing expansion of wind power generation may lead to extremely high levels of price volatility in wholesale electricity markets. Storage technologies, regardless of their specific forms, e.g., pump-storage hydro, large-scale, or distributed batteries, are capable of alleviating the extreme price volatility levels due to their energy usage time shifting, fast-ramping, and price arbitrage capabilities. In this chapter, we present a stochastic bi-level optimization model to find the optimal nodal storage capacities required to achieve a certain price volatility level in a highly volatile energy-only electricity market. The decision on storage capacities is made in the upper-level problem and the operation of strategic/regulated generation, storage, and transmission players is modeled in the lower-level problem using an extended stochastic (Bayesian) Cournot-based game. The South Australia (SA) electricity market, which has recently experienced high levels of price volatility is considered as the case study. Our numerical results indicate that 50% price volatility reduction in SA electricity market can be achieved by installing either 430MWh regulated storage or 530MWh strategic storage. In other words, regulated storage firms are more efficient in reducing the price volatility than strategic storage firms.

In this section, we first discuss the transformations that are revolutionizing the electricity markets. More specifically, storage integration, in large scale or in distributed fashion, and large integration of intermittent renewables are discussed in Australia's electricity market.

22.1.1 Motivation

Electricity generation industry in many countries around the world has experienced a significant transformation from being a centrally coordinated monopoly to a deregulated competitive market, during the last three decades [1]. Moreover, it is expected that the generated electricity will be green, reliable, and affordable in future electricity markets, as shown in Figure 22.1. As a result, the electricity markets have to overcome challenges such as: (1) the over- and undercapacity in the generation, transmission, and distribution sides that are imposing costs on market participants, (2) market power that leads to high electricity prices; (3) integration of intermittent renewables into the network while

Figure 22.1 Three main criteria in designing the future electricity markets.

ensuring the system reliability; and (4) high levels of CO_2 emission and the challenges for decarbonization of electricity markets.

Some of these challenges can be partially addressed by the closure of coal power plants and installing more renewables in the network. Recent studies indicate that the closure of coal power plants reduces the emission in the market but it may lead to high electricity prices [2]. Moreover, the integration of wind and solar in power grids leads to a more clean electricity generation, but it brings high levels of price volatility in the network [3]. These observations imply that these solutions cannot completely address the challenges.

Electricity storage has been proposed as a promising solution for some of the existing challenges in the electricity markets. For example, electricity storage can be used for multiple purposes including the reduction in expensive peaking capacity, managing the intermittency of distributed wind/solar generation, and managing excess generation from base-load coal/nuclear during off-peak times. By providing virtual generation capacity, storage may alleviate existing problems by reducing the impacts of intermittent power generation, market power, and volatility [4].

Given the continuing decrease in battery costs, a large amount of battery storage capacity is expected to be installed in transmission and/or distribution networks in the near future. Moreover, global road map vision indicates significant capacity increase for pump-storage hydro power, from 140GW in 2012 to 400 to 700 W in 2050 [5]. In 2017, the SA government took an initiative to improve the SAs' electricity market structure by installing 100MW (125MWh) electricity storage capacity which significantly improved the quality of regulation and contingency services in the National Electricity Market (NEM) [6].

Motivated by these observations, this chapter presents a framework for mitigating the price volatility in electricity markets by optimal allocation of electricity storage in the networks. In the rest of this section, we discuss the price volatility problem and structure of Australian NEM which has witnessed significant price volatility levels.

Price Volatility Problem

A high level of intermittent wind generation may also result in frequent high prices and high levels of price volatility in electricity markets [7, 8]. High levels of price volatility in a market refers to a situation in which the market prices vary in a wide range. For example, one hundred hours with highest levels of electricity prices resulted in 21% of the annual monetary market share in 2015 in South Australia, which is a highly price volatile region in the NEM market [9]. Price volatility makes the task of price prediction highly uncertain, which consequently imposes large financial risks on the market participants. In the long term, extreme levels of price volatility can lead to undesirable consequences such as the bankruptcy of retailers [10] and market suspension. In a highly volatile electricity market, the participants, such as generators, utility companies, and large industrial consumers, are exposed to a high level of financial risk as well as costly risk management strategies [11]. In some electricity markets, such as the NEM, the market is suspended if the sum of spot prices over a certain period is more than the cumulative price threshold (CPT). A highly volatile market is subject to frequent CPT breaches due to the low conventional capacity and high level of wind variability. Storage, with price arbitrage capability, can resolve the problem of high electricity prices and, consequently, it can prevent high levels of price volatility.

Australia's National Electricity Market

The NEM commenced operation as a wholesale spot market in December 1998. It interconnects five regional market jurisdictions: South Australia, Queensland, Tasmania, New South Wales (including the Australian Capital Territory), and Victoria. In the NEM, electricity is an ideal commodity that is exchanged between producers and consumers through a pool. The market operator must ensure the agreed standards of security and reliability. A high level of reliability across the NEM requires a certain level of reserve. However, when security and reliability are threatened, the market operator is equipped with a variety of tools including demand-side management, load shedding, and reserve trading to maintain the supply and demand balance.

Operating the NEM consists of estimating the electricity demand levels, receiving the bidding offers, scheduling and dispatching the generators, calculating the spot price, and financially settling the market. Electricity demand in a region is forecasted based on different factors, like population, temperature, and sectoral energy consumption in that region. Electricity supply bids (offers) are submitted in three forms of daily bids, rebids, and default bids [12]. Using the rising-price stack, generators are scheduled and dispatched in the market.

22.1.2 Contributions

The current work presents a stochastic optimization framework for finding the required nodal storage capacities in electricity markets with high levels of wind penetration such that the price volatility in the market is kept below a certain level. The contributions of this work are summarized as follows:

1. A bi-level optimization model is developed to find the optimal nodal storage capacities required for avoiding the extreme price volatility levels in a nodal electricity market.
2. In the upper-level problem, the total storage capacity is minimized subject to a price volatility target constraint in each node and at each time.
3. In the lower-level problem, the noncooperative interaction between generation, transmission and storage players in the market is modeled as a stochastic (Bayesian) Cournot-based game with an exponential inverse demand function.
4. The existence of Bayesian Nash equilibrium (Bayes–NE) [13] under the exponential inverse demand function is established for the lower-level problem.

Under the proposed framework, the size of storage devices at two nodes of SA and Victoria (VIC) in NEM is determined such that the market price volatility is kept below a desired level at all times. The desired level of price volatility can be determined based on various criteria such as net revenue earned by the market players, occurrence frequency of undesirable prices, number of CPT breaches [14].

The rest of the chapter is organized as follows. The literature review and related works are discussed in Section 22.2. The system model and the developed bi-level optimization problem are formulated in Section 22.3. The equilibrium analysis of the lower-level problem and the solution method are presented in Section 22.4. The simulation results are presented in Section 22.5. The concluding remarks are discussed in Section 22.6.

22.2 Related Works

The problem of optimal storage operation or storage allocation for facilitating the integration of intermittent renewable energy generators in electricity networks has been studied in [15–22], with total cost minimization objective functions, and in [23–28], with profit maximization goals. However, the price volatility management problem using optimal storage allocation has not been investigated in the literature.

The operation of a storage system is optimized, by minimizing the total operating costs in the network, to facilitate the integration of intermittent renewable resources in power systems in [15]. Minimum (operational/installation) cost storage allocation problem for renewable integrated power systems is studied in [16–18] under deterministic wind models, and in [19] under a stochastic wind model. The minimum-cost storage allocation problem is studied in a bi-level problem in [20, 21], with the upper and lower levels optimizing the allocation and the operation, respectively. The paper [22] investigates the optimal sizing, siting, and operation strategies for a storage system to be installed in a distribution company controlled area. We note that these works only study the minimum cost storage allocation or operation problems, and do not investigate the interplay between the storage firms and other participants in the market.

The paper [23] studies the optimal operation of a storage unit, with a given capacity, which aims to maximize its profit in the market from energy arbitrage and provision of regulation and frequency response services. The paper [24] computes the optimal supply and demand bids of a storage unit so as to maximize the storage's profit from energy arbitrage in the day-ahead and the next 24-hour-ahead markets. The paper [25] investigates the profit maximization problem for a group of independently operated investor-owned storage units that offer both energy and reserve in both day-ahead and hour-ahead markets. In these works, the storage is modeled as a price taker firm due to its small capacity.

The operation of a price-maker storage device is optimized using a bi-level stochastic optimization model, with the lower-level clearing the market and the upper-level maximizing the storage profit by bidding on price and charge/discharge in [26]. The storage size in addition to its operation is optimized in the upper-level problem in [27] when the lower-level problem clears the market. Note that the price bids of market participants other than the storage firm are treated exogenously in these models. The paper [28] also maximizes the day-ahead profit of a load-serving entity that owns large-scale storage capacity, assuming the price bids in the wholesale market as exogenous parameters.

The paper [29] maximizes a large-scale energy storage system's profit considering the storage as the only strategic player in the market. Using Cournot-based electricity market models, the generation and storage firms are considered as strategic players in [30, 31]. However, they do not study the storage sizing problem and the effect of intermittent renewables on the market. Therefore, to the best of our knowledge, the problem of finding optimal storage capacity subject to a price volatility management target in electricity markets has not been addressed before.

22.3 System Model

In this section, we present a bi-level optimization approach for finding the minimum required total storage capacity in the market such that the market price volatility stays within a desired limit at each time.

Consider a nodal electricity market with I nodes. Let $\mathcal{N}_i^{cg}$ be the set of classical generators, such as coal and gas power plants, located in node i and $\mathcal{N}_i^{wg}$ be the set of wind generation firms located in node i. The set of neighboring nodes of node i is denoted by $\mathcal{N}_i$. Because the wind availability is a stochastic parameter, a scenario-based model, with N_w different scenarios, is considered to model the wind availability in the electricity network. The nodal prices in our model are determined by solving a stochastic (Bayesian) Cournot-based game among all market participants, that is, classical generators, wind firms, storage firms, and transmission interconnectors that are introduced in detail in the lower-level problem, given the wind power availability scenarios. The decision variables, feasible region, and objective function for each player in our game model are discussed in Subsection 22.3.3. In a Cournot game, each producer (generator) competes for maximizing its profit, which is defined as its revenue minus its production cost, given the generation of other players. The revenue of each player is its production level times the market price. Also, the market price is a function of the total generation. Following the standard Cournot game models, any player in our model maximizes its objective function given the decision variables of other players (generation, transmission, and storage firms). Considering different wind power availability scenarios with given probabilities makes our game model consistent with the Bayesian game definition. In a Bayesian game, players maximize their expected utility over a set of scenarios with a given probability distribution [13].

22.3.1 Inverse Demand (Price) Function

The two most commonly used inverse demand functions in microeconomics literature are the linear and iso-elastic models [32], e.g., in [31, 33]. Exponential inverse demand function has also been used in energy market models [34]. The inverse demand function of most commodities follows a nonlinear downward sloping price versus demand relation [35] and a linear inverse demand function is just its first-order approximation at an operating price and demand level. The linear function may become invalid when the operating point changes drastically, e.g., when the price plunges from the very high amount of 11,000 $/MWh to low level of 50 $/MWh.

The iso-elastic and exponential functions can more accurately illustrate the price and demand relation. In fact, the exponential function, $p = \alpha' e^{-\beta' q}$, is the modified version of the iso-elastic function, $\ln(p) = \alpha - \beta \ln(q)$ or $p = \tilde{\alpha} e^{-\beta \ln(q)}$ with $\tilde{\alpha} = e^{\alpha}$, which substitutes the logarithmic demand levels with nominal levels. We discuss three reasons privileging the exponential inverse demand function over the iso-elastic. Firstly, the Karush–Kuhn–Tucker (KKT) conditions of the lower-level game become highly nonlinear under the iso-elastic function and it becomes hard to numerically solve them. The derivative of the exponential inverse demand function with respect to demand is $\frac{\partial p}{\partial q} = -\beta' p$, while the derivative of the iso-elastic function respect to demand is $\frac{\partial p}{\partial q} = -\beta p q^{-1}$. Secondly, the exponential function has a finite price feature while the iso-elastic function goes to infinity for small levels of demand. Lastly, the exponential function partially covers the specifications of both linear and iso-elastic functions. Consequently, we use and calibrate exponential inverse demand functions to characterize the price and demand relations in our model.

In electricity market models, the constant coefficients in the inverse demand functions are usually calibrated based on actual price/demand data, p/q, and price elasticity levels, $\epsilon = \frac{\frac{\partial q}{q}}{\frac{\partial p}{p}}$ [35], which are given as input to our model. Given two equations of price-demand function and elasticity function, i.e., $p = f(q)$ and $\epsilon = \frac{\partial q}{\partial p} \frac{p}{q}$, and two unknowns, we can find the both parameters in all three discussed inverse demand functions. For instance, given the price of $p = 50$ $/MWh, demand of $q = 15{,}00$MW and price elasticity of demand $\epsilon = -0.3$, the linear function $p = \frac{650}{3} - \frac{1}{9} q$, the iso-elastic function $\ln(p) = 28.28 - \frac{10}{3} \ln(q)$, and the exponential function $p = 50 e^{\frac{10}{3}} e^{-\frac{1}{450} q}$ can be extracted. Figure 22.2 compares the calibrated linear, exponential, and iso-elastic

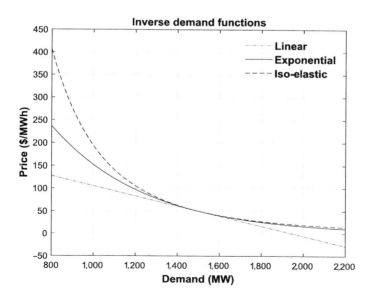

Figure 22.2 Calibrated linear, exponential and iso-elastic inverse demand functions at price 50 $/MWh, demand 15,00MW, and elasticity -0.3.

inverse demand functions. The properties of the exponential function lie between the linear and iso-elastic functions.

22.3.2 Upper-Level Problem

In the upper-level optimization problem, we determine the nodal storage capacities such that a price volatility constraint is satisfied in each node at each time. In this chapter, estimates of variances are used to capture the volatilities [36], i.e., the variance of market price is considered as a measure of price volatility. The variance of the market price in node i at time t, i.e., $\text{Var}(P_{itw})$, can be written as:

$$\text{Var}(P_{itw}) = \mathsf{E}_w\left[(P_{itw}(\boldsymbol{q}_{itw}))^2\right] - (\mathsf{E}_w[P(\boldsymbol{q}_{itw})])^2$$

$$= \sum_w \left(P_{itw}(\boldsymbol{q}_{itw})\right)^2 \Psi_w - \left(\sum_w P_{itw}(\boldsymbol{q}_{itw}) \Psi_w\right)^2 \quad (22.1)$$

where Ψ_w is the probability of scenario w, and $P_{itw}(\boldsymbol{q}_{itw})$ is the market price in node i at time t under the wind availability scenario w, which is a function of the collection of all players' strategies $\boldsymbol{q}_{itw}$, i.e., the decision variables in the lower-level game.

The notion of variance quantifies the *effective* variation range of random variables, i.e., a random variable with a small variance has a smaller effective range of variation when compared with a random variable with a large variance.

Given the price volatility relation (22.1) based on the Bayes–NE strategy collection of all firms $\boldsymbol{q}_{itw}^\star$, the upper-level optimization problem is given by:

$$\min_{\{Q_i^s\}_i} \sum_{i=1}^{I} Q_i^s$$

$$\text{s.t.} \quad Q_i^s \geq 0 \quad \forall i \quad (22.2a)$$

$$\text{Var}(P_{itw}(\boldsymbol{q}_{itw}^\star)) \leq \sigma_0^2 \quad \forall i,t \quad (22.2b)$$

where Q_i^s is the storage capacity in node i, $P_{itw}(\boldsymbol{q}_{itw}^\star)$ is the market price in node i at time t under the wind availability scenario w, and σ_0^2 is the price volatility target. The price volatility of the market is defined as the maximum variance of market price, i.e., $\max_{it} \text{Var}(P_{itw}(\boldsymbol{q}_{itw}^\star))$.

22.3.3 Lower-Level Problem

In the lower-level problem, the nodal market prices and the Bayes–NE strategies of firms are obtained by solving an extended stochastic Cournot-based game between wind generators, storage firms, transmission firms, and classical generators. Our model differs from a standard Cournot game, such that it includes regulated players in addition to strategic players in generation, storage, and transmission levels.

DEFINITION 22.1 *A strategic (price maker) firm decides on its strategies over the operation horizon $\{1, \ldots, N_T\}$ such that its aggregate expected profit, over the operation* horizon, is maximized. However, a regulated (price taker) firm aims to improve the net market value, i.e., the social welfare [37].

The market price in node i at time t under the wind availability scenario w is given by an exponential inverse demand function:

$$P_{itw}(q_{itw}) = \alpha_{it} e^{-\beta_{it}\left(q^{s}_{itw} + \sum_{m \in \mathcal{N}^{wg}_i} q^{wg}_{mitw} + \sum_{n \in \mathcal{N}^{cg}_i} q^{cg}_{nitw} + \sum_{j \in \mathcal{N}_i} q^{tr}_{ijtw}\right)} \quad \forall i,t,w \quad (22.3)$$

where α_{it}, β_{it} are positive real values in the inverse demand function, q^{cg}_{nitw} is the generation strategy of the nth classical generator located in node i at time t under scenario w, q^{wg}_{mitw} is the generation strategy of the mth wind generator located in node i at time t under scenario w, q^{s}_{itw} is the charge/discharge strategy of the storage firm in node i at time t under scenario w, and q^{tr}_{ijtw} is the strategy of transmission firm located between node i and node j at time t under scenario w. The collection of strategies of all firms located in node i at time t under the scenario w is denoted by q_{itw}. Note that the total amount of power supply from the generation and storage firms plus the net import/export, i.e., $q^{s}_{itw} + \sum_{m \in \mathcal{N}^{wg}_i} q^{wg}_{mitw} + \sum_{n \in \mathcal{N}^{cg}_i} q^{cg}_{nitw} + \sum_{j \in \mathcal{N}_i} q^{tr}_{ijtw}$, is equal to the net electricity demand in each node, at each time, and under each scenario, which represents the nodal electricity balance in our model.

In what follows, the variable μ is used to indicate the associated Lagrange variable with its corresponding constraint in the model. The function $P_{itw}(.)$ refers to (22.3).

Wind Generators

The Bayes–NE strategy of the mth wind generator in node i is obtained by solving the following optimization problem:

$$\max_{\{q^{wg}_{mitw}\}_{tw} \geq 0} \sum_w \Psi_w \sum_{t=1}^{N_T} P_{itw}(.) q^{wg}_{mitw} \left(1 - \gamma^{wg}_{mi}\right) + \gamma^{wg}_{mi} \left(\frac{P_{itw}(.)}{-\beta_{it}}\right)$$

s.t. $\quad q^{wg}_{mitw} \leq Q^{wg}_{mitw} \quad : \quad \mu^{wg,max}_{mitw} \quad \forall t, w \quad (22.4a)$

$\quad P_{itw}(.) \leq P^{cap} \quad : \quad \mu^{wg,cap}_{mitw} \quad \forall t, w \quad (22.4b)$

where q^{wg}_{mitw} and Q^{wg}_{mitw} are the generation level and the available wind capacity of the mth wind generator located in node i at time t under scenario w. The parameter P^{cap} represents the price cap in the market, which is, for instance, 11,000 $/MWh in the NEM market. Setting cap price in electricity markets also aims to limit the price levels and price volatility levels. Note that the wind availability changes in time in a stochastic manner, and the wind firm's bids depend on the wind availability. As a result, the nodal prices and decisions of the other firms become stochastic in our model [38].

The mth wind firm in node i acts as a strategic firm in the market if γ^{wg}_{mi} is equal to zero and acts as a regulated firm if γ^{wg}_{mi} is equal to one. The difference between regulated and strategic players corresponds to the strategic price impacting capability. In fact, a regulated firm behaves as a price taker player while a strategic firm behaves as a price maker player.

Classical Generators

Classical generators include coal, gas, hydro, and nuclear power plants. The Bayes–NE strategy of nth classical generator located in node i is determined by solving the following optimization problem:

$$\max_{\{q_{nitw}^{cg}\}_{tw} \geq 0} \sum_w \Psi_w \sum_{t=1}^{N_T} P_{itw}(.) q_{nitw}^{cg} \left(1 - \gamma_{ni}^{cg}\right) - c_{ni}^{cg} q_{nitw}^{cg} + \gamma_{ni}^{cg} \left(\frac{P_{itw}(.)}{-\beta_{it}}\right)$$

s.t.

$$q_{nitw}^{cg} \leq Q_{ni}^{cg} \quad : \quad \mu_{nitw}^{cg,max} \quad \forall t, w \tag{22.5a}$$

$$q_{nitw}^{cg} - q_{ni(t-1)w}^{cg} \leq R_{ni}^{up} \quad : \quad \mu_{nitw}^{cg,up} \quad \forall t, w \tag{22.5b}$$

$$q_{ni(t-1)w}^{cg} - q_{nitw}^{cg} \leq R_{ni}^{dn} \quad : \quad \mu_{nitw}^{cg,dn} \quad \forall t, w \tag{22.5c}$$

$$P_{itw}(.) \leq P^{cap} \quad : \quad \mu_{nitw}^{cg,cap} \quad \forall t, w \tag{22.5d}$$

where q_{nitw}^{cg} is the generation level of the nth classical generator in node i at time t under scenario w, Q_{ni}^{cg} and c_{ni}^{cg} are the capacity and the short-term marginal cost of the nth classical generator in node i, respectively. The constraints (22.5b) and (22.5c) ensure that the ramping limitations of the nth classical generator in node i are always met. The nth classical generator in node i acts as a strategic firm in the market if γ_{ni}^{cg} is equal to zero and acts as a regulated firm if γ_{ni}^{cg} is equal to one.

Storage Firms

Storage firms benefit from price difference at different times to make profit, i.e., they sell the off-peak stored electricity at higher prices at peak times. The Bayes–NE strategy of storage firm located in node i is determined by solving the following optimization problem:

$$\max_{\substack{\{q_{itw}^{dis}, q_{itw}^{ch}\}_{tw} \geq 0 \\ \cdot \{q_{itw}^{s}\}_{tw}}} \sum_w \Psi_w \sum_{t=1}^{N_T} P_{itw}(.) q_{itw}^{s} \left(1 - \gamma_i^{s}\right) - c_i^{s} \left(q_{itw}^{dis} + q_{itw}^{ch}\right) + \gamma_i^{s} \left(\frac{P_{itw}(.)}{-\beta_{it}}\right)$$

s.t.

$$q_{itw}^{s} = \eta_i^{dis} q_{itw}^{dis} - \frac{q_{itw}^{ch}}{\eta_i^{ch}} \quad : \quad \mu_{itw}^{s} \quad \forall t, w \tag{22.6a}$$

$$q_{itw}^{dis} \leq \zeta_i^{dis} Q_i^{s} \quad : \quad \mu_{itw}^{dis,max} \quad \forall t, w \tag{22.6b}$$

$$q_{itw}^{ch} \leq \zeta_i^{ch} Q_i^{s} \quad : \quad \mu_{itw}^{ch,max} \quad \forall t, w \tag{22.6c}$$

$$0 \leq \sum_{k=1}^{t} \left(q_{ikw}^{ch} - q_{ikw}^{dis}\right) \Delta \leq Q_i^{s} : \mu_{itw}^{s,min}, \mu_{itw}^{s,max} \quad \forall t, w \tag{22.6d}$$

$$P_{itw}(.) \leq P^{cap} \quad : \quad \mu_{itw}^{s,cap} \quad \forall t, w \tag{22.6e}$$

where q_{itw}^{dis} and q_{itw}^{ch} are the discharge and charge levels of the storage firm in node i at time t under scenario w, respectively, c_i^{s} is the unit operation cost, η_i^{ch}, η_i^{dis} are the charging and discharging efficiencies, respectively, and q_{itw}^{s} is the net supply/demand of the storage firm in node i. The parameter ζ_i^{ch} (ζ_i^{dis}) is the percentage of storage capacity

Q_i^s, which can be charged (discharged) during period Δ. It is assumed that the storage devices are initially fully discharged. The energy level of the storage device in node i at each time is limited by its capacity Q_i^s. Note that the nodal market prices depend on the storage capacities, i.e., Q_i^ss, through the constraints (22.6b)–(22.6d). This dependency allows the market operator to meet the volatility constraint using the optimal values of the storage capacities. The storage capacity variables are the only variables that couple the scenarios in the lower-level problem. Therefore, each scenario of the lower-lever problem can be solved separately for any storage capacity amount. The storage firm in node i acts as a strategic firm in the market if γ_i^s is equal to zero and acts as a regulated firm if γ_i^s is equal to one.

PROPOSITION 22.1 *At the Bayes–NE of the lower-level game, each storage firm is either in the charge mode or discharge mode at each scenario, i.e., the charge and discharge levels of each storage firm cannot be simultaneously positive at the NE of each scenario.*

Proof We show that the charge and discharge levels of any storage device cannot be simultaneously positive at the NE of the lower game under each scenario. Consider a strategy in which both charge and discharge levels of storage device i at time t under scenario w, i.e., $q_{itw}^{dis}, q_{itw}^{ch}$, are both positive. We show that this strategy cannot be a NE strategy of scenario w as follows. The net electricity flow of storage can be written as $q_{itw}^s = \eta_i^{dis} q_{itw}^{dis} - \frac{q_{itw}^{ch}}{\eta_i^{ch}}$. Let $\bar{q}_{itw}^{dis}$ and $\bar{q}_{itw}^{ch}$ be the new discharge and charge levels of storage firm i defined as $\{\bar{q}_{itw}^{dis} = q_{itw}^{dis} - \frac{q_{itw}^{ch}}{\eta_i^{dis}\eta_i^{ch}}, \bar{q}_{itw}^{ch} = 0\}$ if $q_{itw}^s > 0$, or $\{\bar{q}_{itw}^{dis} = 0, \bar{q}_{itw}^{ch} = q_{itw}^{ch} - q_{itw}^{dis}\eta_i^{dis}\eta_i^{ch}\}$ if $q_{itw}^s < 0$. The new net flow of electricity can be written as $\bar{q}_{itw}^s = \eta_i^{dis} \bar{q}_{itw}^{dis} - \frac{\bar{q}_{itw}^{ch}}{\eta_i^{ch}}$. Note that the new variables $\bar{q}_{itw}^s, \bar{q}_{itw}^{ch}$, and $\bar{q}_{itw}^{dis}$ satisfy the constraints (22.6a–22.6d).

Considering the new charge and discharge strategies $\bar{q}_{itw}^{dis}$ and $\bar{q}_{itw}^{ch}$, instead of q_{itw}^{dis} and q_{itw}^{ch}, the nodal price and the net flow of storage device i do not change. However, the charge/discharge cost of the storage firm i, under the new strategy, is reduced by:

$$c_i^s(q_{itw}^{ch} + q_{itw}^{dis}) - c_i^s(\bar{q}_{itw}^{dis} + \bar{q}_{itw}^{ch}) > 0$$

Hence, any strategy at each scenario in which the charge and discharge variables are simultaneously positive cannot be a NE, i.e., at the NE of the lower game under each scenario, any storage firm is either in the charge mode or discharge mode. □

Transmission Firms

The Bayes–NE strategy of the transmission firm between nodes i and j is determined by solving the following optimization problem:

$$\max_{\{q_{jitw}^{tr}, q_{ijtw}^{tr}\}_{tw}} \sum_w \Psi_w \sum_{t=1}^{N_T} \left(P_{jtw}(.) q_{jitw}^{tr} + P_{itw}(.) q_{ijtw}^{tr} \right)(1 - \gamma_{ij}^{tr}) +$$

s.t.
$$\gamma_{ij}^{tr} \left(\frac{P_{jtw}(.)}{-\beta_{jt}} + \frac{P_{itw}(.)}{-\beta_{it}} \right) \quad (22.7a)$$

$$q_{ijtw}^{tr} = -q_{jitw}^{tr} \quad : \quad \mu_{ijtw}^{tr} \quad \forall t, w \tag{22.7b}$$

$$-Q_{ij}^{tr} \leq q_{ijtw}^{tr} \leq Q_{ij}^{tr} \quad : \quad \mu_{ijtw}^{tr,\min}, \mu_{ijtw}^{tr,\max} \quad \forall t, w \tag{22.7c}$$

$$P_{ktw}(.) \leq P^{cap} \quad : \quad \mu_{kk'tw}^{tr,cap} \quad k, k' \in \{i, j\}, k \neq k' \quad \forall t, w \tag{22.7d}$$

where q_{ijtw}^{tr} is the electricity flow from nodes j to i at time t under scenario w, and Q_{ij}^{tr} is the capacity of the transmission line between node i and node j. The transmission firm between nodes i and j behaves as a strategic player when γ_{ij}^{tr} is equal to zero and behaves as a regulated player when γ_{ij}^{tr} is equal to one. Note that the term $P_{jtw}(.) q_{ijtw}^{tr} + P_{itw}(.) q_{ijtw}^{tr}$ in the objective function of the transmission firm is equal to $(P_{jtw}(.) - P_{itw}(.)) q_{jitw}^{tr}$ which implies that the transmission firm between two nodes makes profit by transmitting electricity from the node with a lower market price to the node with a higher market price. Moreover, the price difference between the paired nodes indicates the congestion on the transmission lines and can be used to set the value of financial transmission rights (FTR) [39] in electricity markets.

Transmission lines or interconnectors are usually controlled by the market operator and are regulated to improve the social welfare in the market. The markets with regulated transmission firms are discussed as electricity markets with transmission constraints in the literature, e.g., see [40–42]. However, some electricity markets allow the transmission lines to act strategically, i.e., to make revenue by trading electricity across the nodes [43].

DEFINITION 22.2 *Social welfare (SW) is equal to the total surplus of consumers and producers in the market. Under any scenario w in our problem, the social welfare can be written as:*

$$SW_w = \sum_{i,t} \int_0^{y_{itw}} P_{itw}(.) \, \partial y_{itw} - \text{Total Cost}_{it} = \sum_{i,t} \frac{P_{itw}(.) - \alpha_{it}}{-\beta_{it}} - \text{Total Cost}_{it}$$

where Total Cost is the total cost of electricity generation, storage, and transmission.

PROPOSITION 22.2 *Maximizing the proposed objective function for any regulated transmission firm is equivalent to best improving the social welfare. They both result in the same Bayes–NE point in the lower-level game.*

Proof We show (1) how the objective function of any transmission player in our model is generalized to be either strategic (profit maximizer) or regulated (social welfare improver), and (2) how maximizing the objective function of any regulated transmission firm is equivalent to best improving the social welfare.

The transmission player maximizes its profit when γ_{ij}^{tr} in (22.7a) is zero, and equivalently improves the social welfare when γ_{ij}^{tr} is one. The derivative of regulated term in (22.7a), respect to q_{ijtw}^{tr}, given the constraint (22.7b), that is, $q_{ijtw}^{tr} = -q_{jitw}^{tr}$, is:

$$\frac{\partial \left(\frac{P_{jtw}(.)}{-\beta_{jt}} + \frac{P_{itw}(.)}{-\beta_{it}} \right)}{\partial q_{ijtw}^{tr}} = P_{itw}(.) - P_{jtw}(.),$$

which is equal to the derivative of the social welfare function respect to q^{tr}_{ijtw}. It shows that based on the first order optimality conditions, maximizing the regulated term is equivalent to improving the social welfare for the transmission player. □

22.4 Solution Approach

In this section, we first provide a game-theoretic analysis of the lower-level problem. Next, the bi-level price volatility management problem is transformed to a single-optimization mathematical problem with equilibrium constraints (MPEC).

22.4.1 Game-Theoretic Analysis of the Lower-Level Problem

To solve the lower-level problem, we need to study the best response functions of firms participating in the market. Then, any intersection of the best response functions of all firms in all scenarios will be a Bayes–NE. In this subsection, we first establish the existence of Bayes–NE for the lower-level problem. Then, we provide the necessary and sufficient conditions that can be used to solve the lower-level problem.

To transform the bi-level price volatility management problem to a single-level problem, we need to ensure that for every vector of storage capacities, i.e., $Q^s = [Q_1^s, \ldots, Q_I^s]^\top \geq \mathbf{0}$, the lower-level problem admits a Bayes–NE. At the Bayes–NE strategy of the lower-level problem, no single firm has any incentive to unilaterally deviate its strategy from its Bayes–NE strategy. Note that the objective function of each firm is quasi-concave in its strategy and the constraint set of each firm is closed and bounded for all $Q^s = [Q_1^s, \ldots, Q_I^s]^\top \geq \mathbf{0}$. Thus, the lower-level game admits a Bayes–NE. This result is formally stated in Proposition 22.3.

PROPOSITION 22.3 *For any vector of storage capacities, $Q^s = [Q_1^s, \ldots, Q_I^s]^\top \geq \mathbf{0}$, the lower-level game admits a Bayes–NE.*

Proof Note that the objective function of each firm is continuous and quasi-concave in its strategy. Also, the strategy space is nonempty, compact, and convex. Therefore, according to theorem 1.2 in [44], the lower-level game admits a Bayes–NE. □

Best Responses of Wind Firms

Let $\mathbf{q}_{-(mi)}$ be the strategies of all firms in the market except the wind generator m located in node i. Then, the best response of the wind generator m in node i to $\mathbf{q}_{-(mi)}$ satisfies the necessary and sufficient KKT conditions ($t \in \{1, \ldots, N_T\}; w \in \{1, \ldots, N_w\}$):

$$P_{itw}(.) + (1 - \gamma^{\text{wg}}_{mi})\frac{\partial P_{itw}(.)}{\partial q^{\text{wg}}_{mitw}} q^{\text{wg}}_{mitw} - \frac{\mu^{\text{wg, max}}_{mitw} + \frac{\partial P_{itw}(.)}{\partial q^{\text{wg}}_{mitw}} \mu^{\text{wg, cap}}_{mitw}}{\Psi_w} \leq 0 \perp q^{\text{wg}}_{mitw} \geq 0 \quad (22.10\text{a})$$

$$q^{\text{wg}}_{mitw} \leq Q^{\text{wg}}_{mitw} \perp \mu^{\text{wg, max}}_{mitw} \geq 0 \quad (22.10\text{b})$$

$$P_{itw}(.) \leq P^{\text{cap}} \perp \mu^{\text{wg, cap}}_{itw} \geq 0 \quad (22.10\text{c})$$

where the perpendicularity sign, $\perp$, means that at least one of the adjacent inequalities must be satisfied as an equality [45].

Best Responses of Classical Generation Firms

The best response of the classical generator n in node i to $\boldsymbol{q}_{-(ni)}$, i.e., the collection of strategies of all firms except for the classical generator n in node i, is obtained by solving the following KKT conditions ($t \in \{1, \ldots, N_T\}; w \in \{1, \ldots, N_w\}$):

$$P_{itw}(.) - c_{ni}^{cg} + (1 - \gamma_{ni}^{cg}) \frac{\partial P_{itw}(.)}{\partial q_{nitw}^{cg}} q_{nitw}^{cg} - \frac{\mu_{nitw}^{cg,max} + \frac{\partial P_{itw}(.)}{\partial q_{nitw}^{cg}} \mu_{nitw}^{cg,cap}}{\Psi_w}$$

$$+ \frac{\mu_{ni(t+1)w}^{cg,up} - \mu_{nitw}^{cg,up} + \mu_{nitw}^{cg,dn} - \mu_{ni(t+1)w}^{cg,dn}}{\Psi_w} \leq 0 \perp q_{nitw}^{cg} \geq 0 \quad (22.11\text{a})$$

$$q_{nitw}^{cg} \leq Q_{ni}^{cg} \perp \mu_{nitw}^{cg,max} \quad (22.11\text{b})$$

$$q_{nitw}^{cg} - q_{ni(t-1)w}^{cg} \leq R_{ni}^{up} \perp \mu_{nitw}^{cg,up} \geq 0 \quad (22.11\text{c})$$

$$q_{ni(t-1)w}^{cg} - q_{nitw}^{cg} \leq R_{ni}^{dn} \perp \mu_{nitw}^{cg,dn} \geq 0 \quad (22.11\text{d})$$

$$P_{itw}(.) \leq P^{cap} \perp \mu_{nitw}^{cg,cap} \geq 0 \quad (22.11\text{e})$$

Best Responses of Storage Firms

To study the best response of the storage firm in node i, let $\boldsymbol{q}_{-i}$ denote the collection of strategies of all firms except for the storage firm in node i. Then, the best response of the storage firm in node i is obtained by solving the following KKT conditions ($t \in \{1, \ldots, N_T\}; w \in \{1, \ldots, N_w\}$):

$$P_{itw}(.) + (1 - \gamma_i^s) \frac{\partial P_{itw}(.)}{\partial q_{itw}^s} q_{itw}^s + \frac{\mu_{itw}^s - \frac{\partial P_{itw}(.)}{\partial q_{itw}^s} \mu_{itw}^{s,cap}}{\Psi_w} = 0 \quad (22.12\text{a})$$

$$\frac{-\eta_i^{dis} \mu_{itw}^s - \mu_{itw}^{dis,max} - \Delta \sum_{k=t}^{N_T} \left(\mu_{ikw}^{s,min} - \mu_{ikw}^{s,max}\right)}{\Psi_w} - c_i^s \leq 0 \perp q_{itw}^{dis} \geq 0 \quad (22.12\text{b})$$

$$\frac{\frac{\mu_{itw}^s}{\eta_i^{ch}} - \mu_{itw}^{ch,max} + \Delta \sum_{k=t}^{N_T} \left(\mu_{ikw}^{s,min} - \mu_{ikw}^{s,max}\right)}{\Psi_w} - c_i^s \leq 0 \perp q_{itw}^{ch} \geq 0 \quad (22.12\text{c})$$

$$q_{itw}^s = \eta_i^{dis} q_{itw}^{dis} - \frac{q_{itw}^{ch}}{\eta_i^{ch}} \quad (22.12\text{d})$$

$$q_{itw}^{dis} \leq \zeta_i^{dis} Q_i^s \perp \mu_{itw}^{dis,max} \geq 0 \quad (22.12\text{e})$$

$$q_{itw}^{ch} \leq \zeta_i^{ch} Q_i^s \perp \mu_{itw}^{ch,max} \geq 0 \quad (22.12\text{f})$$

$$0 \leq \sum_{k=1}^{t} \left(q_{ikw}^{ch} - q_{ikw}^{dis}\right) \Delta \perp \mu_{itw}^{s,min} \geq 0 \quad (22.12\text{g})$$

$$\sum_{k=1}^{t} \left(q_{ikw}^{ch} - q_{ikw}^{dis}\right) \Delta \leq Q_i^s \perp \mu_{itw}^{s,max} \geq 0 \quad (22.12\text{h})$$

$$P_{itw}(.) \leq P^{cap} \perp \mu_{itw}^{s,cap} \geq 0 \quad (22.12\text{i})$$

Best Responses of Transmission Firms

Finally, the best response of the transmission firm between nodes i and j, to $q_{-(ij)}$, i.e., the set of all firms' strategies except those of the transmission line between nodes i and j, can be obtained using the KKT conditions ($t \in \{1, \ldots, N_T\}; w \in \{1, \ldots, N_w\}$):

$$P_{itw}(.) + (1 - \gamma_{ij}^{tr}) \frac{\partial P_{itw}(.)}{\partial q_{ijtw}^{tr}} q_{ijtw}^{tr} + \frac{\mu_{jitw}^{tr} + \mu_{ijtw}^{tr} + \mu_{ijtw}^{tr,min} + \mu_{ijtw}^{tr,max} + \frac{\partial P_{itw}(.)}{\partial q_{ijtw}^{tr}} \mu_{ijtw}^{tr,cap}}{\Psi_w} = 0 \tag{22.13a}$$

$$q_{ijtw}^{tr} = -q_{jitw}^{tr} \tag{22.13b}$$

$$-Q_{ij}^{tr} \le q_{ijtw}^{tr} \perp \mu_{ijtw}^{tr,min} \ge 0 \tag{22.13c}$$

$$q_{ijtw}^{tr} \le Q_{ij}^{tr} \perp \mu_{ijtw}^{tr,max} \ge 0 \tag{22.13d}$$

$$P_{itw}(.) \le P^{cap} \perp \mu_{ijtw}^{tr,cap} \ge 0 \tag{22.13e}$$

22.4.2 The Equivalent Single-Level Optimization Problem

Here, the bi-level price volatility management problem is transformed into a single-level MPEC. To this end, note that for every vector of storage capacities the market price can be obtained by solving the firms' KKT conditions. Thus, by imposing the KKT conditions of all firms as constraints in the optimization problem (22.2), the price volatility management problem can be written as the following single-level optimization problem:

$$\min \sum_{i=1}^{I} Q_i^s \tag{22.14}$$

s.t.

(22.2a–22.2b), (22.10a–22.10c), (22.11a–22.11e), (22.12a–22.12i), (22.13a–22.13e)
$m \in \{1, \ldots, N_i^{wg}\}, n \in \{1, \ldots, N_i^{cg}\}, i, j \in \{1, \ldots, I\}, t \in \{1, \ldots, N_T\}; w \in \{1, \ldots, N_w\}$

where the optimization variables are the storage capacities, the bidding strategies of all firms, and the set of all Lagrange multipliers. Because of the nonlinear complementary constraints, the feasible region is not necessarily convex or even connected. Therefore, increasing the storage capacities stepwise, ΔQ^s, we solve the lower-level problem, which is convex.

REMARK 22.1 *It is possible to convert the equivalent single-level problem (22.14) to a mixed-integer nonlinear problem (MINLP). However, the large number of integer variables potentially makes the resulting MINLP computationally infeasible.*

The MPEC problem (22.14) can be solved using extensive search when the number of nodes is small. For large electricity networks, the greedy algorithm proposed in [46] can be used to find the storage capacities iteratively while the other variables are calculated as the solution of the lower-level problem. In each iteration, the lower-level problem is solved as a mixed complementarity problem (MCP) [47], which is sometimes termed as

Algorithm 22.1 The greedy algorithm for finding the storage allocation
---
while $\max_{it} \text{Var}(P_{itw}(\boldsymbol{q}^\star_{itw})) > \sigma_0^2$ **do**
  iteration=iteration+1
  for $i' = 1 : I$ **do**
    $Q^s_{i'}(\text{iteration}) \leftarrow Q^s_{i'}(\text{iteration} - 1) + \Delta Q^s$
    $Q^s_{-i'}(\text{iteration}) \leftarrow Q^s_{-i'}(\text{iteration} - 1)$
    $\boldsymbol{q}^\star \leftarrow$ Lower-level problem Bayes–NE
    Price Volatility$(i') \leftarrow \max_{it} \text{Var}(P_{itw}(\boldsymbol{q}^\star_{itw}))$
  end for
  $i^* \leftarrow \underset{i}{\text{find}}(\min(\text{Price Volatility}(i)))$
  $Q^s_{i^*}(\text{iteration}) \leftarrow Q^s_{i^*}(\text{iteration} - 1) + \Delta Q^s$
end while
---

rectangular variational inequalities. The optimization solution method is illustrated in Algorithm 22.1. The storage capacity variable is discretized and the increment storage capacity of ΔQ^s is added to the selected node i^* at each iteration of the algorithm. Once the price volatility constraint is satisfied with equality, the optimum solution is found.

Although our greedy algorithm just guarantees a locally optimal storage capacity, we obtained the same results in NEM market using the extensive search.

22.5 Case Study and Simulation Results

In this section, we study the impact of storage installation on price volatility in two nodes of Australia's NEM SA and VIC, with regional pricing mechanism, which sets the marginal value of demand at each region as the regional prices. SA has a high level of wind penetration and VIC has high coal-fueled classical generation. Real data for price and demand from the year 2015 is used to calibrate the inverse demand function in the model. Different types of generation firms, such as coal, gas, hydro, wind, and biomass, with generation capacity (intermittent and dispatchable) of 3.7GW and 11.3GW were active in SA and VIC, respectively, in 2015. The transmission line interconnecting SA and VIC, which is a regulated line, has the capacity of 680MW but currently is working with just 70% of its capacity. The generation capacities in our numerical results are gathered from the Australian Electricity Market Operator's (AEMO's) website (aemo.com.au) and all the prices are shown in Australian dollar.

In our study, we consider a set of scenarios each representing a 24-hour wind power availability profile. To guarantee a high level of accuracy, we do not employ scenario reduction methods [48] and instead consider 365 daily wind power availability scenarios, with equal probabilities, using the realistic data from the year 2015 in different regions of NEM (source of data: AEMO). Figure 22.3 shows the hourly wind power availability in SA. On each box in Figure 22.3, the central mark indicates the median level and the bottom and top edges of the box indicate the 25th and 75th percentiles of wind power availability from the 365 scenarios, respectively. It can be seen that in SA the wind

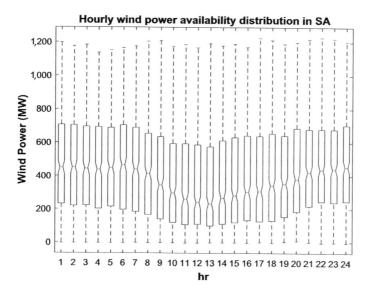

Figure 22.3 SA's Hourly wind power availability distribution in 2015 (the central marks show the median levels and the bottom and top edges of the boxes indicate the 25th and 75th percentiles).

power capacity is about 1,200MW and the wind capacity factor is about 33–42% at different hours.

In what follows, by price volatility we mean the maximum variance of market price, i.e., $\max_{it} \text{Var}(P_{itw}(q^\star_{itw}))$. Also, by the square root of price volatility we mean the maximum standard deviation of market price, i.e., $\max_{it} \sqrt{\text{Var}(P_{itw}(q^\star_{itw}))}$.

22.5.1 One-Region Model Simulations in South Australia

In our one-region model simulations, we first study the impacts of peak demand levels and supply capacity shortage on the standard deviation of hourly electricity prices (or square root of hourly price volatilities) in SA with no storage. Next, we study the effect of storage on the price volatility in SA. Figure 22.4 shows the average and standard deviation of hourly prices for a day in SA (with no storage) for three different cases: (i) a regular-demand day, (ii) a high-demand day, and (iii) a high-demand day with coal-plant outage. An additional load of 1,000 MW is considered in the high-demand case during hours 16, 17, and 18 to study the joint effect of wind intermittency and large demand variations on the price volatility. The additional loads are sometimes demanded in the market due to unexpected high temperatures happening in the region. The coal-plant outage case is motivated by the recent retirement of two coal-plants in SA with total capacity of 770MW [49]. This allows us to investigate the joint impact of wind indeterminacy and low base-load generation capacity on the price volatility.

According to Figure 22.4, wind power fluctuation does not create much price fluctuation in a regular-demand day. The square root of the price volatility in the regular-demand day is equal to 65 $/MWh. Depending on the wind power availability

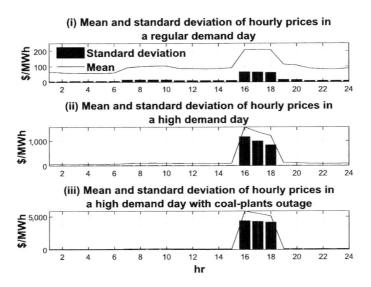

Figure 22.4 Standard deviation and mean of hourly wholesale electricity prices in SA with no storage.

level, the peak price varies from 92 $/MWh to 323 $/MWh, with average of 210 $/MWh, in a regular-demand day. Based on Figure 22.4, the square root of the price volatility in the high-demand day is equal to 1,167 $/MWh. The maximum price in a high-demand day in SA changes from 237 $/MWh to 4,466 $/MWh, with the average of 1,555 $/MWh, because of wind power availability fluctuation. The extra load at peak times and the wind power fluctuation create a higher level of price volatility during a high-demand day compared with a regular-demand day.

The retirement (outage) of coal plants in SA beside the extra load at peak hours increases the price volatility due to the wind power fluctuation. The maximum price during the high-demand day with coal-plant outage varies from 377 $/MWh to the cap price of 11,000 $/MWh, with the average of 5,832 $/MWh. The square root of the price volatility during the high-demand day with coal-plant outage is equal to 4,365 $/MWh. The square root of the price volatility during the high-demand day with coal-plant outage is almost 67 times more than the regular-demand day due to the simultaneous variation in both supply and demand.

Figure 22.5 shows the minimum required (strategic/regulated) storage capacities for achieving various levels of price volatility in SA during a high-demand day with coal-plant outage. The minimum storage capacities are calculated by solving the optimization problem (22.14) for the high-demand day with coal-plant outage case. According to Figure 22.5, a strategic storage firm requires a substantially larger capacity, compared with a regulated storage firm, to achieve a target price volatility level due to the selfish behavior of the storage firms. In fact, the strategic storage firms may sometimes withhold their available capacities and do not participate in the price volatility reduction as they do not always benefit from reducing the price. The price volatility in SA can be reduced

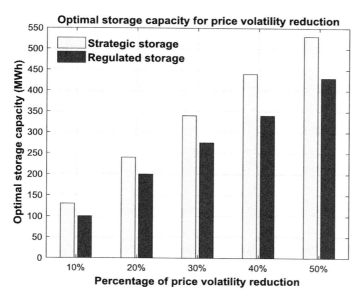

Figure 22.5 Optimal strategic and regulated storage capacity for achieving different price volatility levels in SA region for a high-demand day with coal-plant outage.

by 50% using either 530MWh strategic storage or 430MWh regulated storage. Note that AEMO has forecasted about 500MWh battery storage to be installed in SA until 2035 [50].

According to our numerical results, storage can displace the peaking generators, with high fuel costs and market power, which results in reducing the price level and the price volatility. A storage capacity of 500MWh (or 500MW given the discharge coefficient $\eta^{dis} = 1$) reduces the square root of the price volatility from 4365 $/MWh to 2,692 $/MWh, almost a 38% reduction, during a high-demand day with coal-plant outage in SA.

The behavior of the peak and the daily average prices for the high-demand day with coal-plant outage in SA is illustrated in Figure 22.6. In this figure, the peak price represents the average of highest prices over all scenarios during the day, i.e. $\sum_w \Psi_w(\max_t P_{tw}(q^\star_{tw}))$ and the daily average price indicates the average of price over time and scenarios, i.e., $\frac{1}{N_T} \sum_{tw} P_{tw}(q^\star_{tw})\Psi_w$. Sensitivity analysis of the peak and the daily average prices in SA with respect to storage capacity indicates that high storage capacities lead to relatively low prices in the market. At very high prices, demand is almost inelastic and a small amount of excess supply leads to a large amount of price reduction. According to Figure 22.6, the rate of peak price reduction decreases as the storage capacity increases because large storage capacities lead to lower peak prices, which make the demand more elastic.

Based on Figure 22.6, the impact of storage on the daily average and peak prices depends on whether the storage firm is strategic or regulated. It can be observed that the impacts of strategic and regulated storage firms on the daily peak/average prices are

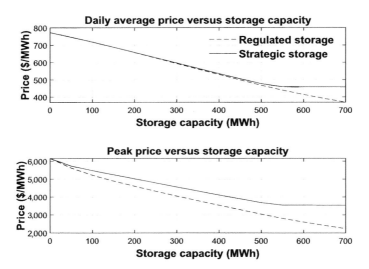

Figure 22.6 Daily peak and average prices in SA versus storage capacity in a high-demand day with coal-plant outage.

almost similar for small storage capacities, i.e., when the storage capacity is smaller than 100MWh (or 100MW given $\eta^{dis} = 1$). However, a regulated firm reduces both the peak and the average prices more efficiently compared with a strategic storage firm as its capacity becomes large. A large strategic storage firm in SA does not use its excess capacity beyond 500MWh to reduce the market price because it acts as a strategic profit maximizer, but a regulated storage firm contributes to the price volatility reduction as long as there is potential for price reduction by its operation.

Figure 22.7 depicts the square root of price volatility versus storage capacity in SA during the high-demand day with coal-plant outage. According to this figure, the price volatility in the market decreases by installing either regulated or strategic storage devices. To reduce the square root of price volatility to 3,350 $/MWh, the required strategic capacity is about 100MWh more than that of a regulated storage. Moreover, a strategic storage firm stops reducing the price volatility when its capacity exceeds a threshold value. In our study, a strategic storage firm does not reduce the square root of price volatility more than 32%, but a regulated firm reduces it by 89%. These observations confirm that regulated storage firms are more efficient than strategic firms in reducing the price volatility.

The impact of the regulated storage firm in reducing the price volatility can be divided into three ranges of initial, efficient, and saturated, as shown in Figure 22.7. In the initial range, an increment in the capacity of the regulated firm slightly reduces the price volatility. Then the price volatility reduces sharply with storage capacity in the second region. Finally, the price volatility reduction gradually stops in the saturated region. This observation implies that although storage alleviates the price volatility in the market, it is not capable to eliminate it completely.

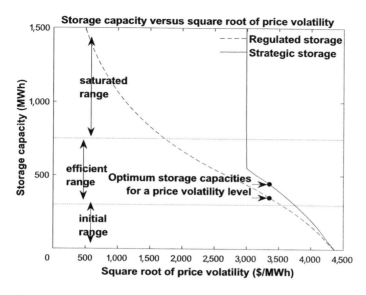

Figure 22.7 Square root of price volatility in SA versus storage capacity during a high-demand day with coal-plant outage.

22.5.2 Two-region Model Simulations in South Australia and Victoria

In the one-region model simulations, we analyzed the impact of storage on the price volatility in SA when the SA–VIC interconnector is not active. In this subsection, we first study the effect of the interconnector between SA and VIC on the price volatility in the absence of storage firms. Next, we investigate the impact of storage firms on the price volatility when the SA–VIC transmission line operates at various capacities. In our numerical results, SA is connected to VIC using a 680MW interconnector which is currently operating with 70% of its capacity, i.e., 30% of its capacity is under maintenance. The numerical results in this subsection are based on the two-node model for a high-demand day with coal-plant outage in SA. To investigate the impact of transmission lines on price volatility, it is assumed that the SA–VIC interconnector operates with 60% and 70% of its capacity.

According to our numerical results, the peak price (the average of the highest prices in all scenarios) in SA is equal to 6,154 $/MWh when the SA–VIC interconnector is completely in outage. However, the peak price reduces to 3,328 $/MWh and 2,432 $/MWh when the interconnector operates at 60% and 70% of its capacity. The square root of price volatility is 4,365 $/MWh, 860 $/MWh, and 614 $/MWh when the capacity of the SA–VIC transmission line is equal to 0%, 60%, and 70%, respectively, which emphasizes the importance of interconnectors in price volatility reduction.

Simulation results show that as long as the interconnector is not congested, the line alleviates the price volatility phenomenon in SA by importing electricity from VIC to SA at peak times. Because the market in SA compared to VIC is much smaller, about three times, the price volatility abatement in SA after importing electricity from VIC is

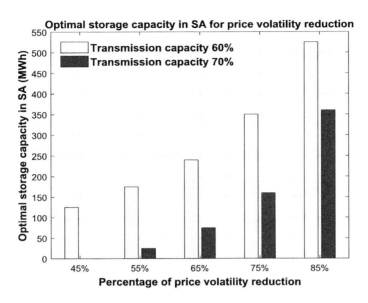

Figure 22.8 Optimal regulated storage capacity versus the percentage of price volatility reduction in the two-node market in a high-demand day with coal-plant outage in SA.

much higher than the price volatility increment in VIC. Moreover, the price volatility reduces as the capacity of transmission line increases.

Figure 22.8 shows the optimum storage capacity versus the percentage of price volatility reduction in the two-node market. According to our numerical results, storage is just located in SA, which witnesses a high level of price volatility as the capacity of the transmission line decreases. According to this figure, the optimum storage capacity becomes large as the capacity of transmission line decreases. Note that a sudden decrease of the transmission line capacity may result in a high level of price volatility in SA. However, based on Figure 22.8, storage firms are capable of reducing the price volatility during the outage of the interconnecting lines.

22.6 Conclusion

High penetration of intermittent renewables, such as wind or solar farms, brings high levels of price volatility in electricity markets. Our study presents an optimization model that decides on the minimum storage capacity required for achieving a price volatility target in electricity markets. Based on our numerical results, the impact of storage on the price volatility in one-node electricity market of SA and two-node market of SA–VIC can be summarized as:

- Storage alleviates price volatility in the market due to the wind intermittency. However, storage does not remove price volatility completely, i.e., storage stops reducing the price volatility when it is not profitable.

- The effect of a storage firm on price volatility reduction depends on whether the firm is regulated or strategic. Both storage types have similar operation behavior and price reduction effect when they possess small capacities. For larger capacities, a strategic firm may underutilize its available capacity and stop reducing the price level due to its profit maximization strategy. However, a regulated storage firm is more efficient in price volatility reduction because of its strategy on improving the social welfare. The price level and volatility reduction patterns observed when storage firms are regulated provide stronger incentives for the market operator to subsidize the storage technologies.
- Both storage devices and transmission lines are capable of reducing the price volatility. High levels of price volatility that may happen due to the line maintenance can be alleviated by storage devices.
- Although many parameters affect the price volatility level of a system, penetration of intermittent wind power generation in a region makes the region or node highly price volatile when a classical generation capacity outage happens or high load level is demanded.

This study can be extended by studying the impact of ancillary services markets [51] and capacity markets [52] on the integration of storage systems in electricity networks and studying the wind correlation analysis to look at volatility reduction effectiveness. The future research can also include the optimal storage siting problem subject to the line congestion constraint to alleviate the congestion problem.

Acknowledgment

This work was supported in part by the Australian Research Council Discovery Project DP140100819.

References

[1] A. G. Kagiannas, D. T. Askounis, and J. Psarras, "Power generation planning: A survey from monopoly to competition," *International Journal of Electrical Power & Energy Systems*, pp. 413–421, 2004.

[2] A. Tishler, I. Milstein, and C.-K. Woo, "Capacity commitment and price volatility in a competitive electricity market," *Energy Economics*, vol. 30, no. 4, pp. 1625–1647, 2008.

[3] J. C. Ketterer, "The impact of wind power generation on the electricity price in Germany," *Energy Economics*, vol. 44, pp. 270–280, 2014.

[4] J. Eyer and G. Corey, "Energy storage for the electricity grid: Benefits and market potential assessment guide," Sandia National Laboratories, Tech. Rep., 2010.

[5] IEA, "Technology roadmap, hydro power," International Energy Agency, Tech. Rep., 2012.

[6] D. McConnell, "A month in, Teslas SA battery is surpassing expectations," AEMC's Consumer Advocacy Panel and Energy Consumers Australia, Tech. Rep., 2018.

[7] D. Wozabal, C. Graf, and D. Hirschmann, "The effect of intermittent renewables on the electricity price variance," *OR Spectrum*, pp. 1–23, 2014.

[8] C.-K. Woo, I. Horowitz, J. Moore, and A. Pacheco, "The impact of wind generation on the electricity spot-market price level and variance: The Texas experience," *Energy Policy*, vol. 39, no. 7, pp. 3939–3944, 2011.

[9] D. Chattopadhyay and T. Alpcan, "A game-theoretic analysis of wind generation variability on electricity markets," *Power Systems, IEEE Transactions on*, vol. 29, no. 5, pp. 2069–2077, Sept. 2014.

[10] S.-J. Deng and S. S. Oren, "Electricity derivatives and risk management," *Energy*, vol. 31, no. 6, pp. 940–953, 2006.

[11] H. Higgs and A. Worthington, "Stochastic price modeling of high volatility, mean-reverting, spike-prone commodities: The Australian wholesale spot electricity market," *Energy Economics*, vol. 30, p. 3172–3185, 2008.

[12] X. Hu, G. Grozev, and D. Batten, "Empirical observations of bidding patterns in Australias National Electricity Market," *Energy Policy*, vol. 33, p. 2075–2086, 2005.

[13] J. C. Harsanyi, "Games with incomplete information played by "bayesian" players, I-III. Part II. Bayesian equilibrium points," *Management Science*, vol. 14, no. 5, pp. 320–334, 1968.

[14] AEMC, "Potential generator market power in the NEM," Australian Energy Market Commission, Tech. Rep. ERC0123, Apr. 26, 2013.

[15] N. Li, C. Ukun, E. M. Constantinescu, J. R. Birge, K. W. Hedman, and A. Botterud, "Flexible operation of batteries in power system scheduling with renewable energy," *IEEE Transactions on Sustainable Energy*, vol. 7, no. 2, pp. 685–696, Apr. 2016.

[16] V. Krishnan and T. Das, "Optimal allocation of energy storage in a co-optimized electricity market: Benefits assessment and deriving indicators for economic storage ventures," *Energy*, vol. 81, pp. 175–188, 2015.

[17] A. Berrada and K. Loudiyi, "Operation, sizing, and economic evaluation of storage for solar and wind power plants," *Renewable and Sustainable Energy Reviews*, vol. 59, pp. 1117–1129, 2016.

[18] W. Qi, Y. Liang, and Z.-J. M. Shen, "Joint planning of energy storage and transmission for wind energy generation," *Operations Research*, vol. 63, no. 6, pp. 1280–1293, 2015.

[19] M. Sedghi, A. Ahmadian, and M. Aliakbar-Golkar, "Optimal storage planning in active distribution network considering uncertainty of wind power distributed generation," *IEEE Transactions on Power Systems*, vol. 31, no. 1, pp. 304–316, Jan. 2016.

[20] L. Zheng, W. Hu, Q. Lu, and Y. Min, "Optimal energy storage system allocation and operation for improving wind power penetration," *IET Generation, Transmission Distribution*, vol. 9, no. 16, pp. 2672–2678, 2015.

[21] J. Xiao, Z. Zhang, L. Bai, and H. Liang, "Determination of the optimal installation site and capacity of battery energy storage system in distribution network integrated with distributed generation," *IET Generation, Transmission Distribution*, vol. 10, no. 3, pp. 601–607, 2016.

[22] Y. Zheng, Z. Y. Dong, F. J. Luo, K. Meng, J. Qiu, and K. P. Wong, "Optimal allocation of energy storage system for risk mitigation of DISCOs with high renewable penetrations," *IEEE Transactions on Power Systems*, vol. 29, no. 1, pp. 212–220, Jan. 2014.

[23] R. Walawalkar, J. Apt, and R. Mancini, "Economics of electric energy storage for energy arbitrage and regulation in New York," *Energy Policy*, vol. 35, no. 4, pp. 2558–2568, 2007.

[24] H. Mohsenian-Rad, "Optimal Bidding, scheduling, and deployment of battery systems in California day-ahead energy market," *IEEE Transactions on Power Systems*, vol. 31, no. 1, pp. 442–453, Jan. 2016.

[25] H. Akhavan-Hejazi and H. Mohsenian-Rad, "A stochastic programming framework for optimal storage bidding in energy and reserve markets," in *Innovative Smart Grid Technologies (ISGT), 2013 IEEE PES*, pp. 1–6, Feb. 2013.

[26] H. Mohsenian-Rad, "Coordinated price-maker operation of large energy storage units in nodal energy markets," *IEEE Transactions on Power Systems*, vol. 31, no. 1, pp. 786–797, Jan. 2016.

[27] E. Nasrolahpour, S. J. Kazempour, H. Zareipour, and W. D. Rosehart, "Strategic sizing of energy storage facilities in electricity markets," *IEEE Transactions on Sustainable Energy*, vol. 7, no. 4, pp. 1462–1472, Oct. 2016.

[28] X. Fang, F. Li, Y. Wei, and H. Cui, "Strategic scheduling of energy storage for load serving entities in locational marginal pricing market," *IET Generation, Transmission Distribution*, vol. 10, no. 5, pp. 1258–1267, 2016.

[29] A. Awad, J. Fuller, T. EL-Fouly, and M. Salama, "Impact of energy storage systems on electricity market equilibrium," *Sustainable Energy, IEEE Transactions on*, vol. 5, no. 3, pp. 875–885, July 2014.

[30] M. Ventosa, R. Denis, and C. Redondo, "Expansion planning in electricity markets: Two different approaches," in *Proceedings of the 14th Power Systems Computation Conference (PSCC), Seville*, 2002.

[31] W.-P. Schill, C. Kemfert et al., "Modeling strategic electricity storage: The case of pumped hydro storage in Germany," *Energy Journal-Cleveland*, vol. 32, no. 3, p. 59, 2011.

[32] M. P. Moghaddam, A. Abdollahi, and M. Rashidinejad, "Flexible demand response programs modeling in competitive electricity markets," *Applied Energy*, vol. 88, no. 9, pp. 3257–3269, 2011.

[33] A. Masoumzadeh, D. Möst, and S. C. Ookouomi Noutchie, "Partial equilibrium modelling of world crude oil demand, supply and price," *Energy Systems*, pp. 1–10, 2016.

[34] P. Graham, S. Thorpe, and L. Hogan, "Non-competitive market behaviour in the international coking coal market," *Energy Economics*, vol. 21, no. 3, pp. 195–212, 1999.

[35] D. Kirschen, G. Strbac, P. Cumperayot, and D. de Paiva Mendes, "Factoring the elasticity of demand in electricity prices," *Power Systems, IEEE Transactions on*, vol. 15, no. 2, pp. 612–617, May 2000.

[36] G. Kirchgässner and J. Wolters, *Introduction to Modern Time Series Analysis*. Berlin: Springer Science & Business Media, 2007.

[37] S. A. Gabriel, A. J. Conejo, J. D. Fuller, B. F. Hobbs, and C. Ruiz, *Complementarity Modeling in Energy Markets*. Vol. 180. New York: Springer Science & Business Media, 2012.

[38] A. Masoumzadeh, E. Nekouei, and T. Alpcan, "Long-term stochastic planning in electricity markets under carbon cap constraint: A Bayesian game approach," in *2016 IEEE Innovative Smart Grid Technologies – Asia (ISGT-Asia)*, pp. 466–471, Nov. 2016.

[39] X. Fang, F. Li, Q. Hu, and Y. Wei, "Strategic CBDR bidding considering FTR and wind power," *IET Generation, Transmission Distribution*, vol. 10, no. 10, pp. 2464–2474, 2016.

[40] J. B. Cardell, C. C. Hitt, and W. W. Hogan, "Market power and strategic interaction in electricity networks," *Resource and Energy Economics*, vol. 19, no. 1–2, pp. 109–137, 1997.

[41] W. W. Hogan, "A market power model with strategic interaction in electricity networks," *The Energy Journal*, vol. 18, no. 4, pp. 107–141, 1997.

[42] E. G. Kardakos, C. K. Simoglou, and A. G. Bakirtzis, "Optimal bidding strategy in transmission-constrained electricity markets," *Electric Power Systems Research*, vol. 109, pp. 141–149, 2014.

[43] AEMO, "An introduction to Australia's national electricity market," Australian Energy Market Operator, Tech. Rep., 2010.
[44] D. Fudenberg and J. Tirole, *Game Theory*. Cambridge, MA: M Press, 1991.
[45] M. C. Ferris and T. S. Munson, "GAMS/PATH user guide: Version 4.3," *Washington, DC: GAMS Development Corporation*, 2000.
[46] M. J. Neely, A. S. Tehrani, and A. G. Dimakis, "Efficient algorithms for renewable energy allocation to delay tolerant consumers," in *Smart Grid Communications (SmartGridComm), 2010 First IEEE International Conference on*. IEEE, pp. 549–554, 2010.
[47] M. C. Ferris and T. S. Munson, "Complementarity problems in GAMS and the PATH solver," *Journal of Economic Dynamics and Control*, vol. 24, no. 2, pp. 165–188, 2000.
[48] J. M. Morales, S. Pineda, A. J. Conejo, and M. Carrion, "Scenario reduction for futures market trading in electricity markets," *IEEE Transactions on Power Systems*, vol. 24, no. 2, pp. 878–888, 2009.
[49] AER, "State of the energy market 2015," Australian Energy Regulator, Tech. Rep., 2015.
[50] AEMO, "Roof-top PV information paper, national electricity forecasting," Australian Energy Market Operator, Tech. Rep., 2012.
[51] D. Chattopadhyay, "Multicommodity spatial Cournot model for generator bidding analysis," *Power Systems, IEEE Transactions on*, vol. 19, no. 1, pp. 267–275, Feb. 2004.
[52] D. Chattopadhyay and T. Alpcan, "Capacity and energy-only markets under high renewable penetration," *Power Systems, IEEE Transactions on*, vol. PP, no. 99, pp. 1–11, 2015.

Index

2003 U.S. Northeastern blackout, 101
2011 San Diego blackout, 101

AC frequency, 371
ACOPF, 363, 368
AC power flow model, 341
AC power generation, 365
active and power injection
   measurement sample, 108
active and reactive power injection
   difference, 109
   net, 108
   variation, 109, 112
ACTIVSg2000, 320, 321, 323, 325, 326, 336
admittance, 366
admittance matrix, 87, 318, 320, 329, 331, 333, 336, 404
AGC, 372
alternating direction method of multipliers (ADMM), 23, 421
ambiguous chance constraint, 390
ancestor nodes, 15
anomaly detection, 156, 158–160, 165, 166
appliance, 407
appliance load monitoring techniques, 231
approximate grid model, 4
asynchronous algorithm, 412
attack
   node-based, 177
   random, 177
   targeted, 177, 179
attack model
   additive, 199, 211
   deterministic , 199
   random, 211
attack vector
   additive, 199
   Bayesian, 199
   centralized, 201
   complementary, 206
   decentralized, 205
   detection, 201
   deterministic, 199
   global, 206
   individual, 205
   information-theoretic, 212
   nondetectable, 201
   random, 211
   with maximum distortion, 205
   with minimum probability of detection, 201
autocorrelation coefficients, 60

bad data, 77
Batch normalization, 58
Bayesian Nash equilibrium, 548
Bayesian risk, 355
Bayesian state estimation, 198
best response
   correspondence, 207
   dynamic, 208
Betti numbers, 180
betweenness centrality
   edge, 177, 186
   node, 177, 186
big power-grid data, 148
Blahut Arimoto algorithm, 249
blockchain, 236
branch-and-bound, 352
Brownian motion, 505
bundle method, 412
bus admittance matrix, 341, 378

capacitors, 458, 459, 462
cascading failure, 101
centralized architecture, 401
central limit theorem, 151
central processor, 401
chance constraint, 381
charging station, 456, 464–466, 479
chemical storage, 458
chi^2 test, 77
Cholesky, 76
Clarke transform, 296
   Clarke matrix, 296
   degrees of freedom, 297
   PCA equivalence, 297
   unbalanced system case, 299
   widely linear model, 305

classical generators, 554
coherence, 348
communication
   one-way, 432, 440
   two-way, 432, 440, 446
communication network, 402
community energy storage, 468, 479
community identification, 333
complex
   simplicial, 180
   Vietoris-Rips, 180
complex power, 341, 404
component
   connected, 178
   giant, 178
compressed sensing, 340
computational complexity, 265
concentration, 179
connectivity loss, 178
consensus, 421
contingency analysis, 101
continuation power flow, 68
control actions, 401
control area, 418
controllable devices, 401
controllable load, 404, 431, 446, 447, 449–453
controllable resource, 406
control inputs, 401
convergence, 410, 411, 413, 415
   linear, 413
convex
   strictly, 432–434, 447
   strongly, 441
convex relaxation, 21, 405
coordinate descent, 349
covariance matrix of PMU data, 392
critical measurements, 77
critical parameters, 91
critical region, 39, 41
cumulative distribution function, 471–474, 479
current, 404
cutoff price, 530–533, 535–537, 541
cutting-plane algorithms, 384
cutting plane method, 412
   disaggregated, 412
cyber data attacks, 271

data
   uncertainty, 183
   variability, 183
data anonymization, 237
data clustering, 278
data compression, 323
data correction, 267
data fitting, 527, 536, 537, 539
data modeling, 148
data obfuscation, 236

data privacy, 275
data recovery, 263
DC approximation to power flows, 370
DCOPF (direct current optimal power flow), 30, 370
decentralized charging algorithm (DCA), 437–440
decision making, 524, 526, 527, 531, 534, 542
demand
   exogenous, 489, 490
   management, 506
   response, 504, 506
demand charge, 459, 464, 465
demand-side management, 406
descendant nodes, 15
detour motif, 185
dictionary, 41
dictionary learning, 41
dictionary learning
   online, 42
differential grid data, 5
differential privacy, 236
discrete Fourier transform, 317
discrete variables, 425
distortion, 201
   average, 207, 208
   excess, 202, 203, 205, 206
   expected, 206, 207
   highest, 207
   mean square error, 227
   minimum, 204
distributed algorithm, 418
distributed dynamic charging, 494
distributed energy resources, 404
distributed feedback control, 432, 446
distribution network, 459, 460, 462, 468
distribution system, 404, 412
disturbances, 401
disutility, 447, 450, 453
dual function, 404, 409, 414
dual method, 408–413
   distributed implementation, 411
dual problem, 404, 411
dual variable, 404, 409
duality gap, 404
duck curve, 28, 45
dynamic
   asymptotically stable, 514
   best-response, 483, 492
   closed-loop, 513–515
   distributed, 492
   macroscopic, 505, 521
   microscopic, 505, 513, 521
   open-loop, 514, 521

economic dispatch
   multiperiod, 33, 35
   single-period, 33, 34

edge weight, 186
effective demand, 476–478
effective secrecy, 214
elastic demand, 406
Electric Reliable Council of Texas (ERCOT), 320
electricity, 455, 456, 459, 460, 463, 464
electricity market, 406, 550
electric vehicle, 404, 407, 483, 491, 497, 502, 504, 506, 507
electrochemical, 457, 458, 462
elimination of variable, 405
empirical mutual information, 254
encryption, 235
energy market, 524–528, 540
energy storage, 404, 407
energy storage systems, 455–465, 467–470, 475–477
equilibrium
  mean-field, 506
  Nash, 483, 487, 506, 519
  robust, 519
Euler digamma function, 226
EUT, 524–527, 531, 532, 535, 537, 539–542
EV charging, 432
expected utility theory, 524, 525, 530

false alarm probability, 90
fault events, 344
financial analytics, 394
Fisher information, 254
fixed frequency modulation, 288
fluid model, 467
Fokker–Planck–Kolmogorov equation, 506
forecast
  point, 28
  probabilistic, 28
forecasting
  error, 484, 485, 491, 495–497, 502
  noise, 496
  signal-to-noise ratio, 490, 491, 496, 502
Frank–Wolfe algorithm, 11
free cumulant, 150
free probability, 150
frequency, 418
frequency control, 446, 450, 452
frequency estimation, 288
  frequency demodulation, 289
  maximum likelihood approach, 292
  smart DFT, 291
frequency filtering, 246
FRP (flexible ramping product), 35
function
  closed, 403
  continuous, 403
  convex, 403
  proper convex, 403

  strictly convex, 411
  strongly convex, 414

game
  aggregative, 488
  coalitional-form, 487
  extensive-form, 487
  formulation, 206
  mean-field, 505, 506, 512, 513
  normal-form, 487
  in normal-form, 206
  ordinal potential, 488
  potential, 207, 208, 488
  solution, 207
  strategic-form, 487, 494
  theory, 483, 487, 505
gather-and-broadcast, 402
Gaussian assumption, 389
Gaussian copula method, 58
Gaussian orthogonal ensemble, 148
Gaussian unitary ensemble, 148
generative adversarial networks, 54
generator, 406
geometry
  global, 178, 188
  local, 178, 187
global positioning satellite, 81
gradient, 403
graph, 15, 176
  ancestor nodes, 15
  descendant nodes, 15
  internal nodes, 15
  Kron reduction, 18
  level sets, 15
  node depth, 15
  terminal nodes, 15
  tree graph, 15
graph Laplacian, 406
graph Laplacian matrix, 13, 342
graph signal processing
  graph Fourier transform, 316
  Laplacian, 315, 316
  graph filters, 317, 319
  graph shift operator, 315
greedy method, 347, 352
grid control, 368
grid game, 527–529, 540, 541
grid operator, 401
grid probing using smart inverters, 13

Hadamard product, 145
Hamilton–Jacobi–Bellman equation, 506, 512
Hamilton–Jacobi–Isaacs equation, 506
Hamiltonian function, 512
Hankel matrix, 262
Holt's method, 511
Holt's model, 508, 509, 511, 519

Index

hybrid state estimator, 82
hypothesis testing, 152, 197, 200, 212, 253

IEEE
   118-bus test system, 102, 116–121, 160, 219
   14-bus test system, 219
   30-bus test system, 219, 221
   33-bus test system, 168
incidence matrix, 406
interior point methods, 391
internal nodes, 15
intrusive load monitoring techniques, 231, 232
inverse demand function, 545, 550
inverter, 404
invisible units detection, 167

Jacobian
   matrix, 171, 198
   measurement matrix, 198, 210, 215, 217, 219
Jensen's inequality, 225

Karush–Kuhn–Tucker conditions, 401
kernel-based learning, 10
kernel learning, 11
Kirchhoff's current law, 144
Kron reduction, 18
Kronecker delta function, 146
Kullback–Leibler divergence, 67, 213

Lagrange multiplier, 89, 409
Lagrangian function, 404
   regularized, 414
Laguerre unitary ensemble, 148, 151
largest normalized residual, 77
Lasso, 340
least absolute value estimator, 80
level sets, 15
leverage point, 80
likelihood ratio, 200, 212
Lindsay–Pilla–Basak method, 217, 219
linear approximation, 405, 408
linear eigenvalue statistics, 144, 151, 152, 168, 169
linear estimation matrix, 198
linear programming, 80
linear quadratic problem, 513
linearized system dynamics, 198
line limit, 369
line outage identification, 115
line temperature, 369
Lipschitz continuous, 414
lithium battery, 456–458, 462, 469, 478
LMP (locational marginal price), 28, 35, 37
   real-time, 35
load management, 431, 453
   centralized, 431
   distributed, 432, 452

load-serving entity, 406
loss aversion, 526, 542
loss function, 55
low-rank matrix, 262
Lyapunov function, 243

magnetic field energy storage, 458
manifold, 418
Marcenko–Pastur law, 149, 151
market
   co-optimized energy and reserve, 37
   day-ahead, 32
   real-time, 33
Markov
   decision process, 484, 497
Markov chain, 463, 465, 466, 470, 473, 475
Markov decision process, 247, 250, 253, 255
MATLAB, 493, 514
MATPOWER, 169, 219, 385
matrix completion, 263
maximum *a-posteriori* probability (MAP) approach, 7
maximum likelihood approach, 5
maximum likelihood estimation
   instantaneous frequency, 293
   frequency estimation, 292
mean-field
   coupling, 511
   equilibrium, 506, 512, 513, 519, 521
   game, 505, 506, 512, 513
   linear-quadratic, 506
mean square error, 198
measurement, 407, 413
measurement-based algorithm, 415
measurement Jacobian, 79
mechanical storage, 457, 462
mechanism design, 505
message passing, 402
method of moments, 150
method of partial correlations, 9
metric
   resilience, 187
   robustness, 176, 184
microgrid, 459, 462, 463
minimax, 56
minimum mean square error, 199
modal transformation, 85
model
   configuration, 185
   Erdos–Renyi, 185
   preferential attachment, 185
   small-world, 185
MOMENTCHI2, 219
Monte Carlo simulation, 30, 65
MSMO (model of real-time system and market operation), 32

Index

multilayer perceptron, 56
multiphase system, 405
multiparametric program
   linear, 33, 39
   quadratic, 33, 39
mutual information, 211, 212, 215, 220, 249

N-transform, 157
Nash equilibrium, 207, 208, 483, 487, 506, 519
national electricity market, 560
natural hazard, 177
net load, 28
network
   complex, 175
   filtration, 180
   motif, 175, 179
   power grid, 176, 180
network admittance matrix, 101, 108
network map, 401, 413, 415
network parameter errors, 88
network topology, 101
Neyman–Pearson Lemma, 212
node
   degree, 177, 186
   strength, 186
node depth, 15
noisy power method, 393
nonintrusive load monitoring techniques, 231, 232
normal distribution, 78

observability, 76
observable islands, 81
Ohm's law, 144
online learning, 432, 440, 441, 446
optimal charging (OC), 433, 434
optimal load control (OLC), 447–452
optimal PMU placement, 325, 331
optimal power flow, 404
   AC, 404, 417
   multi-area, 419
optimal solution
   existence of, 403
optimistic mirror descent (OMD), 441–443, 445
optimization
   centralized, 401
   convex, 403
   decentralized, 403
   distributed, 402
   in-network, 403
   local, 403
   Lyapunov, 425
   nonconvex, 424
   online, 425
optimization problem, 552
orthogonal matching pursuit (OMP), 340, 347

outage detection, 63
outage probability, 463–465, 467, 470, 477, 478

Park transform, 296
   an FM demodulation interpetation, 301
   positive and negative voltage sequences, 301
   sources of bias in unbalanced system conditions, 302
partial differential equation, 506
participant, 524, 526–530, 533, 537, 539–542
participating generators, 372
peer-to-peer communication, 403
percolation limit, 178
performance, 463–465, 467
persistence
   barcode, 181
   diagram, 180, 181
phase imbalance, 345
phase measurement units, 144, 147, 157, 172
phasor, 366
phasor measurement unit (PMU), 74, 102, 111, 112, 116, 121, 261, 312, 340, 408
   synchrophasor, 312
   data, 102
   sample time, 108
photovoltaic system, 404
plug-in electric vehicles, 455, 456, 459, 462–465
PMU (power management unit), 29, 340, 392
Poisson process, 465, 469
polynomial chaos, 391
potential function, 207
power balance equation, 367
power consumption scheduling
   distributed algorithm, 488, 490
   formulation and analysis, 487
   game, 487, 488
   policy, 491
power curtailment, 405
power demand, 101
power flow, 146, 147, 169, 171, 404
   AC, 404, 408
   DC, 406
   limit, 406
power flow equations, 62, 108
power flow problem, 367
power grid, 455, 456, 459, 461, 463–465, 467, 469, 475, 477, 479
power line outages, 343
power network, 432, 446, 449, 452
power quality, 458–461, 468
Price, 434, 437, 438, 440, 441
price of anarchy, 488, 494
price of decentralization, 488
price volatility, 547, 564
pricing function, 442, 443
primal variable, 404, 409

primal-dual algorithm, 448–450
primal-dual gradient, 415
primal-dual gradient method, 418
primal-dual method, 413–418
    measurement-based, 413
    model-based, 413
    regularized, 413
primary response, 372
processor, 401
prospect theory, 524, 525
prosumer, 524–528, 530, 533, 541
proximal term, 412
pseudo-inverse, 379
pseudonyms, 237

Q-learning, 247
quantization, 248
quickest change detection (QCD), 101, 102
    conditional average detection delay (CADD), 105
    CuSum test, 104
    dynamic CuSum (D-CuSum) test, 107
    generalized CuSum test, 114
    generalized D-CuSum test, 114
    generalized Shewhart test, 113
    GLR-CuSum test, 106
    mean time to false alarm (MTFA), 105
    meanshift test, 112
    Shewhart test, 104
    worst-case average detection delay (WADD), 105

R-transform, 149, 150
ramp constraints, 406
random matrix theory, 144
rank, 419
ranked list, 115
rate of change of phase angle, 287
    instantaneous frequency, 287
    maximum likelihood principle, 293
rate-distortion function, 249
real time, 417
real-time bus output under stochastics, 375
receding horizon, 509
    control policies, 510, 511
    equilibrium, 513
    model, 511
    optimization, 511
    problem, 511–513
    strategies, 513
    technique, 508
rechargeable battery, 239, 240
rectified linear units, 57
recursive DFT, 290
    FM demodulation equivalence, 290
    phasor measurement unit (PMU), 290
reference bus, 379

regret minimization, 440–442
reliability function, 179
remedial action scheme, 101
renewable energy, 455, 459, 460, 462, 463
renewable energy source, 239, 240
replica method, 150
residual covariance matrix, 78
residuals, 78
resilience, 176, 188
responding generators, 372
Riccati equation, 515
robust chance-constrained ACOPF, 391
robust equilibrium, 519
robust optimization, 390
robustness, 176
running average, 410, 413

S-transform, 150
saddle point, 414
sample covariance matrix, 223
sampling theory
    graph signal processing, 321, 325
SCADA (supervisory control and data acquisition), 29, 74, 313
scenario, 525–528, 530, 541, 542
scheduling policy, 499
separable optimization problem, 409
setpoint, 407
Shannon lower bound, 249
Sharpe ratio, 394
situational awareness, 101
Slater's constraint qualification, 403
Smart Clarke transform, 305
    algorithm layout, 307
    frequency estimation in ubalanced systems, 306
smart DFT
    frequency estimation, 291
    phasor computation, 293
    window functions, 294
smart grid, 524–528, 541, 542
smart meter privacy, 230, 232
Smart Park transform, 306
    algorithm layout, 307
social welfare, 406
sparse overcomplete representation, 343
stability, 178
star topology, 402, 409, 411
state estimation, 63, 407
    multi-area, 419
state estimator, 101
statistical
    inference, 183
    significance, 183
Stieltjes transform, 149, 150, 157
stochastic, 464, 465, 469, 470, 476, 479
stochastic shortest path problem, 499

storage, 547
storage firm, 554
strategy
   attack, 176
   best-response, 508, 510
   coordination, 506
   mean-field equilibrium, 519, 521
   Nash equilibrium, 519
   optimal scheduling, 501
   stabilizing stochastic charging, 519
   valley-filling, 506
strength, 177
sub-Gaussian random variables, 224
subgradient, 409, 412, 413
supervisory control and data acquisition, 407
swing equation, 372
Symmetrical transform, 300
   a 3-point DFT interpretation, 301
   a spatial DFT interpretation, 301
synchrophasor, 79
synchrophasor technology, 340

target function, 64
technology lock-in, 462
temporal horizon, 524, 542
terminal nodes, 15
test system, 213
   IEEE 118-bus, 160, 219, 221
   IEEE 14-bus, 219
   IEEE 30-bus, 219, 221
   IEEE 33-bus, 168
thermal energy storage, 458, 459
thermal line limits, 377
thermodynamical state, 407
three phase voltages, 287, 289
   a widely linear model, 304
   recursive phasor computation, 289
   voltage sags and their noncircularity diagrams, 302
three-phase state estimation, 85
Tikhonov regularization, 414
time window, 524, 526, 527, 533–537, 539–543
Toeplitz matrix, 219
topology detection, 3, 23
topology identification, 3, 14
transfer learning, 69

transmission system, 406
trapdoor channel, 251
tree graph, 15
trusted third party, 233, 234, 237
two-tailed models, 388

uncertainty-aware DCOPF, 378
uncontrollable inputs, 401
universality principle, 151
utility company, 432, 434, 437, 440, 441
utility function, 431, 434

valley-filling, 433–435
valley-filling algorithm, 491, 493, 494
value-at-risk (aR), 382
variance-aware models, 394
variational inference, 65
voltage, 404
voltage magnitude, 108
voltage phase angle, 101, 102
   difference, 102
   incremental change, 102, 111
   measurements, 102, 108, 121
   variations, 109
voltage stability, 63
vulnerability, 176, 188

wake effect, 408
Wasserstein distance, 56, 182
water-filling algorithm, 242, 250, 493
Weierstrass's theorem, 403
weighted least squares, 75
wholesale price, 545
widely linear model
   three phase voltages, 304
wind farm, 408
wind generator, 553
wind turbine, 408
Wishart
   distribution, 223
   random matrix, 223
Wishart matrix, 148, 151

yaw angle, 408

zero-knowledge proofs, 238
z-score, 185